# THE FUNGI

VOLUME IVA
*A Taxonomic Review with Keys:*
*Ascomycetes and Fungi Imperfecti*

# Contributors to This Volume

G. C. AINSWORTH

RICHARD K. BENJAMIN

J. W. CARMICHAEL

DOROTHY I. FENNELL

W. BRYCE KENDRICK

RICHARD P. KORF

CHARLES L. KRAMER

N. J. W. KREGER-VAN RIJ

E. S. LUTTRELL

E. MÜLLER

B. C. SUTTON

J. A. VON ARX

C. E. YARWOOD

# THE FUNGI

## An Advanced Treatise

*Edited by*

### G. C. AINSWORTH
FORMERLY OF THE COMMONWEALTH MYCOLOGICAL INSTITUTE
KEW, SURREY, ENGLAND

### FREDERICK K. SPARROW
DEPARTMENT OF BOTANY
UNIVERSITY OF MICHIGAN
ANN ARBOR, MICHIGAN

### ALFRED S. SUSSMAN
DEPARTMENT OF BOTANY
UNIVERSITY OF MICHIGAN
ANN ARBOR, MICHIGAN

### VOLUME IVA
A Taxonomic Review with Keys:
Ascomycetes and Fungi Imperfecti

1973

ACADEMIC PRESS   New York   San Francisco   London
A Subsidiary of Harcourt Brace Jovanovich, Publishers

Copyright © 1973, by Academic Press, Inc.
ALL RIGHTS RESERVED.
NO PART OF THIS PUBLICATION MAY BE REPRODUCED OR
TRANSMITTED IN ANY FORM OR BY ANY MEANS, ELECTRONIC
OR MECHANICAL, INCLUDING PHOTOCOPY, RECORDING, OR ANY
INFORMATION STORAGE AND RETRIEVAL SYSTEM, WITHOUT
PERMISSION IN WRITING FROM THE PUBLISHER.

ACADEMIC PRESS, INC.
111 Fifth Avenue, New York, New York 10003

*United Kingdom Edition published by*
ACADEMIC PRESS, INC. (LONDON) LTD.
24/28 Oval Road, London NW1

LIBRARY OF CONGRESS CATALOG CARD NUMBER: 65-15769

PRINTED IN THE UNITED STATES OF AMERICA
80 81 82    9 8 7 6 5 4

To the memory of

GEORGE WILLARD MARTIN

(1886–1971)

# Contents

| | |
|---|---|
| List of Contributors | xi |
| Preface | xiii |
| Contents of Previous Volumes | xv |

## 1. Introduction and Keys to Higher Taxa
*G. C. Ainsworth*

| | | |
|---|---|---|
| I. | Status of Fungi | 1 |
| II. | Circumscription of the Fungi | 2 |
| III. | Taxonomic Arrangements | 3 |
| IV. | Keys to the Higher Taxa | 4 |
| | References | 7 |

## EUMYCOTA

### ASCOMYCOTINA, HEMIASCOMYCETES

## 2. Endomycetales, Basidiomycetous Yeasts, and Related Fungi
*N. J. W. Kreger-van Rij*

| | | |
|---|---|---|
| I. | Introduction | 11 |
| II. | Endomycetales | 12 |
| III. | Basidiomycetous Yeasts | 22 |
| IV. | Yeasts Classified in the Deuteromycotina (Blastomycetes) | 24 |
| | References | 28 |

## 3. Protomycetales and Taphrinales
*Charles L. Kramer*

| | | |
|---|---|---|
| I. | Protomycetales | 33 |
| II. | Taphrinales | 35 |
| | References | 41 |

## ASCOMYCOTINA, PLECTOMYCETES

### 4. Plectomycetes: Eurotiales

*Dorothy I. Fennell*

| | | |
|---|---|---|
| I. | Introduction | 45 |
| II. | Plectomycetes | 45 |
| III. | Eurotiales | 50 |
| IV. | Key | 57 |
| | References | 64 |

## ASCOMYCOTINA, PYRENOMYCETES

### 5. Pyrenomycetes: Erysiphales

*C. E. Yarwood*

| | | |
|---|---|---|
| I. | Erysiphales | 71 |
| II. | Erysiphaceae | 72 |
| III. | Perisporiaceae | 81 |
| | References | 85 |

### 6. Pyrenomycetes: Meliolales, Coronophorales, Sphaeriales

*E. Müller and J. A. von Arx*

| | | |
|---|---|---|
| I. | Introduction | 87 |
| II. | Pyrenomycetes Defined | 87 |
| III. | Meliolales | 89 |
| IV. | Coronophorales | 90 |
| V. | Sphaeriales | 93 |
| | References | 127 |

## ASCOMYCOTINA, LOCULOASCOMYCETES

### 7. Loculoascomycetes

*E. S. Luttrell*

| | | |
|---|---|---|
| I. | Definition of Loculoascomycetes | 135 |
| II. | Outline of Classification | 135 |
| III. | General Characteristics | 136 |
| IV. | Classification | 148 |
| V. | Important Literature | 154 |
| VI. | Synoptic Guide to Orders, Families, and Genera | 155 |

|       |                              |     |
|-------|------------------------------|-----|
| VII.  | Key to Orders                | 161 |
| VIII. | Myriangiales                 | 162 |
| IX.   | Dothideales                  | 165 |
| X.    | Pleosporales                 | 176 |
| XI.   | Hysteriales                  | 194 |
| XII.  | Hemisphaeriales              | 200 |
|       | References                   | 214 |

## ASCOMYCOTINA, LABOULBENIOMYCETES

### 8. Laboulbeniomycetes

*Richard K. Benjamin*

|      |                               |     |
|------|-------------------------------|-----|
| I.   | Introduction                  | 223 |
| II.  | Occurrence                    | 224 |
| III. | Host—Parasite Relationships   | 225 |
| IV.  | Morphology and Development    | 227 |
| V.   | Classification                | 231 |
|      | References                    | 243 |

## ASCOMYCOTINA, DISCOMYCETES

### 9. Discomycetes and Tuberales

*Richard P. Korf*

|       |                           |     |
|-------|---------------------------|-----|
| I.    | Introduction              | 249 |
| II.   | A Note on the References  | 255 |
| III.  | Orders                    | 257 |
| IV.   | Medeolariales             | 257 |
| V.    | Cyttariales               | 258 |
| VI.   | Tuberales                 | 258 |
| VII.  | Pezizales                 | 262 |
| VIII. | Phacidiales               | 277 |
| IX.   | Ostropales                | 281 |
| X.    | Helotiales                | 282 |
|       | References                | 307 |

## DEUTEROMYCOTINA, HYPHOMYCETES

### 10. Hyphomycetes

*W. Bryce Kendrick and J. W. Carmichael*

|     |                                 |     |
|-----|---------------------------------|-----|
| I.  | Introduction                    | 323 |
| II. | Notes on the List of Generic Names | 329 |

|       |                              |     |
|-------|------------------------------|-----|
| III.  | Notes on the Key-Lists       | 331 |
| IV.   | Notes on the Illustrations   | 333 |
| V.    | List of Generic Names        | 334 |
| VI.   | Key-Lists                    | 425 |
|       | Index to Plates              | 503 |

# DEUTEROMYCOTINA, COELOMYCETES

## 11. Coelomycetes

*B. C. Sutton*

|       |                              |     |
|-------|------------------------------|-----|
| I.    | Introduction                 | 513 |
| II.   | Systems of Classification    | 516 |
| III.  | Criteria for Classification  | 524 |
| IV.   | Keys                         | 553 |
|       | References                   | 574 |

|                  |     |
|------------------|-----|
| Author Index     | 583 |
| Taxonomic Index  | 592 |

# List of Contributors

Numbers in parentheses indicate the pages on which the authors' contributions begin.

G. C. Ainsworth, Formerly of the Commonwealth Mycological Institute, Kew, Surrey, England (1)

Richard K. Benjamin, Rancho Santa Ana Botanic Garden, Claremont, California (223)

J. W. Carmichael, Mold Herbarium and Culture Collection, University of Alberta, Edmonton, Alberta, Canada (323)

Dorothy I. Fennell, Northern Regional Research Laboratory, Agricultural Research Service, United States Department of Agriculture, Peoria, Illinois (45)

W. Bryce Kendrick, Department of Biology, University of Waterloo, Waterloo, Ontario, Canada (323)

Richard P. Korf, Plant Pathology Herbarium, Cornell University, Ithaca, New York, (249)

Charles L. Kramer, Division of Biology, Kansas State University, Manhattan, Kansas (33)

N. J. W. Kreger-van Rij, Department of Dermatology, State University of Groningen, Groningen, The Netherlands (11)

E. S. Luttrell, Department of Plant Pathology, University of Georgia, Athens, Georgia (135)

E. Müller, Department of Special Botany, Swiss Federal Institute of Technology, Zurich, Switzerland (87)

B. C. Sutton, Commonwealth Mycological Institute, Kew, Surrey, England (513)

J. A. von Arx, Centraalbureau voor Schimmelcultures, Baarn, The Netherlands (87)

C. E. Yarwood, Department of Plant Pathology, University of California, Berkeley, California (71)

# Preface

Volume IV of this treatise in which an attempt is made to provide generic keys for all the major groups of fungi is unique among multiauthor works. Never before in modern times has such a distinguished group of world specialists on the taxonomy of fungi provided for the professional mycologist such a comprehensive survey of fungal classification at the generic level.

It may be noted that although contributors were asked to conform to a standard pattern, the need for differences in the approaches to some groups—and also the well-known individualism of taxonomists which we were loath to suppress—led to contributions which show a rather wide and, at times, radical variation in treatment. In addition, although contributors to allied groups consulted with one another, a number of genera, and even families, are duplicated by different authors—that is, they appear in the keys of more than one of the higher taxa. Such duplication, as any student of the fungi knows, is, indeed, a reflection of currently unresolved taxonomic problems. These differences and duplications should, however, help rather than hinder the many who will have occasion to consult this volume which is comprised of two parts, IVA and IVB, each part being published separately.

The two parts of Volume IV complete the treatise as planned, and although there is no intention of issuing new editions of any of the volumes the possibility of updating the treatise by one or more supplementary volumes is being considered.

We would like to take this opportunity to thank all the contributors to this treatise for their hard work and for their forbearance with editorial pedantry, and, in conclusion, on behalf of ourselves and every contributor, to thank the staff of Academic Press for their patient and unobtrusive work which has done so much to lighten our labors.

G. C. Ainsworth
F. K. Sparrow
A. S. Sussman

# Contents of Previous Volumes

## VOLUME I
## THE FUNGAL CELL

INTRODUCTION

Historical Introduction to Mycology
G. C. AINSWORTH

Fungal Structure and Organization
C. J. HICKMAN

CELL COMPONENTS

The Cell Wall
JEROME M. ARONSON

Flagella
A. P. KOLE

The Ultrastructure of Fungal Cells
ROYALL T. MOORE

Somatic Nuclei and Forms of Mitosis in Fungi
C. F. ROBINOW AND A. BAKERSPIGEL

Nuclear Behavior during Meiosis
LINDSAY S. OLIVE

Chemical Constituents of the Fungal Cell
1. Elemental Constituents and Their Roles
VIRGIL GREENE LILLY

Chemical Constituents of the Fungal Cell
2. Special Chemical Products
JOHN HOWARD BIRKINSHAW

Carbohydrate Metabolism
1. Glycolysis
HAROLD J. BLUMENTHAL

Carbohydrate Metabolism
2. Tricarboxylic Acid Cycle
D. J. NIEDERPRUEM

Carbohydrate Metabolism
3. Terminal Oxidation and Electron Transport
ARISTID LINDENMAYER

Utilization of Inorganic Nitrogen Compounds and Amino Acids by Fungi
D. J. D. NICHOLAS

Integration of Cellular Metabolism
MARKO ZALOKAR

NUTRITION AND GROWTH OF CELLS

Uptake and Translocation
1. Uptake
ASER ROTHSTEIN

Uptake and Translocation
2. Translocation
E. P. HILL

The Chemical Environment for Fungal Growth
1. Media, Macro- and Micronutrients
VIRGIL GREENE LILLY

The Chemical Environment for Fungal Growth
2. Carbon Sources
D. PERLMAN

The Chemical Environment for Fungal Growth
3. Vitamins and Other Organic Growth Factors
NILS FRIES

The Chemical Environment for Fungal Growth
4. Chemical Inhibition
R. J. W. BYRDE

The Physical Environment for Fungal Growth
1. Temperature
B. J. DEVERALL

The Physical Environment for Fungal Growth
2. *Hydrostatic Pressure*
   RICHARD Y. MORITA

The Physical Environment for Fungal Growth
3. *Light*
   ROBERT M. PAGE

The Physical Environment for Fungal Growth
4. *Effects of Radiation*
   S. POMPER

Kinetics of Fungal Growth
   G. R. MANDELS

The Mechanism of Cellular Extension and Branching
   N. F. ROBERTSON

Growth Rhythms
   STEPHEN JEREBZOFF

Special Growth Techniques (Synchrony, Chemostasis)
   ALLAN CAMPBELL

GENE ACTION
Gene Action
   D. G. CATCHESIDE

*Author Index — Subject Index — Index to Fungi, Lichens, and Actinomycetes*

# VOLUME II
# THE FUNGAL ORGANISM

THE PROTOPLAST
Protoplasts of Fungi
   JULIO R. VILLANUEVA

CELL AGGREGATES
Aggregation of Unicells: Yeasts
   E. O. MORRIS

THE MULTICELLULAR CONDITION
Vegetative Structures
   GILLIAN M. BUTLER

Sporulating Structures in Fungi Imperfecti
   KEISUKE TUBAKI

Fruit Bodies in Ascomycetes
   C. BOOTH

The Hyphal Structure of the Basidiocarp
   ALEXANDER H. SMITH

MECHANISMS OF MORPHOGENESIS
Dimorphism
   ANTONIO H. ROMANO

Morphogenesis in the Myxomycetes
   CONSTANTINE J. ALEXOPOULOS

Organization and Synthesis in the Cellular Slime Molds
   JAMES H. GREGG

Morphogenesis in Aquatic Fungi
   EDWARD C. CANTINO

Morphogenesis in Ascomycetes
   G. TURIAN

Morphogenesis in Basidiomycetes
   W. A. TABER

PHYSIOLOGY OF REPRODUCTION
Sex Hormones in Fungi
   LEONARD MACHLIS

Environmental Influences on Reproduction
   LILIAN E. HAWKER

REPRODUCTION AND INHERITANCE
Life Cycles, Basic Patterns of Sexuality, and Sexual Mechanisms
   JOHN R. RAPER

Mechanisms of Inheritance
1. *Mendelian*
   STERLING EMERSON

Mechanisms of Inheritance
2. *Heterokaryosis*
   ROWLAND H. DAVIS

Mechanisms of Inheritance
3. *The Parasexual Cycle*
   J. A. ROPER

Mechanisms of Inheritance
4. *Extranuclear Inheritance*
   J. L. JINKS

Incompatibility
   KARL ESSER

DISSEMINATION
Spore Release
   C. T. INGOLD

Dispersal
   P. H. GREGORY

Dormancy and Spore Germination
   ALFRED S. SUSSMAN

*Author Index—Subject Index—Index to Fungi, Lichens, and Actinomycetes*

# VOLUME III
## THE FUNGAL POPULATION

ECOLOGY: SAPROBIC FUNGI AND THEIR HABITATS
The Ecology of Terrestrial Fungi
   DAVID PARK

Ecology of Freshwater Fungi
   FREDERICK K. SPARROW, JR.

Saprobic Marine Fungi
   T. W. JOHNSON, JR.

Thermophiles
   RALPH EMERSON

Psychrophiles
   B. J. DEVERALL

ECOLOGY: SYMBIOTIC FUNGI AND THEIR ASSOCIATES
Mycorrhiza
   J. L. HARLEY

ECOLOGY: PARASITIC FUNGI AND THEIR HOSTS
Fungal Parasites of Plants
   B. E. J. WHEELER

Fungal Parasites of Vertebrates
   G. C. AINSWORTH

Fungal Parasites of Invertebrates
1. *Entomogenous Fungi*
   M. F. MADELIN

Fungal Parasites of Invertebrates
2. *Predacious Fungi*
   C. L. DUDDINGTON

Fungi Parasitic on Other Fungi and Lichens
   M. F. MADELIN

ECOLOGY: FUNGI UNDER DOMESTICATION
Fungi under Domestication
   GEORGE SMITH

ECOLOGY: METHODS OF ADJUSTMENT TO THE ENVIRONMENT
Physiological and Biochemical Adjustment of Fungi to Their Environment
   JOHN SAVILLE WAID

Survival of Fungi after Freezing and Desiccation
   PETER MAZUR

Genetical Adjustment of Fungi to Their Environment
   CLAYTON PERSON

ECOLOGY: RESULTS OF ADJUSTMENT IN NATURE
Effects of Adjustment to the Environment on Fungal Form
   PETER K. C. AUSTWICK

Longevity and Survivability of Fungi
   A. S. SUSSMAN

Geographical Distribution of Fungi
   K. A. PIROZYNSKI

The Number of Fungi
  G. C. Ainsworth

**TAXONOMY: TAXONOMIC CRITERIA**

Morphology as a Taxonomic Criterion
  C. G. C. Chesters

Host Specialization as a Taxonomic Criterion
  T. Johnson

Biochemical Differentiation of Taxa with Special Reference to the Yeasts
  J. A. Barnett

Serology as an Aid to Taxonomy
  H. P. R. Seeliger

Genetical and Cytological Aspects of Taxonomy
  Morten Lange

**TAXONOMY: POSSIBLE EVOLUTIONARY PATTERNS**

The Origin and Status of Fungi (with a Note on the Fossil Record)
  G. W. Martin

Possible Interrelationships between Fungal Groups
  D. B. O. Savile

On the Evolution of Fungi
  John R. Raper

*Author Index—Subject Index—Index to Fungi, Lichens, and Actinomycetes*

CHAPTER 1

# Introduction and Keys to Higher Taxa

G. C. AINSWORTH

*Formerly of the
Commonwealth Mycological Institute
Kew, Surrey, England*

---

During recent years, the traditional approaches to the status, circumscription, and taxonomic arrangement of fungi have been seriously questioned. It is customary to classify fungi as plants, if only on the basis that they are not animals, but as Stafleu (1969) writes, "The times in which we recognized two kingdoms, *Regnum vegetabile* and *Regnum animale*, are perhaps surviving only in the subdivision of biology in some of our universities...";  the "seemingly fundamental division in two has proved to be untenable." Where should fungi be classified today?

## I. STATUS OF FUNGI

Dissatisfaction with the division of organisms into two kingdoms is not a new phenomenon. A century ago Haeckel proposed the taxon "Moneres" and subsequently developed a four-kingdom system. Since then there have been a number of proposals for the reorganization of the major taxa. Representative modern examples are the four-kingdom systems of Copeland (1956) and Barkley (1968) and the five-kingdom system of Whittaker (1969). The writings of these authors may be consulted for details of their own and other systems.

There is general agreement that taxonomy should endeavor to reflect phylogeny and that in the virtual absence of a fossil record much of the detailed phylogeny of microorganisms is speculative. However, by supplementing the morphological approach to taxonomy with the results of nutritional and biochemical studies, it is possible to arrive at arrangements which, in general, correspond with current beliefs on possible evolutionary sequences. There is, for example, widespread agreement that the bacteria and blue-green algae, which comprise the kingdom Mychota of Copeland and the kingdom Monera of Barkley and Whittaker, are distinct from other organisms and show primitive characters. Copeland classifies the fungi, with the algal and protozoan groups, in his kingdom Protoctist, mainly in the phylum Inophyta; Barkley's treatment, in which fungi are assigned to the kingdom Protista, mostly as the phylum Mycophyta, is very similar. Whittaker, on the other hand, in an attractive and well-argued arrangement, raises the fungi to the status of a kingdom.

Whittaker's two primitive kingdoms are the prokaryotic Monera and the eukaryotic Protista, and he suggests that it was from Protista-like ancestors that three nutritionally distinct lines have developed: (1) the *plant kingdom*, characterized by photosynthesis, (2) the *animal kingdom*, characterized by ingestive nutrition, and (3) *fungi*, in which nutrition is absorptive.

Recent additional evidence in support of the distinctness of fungi has been derived from the comparative studies by Nolan and Margoliash (1968) on cytochrome c in different organisms. Cytochrome c, a component of the terminal respiratory chain of enzymes in aerobic organisms (see Lindenmayer, 1965), is widely distributed in both plants and animals. It is concluded that the cytochrome c enzymes in all organisms are homologous, that they arise from homologous gene loci, and that each possesses an evolutionary history. From these premises, the interpretation of the experimental findings suggests that the fungi studied form a phylogenetic line distinct from the animal and plant kingdoms. Wheat, the representative of the plant kingdom investigated, was found to be more closely related to man than to fungi from the point of view of its cytochrome c. Whatever the final outcome of such studies, fungi are clearly organisms which, as Martin (1968) concluded in Volume III, "may reasonably be treated as a discrete major taxonomic unit."

## II. CIRCUMSCRIPTION OF THE FUNGI

The circumscription of Fungi is not always well defined. The best-known example of a group of uncertain affinity is that of the slime molds, which has long been claimed by both mycologists (as the Myxomycetes) and zoologists (as the Mycetozoa). Other groups, if less familiar, are of equally uncertain status: the cellular slime molds (Acrasiomycetes), for example, are claimed as Protozoa, a taxon to which the Hydromyxomycetes (including the Labyrinthulales) have been referred.

At the turn of the century, it was a widely held belief that fungi of the class Phycomycetes were derived from algae, a belief reflected in the nomenclature adopted by Clements and Shear (1931) in their "Genera of Fungi" for the orders of Phycomycetes. This hypothesis, which fell into disrepute, has recently been revived in modified form as seen by the classification of the Oomycetes (the only fungal group exhibiting true cellulose in the cell wall) with the algae by both Copeland (1956) and Barkley (1968) and by the German mycologist Kreisel (1969), while the Hyphochytriales (or Hypochytridiomycetes) are also associated with biflagellated algal groups (Heterokontae) by Barkley, Copeland, and Kreisel. In this treatise, following tradition, the Acrasiomycetes and Myxomycetes are included as fungal groups, as are the Oomycetes, but the Hydromyxomycetes are excluded. The recent discovery of biflagellate zoospores (see Mastigomycotina, Volume IVB) in the latter group, however, will necessitate a reconsideration of their status.

## III. TAXONOMIC ARRANGEMENTS

For the past fifty years, most taxonomic arrangements of the fungi have differentiated the plasmodial Myxomycota) from the mycelial (Eumycota) forms, which comprise the bulk of fungi. Three classes of the latter have been universally recognized—Phycomycetes, Ascomycetes, and Basidiomycetes—based on the types of sexually produced spores. These are grouped with the Deuteromycetes, or Fungi Imperfecti, which are characterized by the possession of asexual spores only. Because the class Phycomycetes was patently a miscellaneous assemblage, following the lead of Sparrow (1959), its constituents are with increasing frequency being accommodated in a series of classes: the Chytridiomycetes, Hyphochytridiomycetes, Oomycetes, Zygomycetes, and Trichomycetes. Although it is still held that the Ascomycetes are monophyletic, there have been major adjustments within the group, particularly as a result of the significance now given to uni- and bitunicate asci as taxonomic criteria. For a hundred years, treatment of the basidiomycetes, which include so many of the larger fungi, has been dominated by the Friesian approach which had—and still has—the attraction of differentiating fungi into groups mainly on field characters. More critical microscopic studies of basidiocarp structure and morphogenesis and the recognition that there has been much evolutionary convergence are leading to an abandonment of the Friesian system and the development of more natural groupings. For many basidiomycetes, as for many ascomycetes, the older descriptions are inadequate, and new descriptions have yet to be prepared, so that in many areas the current classification of these fungi is in a state of transition. Some of the other classes recognized in the present account, for example, the Pyrenomycetes and Gasteromycetes, are also clearly heterogeneous, but they are still useful pigeonholes for ordering many common fungi.

In these volumes, the taxonomic arrangement is based on that proposed by Ainsworth (1966) for use in the current edition of the "Dictionary of Fungi" (Ainsworth, 1971). Fungi are treated here as either a separate kingdom or, for the more conservative, as a subkingdom of the plant kingdom, with two divisions, the Myxomycota, for plasmodial forms, and the Eumycota, for nonplasmodial forms which are frequently mycelial. Five subdivisions of the latter are recognized, including the Ascomycotina (for ascomycetes) and the Basidiomycotina (for basidiomycetes), while the imperfect fungi are classified for convenience as the Deuteromycotina, although in a hierarchical classification it is incorrect to equate these fungi with the ascomycetes and the basidiomycetes to which they are subsidiary both taxonomically and by nomenclature. Most lichenized fungi (lichens) have been omitted, as they are to be the subject of a separate book.

## IV. KEYS TO THE HIGHER TAXA

### FUNGI

It is difficult to give a concise diagnostic definition of fungi. The main characteristics of the group are

*Nutrition*: heterotrophic (photosynthesis lacking) and absorptive (ingestion rare).

*Thallus*: on or in the substratum and plasmodial amoeboid or pseudoplasmodial; or in the substratum and unicellular or filamentous (mycelial), the last, septate or nonseptate; typically nonmotile (with protoplasmic flow through the mycelium) but motile states (e.g., zoospores) may occur.

*Cell wall*: well-defined, typically chitinized (cellulose in Oomycetes).

*Nuclear status*: eukaryotic, multinucleate, the mycelium being homo- or heterokaryotic, haploid, dikaryotic, or diploid, the last being usually of limited duration.

*Life cycle*: simple to complex.

*Sexuality*: asexual or sexual and homo- or heterothallic.

*Sporocarps*: microscopic or macroscopic and showing limited tissue differentiation.

*Habitat*: ubiquitous as saprobes, symbionts, parasites, or hyperparasites.

*Distribution*: cosmopolitan.

### KEY TO DIVISIONS OF FUNGI

1. Plasmodium or pseudoplasmodium present . . . . . . . . . . . . . . **Myxomycota** I
1'. Plasmodium or pseudoplasmodium absent, assimilative phase
    typically filamentous . . . . . . . . . . . . . . . . . . . . . . . . . . . **Eumycota** II

# 1. Introduction and Keys to Higher Taxa

## I. Myxomycota

### KEY TO CLASSES OF MYXOMYCOTA

1. Assimilative phase a plasmodium . . . . . . . . . . . . . . . . . . . . 2
1'. Assimilative phase free-living amoebae which unite as a pseudoplasmodium before reproduction . . . . . . . . . . . . . . . . . **Acrasiomycetes**[1]   Vol. IVB
    2(1) Plasmodium forming a network ("net plasmodium") . . . . . . **Labyrinthulales**[2]
    2'(1) Plasmodium not forming a network . . . . . . . . . . . . . . . . . . . . 3
3(2') Plasmodium saprobic, free-living . . . . . . . . . . . **Myxomycetes**   Vol. IVB
3'(2') Plasmodium parasitic within cells of the host plant . . . . . . . . . . . . . . . . . . . **Plasmodiophoromycetes**[3]   Vol. IVB

## II. Eumycota

### KEY TO SUBDIVISIONS OF EUMYCOTA

1. Motile cells (zoospores) present; perfect-state spores typically oospores . . . . . . . . . . . . . . . . . . . . **Mastigomycotina**   III
1'. Motile cells absent . . . . . . . . . . . . . . . . . . . . . . . . . . . . . . 2
    2(1') Perfect state present . . . . . . . . . . . . . . . . . . . . . . . . 3
    2(1') Perfect state absent . . . . . . . . . . . . . . . . . . . **Deuteromycotina**   VII
3(2) Perfect-state spores zygospores . . . . . . . . . . . . . . . . **Zygomycotina**   IV
3'(2) Zygospores absent . . . . . . . . . . . . . . . . . . . . . . . . . . . . . . 4
    4(3') Perfect-state spores ascospores . . . . . . . . . . . . . . . . **Ascomycotina**   V
    4'(3') Perfect-state spores basidiospores . . . . . . . . . . . . **Basidiomycotina**   VI

## III. Mastigomycotina

### KEY TO CLASSES OF MASTIGOMYCOTINA

1. Zoospores posteriorly uniflagellate (flagella whiplash-type) . . . . . . . . . . . . . . . . . **Chytridiomycetes**   Vol. IVB
1'. Zoospores not posteriorly uniflagellate . . . . . . . . . . . . . . . . . . . . . 2
    2(1') Zoospores anteriorly uniflagellate (flagella tinsel-type) . . . . . . . . . . . . . . **Hyphochytridiomycetes**   Vol. IVB
    2'(1') Zoospores biflagellate (posterior flagellum whiplash-type; anterior tinsel-type); cell wall cellulosic . . . . . . . . . . **Oomycetes**   Vol. IVB

---

[1] Excluded from the Myxomycota by Martin and Alexopoulos (1969).
[2] Excluded from this treatment.
[3] Treated as a class of the Mastigomycotina.

## IV. Zygomycotina

### KEY TO CLASSES OF ZYGOMYCOTINA

1. Saprobic or, if parasitic or predacious, having mycelium immersed in host tissue ............................ **Zygomycetes** Vol. IVB
1'. Associated with arthropods and attached to the cuticle or digestive tract by a holdfast and not immersed in the host tissue ............ **Trichomycetes** Vol. IVB

## V. Ascomycotina

### KEY TO CLASSES OF ASCOMYCOTINA

1. Ascocarps and ascogenous hyphae lacking; thallus mycelial or yeastlike ........................ **Hemiascomycetes** p. 9
1'. Ascocarps and ascogenous hyphae present; thallus mycelial ............ 2
    2(1') Asci bitunicate; ascocarp an ascostroma ................... **Loculoascomycetes** p. 133
    2'(1') Asci typically unitunicate; if bitunicate, ascocarp an apothecium ........................... 3
3(2') Asci evanescent, scattered within the astomous ascocarp which is typically a cleistothecium; ascospores aseptate ........ **Plectomycetes** p. 43
3'(2') Asci regularly arranged within the ascocarp as a basal or peripheral layer ................................. 4
    4(3') Exoparasites of arthropods; thallus reduced; ascocarp a perithecium; asci inoperculate ............ **Laboulbeniomycetes** p. 221
    4'(3') Not exoparasites of arthropods ........................ 5
5(4') Ascocarp typically a perithecium which is usually ostiolate (if astomous, asci not evanescent); asci inoperculate with an apical pore or slit ................................ **Pyrenomycetes** p. 69
5'(4') Ascocarp an apothecium or a modified apothecium, frequently macrocarpic, epigean or hypogean; asci inoperculate or operculate .................................... **Discomycetes** p. 247

## VI. Basidiomycotina

### KEY TO CLASSES OF BASIDIOMYCOTINA[4]

1. Basidiocarp lacking and replaced by teliospores (encysted probasidia) grouped in sori or scattered within the host tissue; parasitic on vascular plants ............ **Teliomycetes** Vol. IVB
1'. Basidiocarp usually well-developed; basidia typically organized as a hymenium; saprobic or rarely parasitic ............... 2
    2(1') Basidiocarp typically gymnocarpous or semiangiocarpous; basidia phragmobasidia (**Phragmobasidiomycetidae**) or

[4] If yeastlike, see p. 11.

holobasidia (**Holobasidiomycetidae**); basidiospores
ballistospores . . . . . . . . . . . . . . . . . . **Hymenomycetes** Vol. IVB

2′(1′) Basidiocarp typically angiocarpous; basidia holobasidia;
basidiospores not ballistospores . . . . . . . . . . . **Gasteromycetes** Vol. IVB

### VII. Deuteromycotina

#### KEY TO CLASSES OF DEUTEROMYCOTINA

1. Budding (yeast or yeastlike) cells with or without pseudomycelium characteristic; true mycelium lacking or not well-developed . . . . . . . . . . . . . . . . . . . . . . . . . . **Blastomycetes** p. 24

1′. Mycelium well-developed, assimilative budding cells absent . . . . . . . . . . . 2

    2(1′) Mycelium sterile or bearing spores directly or on special branches (sporophores) which may be variously aggregated but not in pycnidia or acervuli . . . . . . . . **Hyphomycetes** p. 321

    2′(1′) Spores in pycnidia or acervuli . . . . . . . . . . . . **Coelomycetes** p. 511

#### REFERENCES

Ainsworth, G. C. (1966). A general purpose classification for fungi. *Bibl. Syst. Mycol.* No. 1:1–4.

Ainsworth, G. C. (1971). "Ainsworth and Bisby's Dictionary of the Fungi," 6th ed. Commonwealth Mycol. Inst., Kew, Surrey, England.

Barkley, F. A. (1968). "Outline Classification of Organisms," 2nd ed. Hopkins Press, Providence, Massachusetts.

Clements, F. E., and C. L. Shear. (1931). "The Genera of Fungi." Wilson, New York.

Copeland, H. F. (1956). "The Classification of Lower Organisms." Pacific Books, Palo Alto, California.

Kreisel, H. (1969). "Grundzüge eines natürlichen Systems der Pilze." Cramer, Lehre.

Lindenmayer, A. (1965). Carbohydrate metabolism. 3. Terminal oxidation and electron transport. *In* "The Fungi" (G. C. Ainsworth and A. S. Sussman, eds.). Vol. 1, pp. 301–348. Academic Press, New York.

Martin, G. W. (1968). The origin and status of fungi. *In* "The Fungi" (G. C. Ainsworth and A. S. Sussman, eds.), Vol. 3, pp. 635–648. Academic Press, New York.

Martin, G. W., and C. J. Alexopoulos. (1969). "The Myxomycetes," p. 30. Univ. of Iowa Press, Iowa City.

Nolan, C., and E. Margoliash. (1968). Comparative aspects of primary structures of proteins. *Annu. Rev. Biochem.* **37**:727–790.

Sparrow, F. K. (1959). Interrelationships and taxonomy of the aquatic phycomycetes. *Mycologia* **50**:797–813.

Stafleu, F. A. (1969). Biosystematic pathways anno 1969. *Taxon* **18**:485–500

Whittaker, R. H. (1969). New concepts of kingdoms of organisms. *Science* **163**: 150–160.

# Eumycota
# Ascomycotina
# Hemiascomycetes

CHAPTER 2

# Endomycetales, Basidiomycetous Yeasts, and Related Fungi

N. J. W. Kreger-van Rij

*Laboratory of Medical Microbiology*
*State University of Groningen*
*Groningen, The Netherlands*

## I. INTRODUCTION

This chapter covers the Endomycetales (Ascomycotina), most of which are yeasts, and the yeasts in the Basidiomycotina and Deuteromycotina. From this division, it can be seen that the yeasts are not a natural taxonomic entity, although there is a great uniformity of morphology among them. Yeasts are those fungi which, in a stage of their life cycle, occur as single cells, reproducing by budding or fission. As a rule, organisms with plurinucleate cells or producing black pigments are excluded. The distinction of the yeasts from related mycelial fungi is subjective and indicated by practical considerations. Their properties entail a special taxonomic treatment for the differentiation of species in which, because of the scarcity of morphological criteria, physiological characteristics are in general use.

Ascomycetous and basidiomycetous yeasts show, in addition to their different modes of sexual reproduction, other distinguishing features which suggest relationship to genera or species of the Deuteromycotina. Some of these features are the GC content of DNA (Storck and Alexopoulos, 1970), the composition of the cell wall and the capsule (Phaff, 1971), the structure of the cell wall, and the occurrence of ballistospores and dikaryotic mycelium.

In both ascomycetous and basidiomycetous yeasts heterothallic species occur, and several strains of asporogenous species have been found to represent one of the mating types of a perfect species. As conditions for growth of the different stages of life cycles, especially conjugation and sporulation, become better known, it is to be expected that more imperfect forms of yeasts can be transferred to perfect species.

Most of the yeasts and related organisms are saprobes; their habitats are plants (flowers, leaves, bark, wood, and exudates of trees), and animals (insects such as *Drosophila* and bark beetles). They also occur in fresh- and seawater. A few species are parasites and are known as human, animal, or plant pathogens.

The classification of the groups which will be discussed here is shown in the following tabulation:

| | |
|---|---|
| Ascomycotina | |
|   Hemiascomycetes | |
|     Endomycetales | Spermophthoraceae |
| | Ascoideaceae |
| | Endomycetaceae |
| | Saccharomycetaceae |
| Basidiomycotina | Basidiomycetous yeasts |
| Deuteromycotina | |
|   Blastomycetes | Sporobolomycetaceae |
| | Cryptococcaceae |

A recent monograph on yeast taxonomy edited by Lodder (1970) contains more detailed information about most genera mentioned in this chapter. Other literature on yeast taxonomy is found in Wickerham (1951), Kudriavzev (1960), Windisch (1960), Novák and Zsolt (1961), and Tsuchiya et al. (1965). General introductory information about yeasts is given by Phaff et al. (1966). Yeast ecology is surveyed by do Carmo-Sousa (1969), and human and animal pathogens by Gentles and La Touche (1969).

## II. ENDOMYCETALES

### A. Order Endomycetales

Zygotes or single cells are directly transformed into asci. There are different views on the relationships of the Endomycetales. Gäumann (1964) considered them to be primitive ascomycetes, the Prototunicatae (Hemiascomycetes). Bessey (1965) named them Saccharomycetales and regarded them as reduced ascomycetes. Von Arx (1967) classified them together with basidiomycetous and asporogenous orders in a separate class, the Endomycetes. Bartnicki-Garcia (1970) considered the Hemiascomycetes and Euascomycetes as separate offshoots from the same ancestral ascomycetes.

Subdivision of the Endomycetales into families and genera is based on characters of sexual and vegetative reproduction. To the former belong ascus formation preceded by gametangiogamy or somatogamy, formation of ascophores, shape of the ascospores, and number of spores per ascus. Characters of vegetative reproduction are formation of mycelium and arthrospores, budding yeast cells and buds on hyphae, and the occurrence of uni- or plurinucleate cells. In several instances, details of the complete life cycle are not known, and insufficient data are available on other features considered to be of taxonomic importance, such as the composition of the cell wall. The resulting classification of the Endomycetales leaves much to be desired.

Guilliermond's phylogenetic concepts (1909, 1928, 1937) have considerably influenced the differentiation of the Endomycetaceae and the Saccharomycetaceae, and the description of the genus *Endomycopsis* is a consequence of his theories. He assumed a development from the hyphal stage to loose budding cells, with intermediate forms showing bòth types of reproduction, as well as a transition from mycelium to single cells reproducing by fission.

Wickerham (1951) considered that in Guilliermond's system too much value had been attached to the formation of hyphae and budding cells and too little to the shape of the ascospores. Zender (1925) based a classification of genera mainly on the latter property, but it has not found acceptance since it was too one sided and neglected other features. Wickerham (1951) proposed a phylogenetic development from haploid to diploid species. In the genus *Hansenula* (Wickerham, 1970; Wickerham and Burton, 1962), this development agreed with an increased ability to form hyphae and, in the physiological properties, with a decreased dependence on external vitamins and with the capacity to ferment an increased number of sugars.

The number of criteria which may be used for the classification of the yeasts is increasing. Wickerham (1951) extended the number of carbon compounds to be tested for assimilation, so that now 30–40 compounds are used for the description of a species. Secondly, more data are available about the composition of the cell wall and the capsule, as well as about the antigenic structure of the wall. Further, the GC content of DNA is now known for many species, and electron microscopic studies have revealed useful morphological properties. At present, a comparative examination of these features in many species and strains is being made. The results will indicate which of them are of taxonomic value.

**KEY TO FAMILIES OF ENDOMYCETALES**

1. Asci multispored; mycelium abundant . . . . . . . . . . . . . . **Ascoideaceae** p. 15
1'. Asci usually 1- to 8-spored; mycelium absent or present . . . . . . . . . . . . . 2
    2(1') Ascospores needle- or spindle-shaped . . . . . . . **Spermophthoraceae** p. 14
    2'(1') Ascospores not needle- or spindle-shaped . . . . . . . . . . . . . . . . . . 3

3(2') Mycelium usually scanty or lacking; single cells reproducing by
budding or fission . . . . . . . . . . . . . . . . . . . **Saccharomycetaceae** p. 18

3'(2') Mycelium abundant; no single budding cells . . . . . . . **Endomycetaceae** p. 16

## B. Family Spermophthoraceae

Mycelium is present or absent, coenocytic or monokaryotic. Ascospores are needle- or spindle-shaped. Budding yeast cells are present or absent.

This family is characterized by the shape of the spores. The asci are often much larger than the yeast cells.

Species of the general *Spermophthora, Ashbya, Eremothecium,* and *Nematospora* are parasitic on fruit such as cotton bolls, tomatoes, hazel nuts, and citrus. They are transferred by insects.

Guilliermond (1928) described for *Spermophthora* a haploid plurinucleate generation ending up in a sporangium with numerous sickle-shaped spores. After conjugation of the latter, a diploid monokaryotic mycelium arose on which asci with eight to twelve spindle-shaped spores were formed. In the genera *Ashbya* and *Eremothecium,* no conjugation was observed, and Guilliermond (1936) suggested that the spores might be sporangiospores and not ascospores. Do Carmo-Sousa (1970) considered that this might also be true for *Nematospora coryli.* This would mean a close relationship between the genera *Spermophthora, Ashbya, Eremothecium,* and *Nematospora,* all producing sporangiospores—in sharp distinction to *Metschnikowia* and *Coccidiascus,* which have no sporangiospores, but exhibit asci after conjugation. However, Manuel (1938) described conjugation preceding ascus formation in *N. coryli.* In the following key, the spores of *Ashbya, Eremothecium,* and *Nematospora* are provisionally considered to be ascospores.

### KEY TO GENERA OF SPERMOPHTHORACEAE

1. Mycelium with or without budding cells . . . . . . . . . . . . . . . . . . . . . 2

1'. No mycelium, but only budding cells . . . . . . . . . . . . . . . . . . . . . 5

    2(1) Sporangia and asci both formed . . . . . . . . . . . . . . . . . **Spermophthora**
    *S. gossypii* is the single species.

    2'(1) Only asci formed . . . . . . . . . . . . . . . . . . . . . . . . . . . 3

3(2') Budding yeast cells present . . . . . . . . . . . . . . . . . . . . . . . . 4

3'(2') No budding yeast cells present . . . . . . . . . . . . . . . . . . . **Eremothecium**
    *Eremothecium cymbalariae* was the first species of this genus. Routien (1949) classified another species, *E. ashbyi,* described by Guilliermond (1935) in the new genus *Crebrothecium,* slightly differing from *Eremothecium.* It is considered as a synonym of *Eremothecium* here.

    4(3) Conjugation preceding ascus formation . . . . . . . . . . . . . . **Nematospora**
    The single species is *N. coryli.*

    4'(3) No conjugation preceding ascus formation . . . . . . . . . . . . . . **Ashbya**

*Ashbya gossypii* (syn. *Nematospora gossypii*) is the only species described (Guilliermond, 1928; Pridham and Raper, 1950). Strains of this species and of *Eremothecium ashbyi* have the ability to produce riboflavin.

5(1') Ascospores needle-shaped . . . . . . . . . . . . . . . . . . . . . . . **Metschnikowia**

The 5 species recognized by Miller and van Uden (1970) are homo- or heterothallic. In some species chlamydospores occur, and for these the genus *Chlamydozyma* was described by Wickerham (1964). He obtained haploid cells from a diploid culture without spore formation and named this type of reproduction protosexuality. Pitt and Miller (1970) confirmed this observation. These authors (1968) also found that the chlamydospores might change into asci. *Metschnikowia pulcherrima* has been isolated from fruit and flowers, *M. reukaufii* from nectar of flowers. Of the other three species, *M. bicuspidata* is the oldest one; Metschnikoff observed it as a parasite in *Daphnia magna*. Most strains of the three species have been isolated from seawater.

5'(1') Ascospores fusiform, screwed in a helix . . . . . . . . . . . . . . . . **Coccidiascus**

The single species, *C. legeri*, which occurred in the intestines of *Drosophila funebris*, has never been isolated.

## C. Family Ascoideaceae

Mycelium is abundant. The mycelial cells are pluri- or uninucleate. Conidia (blastospores) and arthrospores (oidia) may be formed. Asci are multispored, borne on hyphal tips or arising from gametangia.

This family includes two genera, *Ascoidea* and *Dipodascus*, both with multispored asci, but divergent in the method of ascus formation. Gäumann (1964) classified the genus *Dipodascus* in a separate family, Dipodascaceae. Batra (1959) included in the Dipodascaceae the genus *Helicogonium* (White, 1942). A fourth genus, *Myriogonium*, probably related to the Ascoideaceae, is not classified in a family here. The single species, *M. odontiae* (Cain, 1948), produces croziers and has eight oval spores per ascus. It has been isolated from *Odontia sudans*.

**KEY TO GENERA OF ASCOIDEACEAE**

1. Ascospores round or oval . . . . . . . . . . . . . . . . . . . . . . . **Dipodascus**

Batra (1959) recognized three species, of which *D. albidus* produces plurinucleate mycelial cells and gametangia, and arthrospores. After gametangiogamy, two of the many nuclei fuse and the diploid nucleus, after meiosis and mitotic divisions, yields the ascospores. Korf (1957) studied strains of *D. albidus* indicated as *forma minor. Dipodascus aggregatus* and *D. uninucleatus* (Biggs, 1937) have uninucleate cells and gametangia. *Dipodascus* strains have been isolated from slime flux, soil, frass, and insects.

1'. Ascospores hat-shaped . . . . . . . . . . . . . . . . . . . . . . . . . **Ascoidea**

This genus is characterized by internally proliferating asci. There is no gametangiogamy. Conidia and chlamydospores are formed. The mycelial cells are pluri- or uninucleate. *Ascoidea rubescens* (Walker, 1931, 1935) was the first species described. Other species were described by Batra and Francke-Grosmann (1960, 1964). Strains have been isolated from plant exudates and from ambrosia beetles.

## D. Family Endomycetaceae

Mycelium is abundant; arthrospores and blastospores may be formed. Conjugation is achieved by gametangiogamy or by anastomosis between hyphae. Generally there are one to eight spores per ascus. Erect ascophores may be present.

The Endomycetaceae constitute a small group of genera with few species, probably closely related to the Saccharomycetaceae. The distinction between the two families is not sharp, and Lodder (1970) classified the genus *Endomycopsis* in each family with a question mark.

### KEY TO GENERA OF ENDOMYCETACEAE

1. Mycelium, but no blastospores . . . . . . . . . . . . . . . . . . . . . . . . . . . 2
1′. Mycelium and blastospores . . . . . . . . . . . . . . . . . . . . . . . . . . . . 3

    2(1) Mycelium falling apart into arthrospores . . . . . . . . . . . . . . **Endomyces**
        In *E. magnusii* (Guilliermond, 1909), *E. reessii* (van der Walt, 1959), and *E. candidum* (Butler and Petersen, 1970), gametangiogamy occurs, and the spores are oval. For *E. decipiens* (Guilliermond, 1909), no conjugation was described; the spores are hat-shaped. *Endomyces magnusii* has been isolated from slime flux of an oak, *E. reessii* from retting *Hibiscus cannabis*, *E. candidum* from soil in Puerto Rico, and *E. decipiens* from *Armillaria mellea*.

    2′(1) Mycelium not falling apart into arthrospores . . . . . . . . . . . **Eremascus**
        In the two species *E. albus* and *E. fertilis*, gametangiogamy precedes ascus formation. In *E. albus* the ascus contains 8 almost round spores (Harrold, 1950). Strains have been isolated from contaminated malt extract and mustard powder. Growth on malt agar is enhanced by the addition of 40% sucrose. In *E. fertilis*, young cells are plurinucleate, older cells are uninucleate (Guilliermond, 1909). Eight oval spores are formed per ascus. Strains of this species came from apple jelly and red currants.

3(1′) Erect ascophores bearing asci at the apex . . . . . . . . . . . . . . **Cephaloascus**
        In *C. fragrans* [syn. *Ascocybe grovesii* described by Wells (1954)], conjugation between a gametangium on a hyphal cell with an adjacent cell results in a diploid basal cell from which an erect ascophore arises with asci budding off at the top containing 4 hat-shaped spores (Dixon, 1959; Wilson, 1961; Schippers-Lammertse and Heyting, 1962). Anastomosis between hyphae also occurs. Conidiophores bear conidia in verticils. Strains have been isolated from decayed wood.

    FIG. 1.  *Endomycopsis platypodis*. Round, budding yeast cells and mycelial hyphae.
    FIG. 2.  *Candida parapsilosis*. Pseudomycelium.
    FIG. 3.  *Saccharomycodes ludwigii*. Yeast cells reproducing by bipolar budding.
    FIG. 4.  *Schizosaccharomyces octosporus*. Yeast cells reproducing by splitting.
    FIG. 5.  *Trichosporon cutaneum*. Mycelial hyphae breaking up into arthrospores. Blastospores. are present on the hyphae.
    FIG. 6.  *Rhodosporidium sphaerocarpum*. Dikaryotic mycelium with chlamydospores.
(Photomicrographs by D. C. Dijk.)

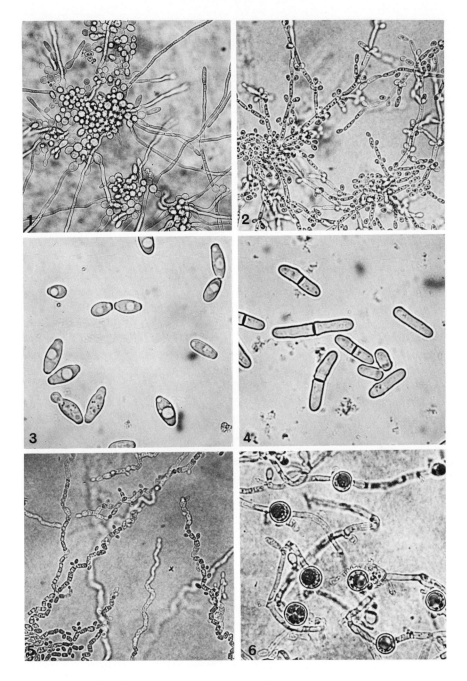

3′(1′) Ascophores bearing asci along the whole length . . . . . . . . . . **Endomycopsella**
   In the single species, *E. phaeospora* (Boedijn, 1960), isolated from fruits of *Flacourtia inermis* in Indonesia, no conjugation was observed.

## E. Family Saccharomycetaceae

Vegetative reproduction occurs by fission or by budding of single cells and, more rarely, by the formation of mycelium. Ascospores are of various shapes, but not needle- or spindle-shaped.

This family comprises most of the ascogenous yeasts. Vegetative reproduction is chiefly by budding of single cells, sometimes adhering in a pseudomycelium. Fission exclusively, as in *Schizosaccharomyces* (Fig. 4), or besides budding, as in *Endomycopsis* (Fig. 1), is exceptional. Two types of budding, multilateral and bipolar (Fig. 3), are distinguished. The moment of conjugation and diploidization in the life cycle varies. In some yeast species, germinating ascospores or the first cells of the haploid stage fuse and give a long diplophase; in others, conjugation between two cells immediately leads to ascus formation of the zygote and the haplophase is long. Transitions occur in which both the diploid and the haploid stage are of longer duration. As a consequence, a yeast culture may consist of diploid or haploid cells, or a mixture of both. Conjugation occurs between cells of the same or different mating type, of, respectively, homothallic or heterothallic strains. Besides isogamous conjugation of cells of the same shape, heterogamous conjugation of cells of different shape, e.g., mother cell and bud, may take place Wickerham (1958) found sexual agglutination, i.e., agglutination of cells of opposite mating type, in several species.

The shape of the ascospores among the Saccharomycetaceae varies considerably, and this is one of the main bases for the differentiation of genera. The spores may be round, oval, hemispherical, oblate-ellipsoidal, reniform, or sickle-shaped. They may have warts, ridges, or narrow or broad ledges which make them hat- or saturn-shaped. The presence of many spores may give the culture a pink or brown appearance. In some species the spores are readily liberated from the ascus upon maturation.

Fermentative ability is found in nearly all yeasts of the Saccharomycetaceae; its vigor varies. Assimilation of nitrate is considered an important characteristic for distinguishing between genera.

**KEY TO SUBFAMILIES AND GENERA OF SACCHAROMYCETACEAE**

1. Vegetative reproduction exclusively by fission . . . . . . . **Schizosaccharomycoideae**
   **Schizosaccharomyces**
   Cultures of *Schizosaccharomyces* spp. contain either single cells or mycelium, which may fall apart into arthrospores. Slooff (1970) recognized four species which all have a high maximum temperature for growth, namely 40°–41°C. Strains of *Schizosaccharomyces* spp. have been isolated from media with a high sugar content, e.g. molasses and cane sugar, and from grapes.

## 2. Yeasts and Related Fungi

1′. Vegetative reproduction exclusively by budding or by budding and fission . . . . . 2

    2(1′) Vegetative reproduction by bipolar budding . . . . . . . . . **Nadsonioideae** 3
    This subfamily (Lodder, 1970), based on bipolar budding, comprises genera which are in other respects rather different. It is therefore questionable whether they are closely related (Phaff, 1970; Kreger-van Rij and Veenhuis, 1971a).

    2′(1′) Vegetative reproduction by multilateral budding or by multilateral budding and fission . . . . . . . . . . . . . . . . . . . . . . . . . . . . . . 6

3(2) Ascospores cap-shaped . . . . . . . . . . . . . . . . . . . . . . . **Wickerhamia**
The single species, *W. fluorescens*, has oval asci which split up in the middle releasing generally one spore. The spores have a ledge which is broad at one side and narrow at the opposite side. A strain of this species has been isolated from dung of a wild squirrel in Japan.

3′(2) Ascospores not cap-shaped . . . . . . . . . . . . . . . . . . . . . . . 4

    4(3′) Conjugation of germinating spores in the ascus . . . . . . . **Saccharomycodes**
    The single species, *S. ludwigii*, has round spores with a narrow ledge. Strains of it come from grape must and wine.

    4′(3′) No conjugation of germinating spores in the ascus . . . . . . . . . . . . 5

5(4′) Ascospores round, warty and dark brown . . . . . . . . . . . . . . . **Nadsonia**
The two species, *N. elongata* and *N. fulvescens*, have been isolated from slime flux. Mother cell and bud conjugate and the two nuclei move into a second bud where, after karyogamy and meiosis, generally one spore develops. The composition of the cell wall is uncommon for ascomycetous yeasts (Phaff, 1970).

5′(4′) Ascospores round, hat-, helmet- or saturn-shaped, smooth or warty, not dark brown . . . . . . . . . . . . . . . . . . . . . . . . . . . . **Hanseniaspora**
Strains of all species have been isolated from grapes and other soft fruit and from *Drosophila*. The ascospores are hat-shaped in *H. valbyensis* and *H. guilliermondii*, round with an equatorial or subequatorial ledge in *H. uvarum*, and round and warty, but without a ledge, in *H. osmophila*. See *Kloeckera*, p. 26.

    6(2′) "Active buds" are formed which may conjugate or change directly into asci . . . . . . . . . . . . . . . . . . . . . . . . . . . **Lipomycetoideae**
    **Lipomyces**
    The peculiar way of ascus formation was first described by Starkey (1946), and later by Slooff (1970) who considered *Zygolipomyces* as a synonym. The spores are smooth, warty or have ridges. Strains of the four species have been isolated from soil.

    6′(2′) No formation of "active buds" . . . . . . . . . . . . **Saccharomycetoideae** 7

7(6′) Cells short-lived on malt agar; strong formation of acetic acid . . . . . . **Dekkera**
The two species described by van der Walt (1970a) are perfect forms of *Brettanomyces* spp. They have hat-shaped spores which are easily liberated from the ascus. Strains have been isolated from beer and wine.

7′(6′) Cells not short-lived on malt agar; no strong formation of acetic acid . . . . . . 8

    8(7′) Mycelium abundant . . . . . . . . . . . . . . . . . . . . . **Endomycopsis**
    Kreger-van Rij (1970) provisionally retained this genus, including 11 species, which forms a heterogeneous combination in the method of ascus formation and the shape of the spores. The spores may be hat-, saturn- or sickle-shaped, round or oval. Other authors (Wickerham, 1970; Boidin *et al.*, 1962, 1964; Abadie *et al.*, 1963) did not agree with this classification and have transferred species to the genera *Hansenula, Pichia, Guilliermondella*, and *Schwanniomyces*. Van der Walt and Scott (1971) changed the

name of the genus to *Saccharomycopsis*. Van der Walt (1972) and von Arx (1972) transferred several *Endomycopsis* species to new genera.

*Endomycopsis lipolytica*, described by Wickerham et al. (1970), is the perfect form of *Candida lipolytica*; it is heterothallic and assimilates hydrocarbons. Kreger-van Rij and Veenhuis (1969) observed septal pores surrounded by a swelling in *E. platypodis* and *E. monospora*. *Endomycopsis* strains have been isolated from various sources, including bark beetles, frass, tan liquid, and slime flux of trees.

 8'(7') Mycelium scarce or lacking . . . . . . . . . . . . . . . . . . . . . . 9

9(8') Assimilation of nitrate . . . . . . . . . . . . . . . . . . . . . . . . . . 10

9'(8') No assimilation of nitrate             12

 10(9) Ascospores round and warty . . . . . . . . . . . . . . . . . . **Citeromyces**
  *Citeromyces matritensis*, isolated from sugar and concentrated sugar solutions, is heterothallic and sexually agglutinative (Wickerham, 1958).

 10'(9) Ascospores not round and warty . . . . . . . . . . . . . . . . . . . . 11

11(10') Ascus thick-walled, ascospores hat-shaped . . . . . . . . . . . . . **Pachysolen**
 Strains of the single species, *P. tannophilus*, have been isolated from tanning liquors. The ascus is a cell from which a tube of variable length has grown out. The tip of the tube contains the protoplast with the spores. The wall of the ascus is very thick and refractile with the exception of the tip, which is thin-walled and from which the spores are easily liberated. Wickerham and Burton (1962) pointed to the similarity of *Pachysolen* and *Hansenula*. In both genera extracellular phosphomannans are produced.

11'(10') Ascus thin-walled, ascospores hat- or saturn-shaped, round or
hemispherical . . . . . . . . . . . . . . . . . . . . . . . . . . . . . . **Hansenula**
 Wickerham (1951, 1970) and Wickerham and Burton (1962) have made studies of this genus and especially of the phylogenetic relationships of the 25 species accepted by Wickerham. He recognized five phylogenetic lines with either homo- or heterothallic species and differing in habitat. Most species are associated with trees and bark beetles; other sources of isolation include fruit and soil. The spores are usually easily liberated from the ascus.

 12(9') No growth on malt agar under normal conditions . . . . . **Saccharomycopsis**
  The single species, *S. uniguttulatus*, occurs in the stomach of rabbits. It requires $CO_2$ in the atmosphere for growth (Richle and Scholer, 1961; Buecher and Phaff, 1970), and it can grow at a pH as low as 2 and a temperature of 30°–40°C. (Phaff, 1970). One to four oval or cylindrical spores are formed per ascus.

 12'(9') Growth on malt agar under normal conditions . . . . . . . . . . . . . 13

13(12') Ascospores oblate-ellipsoidal, dark brown . . . . . . . . . . . . . . **Wingea**
 The single species, *W. robertsii* (syn. *Pichia robertsii*), has been isolated from insects. The spores are not easily liberated from the ascus.

13'(12') Ascospores not oblate-ellipsoidal . . . . . . . . . . . . . . . . . . . . 14

 14(13') Ascospores round or oval with a ledge and warty . . . . . . . . . . . 15

 14'(13') Ascospores not round or oval with a ledge and warty . . . . . . . . . 16

15(14) Conjugation immediately preceding ascus formation . . . . . . . **Schwanniomyces**
 Phaff (1970) recognized four species, all from soil.

15'(14) Conjugation not immediately preceding ascus formation . . . . . . . . . **Pichia**

 16(14') Ascospores hat- or saturn-shaped . . . . . . . . . . . . . . . . . **Pichia**

Boidin et al. (1964, 1965a,b), Pignal and Boidin (1965), and Kreger-van Rij (1964, 1970) described this genus, which includes some 35 species. The ascospores are round, hat-, or saturn-shaped, occasionally warty. Most of them are easily liberated from the ascus. Both homo- and heterothallic species occur, the former generally exhibiting conjugation between mother cell and bud. Many species form pellicles on malt extract. Sources of isolation of *Pichia* spp. are exudates of trees, bark beetles, *Drosophila*, tanning liquid, fruit, and soil.

16'(14') Ascospores round, oval, reniform, or crescentiform . . . . . . . . . . . . 17

17(16') Ascospores reniform or crescentiform . . . . . . . . . . . . . . **Kluyveromyces**
Van der Walt (1956) described the genus for two species, *K. polysporus* and *K. africanus*. He originally supposed a connection between these species with large multispored asci and the genus *Dipodascus*, but later rejected this idea. Other species, originally described as *Saccharomyces*, *Fabospora*, *Zygofabospora*, and *Dekkeromyces* spp., were afterwards accepted in this genus in which van der Walt (1970a) now includes 18 species. Seven of them assimilate and occasionally ferment lactose, and strains of these have been isolated from milk and milk products. Other sources of isolation of *Kluyveromyces* strains are soil, *Drosophila*, slime flux of trees, and seawater. The spores are usually easily liberated from the ascus; they may be round or oval. Several species are interfertile.

17'(16') Ascospores round or oval . . . . . . . . . . . . . . . . . . . . . . 18

18(17') Ascospores warty . . . . . . . . . . . . . . . . . . . . . . . 19

18'(17') Ascospores not warty . . . . . . . . . . . . . . . . . . . . . . 22

19(18) Ascospores oval . . . . . . . . . . . . . . . . . . . . . . **Debaryomyces**
The species of this genus are haploid, and conjugation between mother cell and bud precedes ascus formation. The spores are usually round. Kreger-van Rij (1970) recognized eight species. Van der Walt (1970a) accepted *D. globosus*, the original type of the genus, in *Saccharomyces* (*S. kloeckerianus*). Most strains of *D. hansenii* can assimilate nitrite and creatine, and they have a high salt tolerance. Sources of isolation are cheese, sausages, brined food, and fermenting tobacco. Other species have been isolated from slime flux and soil.

19'(18) Ascospores round . . . . . . . . . . . . . . . . . . . . . . 20

20(19') Fermentation vigorous . . . . . . . . . . . . . . . . . . . . . . **Saccharomyces**
Van der Walt (1970a) recognized 41 species in this genus indicated as *Saccharomyces sensu lato*. Four groups may be distinguished among them. The first one, *Saccharomyces sensu stricto*, includes diploid species, like *S. cerevisiae*. The former genera, *Zygosaccharomyces* and *Torulaspora*, constitute the second and third groups, respectively. Both are mainly haploid. In *Zygosaccharomyces*, conjugation between mother cell and bud often precedes ascus formation. The *Torulaspora* cells produce protuberances on special media. Most *Torulaspora* species have warty spores. According to van der Walt, the fourth group is phylogenetically heterogeneous. In *Saccharomyces*, homo- and heterothallic species occur.

20'(19') Fermentation not vigorous; it may be absent . . . . . . . . . . . . . . 21

21(20') Conjugation immediately preceding ascus formation **Debaryomyces** (see 19 above)

21'(20') No conjugation immediately preceding ascus formation . . . . . . . . . . **Pichia**

22(18') Ascospores oval . . . . . . . . . . . . . . . . . . . . . . 23

22'(18') Ascospores round . . . . . . . . . . . . . . . . . . . . . . 25

23(22) Fermentation slow . . . . . . . . . . . . . . . . . . . . . . . . . . . . **Lodderomyces**
        Strains of the single species, *L. elongisporus*, have been isolated from orange concentrate and from soil. Generally one oval spore is formed per ascus.

23′(22) Fermentation vigorous . . . . . . . . . . . . . . . . . . . . . . . . . . . 24

    24(23′) Spores readily liberated from the ascus . . . **Kluyveromyces** (see 17 above)

    24′(23′) Spores not readily liberated from the ascus . **Saccharomyces** (see 20 above)

25(22′) Fermentation vigorous . . . . . . . . . . . . . . . . . . . . . . . . . . 26

25′(22′) Fermentation weak, slow or absent . . . . . . . . . . . . . . . . . . . **Pichia**

    26(25) Spores readily liberated from the ascus . . . . **Kluyveromyces** (see 17 above)

    26′(25) Spores not readily liberated from the ascus . . . . . . . . . . . . . 27

27(26′) Pellicle formation on malt extract . . . . . . . . . . . **Pichia** (see 15 above)

27′(26′) No pellicle formation on malt extract . . . . . . **Saccharomyces** (see 20 above)

## III. BASIDIOMYCETOUS YEASTS

Budding yeast cells, mycelium with or without clamp connections, and chlamydospores (Fig. 6) are present. Sporidia are formed laterally or terminally on a promycelium from a germinated chlamydospore, or terminal basidiospores are formed on a basidium. Ballistospores may occur. Cells are hyaline, red, or orange.

This group is still small, and the genera included belong to different families. However, since their classification is not always certain, we have refrained from giving a subdivision.

Kluyver and van Niel (1925) presumed a relationship of the genus *Sporobolomyces* with the basidiomycetes based on the resemblance of the discharge of ballistospores in this genus with that of basidiospores in most basidiomycetes. However, convincing evidence for the occurrence of a sexual cycle in yeasts resembling that of fungi belonging to the Heterobasidiomycetes was only given by Nyland (1949) for *Sporidiobolus* and by Banno (1963, 1967) for *Rhodosporidium*. Banno's work has been followed by similar investigations which have resulted in the description of the new genera *Leucosporidium* by Fell *et al.* (1969) and *Aessosporon* by van der Walt (1970b). These authors have given schemes for the alternation of haploid and diploid stages based on crossing experiments, nuclear staining, and quantitative determination of DNA per cell. The three schemes are mentioned below. More than one may occur in a single species. They are

1. Haploid yeast cells of opposite mating type fuse to produce a dikaryotic mycelium with clamp connections. Chlamydospores (teliospores) arise on the mycelium, and in these diploidization occurs. Meiosis takes place upon germination of the chlamydospores with a promycelium from which haploid yeast cells, the sporidia, of different mating type are formed by budding.

2. Yeast cells budding from the promycelium do not conjugate, but from

each of them the whole cycle may develop again (self-sporulating yeast cells). It is presumed that meiosis, upon germination of the chlamydospores, has failed and that the yeast cells are diploid. It is not known how a dikaryotic mycelium develops from diploid cells.

3. No dikaryotic mycelium is formed, but chlamydospores are on uninucleate mycelium without clamp connections. The yeast cells are self-sporulating, and meiosis is supposed to occur in the chlamydospores. In none of these schemes the ballistospores are considered to be basidiospores as proposed by Sainclivier (1952).

*Filobasidium* (Olive, 1968) is a fifth genus belonging to the basidiomycetous yeasts. Large basidia with sessile basidiospores, which are not shot off, are formed on the dikaryotic mycelium with clamp connections. Von Arx (1971, personal communication) found that the species *Leucosporidium capsuligenum* belongs to this genus.

Banno (1967) assigned *Rhodosporidium* to the Ustilaginaceae (or Ustilaginales); Fell *et al.* (1969) classified *Leucosporidium* in the Ustilaginales, but not yet in a family; van der Walt (1970b) placed *Aessosporon* in the Tilletiaceae (Ustilaginales). Nyland (1949) included *Sporidiobolus* in the Heterobasidiomycetes without further classification, but Lodder (1970) retained this genus in the Sporobolomycetaceae, a family of the Fungi Imperfecti. Martin (1963) considered the Sporobolomycetaceae as imperfect Tremellales. Olive (1968) classified *Filobasidium* in a new family (Filobasidiaceae) of the Ustilaginales.

Of the above genera, *Sporidiobolus* shows the greatest variety in cell form: budding yeast cells, ballistospores, dikaryotic mycelium with clamp connections, and chlamydospores. Loss of one or more of these stages might yield organisms which can be identified as *Rhodotorula*, *Sporobolomyces*, or *Rhodosporidium* species. Therefore, a close relationship between these genera seems probable. *Leucosporidium* comprises perfect forms of *Candida* spp. and *Aessosporon* of *Sporobolomyces salmonicolor*. So far, in the genus *Bullera* no perfect stage has been described, but Shadomy (1970) observed mycelium with clamp connections in strains of *Cryptococcus neoformans*. In this connection, it is of interest to note that Slodki *et al.* (1966) found similar physiological properties and extracellular polysaccharides in *Cryptococcus laurentii* and some *Tremella* species. Furthermore, *C. laurentii* closely resembles *Bullera alba*, with the exception of ballistospore formation. In *Trichosporon cutaneum*, which in its physiological properties shows resemblance to *Cryptococcus* spp., Kreger-van Rij and Veenhuis (1971b) observed septal pores of the basidiomycete type. According to Bracker (1967), these are found in the Tremellales, but not in the Ustilaginales. On the other hand, Kreger-van Rij and Veenhuis (1971c) observed this type of pore in *Leucosporidium capsuligenum*, now assigned to *Filobasidium* and considered to belong to the Ustilaginales. *Filobasidium floriforme* also has dolipores. Its haploid form agrees with *Cryptococcus albidus*.

### KEY TO GENERA OF BASIDIOMYCETOUS YEASTS

1. Ballistospores formed . . . . . . . . . . . . . . . . . . . . . . . . . . 2
1'. Ballistospores not formed . . . . . . . . . . . . . . . . . . . . . . . . 3

    2(1) Mycelium with clamp connections . . . . . . . . . . . . . . . . **Sporidiobolus**
    The two species in this genus, *S. johnsonii* and *S. ruinenii*, have been isolated from leaves. These yeasts are red or orange, and they follow cycle 2 (p. 22). Laffin and Cutter (1959a,b) assume that meiosis occurs in the yeast cells and not upon germination of the chlamydospores.

    2'(1) Mycelium without clamp connections . . . . . . . . . . . . . . . . **Aessosporon**
    The life cycle of the single species, *A. salmonicolor*, was described by van der Walt (1970b). It follows cycle 3, but on germination of the chlamydospore the promycelium is not septate and yeast cells are formed at the tip only.

3(1') Cells red or orange . . . . . . . . . . . . . . . . . . . . . . . . **Rhodosporidium**
    The four species described can be differentiated by the shape of the chlamydospores. Most of the strains have been isolated from marine sources. *R. toruloides* exhibits cycles 1 and 2, *R. sphaerocarpum*, 1 and 3 (Fell *et al.*, 1970), and *R. diobovatum*, cycle 1 (Newell and Hunter, 1970). They may be considered as perfect forms of *Rhodotorula glutinis*. The fourth species, *Rhodosporidium malvinellum*, has been described by Fell and Hunter (cf. Fell, 1970b).

3'(1') Cells hyaline . . . . . . . . . . . . . . . . . . . . . . . . . . . . . 4

    4(3') Basidiospores terminal on basidia . . . . . . . . . . . . . . . **Filobasidium**
    Olive (1968) described the first species of this genus, *F. floriforme*, isolated from *Eryanthus giganteus*. The 5 to 8 basidiospores formed on the long erect basidia have a flowerlike appearance. The species described as *Leucosporidium capsuligenum* from cider and sake-moto also belongs to this genus. It has a weak fermentative ability. Both *Filobasidium* species are heterothallic.

    4'(3') Sporidia lateral on a promycelium . . . . . . . . . . . . . . **Leucosporidium**
    Most strains of this genus have been isolated from Antarctic soil and marine waters. They generally have a low maximum temperature of growth, namely ±19°C. The type species, *L. scottii*, is heterothallic, and the sexual cycle represented by schemes 1 and 3 are found in it. Five other *Leucosporidium* spp. accepted by Fell and Phaff (1970) only show the scheme 3 type.

## IV. YEASTS CLASSIFIED IN THE DEUTEROMYCOTINA (BLASTOMYCETES)

There are two families, the Sporobolomycetaceae, characterized by the formation of ballistospores, and the Cryptococcaceae, lacking this characteristic.

### A. *Family Sporobolomycetaceae*

Budding yeast cells and/or mycelium occur with or without clamp connections. Pseudomycelium may occur. Ballistospores are present.

The yeasts of this family, defined by Derx (1930), are considered to be imperfect Heterobasidiomycetes. The genera *Tilletiopsis* and *Itersonilia*, in the description by Derx (1948), have no budding cells and are therefore not included in the yeasts. Since they are classified in the same family with the yeast genera *Sporobolomyces* and *Bullera* they are discussed here. Sowell and Korf (1960) considered *Sporidiobolus* and *Tilletiopsis* as possible synonyms of *Itersonilia*. Strains of all species in this family have been isolated from plant materials and especially from leaves infected with rust and smut.

**KEY TO GENERA OF SPOROBOLOMYCETACEAE**

1. Budding cells formed . . . . . . . . . . . . . . . . . . . . . . . . . . . . 2
1'. No budding cells formed . . . . . . . . . . . . . . . . . . . . . . . . . 3
    2(1) Ballistospores symmetrical; cultures cream-colored to yellowish . . . . **Bullera**
    There are four species of which *B. alba* is the type.

    2'(1) Ballistospores generally asymmetrical; cultures red or pink, exceptionally cream-colored . . . . . . . . . . . . . . . . . . . . . . . . . **Sporobolomyces**
    Phaff (1970) accepted nine species. Mycelium without clamp connections may be formed. The ballistospores are kidney- or sickle-shaped.

3(1') Mycelium with clamp connections . . . . . . . . . . . . . . . . . . **Itersonilia**
    Three species have been described, of which the first one is *I. perplexans*. Sowell and Korf (1960) isolated a mono- and a dikaryotic phase from this species; the first could be obtained from the second, but not vice versa. They observed chlamydospores in both phases. Since a complete life cycle was not yet described for this genus, it is provisionally retained in the Sporobolomycetaceae. Tubaki (1952b) and Olive (1952) both studied *Itersonilia* spp. The ballistospores in this genus are kidney-shaped.

3'(1') Mycelium without clamp connections . . . . . . . . . . . . . . . . . **Tilletiopsis**
    The ballistospores are sickle-shaped. Nyland (1950) described *T. washingtonensis* and *T. minor*. He observed hyaline chlamydospores. Tubaki (1952a) also studied *Tilletiopsis* spp.

## B. Family *Cryptococcaceae*

Budding cells are present. Pseudomycelium (Fig. 2), true mycelium, and arthrospores (Fig. 5) may be formed. Cultures are cream colored, yellow, orange, or red.

This family comprises imperfect forms of ascomycetous and basidiomycetous yeasts. The differentiation in genera is mainly based on morphological and, exceptionally, on physiological properties. Among the latter, the assimilation of inositol very often correlates with the capacity to form starch-like compounds in the capsule. Lack of features of sexual reproduction often impedes a satisfactory classification and results in some very heterogeneous genera.

## KEY TO GENERA OF CRYPTOCOCCACEAE

1. Monopolar budding . . . . . . . . . . . . . . . . . . . . . . . . . . . . **Pityrosporum**
Slooff (1970) recognized three species, two of which, *P. ovale* and *P. orbiculare*, require lipids for growth. They occur on the skin of man and animals. Keddie (1968) considered *Pityrosporum* as a synonym of *Malassezia*. *Malassezia furfur* has a hyphal stage. In *Pityrosporum*, the successive buds leave a collar on the mother cell around the scar plug.

1'. No monopolar budding . . . . . . . . . . . . . . . . . . . . . . . . . . . . 2

    2(1') Bipolar budding . . . . . . . . . . . . . . . . . . . . . . . . . . . . 3

    2'(1') No bipolar budding . . . . . . . . . . . . . . . . . . . . . . . . . . 4

3(2) Fermentative ability . . . . . . . . . . . . . . . . . . . . . . . . . . . **Kloeckera**
Like the yeasts of the genus *Hanseniaspora*, which is the perfect form of *Kloeckera*, most strains have been isolated from fruit and soil. Phaff (1970) accepted four species.

3'(2) No fermentative ability . . . . . . . . . . . . . . . . . . . . . . **Schizoblastosporion**
Strains of the single species, *S. starkeyi-henricii*, have been isolated from soil.

    4(2') Buds formed on stalks . . . . . . . . . . . . . . . . . . . **Sterigmatomyces**
Fell (1966, 1970a) described this genus. A septum is formed in the middle of the stalk separating mother cell and bud. The lamellar structure of the cell wall observed in 2 species resembles that of basidiomycetous and related yeasts; the same holds for bud formation, but the neck of the bud is often very long (Kreger-van Rij and Veenhuis, 1971c). Most strains of *S. halophilus* and *S. Indicus* have been isolated from seawater. *S. elviae* (Sonck and Yarrow, 1969) comes from a human skin infection, and *S. polyborus* (Scott and van der Walt, 1970) from insect frass.

    4'(2') Buds not formed on stalks . . . . . . . . . . . . . . . . . . . . . . . . 5

5(4') Triangular cells occur besides oval ones . . . . . . . . . . . . . . . **Trigonopsis**
*T. variabilis* is the single species. Young cells are oval; older cells are very often triangular. Strains have been isolated from beer and grapes.

5'(4') No triangular cells present . . . . . . . . . . . . . . . . . . . . . . . . 6

    6(5') Most cells lunate . . . . . . . . . . . . . . . . . . . . . . . . . **Selenotila**
Krassilnikov (cf. Kursanova *et al.*, 1954) and Yarrow (1969) each described a species, *S. intestinalis* from the intestines of a butterfly and *S. peltata* from a case of bovine mastitis, respectively.

    6'(5') No lunate cells formed . . . . . . . . . . . . . . . . . . . . . . . . . 7

7(6') Slow growth on malt agar of short-lived cells, generally with strong acetic acid production . . . . . . . . . . . . . . . . . . . . . . . . . . . . . . . . . **Brettanomyces**
Van der Walt (1970a) accepted 7 species, most strains of which had been isolated from beer and wine. The genus is mainly characterized by physiological properties. Ogive-shaped cells may occur.

7'(6') No slow growth on malt agar of short-lived cells which produce acetic acid abundantly . . . . . . . . . . . . . . . . . . . . . . . . . . . . . . . . . 8

    8(7') Formation of true mycelium . . . . . . . . . . . . . . . . . . . . . . . 9

    8'(7') No formation of true mycelium . . . . . . . . . . . . . . . . . . . . . 11

9(8) Formation of arthrospores . . . . . . . . . . . . . . . . . . . . . . **Trichosporon**
Do Carmo-Sousa (1970) recognized three groups among the 7 species described. One

of the groups, including *T. capitatum* and *T. fermentans*, seems to be related to the ascogenous yeasts. *Trichosporon cutaneum* of the second group is probably related to the Basidiomycetes. Do Carmo-Sousa reported the occurrence of asexual endospores in this species. Strains of it have been isolated from white piedra and skin lesions.

9'(8) No formation of arthrospores . . . . . . . . . . . . . . . . . . . . . . . . 10

    10(9') Formation of asexual endospores . . . . . . . . . . . . . . **Oosporidium**
    Do Carmo-Sousa (1970) recognized one species, *O. margaritiferum*, isolated from slime flux of trees. Growth is very slow, and a pink or orange-yellow pigment is formed.

    10'(9') No formation of asexual endospores . . . . . . . . . . . . . . . . **Candida**

11(8') Formation of pseudomycelium . . . . . . . . . . . . . . . . . . . . . . 12

11'(8') Pseudomycelium not formed or rudimentary . . . . . . . . . . . . . . . . 13

    12(11) Cultures red or orange . . . . . . . . . . . . . . . . . . . . **Rhodotorula**

    12'(11) Cultures not red or orange . . . . . . . . . . . . . . . . . . . **Candida**
    This genus comprises imperfect forms of ascomycetous and basidiomycetous yeasts of various genera. Van Uden and Buckley (1970) recognized 81 species, mainly differentiated by physiological properties, and several new species have been described since then. *Candida albicans* is a pathogen, generally recognized by the formation of chlamydospores. Van der Walt (1967, 1969) described a sexual cycle for a strain of *C. albicans* which he transferred to the genus *Syringospora*, considered to belong to the Tulasnellaceae (Tremellales, Heterobasidiomycetes). However, some features of *C. albicans*, such as the GC content of DNA (Nakase and Komagata, 1968; Stenderup and Leth Bak, 1968), and the structure of the cell wall, leave doubt about the correctness of this classification.

13(11') Inositol assimilated . . . . . . . . . . . . . . . . . . . . . . **Cryptococcus**
Phaff and Fell (1970) accepted 17 species, most of which have capsulated cells and produce starchlike compounds. The cultures are often mucous; some of them have a red or orange color. The difference between *Cryptococcus* and *Rhodotorula* lies in the assimilation of inositol (Phaff and Spencer, 1969). *Cryptococcus neoformans* is pathogenic for man. Gordon and Devine (1970) observed endogenous spore formation in a filamentous variant of this species; the spores could not be stained with malachite green. Shadomy (1970) found mycelium with clamp connections in some strains, which strongly suggests the relation of *C. neoformans* to the Heterobasidiomycetes.

13'(11') Inositol not assimilated . . . . . . . . . . . . . . . . . . . . . . . . . 14

    14(13') Cultures cream-colored . . . . . . . . . . . . . . . . . . . . **Torulopsis**
    Van Uden and Vidal-Leiria (1970) recognized 36 species, most of them with fermentative ability. Like *Candida*, *Torulopsis* is heterogeneous, comprising imperfect forms of at least 5 different ascosporogenous genera. Moreover, the distinction from *Candida* is not always clear-cut.

    14'(13') Cultures red, orange or yellow . . . . . . . . . . . . . . . . **Rhodotorula**
    Phaff and Ahearn (1970) accepted 9 species, most of which have capsulated cells and mucous cultures, but which do not produce starchlike compounds. Hasegawa *et al.* (1960) gave a definition of the genus different from the one considered here. Some *Rhodotorula* strains produce both intracellular and extracellular lipids (Stodola *et al.*, 1967). *Rhodotorula glutinis* is the imperfect form of *Rhodosporidium* species and probably also of *Sporidiobolus* species.

## REFERENCES

Abadie, F., M. C. Pignal, and J. L. Jacob. (1963). Les levures à spores verruqueuses. *Bull. Soc. Mycol. Fr.* **79**:16–70.

Banno, I. (1963). Preliminary report on cell conjugation and mycelial stage in *Rhodotorula* yeasts. *J. Gen. Appl. Microbiol.* **9**:249–251.

Banno, I. (1967). Studies on the sexuality of *Rhodotorula. J. Gen. Appl. Microbiol.* **13**:167–196.

Bartnicki-Garcia, S. (1970). Cell wall composition and other biochemical markers in fungal phylogeny. *In* "Phytochemical Phylogeny, Proceedings of the Phytochemical Society, Symposium, 1969" (J. B. Harborne, ed.), pp. 81–103. Academic Press, New York.

Batra, L. R. (1959). A comparative morphological and physiological study of the species of *Dipodascus. Mycologia* **51**:329–355.

Batra, L. R., and H. Francke-Grosmann, (1960). Contributions to our knowledge of ambrosia fungi. I *Ascoidea hylecoeti* sp. nov. (Ascomycetes). *Amer. J. Bot.* **48**:453–456.

Batra, L. R., and H. Francke-Grosmann. (1964). Two new ambrosia fungi—*Ascoidea asiatica* and *A. africana. Mycologia* **56**:632–636.

Bessey, E. A. (1965). "Morphology and Taxonomy of Fungi." Hafner, New York.

Biggs, R. (1937). *Dipodascus uninucleatus. Mycologia* **29**:34–44.

Boedijn, K. B. (1960). On a new genus of the Endomycetaceae. *Mycopathol. Mycol. Appl.* **12**:163–167.

Boidin, J., F. Abadie, J. L. Jacob, and M. C. Pignal. (1962). Les levures à spores réniformes. *Bull. Soc. Mycol. Fr.* **78**:155–203.

Boidin, J., M. C. Pignal, Y. Lehodey, A. Vey, and F. Abadie. (1964). Le genre *Pichia sensu lato* (Première contribution). *Bull. Soc. Mycol. Fr.* **80**:396–438.

Boidin, J., F. Abadie, and Y. Lehodey. (1965a). Le genre *Pichia sensu lato* (Deuxième contribution). *Bull. Soc. Mycol. Fr.* **81**:5–23.

Boidin, J., M. C. Pignal, and M. Besson. (1965b). Le genre *Pichia sensu lato* (Quatrième contribution). *Bull. Soc. Mycol. Fr.* **81**:566–606.

Bracker, C. E. (1967). Ultrastructure of fungi. *Annu. Rev. Phytopathol.* **5**:343–374.

Buecher, E. J., and H. H. Phaff. (1970). Growth of *Saccharomycopsis* Schiönning under continuous gassing. *J. Bacteriol.* **104**:133–137.

Butler, E. E., and L. J. Petersen. (1970). Sexual reproduction in *Geotrichum candidum. Science* **169**:481–482.

Cain, R. F. (1948). *Myriogonium*, a new genus among the simplified Ascomycetes. *Mycologia* **40**:158–167.

Derx, H. G. (1930). Etude sur les Sporobolomycètes. *Ann. Mycol.* **28**:1–23.

Derx, H. G. (1948). *Itersonilia*, nouveau genre de Sporobolomycètes à mycélium bouclé. *Bull. Bot. Gard. Buitenz.* [3] **17**:465–472.

Dixon, P. A. (1959). Life-history and cytology of *Ascocybe grovesii* Wells. *Ann. Bot. (London)* [N.S.] **23**:509–520.

do Carmo-Sousa, L. (1969). Distribution of yeasts in nature. *In* "The Yeasts" (A. H. Rose and J. S. Harrison, eds.), Vol. I, pp. 79–105. Academic Press, New York.

do Carmo-Sousa, L. (1970). *In* "The Yeasts" (J. Lodder, ed.) pp. 440–447, 1161–1166, 1309–1352. North-Holland Publ.. Amsterdam.

Fell. J. W. (1966). *Sterigmatomyces*, a new fungal genus from marine areas. *Antonie van Leeuwenhoek; J. Microbiol. Serol.* **32**:99–104.

Fell, J. W. (1970a). *In* "The Yeasts" (J. Lodder, ed.) pp. 1229–1234. North-Holland Publ., Amsterdam.

Fell, J. W. (1970b). Yeasts with heterobasidiomycetous life cycles. *In* "Recent Trends in Yeast Research" (D. G. Ahearn, ed.), pp. 49–66. Georgia State Univ., Atlanta.

Fell, J. W., and H. J. Phaff. (1970). *In* "The Yeasts" (J. Lodder, ed.) pp. 776–802. North-Holland Publ., Amsterdam.
Fell, J. W., A. C. Statzell, I. L. Hunter, and H. J. Phaff, (1969). *Leucosporidium*, gen. n., the heterobasidiomycetous stage of several yeasts of the genus *Candida*. *Antonie van Leeuwenhoek; J. Microbiol. Serol.* **35**:433–462.
Fell, J. W., H. J. Phaff, and S. Y. Newell. (1970). *In* "The Yeasts" (J. Lodder, ed.) pp. 803–814. North-Holland Publ., Amsterdam.
Gäumann, E. A. (1964). "Die Pilze." Birkhaeuser, Basel.
Gentles, J. C., and C. J. La Touche. (1969). Yeasts as human and animal pathogens. *In* "The Yeasts" (A. H. Rose and J. S. Harrison, eds.), Vol. 1, pp. 107–182. Academic Press, New York.
Gordon, M. A., and J. Devine. (1970). Filamentation and endogenous sporulation in *Cryptococcus neoformans*. *Sabouraudia* **8**:227–234.
Guilliermond, A. (1909). Recherches cytologiques et taxonomiques sur les Endomycétées. *Rev. Gen. Bot.* **21**:353–401.
Guilliermond, A. (1928). Recherches sur quelques ascomycètes inférieurs isolés de la stigmatomycose des graines de cotonnier. Essai sur la phylogénie des Ascomycètes. *Rev. Gen. Bot.* **40**:328–342, 397–414, 474–485, 555–574, 606–624, 690–704.
Guilliermond, A. (1935). Sur un champignon nouveau, parasite des capsules du cotonnier, l'*Eremothecium ashbyi* et ses relations possibles avec le *Spermophthora gossypii* et les Ascomycètes. *C. R. Acad. Sci.* **200**:1556–1558.
Guilliermond, A. (1936). L'*Eremothecium ashbyi*, nouveau champignon parasite des capsules du cotonnier. *Rev. Mycol.* [N.S.] **1**:115–156.
Guilliermond, A. (1937). "La sexualité, le cycle de développement, la phylogénie et la classification des levures." Masson, Paris.
Harrold, C. E. (1950). Studies in the genus *Eremascus*. I. The rediscovery of *Eremascus albus* Eidam and some new observations concerning its life-history and cytology. *Ann. Bot. (London)* [N.S.] **14**:127–148.
Hasegawa, T., I. Banno, and S. Yamauchi. (1960). A taxonomic study on the genus *Rhodotorula*. I. The subgenus *Rubrotorula* nov. subgen. II. The subgenus *Flavotorula* nov. subgen. *J. Gen. Appl. Microbiol.* **5**:200–212; **6**:196–215.
Keddie, F. M. (1968). Tinea versicolor: The electron microscopic morphology of the genera *Malassezia* and *Pityrosporum*. *Proc. Int. Congr. Dermatol., 13th, 1967* pp. 867–872.
Kluyver, A. J., and C. B. van Niel. (1925). Uber Spiegelbilder erzeugende Hefearten und die neue Hefengattung *Sporobolomyces*. *Zentralbl. Bakteriol. Parasitenk. Infektionskr.*, **263**:1–20.
Korf, R. P. (1957). *Dipodascus albidus* forma minor. *Sydowia, Beih.* **1**:285–288.
Kreger-van Rij, N. J. W. (1964). A taxonomic study of the yeast genera *Endomycopsis*, *Pichia* and *Debaryomyces*. Thesis; University of Leiden.
Kreger-van Rij, N. J. W. (1970). *In* "The Yeasts" (J. Lodder, ed.), pp. 129–156, 166–208, 455–554, North-Holland Publ., Amsterdam.
Kreger-van Rij, N. J. W., and M. Veenhuis. (1969). Septal pores in *Endomycopsis platypodis* and *Endomycopsis monospora*. *J. Gen. Microbiol.* **57**:91–96.
Kreger-van Rij. N. J. W., and M. Veenhuis. (1971a). Bipolar budding in yeasts—an electron microscope study. *Antonie van Leeuwenhoek; J. Microbiol. Serol.* **37**:119–130.
Kreger-van Rij, N. J. W., and M. Veenhuis. (1971b). Septal pores in *Trichosporon cutaneum*. *Sabouraudia* **9**:36–38.
Kreger-van Rij. N. J. W., and M. Veenhuis. (1971c). A comparative study of the cell wall of basidiomycetous and related yeasts. *J. Gen. Microbiol.* **68**:87–95.
Kudriavzev, V. I. (1960). "Die Systematik der Hefen." Akademie-Verlag, Berlin.

Kursanova, L. I., N. A. Naumov, N. A. Krassilnikov, and M. V. Gorlenko. (1954). *In* "Opredelitel Nizshikh Rasteni" ("Determination of Lower Plants") (L. I. Kursanova, ed.), Vol. 3, p. 143. State Publ. House "Soviet Science," Moscow.

Laffin, R. J., and V. M. Cutter. (1959a). Investigations on the life cycle of *Sporidiobolus johnsonii*. I. Irradiation and cytological studies. *J. Elisha Mitchell Sci. Soc.* **75**:89–96.

Laffin, R. J., and V. M. Cutter. (1959b). Investigations on the life cycle of *Sporidiobolus johnsonii*. II. Mutants and micromanipulation. *J. Elisha Mitchell Sci. Soc.* **75**:97–100.

Lodder, J. (1970). *In* "The Yeasts" (J. Lodder, ed.), pp. 1–33. North-Holland Publ., Amsterdam.

Manuel, J. (1938). Sur la formation de l'asque de *Nematospora coryli*, après un phénomène sexuel. *C. r. Acad. Sci.* **207**:1241–1243.

Martin, G. W. (1963). Key to the families of Fungi. *In* "Ainsworth and Bisby's Dictionary of the Fungi" (by G. C. Ainsworth) 5th ed., pp. 497–517. Commonwealth Mycol. Inst., Kew, Surrey, England.

Miller, M. W., and N. van Uden. (1970). *In* "The Yeasts" (J. Lodder, ed.) pp. 408–429. North-Holland Publ., Amsterdam.

Nakase, T., and K. Komagata. (1968). Taxonomic significance of base composition of yeast DNA. *J. Gen. Appl. Microbiol.* **14**:345–357.

Newell, S. Y., and I. L. Hunter. (1970). *Rhodosporidium diobovatum* sp. n., the perfect form of an asporogenous yeast (*Rhodotorula* sp.). *J. Bacteriol.* **104**:503–508.

Novák, E. K., and J. Zsolt. (1961). A new system proposed for yeasts. *Acta Bot. Acad. Sci. Hung.* **7**:93–145.

Nyland, G. (1949). Studies on some unusual heterobasidiomycetes from Washington State. *Mycologia* **41**:686–701.

Nyland, G. (1950). The genus *Tilletiopsis*. *Mycologia* **42**:487–496.

Olive, L. S. (1952). Studies on the morphology and cytology of *Itersonilia perplexans* Derx. *Bull. Torrey Bot. Club* **79**:126–138.

Olive, L. S. (1968). An unusual new heterobasidiomycete with *Tilletia*-like basidia. *J. Elisha Mitchell Sci. Soc.* **84**:261–266.

Phaff, H. J. (1970). *In* "The Yeasts" (J. Lodder, ed.) pp. 430–439, 725–732, 756–766, 831–862, 1146–1160. North-Holland Publ., Amsterdam.

Phaff, H. J. (1971). Structure and biosynthesis of the yeast cell envelope. *In* "The Yeasts" (A. H. Rose and J. S. Harrison, eds.), Vol. 2, pp. 135–210. Academic Press, New York.

Phaff, H. J., and D. G. Ahearn. (1970). *In* "The Yeasts" (J. Lodder, ed.) pp. 1187–1223. North-Holland Publ., Amsterdam.

Phaff, H. J., and J. W. Fell. (1970). *In* "The Yeasts" (J. Lodder, ed.), pp. 1088–1145. North-Holland Publ., Amsterdam.

Phaff, H. J., and J. F. T. Spencer. (1969). Improved parameters in the separation of species in the genera *Rhodotorula* and *Cryptococcus*. *Proc. Int. Symp. Yeasts, 2nd, 1966* pp. 59–65.

Phaff, H. J., M. W. Miller, and E. M. Mrak. (1966). "The Life of Yeasts." Harvard Univ. Press, Cambridge, Massachusetts.

Pignal, M. C., and J. Boidin. (1965). Le genre *Pichia sensu lato* (Troisième contribution). *Bull. Soc. Mycol. Fr.* **81**:197–226.

Pitt, J. I., and M. W. Miller. (1968). Sporulation in *Candida pulcherrima*, *Candida reukaufii* and *Chlamydozyma* species: Their relationships with *Metschnikowia*. *Mycologia* **60**:663–685.

Pitt, J. I., and M. W. Miller. (1970). The parasexual cycle in yeasts of the genus *Metschnikowia*. *Mycologia* **62**:462–473.

Pridham, T. G., and K. B. Raper. (1950). *Ashbya gossypii*—its significance in nature and in the laboratory. *Mycologia* **42**:603–623.

Richle, R., and H. J. Scholer. (1961). *Saccharomycopsis guttulata* vom Kaninchen: Kulturelle Eigenschaften und mögliche Bedeutung. *Pathol. Microbiol.* **24**:783–793.

Routien, J. B. (1949). *Crebrothecium ashbyi. Mycologia* **41**:183–185.
Sainclivier, M. (1952). Caryologie des *Sporobolomyces. Bull. Soc. Bot. Fr.* **99**:147–149.
Schippers-Lammertse, A. F., and C. Heyting. (1962). Physiological properties, conjugation and taxonomy of *Cephaloascus fragrans* Hanawa 1920 (syn.: *Ascocybe grovesii* Wells 1954). *Antonie van Leeuwenhoek; J. Microbiol. Serol.* **28**:5–16.
Scott, De. B., and J. P. van der Walt. (1970). Three new yeasts from South African insect sources. *Antonie van Leeuwenhoek; J. Microbiol. Serol.* **36**:389–396.
Shadomy, H. J. (1970). Clamp connections in two strains of *Cryptococcus neoformans. In* "Recent Trends in Yeast Research" (D. G. Ahearn, ed.), pp. 67–72. Georgia State Univ. Atlanta.
Slodki, M. E., L. J. Wickerham, and R. J. Bandoni. (1966). Extracellular heteropolysaccharides from *Cryptococcus* and *Tremella*: A possible taxonomic relationship. *Can. J. Microbiol.* **12**:489–494.
Slooff, W. C. (1970). *In* "The Yeasts" (J. Lodder, ed.), pp. 379–402, 733–755, 1167–1186. North-Holland Publ., Amsterdam.
Sonck, C. E., and D. Yarrow. (1969). Two new yeast species isolated in Finland. *Antonie van Leeuwenhoek. J. Microbiol. Serol.* **35**:172–177.
Sowell, G., and R. P. Korf. (1960). An emendation of the genus *Itersonilia* based on studies of morphology and pathogenicity. *Mycologia* **52**:934–945.
Starkey, R. L. (1946). Lipid production by a soil yeast. *J. Bacteriol.* **51**:33–50.
Stenderup, A., and A. Leth Bak. (1968). Deoxyribonucleic acid base composition of some species within the genus *Candida. J. Gen. Microbiol.* **52**:231–236.
Stodola, F. H., M. H. Deinema, and J. F. T. Spencer. (1967). Extracellular lipids of yeasts. *Bacteriol. Rev.* **31**:194–213.
Storck, R., and C. J. Alexopoulos. (1970). Deoxyribonucleic acid of fungi. *Bacteriol. Rev.* **34**:126–154.
Tsuchiya, T., Y. Fukazawa, and S. Kawakita. (1965). Significance of serological studies on yeasts. *Mycopathol. Mycol. Appl.* **26**:1–15.
Tubaki, K. (1952a). Studies on the Sporobolomycetaceae in Japan. I. On *Tilletiopsis. Nagaoa* **1**:26–31.
Tubaki, K. (1952b). Studies on the Sporobolomycetaceae in Japan. II. On *Itersonilia. Nagaoa* **2**:62–66.
van der Walt, J. P. (1956). *Kluyveromyces*—a new yeast genus of the Endomycetales. *Antonie van Leeuwenhoek; J. Microbiol. Serol.* **22**:265–272.
van der Walt, J. P. (1959). *Endomyces reessii* nov. spec. *Antonie van Leeuwenhoek; J. Microbiol. Serol.* **25**:458–464.
van der Walt, J. P. (1967). Sexually active strains of *Candida albicans* and *Cryptococcus albidus. Antonie van Leeuwenhoek; J. Microbiol. Serol.* **33**:246–256.
van der Walt, J. P. (1969). The genus *Syringospora* Quinquaud emend. *Antonie van Leeuwenhoek; J. Microbiol. Serol.* **35**: Suppl., A1.
van der Walt, J. P. (1970a). *In* "The Yeasts" (J. Lodder, ed.) pp. 157–165, 316–378, 555–718, 863–892. North-Holland Publ., Amsterdam.
van der Walt, J. P. (1970b). The perfect and imperfect stages of *Sporobolomyces salmonicolor. Antonie van Leeuwenhoek; J. Microbiol. Serol.* **36**:49–55.
van der Walt, J. P. (1972). The yeast genus *Ambrosiozyma* gen. nov. (Ascomycetes). *Mycopathol. Mycol. Appl.* **46**:305–315.
van der Walt, J. P. and D. B. Scott. (1971). The yeast genus *Saccharomycopsis* Schiönning. *Mycopathol. Mycol. Appl.* **43**:279–288.
van Uden, N., and H. Buckley. (1970). *In* "The Yeasts" (J. Lodder, ed.) pp. 893–1087. North-Holland Publ., Amsterdam.

van Uden, N., and M. Vidal-Leiria. (1970). *In* "The Yeasts" (J. Lodder, ed.), pp. 1235–1308. North-Holland Publ., Amsterdam.
von Arx, J. A. (1967). "Pilzkunde." Cramer, Lehre.
von Arx, J. A. (1972). On *Endomyces, Endomycopsis* and related yeast-like fungi. *Antonie van Leeuwenhoek; J. Microbiol. Serol.* **38**:289–309.
Walker, L. B. (1931). Studies on *Ascoidea rubescens*. I. History and development. *Mycologia* **23**:51–76.
Walker, L. B. (1935). Studies on *Ascoidea rubescens*. II. Cytological observations. *Mycologia* **27**:102–127.
Wells, D. E. (1954). *Ascocybe*, a new genus of lower ascomycetes. *Mycologia* **46**:37–51.
White, W. L. (1942). A new hemiascomycete. *Can. J. Res., Sect. C.* **20**:389–395.
Wickerham, L. J. (1951). Taxonomy of yeasts. *U.S., Dep. Agr., Tech. Bull.* **1029**, 1–56.
Wickerham, L. J. (1958). Sexual agglutination of heterothallic yeasts in diverse taxonomic areas. *Science* **128**:1504–1505.
Wickerham, L. J. (1964). A preliminary report on a perfect family of exclusively protosexual yeasts. *Mycologia* **56**:253–266.
Wickerham, L. J. (1970). *In* "The Yeasts" (J. Lodder, ed.), pp. 226–315. North-Holland Publ., Amsterdam.
Wickerham, L. J., and K. A. Burton. (1962). Phylogeny and biochemistry of the genus *Hansenula*. *Bacteriol. Rev.* **26**: 382–397.
Wickerham, L. J., C. P. Kurtzman, and A. I. Herman. (1970). Sexual reproduction in *Candida lipolytica*. *Science* **167**:1141.
Wilson, C. M. (1961). A cytological study of *Ascocybe*. *Can. J. Bot.* **39**:1605–1607.
Windisch, S. (1960). Die Hefeartigen Pilze. *In* "Die Hefen. I. Die Hefen in der Wissenschaft" (F. Reiff *et al.*, eds.), pp. 23–178. Verlag Hans Carl, Nürnberg.
Yarrow, D. (1969). *Selenotila peltata* comb. n. *Antonie van Leeuwenhoek; J. Microbiol. Serol.* **35**:418–420.
Zender, J. (1925). Sur la classification des Endomycétacées. *Bull. Soc. Bot. Genève* **17**:272–302.

CHAPTER 3

# Protomycetales and Taphrinales

CHARLES L. KRAMER

*Department of Biology*
*Kansas State University*
*Manhattan, Kansas*

## I. PROTOMYCETALES

Protomycetales are parasitic fungi that attack flowering plants causing lesions and galls, often with extensive color changes. They have no ascocarp. Their intercellular and apparently diploid mycelium develops thick-walled resting spores which either spread throughout the host tissue or form a continuous subepidermal layer. Germination can occur by rupture of the outer wall, thereby allowing the cell membrane to emerge, or by a papillalike extension of the outer wall to form an enlarged spore sac. The multinucleate protoplast of the spore sac becomes oriented in a peripheral layer, while the nuclei apparently undergo reduction division, resulting in the production of four "endospores" per nucleus. After their formation, the several hundred endospores are forcibly discharged in a single mass. These endospores are uninucleate, they bud after ejection and are culturable under artificial conditions in this "yeast state." The endospores or blastospores apparently conjugate to form a diploid mycelium prior to infection.

### A. *Protomycetaceae*

There is but a single family with characteristics as described for the order. Four genera are presently recognized.

**KEY TO GENERA OF PROTOMYCETALES**

1. Mycelium intercellular, forming thick-walled resting spores; spore sac formed by emergence of cell membrane through ruptured outer wall of resting spore, lacking a columellalike partition or septum separating it from the resting spore . . . . . . . . . 2
1'. Mycelium within the layers of the outer wall of the host epidermal cells; thick-walled resting spores lacking; swollen cells of the mycelium immediately developing papilla-

like extensions that become the spore sacs; spore sac with sporangenous outer layer separated by a columellalike wall and septum separating it from the swollen basal cell . . . . . . . . . . . . . . . . . . . . . . . . . . . . . . . . . . . . . . . . . . **Mixia**

Kramer (1958) recognizes 1 species, *M. osmundae* on *Osmunda*, known from Japan and North America.

2(1) Resting spores formed throughout all tissues of localized lesions of the host; endospores formed only after germination of resting spores . . . . . . . . . . . 3

2'(1) Resting spores formed in a continuous subepidermal layer; endospores formed before emergence of spore sac . . . . . . . . . . . . . . . . . . . . . **Taphridium**

Here considered to embrace *Vokartia*; *T. umbelliferarum* on Umbelliferae; 5 species known from Europe, North Africa and Australia.

3(2) Resting spores with rough walls and formed terminally on the mycelium
. . . . . . . . . . . . . . . . . . . . . . . . . . . . . . . . . . . . . . . . . **Protomycopsis**

Eight species recognized on Compositae and Leguminosae; Europe and Asia.

3'(2) Resting spores with smooth walls and formed intercalarily on the mycelium
. . . . . . . . . . . . . . . . . . . . . . . . . . . . . . . . . . . . . . . . . . **Protomyces**

*P. macrosporus* on Umbelliferae is the most widely distributed species of the order, (Plate I, Figs. 1, 2); 15 species known from North and South America, Europe

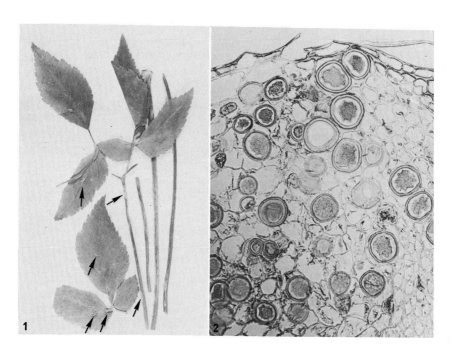

PLATE I

FIG. 1. *Protomyces macrosporus* on *Aegopodium podagraria*. Arrows indicate pustules. ×0.3.

FIG. 2. *P. macrosporus* on *A. podagraria*. Transverse section of stem pustule. × 58.

and Asia; life history studies include those of von Büren (1915, 1922) while Tubaki (1957) has studied these fungi in culture.

## II. TAPHRINALES

Taphrinales are parasitic fungi which attack higher plants and ferns, causing galls on leaves, stems and fruits, as well as hypertrophied and deformed leaves and fruits and leaf spots which are necrotic or slightly hypertrophied and definitely limited. They may also cause close clusters of twigs called witches' brooms. Taphrinales have no ascocarp. Their mycelia have no intercellular or subcuticular, or else it develops within the outer epidermal wall. Haustoria are formed in four species of Taphrinales. The mycelium is perennial in those species which cause witches' brooms; the hyphae are hyaline, and branched, while the mycelial cells are dikaryotic, ultimately forming a subcuticular or subepidermal layer of ascogenous cells. The ascogenous cells are binucleate, and karyogamy occurs within them. The asci arise from ascogenous cells as thin-walled sacs bursting out of the epispore, or by elongation of the ascogenous cell wall; they are primarily clavate to cylindrical in shape. The asci are provided with a basal cell, which is commonly called a stalk cell, cut off from the ascus proper by a septum. Sometimes the stalk cell is lacking. The asci occur in a subcuticular palisade layer or are formed as terminal cells of the septate intercellular hyphae. In some species, the first division of the fusion nucleus is mitotic, furnishing two diploid nuclei—one for the stalk cell, which soon degenerates, and the other for the ascus. In other species which lack stalk cells the first division of the diploid nucleus is meiotic. Meiosis occurs in the young ascus, yielding nuclei for four ascospores or, much more commonly, a subsequent mitosis produces nuclei for eight ascospores. In some species, repeated mitotic division of the haploid nuclei yields many ascosporous nuclei. Ascospores formed in a layer next to the ascus wall are uninucleate, budding before or after ejection from the ascus to form numerous uninucleate blastospores. Ascospores and/or blastospores are forcibly discharged by rupture of the ascus wall at the apex of the ascus; blastospores continue the budding process after ejection and are culturable under artificial conditions in this "yeast state." In one species, ascospores or blastospores conjugate in pairs, giving rise to dikaryotic mycelium; in other species, so far as is known, the dikaryotic condition results from division of the haploid nuclei of the blastospores.

### A. Taphrinaceae

There is but a single monogeneric family with characteristics as described for the order.

*Taphrina*

Ninety-two species are recognized. A list of the more common species and their hosts from North America includes *T. caerulescens* on *Quercus, T. carnea* on *Betula, T. communis* on *Prunus* [plum], *T. cystopteridis* on *Cystopteris, T. deformans* on *Prunus* [peach], *T. farlowii* and *T. flavorubra* on *Prunus* [cherry], *T. occidentalis* on *Alnus, T. polystichi* on *Polystichum, T. populina* on *Populus, T. robinsoniana* on *Alnus, T. sacchari* on *Acer, T. tormentillae* on *Potentilla, T. ulmi* on *Ulmus*, and *T. virginica* on *Ostrya*.

## B. General Characteristics

The order Taphrinales was established by Gäumann (1926) to include the two families Protomycetaceae and Taphrinaceae. However, in 1949 Gäumann proposed a supplementary group of primitive ascomycetes that he called the Synascomycetes, which included the two families Protomycetaceae and Pericystaceae. Members of the Synascomycetes were distinguished by the formation of ascospores in a spore sac, the synascus, which he concluded to be essentially a compound structure equivalent to many asci. Martin, in his "Key to the Families of Fungi" (1961), followed this proposal and in the subclass Hemiascomycetidae established the order Protomycetales as the equivalent of Gäumann's Synascomycetes. Thus, with the establishment of the order Protomycetales, the order Taphrinales was left to include the single family Taphrinaceae. In 1964, Gäumann maintained this separation of the two families, but referred to the ascus wall in the formation of subdivisions of the class Ascomycetes. The order Protomycetales was included in the subclass Prototunicatae in which the ascospores are not actively discharged, while the order Taphrinales was included in the subclass Eutunicatae, group Unitunicatae.

The order Taphrinales is here treated as monogeneric. In the past, the group has been divided in various ways and has included the generic names *Ascomyces, Ascosporium, Exoascus, Magnusiella,* and *Sarcorhophalum*. However, Mix (1949), following the lead of Sadebeck (1893), chose to include all species under the one genus *Taphrina* founded by Elias Fries (1832) based on *T. populina*. This concept has been followed here.

All species of the genus are dimorphic, forming mycelium and asci in their parasitic phase and budding yeast cells in their saprobic phase. In the parasitic phase, species of *Taphrina* develop mycelium within the host plants that gives rise directly to asci without the formation of a fruiting body. The mycelium is intercellular, subcuticular, or located within the outer tangential walls of the epidermal cells of the host. It may be either annual, dying along with the infected parts of the host plant each year or, in species producing witches' brooms, the mycelium may be perennial, overwintering

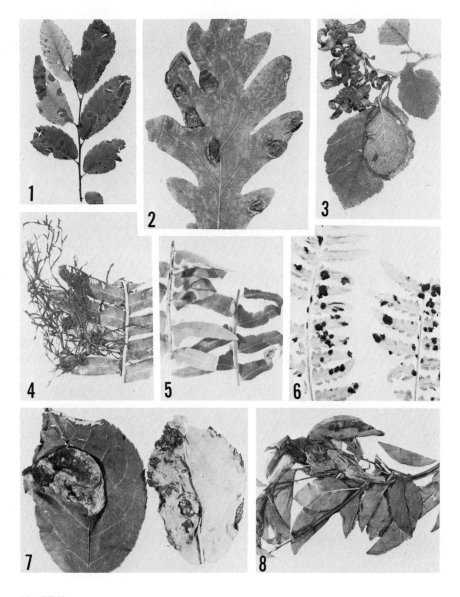

PLATE II

FIG. 1. *Taphrina ulmi* on *Ulmus americana*. ×0.4.
FIG. 2. *T. caerulescens* on *Quercus alba*. ×0.4.
FIG. 3. *T. robinsoniana* on *Alnus rugosa*. ×0.6.
FIG. 4. *T. laurencia* on *Pteris quadriaurita*. ×0.6.
FIG. 5. *T. polystichi* on *Polystichum acrostichoides*. ×0.2.
FIG. 6. *T. californica* on *Dryopteris arguta*. ×0.6.
FIG. 7. *T. confusa* on *Prunus virginiana* var. *melanocarpa*. ×1.
FIG. 8. *T. communis* on *Prunus angustifolia*. ×0.6.

in the woody twigs. In *T. whetzelii*, the mycelium invades the buds of the host and becomes systemic, deforming the whole shoot of the current season.

Symptomatic effects on the host plants vary greatly (Plate II, Figs. 1–8), ranging from simple necrotic lesions to very elaborate galls, twig, flower, and fruit deformations, and witches' brooms. Perhaps the most common type of disease symptom caused by species of *Taphrina* is the unthickened, necrotic leaf spot in which there is little or no hypertrophy of host cells. Cells within the infected tissue soon die and the spots are then characteristically small (Plate II, Fig. 1). However, many species cause thickening of the diseased tissue due to hypertrophy and hyperplasia of the host cells. When this becomes pronounced, it may cause a bulging of the diseased tissue to produce a blisterlike lesion (Plate II, Figs. 2 and 5), or when large portions of the entire leaf blade are involved, a curling effect results (Plate II, Fig. 7). In some cases, when extensive hypertrophy and hyperplasia occur, deformations such as leaf galls (Plate II, Figs. 4 and 6), deformation of twig tips (Plate II, Fig. 8), or enlarged bracts of female catkins of *Alnus* (Plate II, Fig. 3) may result.

In most species, the mycelium forms a compact, subcuticular layer of hyphae that fragment into individual ascogenous cells (Plate III, Figs. 1–3) each of which develops into an ascus (Plate III, Figs. 4–6, 11–12). However, in some, only intercellular mycelium is formed; thus, hyphal branches emerge from between host epidermal cells to form asci directly without forming a subcuticular ascogenous layer (Plate III, Figs. 7–10).

In some species, asci are formed when the outer wall of the ascogenous cells ruptures to allow the inner membrane to emerge and form the ascus wall. In others, the ascogenous cell wall stretches to form the ascus. In many species, a septum is formed across the basal portion of the developing ascogenous cell to cut off a stalk cell at the base and ascus at the apex (Plate

PLATE III
- FIG. 1. *Taphrina confusa*. Ascogenous layer on upper and lower leaf surfaces. ×70.
- FIG. 2. *T. confusa*. Ascogenous layer. ×350.
- FIG. 3. *T. confusa*. Ascogenous layer as seen from surface view. ×350.
- FIG. 4. *T. confusa*. Asci. ×287.
- FIG. 5. *T. pruni*. Asci. ×287.
- FIG. 6. *T. carveri*. Asci. ×287.
- FIG. 7. *T. caerulescens*. Ascus. ×287.
- FIG. 8. *T. populi-salicis*. Asci. ×287.
- FIG. 9. *T. populi-salicis*. Ascus with stalk cell. ×287.
- FIG. 10. *T. populi-salicis*. Young ascus with ascospore nuclei; another with ruptured apex following ascospore discharge. ×287.
- FIG. 11. *T. deformans*. Ascogenous cells and asci. ×287.
- FIG. 12. *T. deformans*. Ascogenous cells. ×287.

# 3. Protomycetales and Taphrinales

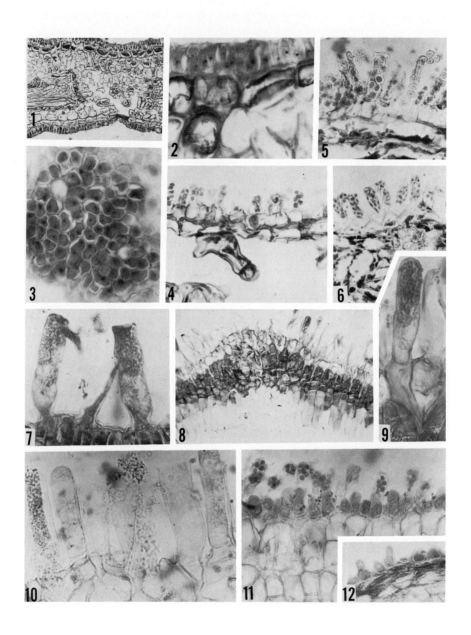

PLATE III

III, Figs. 8 and 9). In other species a septum is not formed; the entire ascogenous cell is converted into an ascus (Plate III, Fig. 7).

The dikaryophase has been reported by Wieben (1927) in *T. epiphylla* to arise from the conjugation of two ascospores (or blastospores) and the development of a dikaryotic hypha from the fusion cell. In other species, it apparently arises through the division of the haploid nucleus of the ascospore (or blastospore) at the time a germ tube is formed prior to penetration of the host plant. The dikaryophase remains evident through the entire vegetative mycelial phase of the organism. Karyogamy occurs in the ascogenous cells prior to ascus formation.

In some species, karyogamy is immediately followed by a mitotic division of the diploid nucleus. One of the daughter nuclei remains in the basal portion of the developing ascogenous cell where it is cut off by a cross wall and soon disintegrates, leaving the stalk cell empty. The second daughter nucleus migrates into the apical portion of the developing ascus. In some species, eight haploid ascospore nuclei are produced in the manner common to most ascomycetes. However, in *T. populi-salicis* and perhaps several other species, there is repeated nuclear division to produce many ascospore nuclei resulting in asci that may be filled with many small ascospores (Plate III, Fig. 9). In still others, the ascus may become filled with many small blastospores formed as a result of budding of the ascospores before discharge (Plate III, Fig. 6). A study of ascus development of several species was done by Kramer (1960).

Ascospores (and/or blastospores) are forcibly ejected by a rupturing of the apex of the unitunicate ascus wall to allow the spores to be discharged in a single mass (Plate III, Fig. 10). These spores develop budding yeast colonies that range from a salmon pink to pinkish-yellow color and give a dry yeastlike appearance to the colony surface.

Interpretation of species has not been consistent among the species of *Taphrina* that occur on different host groups. For example, Mix (1949), following the trend that developed in formulating a concept of the species that occur on the host genus *Acer*, recognized eleven species of *Taphrina* distinguished primarily on host specificity and distribution of the organism on each host species. Morphologically these species are essentially indistinguishable. A contrasting view was followed in the case of *Taphrina* that is known to occur on some fifty species of the host genus *Quercus*. Here a single species, *T. caerulescens*, is recognized, despite the fact that considerable morphological variation occurs within the organism among the various species of *Quercus*. It is hoped that with additional information obtained from serologic, biochemical, nutritional, and cross-inoculation studies, a more sound concept of speciation within the group may be formed.

## REFERENCES

Fries, E. M. (1832). "Systema Mycologicum," Vol. 3, p. 520. Lund.
Gäumann, E. (1926). "Vergleichende Morphologie der Pilze." Fischer, Jena.
Gäumann, E. (1949). "Die Pilze. Grundzuge ihrer Entwicklungsgeschichte und Morphologie." Birkhaeuser, Basel.
Gäumann, E. (1964). "Die Pilze. Grundzuge ihrer Entwicklungsgeschichte und Morphologie." Birkhaeuser, Basel.
Kramer, C. L. (1958). A new genus in the Protomycetaceae. *Mycologia* **50**:916–926.
Kramer, C. L. (1960). Morphological development and nuclear behavior in the genus *Taphrina*. *Mycologia* **52**:295–320.
Martin, G. W. (1961). *In* "Ainsworth and Bisby's Dictionary of the Fungi" (by G. C. Ainsworth), pp. 497–517. Commonwealth Mycol. Inst., Kew, Surrey, England.
Mix, A. J. (1949). A monograph of the genus *Taphrina*. *Univ. Kans. Sci. Bull.* **33**:3–167.
Sadebeck, R. (1893). Die parasitischen Exoasceen, eine Monographie. *Jahrb. Hamburg Wiss., Anst.* **10**:5–110.
Tubaki, K. (1957). Biological and cultural studies of three species of *Protomyces*. *Mycologia* **49**:44–54.
von Büren, G. (1915). Die Schweizerischen Protomycetaceen mit besonderer Berucksichtigung ihrer Entwicklungsgeschichte und Biologie. *Beitr. Kryptogamenflora Schweiz* **5**:1–95.
von Büren, G. (1922). Weitere Untersuchungen über die Entwicklungsgeschichte und Biologie der Protomycetaceen. *Beitr. Kryptogamenflora Schweiz* **5**:1–94.
Wieben, M. (1927). Die Infektion, die Myzeluberwinterung, und die Kopulation bei Exoasceen. *Forsch. Gebiet Pflanzenkr.* **3**:139–176.

# Ascomycotina
# Plectomycetes

CHAPTER 4

# Plectomycetes; Eurotiales

DOROTHY I. FENNELL

*Northern Regional Research Laboratory*
*Agricultural Research Service*
*United States Department of Agriculture, Peoria, Illinois*

## I. INTRODUCTION

Our knowledge of the Ascomycotina has been greatly increased by studies on the structure of the ascus and ascocarp centrum initiated by Miller (1949), continued by Luttrell (1951, 1955), and pursed further by other investigators working on individual families and genera. Emphasis in these studies, however, has been on the Pyrenomycetes, and conclusions regarding the Plectomycetes have been peripheral to their main theme.

Segregation of genera with bitunicate asci produced in unwalled locules as the Loculoascomycetes (Luttrell, 1955) has been widely accepted. No such clear-cut separation between the Plectomycetes, the unitunicate Pyrenomycetes, and certain Discomycetes has been established. Gäumann's (1964) proposed segregation of the subclass Prototunicatae to encompass most Hemiascomycetes, as well as the Ascosphaerales, Aspergillales, Microascales, and Onygenales, represented an attempt to utilize ascus-wall structure in the classification of these fungi, but this thought-provoking scheme was not followed by either Dennis (1968) or Ainsworth (1971), nor is it applied in this volume. Cain's recent (1972) suggestions that the Eurotiaceae should be assigned to the Hypocreales and that the Monascaceae, Onygenaceae, Gymnoascaceae, and Erysiphaceae are derived from the Pezizales illustrate the difficulties encountered in distinguishing between classes.

## II. PLECTOMYCETES

Traditionally, the Plectomycetes have been characterized as producing closed ascocarps in which globose evanescent asci are borne at all levels from ascogenous hyphae ramifying irregularly throughout the central tissue of the fruiting bodies. This definition, based on a majority opinion for each attribute, has not been universally accepted or applied. In fact, evanescence of the ascus wall has been the only characteristic uniformly recognized in all major treatments. As Cain (1956) pointed out, no single character ascribed to these fungi is of any great significance in recognizing relationships and not one is unique to the class. Several characters must be correlated to achieve delimitation of any taxon.

Differences of opinion regarding the relative importance of individual characteristics and combinations of characteristics have resulted in the inclusion of families and orders that fail to conform with current concepts of the class.

Nannfeldt (1932) included the Chaetomiaceae in the Plectomycetes because of their evanescent asci despite the formation of a basal cluster of asci and paraphyses. Miller (1941, 1949) and Gäumann (1952) concurred with this placement. Since then *Chaetomium* and its related genera have been generally recognized as Pyrenomycetes (Moreau, 1953; von Arx and Müller, 1954; Müller and von Arx, 1962; Alexopoulos, 1962; Gäumann, 1964; Dennis, 1968; Ainsworth, 1971).

Miller (1949) classified the Myriangiales in the Plectomycetes. Luttrell (1951), recognizing the significance of the bitunicate asci produced in individual loci by these fungi, transferred the order to the Bitunicatae and later (Luttrell, 1955) to the Loculoascomycetes. This transfer has been accepted in recent works cited.

The Coronophorales, placed in the Plectascales by von Arx and Müller (1954), are now classified as Pyrenomycetes in all major treatments.

Disposition of other families and orders remains tentative and fluid. The peregrinations of the following taxa and the obvious uncertainties regarding their classification emphasize the need for much more information on life cycles and ontogeny.

A detailed ontogenic study of the unique spore cyst that constitutes the ascocarp of *Ascosphaera* (=*Pericystis* Betts) led Spiltoir and Olive (1955) to remove the genus from the Hemiascomycetes and establish the monogenic family Ascosphaeraceae in the Plectascales. In the same year, Luttrell treated the genus as the Pericystales, derived from the Hemiascomycetes in a separate phylogenetic line. Martin (1961) accommodated the family in the Hemiascomycetes, while Alexopoulos (1962) considered it under the Eurotiales. Gäumann (1964) elevated the family to ordinal rank in his Prototunicatae, and the order was arranged under the Plectomycetes by Ainsworth (1971). The developmental process described by Spiltoir and Olive (1955) for *Ascosphaera* is reminiscent of the fragmentation of ascogenous hyphae and the independent proliferation of ascogenous cells described by Andrus and Harter (1933, 1937) and Andrus (1936) for species of *Ophiostoma* (*Ceratocystis*) with evanescent asci. Perhaps the Ophiostomataceae (see below) represent an extension of the phylogenetic line suggested by Luttrell for the Pericystales.

Both C. W. Dodge (1929) and Hawker (1954) classified the hypogean Elaphomycetaceae (*Elaphomyces* and *Mesophellia*) in the Plectascales. This placement was accepted by Luttrell (1951), Alexopoulos (1962), and von Arx (1968). Differences in structure of the ascocarp centrum in the two genera led Dodge to separate them as monogenic tribes. This separation was reflected in the fifth edition of the "Dictionary of the Fungi" (Ainsworth, 1961) where *Elaphomyces* was assigned to the Tuberales and *Mesophellia* to the Lycoperdales. Martin (1961) consigned the Elaphomycetaceae to the Tuberales. Ainsworth (1971) returned the family to the Eurotiales, although the two genera are cited individually as they were in the earlier edition of the Dictionary.

*Microascus* and *Ophiostoma* (*Ceratocystis*), both with irregularly arranged evanescent asci and together comprising the Microascales, are the only genera with ostiolate or beaked ascocarps commonly classified as Plectomycetes (Luttrell, 1951, 1955; von Arx and Müller, 1954; Alexopoulos, 1962; Gäumann, 1964; Ainsworth, 1971). In this volume, Müller and von Arx have transferred *Microascus* (Microascaceae) to the Melanosporaceae in the Sphaeriales, and the Ophiostomataceae to the same order as a separate family. They have also removed from the Plectomycetes certain cleistocarpous genera (e.g., *Kernia*, *Europhium*) which they believe show obvious relationships to ostiolate pyrenomycetous genera. Selected genera with dark ascospores having germ pores (e.g., *Thielavia*) have likewise been transferred to the Sphaeriales. Genera with dark two-celled ascospores, *Testudina* and *Neotestudina*, have been found to possess bitunicate asci and have been segregated by von Arx (1971a) as a new family of Loculoascomycetes.

Miller (1941, 1949) assigned the Erysiphaceae to the Plectascales because of their nonostiolate ascocarps, even though asci are produced in a basal fascicle and discharged forcibly. On the basis of these latter characters, Luttrell (1951) transferred the Erysiphales (Erysiphaceae and Meliolaceae) to the Pyrenomycetes. This treatment, with the two families raised to ordinal rank, was accepted by Moreau (1953) and Alexopoulos (1962). Dennis (1968) and Ainsworth (1971) returned them to the Plectomycetes. In this volume (Chapter 5) they are considered as pyrenomycetous orders.

As viewed in this chapter, the Plectomycetes do not include the Ascosphaerales, Elaphomycetaceae, Microascales, Erysiphales, Meliolales, or the individual genera recently transferred to the Pyrenomycetes and Loculoascomycetes by Müller and von Arx. The Ascosphaerales are considered in a separate line derived from the Hemiascomycetes; the Elaphomycetaceae, in the Tuberales; and the remainder are distributed among the Pyrenomycetes and Loculoascomycetes.

Of previously recognized orders, the Eurotiales, Gymnoascales, and Onygenales remain. The existence of intergrading forms argues against separation of the Gymnoascales and Onygenales at the ordinal level, and they are treated here as families of the Eurotiales. Seven additional families are recognized.

The Plectomycetes remain an unnatural class, but deleting the above taxa permits it to be more closely circumscribed as one order (Eurotiales) than was possible previously. Insofar as is known, the genera here regarded as Plectomycetes share all the following characteristics: ascocarps astomatous; paraphyses lacking; asci irregularly distributed, produced from fertile hyphae ramifying throughout the centrum, typically eight-spored, astipitate, thin-walled, quickly evanescent; ascospores unicellular, without germ pores or germ slits.

Many Plectomycetes have conspicuous and characteristic imperfect states, which are encountered in nature more frequently than perfect states. Notable advances in our understanding of imperfect states and their ontogeny have been made by Hughes (1953), Subramanian (1962), Tubaki (1958, 1963), Barron (1968), Cole and Kendrick (1968, 1969a,b), and Kendrick and Cole (1968a,b). Interpretation of the relationship between developmental patterns demonstrated for various types of conidia, and establishment of an acceptable standardized terminology, were the goals of the First International Specialists' Conference on Criteria and Terminology in the Classification of Fungi Imperfecti (Kendrick, 1971).

Recognition is becoming more widespread that the imperfect state of a fungus is an expression of a genetic character of the same magnitude as the perfect state and is equally indicative of natural relationships (Raper, 1957; Kuehn, 1957; Tubaki; 1958, Apinis, 1964; Nicot and Durand, 1969; Müller, 1971). The correlations drawn by Tubaki (1958) and Müller (1971) between perfect genera and their associated imperfect states are particularly revealing. The apparent unifying character of the Trichocomataceae, as defined recently[1] by Malloch and Cain, is the production of catenate phialospores. An increasing body of evidence supports use of the imperfect state as one criterion in the classification of ascomycetous genera. It has been applied here.

## A. Taxonomic Criteria

The following characteristics are believed to be generically, if not supragenerically, significant, and data regarding them are indispensable to clarification of the phylogeny of these organisms.

[1]CBA/AIBS Meeting, Edmonton, Canada, 1971.

## 1. Ascocarp Initials

The variations observed in these structures have not been recorded fully in the past. Recent studies show a trend toward correction of this inadequacy. Cytological studies are essential to determine the presence or absence of functional or nonfunctional antheridia and the details of plasmogamy and karyogamy. As an example, the ascocarps of the genus *Eupenicillium* are widely accepted as arising as a result of somatogamous copulation between undifferentiated hyphae in the crotch of a loosely branched hyphal tuft. The actual initiation of the perfect state may well be concealed within the preformed sclerotioid structures arising from the "hyphal tufts." Investigation of this aspect of development might well demonstrate differences that correlate with the two "series" now recognized within the genus.

## 2. Ascocarp Peridium

The origin of the peridial hyphae, whether from the archicarp, the surrounding vegetative mycelium, or a combination of the two, is perhaps more significant than the thickness or density of the resulting covering.

## 3. Asci

Within the class, asci have been reported to arise with or without the intervention of croziers and to be borne singly or in chains from the ascogenous hyphae. These would seem to be characters of more than specific importance but have not been so considered in the past. The recent work of Stolk and Samson (1971), who separated *Hamigera* from *Talaromyces*, reflects an awareness of this error.

## 4. Ascospores

The significance of bivalve versus nonbivalve ascospores has been little investigated, but it is believed to be important, at least at the generic level. The former split without swelling at germination; the latter swell and appear to germinate at any point on their circumference.

Genera with septate ascospores are not included among the Plectomycetes in this treatment. While degree of multicellularity may not be particularly significant to classification, its occurrence is.

## 5. Imperfect State

Before the extent to which the imperfect state should be considered in the classification of ascomycetous genera can be fully assessed, the phylogenetic relationship between the various types of conidia must be better understood, and definite proof of connections between more sexual and asexual fructifications must be demonstrated. Accurate identifications of both the perfect and imperfect states are of paramount importance. Means must be found to determine whether so-called conidia are asexual spores or spermatia, hence serving a sexual function. Also, factors influencing the development of the perfect state in pure culture must be investigated.

In my opinion, different imperfect states associated with two otherwise identical perfect forms represent a more profound difference than current widely held beliefs would indicate.

## III. EUROTIALES

### A. General Characteristics

Members of the order occur in soil and on a variety of plant and animal debris, such as wood, dung, hair, feathers, horns, and hooves. They are primarily saprobic, but may be parasitic on plants, and are the causative agents of dermatomycoses and other infections of animals, including man. Many are of importance to both industry and medicine. The majority are among the most common and widespread of all fungi and are responsible for the decomposition of much organic material. A few are rarely encountered. Members of the order have a wide temperature range and may be thermophilic or thermotolerant. Some are xerophilous. Most are homothallic; a few are heterothallic. In general, they grow and produce ascocarps readily in pure culture.

Ascocarps are small, usually spherical, nonostiolate, predominantly sessile but stalked in a few genera, produced on a more or less well-developed septate mycelium, rarely in a stroma. Paraphyses are lacking. Ascogenous hyphae ramify throughout the centrum and bear asci at all levels. Asci are produced singly or in chains from the ascogenous hyphae, with or without the intervention of croziers. They are globose or subglobose, unstalked, usually eight spored, uniformly thin walled, lacking pore or operculum, evanescent, deliquescing to liberate the ascospores within the ascocarp. Ascospores are unicellular, hyaline or colored in pale to dark shades, variously ornamented, without germ pores, frequently of bivalve construction.

Ascocarp development may be initiated through the copulation of two spirally entwined gametangia; through independent development of the ascogonia in the presence of a nonfunctional antheridium or in its absence; through somatogamous copulation between two apparently undifferentiated vegetative cells; but rarely, if ever, by meristogenous growth of a swollen intercalary hyphal cell. Ascogonia are usually free on the mycelium and become surrounded by mycelium arising from the archicarp or from surrounding vegetative hyphae or both. This mycelium differentiates into a

## 4. Plectomycetes; Eurotiales

characterisitc simple astomous ascocarp. Peridia vary from a loose reticulum to a thick parenchymatous layer. Cleistothecia open by weathering or by rupturing from internal pressure created by maturing ascogenous hyphae.

### B. Families

Nine families are recognized. Separation is based on the nature of the ascocarp peridium and the associated imperfect state where known.

*1. Amorphothecaceae* (Parbery, 1969)

This family is monogenic and is included provisionally. The ascocarps are discrete, submerged, or superficial on natural substrate. The peridium is an amorphous melanoid membrane. Hyphae of the ground tissue radiate from a central point, originating from below the gametangia. The asexual spores are blastosporous, *Cladosporium*. Its true relationship may be near the Hemiphacideaceae.

*2. Gymnoascaceae* (Figs. 1-7) (Baranetzky, 1872; Schröter, 1893)

Ascocarps lack a definite peridial wall. The ascogenous hyphae and asci occasionally occur in naked clusters but are usually surrounded and enmeshed by a network of hyphae similar to, or different from, the vegetative hyphae; if different, they are usually colored, thick walled, smooth or variously roughened, forming a net- or bramblelike aggregate with or without modified peripheral appendages. Sexual apparatus is composed of an antheridium surrounded by a coiled ascogonium arising from the same or different hyphae. The antheridium is often nonfunctional. The asci arise from croziers and the ascospores are various. Known imperfect states are aleuriosporous or arthrosporous and include *Chrysosporium, Trichophyton, Microsporum, Oidiodendron*. Racquet mycelium is present. Generic separations within the Gymnoascaceae are based primarily on the degree of modification of peridial hyphae, on the presence or absence of appendages and their nature, and on ascospore characters. Since opinions regarding synonymy in the Gymnoascaceae continue to differ, contested genera are included. The key to this family is arranged to bring like genera together, and the various sections reflect generally the three subfamilies proposed by Apinis (1964): Gymnoascoideae, Arthrodermoideae, and Arachnioideae.

Benjamin (1956) suggested that the Gymnoascaceae may represent an evolutionary blind alley. While this suggestion may be true of the first of these subfamilies, it appears somewhat unlikely for the latter two. In *Ctenomyces*, the unique ctenoid appendages suggest a more highly developed state of the distinctive asymmetrical protruberances seen in the peridial hyphae of *Arthroderma*. On the other hand, the thin prosenchymatous layer of thin-walled hyphae surrounding the ascigerous mass in *Ctenomyces* indicates a

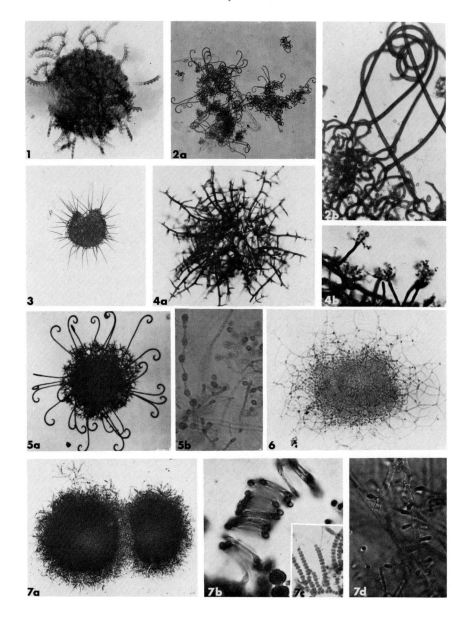

relationship to the Onygenaceae and Eurotiaceae. Derivation of the Gymnoascaceae from the Endomycetales, by way of such genera as *Eremascus*, is suggested by those genera developing clusters of ascigerous hyphae naked in the mycelium or loosely surrounded by inconspicuous scanty wefts of thin-walled hyphae. All bear a marked resemblance to *Byssochlamys*, here keyed in the Eurotiaceae because of its conspicuous *Paecilomyces* imperfect state. These apparent areas of transition to other families in the Eurotiales and elsewhere may, however, represent evolutionary levels in separate phylogenetic lines.

*3. Onygenaceae* (Figs. 8–11) (Fries, 1849; Fischer, 1897)

This family may be considered as cleistocarpous Gymnoascaceae, frequently occurring on keratinic substrates. Details of development of *Onygena*, the genus upon which the family name is based, have not been investigated, and its affinities with the other genera included can only be inferred. With them it shares similar ascocarps, asci, ascospores, and keratinic habitats. Members of the family show the following characters: ascocarp initials (unknown in *Onygena*) consisting of coiled ascogonia; ascocarps usually spherical, stipitate or sessile, red brown to brown, peridium pseudoparenchymatous, with or without appendages; asci arising from croziers. Conidial states are arthrospores or aleuriospores.

For a different analysis of the family the reader is referred to Malloch and Cain (1971), who include *Dichotomomyces, Thermoascus*, and two new monotypic genera, *Ascocalvatia* and *Xynophila*, in their treatment.

*4. Monascaceae* (Figs. 12 and 13) (Schröter, 1897)

The Monascaceae are separated from the Gymnoascaceae by a distinct peridium composed of flattened cells, and from the Eurotiaceae by the arthr-

---

EUROTIALES—GYMNOASCACEAE

FIG. 1. *Ctenomyces serratus*. Mature ascocarp. ×58.

FIG. 2. *Tripedotrichum herbariensis*. (a) Ascocarps, ×28; (b) appendages arising from trichotomous anastomoses of peridial hyphae, ×150.

FIG. 3. *Toxotrichum cancellatum*. Mature ascocarp. ×50.

FIG. 4. *Eidamella deflexa*. (a) Mature ascocarp, ×170; (b) peridial hyphae bearing terminal groups of spiral branchlets, ×325.

FIG. 5. *Myxotrichum*. (a) Mature ascocarp of *M. chartarum*, ×80, (b) imperfect state of *M. uncinatum*, ×415.

FIG. 6. *Gymnoascus reesii*. Mature ascocarp. ×85.

FIG. 7. *Arthroderma curreyi*. (a) Two mature ascocarps, ×53; (b) coiled appendage and one asperulate dumbbell-shaped cell of peridial hyphae, ×650; (c) peridial hyphae composed of dumbbell-shaped cells, ×315; (d) imperfect state, ×725.

Figs. 1, 4, 5, 6, 7 after Benjamin (1956). Figs. 2 and 3 after Orr and Kuehn (1964a,b).

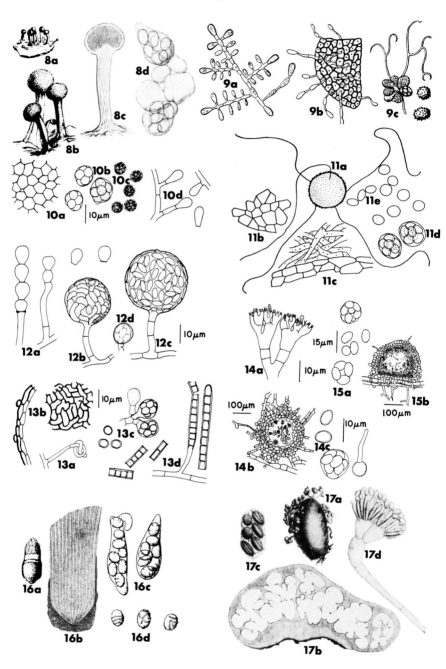

rosporous imperfect state. *Xylogone* is placed here arbitrarily; its true relationship is probably elsewhere.

5. *Thermoascaceae* (Figs. 14 and 15) (Apinis, 1967)

This family is thermophilic, and its cleistothecia are firm, astomous, irregularly globose, superficial, solitary or aggregated and confluent, forming crusts. The peridium is irregular, pseudoparenchymatous, and more or less tomentose. Asexual states are annelloporous or phialosporous. The two genera assigned to this family are considered synonymous by some investigators. Neither is obligately thermophilic, but both are thermophiles as defined by Cooney and Emerson (1964). The ascocarp peridia are clearly different structurally from those of the Eurotiaceae. Ascocarp initials consist of coiled ascogonia without an antheridium, and asci are produced singly from croziers.

EUROTIALES—ONYGENACEAE
FIG. 8. *Onygena equina*. (a) Habit, natural size; (b) stipitate ascocarps, enlarged; (c) longitudinal section through ascocarp; (d) group of asci. ×650.
FIG. 9. *Aphanoascus cinnabarinus*. (a) Conidial stage, ×125; (b) portion of ascocarp wall, ×200, (c) cluster of asci, ×400; (d) mature ascospores, ×600.
FIG. 10. *Anixiopsis fulvescens*. (a) Portion of wall of cleiotothecium; (b) asci; (c) mature ascospores; (d) imperfect state, aleuriospores.
FIG. 11. *Arachnomyces sulphureus*. (a) Cleistothecium with appendages, ×29; (b) surface view of peridium, ×325; (c) cross section through peridium showing exterior tomentose layer of hyphae, ×750; (d) asci, ×750; (e) ascospores, ×750.

MONASCACEAE
FIG. 12. *Monascus ruber*. (a) Imperfect state; (b) and (c) surface and section views of cleistothecium; (d) ascus.
FIG. 13. *Xylogone sphaerospora*. (a) Ascocarp initial; (b) surface and cross section of peridium; (c) asci and ascospores; (d) imperfect state (endogenous arthrospore).

THERMOASCACEAE
FIG. 14. *Dactylomyces thermophilus*. (a) *Polypaecilum* imperfect state; (b) section through cleistothecium; (c) ascus and ascospores (one germinating).
FIG. 15. *Thermoascus aurantiacus*. (a) Ascus and ascospores; (b) section through cleistothecium.

TRICHOCOMATACEAE
FIG. 16. *Trichocoma paradoxa*. (a) Ascocarp, natural size; (b) cross section through ascocarp, schematic; (c) asci with young ascospores, ×650; (d) ascospores, ×650.
FIG. 17. *Penicilliopsis clavariaeformis*. (a) Ascocarps, one-half natural size, on persimmon fruit; (b) cross section through single ascocarp, enlarged; (c) ascus with nearly mature ascospores, much enlarged; (d) *Penicillium*-like imperfect state. ×265.

Figs. 8, 9, 16, 17 after Fischer (1897). Figs. 10, 12, 13 after von Arx, (1970). Fig. 11 after Malloch and Cain (1970a). Figs. 14 and 15 after Apinis (1967).

The thermophilic organisms described as species of *Talaromyces*, with imperfect states suggesting a relationship closer to *Paecilomyces* than to symmetrically biverticillate *Penicillium*, fit the concept of this family except that asci are produced in short chains rather than singly from croziers. Until their generic separation from *Talaromyces* is validated, these species are retained in the Eurotiaceae, but are keyed apart from the genus and as individual species. If the family concept were broadened to include thermotolerant organisms, other genera that now appear to be transitional (*Byssochlamys, Hamigera, Dichotomomyces,* and *Sartorya*) could also be included.

6. *Trichocomataceae* (Figs. 16 and 17) (Fries, 1849; Fischer, 1897)

The ascocarp is a mazaedium with capillitium, or with fertile areas separated into locules by interthecial sterile hyphae. This is probably an artificial grouping. The imperfect state, where known, is *Penicillium* or *Penicillium*-like. *Trichocoma paradoxa*, with a symmetrically biverticillate *Penicillium* imperfect state and horizontally striate ascospores, has been considered conspecific with *Talaromyces luteus*. *Dendrosphaera*, originally figured with bitunicate asci and doubtfully a plectomycete, is included here because of its gross similarity to *Penicilliopsis*. The possibility that *Dendrosphaera* is more closely related to the Elaphomycetaceae should not be overlooked. Ciferri (1957) treated *Trichocoma* and *Dendrosphaera* as monogenic families in the Onygenales and classified the order as transitional between the Plectascales and the Myriangiales.

7. *Eurotiaceae* (Figs. 18–23) (Clements and Shear, 1931)

The ascocarps vary from the soft, poorly delimited, and often confluent ascocarps of *Byssochlamys* and *Talaromyces* to the sclerotioid fructifications of *Eupenicillium* and *Dichlaena*, but in most genera they are true cleistothecia with a more or less firm peridium. In some genera, the peridial layer is prosenchymatous, formed from several layers of interlacing hyphae; in others, it is a well-developed pseudoparenchyma. Sexual apparatus varies from conspicuous spirally entwined gametangia to repeated anastomoses of apparently undifferentiated vegetative hyphae. Asci arise from croziers in some genera, but not in others. Ascospores are commonly, but not consistently, bivalve. The known imperfect states of the Eurotiaceae are represented by catenate conidia borne from phialides and, with few exceptions, fall in the genera *Aspergillus* and *Penicillium*. Separation of genera is based on the structure of the ascocarp peridium, hyphal elements associated with the ascocarps, and the associated imperfect state. *Byssochlamys* is considered to be transitional between the Thermoascaceae and the Eurotiaceae. *Hamigera* may be similarly transitional. *Dichotomomyces* appears to be identical with *Sartorya* except for differences in the associated imperfect states: clearly an *Aspergillus* in *Sartorya*, originally described as aleuriospores but since termed phialospores in *Dichotomomyces*. *Eiona, Amylocarpus,* and *Lilliputia* are arbitrarily placed at the end of this family. Their true relationship is unknown. *Eiona* has an unusual inverted peridial structure similar to that described by Malloch and Cain (1970) for *Mycoarachis* and ascospores with winglike appendages suggestive of *Emericellopsis* in the Pseudeurotiaceae. *Lilliputia*, with ascospores similar to those of *Dendrosphaera* (Trichocomaceae), shows some relationships to the Elaphomycetaceae. The pedunculate asci of *Amylocarpus* have a small thickened apical cap and perhaps would be more correctly classified in the Sphaeriales.

8. Cephalothecaceae (von Höhnel, 1917)

This family is recognized provisionally with a single genus and species. It is separated from other dark-spored genera (see Pseudeurotiaceae) because of its imperfect state, described as penicilloid with phialospores by Chesters (1934), and the villus of septate yellow hyphae that covers the cleistothecia when young. The remaining species of *Cephalotheca*, described with sympodulospores, would fall in *Fragosphaeria* of the Pseudeurotiaceae. Sutures or definite lines of dehiscence in the peridium are conspicuous but not unique to this family.

9. Pseudeurotiaceae (Figs. 24–26) (Malloch and Cain, 1970b)

This family has brown to black ascocarps with thick or thin parenchymatous peridia; ascospores are hyaline to dark, never showing the bivalve construction common in the Eurotiaceae. Conidia of the imperfect state are either gloeoid phialospores (*Cephalosporium, Acremonium, Stilbella*) or sympodulospores (*Sporothrix*). Certain new genera originally described in this family by its authors (Malloch and Cain, 1970b) have been eliminated: *Cryptendoxyla* with a *Chalara* (endogenous phialospores) imperfect state and *Mycoarachis* with two-celled ascospores.

## IV. KEY

### KEY TO FAMILIES AND GENERA OF EUROTIALES

1. Peridium noncellular, composed of dark amorphous material; conidial state: *Cladosporium* (blastospores) . . . . . . . . . . . . . . . . . . . . . . **Amorphothecaceae** p. 51
   **Amorphotheca**
1'. Peridium cellular; conidial state not *Cladosporium* . . . . . . . . . . . . . . . . . 2
    2(1') Imperfect state, where known, arthrospores or aleuriospores . . . . . . . . 3
    2'(1') Imperfect state, where known, phialospores, annellospores or sympodulospores . . . . . . . . . . . . . . . . . . . . . . . . . . . . . . . . 31

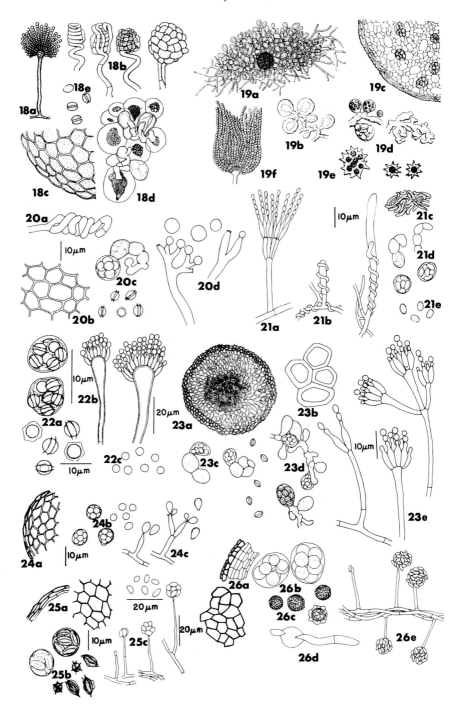

## 4. Plectomycetes; Eurotiales

3(2) Peridium a lax network or coarse bramble of more or less differentiated hyphae, rarely lacking . . . . . . . . . . . . . . . . . . . . . **Gymnoascaceae** p. 51 (Figs, 1–7) 4

3'(2) Peridium prosenchymatous, pseudoparenchymatous, or parenchymatous, with or without appendages; cleistothecia sessile or stalked . . . . . . . . . . . . . . . . . . . . 25

    4(3) Peridial hyphae thick-walled, conspicuously different from vegetative hyphae, mostly with appendages . . . . . . . . . . . . . . . . . . . . . . . . . . . . . . 5

    4'(3) Peridial hyphae thin-walled, more or less similar to vegetative hyphae . . . 19

5(4) Peridial hyphae usually dark, yellow to brown or black, branched and anastomosed to form a coarse bramblelike reticulum; appendages various, usually coarse and rigid . . . . 6

5'(4) Peridial hyphae usually hyaline or lighter colored; reticulum more or less lax; appendages usually not rigid, frequently coiled . . . . . . . . . . . . . . . . . . . . . 13

    6(5) Appendages of one kind . . . . . . . . . . . . . . . . . . . . . . . . . 7

    6'(5) Appendages of more than one kind . . . . . . . . . . . . . . . . . . . 10

### EUROTIALES—EUROTIACEAE

FIG. 18. *Eurotium*. (a) *Aspergillus* imperfect state; (b) ascocarp initial and stages in development of cleistothecium, × 300; (c) peridial cells of mature cleistothecium; (d) asci developing from croziers, × 300; (e) ascospores of *E. amstelodam*, × 500.

FIG. 19. *Emericella*. (a) Cleistothecium surrounded by "hülle" cells, × 150, (b) hülle cells, × 375, (c) cross section through cleistothecium showing peridium (in this immature ascocarp several cell layers thick) and maturing asci, × 200, (d) asci and croziers, × 375, (e) ascus and stellate ascospores of *E. variecolor*; (f) *Aspergillus* imperfect state (phialides biseriate), × 250.

FIG. 20. *Dichotomomyces cejpii*. (a) Ascogonial coil; (b) peridial cells of cleistothecium; (c) asci and ascospores; (d) imperfect state.

FIG. 21. *Talaromyces vermiculatus*. (a) *Penicillium* (Biverticillata-Symmetrica) imperfect state; (b) initials; swollen ascogonia with coiled antheridia; (c) peridial hyphae; (d) asci, produced in chains; (e) ascospores.

FIG. 22. *Sartorya fumigata* var. *glabra*. (a) Asci and ascospores; (b) *Aspergillus* imperfect state; (c) conidia.

FIG. 23. *Eupenicillium*. (a) Cleistothecium, sclerotioid throughout when young, × 75; (b) thick-walled cells of cleistothecium; (c) immature asci produced in chains and mature ascospores of *E. anatolicum*; (d) asci produced singly on ascogenous hyphae in *E. abidjanum*; (e) variety of *Penicillium* imperfect states in this genus.

### PSEUDEUROTIACEAE

FIG. 24. *Pseudeurotium zonatum*. (a) Portion of cleistothecium wall, surface view; (b) asci and ascospores; (c) conidial state, *Sporothrix*.

FIG. 25. *Emericellopsis terricola*. (a) Peridium in section and surface views; (b) asci and ascospores; (c) *Cephalosporium* (*Acremonium*) imperfect state.

FIG. 26. *Hapsidospora irregularis*. (a) Peridium, surface view, × 750, and cross section, × 325; (b) asci, × 750; (c) mature ascospores, × 750; (d) germinating ascospore, × 750; (e) *Cephalosporium* conidial state, × 325.

Figs. 18a, b, d, 19, 23 after Fischer (1897). Figs. 18c and e, 20, 21, 22, 24, 25 after von Arx (1970). Fig. 26 after Malloch and Cain (1970b).

7(6) Appendages ctenoid, uncinate, arising from a sparse weft of orange-brown thick-walled peridial hyphae; ascigerous hyphae enclosed by a thin prosenchymatous layer of thin-walled hyphae . . . . . . . . . . . . . . . . . . . . . . . . . . . . . . **Ctenomyces**

7'(6) Appendages not ctenoid; ascigerous hyphae not enclosed in a layer of thin-walled hyphae . . . . . . . . . . . . . . . . . . . . . . . . . . . . . . . . . . . . . . . 8

   8(7') Appendages unbranched, terminally uncinate, arising from trichotomous anastomoses of peridial hyphae . . . . . . . . . . . . . . . . . . **Tripedotrichum**

   8'(7') Appendages rigid, setalike, branched or unbranched . . . . . . . . . . . . 9

9(8') Appendages unbranched, arising from dichotomous anastomoses of peridial hyphae . . . . . . . . . . . . . . . . . . . . . . . . . . . . . . . . . . . . . **Toxotrichum**

9'(8') Appendages sometimes branched, usually arising from curved pectinate formations in peripheral peridial hyphae . . . . . . . . . . . . . . . . . . . . . . . . . **Pectinotrichum**

   10(6') Appendages with numerous deflexed lateral branches, one or more short hyaline spiral branchlets produced terminally from hyphae surrounding central ascigerous mass . . . . . . . . . . . . . . . . . . . . . . . . . . . . . . . . . . . . . **Eidamella**

   10'(6') Appendages without deflexed lateral branches or short spiral branchlets . 11

11(10') Peridial hyphae with conspicuous knucklelike joints at septa . . . . . **Auxarthron**

11'(10') Peridial hyphae lacking knucklelike joints at septa . . . . . . . . . . . . 12

   12(11') Appendages well-defined, more or less elongate, rigid, simple or branched, apices straight, curved, bent or uncinate . . . . . . . . . . . . . . . . . **Myxotrichum**

   12'(11') Appendages not well-defined; free ends of peripheral hyphae spinelike or more or less elongate with short, apiculate or blunt, straight or curved lateral branches . . . . . . . . . . . . . . . . . . . . . . . . . . . . . . . . . . **Gymnoascus**

13(5') Peridial hyphae uniform, without spinelike branches or other appendages . . . . . . . . . . . . . . . . . . . . . . . . . . . . . . . **Pseudogymnoascus**

13'(5') Peridial hyphae with appendages . . . . . . . . . . . . . . . . . . . . . . 14

   14(13') Appendages simple or branched ends of peridial hyphae with blunt rounded tips . . . . . . . . . . . . . . . . . . . . . . . . . . . . . . . **Neogymnomyces**

   14'(13') Appendages spirally coiled . . . . . . . . . . . . . . . . . . . . . . . 15

15(14') Coiled appendages slender, thin-walled, hyaline, arising terminally or laterally from free ends of peridial hyphae . . . . . . . . . . . . . . . . . . . . . . . . . . . . . 16

15'(14') Coiled appendages robust . . . . . . . . . . . . . . . . . . . . . . . . 18

   16(15) Peridial hyphae composed of irregularly shaped cells, disarticulating at maturity . . . . . . . . . . . . . . . . . . . . . . . . . . . . . . . . **Shanorella**

   16'(15) Peridial hyphae composed of asperulate dumbbell-shaped cells, symmetrical or asymmetrical with protruberances, not disarticulating at maturity . . . . . . . 17

17(16') Cells of peridial hyphae short; ends markedly swollen symmetrically or asymmetrically, thick-walled, spiny; thin-walled and smooth at constriction. Appendages few, all spirals; imperfect state: *Trichophyton* or *Chrysosporium*; homothallic or heterothallic . . . . . . . . . . . . . . . . . . . . . . . . . . . . . . . . . . . . . . . . . . . . **Arthroderma**

17'(16') Cells of peridial hyphae long, symmetrical, slightly swollen at ends, uniformly rough. Appendages mostly spirals, some straight or loosely coiled; imperfect state: *Microsporum*; heterothallic . . . . . . . . . . . . . . . . . . . . . . . . . . . . . . . **Nannizzia**

## 4. Plectomycetes; Eurotiales

18(15') Ascocarps white; coils arising from the cleistothecium or from adjacent vegetative hyphae .......................................................... **Apinisia**

18'(15') Ascocarps light fuscous or reddish fuscous; coils radiating from a common center; heterothallic ................................................ **Ajellomyces**

19(4') Thin-walled peridial hyphae abundant ........................ 20

19'(4') Peridial hyphae sparse or lacking ........................... 21

    20(19) With slender, scimitar-shaped appendages ............ **Spiromastix**

    20'(19) Without appendages; initials large, ascogonium clavate surrounded by densely coiled antheridium; ascospores spherical, thick-walled with irregularly furrowed sheath .......................................................... **Arachnotheca**
                                          Based on *Arachniotus glomeratus*.

21(19') Peridial hyphae sparse, in part thick-walled with enlarged septa; released ascospores remaining in floriate conglomerates ................... **Petalosporus**

21'(19') Peridial hyphae, when present, uniformly thin-walled ............ 22

    22(21') Ascocarp initials consisting of two spirally entwined gametangia of equal size ........................................................ 23

    22'(21') Ascocarp initials not as above ............................ 25

23(22) Asci in chains; ascospores lenticular to oblate, often with equatorial rim or furrow, yellow, or red-brown .............................................. **Arachniotus**

23'(22) Asci in clusters on ascogenous hyphae, from croziers ............ 24

    24(23') Peridium absent, asci single or in pairs; ascospores spherical, 10–12 $\mu$m, spiny, yellow or clear brown, with oil droplet ............... **Eleutherascus**[2]
                                            Based on *Arachniotus lectardii*.

    24'(23') Peridium absent or thin; ascospores spherical, 4–6 (or 8) $\mu$m, thick-walled, reticulate or spiny, yellow to brown, or brown-violet ......... **Amauroascus**

25(22') Initials ringlike, surrounding a central cell; ascospores lenticular or oblate with equatorial band .................................................. **Narasimhella**
                                                               **Rollandina**

25'(22') Initials a series of loose coils; ascospores fusiform, longitudinally striate
.......................................................... **Byssoascus**
                                            Based on *Arachniotus striatosporus*.

    26(3') Cleistothecia small, usually less than 100 $\mu$m; peridium thin, composed of one or two layers of closely interwoven hyphae or flattened hyphal cells
.................... **Monascaceae** p. 53 (Figs. 12 and 13) 27

    26'(3') Cleistothecia larger, sessile or stalked, red-brown to brown; peridium pseudoparenchymatous; appendages present or absent
........................ **Onygenaceae** p. 53 (Figs. 8–11) 28

27(26) Cleistothecia white to red, red-brown, or gray-brown, containing few asci; conidial state: *Basipetospora* (meristem arthrospores) ............... **Monascus**

---

[2]Later considered to be a simple discomycete near *Ascodesmis* by J. A. von Arx (1972, personal communication).

27'(26) Cleistothecia brown; conidial state: endogenous arthrospores . . . . . **Xylogone**

    28(26') Cleistothecia stalked; stalks 1–2 cm high; occurring on horn, hooves, and other keratinous substrates . . . . . . . . . . . . . . . . . . . . . . . . . **Onygena**

    28'(26') Cleistothecia sessile . . . . . . . . . . . . . . . . . . . . . . . . . 29

29(28') Cleistothecia with few long hairlike appendages; ascospores oblate, smooth, red-brown in mass . . . . . . . . . . . . . . . . . . . . . . . . . . . . . . **Arachnomyces**

29'(28') Cleistothecia without appendages . . . . . . . . . . . . . . . . . . . . 30

    30(29') Ascospores pale yellowish brown, lenticular with narrow irregular equatorial crests . . . . . . . . . . . . . . . . . . . . . . . . . . . . . . **Keratinophyton**

    30'(29') Ascospores globose to elliptical without equatorial crests . . . . . . . . 31

31(30') Ascospores red-brown, spinulose . . . . . . . . . . . . . . . . **Aphanoascus**

31'(30') Ascospores pale brown, with alveolate-reticulate wall ornamentation
. . . . . . . . . . . . . . . . . . . . . . . . . . . . . . . . . . . . . . **Anixiopsis**

    32(2') Cleistothecia various, white to yellow or red-brown to light brown; imperfect state: catenate phialospores or annellospores (unknown in two genera of questionable affinity) . . . . . . . . . . . . . . . . . . . . . . . . . . . . . . . . . . 33

    32'(2') Cleistothecia dark brown to black, peridium pseudoparenchymatous, thick or thin; ascospores usually dark, never of bivalve construction; imperfect state: phialospores in gloeoid masses (*Cephalosporium*) or sympodulospores (*Sporothrix*)
. . . . . . . . . . . . . . . . . . **Pseudeurotiaceae** p. 57 (Figs. 24–26) 54

33(32) Thermophilic; cleistothecia rather irregularly shaped, often forming crusts on rich media; development of the red to brown pseudoparenchymatous peridium usually concurrent with ascospore maturation . . . . . **Thermoascaceae** p. 55 (Figs. 14 and 15) 34

33'(32) Not thermophilic; cleistothecia not as above . . . . . . . . . . . . . . . . 35

    34(33) Imperfect state: *Paecilomyces* (phialospores) . . . . . . . . . . **Thermoascus**

    34'(33) Imperfect state: *Polypaecilum* (annellospores) . . . . . . . . . **Dactylomyces**

35(33') Sterile interthecial hyphae present; cleistothecia pedicellate or stipitate . . . . . . . . . . . . . . . **Trichocomaceae** p. 56 (Figs. 16 and 17) 36

35'(33') Without sterile interthecial hyphae; cleistothecia neither pedicellate nor stipitate . . . . . . . . . . . . . . . . . . . . . . . . . . . . . . . . . . 39

    36(35) Cleistothecia pedicellate, rupturing at apex; fertile hyphae forming a brushlike gleba with vertical or near vertical capillitium . . . . . . . . . . . . . . . . 37

    36'(35) Cleistothecia stipitate; fertile areas separated into locules by sterile interthecial hyphae . . . . . . . . . . . . . . . . . . . . . . . . . . . . . . . . . . 38

37(36) Ascospores horizontally striated imperfect state: *Penicilium* . . . . . . **Trichocoma**

37'(36) Ascospores fusoid; imperfect state unknown . . . . . . . . . . **Chaudefaudiella**

    38(36') Ascocarps red-brown, in racemose clusters on stalks 2–3 cm tall, fertile area irregularly multilocular; ascospores with winglike appendages, some with equatorial crests; imperfect state: *Penicillium*-like . . . . . . . . . . . . . . . **Penicilliopsis**

    38'(36') Ascocarps white to yellow, borne terminally on branched stalks to 25 cm high; fertile area peripherally multilocular; ascospores spherical, irregularly echinulate at maturity; imperfect state unknown . . . . . . . . . . . . . . . . **Dendrosphaera**

39'(35') Ascospores hyaline or pale yellow (red to violet in one genus), frequently of bivalve

## 4. Plectomycetes; Eurotiales 63

construction; peridium various, without sutures or lines of
dehiscence . . . . . . . . . . . . . . . . . . . . . **Eurotiaceae** (p. 56) (Figs. 18–23)
   cf. Trichocomataceae of Malloch and Cain, 1971 CBA/AIBS meeting, Edmonton, Canada.

39′(35′) Ascospores brown, not of bivalve construction; cleistochecia black and apically glabrous with conspicuous sutures or lines of dehiscence at maturity, covered with coarse yellow villus when young . . . . . . . . . . . . . . . **Cephalothecaceae** (p. 57)
                                                                           **Cephalotheca**
                                                                            With 1 sp.—*C. sulfurea*.

40(39) Ascocarps sometimes diffuse; peridium composed of loosely interwoven white to bright yellow hyphae, sometimes scanty and inconspicuous . . . . . . . . 41

40′(39) Ascocarps discrete; peridium at maturity composed of one to many layers of flattened isodiametrical or irregular pseudoparenchymatous cells . . . . . . . 45

41(40) Clusters of asci surrounded by scanty wefts of thin hyphae . . . . . . . . . 42

41′(40) Peridium composed of interwoven hyphae, white to light or bright yellow or brownish . . . . . . . . . . . . . . . . . . . . . . . . . . . . . . 43

42(41) Asci developed singly from croziers; ascocarp initial composed of ascogonium coiled about a swollen antheridium; imperfect state: *Paecilomyces*
     . . . . . . . . . . . . . . . . . . . . . . . . . . . . . . **Byssochlamys**

42′(41) Asci produced in chains; ascocarp initial composed of ascogonial coils without antheridium; imperfect state: *Penicillium pallidum* series
     . . . . . . . . . . . . . . . . . . . . . . . . . . . . **Talaromyces emersonii**
                                                                             Thermophilic

43(41′) Asci developed singly from croziers; ascocarp initial a coiled ascogonium without an antheridium; imperfect state: *Penicillium avellaneum* series . . . . . . . . . . **Hamigera**

43′(41′) Asci produced in short chains . . . . . . . . . . . . . . . . . . . . . . 44

44(43′) Imperfect state: *Paecilomyces* or *Penicillium* in other than the Biverticillata-Symmetrica section . . . . . . . . . . . . . . . . . . **Talaromyces thermophilus**
                                                                         **Talaromyces leycettanus**
                                                                                 Thermophilic

44′(43′) Imperfect state: *Penicillium* (Biverticillata-Symmetrica section) . . . . . . . . . . . . . . . . . . . . . . . . . . . . . . . . **Talaromyces**
                                                                               *sensu strictu*

45(40′) Peridium thin, composed of one or few layers of flattened polygonal cells, with or without a loose investiture of hyphal elements . . . . . . . . . . . . . . . . 46

45′(40′) Peridium thick, usually with outer layers of heavy-walled cells and inner layers of thinner walled cells; cleistothecia sometimes sclerotioid . . . . . . . . . . . . . 49

46(45) Cleistothecia without hyphal investiture, usually bright yellow, peridium one cell-layer thick; imperfect state: osmophilic *Aspergillus glaucus* group . . **Eurotium**

46′(45) Cleistothecia loosely invested with sterile hyphae or "hülle" cells . . . . 47

47(46′) Cleistothecia reddish to purple, surrounded by hülle cells; ascospores red to violet or blue; imperfect state: *Aspergillus nidulans* group . . . . . . . . . . . . **Emericella**

47′(46′) Cleistothecia white to slightly yellowish, loosely invested with sterile hyphae; ascospores hyaline to pale yellow with equatorial crests . . . . . . . . . . . . . 48

48(47′) Imperfect state: *Aspergillus fumigatus* group . . . . . . . . . . . . **Sartorya**

48'(47') Imperfect state: phialospores singly or in short chains . . . . . . . . . . . . . . . . . . . . . . . . . . . . . . . . . **Dichotomomyces**

49(45') Cleistothecia sclerotioid, yellow or light brown maturing slowly from the center outward . . . . . . . . . . . . . . . . . . . . . . . . . . . . . . . . . . . . 50

49'(45') Cleistothecia heavy-walled, not sclerotioid . . . . . . . . . . . . . . . . . 52

    50(49) Imperfect state: *Penicillium* (Monoverticillata or Divaricata) . . . . . . . . . . . . . . . . . . . . . . . . . . . **Eupenicillium**
                                                                                           Syn. *Carpenteles*.

    50'(49) Imperfect state: *Aspergillus* . . . . . . . . . . . . . . . . . . . . . . . 51

51(50') Ascospores hyaline, lenticular, with two equatorial crests . . . . **Hemicarpenteles**

51'(50') Ascospores hyaline, subglobose or short ellipsoidal, uniguttulate, without crests . . . . . . . . . . . . . . . . . . . . . . . . . . . . . . . . . . . . **Dichlaena**

    52(49') Cleistothecia brown, coriaceous; ascospores ellipsoid with subterminal radiating gelatinous appendages; of marine origin; imperfect state unknown . . . . . **Eiona**

    52'(49') Cleistothecia cream to yellow later assuming reddish brown tints . . . . 53

53(52') Ascospores globose with slender, rigid raylike surface projections uniguttulate; of marine origin . . . . . . . . . . . . . . . . . . . . . . . . . . . . . . **Amylocarpus**
                                                                                          Syn. *Plectolitus*.

53'(52') Ascospores about 20 μm in diameter, globose, densely warted; not of marine origin; described with *Gliocladium* imperfect state . . . . . . . . . . . . . **Lilliputia**

    54(32') Imperfect state: sympodulospores (*Sporothrix*) . . . . . . . . . . . . . . 55

    54'(32') Imperfect state: phialospores in gloeoid masses (*Cephalosporium, Acremonium, Stilbella*) . . . . . . . . . . . . . . . . . . . . . . . . . . . . . . . . . . . 56

55(54) Peridium heavy; ascospores brown, reniform . . . . . . . . . . . . **Fragosphaeria**

55'(54) Peridium one cell-layer thick; ascospores hyaline then brown, globose or elliptical, smooth . . . . . . . . . . . . . . . . . . . . . . . . . . . . . . . . **Pseudeurotium**

    56(54') Ascospores hyaline, globose, smooth . . . . . . . . . . . . . **Nigrosabulum**

    56'(54') Ascospores dark, variously ornamented . . . . . . . . . . . . . . . . . 57

57(56') Ascospores elliptical with several winglike appendages . . . . . . . **Emericellopsis**

57'(56') Ascospores globose, dark olive green to brown, reticulate . . . . . **Hapsidospora**

### REFERENCES

Ainsworth, G. C. (1961 and 1971). "Ainsworth and Bisby's Dictionary of Fungi," 5th ed. 1961; 6th ed. 1971. Commonwealth Mycol. Inst., Kew, Surrey, England.

Ajello, L. (1968). A taxonomic review of the dermatophytes and related species. *Sabouraudia* **6**:147–159.

Ajello, L., (1971). Sexual reproduction among fungi pathogenic to man. A historical review. *Mykosen* **14**:343–352.

Alexopoulos, C. J. (1962). "Introductory Mycology," 2nd ed., pp. 262–291. Wiley, New York. (1st ed., 1952).

Andrus, C. F. (1936). Cell relations in the perithecium of *Ceratostomella multiannulata*. *Mycologia* **28**:133–153.

Andrus, C. F., and L. L. Harter. (1933). Morphology of reproduction in *Ceratostomella fimbriata*. *J. Agr. Res.* **46**:1059–1078.

## 4. Plectomycetes: Eurotiales

Andrus, C. F., and L. L. Harter. (1937). Organization of the unwalled ascus in two species of *Ceratostomella*. *J. Agr. Res.* **54**:19–46.
Apinis, A. E. (1964). Revision of British Gymnoascaceae. *Mycol. Pap.* **96**:1–56.
Apinis, A. E. (1967). *Dactylomyces* and *Thermoascus*. *Trans. Brit. Mycol. Soc.* **50**:573–582.
Backus, M. P., and P. A. Orpurt. (1961). A new *Emericellopsis* from Wisconsin with notes on other species. *Mycologia* **53**:64–83.
Baranetzky, J. (1872). Entwickelungsgeschichte des *Gymnoascus Reesii*. *Bot. Ztg.* **30**:145–160.
Barron, G. L. (1968). "The Genera of Hyphomycetes from Soil." Williams & Wilkins, Baltimore, Maryland.
Benjamin, C. R. (1955). Ascocarps of *Aspergillus* and *Penicillium*. *Mycologia* **47**:669–687.
Benjamin, R. K. (1956). A new genus of the Gymnoascaceae with a review of the other genera. *El Aliso* **3**:301–328.
Bessey, E. A. (1950). "Morphology and Taxonomy of Fungi." McGraw-Hill (Blakiston), New York.
Boedijn, K. B. (1935a). On the morphology and cytology of *Trichocoma paradoxa*. *Ann. Jard. Bot. Buitenzorg* **44**:243–256.
Boedijn, K. B. (1935b). The genus *Dendrosphaera* in The Netherlands Indies. *Bull. Jard. Bot. Buitenzorg. Ser. III*, **13**:472–477.
Booth, C. (1966). Fruit bodies in Ascomycetes. *In* "The Fungi" (G. C. Ainsworth and A. S. Sussman, eds.), Vol. 2, Chapter 5, p. 133. Academic Press, New York.
Boudier, E. and N. Patouillard. (1900). Note sur deux champignon hypogés. *Bull. Soc. Mycol. Fr.* **16**:144–146.
Brefeld, O. (1891). "Untersuchungen aus dem Gesammtgebiet der Mykologie," Vol. 10, pp. 157–378. Münster, Germany.
Brown, A. H. S. and G. Smith. (1957). The genus *Paecilomyces* Bainier and its perfect stage *Byssochlamys* Westling. *Trans. Brit. Mycol. Soc.* **40**:17–89.
Cain, R. F. (1956). Coprophilous ascomycetes. II. *Can. J. Bot.* **34**:675–687.
Cain, R. F. (1972). Evolution of the fungi. *Mycologia* **64**: 1–14.
Chesters, C. G. C. (1934). Studies on British pyrenomycetes. I. The life histories of three species of *Cephalotheca* Fuck. *Trans. Brit. Mycol. Soc.* **19**:261–279.
Ciferri, R. (1957). *Trichocoma paradoxum* in Santo Domingo and the order Onygenales. *Atti. Ist. Bot. Univ. Lab. Crittogam., Pavia* **14**:1–4.
Clements, F. E., and C. L. Shear. (1931). "The Genera of Fungi." Wilson, New York.
Cole, G. T., and W. B. Kendrick. (1968). Conidium ontogeny in hyphomycetes. The imperfect state of *Monascus ruber* and its meristem arthrospores. *Can. J. Bot.* **46**:987–992.
Cole, G. T., and W. B. Kendrick. (1969a). Conidium ontogeny in hyphomycetes. The annellophores of *Scopulariopsis brevicaulis*. *Can. J. Bot.* **47**:925–929.
Cole, G. T., and W. B. Kendrick. (1969b). Conidium ontogeny in hyphomycetes. The arthrospores of *Oidiodendron* and *Geotrichum*, and the endoarthrospores of *Sporendonema*. *Can. J. Bot.* **47**:1773–1780.
Cooney, D. G., and R. Emerson. (1964). "Thermophilic Fungi. An Account of their Biology, Activities, and Classification." Freeman, San Francisco, California.
Dangeard, P. A. (1907). Recherches sur le développement du périthèce chez les Ascomycètes. *Botaniste* **10**: 1–385.
Davidson, D. E. and M. Christensen. (1971). *Emericellopsis stolkiae* sp. nov. from saline soils in Wyoming. *Trans. Brit. Mycol. Soc.* **57**:385–391.
de Bary, A. (1866). "Morphologie und Physiologie der Pilze, Flechten und Myxomyceten." Engelmann, Leipzig.
Dennis, R. W. G. (1968). "British Ascomycetes." Cramer, Lehre.
Dodge, C. W. (1929). The higher Plectascales. *Ann. Mycol., Berlin* **27**: 145–184.

Emmons, C. W. (1935). The ascocarps of species of *Penicillium*. *Mycologia* **27**:128–150.
Fischer, E. (1897). Plectascineae. *In* "Pflanzenfamilien" (A. Engler and K. Prantl, eds.), I Teil, 1 Abt., pp. 290–320. Engelmann, Leipzig.
Fischer, E. (1900). Nachtrage zu Teil I, Abt. 1. *In* "Pflanzenfamilien" (A. Engler and K. Prantl, eds.), I Teil, 1\*\* Abt., pp. 536–539. Engelmann, Leipzig.
Fries, E. M. (1849). "Summa Vegetabilium Scandinaviae," Sectio posterior. A. Bonnier, Holmiae et Lipsiae.
Gäumann, E. A. (1952). "The Fungi" (transl. by F. L. Wynd.). Hafner, New York.
Gäumann, E. A. (1964). "Die Pilze." Birkhaeuser, Basel.
Hawker, L. E. (1954). British hypogeous fungi. *Phil. Trans. Roy. Soc. London, Ser. B.* **237**:429–546.
Hughes, S. J. (1953). Conidiophores, conidia and classification. *Can. J. Bot.* **31**:577–659.
Kendrick, W. B., ed. (1971). "Taxonomy of Fungi Imperfecti" (Proc. 1st International Specialists' Conference on Criteria and Terminology in the Classification of Fungi Imperfecti). Univ. of Toronto Press, Toronto.
Kendrick, W. B., and G. T. Cole. (1968a). Conidium ontogeny in hyphomycetes. The sympodulae of *Beauvaria* and *Curvularia*. *Can. J. Bot.* **46**:1129–1301.
Kendrick, W. B., and G. T. Cole. (1968b). Conidium ontogeny in hyphomycetes. *Trichothecium roseum* and its meristem arthrospores. *Can. J. Bot.* **47**:345–350.
Kohlmeyer, J. (1968). Dänische Meerespilze (Ascomycetes). *Ber. Deut. Bot. Ges.* **81**:58–60.
Kominami, K., Y. Kobayashi, and K. Tubaki. (1952). Is *Trichocoma paradoxa* conspecific with *Penicillium luteum? Nagaoa* **2**:16–23.
Kuehn, H. H. (1955). Observations on the Gymnoascaceae. I and II. *Mycologia* **47**:533–545 and 878–890.
Kuehn, H. H. (1956). Observations on the Gymnoascaceae. III. *Mycologia* **48**:805–820.
Kuehn, H. H. (1957). Observations on the Gymnoascaceae. IV and V. *Mycologia* **49**:55–67 and 694–706.
Kuehn, H. H. (1958). A preliminary survey of the Gymnoascaceae. I. *Mycologia* **50**:417–439.
Kuehn, H. H. (1959). A preliminary survey of the Gymnoascaceae. II. *Mycologia* **51**:665–692.
Lindau, G. (1899). Ueber Entwickelung and Ernährung von *Amylocarpus encephaloides* Curr. *Hedwigia* **38**:1–19, Tafs. I and II.
Ludwig, F. (1892). "Lehrbuch der niederen Kryptogamen," pp. 263–265. Enke Stuttgart.
Luttrell, E. S. (1951). Taxonomy of the Pyrenomycetes. *Univ. Mo. Stud.* **24**:1–120.
Luttrell, E. S. (1955). The ascostromatic ascomycetes. *Mycologia* **47**:511–532.
Maire, R. (1917). Champignons Nord-Africains nouveaux ou peu connus. *Bull. Soc. Hist. Natur. Afr. N.* **8**:158–159.
Malloch, D., and R. F. Cain. (1970a). The genus *Arachnomyces*. *Can. J. Bot.* **48**:839–845.
Malloch, D., and R. F. Cain. (1970b). Five new genera in the new family Pseudeurotiaceae. *Can. J. Bot.* **48**:1815–1825.
Malloch, D., and R. F. Cain. (1971). New genera of Onygenaceae. *Can J. Bot.* **49**:839–846.
Martin, G. W. (1961). Key to the families of fungi. *In* "Ainsworth and Bisby's Dictionary of the Fungi" (by G. C. Ainsworth), pp. 497–519. Commonwealth Mycol. Inst., Kew, Surrey, England.
Miller, J. H. (1941). The ascomycetes of Georgia. *Plant Dis. Rep., Suppl.* **131**:1–93.
Miller, J. H. (1949). A revision of the classification of the Ascomycetes with special emphasis on the Pyrenomycetes. *Mycologia* **41**:99–127.
Moreau, F. (1953). "Les Champignons, Physiologie, Morphologie, Développement et Systematique," Vol. 2. Chap. 13, *Encycl. Mycol.* Vol. 22/23, pp. 1332–1381. Lechevalier, Paris.

Müller, E. (1971). Imperfect-perfect connections in Ascomycetes. *In* "Taxonomy of Fungi Imperfecti" (W. B. Kendrick, ed.), Chapter 13, p. 184. Univ. of Toronto Press, Toronto.
Müller, E., and J. A. von Arx. (1962). Die Gattungen der didymosporen Pyrenomyceten. *Beitr. Kryptogamenflora Schweiz* **2**:1–922.
Nannfeldt, J. A. (1932). Studien über die Morphologie and Systematik der nichtlichenisierten inoperculaten Discomyceten. *Nova Acta Regiae Soc. Sci. Upsal.* [4] **8**:1–368.
Nicot, J., and F. Durand. (1969). Une espèce remarquable du genre *Arachniotus: A. lectardii* Nicot sp. nov. *Bull. Soc. Mycol. Fr.* **85**:315–320.
Orr, G. F., and H. H. Kuehn. (1963). The genus *Ctenomyces* Eidam. *Mycopathol. Mycol. Appl.* **21**:321–333.
Orr, G. F., and H. H. Kuehn. (1964a). A re-evaluation of *Myxotrichum spinosum* and *M. cancellatum. Mycologia* **56**:473–481.
Orr, G. F., and H. H. Kuehn. (1964b). A new genus of the Gymnoascaceae with dark ascocarps. *Mycologia* **56**:482–487.
Orr, G. F., and H. H. Kuehn. (1971). Notes on Gymnoascaceae. I. A review of eight species. *Mycologia* **63**:191–203.
Padhye, A. A., and J. W. Carmichael. (1971). The genus *Arthroderma* Berkeley. *Can. J. Bot.* **49**:1525–1540.
Parbery, D. G. (1969). *Amorphotheca resinae* gen. nov., sp. nov., the perfect state of *Cladosporium resinae. Aust. J. Bot.* **17**:331–357.
Patouillard, N. (1907). Champignons noveaux du Tonkin. *Bull. Soc. Mycol. Fr.* **23**:69–79.
Raper, K. B. (1957). Nomenclature in *Aspergillus* and *Penicillium. Mycologia* **49**:644–662.
Raper, K. B., and D. I. Fennell. (1965). "The Genus Aspergillus." Williams & Wilkins, Baltimore, Maryland.
Raper, K. B., and C. Thom. (1949). "A Manual of the Penicillia." Williams & Wilkins, Baltimore, Maryland.
Schröter, J. (1893). Die Pilze Schlesiens, in F. Cohn, *Krypt-Fl. Schles.* **3**:210–213.
Schröter, J. (1897). Hemiascineae. *In* "Pflanzenfamilien" (A. Engler and K. Prantl, eds.), I Teil, 1 Abt., pp. 148–149. Engelmann, Leipzig.
Scott, DeB. (1968). The genus *Eupenicillium* Ludwig. *CSIR (Counc. Sci. Ind. Res., S. Afr.), Rep.* No. 272, 150 pp., Pretoria.
Scott, DeB. (1970). *Dichotomomyces cejpii* (Mil'ko) comb. nov. *Trans. Brit. Mycol. Soc.* **55**:313–316.
Spiltoir, C. F., and L. S. Olive. (1955). A reclassification of the genus *Pericystis* Betts. *Mycologia* **47**:238–244.
Stolk, A. C. (1965). Thermophilic species of *Talaromyces* Benjamin and *Thermoascus* Miehe. *Antonie van Leeuwenhoek; J. Microbiol. Serol.* **31**:262–276.
Stolk, A. C., and R. A. Samson. (1971). Studies on *Talaromyces* and related genera. I. *Hamigera* gen. nov. and *Byssochlamys. Persoonia* (Leiden) **6**:341–357.
Stolk, A. C., and DeB. Scott. (1967). Studies on the genus *Eupenicillium* Ludwig. I. Taxonomy and nomenclature of Penicillia in relation to their sclerotioid ascocarpic states. *Persoonia* **4**:391–405.
Subramanian, C. V. (1962). The classification of the Hyphomycetes. *Bull. Bot. Surv. India* **4**:249–259.
Tubaki, K. (1958). Studies on the Japanese Hyphomycetes. V. Leaf and stem group with a discussion of the classification of Hyphomycetes and their perfect stages. *J. Hattori Bot. Lab.* **20**:142–244.
Tubaki, K. (1963). Taxonomic study of Hyphomycetes. *Annu. Rep. Inst. Ferment., Osaka* **1**:25–54.

Turian, G. (1966). Morphogenesis in Ascomycetes. *In* "The Fungi" (G. C. Ainsworth and A. S. Sussman, eds.), Vol. 2, Chapter 11, p. 339. Academic Press, New York.

von Arx, J. A. (1968). "Pilzkunde," pp. 108–115. Cramer, Lehre.

von Arx, J. A. (1970). "Genera of Fungi Sporulating in Pure Culture," pp. 81–96. Cramer, Lehre.

von Arx, J. A. (1971a). Testudinaceae, a new family of Ascomycetes. *Persoonia (Leiden)* **6**:365–369.

von Arx, J. A. (1971b). On *Arachniotus* and related genera of the Gymnoascaceae. *Persoonia* **6**:371–380.

von Arx, J. A., and E. Müller. (1954). Die Gattungen der amerosporen Pyrenomyceten. *Beitr. Kryptogamenflora Schweiz* **2**:1–434.

von Arx, J. A., and T. Nilsson. (1969). *Xylogone sphaerospora*, a new ascomycete from stored pulpwood chips. *Sv. Bot. Tidskr.* **63**:345–348.

von Höhnel, F. (1917). Mykologischen Fragment CLXXIV. Über *Sphaeria fuliginosa* Fries und die Cephalothecaceae. *Ann. Mycol.* **15**:360–363.

Wolf. F. A., and F. T. Wolf. (1947). "The Fungi." Vol. I. Wiley, New York.

# Ascomycotina
# Pyrenomycetes

(FOR KEY TO ORDERS, SEE P. 89)

# CHAPTER 5

# Pyrenomycetes: Erysiphales

C. E. YARWOOD

*Department of Plant Pathology*
*University of California*
*Berkeley, California*

## I. ERYSIPHALES

The Erysiphales, commonly synonymous with the Perisporiales, comprise those fungi with asci arranged as a pallisade within a nonostiolate, dark-colored perithecium borne superficially on the host surface. All forms are parasitic on the above-ground parts of higher plants, are nourished by haustoria, and are obligate parasites. Families within the Erysiphales range from one (Erysiphaceae—Ainsworth, 1963, p. 509) to at least six (Bessey, 1961). Here I will use the common subdivision of the order into two families —the Erysiphaceae, or powdery mildews, with colorless hyphae, and the Perisporiaceae, commonly synonymous with the Meliolaceae, the dark mildews, sometimes incorrectly called the sooty molds.

### A. *Comparison of the Erysiphaceae and the Perisporiaceae*

Except for color, presence or absence of conidiophores, and septation of ascospores, most characters of the Erysiphaceae and Perisporiaceae are similar. Both usually have superficial mycelia and perithecia on the upper and lower surfaces of leaves and on stems. In the Erysiphaceae, the mycelium is held to the leaf surface by appressoria, which are lobed extensions of prostrate hyphal cells. In the Perisporiaceae, the holdfasts are usually two celled and are called capitate hyphopodia. In both families, the haustoria are usually globular in the host epidermal cells and connected to the appressoria or hyphopodia by fine penetration tubes. According to Arnaud (1921) both families are attacked by *Cicinnobolus*, and these are the only groups of fungi attacked by this parasite.

In the Erysiphaceae, conidia are almost universally abundant, whereas in Perisporiaceae they are absent. The ascospores of Erysiphaceae are one celled and colorless, whereas in the Perisporiaceae they mostly have two or more cells and are colored. Septate ascospores of the genera *Schistodes*,

*Astomella*, and *Leucoconis* in the Erysiphaceae are reported from India or Africa (Ainsworth, 1963), but I have not seen these. The Erysiphaceae are more studied and more important, but are clearly less numerous as usually classified, than are the Perisporiaceae. If the species of Erysiphaceae were separated on such small differences as are the Perisporiaceae (Hansford, 1946, 1961, 1963) then there might be as many species of Erysiphaceae as there are of Perisporiaceae. Certainly the conidial states of *Erysiphe polygoni* on *Trifolium pratense, Phaseolus vulgaris, Eschscholtzia californica*, and *Brassica oleracea* var. *capitata* are as different from each other as most species of *Meliola*.

In both families, the original spore from which a colony arises is still identifiable in old colonies as an integral part of the colony. Colonies of Perisporiaceae develop more slowly than those of Erysiphaceae. Asci with less than eight ascospores are common in both families.

The Erysiphaceae extend from the tropics to the Arctic and from below sea level to 4000 m (Hirata, 1966), whereas the Perisporiaceae are primarily tropical. The apparent concentration of Erysiphaceae in temperate areas and their apparent scarcity in tropical areas (Bessey, 1961; Blumer, 1967) may be partly because of the necessity of the perithecial stage for identification. For example, Hansford (1946) writes, "Though the fungi of this family [Erysiphaceae] are extremely common in tropical Africa on a wide range of hosts, no perithecia of any species have hitherto been encountered, so that the writer's collections cannot be referred with certainty to perithecial stages described from other parts of the world." Hirata (1966) also indicates that powdery mildews are moderately abundant in the tropics, and perithecia in several collections of Erysiphaceae are indicated by Viégas (1944).

Erysiphaceae occur predominantly on cultivated plants and are more destructive on cultivated than on wild plants, whereas the Perisporiaceae occur predominantly on wild plants and are rarely injurious.

The relative importance of the Erysiphaceae and the Perisporiaceae may be indicated by the literature concerning each. During the years 1959–1963, in "Review of Applied Mycology," there are 501 indexed items to species of *Erysiphe*, but only three indexed items to species of *Meliola*; these genera are likely the most important and most studied of the Erysiphaceae and Perisporiaceae, respectively.

## II. ERYSIPHACEAE

The first recognition of a powdery mildew as a distinct plant entity may be the naming of *Mucor Erysiphe* on the foliage of *Humulus, Acer, Lamium, Galiopsis*, and *Lithospermum* (Linnaeus, 1753). The Erysiphaceae were perhaps first recognized as a taxonomic group by Leveillé (1851). Major modern taxonomic treatments are by Salmon (1900), Homma (1937), and Blumer (1967). The similarity in the taxonomic treatments by Salmon and Blumer, 67 years apart, indicates that the naming in this group is reasonably stabilized. Other treatments, exclusive of routine treatments in textbooks, are by Brundza (1933), Sawada (1914), Hirata (1966), and Yarwood (1957). The most recent comprehensive bibliography of the group (1462 references) is by Hirata.

The number of genera of Erysiphaceae range up to at least eighteen (Hirata, 1966), according to different authorities; but here I will follow Blumer in restricting the genera to eight.

### KEY TO GENERA OF ERYSIPHACEAE

1. Mycelium superficial . . . . . . . . . . . . . . . . . . . . . . . . . . 2
    2(1) Perithecia with 1 ascus . . . . . . . . . . . . . . . . . . . . . . 3
        3(2) Appendages of perithecia simple . . . . . . . . . . . . . . **Sphaerotheca**
        3'(2) Appendages of perithecia dichotomously branched . . . . . . **Podosphaera**
    2'(1) Perithecia with several asci . . . . . . . . . . . . . . . . . . . 4
        4(2') Appendages of perithecia simple . . . . . . . . . . . . . . . **Erysiphe**
        4'(2') Appendages of perithecia dichotomously branched . . . . . . **Microsphaera**
        4"(2') Appendages of perithecia coiled at tips . . . . . . . . . . . **Uncinula**
1'. Mycelium partly internal . . . . . . . . . . . . . . . . . . . . . . . . 5
    5(1') Appendages of perithecia simple . . . . . . . . . . . . . . . . **Leveillula**
    5'(1') Appendages of perithecia with basal swelling . . . . . . . . . . **Phyllactinia**
1". Perithecia lacking . . . . . . . . . . . . . . . . . . . . . . . . . . **Acrosporium**
                                                                Syn. *Odium* Sacc.

I am omitting some genera described for Africa, Asia, and South America because I do not feel competent to appraise their validity from the literature.

The number of species of Erysiphaceae is given as 60 by Salmon (1900), 125 by Blumer (1967), and 265 by Hirata (1966). Both Salmon and Blumer apparently based their speciation on personal examination of specimens. Blumer split several of Salmon's species but restricted his collection to middle Europe, whereas Salmon examined specimens from throughout the world. Hirata based his speciation on published records from throughout the world and, therefore, likely included several synonyms.

The problem of the validity of genera and species is basically the same in Erysiphaceae as in other groups of plants. *Phyllactinia* is certainly the best defined genus, as its perithecia and conidiophores show no overlapping with other genera. In the other six genera based on perithecia, there is almost a continuous gradation among no appendages, few or short appendages, simple appendages, dichotomously branched appendages, and uncinulate appendages (see Fig. 7), though it is usually easy to decide to which genus a given collection of perithecia should be assigned. Another criticism of the present genera is the inclusion of three radically different conidiophore types in the genus *Erysiphe*. The difference in the conidiophores of *E. graminis*, *E. polygoni*, and *E. cichoracearum* are greater than those existing between the conidiophores of other genera, and each of these conidiophore types might well be the basis of a separate genus.

Some confusion of genera exists between *Erysiphe* and *Microsphaera*. The powdery mildews of bean (U.S. Department of Agriculture 1960), clover (Peterson, 1938), sugar beet (Neuwirth, 1930; Hirata, 1966), and sweet pea (U.S. Department of Agriculture, 1960) are assigned to both *Erysiphe* and *Microsphaera*. That these collections are the same species is indicated by the apparent identity of the conidiophores of *E. polygoni* and *M. alni* and by the similarity of the appendages, except for the tips, which can extend almost to the range of those two genera in a given collection (Blumer, 1967; Fig. 7 of this study).

Erysiphaceae are parasitic on some 7187 host species in 1289 genera, 149 families, and 44 orders of angiosperms (Hirata, 1966). About 90% of these are dicotyledons, and only one species of powdery mildew, *Erysiphe graminis*, is known to attack the monocotyledons. This wide host range for a specific group of foliage pathogens is probably exceeded only by the rusts (Uredinales). Hosts include some 13–38% of the total angiosperms in a given geographic region (Hirata, 1966).

The Erysiphaceae have dark spherical perithecia as well as colorless, septate, uninucleate hyphae, and conidiophores and conidia which appear white on plant surfaces. They occur primarily on the leaves, buds, flowers, and fruits. Other characteristics are the occurrence of haustoria in the epidermal cells of their hosts; their luxuriant development in rain-free seasons; the high water content of their large, turgid, passively liberated, airborne conidia; their diurnal periodicity with respect to several characters; the reversible phototropism of the germ tubes of some species; the compatible association with their natural hosts; the macroscopically visible

FIG. 1. Conidiophores and mycelium of *Sphaerotheca pannosa* on *Prunus persica* [From Tulasne and Tulasne (1861), Plate 3.]

FIG. 2. Conidium types: (A) *Erysiphe graminis* from *Hordeum vulgare* to illustrate elongate conidium and vacuoles. (B) *Erysiphe polygoni* from *Phaseolus vulgaris* to illustrate ovate conidium with vacuoles. (C) *Sphaerotheca fuliginea* from *Cucumis sativus* to illustrate vacuoles and fibrosin bodies. (D) *Sphaerotheca lanestris* from *Quercus agrifolia* to illustrate barrel-shaped conidia. (E) *Phyllactinia corylea* from *Rhamnus californica* to illustrate pointed conidia. (F) *Phyllactinia corylea* from *Quercus agrifolia* to illustrate median constriction of conidia. (Drawn by Margorie Baird Garlin.)

FIG. 3. Stages in the infection process of *Erysiphe polygoni* on resistant and susceptible clones of *Trifolium pratense*. (From Yarwood, 1934.)

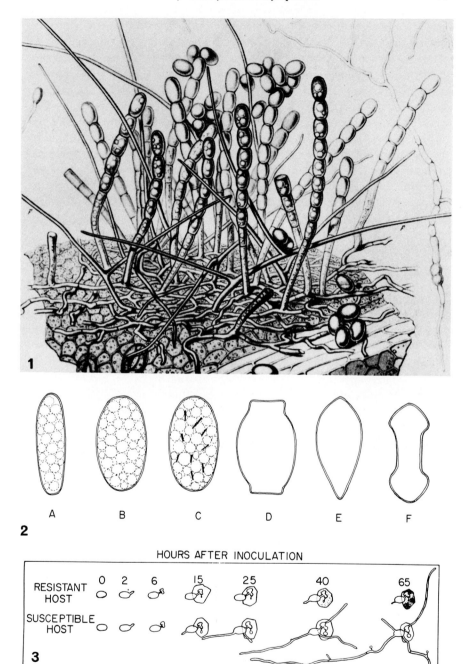

hypersensitive reaction shown by some incompatible host association; the slow injury to their hosts; and their vulnerability to control by fungicides. They are favored by succulent host tissue and by cool, shaded environments. The powdery appearance is caused by the conidiophores and maturing conidia which are formed on erect conidiophores borne singly at right angles to the leaf surface (Fig. 1). The conidiophores are a chain of cells which are shorter but of greater diameter than hyphal cells. With *Erysiphe polygoni, Microsphaera, Uncinula, Leveillula,* and *Phyllactinia,* one conidium is typically formed per conidium per day, and this is slowly and passively abstricted about noon. With *E. graminis, E. cichoracearum, Sphaerotheca,* and *Podosphaera,* several conidia are matured daily. Conidia (Fig. 2) are typically ellipsoidal, with thin walls and numerous conspicuous vacuoles of water. Fibrosin bodies—refractive, straight, or curved rodlike structures—are present in the conidia of *Sphaerotheca* and *Podosphaera.*

With *Erysiphe polygoni* and probably other species, there is an alteration of appressorium-bearing and conidiophore-bearing cells of the hyphae. Probably because of the depression of the leaf surface at the contact of two epidermal cells, there is a slight tendency for the hyphae to follow these intercellular junctions.

Haustoria (Fig. 5) are of two principal types: the globular, formed by most Erysiphaceae, and the digitate, formed by *Erysiphe graminis.* Conidia are short lived. Overwintering is primarily by perithecia or by mycelial infections in buds of perennials. The sexual and asexual life cycle (Fig. 6) is similar to that of many other ascomycetes. Classification is primarily based on the types of appendages (Fig. 7) of the perithecia and the number of asci; these characters are usually clearly expressed, but there is some bothersome overlapping in the characters of the appendages. A major fault with classification by perithecia is that in many areas perithecia are never, or only rarely, formed. In California, of 1164 collections since 1934, only 117 had perithecia. Perithecial formation is also a characteristic of species of mildew, and of host. *Erysiphe aggregata* on alder catkins has been found only as perithecia. *Phyllactinia* is usually found with perithecia (*P. corylea* on *Quercus agrifolia* in the San Francisco Bay area of California is an exception). *Erysiphe polygoni* on *Trifolium longipes* usually has perithecia, but on *T. pratense* it normally does not. I have not found perithecia of *Sphaerotheca fuliginea* on cucurbits.

The perithecium starts as the apposition of two vertical cells at the junction of two prostrate hyphae. These are believed to function as antheridium and oogonium, but details of the fertilization process are controversial (Bessey 1961; Gäumann, 1926). Soon after contact of the sexual cells, other cells grow from the base of the oogonium and surround it. Eventually a wall of several layers is formed around the asci. The outer cells are thick walled, dark brown, and almost devoid of protoplasm, while the inner cells are thin

# 5. Pyrenomycetes: Erysiphales

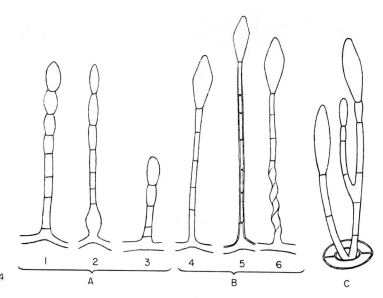

FIG. 4. Conidiophore types: (1) *Erysiphe cichoracearum*; (2) *E. graminis*; (3) *E. polygoni*; (4) *Phyllactinia suffulta*; (5) *P. rigida*; (6) *P. subspiralis*; (C) *Leveillula*. (From Blumer, 1933.)

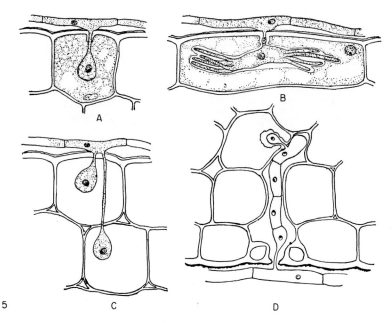

FIG. 5. Haustorium types: (A) *Erysiphe polygoni* (and most species of Erysiphaceae); (B) *Erysiphe graminis*; (C) *Uncinula salicis*; (D) *Phyllactinia corylea*. (From Heald, 1933; adapted from Smith, 1900.)

walled and colorless. In several species, as in *Erysiphe aggregata*, protoplasma-rich nurse cells are characteristic at the base of the asci. Maturing perithecia are at first colorless, then commonly yellowish to brown and, finally, black and opaque. The species *E. trina* is atypical because of its relatively transparent, mature perithecia.

*Typhulochaeta japonica* (Homma, 1937), *Phyllactinia roboris* (Blumer, 1967), and *Erysiphe aggregata* (C. E. Yarwood, 1971, unpublished material) are known only in their perithecial stages. *Erysiphe cichoracearum* (Yarwood,

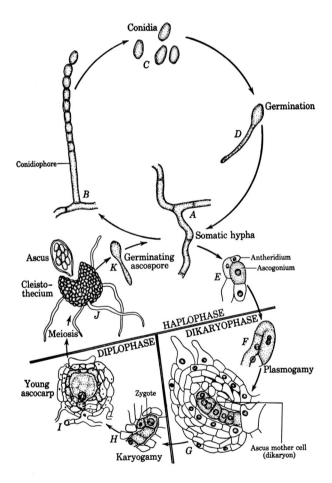

FIG. 6. Life cycle of *Sphaerotheca castagnei*, typical of several Erysiphaceae. (From Alexopoulos, 1952.)

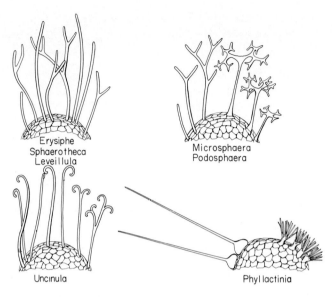

FIG. 7. Types of appendages of Erysiphaceae. (Drawn by Margorie Baird Garlin.)

1957), and *E. graminis* (Hiura, 1962) are believed to be heterothallic. *Sphaerotheca humuli* is homothallic (Homma, 1937).

To permit identification in the absence of perithecia, host indexes (U.S. Department of Agriculture, 1960; Hirata, 1966; Blumer, 1967) are useful. Usually there is only one species of powdery mildew on a given host species or even genus, but occasionally there are several. There are four species: *Erysiphe trina, Microsphaera alni, Sphaerotheca lanestris,* and *Phyllactinia corylea* on *Quercus* species on the Berkeley campus of the University of California, and as many as three species on a single leaf of *Q. agrifolia*. Each is distinct in its gross appearance and in macroscopic characters. As an aid to identification of genera and species in the absence of perithecia, a key based on conidiophore characters has been devised as follows.

### KEY TO ERYSIPHACEAE IN THEIR CONIDIAL STATES

1. Conidiophores usually abundant . . . . . . . . . . . . . . . . . . . . . . 2
　　2(1) Conidia borne single . . . . . . . . . . . . . . . . . . . . . . . . 3
　　　　3(2) Conidiophores some times emerging through stomata . . . . . . . **Leveillula**
　　　　3'(2) Conidiophores on superficial hyphae . . . . . . . . . . . . . . . 4
　　　　　　4(3') Base of conidiophore twisted . . . . . . . . . . . . **Phyllactinia subspiralis**
　　　　　　4'(3') Base of conidiophore straight . . . . . . . . . . . . . . . . . 5
　　　　　　　　5(4') Conidia ellipsoidal . . . . . . . . . . . . . . . . . . . . . 6

6(5) On *Vitis, Salix, Alnus, Parthenocissus, Ulmus, Prunus* .... **Uncinula**
6'(5) On other woody plants ................ **Microsphaera**
6"(5) On herbaceous plants ............... **Erysiphe polygoni**
5'(4') Conidia pointed ....................... **Phyllactinia**
2'(1) Conidia borne in chains .......................... 7
7(2') Base of conidiophore swollen ............. **Erysiphe graminis**
7'(2') Base of conidiophore straight ..................... 8
8(7') Conidia barrel-shaped ............... **Sphaerotheca lanestris**
8'(7') Conidia ellipsoidal ........................ 9
9(8') Conidia with fibrosin bodies ..................... 10
10(9) On herbaceous plants, and on *Rosa* and *Prunus* .... **Sphaerotheca**
10'(9) On other woody plants ................. **Podosphaera**
9'(8') Conidia without fibrosin bodies ........ **Erysiphe cichoracearum**
1'. Conidiophores rare ............................ 11
11(1') Conidia borne singly ................... **Erysiphe trina**
11'(1') Conidia borne in chains .......... **Sphaerotheca humuli** on *Ceanothus*
1". Conidiophores unknown .......................... 12
12(1") On *Quercus* .................... **Phyllactinia roboris**
12'(1") On *Alnus* ..................... **Erysiphe aggregata**

This is based on a moderate acquaintance with Erysiphaceae in California, but limited knowledge of other Erysiphaceae. There is an apparent lack of useful correlation between conidiophore characters and perithecium characters, which indicates that the classification in use does not represent the evolution of the group.

The host range of a morphological species of powdery mildew is usually large, but that of a biological race is usually small. Salmon (1900) lists *Erysiphe polygoni* on 357 host species in 157 genera. Single races, such as those started from a single spore, are sometimes limited to certain clones of a species (Yarwood, 1936), usually to a single host species, commonly to a single genus, and occasionally to several genera and even families (Hammarlund, 1925, 1945). The races attacking a genus are sometimes designated as a biological species, as *E. graminis hordei* with 72 races (Anonymous, 1968).

When conidia of a powdery mildew are dusted on a natural host, germination begins in about 2 hours, appressoria form in about 6 hours, haustoria and secondary hypha become apparent in about 15 hours, and from then on hypal extension continues at about a logarithmic rate (see Fig. 3). Conidiophores are initiated in about 3 days, and the first conidia mature in about 5 days. When conidia are dusted on an immune host, the sequence of events is about the same for the first 15 hours, but then extension of the fungus stops and host necrosis of the penetrated cell becomes apparent in about 24–40 hours. This latter is called the hypersensitive reaction and is characteristic of many incompatible host-parasite associations. The necrosis may or may not be apparent to the unaided eye, and the necrotic area may be limited to a single cell for each conidium, or may involve many cells.

The areas of greatest abundance and of damage due to Erysiphaceae appear from the literature, and my experience, to be California and Israel, both characterized by rain-free summers and intensive agriculture. Treatments which favor succulent growth and large leaves, such as tillage of the soil (when weeds are abundant), shading, induction of water sprouts of woody plants, and fertilization, appear to favor powdery mildew. Liming (Hirata, 1968; Vlamis and Yarwood, 1962) and low soil moisture (Yarwood, 1957) also favor infection.

The most studied, and likely the most important, powdery mildews are *Erysiphe graminis*, *Sphaerotheca fuliginea*, and *Uncinula necator*. On the basis of indexed items in "Review of Applied Mycology" 1964–1968, *E. graminis* is the most important pathogen of barley, but on wheat and oats it is less important than several rusts (*Puccinia* sp.). By the same token, *S. fuliginea* (including *E. cichoracearum* with which it is frequently confused) is the most important pathogen of cucumbers, followed by cucumber mosaic virus. In the same way, *U. necator* is second in importance to *Plasmopara viticola* on grape. In each of the above cases, the pathogen can render the crop worthless, although this rarely happens except under peculiarly favorable environmental conditions.

Because of their superficial position on plants, powdery mildews can be more readily controlled than can most parasitic fungi. Heavy syringing of the host with water (Yarwood, 1957) is moderately effective, but sulfur fungicides are most widely used. For sulfur-sensitive plants, Dinocap (dinitrocaprylphenyl crotonate) is used as a protective and therapeutant. Recently, benomyl [methyl (butylcarbamoyl-2-benzimidazole) carbamate] and other chemicals (Johnston, 1970) have been shown to control mildew when applied to the soil in which plants are growing.

## III. PERISPORIACEAE

The taxonomy of Perisporiaceae is more confusing and unstable than of the Erysiphaceae. For example, of the eighteen genera of Hansford (1946), only two are among the nine genera of Doidge (1917). The number of reported genera range up to at least 45 (Ainsworth, 1963), but here I will confine the genera to those of Hansford (1946, 1961) while excluding his forms with conidial stages. A key to the Perisporiaceae follows.

## KEY TO GENERA OF PERISPORIACEAE

1. Ascospores, two-celled .................................. 2
   2(1) Mycelial setae present ............................. 3
      3(2) Perithecia parenchymatous, apically dehiscent ........... **Balladyna**
      3'(2) Perithecia mucose-diffluent, with meridian hyphae ........ **Linotexis**
   2'(1) Mycelial setae none ............................... 4
      4(2') Perithecia not mucose diffluent ..................... 5
         5(4) Perithecia thick-walled, on radiate mycelial disc ......... **Armatella**
         5'(4) Perithecia thin-walled ........................ 6
            6(5') Perithecia with larviform setae ................. **Wageria**
            6'(5') Perithecia glabrous or with normal setae ......... **Balladynastrum**
            6"(5') Perithecia with hypostroma in stomata .......... **Stomatogene**
      4'(2') Perithecia mucose diffluent ...................... 7
         7(4') Perithecia with meridian hyphae ................ **Parenglerula**
         7'(4') Perithecia without meridian hyphae .............. 8
            8(7') Perithecia sessile ........................ 9
               9(8) Perithecia setose ..................... **Xenostigme**
               9'(8) Perithecia glabrous ................... **Dialacenium**
            8'(7') Perithecia stipitate ...................... **Thrauste**
1'. Ascospores more than two-celled ......................... 10
   10(1') Perithecia applanate ............................ 11
      11(10) No free mycelium ........................... **Actinodothis**
      11'(10) Free mycelium present ....................... **Amazonia**
   10'(1') Perithecia globose ............................ 12
      12(10') Mycelial setae present ........................ **Meliola**
      12'(10') Mycelial setae absent ........................ 13
         13(12') Perithecia with setae ..................... **Irenopsis**
         13'(12') Perithecia with larviform appendages ........... **Appendiculella**
         13"(12') Perithecia with neither setae nor appendages ....... **Asteridella**

Species of Perisporiaceae number about 1840, according to Hansford (1961). About 1814 of these are in the Meliolineae (=Phaeophragmae, three- or more-celled ascospores), and most of these are in the genus *Meliola*. Most members of the subfamily Phaeodidyme (two-celled ascospores) are of doubtful position in the fungi.

The major morphological and taxonomic characters of the Perisporiaceae Figs. 8–12 are as follows.

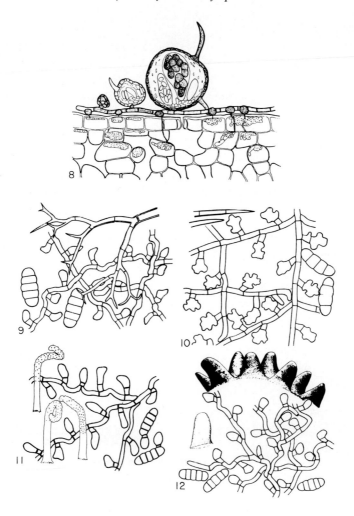

FIG. 8. *Balladynastrum entebeense* on *Mussaenta arcuata* showing superficial mycelium with hyphopodia and haustoria, and perithecia with setae, asci, and ascospores. (From Hansford, 1946.)

FIGS. 9–12. Mycelium, capitate hypopodia, and ascospores of representative Perisporiaceae. (9) *Meliola schimae*, with branched pointed mycelial setae. (10) *Meliola ganglifera*, with pointed mycelial setae. (11) *Irenopsis mikaniae*, with perithecial setae. (12) *Asteridiella venezuelensis*, with surface view of perithecia. (From Hansford, 1963.)

## 1. Mycelium

This is dark and superficial and, in contrast to the Erysiphaceae, its characters are well preserved in dried herbarium specimens. The length, diameter and branching of the hyphal cells are fairly uniform within a species, but vary characteristically between species. Hyphal cells may contain two nuclei.

## 2. Capitate Hyphopodia

These are short, commonly two celled branches from the main hyphae and are closely appressed to the host. The terminal cell is usually swollen and functions as an appressorium from which the haustorium arises.

## 3. Macronate Hyphopodia

These are formed on the hyphae in about the same position, and of similar size, to the capitate hyphopodia, but are one celled, commonly flask shaped, arising at right angles to the leaf, and appearing to have a terminal opening; however, their function is unknown.

## 4. Mycelial Setae

These are characteristic of the genus *Meliola*, and are reminiscent of the perithecial appendages of the Erysiphaceae. Setae are long, erect hyphae formed over the surface of the colony and/or only around the base of the perithecium.

## 5. Perithecia

These are usually globose, may be smooth or verrucose, and usually have two or more layers of cells enclosing the asci. In the genera *Wageria* and *Appenticulella* some of the cells of the outer wall are extended as larviform setae. Sexual fertilization is believed to be similar to the Erysiphaceae.

## 6. Ascospores

These are the most constant structure of the group. They vary from two- to five-celled and are dark. The genus *Anixia* with one-celled ascospores is accepted by Migula, but not by most subsequent reviewers. The ascospores have apparently not been germinated under controlled conditions.

The designation of species characters is commonly by a Beeli (1920) formula of eight figures, which designate qualitative limits of the above characters, as well as quantitative limits for size of spores and mycelial setae.

Even in the absence of perithecia, the Perisporiaceae, like the Erysiphaceae, are usually distinguishable from all other fungi.

For many years it was believed that various Perisporiaceae had imperfect states in such genera as *Helminthosporium, Calonectria,* and *Arthobotryum.* Hansford and others now believe that these imperfect fungi are parasitic on the Perisporiaceae.

Perisporiaceae occur in warm, humid, forested, tropical areas, but may occur in areas with wet seasons alternating with long droughts. They usually are found only on adult leaves.

### REFERENCES

Ainsworth, G. C. (1963). "Ainsworth and Bisby's Dictionary of Fungi," 5th ed. Commonwealth Mycol. Inst., Kew, Surrey, England.

Alexopoulos, C. J. (1952). "Introductory Mycology." Wiley, New York.

Anonymous. (1968). "Annual Report of The Plant Breeding Institute, 1966–1967." Cambridge Univ. Press, London and New York.

Arnaud, G. (1921). Sur les affinités des Erysiphées et des Parodiopsidées. *C. R. Acad. Sci.* **173**: 1394–1396.

Beeli, M. (1920). Note sur le genre *Meliola* Fr. *Bull. Jard. Bot. Brux.* **7**: 89–160.

Bessey, E. A. (1961). "Morphology and Taxonomy of Fungi." Hafner, New York.

Blumer, S. (1933). Die Erysiphaceen Mitteleuropas mit besonderer Berucksichtigung der Schweiz. *Beitr. Kryptogamenflora Schweiz* **7**:1–483.

Blumer, S. (1967). "Echte Mehltaupilze (Erysiphaceae)." Fischer, Jena.

Brundza, K. (1933). Beiträge zur Kenntnis der Erysiphaceen Litanens. *Žemes Ūkio Akad. Matraščio* **2**:107–197.

Doidge, E. M. (1917). South African Perisporiales. *Trans. Roy. Soc. S. Afr.* **6**:713–750.

Gäumann, E. A. (1926). "Vergleichende Morpologie der Pilze." Fischer, Jena.

Hammarlund, C. (1925). Zur Genetik, Biologie und Physiologie einiger Erysiphaceen. *Hereditas* **6**:1–126.

Hammarlund, C. (1945). Beiträge zur Revision einiger imperfekten Mehltau Arten, *Erysiphe polyphaga* nov. sp. *Bot. Notis.* pp. 101–108.

Hansford, C. G. (1946). The foliicolous ascomycetes, their parasites and associated fungi. *Mycol. Pap.* **15**:1–240.

Hansford, C. G. (1961). The Meliolineae, a monograph. *Sydowia*, Beih. **2**:1–806.

Hansford, C. G. (1963). Iconographia Meliolinearum. *Sydowia, Beih.* **5**:1–285.

Heald, F. D. (1933). "Manual of Plant Diseases." McGraw-Hill, New York.

Hirata, K. (1966). "Host Range and Geographical Distribution of the Powdery Mildews" (mimeo.). Niigata Univ., Niigata, Japan.

Hirata, K. (1968). Notes on host range and geographic distribution of the powdery mildew fungi. *Trans. Mycol. Soc. Jap.* **8**:73–88.

Hiura, U. (1962). Hybridization between varieties of *Erysiphe graminis. Phytopathology* **52**: 664–666.

Homma, Y. (1937). Erysiphaceae of Japan. *J. Fac. Agr., Hokkaido Imp. Univ.* **38**:183–461.

Johnston, H. W. (1970). Control of powdery mildew of wheat by soil-applied benomyl. *Plant Dis. Rep.* **54**:91–93.

Leveillé, J. H. (1851). Organisation et disposition méthodique des espèces qui composent le genre *Erysiphe. Ann. Sci. Natur.* [3] **15**:*109–179.*

Linnaeus, C. (1753). "Species Plantarum," Vol. 2, p. 1186. Stockholm.

Neuwirth, F. (1930). *Microsphaera betae* Vanha. *Z. Zuckerind. Cech. Repub.* **55**:75–79.

Peterson, G. A. (1938). Perithecial material of *Erysiphe* and *Microsphaera* on *Trifolium pratense. Mycologia* **30**:299–301.

Salmon, E. S. (1900). A monograph of the Erysiphaceae. *Mem. Torrey Bot. Club* **91**:1–292.
Sawada, K. (1914). On the classification of Erysiphaceae by the conidial stage. (Mostly in Japanese.) *Formosa Agr. Exp. Sta. Spec. Rep.* **9**:1–102. Abstract in *Rev. Appl. Mycol.* **7**:273 (1928).
Smith, G. (1900). The haustoria of the Erysiphaceae. *Bot. Gaz.* **29**:153–184.
Tulasne, L. R., and C. Tulasne. (1861). "Selecta Fungorum Carpologia," Vol. 1 (English translation by W. B. Grove, Oxford Univ. Press (Clarendon), London and New York, 1931.)
U. S. Department of Agriculture. (1960). Index of plant diseases in the United States. *U.S., Dep. Agr., Handb.* **165**:1–531.
Viégas, J. H. (1944). Alguns fungos do Brazil. II. Ascomicitos. *Bragantia* **4**:5–392.
Vlamis, J., and C. E. Yarwood, (1962). Effect of liming of soil on barley powdery mildew. *Plant Dis. Rep.* **46**:886–887.
Yarwood, C. E. (1934). The diurnal cycle of development of *Erysiphe polygoni*. Ph.D. Thesis, University of Wisconsin, Madison.
Yarwood, C. E. (1936). Host range and physiologic specialization of red clover powdery mildew, *Erysiphe polygoni*. *J. Agr. Res.* **52**:659–665.
Yarwood, C. E. (1957). Powdery mildews. *Bot. Rev.* **23**:235–300.

CHAPTER 6

# Pyrenomycetes: Meliolales, Coronophorales, Sphaeriales

E. MÜLLER

*Department of Special Botany*
*Swiss Federal Institute of Technology*
*Zurich, Switzerland*

and

J. A. VON ARX

*Centraalbureau voor Schimmelcultures*
*Baarn, The Netherlands*

## I. INTRODUCTION

Pyrenomycetes cover a series of developmental lines within the Ascomycotina. Unfortunately, the taxon never has been strictly defined or limited. The present classification, based on personal experience and on literature studies, represents only one of several possibilities. For certain groups we have been kindly advised by Prof. Dr. Josef Poelt (Berlin) and Dr. John Krug (Toronto/Zurich). We wish to express our thanks to these colleagues.

## II. PYRENOMYCETES DEFINED

The Pyrenomycetes are defined in this treatment as ascomycetes with ascomata entirely surrounded by a peridial wall and containing unitunicate asci which primarily are arranged in a hymenial layer. In general the ascomata (perithecia) are provided either apically or, rarely, laterally with an opening (ostiole) which is covered inside by hyphalike periphyses. These characteristics differ from those of the Plectomycetes, in which the ascomata (cleistothecia) are astomatous, as well as from those of the Discomycetes, in which the apothecia generally possess an open hymenial layer of asci.

In certain groups of Pyrenomycetes, closely related genera may be either astomatous or ostiolate. Of the genera which are characterized by cleistothecial ascomata, only those related to genera with perithecia will be considered here. Some of these are easily recognised by the size or arrangement of the asci in a hymenial layer as well as by the more complicated form of the ascospores (e.g., septation). In others the arrangement and size of the asci is the same as in typical plectomycetes, the true relationship only being indicated from the corresponding genus with ostiolate ascomata (compare with the introduction to the Sphaeriales).

Pyrenomycetes include a number of different evolutionary lines which are presently placed in several orders. Some of these, such as the Meliolales, the Erysiphales, and the Coronophorales, are well defined. Unfortunately, there is still no general agreement on the definition and delimitation of the others. Therefore, we prefer to arrange the major portion of the Pyrenomycetes in the order Sphaeriales, using a broad definition as given by von Arx and Müller (1954), Müller and von Arx (1962), Munk (1957), and Dennis (1968). Such orders as the Diaporthales, Xylariales and Clavicipitales are easily defined when considering typical members. However, at present it appears to be extremely difficult to place adequately all the pyrenomycetes in any one of these orders, partially because there are too many intermediates for which the true relationship frequently has still to be clarified. Phylogenetically, the Hypocreales, which are characterized by bright and fleshy ascomata, have proved to be a heterogeneous group. The genera classified here belong to a number of different families, especially the Hypocreaceae, Melanosporaceae, Ophiostomataceae, Polystigmataceae, Sphaeriaceae, and Hypomycetaceae. Consequently, it was necessary to place these taxa in more than one of the orders mentioned above.

Pyrenomycetes with bitunicate asci have been excluded. They represent the major portion of the Loculoascomycetes which are considered in Chap. 7.

## A. Outline Classification

Order 1. Erysiphales
  Families: Erysiphaceae, Perisporiaceae (see Chapter 5)
Order 2. Meliolales
  Family: Meliolaceae
Order 3. Coronophorales
  Family: Coronophoraceae
Order 4. Sphaeriales

| | | | |
|---|---|---|---|
| Family A. | Ophiostomataceae | Family F. | Coryneliaceae |
| Family B. | Melanosporaceae | Family G. | Sordariaceae |
| Family C. | Sphaeriaceae | Family H. | Diaporthaceae |
| Family D. | Hypocreaceae | Family I. | Halosphaeriaceae |
| Family E. | Polystigmataceae | Family J. | Diatrypaceae |

# 6. Meliolales, Coronophorales, Sphaeriales

Family K.   Amphisphaeriaceae         Family N.   Clavicipitaceae
Family L.   Xylariaceae               Family O.   Hypomycetaceae
Family M.   Verrucariaceae

## B. Key to Orders

### KEY TO ORDERS OF PYRENOMYCETES

1. Ascomata astomatous or, rarely, possessing a rudimentary ostiole, spherical or hemispherical, developing upon an external mycelium, obligate parasites of higher plants . . . 2
1′. Ascomata ostiolate or, if astomatous, neither developing upon an external mycelium nor obligate parasites of higher plants . . . . . . . . . . . . . . . . . . . . . . . . . 3
    2(1) External mycelium dark, comparatively thick, provided with hyphopodia, asci mainly 2-spored, ascospores dark, typically with several (often 3 to 4) transverse septa . . . . . . . . . . . . . . . . . . . . . . . . . . . . **Meliolales** p. 89
    2′(1) External mycelium hyaline, rarely brown, thin, without hyphopodia, penetrating the host tissue with haustoria, asci 2- to 8-spored, ascospores 1-celled, hyaline . . . . . . . . . . . . . . . . . . . . . . **Erysiphales** (see Chap. 5)
3(1′) Ascomata without a true ostiole, rupturing due to the swelling of large gelatinous cells in the apical region; asci stalked, developing at different levels **Coronophorales** p. 90
3′(1′) Ascomata ostiolate or astomatous, but without an apical layer of gelatinous cells . . . . . . . . . . . . . . . . . . . . . . . . . . . . . **Sphaeriales** p. 93

## III. MELIOLALES

The meliolales are parasitic on higher plants, mostly occurring as epiphytes on leaves. Their mycelium is dark, forming a superficial mat of thick-walled, regularly branched, septate hyphae with hyphopodia, mainly originating from a short lateral cell. The ascomata are superficial on the mycelium, borne as a lateral branch or swelling, spherical or flattened, with a dark peridium composed of thick-walled, relatively large cells. These are either astomatous or, often, with a rudimentary ostiole neck which is glabrous or setose. The asci possess a thin typically evanescent membrane, two- to four- or, rarely, eight-spored at maturity. The ascospores are dark brown, thick walled, smooth, mostly five-celled, rarely one- to three- or four-celled. Paraphyses are often present, while conidial states are absent.

A single family: Meliolaceae.

Most members of the Meliolaceae have been recorded only from warmer geographical areas where they are common on tropical trees and shrubs. They often are characterized as "black mildews," and each species shows a narrow specialization with respect to a particular host species or genus. Unfortunately, they have not as yet been isolated in pure culture. The species of the genera *Meliola*, *Asteridiella*, *Irenopsis*, *Appendiculella* and *Amazonia* were monographed by Hansford (1961, 1965). More than 1800 species are described, distinguished essentially by their host range and by morphological characteristics. Each species is characterized by a "Beeli formula," whereby the morphological features are symbolized by figures. This method has been used successfully by Stevens (1927, 1928).

### KEY TO IMPORTANT GENERA OF MELIOLALES

1. Ascospores 1- or 2-celled at germination, narrowing near the middle . . . . . . **Armatella**
    A. *litseae* and 2 additional species (syn. *Artallendaea*) (von Arx, 1958b; Katumoto, 1962).
1'. Ascospores with 2 to 5 septa . . . . . . . . . . . . . . . . . . . . . . . . . . . . . 2
    2(1') Ascospores 3-celled, with the central cell enlarged, ascomata
        ostiolate . . . . . . . . . . . . . . . . . . . . . . . . . . . . . . . . . . **Diporotheca**
        *D. rhizophila* on roots of *Solanum* (Gordon and Shaw, 1964).
    2'(1') Ascospores mostly composed of 4–5 equal cells . . . . . . . . . . . . . . . . 3
3(2') Ascomata flattened, radiate . . . . . . . . . . . . . . . . . . . . . . . . . **Amazonia**
    A. *psychotriae* and several other species.
3'(2') Ascomata spherical or nearly so, not radiate . . . . . . . . . . . . . . . . . . . . 4
    4(3') Mycelial setae present . . . . . . . . . . . . . . . . . . . . . . . . . **Meliola**
        More than 1000 species especially on tropical plants (Hansford, 1961).
    4'(3') Mycelial setae absent . . . . . . . . . . . . . . . . . . . . . . . . . . . . . 5
5(4') Ascomata glabrous . . . . . . . . . . . . . . . . . . . . . . . . . . . . **Asteridiella**
    A. *solani* and numerous other species (syn. *Irene*).
5'(4') Ascomata setose or possessing appendages . . . . . . . . . . . . . . . . . . . 6
    6(5') Ascomata with setae . . . . . . . . . . . . . . . . . . . . . . . . . . **Irenopsis**
        *I. tortuosa* and other species.
    6'(5') Ascomata with larviform appendages . . . . . . . . . . . . . . **Appendiculella**
        A. *calostroma* and numerous other species.

## IV. CORONOPHORALES

This order is composed of wood-inhabiting ascomycetes which are occasionally parasitic on pyrenomycetous fructifications. The mycelium, which is scanty and easily overlooked, grows within the host tissue. The ascomata are of median size with dark, carbonaceous peridia. They are globose or turbinate, solitary and free or aggregated and then often seated on compact stromatic structures; sometimes the ascomata are surrounded by or seated on hyphal subicula composed of dark brown hyphae. True ostioles are absent, and the ascomata open by disintegration of the apex, the inside of which is bordered by large gelatinous cells forming a mucilaginous mass (Quellkörper). Asci are numerous and generally distinctly stalked, claviform, and irregularly distributed; they usually arise from hyphae to occupy the whole cavity and are unitunicate with a thin membrane and no differentiation at the apex. The ascospores are small to large, one- to multicellular, hyaline to light brown, and often cylindrical and curved, at maturity they fill the interior of the fruiting body as a powdery mass.

A single family: Coronophoraceae

The Coronophoraceae represent a small group of pyrenomycetelike ascomycetes with some characteristics of the Plectomycetes: cleistothecial ascomata opening by a rupture induced by the "Quellkörper," and the arrangement and morphology of the asci. Thus the Coronophorales are clearly distinct from other orders of the Pyrenomycetes. The Coronophoraceae (including the Nitschkeaceae) have been monographed by Fitzpatrick (1923); some genera have been described by von Arx and Müller (1954) and Müller and von Arx (1962).

### KEY TO IMPORTANT GENERA OF CORONOPHORALES

1. Ascomata seated upon or surrounded by a subiculum composed of brown, often dichotomously branched, hyphae . . . . . . . . . . . . . . . . . . . . . . . . . . . . 2
1'. Subiculum not present, ascomata solitary and free or aggregated and often seated on a compact stroma . . . . . . . . . . . . . . . . . . . . . . . . . . . . . . 6
    2(1) Subiculum thin, ascomata aggregated, rarely scattered, ascospores 2-celled (Fig. 1) . . . . . . . . . . . . . . . . . . . . . . . . . . . . **Calyculosphaeria**
        C. *tristis* on old wood (syn. *Sydowinula*).
    2'(1) Subiculum typically well-developed, ascospores 1-celled . . . . . . . . . . 3
3(2') Ascospores provided with filiform appendages at both ends . . . . . . . . . . . . 4
3'(2') Ascospores without appendages . . . . . . . . . . . . . . . . . . . . . . . . 5
    4(3) Asci multispored . . . . . . . . . . . . . . . . . . . . . . **Scortechiniella**
        S. *similis* on bark.
    4'(3) Asci 8-spored . . . . . . . . . . . . . . . . . . . . . . . . **Biciliospora**
        B. *velutina* on old wood.
5(3') Ascomata provided with dark spines . . . . . . . . . . . . . . **Acanthonitschkea**
    A. *argentinensis* and some additional species.
5'(3') Ascomata glabrous . . . . . . . . . . . . . . . . . . . . . . . . . **Tympanopsis**
    T. *euomphala* on dead wood and bark and some additional species (syn. *Coronophorella*, *Euacanthe*, *Fitzpatrickia*, *Scortechinia*, *Teratonema*).
    6(1') Ascospores 1-celled, more or less allantoid . . . . . . . . . . . . . . . . . 7
    6'(1') Ascospores septate, often cylindric and curved . . . . . . . . . . . . . . . 9
7(6) Asci multispored . . . . . . . . . . . . . . . . . . . . . . . . . . . . . . 8
7'(6) Asci 8-spored, ascomata cupulate, mostly aggregated in groups on fructifications of other pyrenomycetes or on basal stromatic structures . . . . . . . . . . . . . . **Nitschkea**
    N. *fuckelii* and a number of additional species primarily on wood or parasitic on other pyrenomycetes (syn. *Coelosphaeria*).

8(7) Ascomata cupulate, mostly aggregated, seated upon basal stromatic structures .................................. **Fracchiaea**
*F. heterogena* and a number of additional species on wood.

8'(7) Ascomata more or less globose, arranged in small or large circular groups, with necks collectively erumpent, rarely solitary, ascospores typically allantoid
.................................................... **Coronophora**
*C. gregaria* and a number of additional species (syn. *Cryptosphaerella*).

9(6') Ascospores 1-septate, hyaline or brown ...................... 10

9'(6') Ascospores 1- to 3-septate, cylindrical, curved, ascomata scattered to densely crowded, turbinate .................................... **Thaxteria**
*T. didyma* and few additional species on old wood (Booth and Müller, 1972).

10(9) Ascospores comparatively large, long fusiform, remaining hyaline .... **Bertia**
*B. moriformis*; monotypic.

10'(9) Ascospores comparatively small, ellipsoidal, becoming brown .... **Gaillardiella**

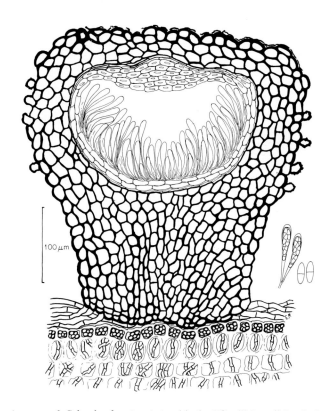

FIG. 1. Ascoma of *Calyculosphaeria tristis* with the "Quellkörper" bordering the apical inside portion. (×250); at the right there are asci and ascospores (×500, ×1000, respectively).

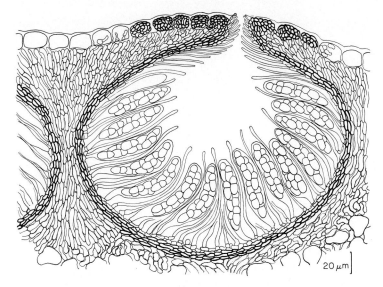

FIG. 2. Ascoma of *Stigmochora controversa* (Polystigmataceae, Sphaeriales) on *Acacia*. The perithecium is embedded in a pseudostroma. (× 350.)

## V. SPHAERIALES

In this order, the ascomata are spherical, hemispherical, or flask-shaped (botuliform). They are mostly ostiolate and rarely astomatous, and have a bright, fleshy or dark membranous or carbonaceous wall; the ostiole is papillate or elongated cylindrical and is provided with periphyses or ostiolar hairs. The ascomata may be solitary or aggregated, free or densely aggregated and connected by stromatic structures, or they may sit on hyphal subicula (Fig. 2). The asci are spherical, clavate, fusiform or cylindrical, with a thin or rather thick single membrane, arranged in a hymenium or irregularly disposed at different levels. The ascospores are one celled or septate, hyaline or colored, sometimes opaque and of very different size; paraphyses are mostly present, often evanescent in an early stage, or they may be absent.

As already mentioned, the Sphaeriales are conceived here in a very broad sense. They include the Sphaeriales *s. str.* (*sensu stricto*), Diaporthales, Xylariales, and Clavicipitales, as well as the Hypocreales and some additional orders segregated from one of the above-mentioned taxa. The major portions of these orders are arranged into widely defined families such as Sphaeriaceae, Diaporthaceae, Xylariaceae, Clavicipitaceae, and Hypocreaceae, and the rest are divided into several additional families.

The most important characteristic distinguishing the families is the mor-

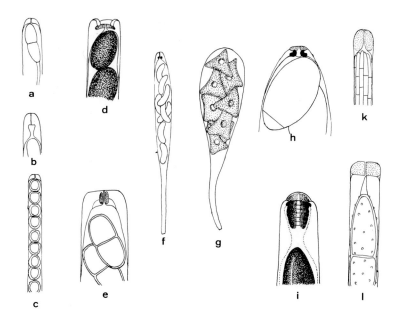

FIG. 3. Ascus types of different Sphaeriales; (a) *Loramyces* (Sphaeriaceae), (b) *Nectria* (Hypocreaceae), (c) *Hypocrea* (Hypocreaceae), (d) *Sordaria* (Sordariaceae), (e) *Melanconis* (Diaporthaceae), (f) *Diatrype* (Diatrypaceae), (g) *Tripospora* (Coryneliaceae), (h) *Chaetapiospora* (Amphisphaeriaceae), (i) *Xylaria* (Xylariaceae), (k) *Cordyceps* (Clavicipitaceae), (l) *Hypomyces* (Hypomycetaceae). ($\times$ 1000.)

phology of the ascus, especially of the ascus apex. Additional suitable characteristics used to define families are the size, structure, and color of the ascospores, and, for example, the presence of germ pores or germ slits, the texture of the peridial wall, the development and the morphology of stromatic structures, and the conidial states formed.

When comparing typical members of the families, different forms and behavior of the ascus apex can be observed. In the Melanosporaceae, Ophiostomataceae, and Halosphaeriaceae, as in the Coryneliaceae, the asci become evanescent before the maturation of the ascospores, and the asci in some cases are spherical or obovate, in others, clavate or cylindrical (Fig. 3). In the Verrucariaceae the ascus wall is persistent, thick and homogeneous.

In the families Sphaeriaceae, Hypocreaceae, and Polystigmataceae, a simple strictly apical pore connects the ascus lumen with the exterior. The ascus wall is apically thickened and forms a more or less distinct plate (Fig. 3). Sometimes this plate increases in thickness downwards. These cases come close to the Diaporthaceae, where the apical porus is surrounded by a ringlike structure which is more refractive and may be colored blue with cotton blue.

Within the Diaporthaceae, we may observe two kinds of behavior of the mature ascus. In one group the asci remain attached to the inner peridial wall forming a well-defined hymenium; in the major portion the ascus base is evanescent so that the asci are easily liberated, filling the whole ascoma.

In the Sordariaceae the ascus apex is funnel-shaped; the base of the funnel is connected with the ascus lumen by a pore which can be surrounded by a cushionlike structure included in the thickened wall (Fig. 3).

The long and narrow asci of the Clavicipitaceae contain strictly filiform ascospores. The apex often is slightly thickened, as in the Sphaeriaceae, provided with a pore and plugged by a large chitinoid body (Fig. 3).

Ringlike structures surrounding an apical pore are also typical for the Xylariaceae, Amphisphaeriaceae, and Diatrypaceae. Contrary to the asci of other families, these structures mostly stain blue with iodine (e.g., Melzer's solution). Considerable morphological differences in these structures may be observed. In the Diatrypaceae, the ascus tip includes just a simple amyloid ring. In the Amphisphaeriaceae, the amyloid structures are similar; a more complicated plug or an extremely thickened ring can be observed and above this structure a chitinoid pulvillus may be present. In other cases the amyloid portion is absent. A complex plug composed of a number of generally amyloid rings or plates are typical for the asci of the Xylariaceae. These also include a chitinoid pulvillus (Fig. 3).

Although the differences in the texture of the peridial wall of ascomata cannot be correlated always to other family characters, there are a few cases in which these may indicate the relationship within a family. Hypocreaceae, Sphaeriaceae, and Polystigmataceae have the same ascus type, but may be differentiated mainly on the structure of the peridial wall. In the Hypocreaceae the wall is bright and fleshy; in the Sphaeriaceae it is brown to dark and membranaceous or carbonaceous; and in the Polystigmataceae the peridial wall is mostly light colored, sometimes hyaline and membranaceous.

Many species of the Sphaeriales include conidial and/or spermatial states in their life cycle. These belong mainly to the Moniliales, less frequently to the Melanconiales or Sphaeropsidales. The conidia are mostly blastoconidia blown out on conidiogenous cells. They have a narrow or broad base due to their kind of development and are born in basipetal succession (phialoconidia) or sympodially.

In some cases certain genera of Pyrenomycetes are connected with certain genera of Deuteromycotina; for example, *Diaporthe* species have *Phomopsis* conidial states (Sphaeropsidales) or *Glomerella* species have a *Colletotrichum* conidial state (Melanconiales). In other cases, pyrenomycetous genera may have conidial states which belong to different genera of Moniliales, but all of them produce their conidia in the same way on phialides. In a number of species chlamydospores are formed. The conidial states of genera with species growing in pure culture are enumerated by von Arx (1970a).

Genera with spherical, ellipsoid or clavate evanescent asci often have been classified within the Plectomycetes. Especially the Ophiostomataceae and some genera of the Melanosporaceae are intermediate between "Pyrenomycetes" and "Plectomycetes." Within these groups a number of genera are characterized by astomatous ascomata. But many of these genera have relatives with ostiolate ascomata. Species belonging to genera with astomatous ascomata often are soil borne or develop deeply immersed in the substratum, whereas species with ostiolate ascomata may be formed within the same species due to the external conditions of growth. The following list gives the relation of some genera with astomatous ascomata to its corresponding genus characterized by ostiolate ascomata.

| Ostiolate | Astomatous |
|---|---|
| *Ceratocystis* | *Europhium* |
| *Melanospora* | *Microthecium* |
| *Chaetomium* | *Chaetomidium* |
| *Lophotrichus* | *Kernia* |
| *Microascus* | *Petriellidium* |
| *Sordaria* | *Boothiella* |
| *Podospora* | *Zopfiella, Tripterospora* |
| *Gelasinospora* | *Anixiella* |
| *Apiosordaria* | *Echinopodospora* |
| *Nectria* | *Heleococcum* |

### KEY TO FAMILIES OF SPHAERIALES

1. Asci evanescent, often spherical, obovate or clavate, ascospores becoming free in the cavity and often discharged in a slimy mass .................... 2
1'. Asci with a persistent membrane, ascospores mostly not liberated in slimy mass, but ejaculated .................... 5
    2(1) Ascomata mostly immersed, ascospores 1-celled or septate, hyaline, often with gelatinous sheaths or appendages (marine fungi, mostly on wood in seawater) .................... **Halosphaeriaceae** p. 116
    2'(1) Ascomata mostly not immersed, ascospores 1-celled (terrestrial, not on wood in seawater) .................... 3
3(2') Ascospores colored, with or without germ pores ...... **Melanosporaceae** p. 98
3'(2') Ascospores without germ pores, mostly hyaline .................... 4
    4(3') Asci spherical, ascospores small, ascomata spherical, with a long beak or astomatous .................... **Ophiostomataceae** p. 98
    4'(3') Asci clavate, mostly with stalks, ripening in succession, ascomata botuliform (parasitic on Podocarpaceae, occasionally on other conifers or other hosts) (Fig. 3a) .................... **Coryneliaceae** p. 108
5(1') Asci with a thick homogeneous membrane (lichenous fungi or growing on lichens) .................... **Verrucariaceae** p. 124

5′(1′) Asci with a thin membrane, often with apical structures or thickenings . . . . . . . . 6
    6(5′) Apical structures of the asci not amyloid (not stained blue by iodine) . . . . . . 7
    6′(5′) Apical structures of the asci amyloid (stained blue by iodine) . . . . . . . . . 13
7(6) Asci without refractive apical caps or rings . . . . . . . . . . . . . . . . . . . . . . 8
7′(6) Asci with refractive apical caps or rings . . . . . . . . . . . . . . . . . . . . . . 11
    8(7) Ascospores mostly dark when ripe, with germ pores (or germ slits), often with a mucous sheath or with appendages, sometimes ornamented
    (Figs. 3d and 4) . . . . . . . . . . . . . . . . . . . . . . . . **Sordariaceae** p. 109
    8′(7) Ascospores without germ pores or germ slits . . . . . . . . . . . . . . . . . 9
9(8′) Ascomata with a thin wall, immersed in the host tissue or in a stroma, ascospores after germination often forming appressoria (mostly parasitic on high plants)
(Fig. 4) . . . . . . . . . . . . . . . . . . . . . . . . . . . **Polystigmataceae** p. 106
9′(8′) Ascomata developing superficially or becoming so, often botryose on a basal stroma
. . . . . . . . . . . . . . . . . . . . . . . . . . . . . . . . . . . . . . . . . 10
    10(9′) Ascomata fleshy, bright, often white, yellow, red, green or blue
    (Fig. 3b,c) . . . . . . . . . . . . . . . . . . . . . . . . . **Hypocreaceae** p. 103
    10′(9′) Ascomata brown or black, developing superficially, ascospores hyaline or bright
    (Fig. 3a) . . . . . . . . . . . . . . . . . . . . . . . . . . **Sphaeriaceae** p. 100

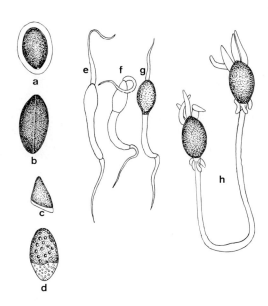

FIG. 4. Different forms of ascospores: (a) *Sordaria*, with germ pore (Sordariaceae), (b) *Xylaria*, with germ slit (Xylariaceae), (c) *Triangularia*, with small appendage cell (Sordariaceae), (d) *Apiosordaria* (Sordariaceae), (e)–(g) *Bombardia*, several stages of development, with primary appendage (originally a cell) and secondary appendages (Sordariaceae), (h) *Zygopleurage* (Sordariaceae). (× 1000.)

Some genera including species with nonamyloid apical structures belong to the Amphisphaeriaceae.

11(7') Asci with an apical ring, mostly visible as 2 strongly refractive bodies, often loosening from the ascogenous cells and embedded in a slimy mass
(Fig. 3e) . . . . . . . . . . . . . . . . . . . . . . . . . . . . . . **Diaporthaceae** p. 111

11'(7') Asci with an apical cap or body, perforated by a narrow canal . . . . . . . . . . 12

12(11') Ascospores fusiform, biapiculate, often verrucose
(Fig. 31) . . . . . . . . . . . . . . . . . . . . . . . . **Hypomycetaceae** p. 127

12'(11') Ascospores filiform (Fig. 3k) . . . . . . . . . . . . **Clavicipitaceae** p. 125

13(6') Asci mostly small, including a simple amyloid ring in the apical portion, ascospores 1-celled, allantoid, colored (sometimes only slightly), ascomata mostly large and aggregated, often connected by stromatic structures (Fig. 3f) . . . . . . . . **Diatrypaceae** p. 118

13'(6') Asci mostly large, amyloid portion a simple ring or a system of rings and, in addition, often with a chitinoid pulvillus (stained by cotton blue), ascospores 1-celled to septate, not allantoid . . . . . . . . . . . . . . . . . . . . . . . . . . . . . . . . . . . . . . . . 14

14(13') Ascospores 1-celled, dark, mostly with a germ slit
(Figs. 3i and 4b) . . . . . . . . . . . . . . . . . . . . . . . **Xylariaceae** p. 121

14'(13') Ascospores hyaline to brown, 1-celled or septate, when dark and 1-celled with germ pores, germ slits rarely occurring (Fig. 3h) . . . **Amphisphaeriaceae** p. 118

## A. Ophiostomataceae

### KEY TO IMPORTANT GENERA OF OPHIOSTOMATACEAE

1. Ascomata dark, ascospores small . . . . . . . . . . . . . . . . . . . . . . . . . . . 2

1'. Ascomata bright, often fleshy . . . . . . . . . . . . . . . . . . . . . . . . . . . . 3

2(1) Ascomata with a long beaklike ostiole . . . . . . . . . . . . . . . . **Ceratocystis**
C. ulmi causes Dutch elm disease; C. fagacearum on oaks; many other species are responsible for wood blueing, often associated with bark beetles = "ambrosia fungi" (Hunt, 1956; Griffin, 1968) (syn. Endoconidiophora, Grosmannia, Linostoma, Ophiostoma, and Rostrella).

2'(1) Ascomata astomatous . . . . . . . . . . . . . . . . . . . . . . . . **Europhium**
E. robustum and 2 additional species (Robinson-Jeffrey and Davidson, 1968).

3(1') Ascomata with a beaklike ostiole . . . . . . . . . . . . . . . **Sphaeronaemella**
S. helvellae on old fructifications of Helvella; S. fimicola on dung (Cain and Weresub, 1957).

3'(1') Ascomata with a papillate ostiolum (parasitic on marine red algae)
. . . . . . . . . . . . . . . . . . . . . . . . . . . . . . curs. cf. **Chadefaudia**, p. 116

## B. Melanosporaceae

(Included are Chaetomiaceae, Achaetomiaceae, Microascaceae, and part of the Thielaviaceae.)

### KEY TO IMPORTANT GENERA OF MELANOSPORACEAE

1. Ascospores depressed, rectangular in outline, truncate at both ends . . . **Ophiostomella**
O. caulincola on plant debris.

## 6. Meliolales, Coronophorales, Sphaeriales

1'. Ascospores not so . . . . . . . . . . . . . . . . . . . . . . . . . . . . . . . . . . 2

   2(1') Ascomata develop on or in an erumpent, dark stroma, ascospores ellipsoidal or reniform, without germ pores, often with a germ slit
   . . . . . . . . . . . . . . . . . . . . . . . . . **Thamnoideae (Xylariaceae)** p. 121

   2'(1') Ascospores without germ slit, ascomata do not develop on an erumpent stroma  3

   3(2') Ascomata immersed in a dark stroma (parasitic on Palmae) . . . . . . . . . . 4

   3'(2') Ascomata not immersed in a stroma . . . . . . . . . . . . . . . . . . . . . 5

      4(3) Ascospores with 2 lateral germ pores . . . . . . . . . . . . . . **Phaeochora**
      *P. steinheilii* on *Chamaerops humilis*; monotypic (von Arx and Müller, 1954; Müller, 1965).

      4'(3) Ascospores without germ pores . . . . . . . . . . . . . . . . . **Serenomyces**
      *S. shearii* on *Serenoa serrulata* and additional species on Palmae (Müller and Dennis, 1965; Müller and Ahmad, 1962).

   5(3') Ascospores at both ends with prominent germ pores, mostly rather large, dark and straight . . . . . . . . . . . . . . . . . . . . . . . . . . . . . . . . . . . . . . . . . . 6

   5'(3') Ascospores not so . . . . . . . . . . . . . . . . . . . . . . . . . . . . . . 9

      6(5) Wall of the ascomata bright . . . . . . . . . . . . . . . . . . . . . . . . . 7

      6'(5) Wall of the ascomata dark . . . . . . . . . . . . . . . . . . . . . . . . . . 8

7(6) Ascomata ostiolate . . . . . . . . . . . . . . . . . . . . . . . . . **Melanospora**
   *M. zamiae* and other species, monographed by Doguet (1955).

7'(6) Ascomata astomatous . . . . . . . . . . . . . . . . . . . . . . . **Microthecium**
   *M. zobelii* and additional species from soil and from fructifications of ascomycetes and basidiomycetes (syn. *Guttularia* and *Lithomyces*) (Udagawa and Cain, 1969a).

   8(6') Ascomata ostiolate, asci clavate . . . . . . . . . . . . . . . . . **Achaetomiella**
   *A. virescens* from soil, monotypic (von Arx, 1970a).

   8'(6') Ascomata astomatous . . . . . . . . . . . . . . . . . . . . . . . . **Thielavia**
   *T. basicola* and additional species (Booth, 1961).

9(5') Ascomata provided with long, branched, coiled, circinate or undulate hairs . . . 10

9'(5') Ascomata without such hairs, smooth or setose . . . . . . . . . . . . . . . . . 14

   10(9) Wall of the ascomata thick, hairs with ampoules (rounded cells), asci cylindrical, conidia borne on denticles mostly present . . . . . . . . . . . . . . **Ascotricha**
   *A. chartarum* and additional species (Ames, 1963; Hawksworth, 1971).

   10'(9) Ampoules and conidia borne on denticles absent . . . . . . . . . . . . . . 11

11(10') Ascospores mostly brown, often biapiculate, with a basal germ pore . . . . . . . 12

11'(10') Ascospores light straw or copper-colored, mostly with 2 inconspicuous germ pores . . . . . . . . . . . . . . . . . . . . . . . . . . . . . . . . . . . . . . . . . . 13

   12(11)Ascomata ostiolate . . . . . . . . . . . . . . . . . . . . . . . . **Chaetomium**
   *C. globosum* and many additional species from soil and plant debris (Ames, 1963; Udagawa and Cain, 1969b).

   12'(11) Ascomata astomatous . . . . . . . . . . . . . . . . . . . . . . **Chaetomidium**
   *C. fimeti* on dung and additional species (Cain, 1961a).

13(11') Ascomata ostiolate . . . . . . . . . . . . . . . . . . . . . . . . . **Lophotrichus**
   *L. ampullus* from soil and plant debris (Ames, 1963).

13'(11') Ascomata astomatous, often irregular, with a thick wall . . . . . . . . . **Kernia**
   *K. nitida* and additional species from soil and dung (Malloch and Cain, 1971).

14(9′) Asci cylindrical, ascospores roundish, with a prominent germ pore . . . . 15
14′(9′) Asci clavate, obovate or spherical . . . . . . . . . . . . . . . . . . 16
15(14) Ascomata ostiolate, dark . . . . . . . . . . . . . . . . . . . . . . **Achaetomium**
       *A. globosum* and few additional species from soil (Rai *et al.*, 1964).
15′(14) Ascomata astomatous, bright . . . . . . . . . . . . . . . . . . . . . **Boothiella**
       *B. tetraspora* in soil, monotypic (syn. *Thielaviella*) (von Arx, 1970a).
16(14′) Ascospores dark, ovoid, with a prominent basal germ pore, ascomata dark, with a long neck . . . . . . . . . . . . . . . . . . . . . . . . . . . . . . . . . **Phaeostoma**
       *P. vitis* on rotten roots (von Arx and Müller, 1954).
16′(14′) Ascospores without a prominent, basal germ pore or not dark . . . . . 17
17(16′) Ascomata ostiolate . . . . . . . . . . . . . . . . . . . . . . . . . . . . . 18
17′(16′) Ascomata astomatous . . . . . . . . . . . . . . . . . . . . . . . . . . . 19
18(17) Ascospores shorter than 7 μm, straw-colored, often curved, conidia not in synnemata . . . . . . . . . . . . . . . . . . . . . . . . . . . . . . . . **Microascus**
       *M. longirostris* and additional species from soil (Barron *et al.*, 1961b).
18′(17) Ascospores longer, reddish-brown . . . . . . . . . . . . . . . . . **Petriella**
       *P. asymmetrica* from soil and additional species (Barron *et al.*, 1961a; Malloch, 1970)
19(17′) Ascomata thin-walled, composed of hyphal elements . . . . . . . **Petriellidium**
       *P. boydii* (Malloch, 1970).
19′(17′) Ascomata thick-walled . . . . . . . . . . . . . . . . . . . . . . . . cf. **Kernia**

## C. Sphaeriaceae

### KEY TO IMPORTANT GENERA OF SPHAERIACEAE

1. Ascomata develop singly or on a hyphal subiculum . . . . . . . . . . . . . . 2
1′. Ascomata develop in or on a stroma . . . . . . . . . . . . . . . . . . . . . 13
    2(1) Ascomata small, with a thin, often mucilaginous wall, ascospores fusiform (parasitic on other fungi, mostly on epiphytic leaf parasites) . . . . . . . . . . . . . . 3
    2′(1) Ascomata mostly with a thick wall (not parasitic on other fungi, mostly saprophytic) . . . . . . . . . . . . . . . . . . . . . . . . . . . . . . . . . . . . . . . 4
3(2) Ascospores 1-celled or with pseudosepta . . . . . . . . . . . . . . . . **Hyaloderma**
       *H. perpusillum* is parasitic on *Asterina* and *Meliola* species on leaves (von Arx, 1958a) (syn. *Epicrea* and *Saccardomyces*).
3′(2) Ascospores 2-celled . . . . . . . . . . . . . . . . . . . . . . . . . . . **Rizalia**
       *R. fasciculata* on mycelium of leaf-parasitic ascomycetes.
       (Ascospores many-celled: *Debaryella, Cryptoleptosphaeria*.)
4(2′) Ascospores 1-celled, asci 8-spored, ascomata setose . . . . . . **Trichosphaeria**
       *T. pilosa* on rotten wood (syn. *Acanthosphaeria* and *Bakeromyces*).
4′(2′) Ascospores septate or filiform or asci with more than 8 ascospores . . . . . 5
5(4′) Ascospores 2-celled, cells separating in an early state, ascomata setose or hairy
. . . . . . . . . . . . . . . . . . . . . . . . . . . . . . . . . . . . . . . **Trichosphaerella**
       *T. decipiens* on rotten wood (Müller and von Arx, 1962) (syn. *Bresadolella, Larseniella, Melanopsammella, Neorehmia, Oplothecium, Oplotheciopsis*).

5′(4′) Ascospores when septate not separating into single cells . . . . . . . . . . . . 6
  6(5′) Ascospores provided with a long, basipetal appendage; (Fig. 3a) (freshwater fungi)
    . . . . . . . . . . . . . . . . . . . . . . . . . . . . . . . . . . . . . . . . . . . . . . . **Loramyces**
    *L. juncicola* on leaves of Gramineae and Cyperaceae (Ingold and Chapman, 1952).
  6′(5′) Ascospores without such an appendage . . . . . . . . . . . . . . . . . . . 7
  7(6′) Ascomata with elongated, cylindrical beaks . . . . . . . . . . . . . . . . . 8
  7′(6′) Ascomata with a papilla or a conical beak . . . . . . . . . . . . . . . . . . 9
    8(7) Ascospores hyaline, 2-celled, beak with peristomal hyphae . . . . . **Klasterskya**
    *K. acuum* on fallen pine needles.
    8′(7) Ascospores brownish, 2-celled, beak without peristomal hyphae   **Rhynchomeliola**
    *R. pulchella* on *Feijoas*; syn. *Ceratostomina*.
  9(7′) Ascomata setose, ascospores mostly 2-celled . . . . . . . . . . . . . . . . **Niesslia**
    *N. exilis* (syn. *Valetoniella*) and other species with *Monocillium* conidial states.
  9′(7′) Ascomata glabrous . . . . . . . . . . . . . . . . . . . . . . . . . . . . . . . . . 10
    10(9′) Ascomata with a thin peridial wall . . . . . . . . . . . . . . . . . . . . . 11
    10′(9′) Ascomata with a thick wall composed of several layers of cells . . . . . . 12
  11(10) Ascomata obpyriform or flask-shaped, ascospores 2-celled, . . . . **Plectosphaerella**
    *P. cucumerina* with a *Fusarium*-like conidial state (Gams and Gerlagh, 1968).
  11′(10) Ascomata globose, ascospores filiform, septate . . . . . . . . . . . . **Tonduzia**
    *T. psychotriae* on *Psychotria*.
    12(10′) Ascospores 3-celled, cylindrical . . . . . . . . . . . . . . . . . . . **Melomastia**
    *M. mastoidea* on dry twigs of different woody plants.
    12′(10′) Ascospores with many septa, fusiform . . . . . . . . . . . . **Saccardoella**
    *S. montellica* and few additional species on bark and twigs (Müller, 1959a).
13(1′) Stromata dark, crustlike, often thin, saprobes on wood . . . . . . . . . . . . 14
13′(1′) Stromata well-developed, dark or bright when crustlike . . . . . . . . . . . 17
  14(13) Ascospores remaining hyaline or becoming slightly yellow . . . . . . . 15
  14′(13) Ascospores brown when ripe . . . . . . . . . . . . . . . . . . . . . . 16
15(14) Ascomata setose, ascospores 2-celled . . . . . . . . . . . . . . . . **Eriosphaeria**
  *E. vermicularia* on wood and some additional species.
15′(14) Ascomata glabrous, ascospores having 2 or more cells, besides ascomata conidiophores
often present . . . . . . . . . . . . . . . . . . . . . . . . . . . . . . . . . . . **Chaetosphaeria**
  *C. innumera*, *C. myriocarpa* and other species with *Catenularia* or *Chloridium* conidial
  states; *C. pomiformis* ( = *Melanopsamma pomiformis*) with a *Stachybotrys*-conidial
  state.
  16(14′) Ascomata setose, ascospores 4-celled with end-cells hyaline   . **Chaetosphaerella**
  *C. phaeostroma* on rotten wood and bark (Booth and Müller, 1972).
  16′(14′) Ascomata glabrous, ascospores 2-celled . . . . . . . . . . . . **Rhynchostoma**
  *R. minutum* on rotten wood.
17(13′) Stromata bright, crustlike (saprobes on bamboo) . . . . . . . . . . . . . . . 18
17′(13′) Stromata dark . . . . . . . . . . . . . . . . . . . . . . . . . . . . . . . . . . 19
  18(17) Ascospores 2-celled . . . . . . . . . . . . . . . . . . . . . . . . . . **Melchioria**
  *M. leucomelaena* on bamboo and some additional species (Hino and Katumoto,
  1968) (syn. *Gaeumannia*).

18′(17) Ascospores multiseptate, ascomata bearing a stellate corona of white hairs round the ostiole . . . . . . . . . . . . . . . . . . . . . . . . . . . . **Ijuhya**
*I. vitraea* on bamboo (von Höhnel, 1912).

19(17′) Stromata erect, clavate or branched, basally immersed in the wood, ascospores brown . . . . . . . . . . . . . . . . . . . . . . . . . . . . . . . . . . . . . . 20

19′(17′) Stromata immersed or erumpent, but not erect . . . . . . . . . . . . . . . 21

    20(19) Ascospores 1-celled . . . . . . . . . . . . . . . . . . cf. **Thamnoideae** p. 121
    Xylariaceae with the genera *Thamnomyces*, *Phylacia*, *Batistia*, and *Wawelia*.

    20′(19) Ascospores 2-celled . . . . . . . . . . . . . . . . . . . . . **Xylobotryum**
    *X. andinum* and additional species on wood (syn. *Melanobotrys*, *Trachyxylaria*, and *Xyloceras*).

21(19′) Ascomata botryose on an erumpent stroma (parasitic on leaves) . . . . . . . 22

21′(19′) Ascomata immersed . . . . . . . . . . . . . . . . . . . . . . . . . . . . . 23

    22(21)Ascospores 1-celled or with pseudosepta . . . . . . . . . . . **Pseudomeliola**
    *P. brasiliensis* and additional species parasitic on leaves (von Arx, 1958a) (syn. *Biotyle*).

    22′(21) Ascospores 2-celled . . . . . . . . . . . . . . . . . . . . . **Rehmiomycella**
    *R. phoradendri* on leaves of *Phoradendron*; monotypic.

23(21′) Stromata on woody substrata, saprobic . . . . . . . . . . . . . . . . . . . 24

23′(21′) Stromata parasitic on lichens or fungi . . . . . . . . . . . . . . . . . . . 29

    24(23) Ascospores brown . . . . . . . . . . . . . . . . . . . . . . . . . . . . 25

    24′(23) Ascospores hyaline . . . . . . . . . . . . . . . . . . . . . . . . . . . 28

25(24)Stromata pustulate, ascomata in groups deeply immersed, ostiole often long . . 26

25′(24) Stromata not pustulate, often reduced to a clypeus . . . . . . . . . . . . 27

    26(25) Ascospores 1-celled . . . . . . . . . . . . . . . . . . . . . . . **Anthostoma**
    *A. decipiens* sensu Eriksson, (1966); many other *Anthostoma* species belong to other genera, e.g., *Anthostomella*.

    26′(25) Ascospores 2-celled . . . . . . . . . . . . . . . . . . . . . . . . . **Valsaria**
    *V. insitiva* on wood and additional species (syn. *Hypoxylonopsis*, *Myrmaecium* and *Pseudothyridaria*).

    27(25′) Ascospores 2-celled . . . . . . . . . . . . . . . . . . . . . . . . . **Endoxylina**
    *E. astroidea* on rotten wood.

    27′(25′) Ascospores more than 2-celled, ascomata often in groups . . . . . . . **Thyridaria**
    *T. incrustans* and additional species on wood (Wehmeyer, 1941b).

    28(24′) Ascospores 1-celled, allantoid, asci with a long stipe and a broad apex, ascomata in circular groups, necks long, collectively erumpent . . . . . . . . **Calosphaeria**
    *C. pulchella* and additional species on rotten wood and on fructifications of other pyrenomycetes; the genus seems to be isolated in its taxonomic position; Munk (1957) erected a family for it.

    28′(24′) Ascospores 1- to many-celled, fusiform, producing ascoconidia, ascomata embedded in an elongate, light-colored stroma . . . . . . . . . . . . **Pleocryptospora**
    *P. bambusae* on bamboo (Reid and Booth, 1969).

29(23′) Ascospores 1-celled or with pseudosepta . . . . . . . . . . . . **Schweinitziella**
    *S. styracum* and additional species parasitic on ascomycete fructifications (von Arx, 1958a) (syn. *Paracesatiella*).

29′(23′) Ascospores 2-celled . . . . . . . . . . . . . . . . . . . . . . . **Rhagodostoma**
      *R. lichenicola* and few additional species parasitic on lichens and fungi (syn. *Castagnella*).

## D. Hypocreaceae

### KEY TO IMPORTANT GENERA OF HYPOCREACEAE[1]

1. Ascomata immersed in a bright fleshy stroma (which may be reduced in some species; in other species the ascomata are protruding, especially in species parasitic on bamboo) . . 2
1′. Ascomata not immersed in a stroma, developing singly superficially or immersed in the host tissue, occasionally botryose on a basal stroma . . . . . . . . . . . . . . . 13
    2(1) Ascospores 1-celled . . . . . . . . . . . . . . . . . . . . . . . . . . . 3
    2′(1) Ascospores multicelled . . . . . . . . . . . . . . . . . . . . . . . . . 5
3(2) Ascospores large, hyaline . . . . . . . . . . . . . . . . . . . . . . . . **Selinia**
      *S. pulchra* on dung (von Arx and Müller, 1955).
3′(2) Ascospores dark . . . . . . . . . . . . . . . . . . . . . . . . . . . . . . 4
    4(3′) Ascospores glabrous . . . . . . . . . . . . . . . . . . . . . **Thuemenella**
      *T. javanica* and some other species on wood (Boedijn, 1964).
    4′(3′) Ascospores ornamented . . . . . . . . . . . . . . . . . . . . **Sarawakus**
      *S. lycogaloides* on wood.
5(2′) Ascospores 2-celled . . . . . . . . . . . . . . . . . . . . . . . . . . . . 6
5′(2′) Ascospores with more than 2 cells, parasitic on bamboo . . . . . . . . . . . 12
    6(5) Ascospores separating in an early stage, asci finally containing 16 1-celled spores
    . . . . . . . . . . . . . . . . . . . . . . . . . . . . . . . . . . . . . . . 7
    6′(5) Ascospores not separating in an early stage . . . . . . . . . . . . . . . 9
7(6) Stromata upright, often clavate . . . . . . . . . . . . . . . . . . . **Podostroma**
      *P. leucopus* and other species on wood and soil (Boedijn, 1934).
7′(6) Stromata crustose, pulvinate or hemispherical . . . . . . . . . . . . . . . . 8
    8(7′) Ascomata immersed, saprobic . . . . . . . . . . . . . . . . . . . **Hypocrea**
      *H. rufa* and a number of additional species with conidial states belonging to the form genera *Trichoderma*, *Gliocladium* (Dingley, 1951b). Doi (1969) separates it into several genera (syn. *Creopus*)
    8′(7′) Ascomata protruding, parasitic on bamboo . . . . . . . . . . . . . . . 11
9(6′) Stromata immersed, broadly conical, ascospores brownish, saprophytic on wood or in soil . . . . . . . . . . . . . . . . . . . . . . . . . . . . . . . . . . **Valsonectria**
      *V. andina* and some additional species.
9′(6′) Stromata develop superficially or become erumpent in an early state . . . . . 10
    10(9′) Saprobic, stromata mostly crustose . . . . . . . . . . . . . . **Hypocreopsis**
      *H. riccioidea* and a number of similar species (syn. *Stilbocrea*, *Myrmaeciella*).
    10′(9′) Parasites of bamboo, stromata mostly tuberose . . . . . . . . . . . . . 11

[1] General reference: Rogerson (1971).

11(10′) Stromata knoblike, often very large, becoming larger with each successive perithecial generation . . . . . . . . . . . . . . . . . . . . . . . . . . . . . . . . . . . **Mycocitrus**
      *M. aurantium* (Möller, 1901; Doi, 1967) (syn. *Shiraiella*).

11′(10′) Ascomata not produced in successive generations, ascospores in some species separating into 1-celled part-spores . . . . . . . . . . . . . . . . . . . . **Peloronectriella**
      *P. sasae* (Doi, 1968).

    12(5′) Ascospores with transverse septa only . . . . . . . . . . . . **Peloronectria**
        *P. vinosa* (Möller, 1901).

    12′(5′) Ascospores with transverse and longitudinal septa . . . . . . . . . . **Shiraia**
        *S. bambusicola*.

13(1′) Ascospores 1-celled, not filiform, asci 8-spored . . . . . . . . . . . . . . . . 14

13′(1′) Ascospores septate or filiform, or asci with more than 8 ascospores . . . . . . . . 17

    14(13) Ascomata immersed in wood in seawater . . . . . . . . . . . . **Halonectria**
        *H. milfordensis* (Gareth-Jones, 1965).

    14′(13) Ascomata not immersed in wood . . . . . . . . . . . . . . . . . . . 15

15(14′) Ascospores with thick, sometimes sculptured walls . . . . . . . . **Neocosmospora**
    *N. vasinfecta* and *N. africana* common on plant debris in warmer areas, occasionally infecting roots of different crops (von Arx, 1955; Udagawa, 1963).

15′(14′) Ascospores with thin walls, ellipsoidal, fusiform or aciculate . . . . . . . . . . 16

    16(15′) Ascospores elliptical . . . . . . . . . . . . . . . . . . . . . **Pseudonectria**
      *P. rousseliana* occurs saprobically on *Buxus*; it has a *Volutella* conidial state (Bezerra, 1963). Additional species may have *Sesquicillium* conidial states.

    16′(15′) Ascospores acicular, pointed at both ends, often with pseudosepta
      . . . . . . . . . . . . . . . (cf. *Hyaloderma* and other genera in the Sphaeriaceae).

17(13′) Ascospores 2-celled, occasionally sprouting within the ascus . . . . . . . . . 18

17′(13′) Ascospores with more than one septum or filiform, often sprouting within the ascus . . . . . . . . . . . . . . . . . . . . . . . . . . . . . . . . . . . . . 24

    18(17) Ascomata astomatous, asci spherical . . . . . . . . . . . . . **Heleococcum**
      *H. aurantiacum* and an additional species from soil.

    18′(17) Ascomata ostiolate, asci elongated . . . . . . . . . . . . . . . . . . 19

19(18′) Ascomata immersed in the host tissue or developing in water . . . . . . . . 20

19′(18′) Ascomata develop superficially on the host and sometimes on subicula or stromata . . . . . . . . . . . . . . . . . . . . . . . . . . . . . . . . . . . . 22

    20(19) Ascomata with a thick double wall . . . . . . . . . . . . . **Hydronectria**
      *H. kriegeriana* on stones in water.

    20′(19) Ascomata with membranaceous walls, often parasitic on fungus fructifications or lichens . . . . . . . . . . . . . . . . . . . . . . . . . . . . . . . . . . 21

21(20′) Ascospores hyaline . . . . . . . . . . . . . . . . . . . . . . . . . **Nectriella**
    *N. robergei* on thalli of lichens; it has an *Illosporium* conidial state, other species on plant debris (syn. *Charonectria*, *Cryptonectriella* and *Pronectria*).

21′(20′) Ascospores brownish, when mature . . . . . . . . . . . . . . . . **Passerinula**
    *P. candida* in fructifications of pyrenomycetes (syn. *Dubitatio*, *Spegazzinula* and *Xenonectriella*).

    22(19′) Ascomata with an elongated neck, ascospores fusiform or acicular
      . . . . . . . . . . . . . . . . . . . . . . . . . . . . . . . . . . **Mycorhynchus**

*M. marchalii* on dung and additional species (Breton and Faurel, 1967). *Rhynchonectria* may be a synonym.

22′(19′) Ascomata with a papillate ostiole .................... 23

23(22′) Ascomata small, usually deep violet or red, wall rather thick, consisting of strongly flattened cells .............................................. **Nectriopsis**
*N. violacea* and *N. candida* are parasitic on myxomycetes.

23′(22′) Ascomata small or of medium size, with a wall composed of rounded or slightly flattened cells (Fig. 3b) ............................................ **Nectria**
*N. cinnabarina* is characterised by densely aggregated ascomata sitting on a well developed stroma; it has a *Tubercularia* conidial state; *N. galligena* has no stroma and its conidial state is *Cylindrocarpon mali*, whereas the conidial state of *N. haematococca* is *Fusarium javanicum* (*F. solani*); many species partly parasitic on higher plants (Booth, 1959, 1960; Dingley, 1951a, 1957; Gerlach, 1970). For synonyms see Müller and von Arx (1962).

24(17′) Ascospores with transverse and longitudinal septa ............ 25

24′(17′) Ascospores with transverse septa only .................. 26

25(24) Ascospores with cilialike appendages at both ends .......... **Ciliomyces**
*C. oropensis* parasitic on lichens.

25′(24) Ascospores without cilia, ascospores often sprouting within the asci or soon after liberation .................................... **Thyrionectria**
*T. patavina* and several additional species (Dingley, 1952; Booth, 1959) (syn. *Pleonectria* and *Thyronectroidea*).

26(24′) Ascomata immersed, developing in marine algae .......... **Orcadia**
*O. ascophylli*; monotypic.

26′(24′) Ascomata not in marine algae ........................ 27

27(26′) Ascospores elliptical or fusiform ........................ 28

27′(26′) Ascospores vermiform, filiform or acicular ................ 30

28(27) Ascomata immersed in the host tissue ............ **Micronectriella**
*M. pterocarpi* parasitic on *Pterocarpus*, *M. stoveri* on banana (Booth, 1964). *Griphosphaeria nivalis* is similar, but the ascus apical apparatus stains blue with iodine; it belongs to the Amphisphaeriaceae.

28′(27) Ascomata more or less superficial ...................... 29

29(28′) Ascomata blue or violet ............................. **Gibberella**
*G. fujikuroi* is used for the production of gibberellin and gibberellic acid; *G. zeae* and *G. saubinetii* are common species, all with *Fusarium* conidial states (Dingley, 1952).

29′(28′) Ascomata white, yellow or reddish .................... **Calonectria**
*C. decora* on bark of different trees, conidial state: *Fusarium ciliatum*; a number of additional species develop as hyperparasites of *Meliola* and other fungi (Hansford, 1946).

30(27′) Ascospores vermiform or filiform ...................... 31

30′(27′) Ascospores acicular or elongated .................... **Debaryella**
*D. hyalina* on fruitbodies of *Valsa*.

31(30) Stroma present, ascospores sprouting within the ascus ...... **Scoleconectria**
*S. cucurbitula* on conifer branches has a *Zythiostroma* conidial state.

31′(30) Stroma absent, ascospores not sprouting within the ascus ...... **Ophionectria**

*O. trichospora*; similar ascomycetes may have bitunicate asci and belong therefore to the Loculoascomycetes.

## E. *Polystigmataceae*

### KEY TO IMPORTANT GENERA OF POLYSTIGMATACEAE

1. Ascomata single or aggregated in groups, not stromatic, mostly immersed in the host tissue, ascospores 1-celled . . . . . . . . . . . . . . . . . . . . . . . . . . . . . . 2
1'. Ascomata immersed in a stroma or pseudostroma, covered by a stromatic clypeus or erumpent and superficial, ascospores 1-celled or septate . . . . . . . . . . . . 4

    2(1) Ascospores comparatively large with a granulose or milky, often slightly colored cytoplasm (saprobic) . . . . . . . . . . . cf. *Physalospora* (Amphisphaeriaceae)

    2'(1) Ascospores comparatively small, translucent, hyaline . . . . . . . . . . . 3

3(2') Ascomata small, spherical, with a dark, parenchymatous wall . . . . . . . **Glomerella**
    *G. cingulata* is the ascigerous state of *Colletotrichum gloeosporioides*, *G. tucumanensis* that of *C. graminicola*; both and few additional species are parasitic on a large number of hosts (von Arx, 1970b).

3'(2') Ascomata usually with a wall composed of flattened cells, darker around the papilla . . . . . . . . . . . . . . . . . . . . . . . . . . . . . . . . . **Plectosphaera**
    *P. bersamae* and some additional species parasitic on leaves; other species are saprobic.

    4(1') Ascospores hyaline or when brown, uniformly colored . . . . . . . . . . . 5

    4'(1') Ascospores dark, with a hyaline equatorial band . . . . . . . . . . . . 27

5(4) Ascomata or stromata develop within the host tissue . . . . . . . . . . . . . 6

5'(4) Ascomata or stromata erumpent . . . . . . . . . . . . . . . . . . . . . 22

    6(5) Pseudostromata bright, without a dark clypeus, developing in a hypertrophy of the host tissue . . . . . . . . . . . . . . . . . . . . . . . . . . . . . . . . . 7

    6'(5) Pseudostromata or stromata dark, at least covered with a dark stromatic clypeus . . . . . . . . . . . . . . . . . . . . . . . . . . . . . . . . . 9

7(6) Ascospores l-celled . . . . . . . . . . . . . . . . . . . . . . . . **Polystigma**
    *P. rubrum* on leaves of *Prunus*, *P. astragali* on *Astragalus* and additional species on Rosaceae and Leguminosae; the pseudostroma stains blue with iodine.

7'(6) Ascospores 2-celled . . . . . . . . . . . . . . . . . . . . . . . . . . . 8

    8(7') Ascospores hyaline . . . . . . . . . . . . . . . . . . . . . . **Polystigmella**
    *P. ussuriensis* on *Prunus*.

    8'(7') Ascospores light brown when mature, ascomata deeply immersed in the host and provided with a long beak . . . . . . . . . . . . . . . . . . . . . **Gibellina**
    *G. cerealis* parasitic on cereals.

9(6') Ascospores l-celled . . . . . . . . . . . . . . . . . . . . . . . . . . 10

9'(6') Ascospores septate or filiform . . . . . . . . . . . . . . . . . . . . . 17

    10(9) Ascospores brown, ascomata immersed in a hard, carbonaceous stroma . . . . . . . . . . . . . . . . . . . . . . . . . . . . . . . **Sphaerodothis**
    *S. arengae* on living leaves of different palmae.

    10'(9) Ascospores hyaline or light yellow . . . . . . . . . . . . . . . . . . 11

11(10′) Pseudostromata, staining dark blue with iodine, covered by a dark stromatic clypeus, light inside ............................................................ **Diachora**
    *D. onobrychidis* on *Lathyrus* and related hosts; has in the parasitic stage *Diachorella* conidial state, ascomata ripening during winter (Ciccarone, 1963).

11′(10′) Pseudostromata not staining blue with iodine ............... 12

    12(11′) Ascospores 1-celled, provided with bristlelike appendages ........ 13

    12′(11′) Ascospores 1-celled or septate or filiform, without appendages ..... 14

13(12) Pseudostromata light-colored, covered by a dark epidermal clypeus ............................................................ **Uropolystigma**
    *U. atro-testaceum* on Malpighiaceae; monotypic.

13′(12) Pseudostromata dark-colored, covered by a dark epidermal clypeus ............................................................ **Schizochora**
    *S. elmeri* on *Ficus* (syn. *Telimenopsis*).

    14(12′) Pseudostromata forming large crusts or galls, erumpent but covered by some portions of the epidermis, beaklike ostioles often furrowed ........... 15

    14′(12′) Pseudostromata immersed, covered by the epidermis of the host .... 16

15(14) Pseudostromata forming galls ...................... **Lohwagia**
    *L. intermedia* on Sapindaceae; monotypic.

15′(14) Pseudostromata forming widespread crusts, covering the host stems ... **Phylleutypa**
    *P. dioscoreae* on *Dioscorea*, *P. wittrockii* on *Linnaea*.

    16(14′) Ascomata entirely covered by the clypeus, ostioles not erumpent; peridial wall light-colored, fleshy ........................................ **Isothea**
        *I. rhytismoides* on *Dryas*; monotypic.

    16′(14′) Ascomata with ostioles erumpent, peridial wall often dark, membranaceous .................................................... **Phyllachora**
        *P. graminis* and numerous species parasitic on Gramineae (Parbery, 1964, 1967) and especially in warmer regions on *Ficus* and many other plants (syn. *Catacauma*, *Endodothella*, *Geminispora*, *Halstedia*, *Metachora*, and *Trabutiella*).

17(9′) Ascospores 2-celled ................................. 18

17′(9′) Ascospores with more than 2 cells or filiform ............... 21

    18(17) Ascospores composed of a large upper cell and a small basal cell ..... 19

    18′(17) Ascospores composed of two more or less equal cells (Fig. 4) .. **Stigmochora**
        *S. controversa* and a few more species on Leguminosae.

19(18) Ascospores brown with a hyaline basal cell ............ **Coccochorella**
    *C. quercicola* on *Quercus*; monotypic.

19′(18) Ascospores hyaline ................................ 20

    20(19′) Stromata subcuticular, hemispherical ............ **Rehmiodothis**
        *R. osbeckiae* on Melastomataceae; monotypic (syn. *Munkiodothis*).

    20′(19′) Stromata (pseudostromata) permeating the whole leaf ..... **Apiosphaeria**
        *A. guaranitica* on Bignoniaceae and a few additional species (syn. *Anisochora* and *Oswaldia*).

21(17′) Ascospores filiform ............................ **Ophiodothella**
    *O. atromaculans* parasitic on leaves of *Lonchocarpus* and few additional species.

21′(17′) Ascospores double-fusiform or cylindric, with several septa ...... **Telimenia**

T. *erythrinae* on *Ficus* and T. *gangranae* on Gramineae (syn. *Telimeniella*).

22(5′) Ascospores 2-celled, ascomata immersed in a bright stroma . . . **Phyllocrea**
  P. *quitensis* parasitic on leaves of *Galium*; monotypic.

22′(5′) Ascospores 1-celled, stroma bright or dark . . . . . . . . . . . . . . . . 23

23(22′) Ascomata single, stroma develops only as a footlike hypostroma . . . . . 24

23′(22′) Ascomata single or in groups immersed in an erumpent stroma . . . . . . . 25

  24(23) Ascomata glabrous, hemispherical . . . . . . . . . . . . . . . **Griggsia**
    G. *cyatheae* on ferns; monotypic.

  24′(23) Ascomata spherical, mostly setose . . . . . . . . . . . . . **Ciferriomyces**
    C. *pulcher* parasitic on *Ilex* and a few additional species.

25(23′) Ascomata single in a stroma with crownlike protuberances . . . . . . **Erikssonia**
  E. *pulchella* parasitic on leaves and a few additional species (syn. *Paidania*, *Periaster* and *Rinia*).

25′(23′) Ascomata botryose on or immersed in a mostly knoblike stroma . . . . . . 26

  26(25′) Ascomata immersed in the stroma, ascospores ellipsoidal, hyaline or brownish . . . . . . . . . . . . . . . . . . . . . . . . . . . . . **Coccodiella**
    C. *arundinariae* and many different species parasitic on leaves of Melastomataceae, Palmae, etc. = *Bagnisiopsis* sensu Petrak (1928), = *Coccostroma* sensu von Arx and Müller (1954), cf. Katumoto (1968); *Camarotella*, *Causalis*, *Coccostromopsis*, *Dothidina*, *Haplostroma*, *Leveillinopsis*, *Phoenicostroma*, *Scolecoccoidea*, *Succinaria* are additional synonyms.

  26′(25′) Ascomata botryose, ascospores elongated fusiform . . . . cf. *Schweinitziella* and *Pseudomeliola* (Sphaeriaceae).

27(4′) Stromata erumpent . . . . . . . . . . . . . . . . . . . . . . . **Pseudothiella**
  P. *hirtellae* parasitic on *Hirtella*; monotypic.

27′(4′) Stromata remaining immersed . . . . . . . . . . . . . . . . . **Phaeochorella**
  P. *parinarii* parasitic on leaves of *Parinarium* and a few additional species.

## F. Coryneliaceae

### KEY TO IMPORTANT GENERA OF CORYNELIACEAE[2]

1. Ascospores stellate with more or less prominent lobes (Fig. 3g) . . . . . . **Tripospora**
    T. *tripos* and few additional species, all parasites on leaves of *Podocarpus*.

1′. Ascospores spherical or ellipsoidal, not lobed . . . . . . . . . . . . . . . . . 2

  2(1′) Ascospores echinulate, rather large, spherical . . . . . . . . . . . . . . 3

  2′(1′) Ascospores smooth . . . . . . . . . . . . . . . . . . . . . . . . . . 4

3(2) Ascocarps apically dehiscent, with one or several clefts . . . . . . . . . . **Corynelia**
    C. *uberata* and some additional species, all parasitic on leaves of *Podocarpus*.

3′(2) Ascocarps not dehiscent with clefts, only with an apical perforation, ascospores very variable in size and ornamentation . . . . . . . . . . . . . . . . **Coryneliospora**
    C. *fructicola* on fruits of *Myrsine*.

[2]Reference: Fitzpatrick (1942).

4(2') Ascospores large, thick-walled, asci 2-spored . . . . . . . . . . **Lagenulopsis**
   *L. bispora* on *Podocarpus*.

4'(2') Ascospores small to medium size, with thinner walls, asci 8-spored, not growing on *Podocarpus* . . . . . . . . . . . . . . . . . . . . . . . . . . . . . . . . . . . . . . . **Caliciopsis**
   *C. pinea* and a number of other species on conifers and various other hosts (Fitzpatrick, 1942; Funk 1963).

## G. Sordariaceae

### KEY TO IMPORTANT GENERA OF SORDARIACEAE

1. Ascospores dark brown, with a germ slit, ascomata glabrous or with setae . . . **Coniochaeta**
   *C. ligniaria* and other species on dead wood, *C. tetraspora* and others in soil, *C. scatigena* on dung.

1'. Ascospores dark brown or with lighter colors, without germ slits, often with one or more germ pores . . . . . . . . . . . . . . . . . . . . . . . . . . . . . . . . . . . . . . . . . . . . . 2

   2(1') Ascospores light-colored when mature (hyaline, yellow or light brown), cylindrical to filiform and transversely septate, rarely allantoid and nonseptate (on wood) . . . 3

   2'(1') Ascospores at least partly dark brown . . . . . . . . . . . . . . . . . . . . 4

3(2) Ascospores filiform, coiled in the ascus, hyaline, breaking into numerous fragments at maturity . . . . . . . . . . . . . . . . . . . . . . . . . . . . . . . . . **Mycomedusiospora**
   *M. flavida* on rotten wood; monotypic (Carroll and Munk, 1964).

3'(2) Ascospores not breaking into fragments, very variable in spore morphology from species to species, ascomata setose . . . . . . . . . . . . . . . . . . . . . . . . . **Lasiosphaeria**
   *L. hirsuta* and many other species on rotten wood; the genus is only imperfectly known and it may be heterogeneous, ascospores vary from allantoid and unicellular to cylindric and septate and they may be provided with small appendages (Munk, 1957; Carroll and Munk, 1964).

   4(2') Ascospores 1-celled or composed of 2 unequal cells, basal cell hyaline, dark cell rarely with a secondary wall . . . . . . . . . . . . . . . . . . . . . . . . . . . . . . . . 5

   4'(2') Ascospores 2- or more-celled, cells sometimes separating . . . . . . . . . 23

5(4) Ascospores with appendages and/or composed of 2 unequal cells . . . . . . . . . 6

5'(4) Ascospores without appendages, often with a mucous sheath or ornamented . . 15

   6(5) Ascomata large, cylindrical, obovoid, rounded above (Fig. 4e–g) . . . **Bombardia**
      *B. fasciculata* on wood.

   6'(5) Ascomata spherical, obpyriform of flask-shaped . . . . . . . . . . . . . . . 7

7(6') Ascospores composed of two generally unequal cells, one dark brown, the other remaining hyaline (often considered as an appendage) . . . . . . . . . . . . . . . . . . . . 8

7'(6') Ascospores 1-celled, with 2 or occasionally many true appendages (not originating from cells) and 1 or 2 germ-pores . . . . . . . . . . . . . . . . . . . . . . . . . . . **Arnium**
   *A. lanuginosum* and other species, mostly on dung (Lundqvist, 1972).

   8(7) Ascospores with upper cell warty or spiny . . . . . . . . . . . . . . . . . . . 9

   8'(7) Ascospores smooth or finely verrucose . . . . . . . . . . . . . . . . . . . 10

9(8) Ascomata ostiolate, dark (Fig. 4d) . . . . . . . . . . . . . . . . . . . **Apiosordaria**
   *A. verruculosa* is a heterothallic soil fungus (von Arx and Gams, 1966).

9′(8) Ascomata astomatous, light-colored (semitransparent) . . . . . . **Echinopodospora**
      *E. jamaicensis* isolated from soil and sugar-cane roots (Robison, 1970).

  10(8′) Ascospores vermiform at first, with filiform appendages at each end, the upper cell developing to a darkened head . . . . . . . . . . . . . . . . **Lasiosordaria**
        *L. lignicola* on rotten wood and other species (= *Ceriophora* sensu Lundqvist, 1972).

  10(8′) Ascospores not cylindric–vermiform at first . . . . . . . . . . . . . . . 11

11(10′) Ascospores with an apical hyaline cell and a basal germ pore . . . . **Anopodium**
      *A. ampullaceum* and some more species on dung (Lundqvist, 1964).

11′(10′) Ascospores with a basal hyaline cell and an apical germ pore . . . . . . . . 12

  12(11′) Ascospores ellipsoidal at first, hyaline cell short and comparatively broad . . 13

  12′(11′) Ascospores distinctly clavate at first, hyaline cell longer than broad, mostly cylindrical, provided with true appendages, occasionally the upper end also provided with appendages . . . . . . . . . . . . . . . . . . . . . . . . . . . . . . . . . 14

13(12) Dark cell of the ascospores broadly fusiform, basal cell small . . . . **Lacunospora**
      *L. stercoraria* and other species on dung (Cailleux, 1968).

13′(12) Dark cell of the ascospores triangular, basal cell broad (Fig. 4c) . . . **Triangularia**
      *T. bambusae* and some other species on wood and dung (Cain and Farrow, 1956; von Arx and Hennebert, 1968).

  14(12′) Ascomata ostiolate . . . . . . . . . . . . . . . . . . . . . . . . . . . **Podospora**
        *P. fimicola* and many additional species on dung and occasionally on other organic substrata (Mirza and Cain, 1969).

  14′(12′) Ascomata astomatous . . . . . . . . . . . . . . . . . . . . . . . . . **Zopfiella**
        *Z. tabulata* on dung develops sometimes an additional septum in the dark upper cell; other species are strictly didymosporous; *Tripterospora* (Cain, 1956; Lundqvist, 1969; Cailleux, 1970) is a similar genus which even may be identical with *Zopfiella*.

15(5′) Ascospores glabrous, with a gelatinous sheath and one germ pore . . . . . . . 16

15′(5′) Ascospores without a gelatinous sheath or with more than one germ pore . . . 17

  16(15) Ascospores provided with a germ pore at the lower end
      (Fig. 3d, Fig. 4a) . . . . . . . . . . . . . . . . . . . . . . . . . . . . . **Sordaria**
        *S. fimicola* and several additional species on dung or in soil (Moreau, 1954; Cailleux, 1971; Lundqvist, 1972).

  16′(15) Ascospores provided with a germ pore at the upper end . . . . . **Apodospora**
        *A. simulans* and few additional species on dung (Cain and Mirza, 1970).

17(15′) Ascospores smooth . . . . . . . . . . . . . . . . . . . . . . . . . . . . . . 18

17′(15′) Ascospores striate, pitted or reticulate . . . . . . . . . . . . . . . . . . . . 20

  18(17) Ascomata large, rounded above; ascospores with two terminal
      germ pores . . . . . . . . . . . . . . . . . . . . . . . . . . . . **Bombardioidea**
        *B. bombardioides* on dung (Moreau, 1954).

  18′(17) Ascomata spherical, obpyriform or flask-shaped . . . . . . . . . . . . . 19

19(18′) Ascospores with one terminal germ pore at the lower end; ascomata setose, with 2-celled, brown conidia . . . . . . . . . . . . . . . . . . . . . . . **Helminthosphaeria**
      *H. clavariarum* on *Clavaria* (Parguey-Leduc, 1961).

19′(18′) Ascospores with 1 terminal and some, excentric germ pores, conidial states lacking
. . . . . . . . . . . . . . . . . . . . . . . . . . . . . . . . . . . . **Fimetariella**

*F. rabenhorstii* on dung (Lundqvist, 1964). *Amphisphaerella* with ascospores provided with equatorial germ-pores belongs to the Amphisphaeriaceae.

20(17') Ascospores striate, provided with one or two germ pores . . . . **Neurospora**
   *N. sitophila* and *N. crassa* are heterothallic with a *Monilia* conidial state; other species are homothallic and have no conidial state.

20'(17') Ascospores pitted or reticulate . . . . . . . . . . . . . . . . . . . . . . 21

21(20') Ascomata ostiolate . . . . . . . . . . . . . . . . . . . . . . . . **Gelasinospora**
   *G. tetrasperma* and some more species on rotten plant material and on dung (Cain, 1950; Cailleux, 1971).

21'(20') Ascomata astomatous . . . . . . . . . . . . . . . . . . . . . . . . . . . . 22

22(21') Asci 2-spored (1- to 3-spored), ascomata dark . . **Copromyces** (*Rechingeriella*)
   *C. bisporus* on dung (Lundqvist, 1967; Petrak, 1970).

22'(21') Asci 8-spored, ascomata bright . . . . . . . . . . . . . . . . . **Anixiella**
   *A. reticulata* (Cain, 1961b).

23(4') Ascospores pitted, ascomata astomatous . . . . . . . . . . . **Diplogelasinospora**
   *D. princeps* on seeds (Cain, 1961b).

23'(4') Ascospores smooth, often with appendages . . . . . . . . . . . . . . . . . 24

24(23') Central part of the ascospores comprises 2 brown cells, appendages at both ends striate . . . . . . . . . . . . . . . . . . . . . . . . . . . . . . . . **Zygospermella**
   *Z. setosa* on dung, and few other species (Lundqvist, 1969).

24'(23') The 2 brown cells of the ascospores are connected by a narrow hyaline central part and become separated when mature (Fig. 4h). . . . . . . . . . **Zygopleurage**
   *Z. zygospora* on dung (Lundqvist, 1969).

## H. Diaporthaceae

### KEY TO IMPORTANT GENERA OF DIAPORTHACEAE

1. Asci cylindrical, remaining attached to the ascogenous cells; ascospores mostly relatively broad, rounded at the ends . . . . . . . . . . . . . . . . . . . . . . . . . . . . 2
1'. Asci fusiform or clavate or nearly cylindrical, loosening from the ascogenous cells at an early stage and lying free in the ascoma; ascospores mostly allantoid, fusiform, pyriform or filiform . . . . . . . . . . . . . . . . . . . . . . . . . . . . . . . . . . . . . . 21

   2(1) Ascospores 1-celled or with pseudosepta . . . . . . . . . . . . . . . . . . 3

   2'(1) Ascospores septate . . . . . . . . . . . . . . . . . . . . . . . . . . . . 5

3 (2) Ascomata solitary . . . . . . . . . . . . . . . . . . . . . . . . . . . . . . . 4

3' (2) Ascomata in circular groups immersed in a yellow or reddish stroma and with long ostiolar necks collectively erumpent; ascospores subcylindrical, curved, large
   . . . . . . . . . . . . . . . . . . . . . . . . . . . . . . . . . . . . . . **Cryptospora**
   *C. suffusa* on *Alnus* and a few additional species (Munk, 1957).

   4(3) Ascomata with a thin wall, asci narrow (mostly on herbaceous stems)
   . . . . . . . . . . . . . . . . . . . . . . . . . . . . . . . . . . . . **Phomatospora**
   *P. berkeleyi* on herbaceous stems and some additional species (von Arx and Müller, 1954).

   4'(3) Ascomata with a thick wall, large (on decaying wood) . . . . . . . . **Endoxyla**
   *E. operculata* and some additional species (Munk, 1965).

5(2′) Ascospores septate in the lower third or near the lower end . . . . . . . . . . . 6
5′(2′) Ascospores with a median cross wall or multiseptate . . . . . . . . . . . . 7
    6(5) Ascospores truncate at both ends, ascomata immersed in an erumpent stroma
. . . . . . . . . . . . . . . . . . . . . . . . . . . . . . . . . . . . . . . . . **Pseudothis**
*P. coccodes* on Leguminosae.
    6′(5) Ascospores rounded, with a small lower cell, ascomata covered by a clypeus
. . . . . . . . . . . . . . . . . . . . . . . . . . . . . . . . . . . . . . . . . **Anisomyces**
*A. nectrioides* parasitic on *Paullinia*, and additional species (syn. *Phaeoapiospora, Stegastroma*).
    7(5′) Ascospores pyriform, 2-celled, ascomata lying horizontally in the attacked leaf tissue, with a lateral beak . . . . . . . . . . . . . . . . . . . . . . . . . . . . . . **Plagiostigme**
*P. couraliae* parasitic on *Couralia*, and few additional species (syn. *Bioporthe*).
    7′(5′) Ascospores mostly ellipsoidal, ascomata with an apical beak or papilla . . . . . 8
        8(7′) Ascospores 2-celled . . . . . . . . . . . . . . . . . . . . . . . . . . . . . 9
        8′(7′) Ascospores with 2 or more septa . . . . . . . . . . . . . . . . . . . . 12
    9(8) Ascomata immersed in the host tissue, stroma absent . . . . . . . . . . **Sydowiella**
*S. fenestrans* common on stems of *Epilobium*.
    9′(8) Ascomata immersed in a stroma or pseudostroma (mostly on bark) . . . . . . . 10
        10(9′) Stroma compact, with a dark margin (bound), pustulate . . . . . **Hercospora**
*H. tiliae* with the conidial state *Rabenhorstia tiliae*.
        10′(9′) Stroma without a dark margin . . . . . . . . . . . . . . . . . . . . 11
    11(10′) Asci with an apical ring, ascospores hyaline or brown (Fig. 3e) . . . . **Melanconis**
*M. stilbostoma* on *Betula* and a number of other species on different trees, with *Melanconium* conidial states (Wehmeyer, 1941a) (syn. *Calospora, Discodiaporthe, Macrodiaporthe, Melanconiella, Neokeissleria, Phaeodiaporthe, Pseudodiaporthe, Pseudovalsella*).
    11′(10′) Asci apically with a spherical body, ascospores large, dark . . . . **Massariovalsa**
*M. sudans* on different species of trees and a few additional species (Wehmeyer, 1941a).
        12(8′) Ascomata solitary or loosely aggregated, immersed in the host tissue or superficial . . . . . . . . . . . . . . . . . . . . . . . . . . . . . . . . . . . . . 13
        12′(8′) Ascomata aggregated and connected by stromatic tissue or immersed in a stroma or pseudostroma, rarely solitary . . . . . . . . . . . . . . . . . . 15
13 (12) Ascospores hyaline or slightly pigmented, transversely septate or filiform . . . 14
13′(12) Ascospores hyaline, transversely and longitudinally septate, sometimes producing blastospores within the ascus . . . . . . . . . . . . . . . . . . . . . . . **Rhamphoria**
Several species (Munk, 1957).
    14(13) Ascomata with a black, hard peridial wall, ostiolum short . . . . . **Zignoëlla**
Several species (Munk, 1957).
    14′(13) Ascomata with a soft peridial wall, ostiolum beaklike . . . . **Ceratosphaeria**
(Munk, 1957).
15(12′) Ascospores filiform . . . . . . . . . . . . . . . . . . . . . . . . . . . . . 16
15′(12′) Ascospores not filiform, transversely septate or dictyosporous . . . . . . . . . 17
    16(15) Ascomata immersed in a stroma, ascospores filiform . . . . . . . . . . **Sillia**
*S. ferruginea* on different trees; monotypic (Munk, 1957).

# 6. Meliolales, Coronophorales, Sphaeriales

16′(15) Ascomata in a pustulate pseudostroma, ascospores long fusiform with narrowed parts, transversely septate . . . . . . . . . . . . . . . . . . . **Vleugelia**
   *V. betulina*; monotypic (Reid and Booth, 1969).

17(15′) Ascomata in a pseudostroma; ascospores dictyosporous, hyaline . . **Dictyoporthe**
   *D. ahmadii* and a few additional species (Müller, 1962).

17′(15′) Ascospores phragmosporous . . . . . . . . . . . . . . . . . . . . . 18

18(17′) Ascospores fusiform or cylindrical and curved, brown; endcells often lighter
   . . . . . . . . . . . . . . . . . . . . . . . . . . . . . . . . . . . . . 19

18′(17′) Ascospores ellipsoidal or fusiform, when brown without light-colored endcells, ascomata in a stromatic disc, divided into ecto- and endostroma . . . . . . . 20

19(18) Ascomata immersed in a stroma . . . . . . . . . . . . . . . . **Melogramma**
   *M. spiniferum* on *Fagus* (Munk, 1957).

19′(18) Ascomata free, sitting on a stromatic crust, with dark, often clavate setae
   . . . . . . . . . . . . . . . . . . . . . . . . . . . . . . . . . **Melanochaeta**
   *M. hemipsila* on rotten wood (Müller *et al.*, 1969); conidial state *Sporoschisma*.

20(18′) Entostroma dark-colored, ascospores not appendaged, conidia borne superficially . . . . . . . . . . . . . . . . . . . . . . . . . . . . . . . **Pseudovalsa**
   *P. lanciformis* and a few additional species on branches of trees (Wehmeyer, 1941a); conidial state *Coryneum*.

20′(18′) Entostroma not dark-colored, ascospores with appendages, at least when young; conidia borne in cavities or locules . . . . . . . . . . . . . **Prosthecium**
   *P. ellipsosporum*, on *Carpinus*, and additional species (Wehmeyer, 1941a); conidial state *Stilbospora*.

21(1′) Ascospores allantoid (sausage-shaped) . . . . . . . . . . . . . . . . . 22

21′(1′) Ascospores not allantoid . . . . . . . . . . . . . . . . . . . . . . . 25

22(21) Ascomata loosely aggregated, immersed in the host tissue, often arranged in circles . . . . . . . . . . . . . . . . . . . . . . . . . . . . . . . . **Enchnoa**
   *E. infernalis* on wood and some additional species.

22′(21) Ascomata immersed in a stroma or a pseudostroma . . . . . . . . . . . 23

23(22′) Ascomata immersed in a dark, pulvinate stroma . . . . . . . . . **Valseutypella**
   *V. tristicha* on *Rosa* (Hubbes, 1960); conidial state *Cytospora*.

23′(22′) Ascomata in a pseudostroma, often arranged in circles . . . . . . . . . 24

24(23′) Pseudostroma distinct, often delineated by dark lines . . . . . . **Leucostoma**
   *L. persoonii* and additional species on trees; conidial state *Cytospora*; including *Valsella* with polysporous asci (Kern, 1957).

24′(23′) Pseudostromata without dark lines, often reduced to clypeuslike structures
   . . . . . . . . . . . . . . . . . . . . . . . . . . . . . . . . . . . **Valsa**
   *V. ceratophora* and many additional species on woody plants; conidial state *Cytospora*.

25(21′) Ascospores 1-celled, not filiform . . . . . . . . . . . . . . . . . . 26

25′(21′) Ascospores septate or filiform . . . . . . . . . . . . . . . . . . . 37

26(25) Ascomata surrounded by a cartilaginous stroma . . . . . . . . . . . . 27

26′(25) Stromata absent or not cartilaginous . . . . . . . . . . . . . . . . 29

27(26) Stroma developing in hypertrophic portions of attacked leaves, covered by a clypeus-

like crust, ascomata with elongated beaks . . . . . . . . . . . . . . . **Mamianiella**
   *M. coryli* parasitic on *Corylus* (von Arx and Müller, 1954).
27'(26) Stromata not developing in the whole depth of a leaf, mostly sclerotiumlike . 28
   28(27') Stromata enclosing a single ascoma . . . . . . . . . . . . . . . **Heteropera**
      *H. borealis* (von Arx and Müller, 1954) (syn. *Cryptonectriopsis* and *Paramazzantia*).
   28'(27') Stromata flat or elongate, mostly containing several ascomata . . **Mazzantia**
      *M. galii* on *Galium* (von Arx and Müller, 1954) (syn. *Clypeocarpus*).
29(26') Asci 16-spored, ascospores often pseudoseptate . . . . . . . . . . . . . . 30
29'(26') Asci 4–8-spored . . . . . . . . . . . . . . . . . . . . . . . . . . . . . 31
   30(29) Ascomata large, with conical or short cylindrical beaklike ostioles, covered by a clypeus . . . . . . . . . . . . . . . . . . . . . . . . . . . . . . . **Ditopella**
      *D. ditopa* on *Alnus* monotypic (von Arx and Müller, 1954).
   30'(29) Ascomata small, with a long cylindrical beak, without stroma . . . **Rehmiella**
      *R. alpina* on *Alchemilla* (von Arx and Müller, 1954).
31(29') Ascomata with long beaks, stroma absent or present . . . . . . . . . . . . 32
31'(29') Ascomata with short beaks, mostly immersed in a stroma . . . . . . . . . 34
   32(31) Ascomata solitary, free . . . . . . . . . . . . . . . . . . . . . **Gnomoniella**
      *G. tubaeformis*, on *Alnus*, and a few additional species.
   32'(31) Ascomata in groups, free or connected by stromatic structures . . . . . 33
33(32') Ascomata embedded at the same level in the host tissue, beaks collectively erumpent; ascospores broadly fusiform, large . . . . . . . . . . . . . . . . . . . **Cryptosporella**
   *C. hypodermia* on *Ulmus*.
33'(32') Ascomata embedded at different levels in a cushionlike stroma, beaks individually erumpent; ascospores ellipsoid . . . . . . . . . . . . . . . . . **Pseudocryptosporella**
   *P. polylepidis* on *Polylepis*; monotypic (Reid and Booth, 1969).
   34(31') Ascomata covered by a clypeus, ascospore content at the ends granular, dense and easily staining with cotton-blue; in the middle portion glassy and not staining . . . . . . . . . . . . . . . . . . . . . . . . . . . . . . . . . . . **Bagcheea**
      *B. albomaculans* parasitic on leaves of *Castanopsis*; monotypic (Katumoto, 1965).
   34'(31') Ascospores with homogeneous cytoplasm . . . . . . . . . . . . . . . . 35
35(34') Ascomata small (less than 150 μm in diameter) mostly on leaves . . . . . . . 36
35'(34') Ascomata larger, mostly on stems . . . . . . . . . . . . . . . **Diaporthopsis**
   *D. angelicae*, on Umbelliferae, and a few additional species.
   36(35) Ascomata covered by a clypeus . . . . . . . . . . . . . **Sphaerognomonia**
      *S. carpini* on dead leaves of *Carpinus* (syn. *Apiosporopsis*).
   36'(35) Ascomata surrounded by a cellular stroma . . . . . . . . . . . **Diplacella**
      *D. marayensis* parasitic on leaves of *Paullinia*; monotypic.
37(25') Ascospores 2-, rarely 4-celled, mostly fusiform, pyriform or ellipsoid . . . . 38
37'(25') Ascospores filiform, vermiform or double-fusiform, at least 8 times as long as broad . . . . . . . . . . . . . . . . . . . . . . . . . . . . . . . . . . . . 51
   38(37) Ascomata without a stroma; mostly on leaves . . . . . . . . . . . . . . . 39
   38'(37) Ascomata immersed in a stroma or growing in valsoid groups or covered by a clypeus . . . . . . . . . . . . . . . . . . . . . . . . . . . . . . . . . . . 42

## 6. Meliolales, Coronophorales, Sphaeriales

39(38) Ascomata flattened, lying horizontally, with lateral beaks . . . . . . . . . . 40
39'(38) Ascomata upright with an apical beak . . . . . . . . . . . . . . . . . . 41
    40(39) Ascospores septate near the lower end . . . . . . . . . . . . **Plagiostomella**
    *P. carpinicola* on leaves of *Carpinus*.
    40'(39) Ascospores septate in or near the middle . . . . . . . . . . . . **Plagiostoma**
    *P. euphorbiae*, on stems of *Euphorbia*, and some additional species (syn. *Chalcosphaeria*, *Gnomonina* and *Laestadia*).
41(39') Ascospores septate in the lower third portion . . . . . . . . . . . **Apiognomonia**
    *A. errabunda* (= *Gnomonia platani*) with a *Discula* conidial state on leaves of *Platanus*, *Tilia* and other trees.
41'(39') Ascospores septate in or near the middle or 4-celled . . . . . . . . . . **Gnomonia**
    *G. fructicola* is parasitic on strawberries and has a *Zythia*-conidial state; many additional species develop on dead leaves; some species have a *Discula* or a *Cylindrosporella* conidial state.
    42(38') Ascomata immersed in a pseudostroma developing in (often hypertrophic) portions of leaves . . . . . . . . . . . . . . . . . . . . . . . . . . . . . . . . . 43
    42'(38') Ascomata immersed in a stroma or in a pseudostroma (on twigs or on woody stems) . . . . . . . . . . . . . . . . . . . . . . . . . . . . . . . . . . . . 46
43(42) Stromata erumpent, ascospores septate near the middle . . . . . . **Phylloporthe**
    *P. vernoniae* on living leaves of *Vernonia*.
43'(42) Stromata not erumpent . . . . . . . . . . . . . . . . . . . . . . . . . . 44
    44(43') Ascomata depressed, with lateral beaks . . . . . . . . . . . . . **Hypospilina**
    *H. bifrons* on *Quercus*; *H. salicina* has 1-celled ascospores.
    44'(43') Ascomata spherical, with apical beaks . . . . . . . . . . . . . . . . . 45
45(44') Beaks elongated, stroma bright inside . . . . . . . . . . . . . . . **Mamiania**
    *M. fimbriata* on *Carpinus*.
45'(44') Ascomata with short cylindrical beaks . . . . . . . . . . . . . . . . **Lambro**
    *L. insignis* on living leaves of *Sterculia*; *L. ulmea* on *Ulmus*.
46(42') Ascospores septate near the lower end . . . . . . . . . . . . . . . . . . 47
46'(42') Ascospores with a septum in or near the middle . . . . . . . . . . . . . 48
47(46) Ascomata immersed in a dark, pulvinate stroma . . . . . . . . . . **Anisogramma**
    *A. virgultorum* parasitic on *Betula* (syn. *Apioporthe*).
47'(46) Ascomata immersed in the host tissue and surrounded by a
pseudostroma . . . . . . . . . . . . . . . . . . . . . . . . . . . . . . . **Apioporthella**
    *A. bavarica* on *Alnus*.
    48(46') Stroma light, mostly yellow, fleshy . . . . . . . . . . . . . . . . **Endothia**
    *E. parasitica* is a parasite of *Castanea*; other species are saprobes (Kobayashi and Ito, 1956).
    48'(46') Stroma not fleshy when present . . . . . . . . . . . . . . . . . . . . 49
49(48') Ascomata immersed in a dark, pulvinate stroma . . . . . . . . . . **Diaporthella**
    *D. aristata* parasitic on twigs of *Betula*.
49'(48') Ascomata immersed in the host tissue and mostly surrounded by a
pseudostroma . . . . . . . . . . . . . . . . . . . . . . . . . . . . . . . . . . 50
    50(49') Pseudostroma distinct, often delimited by dark lines; conidial state:

*Phomopsis* .................................................. **Diaporthe**
   *D. eres* and a large number of other species on twigs of trees and shrubs and on stems of herbaceous plants (Wehmeyer, 1933).

   50'(49') Pseudostroma without dark lines, often reduced to clypeuslike structures; conidial states: *Discella, Chondroplea* ...................... **Cryptodiaporthe**
   *C. populea* parastic on *Populus* (Wehmeyer, 1933; Butin, 1958).

51(37') Ascospores double-fusiform, with a narrow middle portion ........... 52

51'(37') Ascospores not narrowed in the middle ...................... 53

   52(51) Parasitic on leaves, wall of the ascomata light .......... **Diatractium**
   *D. cordiana* on *Cordia*.

   52'(51) Saprobic on leaves or twigs, wall of the ascomata dark, or ascomata immersed in a pseudostroma ............................................ **Pleuroceras**
   *P. cryptoderis* on *Populus* leaves (syn. *Cryptoderis*).

53(51') Ascospores filiform, with a somewhat swollen upper cell, ascomata with an apical beak ............................................................ **Ophiognomonia**
   *O. pleurostyla*, on *Salix*, and a few additional species.

53'(51') Upper cell of the ascospores not swollen ...................... 54

   54(53') Ascomata with a thin wall, covered by a clypeus or immersed in a pseudostroma .......................................................... 55

   54'(53') Ascomata with a thick wall, immersed in the host tissue or erumpent .... 56

55(54) Ascomata with a lateral beak ...................... **Linospora**
   *L. capreae*, on *Salix*, and some additional species (Schrantz, 1960).

55'(54) Ascomata with an apical papilla ...................... **Linocarpon**
   *L. pandani* on *Pandanus*.

   56(54') Ascomata with a lateral beak .................. **Plagiosphaera**
   *P. moravica* on rotten stems of *Urtica*.

   56'(54') Ascomata with a (short) cylindrical, apical beak ..... **Gaeumannomyces**
   *G. graminis* (= *Ophiobolus graminis*) parasitic on roots of cereals and grasses, and some other species (Schrantz, 1960; Skou, 1968; Walker, 1972).

## I. Halosphaeriaceae

### KEY TO IMPORTANT GENERA OF HALOSPHAERIACEAE[3]

1. Asci spherical or broadly clavate; ascospores 1-celled, small (parasitic on red algae)
   .................................................. **Chadefaudia**
   syn. *Mycophycophila*; *M. corallinarum* (Crouan and Crouan) Kohlm. = *Chadefaudia corallinarum* (Crouan and Crouan) comb. nov. Basionym: *Sphaeria corallinarum* Crouan and Crouan in Fl. Finistère Paris, p. 24, 1867.

1'. Asci clavate or cylindrical, not spherical (mostly saprobic) ............. 2

   2(1') Ascospores without appendages, at most with an irregular gelatinous sheath ............................................................ 3

---

[3]General references: Johnson and Sparrow (1961); Kohlmeyer and Kohlmeyer (1964–1969).

## 6. Meliolales, Coronophorales, Sphaeriales

2′(1′) Ascospores provided with appendages inserted apically and/or laterally . . . . 5

3(2) Ascospores 1-celled, surrounded by a gelatinous, irregular or undulate
sheath . . . . . . . . . . . . . . . . . . . . . . . . . . . . . . . . . **Samarosporella**
    *S. pelagica* (Johnson and Sparrow, 1961); monotypic.

3′(2) Ascospores 2-celled . . . . . . . . . . . . . . . . . . . . . . . . . 4

    4(3′) Ascospores surrounded by a gelatinous sheath . . . . . . **Didymosamarosporella**
        *D. euryhyalina*; monotypic.

    4′(3′) Ascospores without a gelatinous sheath . . . . . . . . . . . . . **Lignincola**
        *L. laevis*; monotypic.

5(2′) Ascospore appendages predominantly ciliate or rigid, inserted apically and/or
laterally . . . . . . . . . . . . . . . . . . . . . . . . . . . . . . . . . 6

5′(2′) Ascospore appendages predominantly gelatinous or slimy . . . . . . . . . . . 9

    6(5) Ascomata growing superficially on a crusty subiculum, more seldom with the lower portion immersed; ascospores 1-celled to multicelled, appendages at least partly ciliate . . . . . . . . . . . . . . . . . . . . . . . . . . . . . **Corollospora**
        *C. maritima* and several additional species, mainly differentiated by ascospore characteristics (syn. *Arenariomyces, Peritrichospora*).

    6′(5) Ascomata normally immersed in the host tissue . . . . . . . . . . . . . . . 7

7(6′) Ascospores 1-celled, with apical and lateral tufts of cilialike hairs . . . . **Nautosphaeria**
    *N. cristaminuta*; monotypic.

7′(6′) Ascospores septate . . . . . . . . . . . . . . . . . . . . . . . . . . . 8

    8(7′) Ascospores 2-celled . . . . . . . . . . . . . . . . . . . . . . . **Halosphaeria**
        *H. appendiculata* and several additional species (syn. *Antennospora, Halosphaeriopsis*).

    8′(7′) Ascospores multicelled, appendages radiating at one or both ends
    . . . . . . . . . . . . . . . . . . . . . . . . . . . . . . . . . **Torpedospora**
        *T. radiata* and several additional species.

9(5′) Ascospores with gelatinous caps at both ends . . . . . . . . . . . . . . . 10

9′(5′) Ascospores with gelatinous, hamate or straight appendages . . . . . . . . . 12

    10(9) Ascospores 1-celled, ellipsoidal, vermiform or filiform . . . . . . . . . . 11

    10′(9) Ascospores 2-celled . . . . . . . . . . . . . . . . . . . . . . **Lentescospora**
        *L. submarina*; monotypic.

11(10) Ascospores vermiform or filiform, gelatinous caps cell-like . . . . . . **Lulworthia**
    *L. fucicola* and several additional species (syn. *Halophiobolus*).

11′(10) Ascospores ellipsoidal . . . . . . . . . . . . . . . . . . . . . . **Haloguignardia**
    *H. decidua* and several additional species (Cribb and Cribb, 1956).

    12(9′) Ascospores 2-celled . . . . . . . . . . . . . . . . . . . . . . . . . 13

    12′(9′) Ascospores multicelled . . . . . . . . . . . . . . . . . . . . . . . . 14

13(12) Wall of ascomata bright and soft, consisting of relatively large cells, ascospores with apical appendages, appearing delicately striped under phase contrast . . . **Remispora**
    *R. maritima* and some additional species (syn. *Palomyces*).

13′(12) Wall of ascomata dark and brittle, consisting of narrow, thick-walled cells; ascospores with apical, or apical and lateral appendages, appearing delicately striped under phase contrast . . . . . . . . . . . . . . . . . . . . . . . . . . . . . . . **Ceriosporopsis**
    *C. halima* and few additional species.

14(12′) Ascospores filiform, appendages delicate, often inflated . . . . . . . **Lindra**
L. *inflata*; monotypic.

14′(12′) Ascospores more or less broadly fusiform, appendages inserted at the ends or slightly laterally . . . . . . . . . . . . . . . . . . . . . . . . . . . . **Haligena**
H. *elaterophora*.

## J. Diatrypaceae

### KEY TO IMPORTANT GENERA OF DIATRYPACEAE

1. Stromata formed entirely by fungus elements (true stromata), disclike, spreading or erumpent, but delimited . . . . . . . . . . . . . . . . . . . . . . . . . . . 2

1′. Stromata formed by fungus elements and host tissue (pseudostromata), not well delimited, reduced or absent . . . . . . . . . . . . . . . . . . . . . . . . . . . . . 3

    2(1) Asci 8-spored (Fig. 3f) . . . . . . . . . . . . . . . . . . . . . . . **Diatrype**
    D. *disciformis* and other species, often difficult to differentiate.

    2′(1) Asci polysporous . . . . . . . . . . . . . . . . . . . . . . . . . **Diatrypella**
    D. *verrucaeformis* and a number of additional species which are difficult to differentiate.

3(1′) Pseudostromata erumpent, often pustulate . . . . . . . . . . . . . . . . . . . 4

3′(1′) Pseudostromata not erumpent, entirely included in the host, often reduced to black clypeal zone or even absent, ascomata solitary or in small groups, embedded in the host . . . . . . . . . . . . . . . . . . . . . . . . . . . . . . . . . **Cryptosphaeria**
C. *millepunctata*.

    4(3) Ascomata arranged in circular groups, ostioles long, collectively erumpent in the center of covering disc . . . . . . . . . . . . . . . . . . . . . . . . . . . 5

    4′(3) Ascomata solitary or in small groups with ostioles not collectively erumpent . . . . . . . . . . . . . . . . . . . . . . . . . . . . . . . **Eutypa**
    E. *lata* on hard wood, and many other species.

5(4) Ascomata numerous, forming a dense cluster, embedded in a marked pseudostromatic pustule . . . . . . . . . . . . . . . . . . . . . . . . . . . . . . . **Eutypella**

5′(4) Ascomata not numerous, only weakly erumpent; pseudostromata often reduced and delimited by dark lines . . . . . . . . . . . . . . . . . . . . . . . . **Quaternaria**
Q. *persoonii* on *Fagus* with *Libertella* conidial state.

## K. Amphisphaeriaceae

### KEY TO IMPORTANT GENERA OF AMPHISPHAERIACEAE

1. Ascospores double-fusiform, narrowed in the center, with one septum . . . . **Vialaea**
V. *insculpta* on *Ilex* (Schrantz, 1960).

1′. Ascospores ellipsoidal, ovoid or fusiform . . . . . . . . . . . . . . . . . . . . 2

    2(1′) Ascospores 1-celled . . . . . . . . . . . . . . . . . . . . . . . . . . . 3

    2′(1′) Ascospores septate . . . . . . . . . . . . . . . . . . . . . . . . . . . 12

3(2) Ascospores hyaline or bright . . . . . . . . . . . . . . . . . . . . . . . . . 4

3′(2) Ascospores brown or dark . . . . . . . . . . . . . . . . . . . . . . . . . 10

4(3) Ascospores longer than 20 μm, with a granular or milky content; ascomata glabrous
or setose, with a fleshy wall ......................... **Physalospora**
  *P. alpestris* on *Carex*, *P. rhododendri* on *Rhododendron* and some additional species
  (von Arx and Müller, 1954; Barr, 1970) (syn. *Acanthorhynchus, Benedekiella, Pseudoguignardia, Pseudophysalospora*, and *Trichophysalospora*).

4'(3') Ascospores smaller or translucent ...................... 5

5(4') Ascomata immersed in the host tissue, arranged around a collective
ostiolum ............................................. **Myelosperma**
  *M. tumidum* on *Cocos nucifera*; monotypic.

5'(4') Ascomata not arranged around a collective ostiolum ............... 6

  6(5') Ascomata with an often lateral ostiolum ............. **Xylochora**
    Two species on wood.

  6'(5') Ascomata with an apical ostiolum ...................... 7

7(6') Ascospores rostrate ......................... **Urosporella**
  *U. americana* on old stems, *U. coccifera* on *Quercus* (Barr, 1966).

7'(6') Ascospores not rostrate ............................ 8

  8(7') Ascospores attenuated and curved near the base, ascomata often stromatic . . 17

  8'(7') Ascospores not curved at the base, ascomata not stromatic ......... 9

9(8') Ascomata with a wall composed of enlarged cells ......... **Anisostomula**
  *A. cookeana* on dead leaves of *Quercus*.

9'(8') Ascomata with a wall composed of bright, flattened cells ....... **Hyponectria**
  *H. buxi* on leaves of *Buxus*; monotypic.

  10(3') Ascospores with equatorially arranged germ pores; asci amyloid or not
    .................................... **Amphisphaerella**
    *A. xylostei* on *Lonicera* and a few additional species on branches (Eriksson, 1966).

  10'(3') Germ pores when present not equatorial ................... 11

11(10') Ascospores with a basal, appendagelike cell and an apical germ pore ..... 15

11'(10') Ascospores fusiform, with mucilaginous appendages at both ends . . **Pemphidium**
  *P. nitidum* on palms and bamboo; monotypic.

  12(2') Ascospores 2-celled ........................... 13

  12'(2') Ascospores at maturity at least partly with 2 or more septa ......... 26

13(12) Ascospores septate near the lower end ..................... 14

13'(12) Ascospores septate near the middle ..................... 18

  14(13) Ascospores brownish or dark when ripe .................. 15

  14'(13) Ascospores hyaline ............................ 16

15(11,14) Ascomata setose, asci amyloid (Fig. 3h) ............. **Chaetapiospora**
  *C. islandica*, on *Dryas*, and additional species.

15'(11,14) Ascomata glabrous, asci not amyloid; ascospores with an apical germ pore, the
large upper cell brown and the small basal cell hyaline ............ **Entosordaria**
  *E. perfidiosa* on bark of *Acer*.

  16(14') Ascomata depressed, with a lateral ostiolum .......... **Apiothyrium**
    *A. arcticum* on *Diapensia*; monotypic.

  16'(14') Ascomata spherical, with an apical ostiolum ............... 17

17(16′) Ascomata united in an often striate stroma, asci mostly not amyloid, ascospore septum often formed late or absent . . . . . . . . . . . . . . . . . . . . **Apiospora**
 A. *montagnei*, on Gramineae, and additional species.
17′(16′) Ascomata not united in a stroma, asci mostly amyloid . . . . . . **Pseudomassaria**
 P. *chondrospora*, on bark of *Tilia*, and additional species (Barr, 1964).
 18(13′) Ascospores hyaline . . . . . . . . . . . . . . . . . . . . . . . . 19
 18′(13′) Ascospores pigmented, mostly brown . . . . . . . . . . . . . . . . 21
19(18) Ascospores not rostrate or attenuated . . . . . . . . . . . . . . . **Lejosphaerella**
 L. *praeclara*, on *Vaccinium*, and additional species.
19′(18) Ascospores attenuated at both ends . . . . . . . . . . . . . . . . . . . 20
 20(19′) Ascomata spherical, with an apical ostiolum . . . . . . . . . . **Ceriospora**
  C. *dubyi*, on hops, and additional species.
 20′(19′) Ascomata depressed, with a lateral, curved ostiolum . . . . . . . . **Oxydothis**
  O. *grisea* and other species on Palmae, Musaceae, and Gramineae (Müller and von Arx, 1962).
21(18′) Ascospores thick-walled, smooth, without germ pores, often with a mucilaginous sheath . . . . . . . . . . . . . . . . . . . . . . . . . . . . . . . . . . . **Amphisphaeria**
 A. *umbrina* on bark of hardwood and a number of additional species (syn. *Massariella*, *Massariopsis*, and *Phorcys*).
21′(18′) Ascospores with germ pores or apical caps or appendages or ridges . . . . . 22
 22(21′) Ascomata depressed, mostly hemispherical . . . . . . . . . . . . . . 23
 22′(21′) Ascomata spherical or nearly so . . . . . . . . . . . . . . . . . . . 24
23(22) Ascospores with gelatinous appendages . . . . . . . . . . . . . . . . **Seynesia**
 S. *erumpens* on Palmae and bamboo; monotypic (syn. *Steganopycnis*).
23′(22) Ascospores with longitudinal ridges . . . . . . . . . . . . . . . . . **Roussoëlla**
 R. *hysterioides* on bamboo and Palmae.
 24(22′) Ascospores with appendages at both ends . . . . . . . . . . . . **Ceriophora**
 24′(22′) Ascospores with germ pores or apical caps . . . . . . . . . . . . . . . 25
25(24′) Ascospores with ridges and apical caps . . . . . . . . . . . . . . . . . **Cainia**
 C. *graminis* on grasses, C. *desmazieri* on *Spartium*.
25′(24′) Ascospores with 2 germ pores . . . . . . . . . . . . . . . . . . . **Cainiella**
 C. *johansonii*, on *Dryas*, and other species (Barr, 1959).
 26(12′) Ascospores hyaline . . . . . . . . . . . . . . . . . . . . . . . . 27
 26′(12′) Ascospores pigmented . . . . . . . . . . . . . . . . . . . . . . . 31
27(26) Ascomata immersed in an elongated stroma composed of a bright plectenchyma with a dark crust . . . . . . . . . . . . . . . . . . . . . . . . . . . . . . . 28
27′(26) Stroma absent or not compact . . . . . . . . . . . . . . . . . . . . . . 29
 28(27) Ascospores fusiform . . . . . . . . . . . . . . . . . . . . . . . **Monographus**
  M. *aspidiorum* on ferns (Obrist, 1959).
 28′(27) Ascospores ellipsoid . . . . . . . . . . . . . . . . . . . . . . . **Exarmidium**
  E. *hysteriiforme* on wood.
29(27′) Asci not amyloid . . . . . . . . . . . . . . . . . . . . . . . . **Griphosphaerioma**
 G. *kansense* on *Symphoricarpos* has a *Labridella* conidial state (Shoemaker, 1963).

29'(27') Asci amyloid . . . . . . . . . . . . . . . . . . . . . . . . . . . . 30

    30(29)') Ascomata immersed in an often erumpent stroma . . . . . . . **Discostroma**
        *D. massarinum* on *Ribes* has a *Seimatosporium* conidial state (Shoemaker and Müller, 1964) (syn. *Clathridium* Berlese non Sacc.).

    30'(29') Ascomata immersed in the host tissue, often covered by a clypeus and surrounded by intramatrical hyphae . . . . . . . . . . . . . . . . . . . . . **Griphosphaeria**
        *G. corticola* on *Rosa* and *G. nivalis* on Gramineae with a *Fusarium* conidial state (syn. *Monographella* and *Griphosphaerella*).

31(26') Ascospores with hyaline or bright end cells . . . . . . . . . . . . . . . . 32

31'(26') All cells of the ascospores pigmented . . . . . . . . . . . . . . . . . . . 33

    32(31) Asci mostly not amyloid . . . . . . . . . . . . . . . . . . . . . . **Broomella**
        Four species on *Clematis* (Shoemaker and Müller, 1963); conidial state *Pestalotia*.

    32'(31) Asci amyloid . . . . . . . . . . . . . . . . . . . . . . . . . **Chitonospora**
        *C. ammophila* on Gramineae.

33(31') Ascospores pitted by germ pores . . . . . . . . . . . . . . . . **Blogiascospora**
    *B. marginata* on *Rosa* has a *Seiridium* conidial state (Shoemaker *et al.*, 1966).

33'(31') Ascospores not pitted by germ pores . . . . . . . . . . . . . . . . . . . 34

    34(33') Ascomata immersed in an erumpent stroma . . . . . . . . . **Phragmodiscus**
        *P. arundinariae* and other species on bamboo.

    34'(33') Stroma absent or not erumpent . . . . . . . . . . . . . . . . . . . . . 35

35(34') Ascomata united in valsoid groups, ascospores muriform . . . . . . . . . . . . . . . . . . . . **Mycothyridium** (E. Müller nom. nov.)
    [for *Thyridium* Fuck., *Symb. Mycol.* p. 195, 1869 (non Mitt., 1868, mosses) *Mycothyridium vestitum* (Fr.) *comb. nov.* basionym: *Sphaeria vestita* Fr., *Syst. Mycol.* 2:410, 1823].

35'(34') Ascomata not united in valsoid groups, mostly immersed, often covered by clypeuslike structures . . . . . . . . . . . . . . . . . . . . . . . . . . . . . . . . **Lepteutypa**
    *L. fuckelii* on bark of *Tilia*; including *Lepteutypella*, *Hymenopleella*, *Diapleella*, *Trematomyces*.

## L. Xylariaceae

### KEY TO IMPORTANT GENERA OF XYLARIACEAE[4]

1. Asci with a persistent wall and with amyloid apical structures, mostly cylindrical, ascomata always ostiolate . . . . . . . . . . . . . . . . . . . . true **Xylariaceae** 5

1'. Asci evanescent, spherical, clavate, occasionally cylindrical; ascomata astomatous or ostiolate and ascospores discharged in a slimy mass . . . . . . . . . . **Thamnoideae** 2

    2(1') Stromata erect, dendroid or threadlike; ascomata ostiolate . . . . . . . . . 3

    2'(1') Stromata not dendroid or threadlike; ascomata astomatous . . . . . . . . . 4

3(2) Ascomata with often long, glabrous necks, ascospores protruding in a slimy mass, without a germ slit . . . . . . . . . . . . . . . . . . . . . . . . . . . . . **Thamnomyces**
    *T. chamissonis* and some additional species, all tropical (Dennis, 1957).

---

[4]General references: von Arx and Müller (1954); Munk (1957); Martin (1967).

3'(2) Ascomata with shorter, ciliate necks, ascospores with a germ slit . . . . . **Wawelia**
*W. regia* on dung, Europe (Müller, 1959b).

    4(2') Stromata turbinate or headlike, with a footlike basal portion, containing a number of elongated cavities . . . . . . . . . . . . . . . . . . . . . . . . . . **Phylacia**
*P. globosa* and some additional species, all tropical (syn. *Henningsinia*) (Dennis, 1957; Martin, 1969b).

    4'(2') Stromata small, stalked, containing a single cavity; wall with sutures; ascospores small, pale . . . . . . . . . . . . . . . . . . . . . . . . . . . . **Batistia**
*B. annulipes*; monotypic (Ciferri, 1958).

5(1) Ascomata single or aggregated, free or surrounded by a hyphal subiculum or immersed in the host tissue and covered by a small stromatic clypeus . . . . . . . . . . . . 6

5'(1) Ascomata several to many, embedded in a stroma or sitting on a conspicuous erect or wirelike stroma consisting entirely of fungus tissue . . . . . . . . . . . . . . . . 9

    6(5) Ascomata superficial . . . . . . . . . . . . . . . . . . . . . . . . . . . 7

    6'(5) Ascomata immersed in the host tissue . . . . . . . . . . . . . . . . . . . 8

7(6) Ascomata often surrounded by a hyphal subiculum, without protuberances, ascospores sometimes provided with an apical gelatinous appendage . . . . . . . . . . **Rosellinia**
*R. aquila* on rotten wood and many additional species; *R. necatrix* on roots of *Vitis* and *R. quercina* on roots of *Quercus* are parasites (syn. *Pleosporopsis* and *Sphaeropyxis*).

7'(6) Ascomata laterally provided with conical protuberances . . . . . . **Stilbohypoxylon**
*S. moelleri*; monotypic.

    8(6') Ascomata dispersed or effuse, occasionally covered by a weakly developed stromatic clypeus . . . . . . . . . . . . . . . . . . . . . . . . . . . . . . . **Anthostomella**
*A. punctulata* on dead leaves of *Carex* and many additional species on different hosts (Martin, 1969c) (syn. *Paranthostromella, Phaeophomatospora, Myconeesia*, and *Phaeaspis*).

    8'(6') Ascomata arranged in valsoid groups, ostioles collectively erumpent . . . . . . . . . . . . . . . . . . . . . . . . . . . . . . **Lopadostoma**
*L. gastrinum* and *L. turgidum* on bark of hardwood; a few additional species (Martin, 1969c).

9(5') Stromata composed of a light-colored prosenchyma, occasionally covered by a dark crust . . . . . . . . . . . . . . . . . . . . . . . . . . . . . . . . . . 10

9'(5') Stromata dark throughout . . . . . . . . . . . . . . . . . . . . . . . . . 20

    10(9) Ascomata monostichous, often developing near the periphery of the stromata  11

    10'(9) Ascomata polystichous, completely embedded in the stroma, often at the base of the stromata . . . . . . . . . . . . . . . . . . . . . . . . . . . . . 18

11(10) Stromata more or less erect or stipitate, stalks sometimes embedded in the substrate . . . . . . . . . . . . . . . . . . . . . . . . . . . . . . . . . 12

11'(10) Stromata hemispherical, forming galls or widely dispersed crusts . . . . . . . 14

    12(11) Stromata occurring on dung . . . . . . . . . . . . . . . . . . . . . . . 13

    12'(11) Stromata on plants or debris or originating from termitaria, erect clavate, branched or with head-like fertile apex (Fig. 3i; 4b) . . . . . . . . . **Xylaria** nom. cons.
*X. hypoxylon* and many additional species on old wood of different trees, on old fruits or on termitaria; some authors are using the genus name *Xylosphaera* (Martin, 1970). For synonyms, see von Arx and Müller (1954).

13(12) Stromata flat-topped, with white crust; ascomata only beneath the flat upper surface . . . . . . . . . . . . . . . . . . . . . . . . . . . . . . . . . . **Poronia**
P. punctata and P. oedipus.

13'(12) Stromata more or less clavate, surface crust brown; ascomata often protruding, arranged not only beneath the upper surface . . . . . . . . . . . . . . . **Podosordaria**
P. mexicana and some other species (Martin, 1970).

    14(11') Stromata with a gelatinous interior . . . . . . . . . . . . . . . **Entonaema**
    E. lignescens on wood.

    14'(11') Stromata not gelatinous . . . . . . . . . . . . . . . . . . . . . . . . . 15

15(14') Stromata sessile, sitting broadly on the substrate . . . . . . . . . . . . . . 16

15'(14') Stromata indistinctly stipitate, flattened or subglobose . . . . . . . . . . . 17

    16(15) Stromata with a light-colored surface . . . . . . . . . . . . **Sarcoxylon**
    S. compunctatum on wood.

    16'(15) Stromata with dark-colored surface . . . . . . . . . . . . . . . . **Penzigia**
    P. cranioides and some additional species on wood.

17(15') Stromata small or forming widespread crust, flat-topped . . . . . . **Kretzschmaria**
K. clavus, K. deusta (= Ustulina vulgaris) on rotten wood and some additional species (Martin, 1970).

17'(15') Stromata not flat-topped, small, subsessile, subglobose . . . . . . . **Hypoxylina**
H. umbilicata and a few additional species.

    18(10') Flesh of the stroma colored, fertile portions continuous . . . . . **Peridoxylon**
    P. petersii on wood; Martin (1969b) considers Nummulariola a synonym.

    18'(10') Flesh of the stroma white . . . . . . . . . . . . . . . . . . . . . . . . 19

19(18') Stromata with reticulate fertile portions, enclosing sterile portions, ascospores comparatively large . . . . . . . . . . . . . . . . . . . . . . . . . . . . . . **Engleromyces**
E. goetzii on wood.

19'(18') Reticulate fertile portions absent in the stroma, ascospores comparatively small . . . . . . . . . . . . . . . . . . . . . . . . . . . . . . . . . . **Sarcostromella**
    S. polysticha on wood (Boedijn, 1959).

    20(9') Stromata usually immersed in the substratum, occasionally reduced . . . . . . . . . . . . . . . . . . . . . . . . . . . . . . . . . . **Hypocopra**
    H. fimicola and some additional species on dung.

    20'(9') Stromata more or less superficial . . . . . . . . . . . . . . . . . . . . 21

21(20') Stromata subglobose to globose, showing concentric zones in section . . **Daldinia**
D. concentrica on wood (Martin, 1969b).

21'(20') Stromata without concentric zones . . . . . . . . . . . . . . . . . . . . 22

    22(21') Ascomata monostichously arranged in the stroma . . . . . . . . . . . 23

    22'(21') Ascomata polystichous . . . . . . . . . . . . . . . . . . . . . . . . **Bolinia**
    B. tubulina on wood; Martin (1969b) considers Nummulariola a synonym.

23(22) Stromata erect, cylindrical, with elongated cylindrical ascomata embedded in a single layer beneath an apical disc . . . . . . . . . . . . . . . . . . . . . . **Camillea**
C. lepriaeri and a few additional species on wood; Martin (1969b) considers Nummulariola a synonym.

23'(22) Stromata crustlike to hemispherical or globose . . . . . . . . . . . . . . 24

    24(23') Ascomata sunken in the base of the stromata and provided with

long necks . . . . . . . . . . . . . . . . . . . . . . . . . . . . . . . . . **Nummulariola**
*N. bulliardii* on wood of *Fagus* and *Alnus*; *Nummulariola* has to replace the illegitimate name *Nummularia* (Martin, 1969b).

24'(23') Ascomata embedded in the periphery of the stroma, provided with short necks . . . . . . . . . . . . . . . . . . . . . . . . . . . . . . . . . . . 25

25(24') Ascomata elongate cylindrical, often extremely long (5 mm), easily separating . . . . . . . . . . . . . . . . . . . . . . . . . . . . . . . . . **Camarops**
*C. polyspermum* and some additional species on wood; Martin (1969b) considers *Nummulariola* a synonym.

25'(24') Ascomata semiglobose to angular, not separating . . . . . . . . . . **Hypoxylon**
*H. fragiforme* ( = *H. coccineum*) on wood of *Fagus* and a large number of additional species; monographed by Miller (1961) and Martin (1968a,b, 1969a,b) conidial states (as in some other genera of Xylariacea) *Nodulisporium* and *Geniculisporium*.

## M. Verrucariaceae

### KEY TO IMPORTANT GENERA OF VERRUCARIACEAE

1. Ascospores 1-celled (all lichenized) . . . . . . . . . . . . . . . . . . . . . . . 2
1'. Ascospores septate, lichenized or rarely wood-inhabiting species or hyperparasites . . 8
   2(1) Asci polysporous; ascospores minute; lichen thalli crustose or scaly
   . . . . . . . . . . . . . . . . . . . . . . . . . . . . . . . . . . . . **Trimmatothele**
   *T. maritima* on rocks or stones (syn. *Trimmatothelopsis*)

   2'(1) Asci with 8 ascospores . . . . . . . . . . . . . . . . . . . . . . . . 3
3(2') Lichen thallus foliose-scaly . . . . . . . . . . . . . . . . . . . . . . . . . 4
3'(2') Lichen thallus crustose or areolate . . . . . . . . . . . . . . . . . . . . . 5
   4(3) Ascomata with markedly developed involucrellum (carbonaceous portion formed by the lichen thallus covering or enclosing the ascoma) . . . . . **Involucrocarpon**
   *I. lachneum* and a few additional species.

   4'(3) Involucrellum absent . . . . . . . . . . . . . . . . . . . . **Dermatocarpon**
   *D. fluviatiale* and about 50 or more species.

5(3') Ascomata light-colored, involucrellum absent . . . . . . . . . . . . **Leucocarpia**
   *L. biatorella* on limestone.

5'(3') Ascomata more or less carbonaceous . . . . . . . . . . . . . . . . . . . . 6
   6(5') Involucrellum absent or only weakly developed . . . . . . . . . **Amphoridium**
   *A. calcisedum*.

   6'(5') Involucrellum well-developed . . . . . . . . . . . . . . . . . . . . . . 7
7(6') Involucrellum radially split . . . . . . . . . . . . . . . . . . . **Protobagliettoa**
   *P. parmigera* and a few additional species.

7'(6') Involucrellum not radially split . . . . . . . . . . . . . . . . . . . **Verrucaria**
   *V. calciseda* and numerous additional species.

   8(1') Ascospores with transverse septa only . . . . . . . . . . . . . . . . . 9
   8'(1') Ascospores with transverse and longitudinal septa . . . . . . . . . . 14

9(8) Ascospores strictly 2-celled; nonlichenized species . . . . . . . . . . . . . . 10
9'(8) Ascospores 2- to many-celled . . . . . . . . . . . . . . . . . . . . . . 11
    10(9) Ascospores brown, asci with 4, 8 or more ascospores . . . . . . **Tichothecium**
        *T. pygmaeum* and a few additional species.
    10'(9) Ascospores hyaline, asci 8-spored . . . . . . . . . . . . . . . . **Pharcidia**
        *P. epicymatia* and some additional species.
11(9') Lichen thalli scaly . . . . . . . . . . . . . . . . . . . . . . . . . . . 12
11'(9') Lichen thalli crustose . . . . . . . . . . . . . . . . . . . . . . . . . 13
    12(11) Ascomata without involucrellum . . . . . . . . . . . . . . . **Placidiopsis**
        *P. grappae* and a few additional species.
    12'(11) Ascomata with a well-developed involucrellum . . . . . . . **Paraplacidiopsis**
        *P. crenulata* and a few additional species.
    13(11') Ascomata without or with a weakly developed involucrellum . . . . **Involucrothele**
        *I. dufourii* and additional species.
    13'(11') Ascomata with a well-developed involucrellum . . . . . . . . . . . **Thelidium**
        *T. decipiens* and additional species.
    14(8') Lichen thalli minutely scaly . . . . . . . . . . . . . . . . . . . **Agonimia**
        *A. tristicula*.
    14'(8') Lichen thalli crustose . . . . . . . . . . . . . . . . . . . . . . . . 15
    15(14') Involucrellum absent or only weakly developed . . . . . . . . **Amphoroblastia**
        *A. quinqueseptata* and a few additional species.
    15'(14') Involucrellum well developed . . . . . . . . . . . . . . . . . . . . 16
    16(15') Involucrellum radially split . . . . . . . . . . . . . . . . . . . **Bagliettoa**
        *B. inaequata* and a few additional species.
    16'(15') Involucrellum not radially split . . . . . . . . . . . . . . . . . **Polyblastia**
        *P. intercedens* and additional species.

## N. Clavicipitaceae

### KEY TO IMPORTANT GENERA OF CLAVICIPITACEAE[5]

1. Parasitic on Gramineae, including bamboo; stromatic structures always present . . . . 2
1'. Not parasitic on Gramineae, stromatic or not stromatic . . . . . . . . . . . . 11
    2(1) Ascomata aggregated in small, stromatical groups, surrounded by a hyaline subiculum . . . . . . . . . . . . . . . . . . . . . . . . . . . . . . . . . . . **Oomyces**
        *O. carneoalbus* and a number of additional species (syn. *Coscinaria*).
    2'(1) Stromatic structures mostly predominant . . . . . . . . . . . . . . . . 3
3(2') Ascomata develop on or in a bright, crustose stroma encircling stems of grasses
    . . . . . . . . . . . . . . . . . . . . . . . . . . . . . . . . . . . **Epichloë**
    *E. typhina* on a number of different graminaceous hosts; and several other species.
3'(2') Ascomata deeply immersed or stromata not crustose . . . . . . . . . . . . . 4

[5] General reference: Rogerson (1971).

4(3') Stromata developing as an apical head on an erect stalk, arising from a dark sclerotium ................................................ **Claviceps**
   *C. purpurea* a parasite on different species of Gramineae and a number of additional species (syn. *Balansiella*).

4'(3') Fertile stromata not stalked or not arising from a dark sclerotium ...... 5

5(4') Stromata large, bright, tuberose or flattened; ascospores often separating into single cells, growing on bamboo ........................................ 6

5'(4') Stromata often dark, irregularly spherical, effuse or crustose, rarely stalked; ascospores mostly not separating ....................................... 9

    6(5) Ascospores brownish when mature .................... **Konradia**
   *K. bambusina*.

    6'(5) Ascospores remaining hyaline ........................... 7

7(6') Stromata flattened, similar to *Polyporus*; ascomata immersed in the lower part of the stroma, which is sterile in the upper part ............... **Ascopolyporus**
   *A. polychrous*.

7'(6') Stromata tuberose, not flattened ............................ 8

    8(7') Ascomata densely immersed in a spherical stroma ......... **Mycomalus**
   *M. bambusina* with large stromata.

    8'(7') Ascomata scattered, often protruding ................ **Dussiella**
   *D. tuberiformis* (syn. *Echinodothis*).

9(5') Conidial fructifications absent; fertile stromata often stalked ...... **Balansiopsis**
   *B. gaduae* and some additional species.

9'(5') Conidial fructifications present in young stromata ............... 10

    10(9') Conidial fructifications dark .................... **Ustilaginoidea**
   *U. oryzae* and additional species with a smutlike conidial state.

    10'(9') Conidial fructifications bright ..................... **Balansia**
   *B. claviceps* and a number of additional species on various Gramineae (syn. *Atkinsonella*) (Diehl, 1950).

11(1') Stromata present (mostly parasitic on insects or on other fungi) ....... 12

11'(1') Stromata absent (mostly saprobic on plants or plant debris)     15

    12(11) Ascospores separating into part-spores within the ascus ......... 13

    12'(11) Ascospores not separating into part-spores, occasionally separating outside the asci ............................................. 14

13(12) Entomogenous, stromata formed as byssoid subiculum ........ **Torrubiella**
   *T. aranicida* and several additional species (syn. *Globulina*).

13'(12) Entomogenous; stromata fleshy or horny, often completely smothering the host, ascomata immersed ............................... **Hypocreella**
   *H. discoidea* and a number of additional species (syn. *Fleischeria, Hypocreophis, Moelleriella*).

    14(12') Entomogenous; stromata formed as byssoid subiculum ..... **Podonectria**
   *P. coccicola* and a few additional species.

    14'(12') Stromata headlike or clavate, stalked, arising from endosclerotial structures in insects or fungus fructifications (Fig. 3k) ............... **Cordyceps**
   *C. militaris* and additional species on insects. *C. canadensis* and *C. ophioglossoides* on *Elaphomyces* spp. (syn. *Ophiocordyceps*).

15(11′) Fungicolous, ascomata superficial, ascospores 1-celled . . . . . . . . . . **Barya**
   *B. parasitica* and a few additional species.
15′(11′) Not fungicolous, mostly saprobic on plants or on plant debris; ascospores septate . . . . . . . . . . . . . . . . . . . . . . . . . . . . . . . . . . . . . 16
   16(15′) Ascomata more or less elongated cylindrical, only basally immersed in the host tissue, with an apical porus . . . . . . . . . . . . . . . . . . . **Acrospermum**
      *A. compressum* and other species on plant debris.
   16′(15′) Ascomata turbinate, lageniform to vermiform, mostly with an elongated ostiolum . . . . . . . . . . . . . . . . . . . . . . . . . . . . . . . . . . . 17
17(16′) Ascomata terricolous, hyaline . . . . . . . . . . . . . . . . . . . . . **Romanoa**
   *R. terricola* (Thirumalachar, 1955).
17′(16′) Ascomata immersed in the host tissue, more or less colored . . . **Acrospermoides**
   *A. subulata* and few additional species.

## O. Hypomycetaceae

### KEY TO IMPORTANT GENERA OF HYPOMYCETACEAE

1. Ascospores 1-celled . . . . . . . . . . . . . . . . . . . . . . . . . . . . **Peckiella**
   *P. viridis*, on Russulaceae, and a few other species.
1′. Ascospores 2-celled . . . . . . . . . . . . . . . . . . . . . . . . . . . . . . . . . . 2
   2(1′) Ascospores enclosed in a stout epispore, not breaking into their component cells . . . . . . . . . . . . . . . . . . . . . . . . . . . . . . . . . . . . . . . 3
   2′(1′) Ascospores thin-walled, readily breaking into their component cells . . . . . . . . . . . . . . . . . . . . . . . . . . . . . . . . . . . . . **Arachnocrea**
      *A. papyracea* (Moravec, 1956).
3(2) Ascomata small, with subulate ostioles; asci clariform; ascospores with a smooth epispore . . . . . . . . . . . . . . . . . . . . . . . . . . . . . . . **Pyxidiophora**
   *P. asterophora* with *Chalara*-like endoconidial state.
3′(2) Ascomata medium-sized, with short, cylindrical or papillate ostioles; asci cylindrical; ascospores smooth or with sculptured epispore (Fig. 31) . . . . . . . . . **Hypomyces**
   *H. lactifluorum* and other species with *Cladobotryum* and *Sepedonium* conidial states.

### REFERENCES

Ames, L. M. (1963). A monograph of the Chaetomiaceae. *U.S. Army Res. Develop. Ser.* **2**:1–125.
Barr, M. E. (1959). Northern Pyrenomycetes. I. Canadian Eastern Arctic. *Contrib. Inst. Bot. Univ. Montreal* **73**:1–98.
Barr, M. E. (1964). The genus *Pseudomassaria* in North America. *Mycologia* **56**:690–693.
Barr, M. E. (1966). Observations on *Urosporella*. *Mycologia* **58**:841–862.
Barr, M. E. (1970). Some amerosporous ascomycetes on Ericaceae and Empetraceae. *Mycologia* **62**:377–394.
Barron, G. L., R. F. Cain, and J. C. Gilman. (1961a). A revision of the genus *Petriella*. *Can. J. Bot.* **39**:837–845.
Barron, G. L., R. F. Cain, and J. C. Gilman. (1961b). The genus *Microascus*. *Can. J. Bot.* **39**:1609–1631.
Bezerra, J. L. (1963). Studies on *Pseudonectria rousseliana*. *Acta Bot. Neer.* **12**:58–63.

Boedijn, K. B. (1934). The genus *Podostroma* in the Netherlands Indies. *Bull. Jard. Bot. Buitenz.* [3] **13**:269–275.
Boedijn, K. B. (1959). On a new family of the Sphaeriales. *Persoonia (Leiden)* **1**:15–19.
Boedijn, K. B. (1964). The genus *Thuemenella* with remarks on Hypocreaceae and Nectriaceae. *Persoonia (Leiden)* **3**:1–7.
Booth, C. (1959). Studies of the Pyrenomycetes. IV. *Nectria* (Part I). *Mycol. Pap.* **73**:1–115.
Booth, C. (1960). Studies of the Pyrenomycetes. V. Nomenclature of some fusaria in relation to their nectroid perithecial states. *Mycol. Pap.* **74**:1–16.
Booth, C. (1961). Studies on Pyrenomycetes. VI. *Thielavia* with notes of some allied genera. *Mycol. Pap.* **83**:1–15.
Booth, C. (1964). Studies of the Pyrenomycetes. VII. *Mycol. Pap.* **94**:1–16.
Booth, C., and E. Müller. (1972). On the taxonomic position of *Sphaeria phaeostroma* Fuck. *Trans. Brit. Mycol.* **58**:73–77.
Breton, A., and L. Faurel. (1967). Etude des affinités du genre *Mycorhynchus* Sacc. et description de plusieurs espèces nouvelles. *Rev. Mycol.* **32**:229–258.
Butin, H. (1958). Ueber die auf *Salix* und *Populus* vorkommenden Arten der Gattung *Cryptodiaporthe* Petr. *Phytopathol. Z.* **32**:399–415.
Cailleux, R. (1968). Champignons stercoraux de République Centreafricaine. *Cah. Maboke* **6**:91–98.
Cailleux, R. (1970). Champignons stercoraux de République Centreafricaine. IV. *Tripterospora. Cah. Maboke* **8**:5–16.
Cailleux, R. (1971). Recherches sur la mycoflore coprophile centrafricaine. Les genres *Sordaria, Gelasinospora, Bombardia. Bull. Soc. Mycol. Fr.* **87**:461–626.
Cain, R. F. (1950). Studies of coprophilous ascomycetes. 1. *Gelasinospora. Can. J. Res., Sect. C* **28**:566–576.
Cain, R. F. (1956). Studies of coprophilous ascomycetes. 4. *Tripterospora*, a new cleistocarpous genus in a new family. *Can. J. Bot.* **34**:699–710.
Cain, R. F. (1961a). Studies of soil fungi. III. New species of *Coniochaeta, Chaetomidium* and *Thielavia. Can. J. Bot.* **39**:1231–1239.
Cain, R. F. (1961b). *Anixiella* and *Diplogelasinospora*, two genera with cleistothecia and pitted ascospores. *Can. J. Bot.* **39**:1667–1677.
Cain, R. F., and W. M. Farrow. (1956). Studies of coprophilous ascomycetes. III. The genus *Triangularia. Can. J. Bot.* **34**:689–697.
Cain, R. F., and J. H. Mirza. (1970). *Apodospora*, a new genus of the Sordariaceae. *Can. J. Bot.* **48**:891–896.
Cain, R. F., and L. K. Weresub. (1957). Studies of coprophilous ascomycetes. V. *Sphaeronaemella fimicola. Can. J. Bot.* **35**:119–131.
Carroll, G. C., and A. Munk. (1964). Studies on lignicolous Sordariaceae. *Mycologia* **56**:77–98.
Ciccarone, A. (1963). Osservazioni sistematiche su *Diachorella onobrychidis* (DC) v. Höhn. *Phytopathol. Mediterr.* **2**:239–250.
Ciferri R. (1958). *Batistia*, a new genus of the Cephalotheca family. *Atti 1st. Bot. Univ. Lab. Crittogam., Pavia* [5] **15**:160–170.
Cribb, A. B., and J. W. Cribb. (1956). Marine fungi from Queensland. II. *Univ. Queensl. Pap.* **3**:97–105.
Dennis, R. W. G. (1957). Further notes on tropical American Xylariaceae. *Kew Bull.* **12**:297–333.
Dennis, R. W. G. (1968). "British Ascomycetes." Cramer, Weinheim.
Diehl, W. W. (1950). *Balansia* and the Balaniae in America. *U.S., Dep. Agr., Agr. Monogr.* **4**:1–82.

Dingley, J. M. (1951a). The Hypocreales of New Zealand. II. The genus *Nectria*. *Trans. Roy. Soc. N. Z.* **79**:177–202.
Dingley, J. M. (1951b). The Hypocreales of New Zealand. III. The genus *Hypocrea*. *Trans. Roy. Soc. N. Z.* **79**:323–337.
Dingley, J. M. (1952). The Hypocreales of New Zealand. IV. The genera *Calonectria*, *Gibberella* and *Thyronectria*. *Trans. Roy. Soc. N. Z.* **79**:403–411.
Dingley, J. M. (1957). Life history studies of New Zealand species of *Nectria* Fr. *Trans. Roy. Soc. N. Z.* **84**:467–477.
Doguet, G. (1955). Le genre *Melanospora*. *Botaniste* **39**:1–313.
Doi, Y. (1967). A revision of Hypocreales with cultural observations. II. On *Mycocitrus phyllostachidis* (Syd.) Doi, a perfect state of *Cephalosporium*. *Bull. Nat. Sci. Mus., Tokyo* **10**:31–36.
Doi, Y. (1968). Revision of the Hypocreales with cultural observations. I. The genus *Peloronectriella* with a note on bambusicolous Hypocreales with large persistent stroma. *Bull. Nat. Sci. Mus., Tokyo* **11**:179–184.
Doi, Y. (1969). Revision of the Hypocreales with cultural observations. IV. The genus *Hypocrea* and its allies in Japan. (1) General part. *Bull. Nat. Sci. Mus., Tokyo* **12**:693–724.
Eriksson, O. (1966). On *Anthostomella* Sacc., *Entosordaria* (Sacc.) v. Höhn. and some related genera (Pyrenomycetes). *Sv. Bot. Tidskr.* **60**:315–324.
Fitzpatrick, H. M. (1923). Monograph of the Nitschkeaceae. *Mycologia* **15**:23–67.
Fitzpatrick, H. M. (1942). Revisionary studies in the Coryneliaceae. I and II. *Mycologia* **34**:464–514.
Funk, A. (1963). Studies in the genus *Caliciopsis*. *Can. J. Bot.* **41**:503–543.
Gams., W., and M. Gerlagh. (1968). Beiträge zur Systematik und Biologie von *Plectosphaerella cucumeris* und der zugehörigen Nebenfruchtform. *Persoonia (Leiden)* **5**:177–188.
Gareth-Jones, E. B. (1965). *Halonectria milfordensis* gen. et spec. nov., a marine pyrenomycete on submerged wood. *Trans. Brit. Mycol. Soc.* **48**:287–290.
Gerlach, W. (1970). Suggestions to an acceptable *Fusarium* system. *Ann. Acad. Sci. Fenn., Ser. A4* **168**:37–43.
Gordon, C. C., and C. G. Shaw. (1964). Ascocarpic development in *Diporotheca rhizophila*. *Can. J. Bot.* **42**:1525–1530.
Griffin, H. D. (1968). The genus *Ceratocystis* in Ontario. *Can. J. Bot.* **46**:689–718.
Hansford, C. G. (1946). The foliicolous ascomycetes, their parasites and associated fungi. *Mycol. Pap.* **15**:1–240.
Hansford, C. G. (1961). The Meliolineae, a monograph. *Sydowia, Beih.* **2**:1–806.
Hansford, C. G. (1965). Iconographia Meliolinearum. *Sydowia, Beih.* **5**:I–CCLXXXIV.
Hawksworth, D. L. (1971). A revision of the genus *Ascotricha* Berk. *Mycol. Pap.* **126**:1–28.
Hino, I., and K. Katumoto. (1968). The genus *Melchiora* in Japan. *J. Jap. Bot.* **42**:26–32.
Hubbes, M. (1960). Untersuchungen über die Valsaceengattung *Valseutypella* v. H. *Phytopathol. Z.* **39**:389–400.
Hunt, J. (1956). Taxonomy of the genus *Ceratocystis*. *Lloydia* **19**:1–58.
Ingold, C. T., and B. Chapman. (1952). Aquatic ascomycetes: *Loramyces juncicola* Weston and *L. macrospora* n. sp. *Trans. Brit. Mycol. Soc.* **35**:268–272.
Johnson, T. W., and F. K. Sparrow. (1961). "Fungi in Oceans and Estuaries". Cramer, Weinheim.
Katumoto, K. (1962). On the genus *Artallendea* Bat. et Maia. *Bull. Fac. Agr., Yamaguti Univ.* **13**:291–296.
Katumoto, K. (1965). On the genus *Bagcheea* in Japan. *Bull. Fac. Agr., Yamaguti Univ.* **16**:615–622.
Katumoto, K. (1968). On the genus *Coccodiella* Hara. *J. Jap. Bot.* **43**:277–284.

Kern, H. (1957). Untersuchungen über die Umgrenzung der Arten in der Ascomycetengattung *Leucostoma. Phytopathol. Z.* **30**:149–180.

Kobayashi, Y., and K. Ito. (1956). Notes on the genus Endothia in Japan. I. Species of *Endothia* collected in Japan. *Bull. Govt. Forest. Exp. Sta.* **92**:91–97.

Kohlmeyer, J., and E. Kohlmeyer. (1964–1969). "Icones Fungorum Maris." Cramer, Weinheim and Lehre.

Lundqvist, N. (1964). *Anopodium*, a new genus of coprophilous pyrenomycetes with apically pedicellate spores. *Bot. Notis.* **117**:355–365.

Lundqvist, N. (1967). On spore ornamentation in the Sordariaceae, exemplified by the new cleistocarpous genus *Copromyces. Ark. Bot.* **6**:327–337.

Lundqvist, N. (1969). *Tripterospora* (Sordariaceae s. lat., Pyrenomycetes). *Bot. Notis.* **122**:589–603.

Lundquvist, N. (1972). Nordic Sordariaceae. *Symb. Bot. Upsal.* **20**, 1:1–374.

Malloch, D. (1970). New concepts in the Microascaceae illustrated by two new species. *Mycologia* **62**:727–740.

Malloch, D., and R. F. Cain. (1971). The genus *Kernia. Can. J. Bot.* **49**:855–867.

Martin, P. (1967). Studies in the Xylariaceae. I. New and old concepts. *J. S. Afr. Bot.* **33**:205–240.

Martin, P. (1968a). Studies in the Xylariaceae. III. *J. S. Afr. Bot.* **34**:153–199.

Martin, P. (1968b). Studies in the Xylariaceae. IV. *J. S. Afr. Bot.* **34**:303–330.

Martin, P. (1969a). Studies in the Xylariaceae. V. *J. S. Afr. Bot.* **35**:149–206.

Martin, P. (1969b). Studies in the Xylariaceae. VI. *Daldinia*, *Nummulariola* and their allies. *J. S. Afr. Bot.* **35**:267–320.

Martin, P. (1969c). Studies in the Xylariaceae. VII. *Anthostomella* and *Lopadostoma. J. S. Afr. Bot.* **35**:393–410.

Martin, P. (1970). Studies in the Xylariaceae. VIII. *Xylaria* and its allies. *J. S. Afr. Bot.* **36**:73–138.

Miller, J. H. (1961). "A Monograph of the World Species of *Hypoxylon*." Univ. of Georgia Press, Athens.

Mirza, J. H., and R. F. Cain. (1969). Revision of the genus *Podospora. Can. J. Bot.* **47**:1999–2048.

Möller, A. (1901). Phycomyceten und Ascomyceten. *Bot. Mitt. Tropen* **9**:1–319.

Moravec, Z. (1956). *Arachnocrea*, un genre nouveau de la famille des Néctriacées. *Bull. Soc. Mycol. Fr.* **72**:160–166.

Moreau, C. (1954). Les genres *Sordaria* and *Pleurage. Encycl. Mycol.* **25**:1–330.

Müller, E. (1959a). Pilze aus dem Himalaya. II. *Sydowia* **12**:160–184.

Müller, E. (1959b). Ueber die Stellung der Ascomycetengattung *Wawelia* Namyslowsky. In "Omagiu lui Traian Săvulescu," pp. 515–518. Editura Acad. Rep. Pop. Romîne.

Müller, E. (1962). Ueber einige Ascomyceten aus Südfrankreich. *Sydowia* **15**:84–91.

Müller, E. (1965). Beobachtungen an Ascomyceten. *Sydowia* **18**:86–105.

Müller, E., and S. Ahmad. (1962). Ueber einige neue oder bemerkenswerte Ascomyceten aus Pakistan. V. *Biologia (Lahore)* **8**:151–162.

Müller, E., and R. W. G. Dennis. (1965). Fungi venezuelani. VIII. Plectascales, Sphaeriales, Loculoascomycetes. *Kew Bull.* **19**:357–386.

Müller, E., and J. A. von Arx. (1962). Die Gattungen der didymosporen Pyrenomyceten. *Beitr. Kryptogamanflora Schweiz* **11**:1–922.

Müller, E., J. Harr, and P. Sulmont. (1969). Deux ascomycètes dont le stade conidien présente des conidies phaeophragmiées endogènes. *Rev. Mycol.* **33**:369–378.

Munk, A. (1957). Danish Pyrenomycetes. *Dan. Bot. Ark.* **17**:1–149.

Munk, A. (1965). On some species of *Endoxyla* recently found in Denmark. *Bot. Tidsskr.* **61**:56–70.

Obrist, W. (1959). Untersuchungen über einige dothideale Gattungen. *Phytopathol. Z.* **35**:357–388.
Parbery, D. G. (1964). Studies on graminicolous species of *Phyllachora* Fuckel. I, II, and III. *Aust. J. Bot.* **11**:117–151.
Parbery, D. G. (1967). Studies on graminicolous species of *Phyllachora* Nke. in Fuck. V. A taxonomic monograph. *Aust. J. Bot.* **15**:271–375.
Parguey-Leduc, A. (1961). Etude des asques et du développement de l'*Helminthosphaeria clavariarum* (Desm.) Fuck. ap. Munk. *Bull. Soc. Mycol. Fr.* **77**:15–33.
Petrak, F. (1928). Ueber *Bagnisiopsis* und verwandte Gattungen. *Hedwigia* **68**:251–290.
Petrak, F. (1970). Ergebnisse einer Revision der Grundtypen verschiedener Gattungen der Ascomyceten und Fungi Imperfecti 133. *Unamunoa. Sydowia* **23**:268–269.
Rai, J. N., V. P. Tewari, and K. G. Mukerji. (1964). *Achaetomium*, a new genus of ascomycetes. *Can J. Bot.* **42**:693–697.
Reid, J., and V. C. Booth. (1969). Some species segregated from the genera *Cryptospora, Cryptosporella* and *Sillia. Can. J. Bot.* **47**:1055–1060.
Robinson-Jeffrey, R. C., and R. W. Davidson. (1968). Three new *Europhium*-species with *Verticicladiella* imperfect states on blue stained pine. *Can. J. Bot.* **46**:1523–1527.
Robison, B. M. (1970). A new cleistocarpous genus *Echinopodospora* with two new species. *Trans. Brit. Mycol. Soc.* **54**:318–322.
Rogerson, C. T. (1971). The hypocrealean fungi (Ascomycetes, Hypocreales). *Mycologia* **62**:865–910.
Schrantz, J. B. (1960). Recherches sur les pyrenomycètes de l'ordre des Diatrypales sensu Chadefaud. *Bull. Soc. Mycol. Fr.* **76**:305–407.
Shoemaker, R. A. (1963). Generic correlations and concepts: *Griphosphaerioma* and *Labridella*. *Can. J. Bot.* **41**:1419–1423.
Shoemaker, R. A., and E. Müller. (1963). Generic correlations and concepts: *Broomella* and *Pestalotia. Can. J. Bot.* **41**:1235–1244.
Shoemaker, R. A., and E Müller. (1964). Generic correlations and concepts: *Clathridium* (= *Griphosphaeria*) and *Seimatosporium* (= *Sporocadus*). *Can. J. Bot.* **42**:403–410.
Shoemaker, R. A., E. Müller, and G. Morgan-Jones. (1966). Fuckel's *Massaria marginata* and *Seiridium marginatum* Nees ex Steudel. *Can. J. Bot.* **44**:247–254.
Skou, J. P. (1968). Studies on the take-all fungus *Gaeumannomyces graminis* I. *Yearb. Roy. Vet. Agr. Coll. København 1968* pp. 109–116.
Stevens, F. L. (1927). The Meliolineae. I. *Ann. Mycol.* (*Berlin*) **25**:405–469.
Stevens, F. L. (1928). The Meliolineae. II. *Ann. Mycol.* (*Berlin*) **26**:165–383.
Thirumalachar, M. J. (1955). *Romanoa*, a new genus of soil fungus with antibacterial activity. *Rend. Ist. Super. Sanita* **17**:206–212.
Udagawa, S. (1963). *Neoscosmospora* in Japan. *Trans. Mycol. Soc. Jap.* **4**:121–125.
Udagawa, S., and R. F. Cain. (1969a) Notes on the genus *Microthecium. Can. J. Bot.* **47**:1915–1933.
Udagawa, S., and R. F. Cain. (1969b). Some new or noteworthy species of the genus *Chaetomium. Can. J. Bot.* **47**:1939–1951.
von Arx, J. A. (1955). Ein neuer Ascomycet aus Afrika. *Antonie van. Leeuwenhoek; J. Microbiol. Serol.* **21**:161–165.
von Arx, J. A. (1958a). Ueber einige Ascomyceten aus Südamerika. *Acta Bot. Neer.* **7**:503–518.
von Arx, J. A. (1958b). Die systematische Stellung der Ascomycetengattung *Armatella* Theiss. et Syd. *Fungus* (*Wageningen*) **28**:1–3.
von Arx, J. A. (1970a). "The Genera of Fungi Sporulating in Pure Culture." Cramer, Lehre.
von Arx, J. A. (1970b). "A Revision of Fungi Classified as *Gloeosporium*." Cramer, Lehre.

von Arx, J. A., and W. Gams. (1966). Ueber *Pleurage verruculosa* und die zugehörige *Cladorhinium*-Konidienform. *Nova Hedwgia* **13**:199–208.

von Arx, J. A., and G. L. Hennebert. (1968). *Triangularia mangenotii* nov. spec. *Bull. Soc. Mycol. Fr.* **84**:423–426.

von Arx, J. A., and E. Müller. (1954). Die Gattungen der amerosporen Pyrenomyceten. *Beitr. Kryptogamenflora Schweiz* **11**:1–434.

von Arx, J. A., and E. Müller. (1955). Ueber die Gattungen *Selinia* und *Seliniella* nov. gen. und ihre phylogenetische Bedeutung. *Acta Bot. Neer.* **4**:116–125.

von Höhnel, F. (1912) Fragmente zur Mykologie 762. *Iyuhya vitrae* Starb. var. *javanica* v. Höhn, *Sitzungsber. Akad. Wiss. Wien, Math.-Naturw. Kl. (Abt.* 1) **121**:380–381.

Walker, J. (1972). Type studies on *Gaeumannomyces graminis* and related fungi. *Trans. Brit. Mycol. Soc.* **58**:427–457.

Wehmeyer, L. E. (1933). The genus *Diaporthe. Univ. Mich. Stud.* **9**:1–349.

Wehmeyer, L. E. (1941a). A revision of *Melanconis, Pseudovalsa, Prosthecium* and *Titania, Univ. Mich. Stud.* **14**:1–161.

Wehmeyer, L. E. (1941b). The genus *Thyridaria* (Pyrenomycetes). *Lloydia* **4**:241–261.

# Ascomycotina
# Loculoascomycetes

# CHAPTER 7

# Loculoascomycetes

E. S. LUTTRELL

*Department of Plant Pathology*
*University of Georgia*
*Athens, Georgia*

## I. DEFINITION OF LOCULOASCOMYCETES

In this class the asci are bitunicate. The ascocarp is an ascostroma with the asci individually and irregularly distributed in the stromal tissue or grouped in locules; the ascostroma then becomes a perithecioid, or less commonly an apothecioid, pseudothecium. The pseudothecia are separate, grouped on a common basal stroma in which they are more or less immersed, or else they are completely immersed and appear as unwalled locules in a multiloculate stroma. Usually the ascocarps are dark, although in some genera they are bright colored. They may be entirely superficial, erumpent, or immersed in the substratum. The centrum structure is variable, but generally distinguishable from types characteristic of Pyrenomycetes; it is most commonly composed of asci interspersed with persistent pseudoparaphyses or of fascicles of aparaphysate asci in disintegrating centrum tissue. The ostioles are lysigenous or, less commonly, schizogenous and periphysate in perithecioid forms; apothecioid forms open by splitting. Usually the ascospores are septate, although they are amerosporous in a relatively few genera. Loculoascomycetes appear as superficial epiphytes, parasites, or hyperparasites of superficial fungi and insects, as internal parasites fruiting on green leaves and stems, as parasites fruiting on dead leaves and stems, or as saprotrophs on dead leaves, herbaceous stems, wood, dung, and plant debris.

## II. OUTLINE OF CLASSIFICATION

Order 1. Myriangiales, p. 162
    Family A.    Atichiaceae
    Family B.    Myriangiaceae
    Family C.    Saccardiaceae
    Family D.    Saccardinulaceae

Order 2. Dothideales, p. 165
    Family A.    Trichothyriaceae
    Family B.    Chaetothyriaceae
    Family C.    Parodiopsidaceae
    Family D.    Englerulaceae

| | | | |
|---|---|---|---|
| Family E. | Pseudosphaeriaceae | Family B. | Arthoniaceae |
| Family F. | Capnodiaceae | Family C. | Opegraphaceae |
| Family G. | Dothideaceae | Family D. | Phillipsiellaceae |
| Family H. | Dothioraceae | Family E. | Patellariaceae |
| Order 3. Pleosporales, p. 176 | | Family F. | Lecanactidaceae |
| Family A. | Dimeriaceae | Order 5. Hemisphaeriales, p. 200 | |
| Family B. | Venturiaceae | Family A. | Microthyriaceae |
| Family C. | Mesnieraceae | Family B. | Trichopeltinaceae |
| Family D. | Botryosphaeriaceae | Family C. | Munkiellaceae |
| Family E. | Lophiostomataceae | Family D. | Micropeltidaceae |
| Family. F. | Sporormiaceae | Family E. | Asterinaceae |
| Family G. | Pleosporaceae | Family F. | Brefeldiellaceae |
| Family H. | Mycoporaceae | Family G. | Aulographaceae |
| Order 4. Hysteriales, p. 195 | | Family H. | Parmulariaceae |
| Family A. Hysteriaceae | | Family I. | Stephanothecaceae |
| Subfamily 1. Hysterioideae | | Family J. | Schizothriaceae |
| Subfamily 2. Lophioideae | | Family K. | Leptopeltidaceae |

## III. GENERAL CHARACTERISTICS

### A. Bitunicate Asci

The primary character of the Loculoascomycetes is the bitunicate ascus. The ascus wall consists of two layers which are separable in the normal course of ascospore discharge: a thin, inextensible outer layer, or ectoascus, and a thick, extensible inner layer, or endoascus. The very young ascus is thin walled (Figs. 1A, 2C). During most of its development, however, the endoascus is conspicuously thickened, especially toward the apex where it is indented or partially penetrated by a tubular channel forming a subapical chamber into which the protoplast extends (Figs. 1C, D, E, F, G, H, I; 2C; 3C). This marks the pore through which the ascospores are later ejected. Although the endoascus becomes thinner as the ascus matures, the subapical chamber characteristically remains visible as an indentation in the thicker apex (Figs. 2A, C, D; 3A). As the ascus swells prior to dehiscence the entire wall appears uniformly thin.

At dehiscence the bitunicate structure becomes apparent. The ectoascus splits at the apex; in some species it ruptures circumscissally, throwing off the apex as a thimble-shaped cap (Fig. 1D). The endoascus expands as a tubular extension which doubles or triples the original length of the ascus (Figs. 1B, D, G, H; 2C, D). The margin of the ruptured ectoascus is visible as a faint line around the ascus below a slight or definite bulge in the freed endoascus. The wall of the expanded endoascus is uniformly thin. The ascospores move up in the endoascus. One advances into the elastic pore in the tip and hangs there briefly at its widest diameter (Fig. 2D). Spores constricted at the septa are trapped momentarily at a constriction (Fig. 1B). As the first spore is shot

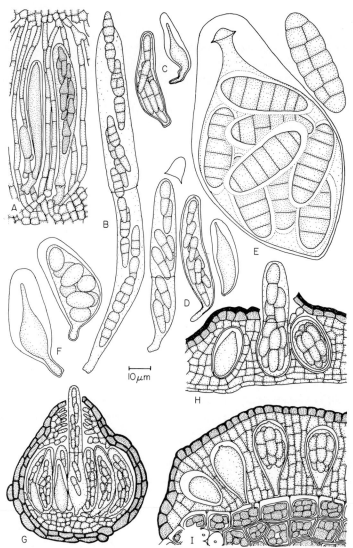

FIG. 1. (A) *Letendraea padouk*, young and mature asci among pseudoparaphyses. (B) *Dothiora schizospora*, ascus with expanded endoascus in process of spore discharge. (C) *Scirrhia acicola*, young and mature asci. (D) *Stomiopeltis* sp., young and mature asci and ascus with tip of ectoascus thrown off as thimble-shaped cap by expanded endoascus. (E) *Leptosphaerulina americana*, ascus with maturing spores and mature ascospore. (F) *Ellisiodothis inquinans*, young and mature asci. [From *Amer. J. Bot.* **35**:62, Fig. 34 (1948).] (G) *Mycosphaerella* sp., section of mature perithecioid pseudothecium with fascicle of aparaphysate asci and periphysate ostiole. (H) *Schizothyrium pomi*, section of part of dimidiate-scutate ascocarp, one ascus with expanded endoascus emerging through split in the shield. (I) *Leptopeltopsis* sp., section of dimidiate-scutate ascocarp on surface of host leaf with asci individually distributed in the stromatic tissue of vertically arranged cells; internal mycelium massed in host cells, forming pseudostroma (hypostroma).

away, the second spore moves into the pore and, as pressure rebuilds, is in turn shot away, the ascus tip recoiling slightly at each discharge. The whole complement of spores may be ejected in rapid succession, or there may be longer pauses between spores or between groups of spores. After all spores are discharged, pressure is rapidly released and the ascus shrinks. The endoascus swells until only a narrow lumen remains, and the subapical chamber is again evident. The ectoascus is now readily apparent as a wrinkled collar constricting the endoascus.

In many Loculoascomycetes, such as *Sporormia* spp., the entire process of ascus dehiscence and ascospore discharge occurs normally in detached asci or in sections of ascocarps in water mounts and may be observed easily under the microscope. In others, ascus dehiscence and elongation of the endoascus occur, but turgor is gradually lost and the spores are trapped in the swollen endoascus walls (Fig. 3C). In *Botryosphaeria* spp., the endoascus begins to swell and dissolve as soon as it emerges from the ectoascus, and the ascospores have often been described as oozing from the ascus in a gelatinous sheath. In such species, placing single ascocarps or a bit of substrate bearing ascocarps on squares of moistened filter paper in the lids of petri dishes and inverting them over an agar plate will demonstrate that the ascospores are forcibly discharged. The irregular pattern of the groups of eight ascospores on the agar resulting from the discharge of spores singly, and with greater or lesser intervals between discharges, is presumptive evidence that the ascus is bitunicate. Close groupings of eight ascospores in contact is characteristic of unitunicate asci, in which spore discharge usually is simultaneous. Even in unitunicate asci with narrow pores, such as those of the Clavicipitaceae, which discharge their spores successively, the spores follow one another quickly and smoothly without the pause and readjustment of the ascus between spore discharges typical of bitunicate asci. The spores consequently follow more nearly similar trajectories and usually are more closely grouped on the agar. Spore patterns may be more easily recognized if the plates are examined after only a few asci have discharged. On a thin clear agar spores may be observed by inverting the plate on the microscope stage and focusing on the surface of the agar through the bottom of the dish. The target area should be marked on the bottom of the dish so that the lid bearing the ascocarps can be rotated.

Ascospore discharge often may be observed directly under the microscope or stereomicroscope by mounting moistened whole ascocarps or substrate bearing ascocarps on dry slides or at the edges of moistened squares of filter paper so that the ostioles or hymenial surfaces can be focused on in profile. If a number of mature asci are present, spore discharge may be observed during the short period before the mounted ascocarps become too dry.

A characteristic structure of most bitunicate asci is the "nasse" (Chade-

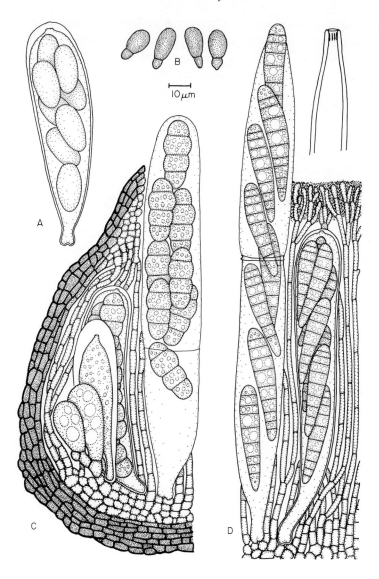

FIG. 2. (A) *Botryosphaeria dothidea*, mature ascus. (B) *Apiosporina morbosa*, greenish apiosporous ascospores. (C) *Leptosphaeria* sp., section of part of perithecioid pseudothecium with asci in various stages of development among pseudoparaphyses. (D) *Lecanidion atratum*, portion of hymenium from section of apothecioid pseudothecium with asci among paraphysoidal hyphae whose branching tips form an epithecium; upper right, tip of expanded endoascus after spore discharge showing rod-shaped elements of the "nasse."

faud, 1960) formed in the peripheral layers of the cytoplasm against the walls of the subapical chamber. This usually has the form of four vertically arranged, refringent rods (Fig. 2D) staining with cotton blue. The rods may branch and anastomose to form a network. In a few genera of Loculoascomycetes, such as *Pyrenophora* and *Leptosphaerulina*, the subapical chamber of the ascus has a flangelike extension into the lateral walls of the endoascus and at some stages of ascus development is characteristically hat shaped or trilobate in optical longitudinal section (Figs. 1E, 3C). In *Pyrenophora* and *Wettsteinina*, a nonamyloid annulus staining with blue-black inks and congo red is differentiated within the endoascus wall surrounding the subapical chamber (Eriksson, 1967a). A refractive apical apparatus is present also in *Didymosphaeria rhizophorae* (Kohlmeyer and Kohlmeyer, 1967, Table 62) and in *Otthia pruni* (Munk, 1957: 403). Generally the presence of more complex apical structures, such as the refractive, chitinoid annulus of the Diaporthaceae and Sordariaceae, the double amyloid annulus of the Xylariaceae, and the caplike apical plug of the Clavicipitaceae, or a positive reaction to iodine are characters that indicate exclusion from the Loculoascomycetes. The complexly structured asci of *Melogramma spiniferum* and of *Helicascus*, which in other characters seem related to Euascomycetes,[1] however, have been described as bitunicate (Doguet, 1960; Kohlmeyer, 1969).

Despite the peripheral uncertainties and difficulties that have been mentioned, the bitunicate structure of the ascus is the most useful single character in placing a fungus in the Loculoascomycetes. Structure of mature and immature asci alone, even in long-dried herbarium specimens, is usually sufficient for identification. In questionable cases the normal process of ascospore discharge must be observed and the structure of the ascus in all stages of development must be worked out. This has been done for too few species.

### B. Ascostromatic Ascocarps

Historically the Loculoascomycetes (Luttrell, 1955), first as the Dothidiineae of Theissen and Sydow (1918) and later as the Ascoloculares of Nannfeldt (1932), have been distinguished from Euascomycetes primarily on the basis of the ascostromatic nature of the ascocarp. In Euascomycetes[1] the ascogonia, whether free or in a stroma, become surrounded by an envelope of hyphae which develops into the peridium of the ascocarp. In Loculoascomycetes the ascogonia appear within a stromatic initial which develops directly into the ascocarp, and the asci lie in unwalled locules in the stroma. The ascostromatic nature of the ascocarp is most clearly apparent, and was

---

[1] A taxon covering Pyrenomycetes, Laboulbeniomycetes, and Discomycetes as delimited in this volume.

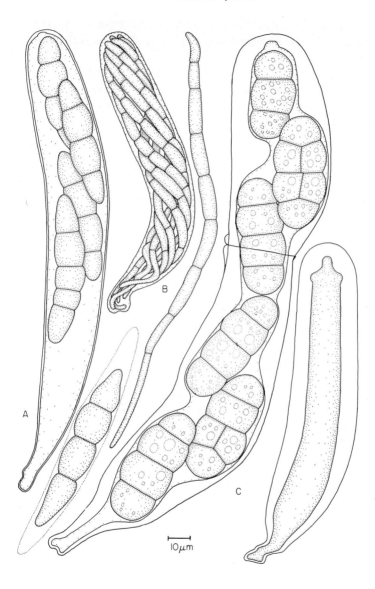

FIG. 3. (A) *Trichometasphaeria (Keissleriella) turcica*, mature ascus and single discharged ascospore with tubular gelatinous sheath. [From *Phytopathology* **48**:283, Fig. 142 (1958).] (B) *Cochliobolus* perfect state of *Bipolaris micropa*, mature ascus with helically coiled, filiform ascospores and single ascospore. (C) *Pyrenophora bromi*, immature ascus and mature ascus with partially expanded endoascus collapsed before discharge, trapping spores in the swollen wall.

earliest recognized, in the Myriangiaceae, in which the asci are individually distributed in the stroma (Fig. 4A), and in the compound Dothideaceae, in which the multiloculate stroma contains groups of asci in locules which obviously are not separated from the surrounding stromata by peridia (Fig. 5A). When the ascocarps are uniloculate, the surrounding stroma may be reduced to a thin wall and the globose or flask-shaped, solitary pseudothecium (Figs. 1G, 7B, 8B) then is difficult to distinguish from the solitary perithecium of the Pyrenomycetes. Tissue types occasionally are helpful since there is a tendency among Loculoascomycetes to produce parenchymatous stromal tissues. This is expressed in the vertically oriented rows of angular cells in the ascostromata of many of the Dothideaceae and Leptopeltidaceae (Fig. 1I) and in the origin of the ascocarp initial through three-dimensional division of a single cell, or a few cells, in fungi such as *Sporormia*. Observation of the ascocarp initials sometimes may be made without great difficulty in species which fruit readily in agar cultures. Even intensive ontogenetic studies may prove equivocal, however, since the stromatic initial of the Loculoascomycetes occasionally is severely reduced.

Although the ascostromatic ascocarp is a fundamental character of the Loculoascomycetes, it is not an exclusive character. A meristogenous origin of the ascocarp similar to that in *Sporormia* occurs in some fungi which are generally classified in the Euascomycetes (Barr, 1956). It is further possible that a similar ascostromatic structure has been arrived at in some Euascomycetes by reduction of peridia. Although peridia are clearly evident in stromatic Pyrenomycetes in the Xylariaceae, they are indistinct in many of the Hypocreaceae and Clavicipitaceae and appear to be lacking entirely in the Coronophoraceae and in some discomycetous fungi such as the Phacidiaceae. The Coronophoraceae, Phacidiaceae, and similar groups of Euascomycetes, however, have other distinctive characters by which they may be excluded from the Loculoascomycetes, and the presence of an ascostromatic ascocarp is a useful criterion when this structure can be inferred from mature specimens.

## C. Diagnostic Types of Internal Structure

Although no single type of ascocarp structure is an index to all Loculoascomycetes, the structural types found in Loculoascomycetes generally are distinct from those characterizing the various groups of the Euascomycetes. Three major types, exhibited in the Myriangiales, Dothideales, and Pleosporales, distinguish the majority of Loculoascomycetes.

In the Myriangiales the asci are scattered individually and usually at several levels in relatively unaltered stromal tissue (Fig. 4A). The ascostromata are crustose, pulvinate, discoid, or subglobose. The asci are distributed throughout the stroma or in restricted fertile regions. Asci are globose and

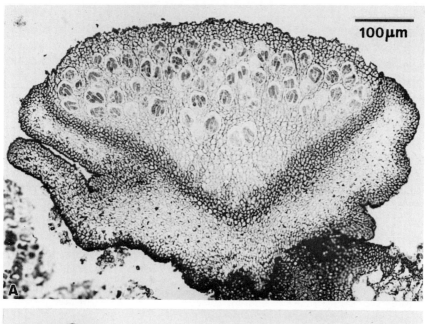

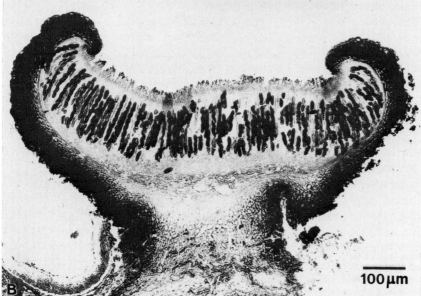

FIG. 4. (A) *Myriangium duriaei*, section of fertile apothecioid conceptacle on basal stroma with globose asci individually distributed at various levels in the stromal parenchyma. (B) *Rhytidhysterium rufulum*, section of mature apothecioid pseudothecium with tips of paraphysoidal hyphae forming an epithecium above the hymenium of cylindrical asci.

thick walled. They expand individually through the softened or crumbling stromal tissue to discharge their spores above the surface. There is little probability of confusing these stromatic ascocarps with the more regular perithecia of the Plectomycetes in which the evanescent, thin-walled asci likewise are irregularly distributed in the sterile tissues of the centrum.

In the Dothideales the ovate to short-cylindrical asci typically are in fascicles in small perithecioid locules in uni- or multiloculate ascostromata (Figs. 1G, 5A, 6A). They push up as a group into disintegrating centrum parenchyma and are aparaphysate. The locules open at the apex through ostioles which are lysigenous (Fig. 5A) or, less commonly, schizogenous and periphysate (Fig. 1G). The periphyses may be abundant, and in some species such as *Scirrhia acicola* (Fig. 6B), *Mycosphaerella ascophylli* (Kohlmeyer and Kohlmeyer, 1968, Table 76), *Metacapnodium juniperi* (Corlett, 1970), and *Pleurisperma dahlbergiae* (Sivanesan, 1970a), periphysislike hyphae (lateral paraphyses, extrahymenial paraphyses) projecting inward from the lateral walls of the pseudothecium occupy most of the space above the developing fascicle of asci. The perithecial centrum of the Diaporthales (Euascomycetes) is similarly parenchymatous; but the more numerous, unitunicate asci have refractive annuli in their apexes, arise in an extended wall layer, and become detached and fill the perithecial cavity. In variations of the dothideaceous type, as in the Pseudosphaeriaceae, the asci are relatively large and few (Fig. 7A). The sterile centrum parenchyma cells may be compressed or may elongate between the maturing asci and remain more or less evident as interascicular strands (paraphysoids) at maturity (Fig. 7B). In other variations, as in the Dothioraceae and in genera of the Asterinaceae, the ascocarps tend to be apothecioid. The locules are laterally extended and enclose a disk-shaped or elongated layer of asci (Fig. 5B) which are exposed by rupture of the overlying stromal tissue. As in the perithecioid forms, remnants of the sterile centrum tissues may be more or less persistent as strands between the mature asci.

In the Pleosporales the clavate- to long-cylindrical asci typically occur in broad basal layers in large perithecioid locules in uni- or less commonly, multiloculate ascostromata (Figs. 2C, 8B, 9A). The ostioles are lysigenous or, commonly, schizogenous and periphysate. The asci are interspersed with pseudoparaphyses. Even at maturity the attachment of the pseudoparaphyses at the apex of the locule can be established (Figs. 1A, 2C, 8B, 9B), and

FIG. 5. (A) *Dothidea puccinioides*, section of perithecioid locule immersed in pulvinate, multiloculate, parenchymatous ascostroma with fascicle of aparaphysate, short-cylindrical asci arising from hemispherical placenta in base of locule. [From *Amer. J. Bot.* **38**:464, Fig. 15 (1951).] (B) *Dothiora schizospora*, section of apothecioid pseudothecium with broad layer of young asci arising in the parenchymatous tissue of the disc-shaped locule. [From *Mycologia* **52**:68, Fig. 5 (1960).]

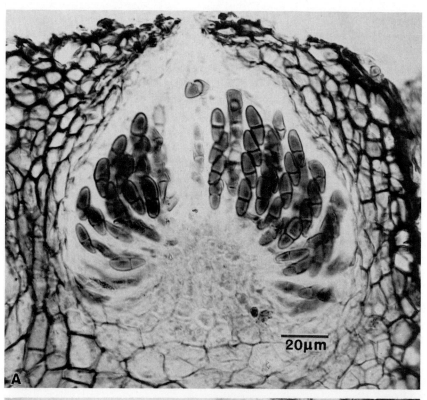

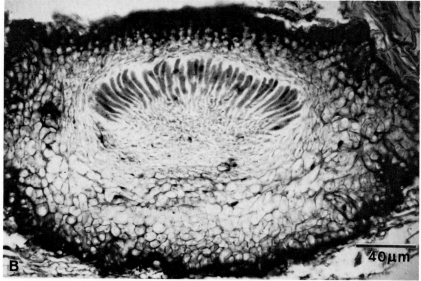

this will distinguish them from the paraphyses of Euascomycetes. In some of the smaller pleosporaceous fungi the pseudoparaphyses are sparse and the asci are relatively few. In squash mounts the asci may be squeezed out in a basally bound group, while the apically attached pseudoparaphyses remain within the ascocarp. Sections may be necessary to demonstrate the presence of pseudoparaphyses. Sections of younger ascocarps, in which the immature asci may be observed arising among preexisting pseudoparaphyses (Fig. 8A) attached at both the top and bottom of the locule, are helpful in distinguishing pseudoparaphyses from the paraphyses of Euascomycetes and from the interascicular strands of some dothideaceous fungi. Apothecioid variations occur in the Hysteriaceae in which the pseudoparaphysate asci develop in extended flat layers in the elongated ascocarps and are exposed by a longitudinal slit in the overlying stromal tissue.

A fourth structural type is represented by the Patellariaceae, in which the ascocarps are essentially identical with apothecia. In *Rhytidhysterium rufulum* ascogonia arise in a stroma. A palisade of vertically oriented paraphysoidal hyphae attached at both base and apex develops across the upper part of the stroma, and asci grow up among the paraphysoidal hyphae. The overlying stromal tissue splits and, as the ascocarp expands to its mature apothecial shape (Fig. 4B), the apical ends of the paraphysoidal hyphae separate from the overlying stromal tissue. Their freed tips regenerate and produce numerous short branches with swollen tips which form an epithecium over the asci. At maturity there is little to distinguish these paraphysoidal hyphae from the paraphyses of Discomycetes except their tendency to produce lateral branches at various points along their length. In other, possibly related, forms such as *Pseudoscypha* (Reid and Pirozynski, 1966) the paraphysoidal hyphae arise from the surface of a flat basal stroma or subiculum. Their tips are free from the beginning, and at present there is no basis for distinguishing them from paraphyses. Presence of the bitunicate ascus is the primary reason for including these fungi in the Loculoascomycetes, and ascus structure itself has been inadequately studied in this group.

## D. Septate Ascospores

Unicellular ascospores occur in relatively few genera of Loculoascomycetes. If the ascospores are unicellular the odds are perhaps nine to one that the genus belongs in the Euascomycetes and are even higher if the spores are globose or are provided with germ slits or pores or are conspicuously ornamented. None of the Loculoascomycetes has allantoid spores. If the ascospores are one septate, the odds are approximately two to one that the

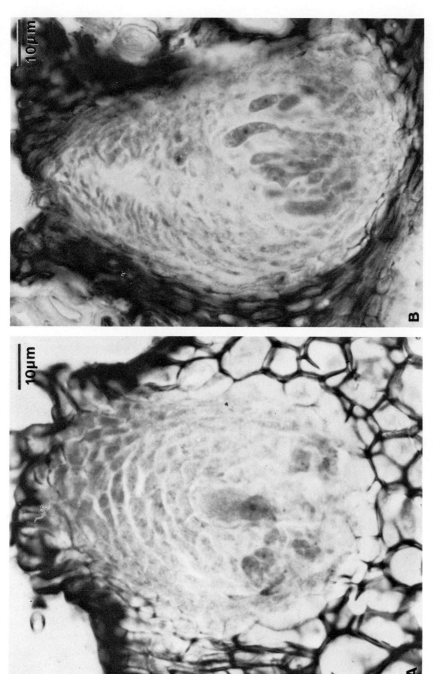

FIG. 6. (A) *Cymadothea trifolii*, section of perithecioid locule partially immersed in flat parenchymatous stroma with ascogenous cells and young ascus in the parenchymatous centrum tissue. (B) *Scirrhia acicola*, section of perithecioid lobe from basal stroma with fascicle of young asci in locule filled with periphysislike hyphae growing inward from lateral walls.

genus is in the Loculoascomycetes. If the ascospores are phragmosporous or dictyosporous, the odds are nine to one or higher that the genus is in the Loculoascomycetes.

## IV. CLASSIFICATION

The Myriangiales has persisted through all classifications, and this order is maintained for the small group of Loculoascomycetes with globose asci distributed singly within the tissue of ascostromatic ascocarps. These are usually superficial on living leaves and young stems. The great majority of the Loculoascomycetes commonly encountered on wood, and herbaceous stems, and dead leaves produce perithecioid pseudothecia which may be solitary, (Figs. 1G, 7B, 8B), grouped on a stroma (Figs. 6A, 9A), or immersed in a common stroma in which they appear as locules (Fig. 5A). Most of these have relatively large locules and pseudoparaphysate asci (Figs. 2C, 8B) and are placed in the order Pleosporales. A less numerous group with relatively small locules containing fascicles of aparaphysate asci (Figs. 1G, 5A) is included in the old order Dothideales. With few exceptions, Loculoascomycetes with apothecioid pseudothecia (Fig. 4B) as a matter of practical convenience are lumped in the order Hysteriales. Although the Hysteriales is limited to forms in which the pseudothecia are definitely discomycete-like, it is probably heterogeneous. The affinities of the Hysteriaceae with the Loculoascomycetes have long been recognized. Except for the elongated and slitlike opening of the ascocarp, their structure is similar to that of the Pleosporales, and they might well be included in this order. Although its relationships remain in doubt, the discomycete family Patellariaceae (Figs. 2D, 4B) is provisionally included in the Hysteriales. The very few species in the Dothioraceae, despite the disklike hymenium (Fig. 5B) exposed by rupture of the overlying stromal tissue of the pulvinate ascocarp, are fundamentally so similar to the Dothideaceae that they are separated from other apothecioid forms and included in the Dothideales. Apothecioid forms that are clearly myriangiaceous in internal structure (Fig. 4A) are similarly retained in the Myriangiales.

There remains a number of Loculoascomycetes which occur primarily on living leaves, fruits, and young stems as epiphytes, ectocommensals, parasites, and hyperparasites and which produce small superficial ascocarps and commonly superficial mycelium. They are separated into a large number genera, most of which occur in tropical and subtropical regions. These present the most difficult problems of classification chiefly because of the small size of their ascocarps and lack of study of their development. Most of these

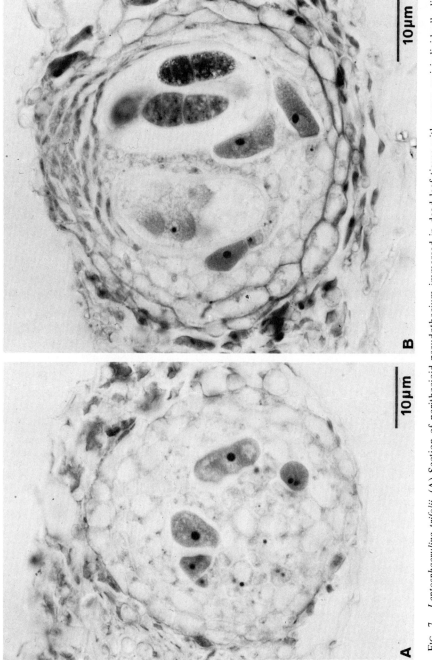

FIG. 7. *Leptosphaerulina trifolii.* (A) Section of perithecioid pseudothecium immersed in dead leaf tissue with young asci individually distributed in the parenchymatous centrum tissue. (B) Section of older pseudothecium with maturing asci partially separated by elongated strands of centrum tissue.

fungi have classically been placed in the old orders Perisporiales and Hemisphaeriales. Those in the Hemisphaeriales are characterized by dimidiate pseudothecia flattened against the host surface and covered over the upper exposed surface by a shield (Figs. 1H, I; 9B). They show variations in internal structure ranging through those characteristic of the Myriangiales, Dothideales, Pleosporales, and Hysteriales; and it seems certain, as suggested by Luttrell (1951), that they must ultimately be distributed among these orders. Nevertheless, the dimidiate-scutate structure of the ascocarp, although undoubtedly a response to the superficial habit, is not a necessary response, and its theoretical significance cannot be entirely ignored. Furthermore, it is a practical character which sets apart a large group of superficial forms for convenience of identification. The order Hemisphaeriales, therefore, is provisionally maintained. It is divided into a number of narrowly delimited families to facilitate possible future distribution among other orders.

Of the superficial forms with globose ascocarps formerly placed in the Perisporiales, the Dimeriaceae, although small, apparently have pseudoparaphysate centra. They intergrade with the Venturiaceae and have been placed alongside this family in the Pleosporales. The rest of the families containing these fungi show closer affinities with the Dothideales on the one hand, and with reduced forms of the Myriangiales on the other. The asci are broad clavate to short cylindrical and are relatively large and few in the small pseudothecia. They may be partially separated by strands of sterile tissue. Because of their perithecioid pseudothecia, these fungi are included in the Dothideales in the families Parodiopsidaceae, Englerulaceae, and Capnodiaceae. A heterogenous group of fungi, largely parasitic on green leaves with superficial, more or less apothecioid ascocarps, is placed in the Hysteriales as the family Phillipsiellaceae. Phillipsiellaceae is used for this group instead of Schizothyriaceae (Müller and von Arx, 1962) because *Schizothyrium* has the same myriangiaceous internal structure (Fig. 1H) that characterizes the Stephanothecaceae. The Schizothyriaceae, as typified by *Schizothyrium*, therefore is retained alongside the Stephanothecaceae in the Hemisphaeriales.

The lichen family Mycoporaceae is included in the Pleosporales and the lichen families Arthoniaceae (Fig. 9C), Opegraphaceae, and Lecanactidaceae in the Hysteriales although no attempt has been made to give a comprehensive treatment of their genera. These families are mentioned, first, to indicate the affinities of a number of lichen fungi with the Loculoascomycetes and, second, because the affinities often are so close that it is questionable whether lichenized and nonlichenized species should even be segregated into separate genera.

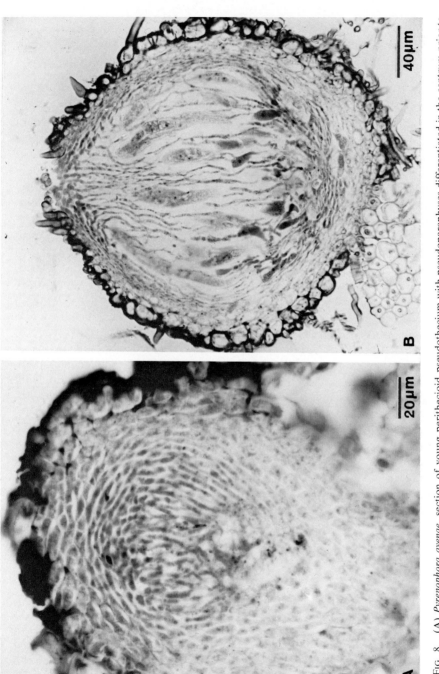

Fig. 8. (A) *Pyrenophora avenae*, section of young perithecioid pseudothecium with pseudoparaphyses differentiated in the centrum prior to formation of asci. (B) *Trichometasphaeria* (*Keissleriella*) *turcica*, section of mature, perithecioid pseudothecium with cylindrical asci among persistent pseudoparaphyses. [From *Amer. J. Bot.* **51**:216, Fig. 19 (1964).]

This classification is basically similar to that of Müller and von Arx (1962; von Arx and Müller, 1954; von Arx, 1963). The treatment of families genera, and keys is based on their monographs as far as possible and suffers when it extends beyond them. Müller and von Arx, however, recognized only three orders: the Myriangiales in the traditional sense, the Pseudosphaeriales for all perithecioid forms, and the Dothiorales for all apothecioid forms. The Pseudosphaeriales, as they defined it, corresponds essentially to the Pleosporales since the central family is the Pleosporaceae. The type genus of the Pseudosphaeriaceae, *Pseudosphaeria* (as a synonym of *Wettsteinina*), was included in the Pleosporaceae, and Pseudosphaeriaceae became a synonym of Pleosporaceae. Also included in the Pseudosphaeriales was the central family of the Dothideales, the Dothideaceae, for which the name Mycosphaerellaceae was employed. The dothideaceous families Capnodiaceae, Chaetothyriaceae, and Trichothyriaceae joined the Dothideaceae (Mycosphaerellaceae) in the Pseudosphaeriales. The Pseudosphaeriales also contained the families of Hemisphaeriales with hemiperithecioid ascocarps, the Microthyriaceae and Micropeltidaceae.

The Dothiorales (von Arx and Müller; 1954; Müller and von Arx, 1962) was proposed in place of the older ordinal name Hysteriales. This order included the Hysteriaceae and Patellariaceae as the central groups of saprotrophs. A group of leaf parasites with more or less apothecioid ascocarps formed the Schizothyriaceae, which was greatly expanded to include genera unrelated to the type. With the removal of *Schizothyrium*, which is similar in structure to the genera *Pycnoderma*, *Pycnodermina*, and *Stephanotheca* subsequently placed in the Myriangiales by von Arx (1963), the Schizothyriaceae of Müller and von Arx (1962) corresponds essentially to the Phillipsiellaceae. Because of the more or less clavate asci and dehiscence of the ascocarp by a broad pore or by disintegration or splitting, the Englerulaceae and the Parodiopsidaceae (with the new name Perisporiopsidaceae) were included with the apothecioid forms in the Dothiorales. Included also were families of the Hemisphaeriales, in which the ascocarp was considered apothecioid, the Asterinaceae, Brefeldiellaceae, Leptopeltidaceae, and Parmulariaceae; the family Dothioraceae from the Dothideales; and the family Atichiaceae from the Myriangiales.

FIG. 9. (A) *Apiosporina morbosa*, section of mature perithecioid pseudothecia seated on crustose basal stroma, with centrum of cylindrical asci and persistent pseudoparaphyses. (B) *Ellisiodothis smilacis*, section of superficial, dimidiate-scutate pseudothecium with mature asci and persistent pseudoparaphyses in locule. [From *Amer. J. Bot.* **31**:642, Fig. 12 (1944).] (C) *Stirtonia* sp., section of apothecioid pseudothecium in crustose thallus on phellem of twig of *Prunus serotina* with young asci scattered in the tangle of hyphae forming the ascocarp.

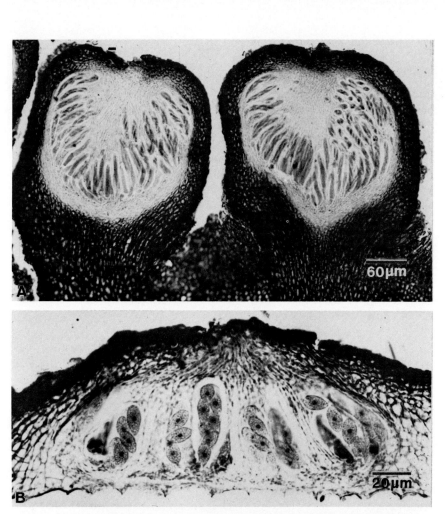
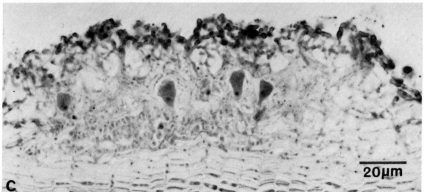

Chadefaud (1960) and Parguey-Leduc (1966–1967) adopted, with some modifications, this system of Müller and von Arx, placing primary emphasis on ascus shape in recognizing three orders: Myriangiales with globose asci, Pleosporales with asci broadest in the lower part, and Dothiorales with asci broadest toward the apex.

Authors of family names may be found in Cooke and Hawksworth (1970). Note, however, that the family name Parodiopsidaceae was validly published by Toro (1952), and Perisporiopsidaceae (Müller and von Arx, 1962) is an invalid name for the same family. Aulographaceae (Arnaud) stat. nov. is based on Arnaud's (1930: 285) tribe Aulographeés. Munkiellaceae (Theissen & Sydow) stat. nov. is based on Theissen and Sydow's (1917: 400) subfamily Munkielleae to replace the family name Stigmateaceae, which, with the transfer of *Stigmatea* to the Venturiaceae, becomes a prior synonym of Venturiaceae.

## V. IMPORTANT LITERATURE

The indispensible critical treatments of genera of Loculoascomycetes are "Die Gattungen der Amerosporen Pyrenomyceten" by von Arx and Müller (1954) and "Die Gattungen der Didmosporen Pyrenomyceten" by Müller and von Arx (1962). These volumes are profusely illustrated, and the keys to families and genera have been translated into English (von Arx and Müller, 1969). These monographs do not cover the phragmosporous and dictyosporous genera. They are supplemented, however, by von Arx's (1963) "Die Gattungen der Myriangiales." Useful general treatments of Loculoascomycetes are included in "British Ascomycetes" by Dennis (1968) and in "Danish Pyrenomycetes" by Munk (1957). Comprehensive monographs of species have appeared in "Monografia dos Fungos Micropeltaceae" by Batista (1959b), "The Chaetothyriales" by Batista and Ciferri (1962), and "Capnodiales" by Batista and Ciferri (1963), but these treatments are difficult to use and are less dependable than might be desired. Zogg's (1962) "Die Hysteriaceae S. Str. und Lophiaceae" and Margaret Barr's (1968) "The Venturiaceae in North America" are excellent guides to these families. Monographs of genera are mostly lacking, Wehmeyer's (1961) "A World Monograph of the Genus *Pleospora* and its Segregates," Mirza's (1968) "Taxonomic Investigations on the Ascomycetous Genus *Cucurbitaria*," Hedjaroude's (1968) "Études Taxonomiques sur les Phaeosphaeria et leurs Formes Voisines" and Reynolds' (1971) "The Sooty Mold Ascomycete Genus *Limacinula*" being notable exceptions. Available monographs of genera are cited under the appropriate genera in the keys. References to descriptions and, whenever possible, to illustrations of representative species are also given under each genus in the keys. Some of the most detailed illustrations of these fungi, primarily of Hemisphaeriales, have been published by Arnaud (1918, 1921, 1923, 1925, 1930, 1931).

## VI. SYNOPTIC GUIDE TO ORDERS, FAMILIES, AND GENERA

### A. Ascospores

*i. Ascospores nonseptate.* Myriangiales, Saccardinulaceae—*Hyalotheles, Piedraia.* Dothideales, Parodiopsidaceae–*Cleistosphaeria, Pilgeriella*; Pseudosphaeriaceae—*Montagnellina*; Dothideaceae—*Auerswaldia, Columnosphaeria, Guignardia, Plurisperma, Vestergrenia*; Dothioraceae—*Bagnisiella, Hypnotheca.* Pleosporales, Mesnieraceae—*Mesniera*; Botryosphaeriaceae—*Auerswaldiella, Botryosphaeria, Pyrenostigme*; Sporormiaceae—*Semidelitschia.* Hysteriales, Hysteriaceae—*Farlowiella.* Hemisphaeriales, Microthyriaceae—*Ellisiodothis, Myiocopron*; Munkiellaceae—*Microdothella, Parastigmatea, Trabutia*; Micropeltidaceae—*Dictyothyrina*; Leptopeltidaceae—*Moeszopeltis.*

*ii. Ascospores nonseptate, brown with a pale-colored equatorial band.* Hemisphaeriales, Munkiellaceae—*Entopeltis, Vizella*; Parmulariaceae—*Blasdalea.*

*iii. Ascospores 1-septate.* Myriangiales, Atichiaceae—all; Saccardinlaceae—*Micularia.* Dothideales, Trichothyriaceae—nearly all; Chaetothyriaceae—*Akaropeltis, Macrocallis*; Parodiopsidaceae—nearly all, Englerulaceae—all; Dothideaceae—nearly all; Capnodiaceae—*Calyptra, Capnobatista, Capnogonium, Laterotheca*; Pleosporales, Dimeriaceae—nearly all; Venturiaceae—all; Botryosphaeriaceae—*Neodeightonia*; Lophiostomataceae—*Ostropella, Byssolophis, Lophiosphaera*; Sporormiaceae—*Delitschia, Trichodelitschia*; Pleosporaceae—*Cercidospora, Didymosphaeria, Microthelia, Otthia, Paraliomyces, Phaeospora, Tomasellia*; Mycoporaceae—*Leiophloea, Sporoschizon.* Hysteriales, Hysteriaceae—*Actidium, Glonium*; Phillipsiellaceae—most; Patellariaceae—many. Hemisphaeriales, Microthyriaceae—most; Trichopeltinaceae—most; Munkiellaceae—*Hormotheca, Munkiella*; Micropeltidaceae—many; Asterinaceae—most; Brefeldiellaceae—all; Aulographaceae—all; Parmulariaceae—nearly all; Stephanothecaceae—*Campoa, Protothyrium*; Schizothriaceae—*Leptophyma, Schizothyrium*; Leptopeltidaceae—most.

*iv. Ascospores 0- to 1-septate, thick-walled with sculptured epispore.* Pleosporales, Mesnieraceae—*Mesniera, Stegasphaeria.*

*v. Ascospores 1-septate, apiosporous.* Dothideales, Dothideaceae—*Achorodothis, Lasiobotrys, Omphalospora.* Pleosporales, Venturiaceae—*Apio-*

*sporina, Atopospora, Botryostroma, Coccoidea, Coleroa, Crotone, Platychora*; Pleosporaceae—*Didymopleela, Pteridospora*. Hysteriales, Hysteriaceae—*Farlowiella*; Phillipsiellaceae—*Coccodothis*. Hemisphaeriales, Munkiellaceae—*Munkiella*; Parmulariaceae—*Inocyclus, Polycyclus*.

    *vi. Ascospores 1- to several-septate, variable.* Dothideales, Trichothyriaceae—*Actinopeltis, Trichothyrium*; Parodiopsidaceae—*Perisporiopsis;* Pseudosphaeriaceae—*Wettsteinina*; Dothideaceae—*Plowrightia*. Pleosporales, Venturiaceae—*Eudarluca*; Pleosporaceae—*Herpotrichia, Herpotrichiella, Keissleriella, Massarina*. Hysteriales, Arthoniaceae—*Arthonia, Stirtonia*.

    *vii. Ascospores 1-septate, large, brown, with 2 additional septa cutting off tiny, apiculate cells at either end.* Pleosporales, Pleosporaceae—*Caryospora*.

    *viii. Ascospores phragmosporous or dictyosporous.* Myriangiales—nearly all. Dothideales, Chaetothyriaceae—most; Capnodiaceae—nearly all; Dothideaceae—*Phaeodothiora, Pringsheimia, Sphaerulina*; Dothioraceae—nearly all. Pleosporales, Dimeriaceae—*Aphanostigme*; Lophiostomataceae—most; Sporormiaceae—most; Pleosporaceae—most; Mycoporaceae—most. Hysteriales, Hysteriaceae—nearly all; Opegraphaceae—all; Phillipsiellaceae—*Annajenkinsia, Protoscypha*; Patellariaceae—many; Lecanactidaceae—all. Hemisphaeriales, Microthyriaceae—*Actinomyxa, Caenothyrium, Halbaniella, Micropeltopsis, Phragmothyrium, Platypeltella, Yatesula*; Trichopeltinaceae—*Trichopeltum*; Munkiellaceae—*Melanochlamys, Stigmatodothis*; Micropeltidaceae—most; Stephanothecaceae—most; Schizothyriaceae—many; *Leptopeltidaceae—Dothiopeltis, Pycynothyrium*.

    *ix. Ascospores phragmo-dictyosporous, variable.* Myriangiales, Myriangiaceae—*Bitancourtia, Elsinoë*; Saccardinulaceae—*Molleriella, Saccardinula, Xenodium*. Dothideales, Pseudosphaeriaceae—*Leptosphaerulina*. Pleosporales, Pleosporaceae—*Karstenula, Pyrenophora*.

    *x. Ascospores scolecosporous.* Dothideales, Parodiopsidaceae—*Ophioparodia*. Pleosporales, Lophiostomataceae—*Lophionema*; Pleosporaceae—*Acanthophiobolus, Cochliobolus* (helically coiled), *Dolabra, Entodesmium, Leptosphaeria, Leptospora* (helically coiled at apex), *Nodulosphaeria, Ophiobolus, Phaeosphaeria, Plejobolus, Podonectria, Rosenscheldia, Tubeufia*; Mycoporaceae—*Leptorhaphis*. Hysteriales, Hysteriaceae—*Glyphium, Lophium*. Hemisphaeriales, Micropeltidaceae—*Scolecopeltis*; Schizothyriaceae—*Kerniomyces*.

    *xi. Ascospores with germ slits.* Pleosporales, Sporormiaceae—all but *Trichodelitschia*.

    *xii. Ascospores with germ pores.* Pleosporales, Pleosporaceae—*Pontoporeia*; Sporormiaceae—*Trichodelitschia*. Hysteriales, Patellariaceae—*Eutryblidiella*.

*xiii. Ascospores tapering to filiform extensions at 1 or both ends.* Myriangiales, Saccardinulaceae—*Piedraia*. Pleosporales, Pleosporaceae—*Rebentschia*. Hemisphaeriales, Microthyriaceae—*Caudella*; Micropeltidaceae—*Stomiotheca*.
*xiv. Ascospores with gelatinous sheaths.* Dothideales, Pseudosphaeriaceae—*Wettsteinina*. Pleosporales, Sporormiaceae—all; Pleosporaceae—*Asteromassaria, Julella, Keissleriella, Massaria, Massarina, Peltosphaeria, Pleomassaria, Pteridospora, Pyrenophora, Stigmatomassaria, Trichometasphaeria*.
*xv. Ascospores with slimy appendages.* Pleosporales, Lophiostomataceae—*Lophiosphaeria, Lophiostoma, Lophiotrema*; Pleosporaceae—*Herpotrichia diffusa, Nodulosphaeria, Paralomyces*.
*xvi. Ascospores with cilia.* Hysteriales, Patellariaceae—*Banhegyia*. Hemisphaeriales, Microthyriaceae—*Microthyrium macrosporum*.

## B. Asci

*i. Asci globose to broad oblong or clavate (less than 3 times as long as broad).* Myriangiales—all. Dothideales, Parodiopsidaceae—all; Englerulaceae—all. Hysteriales, Arthoniaceae—all; Phillipsiellaceae—*Coccodothis, Plochmopeltis*. Hemisphaeriales, Asterinaceae—most; Brefeldiellaceae—most; Parmulariaceae—many, e.g., *Aulacostroma, Ferrarisia, Rhipidocarpon*; Stephanothecaceae—most; Schizothyriaceae—most; Leptopeltidaceae—all.
*ii. Asci tending to be wider toward the base, ovate or obclavate, sessile or abruptly short-stipitate.* Dothideales—Trichothyriaceae, Chaetothyriaceae, Pseudosphaeriaceae (short ovate), Capnodiaceae, Dothideaceae. Pleosporales—Dimeriaceae, Venturiaceae. Hemisphaeriales—Microthyriaceae, Munkiellaceae, Micropeltidaceae.
*iii. Asci tending to be wider toward the apex, long cylindrical or clavate, tapering to the stipe.* Dothideales—Dothioraceae. Pleosporales—Botryosphaeriaceae, Lophiostomataceae, Sporormiaceae, Pleosporaceae. Hysteriales—Hysteriaceae, Opegraphaceae, Phillipsiellaceae, Patellariaceae, Lecanactidaceae. Hemisphaeriales—Parmulariaceae.
*iv. Asci globose, on stalks of varying lengths.* Dothideales, Dothideaceae—*Vestergrenia*.
*v. Asci polysporous.* Myriangiales, Saccardinulaceae—*Xenodium*. Dothideales, Capnodiaceae—*Capnodaria* (= *Capnodium*); Dothideaceae—*Delphinella, Plurisperma*; Dothioraceae—*Endodothiora, Keisslerina, Sydowia*. Pleosporales, Sporormiaceae—*Preussia* (by separation of phragmosporous spore cells), *Trichodelitschia* (true or by separation of didymosporous spore cells); Pleosporaceae—*Capronia*. Hemisphaeriales, Asterinaceae—*Anariste* (by separation of didymosporous spore cells);

Parmulariaceae—*Cyclostomella* (by separation of didymosporous spore cells). Species with polysporous asci may occur in other genera since this character usually is not employed as a generic distinction. Four-spored asci occur in species such as *Dothidea puccinoides*. Asci in other genera such as *Cochliobolus* and *Keissleriella* contain 1–8 spores.

### C. Ascocarps

*i. Ascocarps bright-colored.* Myriangiales, Myriangiaceae—*Cookella, Uleomyces*; Saccardiaceae—*Dictyonella*. Dothideales, Dothideaceae—*Cerodothis*. Pleosporales, Venturiaceae—*Allonecte, Letendraea, Placocrea*; Pleosporaceae—*Podonectria, Tubeufia*. Hysteriales, Phillipsiellaceae—*Annajenkinsia, Gymnoascopsis*, Hemisphaeriales, Stephanothecaceae—*Pycnoderma*; Schizothriaceae—*Leptophyma*.

*ii. Ascocarp gelatinous.* Myriangiales, Atichiaceae—all. Hemisphaeriales, Schizothyriaceae—*Leptophyma*.

*iii. Ascocarp dissolving into slime mass.* Dothideales, Englerulaceae—all. Hysteriales, Phillipsiellaceae—*Asterotexis, Henningsiella*. Hemisphaeriales, Microthyriaceae—*Actinomyxa*; Asterinaceae—*Asterina, Clypeolella*.

*iv. Ascocarp lenticular, with upper and lower plates of radiate structure.* Dothideales, Trichothyriaceae—all.

*v. Ascocarp apothecioid.* Myriangiales, Myriangiaceae—*Myriangium*; Saccardiaceae—many. Dothideales, Dothideaceae—*Mycosphaerella pneumatophorae*; Dothioraceae—all. Pleosporales, Venturniaceae—*Pseudoparodia*; Hysteriales—nearly all. Hemisphaeriales, Asterinaceae—many; Parmulariaceae—all; Schizothyriaceae—*Kerniomyces, Schizothyrium*.

*vi. Ascocarp elongated or with elongated locules opening by a slit.* Pleosporales, Lophiostomataceae—*Byssolophis*. Hysteriales, Hysteriaceae—all; Opegraphaceae—nearly all. Hemisphaeriales, Asterinaceae—many, e.g., *Aulographina, Lembosia, Morenoina*; Parmulariaceae—all; Aulographaceae—all; Stephanothecaceae—*Campoa, Protothyrium*; Schizothyriaceae—*Kerniomyces, Schizothrium*.

*vii. Ascocarp with laterally compressed beak and slitlike ostiole.* Pleosporales, Lophiostomataceae—all.

### D. Size of Ascocarp

*i. Ascocarps very large (1–5 mm), solitary.* Myriangiales, Myriangiaceae—*Myriangium, Uleomyces* (some), *Elsinoë* (some), Saccardiaceae—*Angatia, Dictyonella* (some). Dothideales, Dothioraceae—most. Hysteriales, Hysteriaceae—all; Patellariaceae—all.

*ii. Ascocarps very large, compound (often with small locules).* Dothideales, Dothideaceae—*Auerswaldia, Coccoidella, Dothidea, Microcyclus*. Pleosporales, Venturiaceae—*Auerswaldiella, Coccoidea, Eudarluca, Placocrea*,

*Platychora, Polyrhizon, Uleodothis*; Pleosporaceae—*Rhopographus*. Hysteriales, Phillipsiellaceae—*Asterotexis, Englerodothis*. Hemisphaeriales, Mycrothyriaceae—*Dothidella*; Munkiellaceae—*Munkiella*; Parmulariaceae—most.

*iii. Ascocarps fused laterally, forming extensive flat crusts.* Hemisphaeriales, Microthyriaceae—*Ellisiodothis, Myiocopron, Palawania, Polycyclinopsis, Xenostomella*; Leptopeltidaceae—many, e.g., *Leptopeltopsis*.

*iv. Ascocarps large (400–1000 μm).* Myriangiales, Myriangiaceae—*Anhellia* (some), *Bitancourtia*; Saccardiaceae—*Saccardia* (some). Dothideales, Pseudosphaeriaceae—*Wettsteinina* (some). Pleosporales, Sporormiaceae—most; Lophiostomataceae—most; Pleosporaceae—most. Hemisphaeriales, Microthyriaceae—*Caudella, Cyclotheca*; Asterinaceae—*Echidnodella, Symphaster*; Stephanothecacea—many; Schizothriaceae—*Schizothyrium* (some).

*v. Ascocarps middle-sized (200–350 μm).* Myriangiales, Myriangiaceae—many; Saccardiaceae—many. Pleosporales, Venturiaceae—most; Mesnieraceae—all; Sporormiaceae—many, Pleosporaceae—many. Hysteriales, Phillipsiellaceae—most. Hemisphaeriales—most except for Parmulariaceae.

*vi. Ascocarps small (40–150 μm).* Myriangiales, Saccardinulaceae—all. Dothideales—nearly all except Dothioraceae. Pleosporales, Dimeriaceae—most; Venturiaceae—some. Hemisphaeriales—many in Microthyriaceae, Micropeltidaceae, and Asterinaceae.

### E. Mycelium

*i. Mycelium a membrane of radiating or parallel hyphae on leaves or young stems.* Myriangales, Saccardinulaceae—*Saccardinula*. Dothideales, Trichothyriaceae—all; Capnodiaceae—*Trichopeltheca*. Pleosporales, Venturiaceae—*Venturia asteromorpha*. Hemisphaeriales, Trichopeltinaceae—all; Munkiellaceae—*Hormotheca*; Brefeldiellaceae—all.

*ii. Mycelium forming a dark superficial net on leaves and young stems.* Dothideales, Parodiopsidaceae—many; Englerulaceae—many. Pleosporales, Dimeriaceae—most. Hemisphaeriales, Microthyriaceae—many; Micropeltidaceae—many; Asterinaceae—all.

*iii. Superficial mycelium with hyphopodia.* Dothideales, Parodiopsidaceae—*Balladyna*; Englerulaceae—*Paraenglerula, Rhytidenglerula, Schniffnerula, Thrauste*. Hemisphaeriales, Microthyriaceae—*Asterinema, Caudella, Maublancia, Platypeltella, Xenostomella*; Asterinaceae—many; Brefeldiellaceae—*Pycnocarpon*.

*iv. Mycelium forming thick, widespread, black mats easily stripped from leaves and stems (sooty molds).* Dothideales, Capnodiaceae—most.

*v. Mycelium forming a pale superficial pellicle beneath which subglobose pseudothecia develop.* Dothideales, Chaetothyriaceae—all.

*vi. Mycelium forming a brown felt over conifer seedlings.* Pleosporales, Pleosporaceae—*Herpotrichia.*

*vii. Mycelium forming a brown to black subiculum on bark and wood.* Pleosporales, Lophiostomataceae—*Byssolophis*; Pleosporaceae—*Herpotrichia, Herpotrichiella, Thaxteria, Tubeufia*, Hysteriales, Hysteriaceae—*Glonium.*

*viii. Mycelium forming a lichenlike thallus.* Pleosporales, Pleosporaceae—*Cucurbidothis.*

### F. Habit

*i. Lichenized.* Pleosporales, Mycoporaceae—all; Pleosporaceae—*Didymosphaeria* (some); *Microthelia* (some), *Tomasellia*. Hysteriales, Arthoniaceae—all; Opegraphaceae—all; Patellariaceae—*Buellia* (some), *Karschia* (some); Lecanactidaceae—all.

*ii. Parasitic on lichens.* Pleosporales, Dimeriaceae—*Echinothecium*; Pleosporaceae—*Cercidospora, Didymosphaeria* (some), *Phaeospora*. Hysteriales, Patellariaceae—*Abrothallus, Buellia* (some).

*iii. Parasitic on mosses.* Pleosporales, Dimeriaceae—*Lizonia.*

*iv. Hyperparasitic on superficial fungi.* Myriangiales, Myriangiaceae—*Cookella, Uleomyces*. Dothideales, Trichothyriaceae—all. Pleosporales, Dimeriaceae—*Dimerina, Dimerium, Phaeodimeriella*; Venturiaceae—*Eudarluca* (in and around rust pustules); Pleosporaceae—*Tubeufia*; Hemisphaeriales, Trichopeltinaceae—all; Stephanothecaceae—*Pycnoderma.*

*v. Hyperparasitic on scale insects.* Myriangiales, Myriangiaceae—*Myriangium*; Saccardiaceae—*Angatia*. Pleosporales, Pleosporaceae—*Podonectria.*

*vi. Superficial epiphytes on insect honey dew.* Myriangiales, Atichiaceae—all. Dothideales, Capnodiaceae—all.

*vii. On human hair.* Myriangiales, Saccardinulaceae—*Piedraia.*

*viii. Fruiting on green leaves or young stems.* Myriangiales—most. Dothideales, Chaetothyriaceae—most; Trichothyriaceae—most; Parodiopsidaceae—most; Englerulaceae—most; Capnodiaceae—most; Dothideaceae—*Achorodothis, Cerodothis, Coccoidella, Lasiobotrys, Microcylus, Rhizogene*. Pleosporales, Dimeriaceae—most; Venturiaceae—most; Mesnieraceae—all; Pleosporaceae—*Phragmocauma, Muellerites, Tomasellia*. Hysteriales, Phillipsiellaceae—most. Hemisphaeriales—most. (Compare with black mildews in Meliolaceae, Euascomycetes.)

*ix. Fruiting on wood, bark, dead herbaceous stems, and dead leaves.* Dothideales, Pseudosphaeriaceae—most; Dothideaceae—most; Dothioraceae—most. Pleosporales, Venturiaceae—*Didymella, Licopolia, Venturia*; Botryosphaeriaceae—all; Lophiostomataceae—all. Pleosporaceae—most. Hysteriales, Hysteriaceae—all; Patellariaceae—all. Hemisphaeriales, Microthyriaceae—*Arnaudiella, Ellisiodothis, Microthyrium, Myiocopron, Palawania*; Munkiellaceae—*Microdothella.*

*x. On dung.* Pleosporales, Sporormiaceae—all.
*xi. Marine.* Dothideales, Dothideaceae—*Mycosphaerella*. Pleosporales, Venturiaceae—*Didymella*; Pleosporaceae—*Didymosphaeria, Halotthia, Herpotrichiella, Keissleriella, Leptosphaeria, Microthelia, Paraliomyces, Phaeosphaeria, Pleospora, Pontoporeia, Thallassoascus, Trematosphaeria*. Hysteriales, Patellariaceae—*Banhegyia, Buellia*.
*xii. Of economic importance on crop and ornamental plants.* Myriangiales, Myriangiaceae—*Elsinoë*, spot anthracnoses of many plants, *E. ampelina* on *Vitis*. Dothideales, Pseudosphaeriaceae—*Leptosphaerulina*, leaf spots, *L. trifolii*, pepper spot of *Trifolium*. Capnodiaceae—various genera, sooty molds of ornamentals. Dothideaceae—*Cymadothea*, leaf spots, *C. trifolii*, sooty blotch of *Trifolium*; *Guignardia*, leaf spots and fruit rots, *G. bidwellii*, black rot of *Vitis*; *Microcyclus*, leaf spots, *M. ulei*, leaf blight of *Hevea*; *Mycosphaerella*, numerous leaf spots, *M. musicola*, sigatoka disease of *Musa*; *Scirrhia*, leaf blights, *S. acicola*, brown spot of *Pinus*. Pleosporales, Venturiaceae—*Apiosporina*, hypertrophies, *A. morbosa*, black knot of *Prunus*; *Phaeocryptopus*, needle blights, *P. gaeumannii* on *Pseudotsuga*; *Venturia*, leaf spots and fruit rots, *V. inaequalis*, scab of *Malus*. Botryosphaeriaceae—*Botryosphaeria*, stem cankers, twig blights, fruit rots, *B. corticis* on *Vaccineum*. Pleosporaceae—*Cochliobolus*, leaf spots and blights, *C. victoriae*, Victoria blight of *Avena*; *Herpotrichia*, felt blight of conifers; *Phaeosphaeria*, leaf spots, stem rots, *P. nodorum*, glume blotch of *Triticum*; *Pleospora*, leaf spots, *P. herbarum* on *Medicago*; *Pyrenophora*, leaf spots and blights, *P. bromi* on *Bromus*; *Trichometasphaeria (Keissleriella)*, leaf spots and blights, *T. turcica* on *Zea*. Hemisphaeriales, Micropeltidaceae—*Stomiopeltis* sp., olive blotch of *Rubus*. Schizothyriaceae—*Schizothyrium*, fly speck, *S. pomi* on *Malus*.

## VII. KEY TO ORDERS

### KEY TO ORDERS OF LOCULOASCOMYCETES

1. Ascocarps dimidiate-scutate, with inverse, basipetal development, superficial or rarely subcuticular, on leaves, fruits, and young stems
 . . . . . . . . . . . . . . . . . . . . **Hemisphaeriales (Microthyriales)** p. 202
1'. Ascocarps not dimidiate-scutate, perithecioid, apothecioid, or pulvinate, with acropetal or centrifugal development, superficial to commonly erumpent or innate, on various substrates . . . . . . . . . . . . . . . . . . . . . . . . . . . . 2
    2(1') Asci globose to broad clavate, singly and irregularly distributed in the tissue of a parenchymatous or plectenchymatous ascostroma or gelatinous thallus . . . . . . . . . . . . . . . . . . . . . . . . **Myriangiales** p. 162
    2'(1') Asci ovoid to commonly clavate or cylindrical, grouped in locules in pseudothecia; pseudothecia separate, grouped on a stroma or immersed in crustose to pulvinate stromata in which they appear as locules . . . . . . . . . . . . . . . . . 3
3(2') Asci aparaphysate, in fascicles in small perithecioid locules or rarely in broad discoid layers . . . . . . . . . . . . . . . . . . . . . . . . . . . **Dothideales** p. 165

3′(2′) Asci interspersed with pseudoparaphyses or with paraphysoidal hyphae which branch at the tips to form an epithecium, in usually middle-sized to large locules ...... 4

4(3′) Pseudothecia perithecioid ............... **Pleosporales** p. 176

4′(3′) Pseudothecia apothecioid or lirelliform, rarely reduced to a flat basal stroma from which the asci and paraphysoidal hyphae arise ......... **Hysteriales** p. 194

## VIII. MYRIANGIALES

The Myriangiales comprises a relatively small number of Loculoascomycetes occurring as epiphytes, parasites, or hyperparasites on superficial fungi or scale insects on living leaves and stems. These fungi are mostly tropical or subtropical. Most common in temperate zones are species of *Myriangium* on scale insects and the many species of *Elsinoë* causing spot anthracnoses of stems, leaves, and fruits of various plants. In *Elsinoë* and *Bitancourtia* the ascocarps are intra- or subepidermal and are only partially erumpent. In all others they are superficial except for a basal foot, or are entirely superficial. They occur in the form of extensive crusts or large to small, pulvinate, subglobose, or discoid ascostromata. The most complex structures are produced by *Myriangium* spp., in which the asci are limited to fertile regions in short-stalked, apotheciumlike conceptacles covering small basal stromata.

Typical Myriangiales are distinguished from all other fungi of similar habit by internal structure. The asci are globose, and the ascospores are usually phragmosporous or dictyosporous. The asci are individually distributed in the parenchymatous tissue of the ascostroma, and in typical forms in the Myriangiaceae are irregularly scattered at several levels (Fig. 4A). In the Saccardiaceae the asci are restricted to a single irregular layer in a discoid to crustose stroma. The stromal tissues may be compressed between the asci and show a tendency to become paraphysoidal. The Saccardiaceae intergrade with the Phillipsiellaceae of the Hysteriales. Lichen fungi in the Arthoniaceae (Fig. 9C) likewise are similar to the Saccardiaceae and might be as well placed in the Myriangiales as in the Hysteriales. Genera of the Stephanothecaceae, Schizothyriaceae (Fig. 1H), and Leptopeltidaceae (Fig. 1I) of the Hemisphaeriales are distinguished only by their flattened, dimidiate-scutate ascocarps.

Genera of the Saccardinulaceae have small subglobose ascostromata occurring on leaf hairs and glandular scales. The asci are in small clusters, and the interascicular stromal tissue may be partially disintegrated. The structure of the ascocarp differs only slightly from that in the Parodiopsidaceae and Englerulaceae of the Dothideales. Appended to the Saccardinulaceae is the only loculoascomycete of medical interest, *Piedraia*. *Piedraia hortai* may form tiny ascostromata on human hair in the humid tropics.

The Atichiaceae are possibly unrelated to the other families in the order but have generally been appended to the Myriangiales. They are easily distinguished regardless of their location in the taxonomic system. The globose to lobed, grayish, gelatinous thalli grow in insect honey dew on leaves and stems, often in association with sooty molds, and can be recognized even in the absence of asci by their peculiar multicellular propagulae.

**KEY TO FAMILIES AND IMPORTANT GENERA OF MYRIANGIALES**

1. Thallus a globose or lobed body of toruloid hyphae in a gelatinous matrix bearing multicellular asexual propagulae; asci broad clavate, embedded individually in a single layer in the periphery of the thallus; ascospores 1-septate, hyaline to pale brown; saprotrophic in insect secretions on leaves and stems . . . . . . . . . . . . . . . . **Atichiaceae** p. 163

1'. Thallus a mycelium on which pseudoparenchymatous or plectenchymatous ascostromata are produced; mycelium sometimes reduced and thallus almost entirely stromatic; ascostroma superficial or erumpent; parasites on leaves or stems or hyperparasites on scale insects or fungi . . . . . . . . . . . . . . . . . . . . . . . . . . . . . . . . . . . . . 2

    2(1') Asci irregularly distributed in several layers in a pulvinate ascostroma or in apothecioid fertile regions of the stroma, typically globoid . . . **Myriangiaceae** p. 163

    2'(1') Asci in a single irregular layer or group in a pulvinate to subglobose or discoid ascostroma, globoid to broad clavate . . . . . . . . . . . . . . . . . . . . . 3

3(2') Ascostroma discoid or flat crustose; ascospores dictyosporous.
. . . . . . . . . . . . . . . . . . . . . . . . . . . . . . . . . . . **Saccardiaceae** p. 164
3'(2') Ascostroma small pulvinate to irregularly subglobose; on plant leaf hairs, one on human hair . . . . . . . . . . . . . . . . . . . . . . . . . . **Saccardinulaceae** p. 164

## A. Atichiaceae

1. Thallus usually lobed; asci in specialized disk-shaped protusions from the surface; propagulae triangular, branched structures resembling staurospores, formed in cavities in the thallus . . . . . . . . . . . . . . . . . . . . . . . . . . . . . . . . . . . . . . **Atichia**
(Boedijn, 1961; Müller and von Arx, 1962: 229, 230, icon.; Arnaud, 1925, Pl. IX, X; Verona and Benedek, 1968, C-319.)

1'. Thallus semiglobose, often fused into lobed masses; asci in undifferentiated peripheral layers; propagulae globose, multicellular, formed on surface of the thallus . . **Phycopsis**
(Boedijn, 1961; Müller and von Arx, 1962: p. 227, consider *Phycopsis* synonymous with *Atichia*; Verona and Benedek. 1968, C-319.)

## B. Myriangiaceae

1. Ascostroma complex; asci borne in differentiated, cupulate fertile portions arising on stalks from a basal stroma; on stems or rarely leaves, usually parasitic on scale insects; ascospores dictyosporous, hyaline . . . . . . . . . . . . . . . . . . . . . . . . . . . . . **Myriangium**
(von Arx, 1963: 424; Miller, 1938, 1940; Dennis, 1968: 354, Pl. XXXVI C.)

1'. Ascostroma simple . . . . . . . . . . . . . . . . . . . . . . . . . . . . . . . . . 2

    2(1') Ascostroma innate, intra- or subepidermal, partially erumpent at maturity, small-pulvinate to crustlike; ascospores 3-septate with longitudinal septa in some cells, hyaline to yellowish . . . . . . . . . . . . . . . . . . . . . . . . . . . . . **Elsinoë**
(von Arx, 1963: 442, icon.; Dennis, 1968: 355, Pl. XXXVI B.)

    2'(1') Ascostroma erumpent or superficial . . . . . . . . . . . . . . . . . . . . . 3

3(2') Ascostroma erumpent, usually dark-colored; parasitic on leaves or stems ..... 4
3'(2') Ascostroma superficial, usually bright-colored; hyperparasites on fungi ..... 6
    4(3) Ascostroma attached across entire base, pulvinate, erumpent through epidermis; ascospores 3-septate, with longitudinal septa in some cells ...... **Bitancourtia**
    (von Arx, 1963; Thirumalachar and Jenkins, 1953.) Close to *Elsinoë*.
    4'(3) Ascostroma globose to discoid, tapering basally to a narrow foot; ascospores dictyosporous, hyaline to pale brown ..... 5
5(4') Ascostroma globoid with tapering foot seated in leaf stoma ...... **Diplotheca**
    (von Arx, 1963; Dodge, 1939, icon. as *Stevensea*.)
5'(4') Ascostroma discoid to hemiglobose, with foot erumpent through epidermis from intraepidermal hypostroma ..... **Anhellia**
    (von Arx, 1963: 435, icon; Pl. 13, Fig. 1; Verona and Benedek, 1965, C-191.) *Butleria* (von Arx, 1963: 436) is probably a synonym.
    6(3') Ascospores dark brown, dictyosporous; ascostroma white fleshy .... **Cookella**
    (von Arx, 1963: 459, icon.)
    6'(3') Ascospores hyaline or rust red, dictyosporous or rarely phragmosporous; ascostroma bright to dark rust-red ..... **Uleomyces**
    (von Arx, 1963, Pl. 3, Fig. 2; Pl. 4.)

## C. Saccardiaceae

1. Ascostroma differentiated into a fertile discoid portion connected by a tapering foot to a superficial basal stroma; interfascicular tissue paraphysoidal; ascospores hyaline; parasitic on scale insects ..... **Angatia**
    (von Arx, 1963: 430, icon.; Verona and Benedek, 1966, C-231.) Like *Myriangium* in habit but quite different in arrangement of asci.
1'. Ascostroma seated on a superficial mycelium ..... 2
    2(1') Ascospores brown; ascostromata bright to dark; superficial mycelium red-brown to brownish ..... **Dictyonella**
    (von Arx, 1963: 461, 462, icon.; Verona and Benedek, 1965, C-187.)
    2'(1') Ascospores hyaline ..... 3
3(2') Superficial mycelium hyaline or bright colored ..... **Saccardia**
    (von Arx, 1963: 463; Ellis and Everhart, 1892, Pl. 9.)
3'(2') Superficial mycelium brown ..... **Byssogene**
    (von Arx, 1963: 465.)

## D. Saccardinulaceae

1. Ascostromata on human hair; asci ellipsoid, single or in small groups; ascospores nonseptate, hyaline, fusoid, tapering at each end to a short filiform extension ... **Piedraia**
    (von Arx, 1963: 472, icon.)
1'. Ascostromata on leaf hairs and glandular scales of plants ..... 2
    2(') Ascospores nonseptate, globose, hyaline to pale brown; ascostromata globoid-pulvinate, containing a slime mass in which the asci are embedded .. **Hyalotheles**
    (von Arx, 1963: 542; Müller and von Arx, 1962: 827, icon.)
    2'(1') Ascospores septate ..... 3

3(2′) Ascospores 1-septate, hyaline . . . . . . . . . . . . . . . . . . . . . . . **Micularia**
    (von Arx, 1963; Boedijn, 1961: 68 icon.)
3′(2′) Ascospores 3-to-5-septate, often with longitudinal septa in some cells, hyaline to pale brown . . . . . . . . . . . . . . . . . . . . . . . . . . . . . . . . . . . . . . . . 4
    4(3′) Ascostromata grouped on a radiate, superficial, cellular membrane
. . . . . . . . . . . . . . . . . . . . . . . . . . . . . . . . . . . . . . . . **Saccardinula**
        (von Arx, 1963; Arnaud, 1925, Pl. VIII.)
    4′(3′) Radiate membrane lacking . . . . . . . . . . . . . . . . . . . . . . . . . 5
5(4′) Asci 4- to 8-spored . . . . . . . . . . . . . . . . . . . . . . . . . . . . . . **Molleriella**
    (von Arx, 1963: 447 icon.; Boedijn, 1961: 66, icon.; Arnaud, 1925, Pl. VI.)
5′(4′) Asci many-spored . . . . . . . . . . . . . . . . . . . . . . . . . . . . . . . **Xenodium**
    (von Arx, 1963: 451.)

## IX. DOTHIDEALES

Loculoascomycetes producing fascicles of aparaphysate, obclavate to short cylindrical asci in very small, ostiolate, spherical locules in multiloculate, pulvinate to crustose ascostromata (Fig. 5A) or in small perithecioid pseudothecia (Fig. 1G) are included in the central family of the Dothideales, the Dothideaceae. The asci develop synchronously and push up as a group into disintegrating centrum parenchyma. The ostioles are lysigenous or schizogenous and periphysate. These fungi are common and widely distributed on dead leaves and stems. Many that mature on dead leaves cause leaf spot diseases with which their various conidial states are associated. A few produce mature pseudothecia on green leaves or stems or in necrotic lesions on living leaves. At one extreme there is a tendency for the asci to become long cylindrical and more numerous and to form broader, disk-shaped layers. This tendency culminates in the Dothioraceae in apothecioid forms (Figs. 5B, 1B) in which the overlying stroma is ruptured and thrown back to expose the disk of asci. These are distinguished from apothecioid forms in the Hysteriales by the lack of pseudoparaphyses among the asci. At the other extreme the asci become fewer, larger, and asynchronous in development (Fig. 6A). In the Pseudosphaeriaceae the asci are saccate (Fig. 1E) and arise individually in the centrum tissue (Fig. 7A). Even at maturity they may be more or less separated by remnant strands of centrum tissue (Fig. 7B).

Generally the Dothideaceae are easily distinguished from the Pleosporales with their large locules and pseudothecia, persistent pseudoparaphyses, long cylindrical asci, and many-septate ascospores (Figs. 2C, 8B) by the small locules and pseudothecia, the fasciculate, aparaphysate asci, the obclavate to ovate shape of the asci and the usually one-septate ascospores (Fig. 1G). However, in the Dimeriaceae and Venturiaceae of the Pleosporales the ascospores are typically one-septate and the pseudothecia may be small, with the asci reduced in number and pseudoparaphyses sparse. Furthermore, in some Dothideaceae, strands of centrum tissue may persist among the asci and in others the centrum may be largely filled with lateral paraphyses (Fig. 6B). Consequently, uncertainty exists in the placement of borderline genera, and fungi with small pseudothecia, particularly those superficial on leaves, must be sought in both orders. Even in typical genera of the Pleosporaceae, species with small, reduced ascocarps occur. The genera *Leptosphaerulina* (Fig. 1E) and *Wettsteinina* of the Pseudosphaeriaceae resemble *Pyrenophora* (Fig. 3C) of the Pleosporaceae in ascus and ascospore structure and might be interpreted as reduced pleosporaceous fungi.

Grouped around the Dothideaceae are several small and distinctive families of leaf and stem inhabiting, chiefly, tropical and subtropical fungi with small, superficial, perithecioid pseudothecia: the Parodiopsidaceae, Englerulaceae, Trichothyriaceae, Chaetothyriaceae, and Capnodiaceae. These must be distinguished from fungi similar in habit in the Dimeriaceae and Venturiaceae of the Pleosporales, the Phillipsiellaceae of the Hysteriales, the Saccardinulaceae of the Myriangiales, and the black mildews in the Meliolaceae of the Pyrenomycetes. The Capnodiaceae and Chaetothyriaceae have ascocarps apparently similar in internal structure to those of the Dothideaceae. The Capnodiaceae are distinguished by habit. These are sooty mold fungi growing epiphytically on insect honey dew and producing dark wooly mats or spongy masses of mycelium easily stripped from the substrate. They have a variety of pycnidial and hyphomycetous conidial states. Many species are common in temperate regions, but often only the conidial states are present on the mats. In the Chaetothyriaceae, the hyaline to brown mycelium forms a thin pellicle. The globose to flattened pseudothecia develop beneath the pellicle, and the upper surface of the pseudothecium is fused with the overlying pellicle. This structure is easier described than determined. Consequently, confusion persists in the distribution of fungi between the Capnodiaceae and Chaetothyriaceae and also in the delimitations of genera within each of these families. Genera of Chaetothyriaceae with flattened pseudothecia also may be confused with Hemisphaeriales.

The Parodiopsidaceae, Englerulaceae, and Trichothyriaceae resemble the Pseudophaeriaceae in internal structure, with the asci more or less separated by remnants of centrum tissue. The Parodiopsidaceae and Englerulaceae are distinguished from Dothideaceae by their very few, broad clavate to oblong asci and globose pseudothecia which open in the Parodiopsidaceae by disintegration at the apex to form a broad pore and in the Englerulaceae by dissolution of most of the ascocarp into a slime mass.

Although the Trichothyriaceae have been confused with Hemisphaeriales, they are distinguished from all ascomycetes by their lenticular pseudothecia with both upper and lower walls formed by plates of radially arranged cells.

### KEY TO FAMILIES AND IMPORTANT GENERA OF DOTHIDEALES

1. Ascocarp a perithecioid or lenticular pseudothecium containing a relatively few asci arising individually and separated by strands of interascicular stromal tissue which may be more or less persistent at maturity; pseudothecium opening by an apical pore or disintegrating into a slime mass; mostly superficial on leaves, rarely innate . . . . . . . 2

1'. Ascocarp a perithecioid, or rarely apothecioid, pseudothecium containing asci in araphysate fascicles in small locules or rarely in extended discoid layers; pseudothecia separate and usually innate (except in sooty mold fungi in Capnodiaceae) or grouped on or immersed in a crustose to pulvinate stroma . . . . . . . . . . . . . . . . . . . . . . . . . . . 6

    2(1) Pseudothecium lenticular, composed of upper and lower radiate plates or beneath a membranous pellicle and depressed globose to globose . . . . . . . . . . . . 3

    2'(1) Pseudothecium neither lenticular nor beneath a mycelial membrane, containing a few, relatively large, ovate to broad-clavate, erect asci . . . . . . . . . . . . 4

3(2) Pseudothecium lenticular, composed of upper and lower, radiately structured plates fused at the periphery, arising from the surface of the superficial mycelium or at the tip of a single stalk hypha; asci semiprostrate, with their tips directed toward a pore in the center of the upper plate; ascospores usually 1-septate, hyaline, rarely with additional transverse septa or rarely brown; hyperparasites on superficial fungi
. . . . . . . . . . . . . . . . . . . . . . . . . . . . . . **Trichothyriaceae** p. 168

3'(2) Pseudothecium depressed globose or strongly flattened, beneath, and in the apical region fused with, a covering shieldlike membrane developed from the superficial mycelium; mycelium usually pale brown and inconspicuous, rarely dark, asci erect or converging toward the apical pore; ascospores usually phragmosporous or dictyosporous, rarely 1-septate; ectocommensals on the cuticle of vascular plants . **Chaetothyriaceae** p. 168

    4(2') Pseudothecia superficial; parasites on living leaves . . . . . . . . . . . . . 5

    4'(2') Pseudothecia innate, usually in dead leaf or herbaceous stem tissue
. . . . . . . . . . . . . . . . . . . . . **Pseudosphaeriaceae** p. 171

5(4) Pseudothecia remaining intact, opening by a broad pore or crumbling at the apex
. . . . . . . . . . . . . . . . . . . . . . . . . **Parodiopsidaceae** p. 169

5'(4) Pseudothecia disintegrating into a slimy mass around the asci at maturity, ascospores 1-septate . . . . . . . . . . . . . . . . . . . . . . . . . **Englerulaceae** p. 170

    6(1') Pseudothecia perithecioid, opening by an apical ostiole which is frequently lined with periphyses; asci grouped in a fascicle . . . . . . . . . . . . . . . . . . . . 7

    6'(1') Pseudothecia apotheciod; asci in a continuous flat disk, exposed by rupture of the overlying stromal tissue of the pulvinate, orbicular to oblong, erumpent, ascocarp . . . . . . . . . . . . . . . . . . . . . . . . **Dothioraceae** p. 175

7(6) Pseudothecia on a dark, wide-spreading, superficial mycelium or subiculum; mycelium often composed of moniliform or agglutinated hyphae, usually easily stripped from the epidermis but penetrating and adhering to phellem of older branches; with a variety of conidial states; sooty mold fungi saprotrophic in insect honeydew on leaves and stems . . . . . . . . . . . . . . . . . . . . . . . . . . **Capnodiaceae** p. 171

7′(6) Pseudothecia innate or erumpent or on or immersed in a crustose to pulvinate, superficial, erumpent, or innate stroma . . . . . . . . . . . . . . **Dothideaceae** p. 173

## A. Trichothyriaceae

1. Mycelium forming membranous bands of parallel hyphae; ascospores rarely more than 1-septate or brown . . . . . . . . . . . . . . . . . . . . . . . . . . . **Trichothyrium**
(Tim, 1971; Müller and von Arx, 1962: 556, icon.; Arnaud, 1918, Pl. XXVI; Hughes, 1953; Verona and Benedek, 1968, C-321.) *Trichothyriopsis* may be used for species with 2-septate ascospores and *Trichothyriella* or *Trichothyriomyces* for those with brown ascospores (Müller and von Arx, 1962: 555).

1′. Superficial mycelium of separate hyphae, inconspicuous . . . . . . . . . . . . . . 2

    2(1′) Pseudothecia glabrous . . . . . . . . . . . . . . . . . . . . . . **Trichothyrina**
    (Müller and von Arx, 1962: 558.)

    2′(1′) Pseudothecia with hyphal hairs or setae around the pore . . . . . . . . . . 3

3(2′) Pseudothecia with hypal hairs . . . . . . . . . . . . . . . . . . . **Trichothyrinula**
    (Müller and von Arx, 1962: 560.)

3′(2′) Pseudothecia setose, ascospores often becoming several septate . . . . **Actinopeltis**
(Arnaud, 1918, icon. as *Mycoangloisia* Pl. XXVII.) *Mycoangloisia* with 1-septate ascospores and *Actinopeltis* with phragmosporous ascospores were considered congeneric by Müller and von Arx (1962: 561).

## B. Chaetothyriaceae

1. Ascospores 1-septate, hyaline to pale brown or pale olivaceous, ellipsoid . . . . . . . 2

1′. Ascospores more than 1-septate, ellipsoid, fusoid, to long cylindrical . . . . . . . . 3

    2(1) Mycelium setose over or beyond pseudothecia . . . . . . . . . . . **Microcallis**
    (Müller and von Arx, 1962: 552, icon.; Verona and Benedek, 1961, C-63.)

    2′(1) Mycelium lacking setae . . . . . . . . . . . . . . . . . . . . . . **Akaropeltis**
    (Müller and von Arx, 1962: 553.)

3(1′) Ascospores with transverse septa . . . . . . . . . . . . . . . . . . . . . . . . . 4

3′(1′) Ascospores with transverse and longitudinal septa . . . . . . . . . . . . . . . . 8

    4(3) Ascospores ellipsoid to cylindrical . . . . . . . . . . . . . . . . . . . . . 5

    4′(3) Ascospores filiform, hyaline . . . . . . . . . . . . . . . . . . . . . . . . 7

5(4) Ascospores hyaline . . . . . . . . . . . . . . . . . . . . . . . . . . . . . . . . 6

5′(4) Ascospores brown, mycellium lacking setae . . . . . . . . . . . . **Skoteinospora**
(Batista and Ciferri, 1962: 103, icon.) The mycelium may be dark and conspicuous.

    6(5) Mycelium setose over or beyond pseudothecia . . . . . . . . . **Chaetothyrium**
    (Hansford, 1946: 146, icon.; Dennis, 1968: 368, Pl. XXXVI E; Verona and Benedek, 1961, C-46.)

    6′(5) Mycelium lacking setae . . . . . . . . . . . . . . . . . . . . . . . **Ciferriusia**
    (Batista and Ciferri, 1962: 18, 19, 20–21, icon.)

    7(4′) Mycelium setose over or beyond pseudothecia . . . . . . . . . . . **Actinocymbe**
    (Hansford, 1946: 159.)

    7′(4′) Mycelium lacking setae . . . . . . . . . . . . . . . . . . . . . . . **Limaciniella**
    (Batista and Ciferri, 1962: 22–23.) It is doubtful that the elongated shape of the ascospores is an adequate basis for separating *Limaciniella* from *Skoteinospora*.

## 7. Loculoascomycetes

8(3') Ascospores hyaline . . . . . . . . . . . . . . . . . . . . . . . . . 9
8'(3') Ascospores brown, mycelium lacking setae . . . . . . . . . **Phaeosaccardinula**
  (Hansford, 1946: 155, 156, icon.)
9(8) Mycelium setose over or beyond pseudothecia . . . . . . . . . . . **Treubiomyces**
  (Batista and Ciferri, 1962: 29, icon.)
9'(8) Mycelium lacking setae . . . . . . . . . . . . . . . . . . . . . . . **Ainsworthia**
  (Batista and Ciferri, 1962: 5–6, 7, icon.) *Ainsworthia* Batista & Ciferri, 1962; non Bossier, 1844 (Umbelliferae) is probably synonymous with *Phaeosaccardinula*.

### C. Parodiopsidaceae

1. Ascospores nonseptate, hyaline . . . . . . . . . . . . . . . . . . . . . . . 2
1'. Ascospores septate . . . . . . . . . . . . . . . . . . . . . . . . . . . . 3
  2(1) Pseudothecia on a superficial, hyphopodiate mycelium . . . . . **Cleistosphaeria**
    (von Arx and Müller, 1954: 82; icon.; Arnaud, 1923, Pl. 1.)
  2'(1) Pseudothecia on small thin parenchymatous or plectenchymatous stromata in superficial mycelium, usually with erect hairs or setae on stromata and mycelium
    . . . . . . . . . . . . . . . . . . . . . . . . . . . . . . . . . . . **Pilgeriella**
    (von Arx and Müller, 1954: 80, icon.; Verona and Benedek, 1966, C-229.)
3(1') Ascospores 1-septate (sometimes more than 1-septate in *Perisporiopsis*) . . . . . . 4
3'(1') Ascospores more than 1-septate . . . . . . . . . . . . . . . . . . . . . . . 13
  4(3) Pseudothecia subcuticular, adnate above to raised cuticle, on dendritic subcuticular mycelium; ascospores small (less than 30 μm), dark brown . . . . . . . . **Jaffuela**
    (Müller and von Arx, 1962: 29, icon.) Perhaps better placed in the Dothideaceae.
  4'(3) Pseudothecia superficial, often connected with an innate hypostroma or mycelium; superficial mycelium usually present . . . . . . . . . . . . . . . . . . . . . 5
5(4') Pseudothecia grouped on small basal stromata; superficial mycelium, if present, lacking hyphopodia . . . . . . . . . . . . . . . . . . . . . . . . . . . . . . 6
5'(4') Pseudothecia not on basal stroma, scattered or grouped on superficial mycelium; except in *Balladyna*, hyphopodia lacking . . . . . . . . . . . . . . . . . . . . . . . 8
  6(5) Superficial mycelium present; pseudothecia on small basal stroma anchored over stomata; ascospores brown . . . . . . . . . . . . . . . . . . . **Stomatogene**
    (Müller and von Arx, 1962: 183, icon.)
  6'(5) Superficial mycelium lacking, stromata erumpent . . . . . . . . . . . . . 7
7(6') Basal stromata erumpent from a subcuticular mycelial membrane; ascospores 1-septate, brown . . . . . . . . . . . . . . . . . . . . . . . . . **Kusanobotrys**
  (Müller and von Arx, 1962: 180, icon.)
7'(6') Basal stroma immersed in leaf tissue, erumpent at apex; ascospores initially 1-septate, hyaline, becoming phragmosporous and brown . . . . . . . . . . . **Chevalieropsis**
  (Müller and von Arx, 1962: 178; Arnaud, 1921, Plate III, as *Chevalieria*.)
  8(5') Pseudothecium borne on erect stalk cell or hypha; ascospores small, less than 30 μm . . . . . . . . . . . . . . . . . . . . . . . . . . . . . . . . . . 9
  8'(5') Pseudothecium seated in the mycelium . . . . . . . . . . . . . . . . . 11
9(8) Innate mycelium intracellular, forming hypostroma from which superficial mycelium is erumpent; ascospores small, less than 15 μm, becoming brownish . . . . . . . **Alina**
  (Müller and von Arx, 1962: 182, icon.)

9'(8) Innate mycelium lacking or sparse, intercellular, usually forming haustoria; ascospores middle-sized, brown . . . . . . . . . . . . . . . . . . . . . . . . . . . . 10

    10(9') Superficial mycelium with lateral hyphopodia . . . . . . . . . . **Balladyna**
    (Müller and von Arx, 1962: 184, icon.; Arnaud, 1918, Pl. XLII.)

    10'(9') Superficial mycelium lacking hyphopodia . . . . . . . . . . . **Dysrhynchis**
    (Müller and von Arx, 1962: 190, icon.; Verona and Benedek, 1966, C-242.)

11(8') Pseudothecia fused laterally into a crust, seated on a thick dark subiculum; ascospores small, less than 30 μm, brown with a light band around each cell . . . . . **Neoparodia**
(Müller and von Arx, 1962: 177.)

11'(8') Pseudothecia separate on the superficial mycelium; ascospores large, brown . . . 12

    12(11') Hypostroma lacking; innate mycelium sparse, intercellular, connected with superficial mycelium by stomatopodia . . . . . . . . . . . . . . . **Perisporiopsis**
    (Müller and von Arx, 1962: 172, icon.; Arnaud, 1918, Fig. 11; as *Parodiopsis* Pl. III, IV.)

    12'(11') Hypostroma present, intracellular, erumpent to form superficial mycelium . . . . . . . . . . . . . . . . . . . . . . . . . . . . . . . . . . . . . **Scolionema**
    (Müller and von Arx, 1962: 176.)

13(3') Ascospores long-fusoid, pale brown; superficial mycelium setose, forming stomatopodia and internal, intercellular nonhaustoriate mycelium . . . . . . . . . . . **Nematostigma**
(Hansford, 1946: 21, 22, icon.; Verona and Benedek, 1968, C-314.)

13'(3') Ascospores cylindric-filiform, subhyaline to pale brown; superficial mycelium lacking spines and hyphopodia . . . . . . . . . . . . . . . . . . . . . . . . . **Ophioparodia**
(Petrak and Ciferri, 1932: 223.)

## D. Englerulaceae

1. Superficial mycelium lacking hyphopodia . . . . . . . . . . . . . . . . . . . 2

1'. Mycelium with lateral hyphopodia . . . . . . . . . . . . . . . . . . . . . . . 3

    2(1) Pseudothecia seated on the superficial mycelium, wall 1-cell thick, early disintegrating into slime mass; mycelium penetrating stomata and forming haustoria or a sparse internal mycelium; conidia lacking or in pycnidia . . . . . . . . . . . **Englerula**
    (Müller and von Arx, 1962: 155, icon.; Verona and Benedek, 1961, C-76.)

    2'(1) Pseudothecia seated singly or in small groups on flat stromata formed by superficial mycelium, globoid, wall composed of several layers of small cells; conidia on long, stout conidiophores from basal stromata . . . . . . . . . . . . . . . **Allosoma**
    (Müller and von Arx, 1962: 222, icon.; Verona and Benedek, 1965, C-190.)

3(1') Supporting hyphae arising from the basal cushion and curving around slime mass containing 1–6 asci . . . . . . . . . . . . . . . . . . . . . . . . . **Parenglerula**
(Müller and von Arx, 1962: 164, icon.; Arnaud, 1918, as *Englerulaster* Pl. XXXVIII; Verona and Benedek, 1967, C-286.)

3'(1') Supporting hyphae lacking . . . . . . . . . . . . . . . . . . . . . . . . . 4

    4(3') Pseudothecium at the tip of a single erect stalk cell, usually containing a single ascus . . . . . . . . . . . . . . . . . . . . . . . . . . . . . . . . . **Thrauste**
    (Müller and von Arx, 1962: 167, icon.; Verona and Benedek, 1968, C-339.)

    4'(3) Pseudothecium seated on the mycelium or on the substrate, not on stalk cell . . 5

5(4') Mycelium mostly thin and delicate; pseudothecium with a narrowed parenchymatous base; conidia lacking or nonseptate and in pycnidia . . . . . . . . . . **Rhytidenglerula**

(Müller and von Arx, 1962: 157, icon.; Verona and Benedek, 1967, as *Dialacenium* C-285; 1968, C-340.)

5'(4') Mycelium stout and thick-walled; pseudothecium with broad flattened base, dissolving almost entirely; conidia on mycelium, elongated, curved, 2–3 septate, or rounded, dark, and multicellular . . . . . . . . . . . . . . . . . . . . . . . . . . . . . . . . **Schiffnerula**
(Müller and von Arx, 1962: 161, icon.; Verona and Benedek, 1968, C-318.)

## E. Pseudosphaeriaceae

1. Ascospores nonseptate, small, hyaline . . . . . . . . . . . . . . . . **Montagnellina**
(von Arx and Müller, 1954: 61, icon.)

1'. Ascospores septate . . . . . . . . . . . . . . . . . . . . . . . . . . . . . . . . . 2

2(1') Ascospores dictyosporous, with 3 or more cross septa and a longitudinal septum in one or more cells, some spores phragmosporous and usually so in *L. trifolii*, hyaline . . . . . . . . . . . . . . . . . . . . . . . . . . . . . . . **Leptosphaerulina**
(Graham and Luttrell, 1961; Dennis, 1968: 357, Fig. 21 E, as *Pseudoplea*.)

2'(1') Ascospores 1-several-septate, hyaline . . . . . . . . . . . . . . . . . . . . 3

3(2') Ascospores usually 1-septate, shorter than 30 μm; pseudothecia small, at most 120 μm diam., depressed globose . . . . . . . . . . . . . . . . . . . . . . . **Monascostroma**
(Müller and von Arx, 1962: 273, icon.)

3'(2') Ascospores usually phragmosporous, large; pseudothecia larger, often short conical toward apex . . . . . . . . . . . . . . . . . . . . . . . . . . . . . . . . . **Wettsteinina**
(Müller and von Arx 1962: 270, icon.; Dennis, 1968: 357, Fig. 10 G; Verona and Benedek, 1968, C-343.)

## F. Capnodiaceae

1. Ascospores, 1-septate . . . . . . . . . . . . . . . . . . . . . . . . . . . . . . . . 2

1'. Ascospores more than 1-septate . . . . . . . . . . . . . . . . . . . . . . . . . . 6

2(1) Ascospores brown; mycelium cottony, with moniliform hyphae; pseudothecia globose, glabrous, sessile . . . . . . . . . . . . . . . . . . . . . **Archaetobotrys**
(Batista and Ciferri, 1963: 50, icon.) *Kusanotheca* (Batista and Ciferri, 1963: 117) is possibly synonymous.

2'(1) Ascospores hyaline . . . . . . . . . . . . . . . . . . . . . . . . . . . . . . . . 3

3(2') Pseudothecia on tips of single erect hyphae, lacking setae . . . . . . **Capnogonium**
(Batista and Ciferri, 1963: 105.) A questionable genus possibly congeneric with *Calyptra*.

3'(2') Pseudothecia not on tips of single erect hyphae . . . . . . . . . . . . . . . . 4

4(3') Mycelium with erect, monopodially branched, sterile hyphae and erect or decumbent synnemata bearing globose pseudothecia and pycnidia . . . . . . . . . **Laterotheca**
(Batista and Ciferri, 1963: 119, icon.) A questionable genus possibly based on discordant elements.

4'(3') Erect sterile hyphae lacking . . . . . . . . . . . . . . . . . . . . . . . . . . 5

5(4') Pseudothecia setose . . . . . . . . . . . . . . . . . . . . . . . . . . **Capnobatistia**
(Batista and Ciferri, 1963: 75, icon.)

5'(4') Pseudothecia lacking setae . . . . . . . . . . . . . . . . . . . . . . . . **Calyptra**
(Batista and Ciferri, 1963, icon. p. 74). *Calyptra* resembles *Scorias* except for its

1-septate ascospores. *Capnocrinum* (Batista and Ciferri, 1963: 78; Verona and Benedek, 1965, C-178), is possibly synonymous.

6(1') Ascospores phragmosporous . . . . . . . . . . . . . . . . . . . . . . . 7

6'(1') Ascospores dictyosporous . . . . . . . . . . . . . . . . . . . . . . . 13

7(6) Ascospores hyaline . . . . . . . . . . . . . . . . . . . . . . . . . . . 8

7'(6) Ascospores brown . . . . . . . . . . . . . . . . . . . . . . . . . . . 10

    8(7) Pseudothecia bearing warty outgrowths or tapering, spinose setae . . . . . . 9

    8'(7) Pseudothecia lacking warts and setae, glabrous or with hyphal outgrowths, globose to elliptical, sessile, stalked, or on agglutinated synnemata . . . . . . . . **Scorias** (Batista and Ciferri 1963: 190, icon.; Ellis and Everhart 1892, Pl. 10.) Possible synonyms are *Antennella* (Batista and Ciferri 1963: 58); *Antennelina* (Batista and Ciferri 1963: 60; Verona and Benedek 1965, C-181); *Capnociferria* (Batista and Ciferri 1963: 76; Verona and Benedek, 1965, C-178); *Hyalocapnias* (Batista and Ciferri, 1963: 114; Verona and Benedek, 1965, C-182); *Uloseia* (Batista and Ciferri, 1963: 224); and *Xystozukalia* (Batista and Ciferri, 1963: 225; Verona and Benedek, 1965, C-183).

9(8) Pseudothecia with warty, parenchymic outgrowths, especially near the apex
. . . . . . . . . . . . . . . . . . . . . . . . . . . . . . . . . . . . . . . **Blastocapnias**
(Batista and Ciferri, 1963: 68–69, icon.)

9'(8) Pseudothecia setose . . . . . . . . . . . . . . . . . . . . . . . . **Trichomerium**
(Batista and Ciferri, 1963: 198, icon.; Verona and Benedek, 1964, C-161.) Possible synonyms *Antennellopsis* (Batista and Ciferri, 1963: 61; Verona and Benedek, 1965, C-181); *Chaetopotius* (Batista and Ciferri, 1962: 110).

10(7') Pseudothecia on erect hyphae . . . . . . . . . . . . . . . . . . . . . 11

10'(7') Pseudothecia not on erect hyphae, sessile or on broad stipes in the mycelium
. . . . . . . . . . . . . . . . . . . . . . . . . . . . . . . . . . . . . . . 12

11(10) Pseudothecia terminal on short, simple, erect hyphae; with long, simple, reticulately sculptured hyphae bearing lateral helicoid conidia . . . . . . . . . . . . **Brooksia**
(Deighton and Pirozynski, 1966: 4, 6, icon.; Verona and Benedek, 1967, C-283)

11'(10) Pseudothecia lateral on erect branched hyphae . . . . . . . . . . **Acrogenotheca**
(Batista and Ciferri, 1963: 53–55, icon.; Deighton and Pirozynski, 1966: 4, 6, icon.; Hughes, 1967b: 506–513, icon.; Verona and Benedek, 1965, C-184.)

    12(10') Pseudothecia with tapering spinose setae . . . . . . . . . . . . . **Setella**
Syn. *Morfea* (Batista and Ciferri, 1963: 148, 149, icon.; Verona and Benedek, 1965, C-179.)

    12'(10') Pseudothecia lacking setae, glabrous or with hyphal outgrowths, sessile, stalked, or on agglutinated synnemata . . . . . . . . . . . . . . . . . . . . . **Limacinia**
Possible synonyms: *Aithalomyces, Metacapnodium, Phaeocapnias, Ophiocapnocoma, Limacinia* has been generally (Barr, 1955: 503, icon.) and, perhaps incorrectly, employed for species distinguished from *Scorias* by their brown ascospores. Intergradations would make it difficult to separate species with long-cylindrical ascospores in *Ophiocapnocoma* (Batista and Ciferri, 1963: 166; Hughes, 1967a; Verona and Benedek, 1965, C-180) from those with ellipsoid to fusoid ascospores. Hughes, fide Corlett, 1970, apparently would divide this group among *Metacapnodium* (Dennis, 1968: 367, Fig. 30 H) with mycelial hyphae tapering and *Aithalomyces* with mycelial hyphae cylindrical and uniform in diameter.

13(6') Ascospores hyaline . . . . . . . . . . . . . . . . . . . . . . . . . . . 14

13'(6') Ascospores brown . . . . . . . . . . . . . . . . . . . . . . . . . . . 15
    14(13) Pseudothecia sessile, collapsing at the apex when mature and dry
    . . . . . . . . . . . . . . . . . . . . . . . . . . . . . . . . . . . . . **Limacinula**
    (Batista and Ciferri, 1963: 134; Verona and Benedek, 1967, C-302; Reynolds, 1971.)
    14'(13) Pseudothecia stalked, not collapsing . . . . . . . . . . . **Paracapnodium**
    (Batista and Ciferri, 1963: 172, icon.)
15(13') Mycelium forming an orbicular to branched-elongated, membranous, setose thallus of radially arranged hyphae; pseudothecia subglobose, seated on the thallus; ascospores mucronate at one or both ends . . . . . . . . . . . . . . . . . . . . . **Trichopeltheca**
    (Hughes, 1965; 330, icon.)
15'(13') Mycelium forming a wooly mat, pseudothecia sessile or stalked . . . . . . . 16
    16(15') Pseudothecia with tapering, spinose setae . . . . . . . . . . . **Capnophaeum**
    (Batista and Ciferri, 1963: 109, icon.)
    16'(15') Pseudothecia not setose, glabrous or bearing hyphal outgrowths
    . . . . . . . . . . . . . . . . . . . . . . . . . . . . . . . . . . . . . **Capnodium**
    (Batista and Ciferri, 1963: 103, icon.; Arnaud, 1925: 661, Fig. 8; Verona and Benedek, 1964, C-160; Dennis, 1968: 367, Fig. 17 E; Ellis and Everhart, 1892, Pl. 10.) Possible synonyms are *Capnocrinum* (Batista and Ciferri, 1963: 78; Verona and Benedek, 1965, C-178) and *Capnodaria* (Batista and Ciferri, 1963: 80). *Naetrocymbe* (Batista and Ciferri, 1963: 155; Verona and Benedek, 1964, C-170) is a questionable genus of uncertain position.

## G. Dothideaceae

1. Ascospores nonseptate . . . . . . . . . . . . . . . . . . . . . . . . . . . . . . 2
1'. Ascospores septate . . . . . . . . . . . . . . . . . . . . . . . . . . . . . . . 6
    2(1) Asci broad-clavate to ovoid, on long, slender stalks of varying lengths and standing at differing heights in the locule; pseudothecia separate, innate in leaf tissue; ascospores brown . . . . . . . . . . . . . . . . . . . . . . . . . . . . . **Vestergrenia**
    (von Arx and Müller, 1954: 76, 77, icon.; Verona and Benedek, 1966, C-230.)
    2'(1) Asci short-stalked, clavate . . . . . . . . . . . . . . . . . . . . . . . . . 3
3(2') Ascospores brown; pseudothecia immersed locules in a pulvinate, erumpent stroma connected by a narrowed basal foot to an innate hypostroma; saprotrophic on bark or wood . . . . . . . . . . . . . . . . . . . . . . . . . . . . . . . . . . **Auerswaldia**
    (von Arx and Müller, 1954: 63, 64, icon.; Verona and Benedek, 1970, C-489.)
3'(2') Ascospores hyaline to pale brown; pseudothecia separate . . . . . . . . . . . . 4
    4(3') Pseudothecia with ostiole periphysate and centrum filled with lateral paraphyses; asci polysporous . . . . . . . . . . . . . . . . . . . . . . . . . . . **Plurisperma**
    (Sivanesan, 1970a, Pl. 47.)
    4'(3') Pseudothecia with ostiole lacking periphyses and centrum parenchymatous, asci 8-spored . . . . . . . . . . . . . . . . . . . . . . . . . . . . . . . . . . . . 5
5(4') Asci arising from the base of the locule; pseudothecia usually in living or dead leaves or herbaceous stems . . . . . . . . . . . . . . . . . . . . . . . . . . **Guignardia**
    (von Arx and Müller, 1954: 44, 50, 51, icon.; Verona and Benedek, 1966, C-248; Reusser, 1964.)
5'(4') Asci arising from a cylindrical placenta projecting upward from the base of the locule;

on woody stems . . . . . . . . . . . . . . . . . . . . . . . . . . **Columnosphaeria**
(Munk, 1957: 473, icon.) von Arx and Müller (1954: 44, 50) considered *Columnosphaeria* a synonym of *Guignardia*.

6(1') Ascospores 1-septate . . . . . . . . . . . . . . . . . . . . . . . . . . . . 7

6'(1') Ascospores more than 1-septate . . . . . . . . . . . . . . . . . . . . 20

7(6) Pseudothecia separate; ascospores septate near the middle; hyaline, or finally brownish . . . . . . . . . . . . . . . . . . . . . . . . . . . . . . . . . . . . 8

7'(6) Pseudothecia seated on or immersed in a stroma . . . . . . . . . . . . . . . . . 9

    8(7) Asci 8-spored; pseudothecia small, immersed in host tissue (usually on dead leaves)
. . . . . . . . . . . . . . . . . . . . . . . . . . . . . . . . . . . . . **Mycosphaerella**
(Müller and von Arx, 1962: 356, icon.; Verona and Benedek, 1962, C-90; Kohlmeyer and Kohlmeyer, 1967, Table 65; 1968, Table 76; Dennis, 1968: 362, Fig. 21 A, B, C.)

    8'(7) Asci polysporous; pseudothecia larger, erumpent, globose or pulvinate and laterally expanded . . . . . . . . . . . . . . . . . . . . . . . . . . . . . . . **Delphinella**
Müller and von Arx (1962: 26, icon.) united in *Delphinella* species with broad disk-shaped locules and those with spherical locules usually placed in *Rehmiellopsis*. (Waterman, 1945: 320, 322, 323, icon. as *Rehmiellopsis*; Dennis, 1968: 366, icon. as *Rehmiellopsis* Pl. XXXVII C.)

9(7') Stroma immersed in substrate . . . . . . . . . . . . . . . . . . . . . . . 10

9'(7') Stroma erumpent or superficial . . . . . . . . . . . . . . . . . . . . . . . 15

    10(9) Ascospores septate near the middle, hyaline to pale yellowish . . . . . . . 11

    10'(9) Ascospores septate in the lower third, hyaline . . . . . . . . . . . . . 14

11(10) Stroma subcuticular, containing small locules with few asci . . . . . . **Euryachora**
(Müller and von Arx, 1962: 387; Müller and Ahmad, 1955: 237, icon.)

11'(10) Stroma intraepidermal or deeper . . . . . . . . . . . . . . . . . . . . . . 12

    12(11') Stromata thin, orbicular or irregular crusts on which erumpent pseudothecia are seated . . . . . . . . . . . . . . . . . . . . . . . . . . . . . **Cymadothea**
(Wolf, 1935: 70, icon.)

    12'(11') Stroma elongated, elliptical to linear . . . . . . . . . . . . . . . . . 13

13(12') Stroma dark-colored, with partially erumpent uniloculate pseudothecia along the apex . . . . . . . . . . . . . . . . . . . . . . . . . . . . . . . . . . . **Scirrhia**
(Müller and von Arx, 1962: 380, icon.; Obrist, 1959: 372, 374, 376, 378, icon.; Verona and Benedek, 1963, C-153; Dennis, 1968: 360, Fig. 22 J.)

13'(12') Stroma bright yellow, waxy; locules deeply immersed, with long ostioles; on green leaves . . . . . . . . . . . . . . . . . . . . . . . . . . . . . . . . . **Cerodothis**
(Muthappa, 1969: 738, icon.; Verona and Benedek, 1970, C-488.)

    14(10') Stroma subcuticular, crustlike, containing small immersed locules with few asci . . . . . . . . . . . . . . . . . . . . . . . . . . . . . . . . . **Omphalospora**
(Müller and von Arx, 1962: 389, icon.)

    14'(10') Stroma intraepidermal and deeper, formed by intracellular hyphae in hypertrophied parts of living leaves; locules larger, immersed . . . . . . **Achorodothis**
(Müller and von Arx, 1962: 391, icon.)

15(9') Stroma superficial, connected with a subcuticular hypostroma . . . . . . . . . 16

15'(9') Stroma erumpent from deeper in the substrate, pulvinate with immersed locules
. . . . . . . . . . . . . . . . . . . . . . . . . . . . . . . . . . . . . . . . . 17

16(15) Stroma sclerotiumlike, with separate pseudothecia attached around margins, connected to hypostroma by erect hyphae; ascospores septate in upper third, brown . . . . . . . . . . . . . . . . . . . . . . . . . . . . . . . . . . . . . . . **Lasiobotrys**
(Müller and von Arx, 1962; 462, icon.; Verona and Benedek, 1968: C-333; Dennis, 1968: 379, Pl. XXXVIII A.)

16'(15) Stroma thick, disk-shaped, with locules immersed in margins and horizontally oriented toward periphery, covered with setae or hyphal hairs, connected to hypostroma by a central foot; ascospores septate in middle, hyaline to pale brownish . . . . . . . . . . . . . . . . . . . . . . . . . . . . . . . . . . . . . . . . . . . . . . . . . **Rhizogene**
(Müller and von Arx, 1962: 640, icon.)

17(15') Stroma broadly erumpent, attached over entire base; on stems . . . . . . . . 18

17'(15') Stroma tapering basally to centrally attached foot; on leaves . . . . . . . . 19

    18(17) Ascospores brown . . . . . . . . . . . . . . . . . . . . . . . . . . . **Dothidea**
    (Müller and von Arx, 1962: 385, icon.; Verona and Benedek, 1963, C-144; 1968, C-336; Dennis, 1968: 358, Fig. 15 K, 18 C.)

    18'(17) Ascospores hyaline . . . . . . . . . . . . . . . . . . . . . . . . . . **Plowrightia**
    (Dennis, 1968: 362, Fig. 15 L.) Loeffler (1957) considered *Plowrightia* synonymous with *Dothidea*.

19(17') Ascospores hyaline . . . . . . . . . . . . . . . . . . . . . . . . . . . **Microcyclus**
(Müller and von Arx, 1962: 371, 372, 373, icon.)

19'(17') Ascospores yellowish to pale brown . . . . . . . . . . . . . . . **Coccoidella**
(Müller and von Arx, 1962: 450, 452, icon.; Verona and Benedek, 1969, C-366.)

    20(6') Ascospores phragmosporous, hyaline; pseudothecia separate, immersed in substrate; asci fasciculate . . . . . . . . . . . . . . . . . . . . . . . . **Sphaerulina**
    (Dennis, 1968: 365, Fig. 21 F; Verona and Benedek, 1969, C-348; Ellis and Everhart, 1892, Pl. 27.)

    20'(6') Ascospores dictyosporous; pseudothecia separate, immersed in substrate; asci on a pulvinate placenta . . . . . . . . . . . . . . . . . . . . . . . . . . 21

21(20') Ascospores hyaline . . . . . . . . . . . . . . . . . . . . . . . . . **Pringsheimia**
(Müller, 1957: 459, icon.; Müller and von Arx, 1950: 379, icon.; Verona and Benedek, 1970, C-465; Dennis, 1968, as *Saccothecium*: 365, Fig. 21 D.)

21'(20') Ascospores brown . . . . . . . . . . . . . . . . . . . . . . . . . **Phaeodothiora**
(Petrak, 1948b.)

## H. Dothioraceae

1. Ascospores nonseptate, hyaline . . . . . . . . . . . . . . . . . . . . . . . . . . . 2

1'. Ascospores septate, hyaline . . . . . . . . . . . . . . . . . . . . . . . . . . . . 3

    2(1) Pseudothecia erumpent, pulvinate, parenchymatous in upper part and containing a single broad layer of asci separated by chains of vertically arranged cells; ascus tips released by crumbling of overlying stroma; on dead stems . . . . **Bagnisiella**
    (von Arx and Müller, 1954: 26; Theissen and Sydow, 1915, icon. Pl. II, Fig. 9, 10.) Although *Bagnisiella* has often figured in discussions of phylogeny, its structure is poorly understood. Species of *Botryosphaeria* and *Auerswaldia* (Pleosporales) have been referred erroneously to *Bagnisiella*. Species of *Bagnisiella* described by Tilak (1963) and Tilak and Rao (1966) possibly belong in *Auerswaldia*.

    2'(1) Pseudothecia subepidermal to erumpent, crateriform, made up of rows of vertically arranged cells; entire upper surface of stroma thrown off to expose the disk-shaped layer of aparaphysate asci; on leaf lesions . . . . . . . . . . . . . . **Hypnotheca** (Tommerup, 1970: 470, 472, icon., Pl. 28.) *Hypnotheca* seems most closely related in its internal structure and its parasitic habit to *Moeszopeltis*, Leptopeltidaceae, Hemisphaeriales.

3(1′) Asci containing 8 spores . . . . . . . . . . . . . . . . . . . . . . . . . . . . . . 4

3′(1′) Asci containing 16 or more spores . . . . . . . . . . . . . . . . . . . . . . . 5

    4(3) Ascospores phragmosporous . . . . . . . . . . . . . . . . . . **Leptodothiora** (Müller, 1955: 218.)

    4′(3) Ascospores dictyosporous . . . . . . . . . . . . . . . . . . . . . . . . **Dothiora** (Müller and von Arx, 1950: 377, icon.; Luttrell, 1960: 70, icon.; Verona and Benedek, 1970, C-486.)

5(3′) Ascospores phragmosporous . . . . . . . . . . . . . . . . . . . . . . . **Sydowia** (Butin, 1963: 115–116, icon.)

5′(3′) Ascospores dictyosporous . . . . . . . . . . . . . . . . . . . . . . . . . . . . . 6

    6(5′) Saprotrophic on stems . . . . . . . . . . . . . . . . . . . . . **Keisslerina** (Müller and von Arx, 1950: 378, icon.)

    6′(5′) Parasitic on *Dothidea collecta* and immersed in the stroma of the host fungus . . . . . . . . . . . . . . . . . . . . . . . . . . . . . . . . . . . . . . . . . . . . . **Endodothiora** (Petrak, 1929: 345.)

## X. PLEOSPORALES

The Pleosporales contains the great mass of common Loculoascomycetes on dead leaves and herbaceous stems, tree branches, and wood as well as many on green leaves and stems. The majority is lumped into one large family, the Pleosporaceae. These have middle-sized to large perithecioid pseudothecia containing cylindrical asci among persistent pseudoparaphyses (Figs. 1A, 2C, 8A, 8B, 9A). The ascospores are commonly phragmosporous or dictyosporous. The pseudothecia are usually solitary but may be gregarious or seated on a basal stroma in which they are partially immersed. There are a few genera with immersed locules in multiloculate stromata. The ascocarps are immersed in the substratum, erumpent, or almost entirely superficial. Some genera such as *Pyrenophora* (Figs. 3C, 8A) and *Cochliobolus* (Fig. 3B) with pseudothecia occurring on dead leaves and stems cause leaf spot and blight diseases with which their conidial states are associated.

A few small families are segregated from the Pleosporaceae, but the basis for their separation is questionable. The Lophiostomataceae have solitary pseudothecia immersed in wood and are distinguished by their laterally compressed beaks with slitlike ostioles. The Sporormiaceae contains a few genera found mostly on dung and characterized by dark ascospores with germ slits or pores. They are distinguished from the Sordariaceae (Pyrenomycetes) which share this habitat by the bitunicate asci and pseudoparaphyses. The Botryosphaeriaceae are separated primarily on the basis of their large, ovoid to oblong, usually hyaline, nonseptate ascospores (Fig. 2A). This seems an inadequate basis for recognition of a family. However, it is so unusual a spore type among Loculoascomycetes that fungi with nearly identical ascospores in *Guignardia* of the Dothideales and *Ellisiodothis* (Fig. 1F), *Myiocopron*, *Microdothella*, *Parastigmatea*, and *Trabutia* of the Hemisphaeriales must be considered for possible relationships with the Botryosphaeriaceae.

Most of the Pleosporales occurring on living leaves are in the families Venturiaceae and Dimeriaceae. The ascocarps are middle sized to small and are usually superficial. The ascospores are almost always one-septate. The Dimeriaceae are hyperparasites, ectocommensals, or parasites with a mycelium which is entirely superficial or may penetrate the host to a limited extent. In the Venturiaceae the mycelium is primarily internal, and if any superficial mycelium is present, it arises from an innate hypostroma. Pseudothecia are superficial, erumpent, or rarely immersed. The Venturiaceae are further characterized by the tendency of the ascospores to be apiosporous (Fig. 2B) and greenish, gray, or yellowish rather than hyaline or deep brown. Genera with similar ascospores in the Munkiellaceae (Hemisphaeriales) are very closely related to the Venturiaceae. The line between the Venturiaceae and those Pleosporaceae with smaller ascocarps and one-septate ascospores is not entirely clear, and the distinction between the Venturiaceae and Dimeriaceae is arbitrary. Venturiaceae with very small pseudothecia also are not easily separable from some genera of Dothideaceae (Dothideales).

The Mesnieraceae accommodates two genera of leaf parasites with immersed, subhyaline pseudothecia indistinctly delimited from the surrounding host tissue and nonseptate or one septate ascospores with dark, thick, sculptured walls. These fungi are similar in habit to genera such as *Physalospora* in the Phyllachoraceae (Sphaeriales).

**KEY TO FAMILIES AND IMPORTANT GENERA OF PLEOSPORALES**

1. Pseudothecia small, mostly superficial on green leaves or stems, rarely innate or maturing in dead tissue; ascospores 1-septate, hyaline to yellowish, green, gray, or brown; parasites, hyperparasites or epiphytic commensals . . . . . . . . . . . . . . . . . . . . . 2
1'. Pseudothecia middle-sized to large, superficial, erumpent, or innate or in or on a stroma, mostly on dead stems, wood, and dung; ascospores nonseptate to 1-septate or most commonly many-septate, hyaline to dark brown . . . . . . . . . . . . . . . . . . . . . . 4
    2(1) Pseudothecia arising on a superficial mycelium; mycelium entirely superficial or, if partially innate, inconspicuous and not forming stromatic masses in host leaf; hyperparasites, epiphytic commensals, or parasites on green leaves; ascospores hyaline . . . . . . . . . . . . . . . . . . . . . . . . . . . . . . **Dimeriaceae** p. 178
    2'(1) Pseudothecia erumpent or immersed, or, if superficial, on an immersed hypostroma or mycelium arising from an immersed hypostroma and superficial mycelium otherwise

lacking; parasites on green leaves or less commonly stems, sometimes maturing in
dead tissue . . . . . . . . . . . . . . . . . . . . . . . . . . . . . 3
3(2') Ascospores with thin, smooth walls, 1-septate, yellow, greenish or brownish; pseudothecial
wall composed of distinct, dark brown cells . . . . . . . . . . **Venturiaceae** p. 179
3'(2') Ascospores with thick walls and sculptured, furrowed, or warty epispore, usually dark
brown; pseudothecia with fleshy, subhyaline to yellowish walls indistinctly delimited from
the living tissue of the host leaf in which they are immersed . . . **Mesnieraceae** p. 183
    4(1') Ascospores nonseptate, (becoming 1-septate in *Neodeightonia*) ovoid to ellipsoid,
hyaline or rarely brown . . . . . . . . . . . . . . . **Botryosphaeriaceae** p. 183
    4'(1') Ascospores 1- many-septate . . . . . . . . . . . . . . . . . . . . 5
5(4') Beak of pseudothecium laterally compressed and ostiole slit-like; saprotrophic
on wood . . . . . . . . . . . . . . . . . . . . . . . . . . **Lophiostomataceae** p. 183
5'(4') Beak of pseudothecium not compressed; ostiole round . . . . . . . . . . . . 6
    6(5') Ascospores with germ slits or germ pores, dark brown, with gelatinous sheath;
chiefly on dung . . . . . . . . . . . . . . . . . . . . . . **Sporormiaceae** p. 184
    6'(5') Ascospores without germ slits or pores . . . . . . . . . . . . . . . 7
7(6') Chiefly on wood and dead herbaceous stems, parasites or saprotrophs
. . . . . . . . . . . . . . . . . . . . . . . . . . . . . . . . . . **Pleosporaceae** p. 184
7'(6') Symbiotic with algae, forming usually crustose lichens . . . **Mycoporaceae** p. 193

## A. Dimeriaceae

1. On vascular plants . . . . . . . . . . . . . . . . . . . . . . . . . . . . 3
1'. On lichens or mosses . . . . . . . . . . . . . . . . . . . . . . . . . 2
    2(1') Parasitic on mosses; pseudothecia superficial on a dark, sparse, erumpent, straight-
walled mycelium; pseudothecial wall several cells thick; ascospores
brownish . . . . . . . . . . . . . . . . . . . . . . . . . . . . . . . **Lizonia**
(Müller and von Arx, 1962: 501, icon.; Verona and Benedek, 1965, C-209; Dennis,
1968: 368, Pl. XXXVII I.)
    2'(1') Parasitic on lichens; mycelium superfical; hyphae dark, short-celled and becom-
ing moniliform; pseudothecial wall of a single layer of rounded cells, bearing hyphal
hairs; ascospores hyaline . . . . . . . . . . . . . . . . . . . . **Echinothecium**
(Müller and von Arx, 1962: 505, icon.; Verona and Benedek, 1966, C-233.)
3(1) Leaf parasites with an innate mycelium or with stomatopodia or haustoria . . . . 4
3'(1) Epiphytic commensals or hyperparasites without stomatopodia or haustoria . . . 7
    4(3) Ascospores 1-septate . . . . . . . . . . . . . . . . . . . . . . . 5
    4'(3) Ascospores phragmosporous, hyaline . . . . . . . . . . . . . **Aphanostigme**
(Müller, 1965: 101, icon.)
5(4) Superficial mycelium abundant, forming stomatopodia or appressoria; pseudothecium
very small, with erect hyphal hairs; ascospores hyaline or pale gray . . . . . . **Eumela**
(Müller and von Arx, 1962: 499, icon.)
5'(4) Superficial mycelium sparse; innate mycelium intracellular; pseudothecia larger than
40 μm in diameter; ascospores hyaline to brown . . . . . . . . . . . . . . . . 6
    6(5') Pseudothecia setose . . . . . . . . . . . . . . . . . . . . . **Dimeriella**

Farr (1965) maintained the genus *Dimeriella*; Müller and von Arx (1962) considered it a synonym of *Wentiomyces*.

6'(5') Pseudothecia glabrous .................................................. **Episphaerella**
(Müller and von Arx, 1962: 498, icon.) According to Farr (1965, 1966), presence or absence of setae is a character of questionable value and *Episphaerella* probably should be considered a synonym of *Dimeriella*.

7(3') Pseudothecia seated on the surface of the mycelium, globose, usually small, smooth or setose; wall usually 1 cell thick; conidial state, when present, phomoid; hyperparasites ............................................................................. 8

7'(3') Pseudothecia nested in the mycelium or on the tips of stromatic columns, globose or slightly depressed, small to middle-sized, with hyphal hairs or appendages or setae, seldom completely glabrous; wall of several layers of cells; epiphytes ............ 10

    8(7) Pseudothecia spinose ................................................ **Phaeodimeriella**
    (Müller and von Arx, 1962: 480, icon.)

    8'(7) Pseudothecia glabrous above ................................................ 9

9(8') Ascospores remaining hyaline ................................................ **Dimerina**
(Müller and von Arx, 1962: 481.)

9'(8') Ascospores becoming more or less dark brown ................. **Dimerium**
(Müller and von Arx, 1962: 478, icon.)

    10(7') Pseudothecia single at the tips of stromatic columns arising from the superficial mycelium, glabrous; ascospores brown, septate above the middle . . . **Pododimeria**
    (Müller and von Arx, 1962: 502, icon; Verona and Benedek, 1965, C-202.)

    10'(7') Pseudothecia nested in the mycelium ..................................... 11

11(10') Mycelium and pseudothecia on the leaf hair mat of living or dead leaves; pseudothecia glabrous above; ascospores hyaline ................................. **Eudimeriolum**
(Müller and von Arx, 1962: 495.)

11'(10') On the surface of living leaves; pseudothecia usually with setae or hyphal hairs ........................................................................... 12

    12(11') Ascospores becoming yellowish, greenish, or brown ......... **Epipolaeum**
    (Müller and von Arx, 1962: 487, 489, icon.; Farr, 1966: 240, icon.; Shoemaker, 1965.)

    12'(11') Ascospores remaining hyaline ................................. **Wentiomyces**
    (Müller and von Arx, 1962: 492, icon.; Verona and Benedek, 1966, C-226.) Farr (1965, 1966) considered *Wentiomyces* a nomen dubium and employed *Epipolaeum* for species with hyaline as well as colored ascospores.

## B. Venturiaceae

1. Pseudothecia superficial or erumpent ................................................ 2

1'. Pseudothecia innate, immersed in the substrate or only barely erumpent at the apex ............................................................................. 27

    2(1) Pseudothecia superficial, on a superficial mycelium or stroma or on a stroma arising from an innate hypostroma or mycelium ................................. 3

    2'(1) Pseudothecia erumpent from beneath the cuticle or epidermis ........ 21

3(2) Pseudothecia on a superficial mycelium or stroma ............................. 4

3'(2) Pseudothecia on a stroma connected by a narrowed basal foot to an innate hypostroma or mycelium or under an umbrellalike shield ................................. 12

4(3) Pseudothecia dark, turbinate, densely crowded in extensive crusts over a superficial mycelium or pseudostroma on hypertrophied host leaves or stems; ascospores septate near the lower end, hyaline, becoming greenish or brownish . . . . . **Apiosporina**
(Müller and von Arx, 1962: 466, icon.)

4'(3) Pseudothecia scattered or in small groups or on or immersed in a crustlike superficial stroma; ascospores septate in the middle or upper third . . . . . . . . . . . 5

5(4') Pseudothecia dark . . . . . . . . . . . . . . . . . . . . . . . . . . . . . . 6

5'(4') Pseudothecia bright-colored, yellow to red or reddish brown . . . . . . . . . . 11

   6(5) Pseudothecia on a superficial mycelium . . . . . . . . . . . . . . . . . . 8

   6'(5) Pseudothecia on or in a stroma . . . . . . . . . . . . . . . . . . . . . . . 7

7(6') Pseudothecia immersed locules in a crustlike superficial stroma connected at many points with an intraepidermal hypostroma and extending at the margins into a radiately structured membrane; ascospores septate in the middle, brown, with a finely warted exospore . . . . . . . . . . . . . . . . . . . . . . . . . . . . . . . . . . . **Polyrhizon**
(Müller and von Arx, 1962: 321, icon.)

7'(6)') Pseudothecia on the surface of a crustlike stroma on the surface of dead leaves; internal mycelium lacking; ascospores septate in the middle, brownish . . . . . . . . . **Licopolia**
(Müller and von Arx, 1962: 332, icon.)

   8(6) Setae present on superficial mycelium and on pseudothecia; ascospores becoming dark brown . . . . . . . . . . . . . . . . . . . . . . . . . . . . . . . **Acantharia**
(Müller and von Arx, 1962: 438, icon.)

   8'(6) Setae lacking on mycelium, present or absent on pseudothecia . . . . . . . . 9

9(8') Pseudothecia on a dense, closely appressed subiculum; ascospores bright brown . . . . . . . . . . . . . . . . . . . . . . . . . . . . . . . . . . . **Metacoleroa**
(Müller and von Arx, 1962: 442, icon.)

9'(8') Pseudothecia on a loose hyphal subiculum or on small scattered basal stromata; ascospores green to olivaceous . . . . . . . . . . . . . . . . . . . . . . . . . . . . . 10

   10(9') Pseudothecia opening by a round ostiole . . . . . . . . . . . . **Antennularia**
(Müller and von Arx, 1962: 430, icon.)

   10'(9') Pseudothecium opening widely by disintegration of the apical region and becoming apothecioid . . . . . . . . . . . . . . . . . . . . . . . . . . . . . **Pseudoparodia**
(Müller and von Arx, 1962: 441, icon.)

11(5') Pseudothecia separate, yellowish; ascospores brown; hyperparasites on fungi . . . . . . . . . . . . . . . . . . . . . . . . . . . . . . . . . . . . **Letendraea**
(Müller and von Arx, 1962: 317, icon.; Verona and Benedek, 1968, C-334; Dennis, 1968: 386, Pl. XXXI K.)

11'(5') Pseudothecia seated on a superficial crustlike stroma anchored at several points in the stomata; stroma bright yellow to red or finally reddish brown; ascospores hyaline . . . . . . . . . . . . . . . . . . . . . . . . . . . . . . . . . . . . . . . . **Placocrea**
(Müller and von Arx: 1962, 368, icon.)

   12(3') Pseudothecia grouped under an umbrellalike shield arising from a central stromatic column connected with a hypostroma and extending marginally into hyphal strands; ascospores hyaline . . . . . . . . . . . . . . . . . . . . . . . . . . . . **Gilletiella**
(Müller and von Arx, 1962: 322, icon.)

   12'(3') Pseudothecia on or in a pulvinate stroma narrowed to a central foot connected with

a hypostroma; stroma reduced to a small basal foot when it bears a single
pseudothecium .................................... 13
13(12') Pseudothecia immersed locules in a pulvinate stroma ............ 14
13'(12') Pseudothecia free or mostly free, grouped or single on the surface of the
stroma ........................................ 17
    14(13) Pseudothecia vertically oriented ...................... 15
    14'(13) Pseudothecia horizontally oriented with ostioles opening laterally around margins of the stroma; stroma delicate with narrow central foot in a stoma and superficial mycelium arising from its margins, ascospores pale brownish .... **Trichodothella**
        (Müller and von Arx, 1962: 458, icon.)
15(14) Pseudothecia in concentric zones around sterile center of stroma; superficial mycelium extending out from margins of stroma; ascospores hyaline, finally grayish or brownish green ................................ **Trichodothis**
    (Müller and von Arx, 1962: 456, 457, icon.)
15'(14) Pseudothecia uniformly distributed throughout stroma; superficial mycelium lacking ....................................... 16
    16(15') Ascospores septate near the lower end, yellowish brown ...... **Coccoidea**
        (Müller and von Arx, 1962: 454, icon.)
    16'(15') Ascospores septate near the middle, hyaline ........... **Uleodothis**
        (Müller and von Arx, 1962: 394–395, icon.)
17(13') Pseudothecia bright-colored, reddish brown, covered with hyaline setae, grouped or single on the reduced basal stroma; ascospores hyaline, septate near the middle ........................................ **Allonecte**
    (Müller and von Arx, 1962: 319, icon.; Verona and Benedek, 1970, C-464.)
17'(13') Pseudothecia dark ................................ 18
    18(17') Pseudothecia borne singly on a reduced stromatic base ......... 19
    18'(17') Pseudothecia in groups on small stromata; ascospores septate near the middle, yellowish, greenish, or brownish ........................ 20
19(18) Pseudothecia small, up to 150 μm in diameter, tapering to a narrow foot seated in a stoma, glabrous; ascospores septate at or below the middle, greenish or bright brown
................................................. **Phaeocryptopus**
    (Müller and von Arx, 1962: 444, icon.; Verona and Benedek, 1967, C-266; Dennis, 1968: 378, Pl. XXXVIII B.)
19'(18) Pseudothecia large, thick-walled, on a broad stromatic base arising from a hypostroma; ascospores large, brown, faintly striate .................... **Parodiella**
    (Müller and von Arx, 1962: 330, icon.; Verona and Benedek, 1960, C-28.)
    20(18') Pseudothecia small, mostly less than 100 μm in diameter, usually glabrous, on a columnar or platelike basal stroma .................. **Xenomeris**
        (Müller and von Arx, 1962: 446, icon.)
    20'(18') Pseudothecia larger, glabrous or often setose, on a cushion-shaped stroma
................................................. **Gibbera**
        (Müller and von Arx, 1962: 421; Verona and Benedek, 1967, C-297; Dennis, 1968: 377, Fig, 21 I, 21 L.)
21(2') Pseudothecia erumpent from beneath cuticle ................. 22
21'(2') Pseudothecia erumpent from epidermis or deeper ............. 24

22(21) Ascospores septate near the lower end, hyaline to pale brown; pseudothecia seated on a subcuticular hypostroma . . . . . . . . . . . . . . . . **Botryostroma**
(Müller and von Arx, 1962: 463, icon.)

22'(21) Ascospores septate near the middle or in the upper or lower third, pale yellow, green, or brown . . . . . . . . . . . . . . . . . . . . . . . . . . . . . . . . 23

23(22') Pseudothecia seated on a crustlike subcuticular basal stroma, often setose . . . . . . . . . . . . . . . . . . . . . . . . . . . . . . . . . . . . . . **Coleroa**
(Müller and von Arx, 1962: 414, icon.; Verona and Benedek, 1959, C-4; Dennis, 1968: 376, Fig. 21 J, 21 K.) *Hormotheca* in the Munkiellaceae, Hemisphaeriales, is similar to *Coleroa*.

23'(22') Pseudothecia immersed in a pulvinate subcuticular stroma, glabrous . . **Atopospora**
(Müller and von Arx, 1962: 472, icon.; Dennis, 1968: 361, as *Euryachora*, Fig. 22 G.)

24(21') Pseudothecia immersed locules in a pulvinate stroma . . . . . . . . . . 25

24'(21') Pseudothecia separate or on a basal crustlike stroma . . . . . . . . . 26

25(24) Ascospores septate in lower third, brownish; on hypertrophied leaf tissue
. . . . . . . . . . . . . . . . . . . . . . . . . . . . . . . . . . . . . . . . . **Crotone**
(Müller and von Arx, 1962: 468, icon.)

25'(24) Ascospores septate near the lower end, hyaline but greenish in mass; host tissue not hypertrophied . . . . . . . . . . . . . . . . . . . . . . . . . . . . . . . **Platychora**
(Müller and von Arx, 1962: 470, icon.; Dennis, 1968: 361, as *Euryachora*, Fig. 22 D.)

26(24') Ascospores greenish-brown, septate in or above the middle; pseudothecia often setose . . . . . . . . . . . . . . . . . . . . . . . . . . . . . . . . . **Stigmatea**
(Müller and von Arx, 1962: 412, icon.; Verona and Benedek, 1970, C-440.)

26'(24') Ascospores hyaline to pale yellow or brown, septate in the middle; pseudothecia glabrous . . . . . . . . . . . . . . . . . . . . . . . . . . . . . . . . **Rosenscheldiella**
(Müller and von Arx, 1962: 377, icon.; Verona and Benedek, 1969, C-413.)

27(1') Pseudothecia basally or wholly immersed in a globose to orbicular-pulvinate stroma, ascospores septate at or slightly below the middle, sometimes becoming phragmosporous; hyperparasites in and around pustules of Uredinales . . . . . . . . . . . . . **Eudarluca**
(Müller and von Arx, 1962: 313, icon.; Eriksson, 1966, icon. Figs. 2–5; 1967b.)

27'(1') Pseudothecia separate . . . . . . . . . . . . . . . . . . . . . . . . . . . . 28

28(27') Ascospores hyaline . . . . . . . . . . . . . . . . . . . . . . . . **Didymella**
(Müller and von Arx, 1962: 362; Corbaz, 1957, icon.; Holm, 1953; Verona and Benedek, 1967, C-296; Kohlmeyer and Kohlmeyer, 1968, Table 82.)

28'(27') Ascospores colored . . . . . . . . . . . . . . . . . . . . . . . . . . . . . 29

29(28') Ascospores brown, septate in the middle, cylindrical or fusiform, relatively long; pseudothecia glabrous; conidia in pycnidia . . . . . . . . . . . . . . **Teratosphaeria**
(Müller and von Arx, 1962: 316, icon.; Verona and Benedek, 1966, C-232.)

29'(28') Ascospores greenish-yellow or rarely pale olive-brown, oblong, septate in the upper or lower third; pseudothecia often setose at the apex; conidial states in Hyphomycetes . . . . . . . . . . . . . . . . . . . . . . . . . . . . . . . . **Venturia**
(Müller and von Arx, 1962: 403, icon.; Menon, 1956; Nüesch, 1960; Dennis, 1968: 375. Fig. 21 G, 21 H.) *Endocoleroa*, Petrak, 1968, seems indistinguishable from *Venturia*.

## C. Mesnieraceae

1. Ascospores nonseptate . . . . . . . . . . . . . . . . . . . . . . . . . . . . . . **Mesniera**
   (von Arx and Müller, 1954: 107, icon.; Moreau and Moreau, 1955: 115, 117, icon.)
1'. Ascospores 1-septate . . . . . . . . . . . . . . . . . . . . . . . . . . . . **Stegasphaeria**
    (Müller and von Arx, 1962: 344, icon.)

## D. Botryosphaeriaceae

1. Pseudothecia superficial, connected with an innate hypostroma; leaf parasites . . . . 2
1'. Pseudothecia immersed or erumpent; usually on woody stems . . . . . . . . . . . 3
    2(1) Pseudothecia small, simple, globose, spinose or hairy . . . . . . **Pyrenostigme**
         (von Arx and Müller, 1954: 70; Petrak, 1948a: 71, as *Chaetomelanops*, icon.)
    2'(1) Pseudothecia immersed in a large pulvinate stroma connected by a broad or narrow basal foot with an innate hypostroma; ascospores hyaline or colored
         . . . . . . . . . . . . . . . . . . . . . . . . . . . . . . . . . . . . . . . **Auerswaldiella**
         (von Arx and Müller, 1954; 66, icon.; Verona and Benedek, 1970, C-485.)
3(1') Ascospores nonseptate, hyaline or rarely brown, pseudothecia single, laterally fused, or more or less immersed in a pulvinate stroma . . . . . . . . . . . **Botryosphaeria**
    (von Arx and Müller, 1954: 36, icon.; Verona and Benedek, 1965, C-172, 1969, C-354; Dennis, 1968: 373, Pl. XXXVIII C, E, J; Ellis and Everhart, 1892, Pl. 36.)
    *Astrocystis* in the sense of Kar and Maity, 1970b, seems identical with *Botryosphaeria*.
3'(1') Ascospores becoming 1-septate and brown . . . . . . . . . . . . . **Neodeightonia**
    (Punithalingham, 1969: 18, 22, 23, icon.) *Neodeightonia* seems to represent a *Botryosphaeria* in which the ascospores ultimately become 1-septate.

## E. Lophiostomataceae

1. Ascospores 1-septate . . . . . . . . . . . . . . . . . . . . . . . . . . . . . . . . . 2
1'. Ascospores more than 1-septate . . . . . . . . . . . . . . . . . . . . . . . . . . . 4
    2(1) Pseudothecia surrounded by a subiculum, ellipsoid, opening by a longitudinal slit; ascospores brown . . . . . . . . . . . . . . . . . . . . . . . . **Byssolophis**
         (Müller and von Arx, 1962: 341–342, icon.; Dennis, 1968; 381, Fig. 15 H.) The relationship of *Byssolophis* to other Lophiostomataceae is questionable, but it is hardly as well accommodated in the Hysteriaceae.
    2'(1) Subiculum lacking; pseudothecia globose with a compressed beak . . . . . . 3
3(2') Ascospores remaining hyaline, with or without appendages at the ends
    . . . . . . . . . . . . . . . . . . . . . . . . . . . . . . . . . . . . . . . . **Lophiosphaera**
    (Müller and von Arx, 1962: 335, icon.)
3'(2') Ascospores becoming brown . . . . . . . . . . . . . . . . . . . . . . . **Ostropella**
    (Müller and von Arx, 1962: 337–338, icon.)
    4(1') Ascospores ellipsoid to fusoid . . . . . . . . . . . . . . . . . . . . . . . . . 5
    4'(1') Ascospores filiform . . . . . . . . . . . . . . . . . . . . . . . . . . **Lophionema**
         (Chesters and Bell, 1970; Ellis and Everhart, 1892, Pl. 25.)
5(4) Ascospores phragmosporous . . . . . . . . . . . . . . . . . . . . . . . . . . . . 6
5'(4) Ascospores dictyosporous . . . . . . . . . . . . . . . . . . . . . . . . . . . . . 7

6(5) Ascospores hyaline, with or without appendages at the ends . . . **Lophiotrema**
(Dennis, 1968: 380, Fig, 13 A; Eriksson, 1967b.)

6′(5) Ascospores brown, with or without appendages at the ends . . . **Lophiostoma**
(Chesters and Bell, 1970: 41, icon.; Ellis and Everhart, 1892, Pl. 25; Eriksson, 1967b; Dennis, 1968: 381, Fig. 13 B.) Chesters and Bell would place all species with 1- several-septate, hyaline to brown ascospores in the single genus *Lophiostoma*.

7(5′) Ascospores hyaline . . . . . . . . . . . . . . . . . . . . . . . . . **Lophidiopsis**

7′(5′) Ascospores brown . . . . . . . . . . . . . . . . . . . . . . . . . **Platystomum**
(Dennis, 1968, as *Lophidium*: 381, Fig. 13 C; Ellis and Everhart, 1892, as *Lophidium*, Pl. 25.) Chesters and Bell, 1970, would include all species with hyaline to brown, dictyosporous ascospores in the single genus *Platystomum*.

## F. Sporormiaceae

1. Ascospores nonseptate, ovoid to oblong, with a germ slit extending the length of the spore . . . . . . . . . . . . . . . . . . . . . . . . . . . . . . . **Semidelitschia**
(Cain and Luck-Allen, 1969: 583, icon.)

1′. Ascospores septate . . . . . . . . . . . . . . . . . . . . . . . . . . . 2

    2(1′) Ascospores 1-septate . . . . . . . . . . . . . . . . . . . . . . 3

    2′(1′) Ascospores phragmosporous, with germ slits in each cell . . . . . . . . . . 4

3(2) Ascospores with germ slits the length of each cell . . . . . . . . . . . . **Delitschia**
(Müller and von Arx, 1962: 348, icon.; Verona and Benedek, 1963, C-120; Dennis, 1968: 412, Fig. 20 L; Ellis and Everhart, 1892, Pl. 17.)

3′(2) Ascospores with a germ pore at each end, often separating into two cells . . . . . . . . . . . . . . . . . . . . . . . . . . . . . . . . . **Trichodelitschia**
(Müller and von Arx, 1962: 350, icon.; Dennis, 1968: 412, Pl. XXXVII J.)

    4(2′) Pseudothecia ostiolate . . . . . . . . . . . . . . . . . . . . . . **Sporormia**
(Dennis, 1968: 411, Pl. XXXVII L; Verona and Benedek, 1963, C-119; Ellis and Everhart, 1892, Pl. 18.)

    4′(2′) Pseudothecia lacking an ostiole, globose, closed; ascospores sometimes separating into their constituent cells and asci appearing polysporous . . . . . . . . **Preussia**
(Cain, 1961: 1161–1162, icon.; Verona and Benedek, 1963, C-128.)

## G. Pleosporaceae

1. Parasitic on lichens . . . . . . . . . . . . . . . . . . . . . . . . . . . 2
1′. On vascular plants . . . . . . . . . . . . . . . . . . . . . . . . . . . 4

    2(1) Ascospores 1-septate; pseudothecia globose, separate, immersed, middle-sized . . . . . . . . . . . . . . . . . . . . . . . . . . . . . . . . . . . . . 3

    2′(1) Ascospores phragmosporous, brown; pseudothecia separate, immersed . . . . . . . . . . . . . . . . . . . . . . . . . . . . . . . . . . . **Phaeospora**
(Dennis, 1968: 371, Pl. XXXVII F.)

3(2) Ascospores hyaline, not constricted at the septum . . . . . . . . . . **Cercidospora**
(Müller and von Arx, 1962: 392, icon.)

3′(2) Ascospores brown, slightly constricted at the septum . . . . . . . **Didymosphaeria**
(Müller and von Arx, 1962: 291, icon.; Scheinpflug, 1958.)

    4(1′) Ascospores 1-septate (additional septa in apiculate tips in *Caryospora*) . . . . 5

## 7. Loculoascomycetes

4'(1') Ascospores more than 1-septate (1- several-septate in *Herpotrichia*, *Herpotrichiella*, *Keissleriella*, and *Massarina*) . . . . . . . . . . . . . . . . . . . . . . . 16

5(4) Pseudothecia superficial or with a broad flat base partially embedded in the substrate
. . . . . . . . . . . . . . . . . . . . . . . . . . . . . . . . . . . . . . 6

5'(4) Pseudothecia immersed in the substrate . . . . . . . . . . . . . . . . . . . . 11

    6(5) Pseudothecia entirely superficial or with a small immersed foot . . . . . . . . 7

    6'(5) Pseudothecia with a broad flat base partially embedded in the substrate . . . . 9

7(6) Pseudothecia entirely superficial, single or usually grouped on a crustose stroma, often laterally fused; ascospores hyaline to brown . . . . . . . . . . . . . . . **Otthia**
(Müller and von Arx, 1962: 274, 277, icon.; Dennis, 1968: 415, Fig. 18 H; Scheinpflug, 1958; Verona and Benedek, 1968, C-323; Ellis and Everhart, 1892, Pl. 26.) Species with hyaline ascospores may be segregated in *Otthiella* (Munk, 1957: 402–403.)

7'(6) Pseudothecia superficial, with a small immersed foot; marine, on wood . . . . . . 8

    8(7') Pseudothecia on a long, stalklike base, with periphysate ostiole; asci in a wall layer; ascospores hyaline to pale gray, thin-walled, rounded at the tips, strongly constricted at the septum . . . . . . . . . . . . . . . . . . . . . . . . . . **Thallassoascus**
(Kohlmeyer and Kohlmeyer, 1964, Table 29.)

    8'(7') Pseudothecia sessile, on a weak hypostroma; ostiole lacking; asci over the surface of a large hemispherical placenta projecting from the base of the locule; ascospores with a thick, 2-layered wall and dark brown epispore, with germ pores and often pseudosepta in the acuminate tips . . . . . . . . . . . . . . . . . . **Pontoporeia**
(Kohlmeyer and Kohlmeyer, 1964, Table 22.)

9(6') Ascospores large (longer than 30 μm), broad-fusoid with acuminate tips and additional septa cutting off tiny cells at either end, dark brown; pseudothecia conical
. . . . . . . . . . . . . . . . . . . . . . . . . . . . . . . . . . . . . **Caryospora**
(Müller and von Arx, 1962: 279, icon.; Dennis, 1968: 405, Fig. 20 F; Ellis and Everhart, 1892, Pl. 24; Verona and Benedek, 1970, C-471.)

9'(6') Ascospores smaller, 1-septate, fusoid to ellipsoid, brown to green; pseudothecia flattened hemispherical . . . . . . . . . . . . . . . . . . . . . . . . . . . . . 10

    10(9') Pseudothecial wall parenchymic, composed of oblong to isodiametric cells
. . . . . . . . . . . . . . . . . . . . . . . . . . . . . . . . . . . . **Microthelia**
(Müller and von Arx, 1962; 284, 258, 287, icon.; Kohlmeyer and Kohlmeyer, 1967, Table 64; Verona and Benedek, 1969, C-367.)

    10'(9') Pseudothecial wall plectenchymic, composed of irregularly tangled hyphae mixed with substrate cells . . . . . . . . . . . . . . . . . . . . . . . . . . . **Halotthia**
(Kohlmeyer and Kohlmeyer, 1964, Table 33.)

11(5') Pseudothecia subcuticular or grouped beneath a subcuticular stroma, flattened
. . . . . . . . . . . . . . . . . . . . . . . . . . . . . . . . . . . . . . . . 12

11'(5') Pseudothecia more deeply immersed, single, globose . . . . . . . . . . . . . 13

    12(11) Ascospores brown; pseudothecia single, under a shield extending at the periphery into a hyphal membrane . . . . . . . . . . . . . . . . . . . . . . **Mycomicrothelia**
(Müller and von Arx, 1962: 326, icon.)

    12'(11) Ascospores hyaline; pseudothecia grouped and often laterally fused under a stromatic shield . . . . . . . . . . . . . . . . . . . . . . . . . . . . **Tomasellia**
(Müller and von Arx, 1962: 328, icon.)

13(11′) Ascospores septate near the lower end, brown . . . . . . . . . . . . . . . 14
13′(11′) Ascospores septate in the middle . . . . . . . . . . . . . . . . . . . . . 15
    14(13) Ascospores with a gelatinous sheath; pseudothecia large, thick-walled
    . . . . . . . . . . . . . . . . . . . . . . . . . . . . . . . . . . . . **Pteridospora**
    (Müller and von Arx, 1962: 282, icon.; Dennis, 1968: 407, Fig. 30 D.)
        14′(13) Ascospores without gelatinous sheath; pseudothecia middle-sized, thin-walled . . . . . . . . . . . . . . . . . . . . . . . . . . . . . . . . **Didymopleela**
        (Müller and von Arx, 1962: 294, icon.)
    15(13′) Ascospores brown, lacking a gelatinous sheath . . . . . . . . . **Didymosphaeria**
    (Müller and von Arx, 1962: 290, icon.)
    15′(13′) Ascospores hyaline with a slimy appendage near the septum; partially or wholly immersed in wood in seawater . . . . . . . . . . . . . . . . . . . . **Paraliomyces**
    (Müller and von Arx, 1962: 310–311; Kohlmeyer and Kohlmeyer, 1964, Table 20; Verona and Benedek, 1966, C-227.)
        16(4′) Ascospores phragmosporous to scolecosporous, brown, pale yellowish-brown or hyaline . . . . . . . . . . . . . . . . . . . . . . . . . . . . . . . . . . . . 17
        16′(4′) Ascospores dictyosporous, longitudinal septa sometimes present in only a few cells or even entirely lacking in some species of *Cucurbidothis* and *Karstenula*, brown to sometimes hyaline . . . . . . . . . . . . . . . . . . . . . . . . . . . . 51
17(16) Pseudothecia light-colored, pink, ochraceous, or yellow and becoming brown; ascospores filamentous or acicular, hyaline . . . . . . . . . . . . . . . . . . . 18
17′(16) Pseudothecia black . . . . . . . . . . . . . . . . . . . . . . . . . . . . 19
    18(17) Pseudothecia gregarious on a loose subiculum, hairy or glabrous; ascospores filiform, hyaline, spirally arranged in cylindrical asci; on scale insects . . **Podonectria**
    (Dennis, 1968: 420, Fig. 19 D.)
    18′(17) Pseudothecia separate; ascospores acicular to filamentous; on plant stems or hyperparasitic on other fungi . . . . . . . . . . . . . . . . . . . . . . **Tubeufia**
    (Dennis, 1968: 421, Fig. 98; Munk, 1957: 474, icon. as *Ophionectria*.) *Ophionectria* may be maintained as a separate genus distinguished by lack of the ring of apical appendages around the ostiole that is typical of *Tubeufia*.
19(17′) Ascospores scolecosporous, mostly hyaline . . . . . . . . . . . . . . . . 20
19′(17′) Ascospores ellipsoid, fusoid, to long cylindric (scolecosporous in many species of *Nodulosphaeria* and some species of *Entodesmium*, *Phaeosphaeria*, and *Leptosphaeria*.)
    . . . . . . . . . . . . . . . . . . . . . . . . . . . . . . . . . . . . . . 26
    20(19) Pseudothecia solitary, ellipsoid, immersed, prostrate with a short, broad neck at one end; asci cylindric, more or less parallel with the surface of the substratum; ascospores hyaline, filiform, fragmenting into part spores; on herbacous stems . . . . . . . . . . . . . . . . . . . . . . . . . . . . . . . . . . . **Plejobolus**
    (Eriksson, 1967b.)
    20′(19) Pseudothecia erect, with apical ostiole; asci perpendicular to surface of the substrate . . . . . . . . . . . . . . . . . . . . . . . . . . . . . . . . . 21
21(20′) Ascospores long-acicular, divided by two septa into three unequal cells, hyaline to pale brownish; pseudothecia densely crowded on a well-developed stroma, composed of a scleroplectenchymatous tissue; parasitic on Labiatae . . . . . . . . **Rosencheldia**
    (Holm, 1968: 220–222, 224, icon. Plate I c.) Holm recognized the affinities of *Rosencheldia* with *Leptosphaeria* but maintained it as a distinct genus primarily

on the basis of spore type. He transferred the common *R. heliopsidis* on stems of Compositae to *Leptosphaeria*.

21'(20') Ascospores filiform, with numerous, often inconspicuous septa . . . . . . . 22
    22(21') Ascospores with a deep constriction in the middle where they readily dissociate into two parts, with a swollen cell above and below the constriction, pale yellowish brown, parallel in the ascus; pseudothecia more or less erumpent, with distinct beaks and wall of uniform thickness composed of small rounded cells; on dead stems of dicots, chiefly Compositae . . . . . . . . . . . . . . . . . . . . . . . . . . **Ophiobolus**
    (Dennis, 1968: 419, Fig. 30 A; Müller, 1952: 324, Fig. 5; 316, Fig. 2 N; Holm, 1957: 73–77, 176, Fig. 5 I, J.) Müller (1952) followed the classical treatment in which fungi of this type on herbaceous stems with scolecosporous ascospores are referred to *Ophiobolus*, those with ellipsoid to fusoid, phragmosporous ascospores to *Leptosphaeria*. Holm (1957) limited *Ophiobolus* to species with a median constriction and two adjacent swollen cells and transferred many species to *Nodulosphaeria*, *Leptosphaeria*, and *Entodesmium*, genera which contain mostly phragmosporous species.
    22'(21') Ascospores of uniform diameter: lacking deep constriction and swollen cells . . . . . . . . . . . . . . . . . . . . . . . . . . . . . . . . . . . . . . . 23
23(22') Ascospores more or less spirally coiled within the ascus; pseudothecia similar to those of *Ophiobolus* . . . . . . . . . . . . . . . . . . . . . . . . . . . . 24
23'(22') Ascospores straight, parallel, hyaline; pseudothecia superficial, wedge-shaped, with the base narrowed to a sterile stipe; on tree cankers . . . . . . . . . . . . **Dolabra**
    (Booth and Ting, 1964: 236, icon.; Verona and Benedek, 1965, C-215.)
    24(23) Pseudothecia bearing sparse setae; on plant debris . . . . . **Acanthophiobolus**
    (Dennis, 1968: 420, Fig, 16 H.)
    24'(23) Pseudothecia lacking setae . . . . . . . . . . . . . . . . . . . . . 25
25(24') Producing a conidial state in *Bipolaris* (*Helminthosporium*) and pseudothecia often covered with conidiophores; on Gramineae . . . . . . . . . . . . . **Cochliobolus**
    (Drechsler, 1934.)
25'(24') Lacking *Bipolaris* conidial state; pseudothecia often laterally compressed, scattered over reddish patches on the substrate; on herbaceous stems of dicots and rarely on wood . . . . . . . . . . . . . . . . . . . . . . . . . . . . . . . . . . . **Leptospora**
    (Dennis, 1968: 418, Pl. XXIX C; Holm, 1957: 103–105; Müller, 1952, icon. as *Ophiobolus*: 334, Fig. 8: 326, Fig. 6 K.)
    26(19') Pseudothecia on a superficial mycelium or subiculum (surface merely covered with hyphal hairs in *Herpotrichia mutabilis*) or on an erumpent, footlike basal stroma . . . . . . . . . . . . . . . . . . . . . . . . . . . . . . . . . . . . . . . 27
    26'(19') Pseudothecia on or in the substrate, sometimes united by an innate stroma . . . . . . . . . . . . . . . . . . . . . . . . . . . . . . . . . . . . . . . 31
27(26) Pseudothecia on a dark subiculum . . . . . . . . . . . . . . . . . . . . 28
27'(26) Pseudothecia on a basal stroma . . . . . . . . . . . . . . . . . . . . . 30
    28(27) Ascospores 3-septate, brown with pale end cells . . . . . . . . . . **Thaxteria**
    (Dennis, 1968: 410, Fig. 16 B.)
    28'(27) Ascospores 1-several-septate, pale or finally brown . . . . . . . . . . . 29
29(28') Pseudothecia middle-sized to large, glabrous or with hyphal hairs, ascospores remaining 1-septate or with additional septa inserted . . . . . . . . . . . . . . . **Herpotrichia**

(Müller and von Arx, 1962: 303, 305, 306, 307, 310, icon.; Bose, 1961: 195, 196, 199, 201, 203, 205, 207, 209, 210, icon.; Ellis and Everhart, 1892; Dennis, 1968: 410, Fig. 16 A; Verona and Benedek, 1965, C-176.) Commonly on wood and branches but *H. coulteri* and *H. juniperi* may cause a felt blight of living shoots of conifers. Müller and von Arx included in *Herpotrichia* the type species of *Pseudotrichia* in which the hairy pseudothecia are aggregated on old stromata of pyrenomycetes and a subiculum is lacking.

29′(28′) Pseudothecia small, membranous, spinose; ascospores 1- several-septate, pale grayish brown . . . . . . . . . . . . . . . . . . . . . . . . . . . . . **Herpotrichiella**
(Müller and von Arx, 1962: 311; Kohlmeyer and Kohlmeyer, 1964, Tab. 17; Verona and Benedek, 1962, C-93.) Munk (1957: 438–441) placed *Herpotrichiella* together with the dictyosporous genera *Dictyotrichiella*, *Capronia*, and *Berlesiella* in a separate family, the Herpotrichiellaceae.

30(27′) Ascospores deeply constricted at the middle septum and each of the two halves 4-septate, pale brown; on woody branches of *Viscum* . . . . . . . . . . **Gibberidea**
(Holm, 1968: 218–219, icon.; Pl. 1a,b.)

30′(27′) Ascospores 3-septate, fusoid, brown; basal stroma erumpent from living leaves of *Juniperus* . . . . . . . . . . . . . . . . . . . . . . . . . . . . . . . . **Muellerites**
(Holm, 1968: 231, icon. *M. juniperi* as *Gibberidea juniperi* Plate III c; Müller and von Arx, 1955: 359, 363, icon. *Gibberidea juniperi* Fig. 2, Fig. 3 d; Casagrande, 1969; Verona and Benedek, 1970, C-466.)

31(26′) Ascospores large, with conspicuous gelatinous sheath; pseudothecia immersed in woody stems . . . . . . . . . . . . . . . . . . . . . . . . . . . . . . . . . . . . 32

31′(26′) Ascospores smaller, lacking a gelatinous sheath (except in species of *Keissleriella*); on woody or herbaceous stems, or leaves . . . . . . . . . . . . . . . . . . . 35

32(31) Ascospores hyaline, 1- several-septate . . . . . . . . . . . . . . **Massarina**
(Dennis, 1968: 406, Fig. 23 A; Bose, 1961: 158, 159, 167, 168, 169, 171, 174, 178, icon.; Verona and Benedek 1965 C-171.)

32′(31) Ascospores brown . . . . . . . . . . . . . . . . . . . . . . . . . . . . 33

33(32′) Pseudothecia remaining immersed . . . . . . . . . . . . . . . . . . . . 34

33′(32′) Pseudothecia erumpent in clusters through bark, often united on a stromatic base; ascospores smooth; with a *Scolecosporium* imperfect state . . . . . . . **Asteromassaria**
(Dennis, 1968: 408, Fig. 20 I; Munk, 1957: 411.)

34(33) Ascospores punctate; pseudothecia separate . . . . . . . . **Stigmatomassaria**
(Munk, 1957: 417.)

34′(33) Ascospores with a smooth wall; pseudothecia separate or clustered and united in a rudimentary stroma . . . . . . . . . . . . . . . . . . . . . . . . . . . **Massaria**
(Munk, 1957: 418; Dennis, 1968: 407, Fig. 23 F, 23 H; Ellis and Everhart, 1892, Pl. 29; Verona and Benedek, 1962, C-92.)

35(31′) Pseudothecia in a stroma . . . . . . . . . . . . . . . . . . . . . . . . . 36

35′(31′) Pseudothecia separate . . . . . . . . . . . . . . . . . . . . . . . . . . . 38

36(35) Pseudothecia in an orbicular, plectenchymatous stroma erumpent through bark of woody stems; ascospores ellipsoid to fusoid, brown 3- to many-septate . . . . . . . . . . . . . . . . . . . . . . . . . . . . . . . . . . . . . . . **Thyridaria**
(Munk, 1957:414; Dennis, 1968:409, Pl. XXXIX B; Wehmeyer, 1941.)

36′(35) Pseudothecia immersed locules in elongated parenchymatous stromata; on leaves of ferns .................................... 37

37(36′) Locules small (less than 200 μm) in 1- several rows in extensive flat, subepidermal stromata of round to angular cells tending to be arranged in vertical rows; ascospores fusoid. hyaline to brown, often with gelatinous appendages at the tips . . **Rhopographus** (Dennis, 1968: 360, Fig. 22 F; Obrist, 1959: 368, icon.; Munk, 1957: 467, icon.)

37′(36′) Locules large (300–500 μm) in a single row in an elongated parenchymatous stroma immersed several cell layers beneath the epidermis; ascospores fusoid, hyaline ................................. **Dangeardiella**
(Obrist, 1959:378, 380, 381, icon.)

    38(35′) Ascospores with a taillike hyaline, basal portion, clavate-fusoid, several septate and slightly constricted at the septa; on herbaceous stems ...... **Rebentischia** (Dennis, 1968: 402, Pl. XXXIX I; Verona and Benedek, 1969, C-422; Müller, 1950: 308–310.)

    38′(35′) Ascospores not tapering to a taillike extremity ............. 39

39(38′) Ascospores consistently 2-septate .................. 40

39′(38′) Ascospores 3- many-septate, rarely 1-septate ............. 42

    40(39) Ascospores with approximately equal cells, hyaline; on wood
................................ **Melomastia**
(Dennis, 1968: 406, Fig. 23 D; Verona and Benedek, 1967, C-291; Munk, 1957: 409.)

    40′(39) Ascospores with unequal cells; on herbaceous stems .......... 41

41(40′) Ascospores with 2 septa toward base and anterior cell making up at least half of spore, oblong to fusoid, hyaline to pale brown ............... **Buergenerula** (Müller, 1950: 307, icon.; Dennis, 1968: 387, Fig. 10 H.)

41′(40′) Ascospores with a septum near either end cutting off two small terminal cells and a larger central cell, hyaline .................. **Leptoguignardia** (Müller, 1955: 217, icon.)

    42(39′) Ascospores ellipsoidal, brown, with rounded hyaline end cells . . **Passeriniella** (Dennis, 1968: 403, Fig. 25N; Verona and Benedek, 1969, C-422.)

    42′(39′) Ascospores concolorous, without rounded hyaline end cells (tips may be pale in fusoid ascospores of *Trematosphaeria*) ................. 43

43(42′) Pseudothecial wall composed of a scleroplectenchyma of isodiametric, thick-walled cells, stout, often strongly thickened laterally at the base; pseudothecial beak short or lacking; ascospores fusoid to rarely scolecosporous, 3- to many-septate, brownish yellow to hyaline; mostly on stems of herbaceous dicots ............ **Leptosphaeria**
(Holm, 1957: 15–50; 172, Fig. 1 a–ee; 173, Fig. 2 a–i; Müller, 1950: 217, 225, icon. *L. salebricola; L. doliolum*, 243, 235; *L. nitschkei*, 237, 235; *L. napelli*, 245, 235; Dennis, 1968: 389, Pl. XXXIX D, G, Fig. 16 G, J., Fig. 24 A–I, X–Z, AA, Fig. 25 C, E, H, I, K, L, M; Kohlmeyer and Kohlmeyer, 1964, Table 18, 34–37, 1965, Table 51; Verona and Benedek, 1960, C-33.) *Metasphaeria*, with hyaline ascospores, is generally considered a synonym of *Leptosphaeria* although it is occasionally used provisionally for species not readily accomodated in *Leptosphaeria*, (Kohlmeyer and Kohlmeyer, 1967, Pl. 64). Traditionally, *Leptosphaeria* has been used for species on herbaceous stems with fusoid ascospores and *Ophiobolus* for those with scolecosporous ascospores, while similar species on wood have been placed in

*Melanomma* and *Trematosphaeria*. Müller's monograph (1950) followed the traditional concept of *Leptosphaeria* and Dennis (1968) conserved this treatment for practical purposes. Holm (1957) in a revision based on a complex of characters, including pseudothecial wall structure as well as ascospore structure, segregated from *Leptosphaeria* several genera such as *Entodesimum, Phaeosphaeria*, and *Nodulosphaeria*. A few species with scolecosporous ascospores such as *L. tanaceti* were included in *Leptosphaeria*. Some species on wood were transferred from *Melanomma* to *Leptosphaeria* and some on herbaceous stems from *Leptosphaeria* to *Melanomma*. Holm (1968, p. 223) established the subgenus *Syncarpella* for a group of species on Compositae with a tendency to form aggregates of pseudothecia, this line culminating in species with the pseudothecia grouped on a basal stroma as in *Leptosphaeria* (*Rosenscheldia*) *heliopsidis*.

43′(42′) Pseudothecial wall usually not scleroplectenchymatous (cells thin-walled except at the periphery), of uniform thickness or sometimes thickened at the apex . . . . . . . . 44

    44(43′) Beak of pseudothecium well developed, usually with a periphysate ostiole; pseudothecial wall composed of small rounded cells, ascospores fusoid to often scolecosporous . . . . . . . . . . . . . . . . . . . . . . . . . . . . . . . . . . . . . . 45

    44′(43′) Beak of pseudothecium short or lacking; ascospores oblong to fusoid . . 46

45(44) Ascospores with a single swollen cell, fusoid to scolecosporous, often with small gelatinous appendages at each end, hyaline to olivaceous; almost exclusively on stems of herbaceous dicots, primarily Compositae . . . . . . . . . . . . . . . **Nodulosphaeria**
    (Holm, 1957: 78–102, 174, Fig. 3g–z; Müller, 1952: 321, 316, icon. *N. fructicum*, as *Ophiobolus fructicum*; Müller and Ahmad, 1958: 28; Dennis, 1968: 394, Fig. 24 J; Holm, 1961.) *Nodulosphaeria* comprises most of the species traditionally placed in *Ophiobolus*.

45′(44) Ascospores lacking swollen cells, fusoid to rarely scolecosporous, 4- to many-septate, in some species fragmenting at the septa; pseudothecial beaks very long; on stems of herbaceous Leguminosae . . . . . . . . . . . . . . . . . . . . . . . . . . **Entodesmium**
    (Holm, 1957: 132–137, 174, Fig. 3a–f; Müller, 1952: 328, 326, icon. *E. mayorii*, as *Ophiobolus mayorii*; Dennis, 1968: 399–400, Fig. 25B.)

    46(44′) Ostiole lined with unicellular, dark brown or rarely hyaline setae and often with setae surrounding the pore; pseudothecia immersed, with wall somewhat thickened and clypeate at the apex and beaks lacking; ascospores hyaline, 1-septate or usually several-septate, often with a thin gelatinous sheath . . . . . . **Keissleriella**
    (Bose, 1961: 181, 182, 184, 185, 187, 189, icon.; Müller and von Arx, 1962: 299–300, icon.; Eriksson, 1967b; Dennis, 1968: 388, Fig. 20 G, Fig. 25 R, S. Q. U; Kohlmeyer and Kohlmeyer, 1965, Table 50; Verona and Benedek, 1964, C-167.) Bose (1961) considered *Trichometasphaeria*, with phragmosporous ascospores, synonymous with *Keissleriella*, with 1-septate ascospores. Species of *Trichometasphaeria* (Luttrell, 1963: 283, icon.) described from culture as perfect states of *Bipolaris* spp. have not been transferred to *Keissleriella* and are provisionally retained in *Trichometasphaeria*.

    46′(44′) Ostiole not lined with setae; ascospores brown . . . . . . . . . . . . 47

47(46′) Pseudothecia immersed in living leaves, irregular in outline, with the wall thickened at the apex, separate or often gregarious and united by a hyphal or cellular stroma; ascospores broad fusoid, large and dark brown . . . . . . . . . . . **Phragmocauma**
    (Müller and Dennis, 1965: 377, icon.)

47′(46′) Pseudothecia on dead herbaceous stems or wood . . . . . . . . . . . . . . 48

# 7. Loculoascomycetes

48(47′) Pseudothecial wall thick; ascospores ellipsoid to fusoid; mostly on wood .................................................... 49

48′(47′) Pseudothecial wall thin; ascospores short cylindric, usually with 1 cell swollen; on herbaceous stems, usually monocots .................... 50

49(48) Pseudothecial wall of numerous layers of cells, variable in shape and size and heavily and irregularly pigmented; pseudothecia partially or wholly immersed; ascospores fusoid, brown, usually with paler tips ................... **Trematosphaeria**
(Dennis, 1968: 404, Fig. 16 D, Fig. 250; Müller and Dennis, 1965: 376, icon.; Müller and Ahmad, 1958: 29, icon.; Kohlmeyer and Kohlmeyer, 1968, Table 77; Verona and Benedek, 1969, C-347; Ellis and Everhart, 1892, Pl. 22.) In Holm's revision (1957) *Trematosphaeria* included some species on herbaceous stems formerly placed in *Leptosphaeria*.

49′(48) Pseudothecial wall of numerous layers of cells, variable in shape and size and rarely immersed, globose, often depressed; ascospores ellipsoid to cuneiform or almost clavate, usually 3-septate; more or less dark olive-brown; mostly on wood
..................................................... **Melanomma**
(Dennis. 1968: 409, Fig. 16 C; 400, Fig. 35 D; 401, Fig. 25 F, G; Verona and Benedek, 1960, C-26; Ellis and Everhart, 1892, Pl. 21; Munk, 1957: 408.) In Holm's revision (1957) *Melanomma* included some species on herbaceous stems formerly referred to *Leptosphaeria*.

50(48′) Ascospore with smooth epispore, fusiform to cylindric, often with 1 cell swollen and part anterior to swollen cell shorter, yellow-olivaceous; conidial states in *Hendersonia*, *Septoria*, or *Phaeoseptoria*; on monocots, primarily Gramineae
..................................................... **Phaeosphaeria**
(Hedjaroude, 1968; Holm, 1957; 106–131, 175, Fig. 41 ff; Müller, 1950: 217 *P. sowerbyi*, as *Leptosphaeria sowerbyi*; Dennis, 1968: 394–398, Fig. 24 K–W; Kohlmeyer and Kohlmeyer, 1965, Table 55.)

50′(48′) Ascospores often with punctate epispore, dark yellowish-brown, oblong to short cylindric, 3- to several-septate, with swollen cell near the base and posterior part shorter; conidial state *Coniothyrium*; on monocots and dicots . . **Paraphaeosphaeria**
(Hejaroude, 1968: 97–100; Eriksson, 1967 b; Dennis, 1968, 399; Fig. 24 BD, Fig. 25 A.)

51(16′) Ascospores small, greenish-gray; pseudothecia small, with spines at least around the pore; on rotten wood or old stromata of Pyrenomycetes (Herpotrichiellaceae) ... 52

51′(16′) Ascospores large, usually brown, rarely pale yellowish-brown to hyaline ... 54

52(51) Pseudothecia densely gregarious on old stromata of Pyrenomycetes; asci 8-spored .......................................... **Berlesiella**
(Dennis, 1968: 370, Fig. 16 E; Munk, 1957: 441.)

52′(51) Pseudothecia scattered on rotted wood ................ 53

53(52′) Asci 8-spored ............................. **Dictyotrichiella**
(Munk, 1957: 440; Verona and Benedek, 1968, C-315, 1969, C-382.)

53′(52′) Asci 16-spored ............................ **Capronia**
(Munk, 1957: 440, icon.) *Berlesiella* and *Dictyotrichiella* probably should be united under *Capronia*. Munk (1957: 438–441) placed these genera with *Herpotrichiella* in his family Herpotrichiellaceae.)

54(51′) Pseudothecia on dead herbaceous stems, single, scattered, large, partially to wholly erumpent .................................... 55

54′(51′) Pseudothecia on wood or bark, separate or grouped on or in a stroma  57

55(54) Ascospores definitely flattened in 1 plane . . . . . . . . . . . . **Clathrospora**
(Wehmeyer, 1961; Verona and Benedek, 1963, C-127.) Erikson (1967a) included in *Clathrospora* Wehmeyer's genus *Platyspora* and Wehmeyer's subgenus *Platysporioides* from the genus *Pleospora*. He recognized 3 subgenera: *Clathrospora*, ascospores with 2–7 rows of longitudinal septa, with additional transverse or "B" septa in the segments cut out by primary or "A" septa at constrictions, smooth, with thin septa, inequilateral; *Platyspora*, ascospores with one row of longitudinal septa, lacking "B" septa, punctate, with thin septa, straight; *Platysporioides*, ascospores with one row of longitudinal septa, lacking "B" septa, smooth, with thick septa, inequilateral, less strongly flattened, darker.

55′(54) Ascospores not flattened, circular in cross section . . . . . . . . . . . . . . 56

56(55′) Ascospores large, hyaline to pale brownish, oblong with almost straight sides and broadly rounded ends 3- 5- 7-septate with a single longitudinal septum in 1 to all cells; pseudothecia large, usually setose and often bearing conidiophores; conidial state in *Drechslera* (*Helminthosporium*) . . . . . . . . . . . . . . . . . . . . . . . **Pyrenophora**
(Dennis, 1968: 414, Pl. XXXIX J; Eriksson, 1967a; Ammon, 1963: 257, 269, icon.; Wehmeyer, 1961; Verona and Benedek, 1961, C-49.)

56′(55′) Ascospores generally smaller, those over 50 μm more than 7-septate, commonly dark brown . . . . . . . . . . . . . . . . . . . . . . . . . . . . . . . . . **Pleospora**
(Wehmeyer, 1961; Dennis, 1968: 413, Pl. XXXIX F; Eriksson, l967a; Kohlmeyer and Kohlmeyer, 1964, Table 21, 40; Verona and Benedek, 1959, C-3.)

57(54′) Pseudothecia in or on a stroma . . . . . . . . . . . . . . . . . . . . . . . 58

57′(54′) Pseudothecia separate, scattered or gregarious . . . . . . . . . . . . . . . 61

58(57) Pseudothecia immersed in small erumpent stromata . . . . . . . . . . . 59

58′(57) Pseudothecia on the surface of a flat stroma . . . . . . . . . . . . . . . 60

59(58) Pseudothecia in disk-shaped stromata with beaks collectively
erumpent . . . . . . . . . . . . . . . . . . . . . . . . . . . . . . . . . . . . **Fenestella**
(Dennis, 1968: 416, Fig. 23 G; Verona and Benedek, 1961, C-65; Ellis and Everhart, 1892, Pl. 35.)

59′(58) Pseudothecia in 2–3 rows in elongated, oblong stromata . . . . . . . **Dictyodothis**
(Müller and Ahmad, 1962: 156, 161, icon.)

60(58′) Pseudothecia clustered on a uniform, pseudoparenchymatous basal stroma; pseudothecial walls pseudoparenchymatous . . . . . . . . . . . . . **Cucurbitaria**
(Mirza, 1968; Dennis, 1968: 414, Pl. XXXIX E; Verona and Benedek, 1967, C-274; Ellis and Everhart, 1892, Pl. 26; Arnaud, 1925: 660, Fig. 6.)

60′(58′) Pseudothecia scattered on a differentiated basal stroma resembling a foliose lichen thallus, black on the upper side, white internally and below, upper layer a scleroplectenchyma, lower layer a loose prosenchyma; pseudothecial wall scleroplectenchymatous; ascospores sometimes lacking longitudinal septa . . . **Cucurbidothis**
(Holm, 1967: 451, icon. Pl. I, II; Casagrande, 1969; Verona and Benedek, 1970, C-473.)

61(57′) Pseudothecia more or less superficial, small, smooth, gregarious . . . **Teichospora**
(Dennis, 1968: 416, Fig. 16 K.)

61′(57′) Pseudothecia immersed in substrate . . . . . . . . . . . . . . . . . . . . . 62

62(61') Ascospores very large (more than 50 µm), brown, with gelatinous sheath; pseudothecia large, solitary . . . . . . . . . . . . . . . . . . . . . . . . 63
62'(61') Ascospores smaller, lacking gelatinous sheath . . . . . . . . . . . . 65
63(62) Ascospores with few longitudinal septa . . . . . . . . . . . . . **Pleomassaria**
(Dennis, 1968: 417, Fig. 23 E; Ellis and Everhart, 1892, Pl. 30.)
63'(62) Ascospores with numerous longitudinal septa . . . . . . . . . . . . . . . . 64
    64(63') Pseudothecial wall thin, of flattened cells . . . . . . . . . . . **Peltosphaeria**
    (Kern, 1959: 281, icon.)
    64'(63') Pseudothecial wall very thick, of isodiametric cells; asci often with 1–4 ascospores . . . . . . . . . . . . . . . . . . . . . . . . . . . . . . . . . . . . . **Julella**
    (Kern, 1959: 281, 282, icon.)
65(62') Pseudothecia gregarious beneath bark, surrounded by dense brown hyphal web, with beaks erumpent individually through pores in the bark; ascospores sometimes lacking vertical septa . . . . . . . . . . . . . . . . . . . . . . . . . . **Karstenula**
(Dennis, 1968: 417, Pl. XXXIX A.)
65'(62') Pseudothecia scattered, immersed in wood . . . . . . . . . . . . . . **Thyridium**
(Dennis, 1968: 418, Fig. 20 A; Verona and Benedek, 1961, C-65; Müller and Ahmad, 1958: 31–32 icon. as *Xylosphaeria*.) According to Dennis (1968) nomenclatural problems are unresolved and neither *Thyridium* nor *Xylosphaeria* is established as a legitimate generic name for these fungi.)

## H. Mycoporaceae

1. Pseudothecia separate or converging at the beaks, ascospores usually hyaline . . . . 2
1'. Pseudothecia fused laterally into a compound ascocarp with the locules only partially separated by sterile tissue but each locule opening by a pore; ascospores usually brown . . . . . . . . . . . . . . . . . . . . . . . . . . . . . . . . . . . . . . . . . 9
    2(1) Ascospores 1-septate . . . . . . . . . . . . . . . . . . . . . . . . . . . . 3
    2'(1) Ascospores more than 1-septate . . . . . . . . . . . . . . . . . . . . . . 4
3(2) Mature ascospores dissociating into two cells . . . . . . . . . . . . . **Sporoschizon**
(Riedl, 1961.)
3'(2) Ascospores not dissociating into separate cells . . . . . . . . . . . . . **Leiophloea**
(Riedl, 1961.)
    4(2') Ascospores phragmosporous . . . . . . . . . . . . . . . . . . . . . . . . 5
    4'(2') Ascospores dictyosporous . . . . . . . . . . . . . . . . . . . . . . . . . 8
5(4) Ascospores brown . . . . . . . . . . . . . . . . . . . . . . . . . . **Microthelia**
(von Keissler, 1938: 24, 25, icon.)
5'(4) Ascospores hyaline . . . . . . . . . . . . . . . . . . . . . . . . . . . . . . 6
    6(5') Cells of ascospores rounded or lenticular . . . . . . . . . . . **Pseudopyrenula**
    6'(5') Cells of ascospores cylindric or cubic . . . . . . . . . . . . . . . . . . . 7
7(6') Ascospores elliptical to oblong . . . . . . . . . . . . . . . . . . . . **Arthopyrenia**
(von Keissler, 1938: 64, 108, 130, icon.)
7'(6') Ascospores acicular . . . . . . . . . . . . . . . . . . . . . . . . . **Leptorhaphis**
(von Keissler, 1938: 239, 242, icon.)

8(4′) Pseudothecia erect, separate .................... **Microglaena**
    (Morgan-Jones and Swinscow, 1965: 48, icon.) The algal host is *Protococcus*. *Polyblastiopsis* (Riedl, 1961) is similar but has *Trentepohlia* as the host.

8′(4′) Pseudothecia arranged in a circle with necks converging to a common center or fused ........................................... **Parmentaria**

9(1′) Ascospores 1-septate ........................ **Mycoporellum**
    (Riedl, 1961; von Keissler, 1938: 472, icon.)

9′(1′) Ascospores more than 1-septate .................... 10

10(9′) Ascospores phragmosporous ................... **Mycoporopsis**
    (Riedl, 1961.)

10′(9′) Ascospores dictyosporous ..................... **Dermatina**
    (Riedl, 1963; Dennis, 1968: 355, Fig. 19 B; von Keissler, 1938: 471, icon.) *Mycoporum* is a synonym of *Dermatina*.

## XI. HYSTERIALES

The Hysteriales is a long-established order for the single family Hysteriaceae whose members are common on the surface of dead woody branches and on bare wood. The Hysteriaceae are easily recognized by their distinctive boat-shaped to linear, carbonaceous pseudothecia opening by a longitudinal slit and becoming apothecioid when moistened. They superficially resemble the ascocarps of the Hypodermataceae (Phacidiales, Euascomycetes). In cross section the pseudothecia look much the same as the perithecioid pseudothecia of the Pleosporales, with long cylindrical asci among persistent pseudoparaphyses. Groups of lichen fungi with somewhat similar pseudothecia have generally been considered related to the Hysteriaceae. Most of these are included in the family Opegraphaceae. Although a few have fruticose thalli (*Roccella*), most form thin crustose thalli appearing as grayish stains on the bark of trees in which the linear, often curved or branched pseudothecia stand out as etched black lines. In the Arthoniaceae the pseudothecia are irregularly rounded or stellate and in the simplest forms are hardly more than a tangled mass of hyphae in which the globoid asci are embedded (Fig. 9C). Structure here seems on the level of that in the Myriangiales. The Lecanactidaceae have rounded, distinctly apotheciumlike pseudothecia.

Two obviously unrelated families, the Patellariaceae and the Phillipsiellaceae, are included in the Hysteriales as a matter of convenience in identification because their ascocarps are apothecioid (Fig. 4B). They should constitute a separate order. The Patellariaceae is a family traditionally classified among the inoperculate Discomycetes in which the asci have proved to be bitunicate. Most of these fungi produce leathery, brown to black, long-lived, saucer-shaped ascocarps on the surface of dead woody branches and bare wood. Some are lichenized and some are lichen parasites. The disk-shaped hymenium is composed of cylindrical asci interspersed with paraphysoidal hyphae which branch and swell at the tips to form an epithecium over the asci (Fig. 2D, 4B). The Patellariaceae have been considered to be closely related to the discomycetous lichen fungi, and species saprotrophic on wood, species forming lichens, and species parasitizing lichens have even been placed in the same genus (*Buellia, Melaspilea*). The primary diagnostic character is the bitunicate ascus (Fig. 2D), and determination of ascus type is critical since the unitunicate asci of many lichen fungi are thick walled and multilayered.

The Phillipsiellaceae comprises a heterogenous group of fungi fruiting on green leaves and producing superficial or erumpent, more or less apothecioid pseudothecia. Some, such as *Annajenkinsia, Epibelonium*, and *Pseudodiscus*, resemble the Patellariaceae in structure. In some, such as *Pseudoscypha* and *Plochmopeltis*, the ascocarp is hardly more than a hymenium of asci and paraphysoidal hyphae borne on the surface of a basal stroma or a basal hyphal weft. Others, such as *Chaetoscutula, Gymnoascopsis, Henningsiella*, and *Asterotexis*, are placed in the Phillipsiellaceae because they are not better accomodated elsewhere. In making identifications, fungi similar in habit in the Myriangiales and Hemisphaeriales (especially Parmulariaceae and Leptopeltidaceae) should be included in the search.

For purposes of identification the Dothioraceae also might be placed in the Hysteriales. Their asci, however, stand side by side in more or less extensive layers (Fig. 5B) in the ascostroma and at most are separated only by remnant strands of stromal tissue. The basic relationships of the Dothioraceae, therefore, appear to be with the Dothideales (Fig. 5A).

### KEY TO FAMILIES AND IMPORTANT GENERA OF HYSTERIALES

1. Pseudothecia horizontally elongated, oblong, boat-shaped, linear and sometimes branched, or resembling a bivalve mollusk shell (round to elongated in some lichen genera), opening by a narrow longitudinal slit; asci interspersed with pseudoparaphyses . . . . . . . 2
1'. Pseudothecia orbicular, crustose to typically apothecioid and opening widely by crumbling or splitting; asci interspersed with paraphysoidal hyphae which are often enlarged and branched at the tips to form an epithecium . . . . . . . . . . . . . . . . . . . . 4

    2(1) Saprotrophs on tree branches and wood; pseudothecia black, carbonaceous, partially immersed at the base, superficial, or on a subiculum . . . . . **Hysteriaceae** p. 196

    2'(1) Symbionts with algae, forming crustose or rarely fruticose lichens . . . . . . 3

3(2') Pseudothecia lacking an exciple, sometimes hardly more than a layer of asci embedded in localized, irregularly rounded to stellate fertile areas of the thallus; thallus crustose . . . . . . . . . . . . . . . . . . . . . . . . . . . **Arthoniaceae** p. 197

3'(2') Pseudothecia with a well-developed exciple, elliptical to linear (rounded in some species of *Roccella*); thallus crustose or rarely fruticose . . . . **Opegraphaceae** p. 197

    4(1') Superficial parasites, or rarely hyperparasites, on leaves; pseudothecia small, orbicular, crustlike to apothecioid, sometime reduced to a basal stroma or hyphal weft bearing a discoid layer of asci interspersed with paraphysoidal hyphae or embedded in a slime mass . . . . . . . . . . . . . . . . . **Phillipsiellaceae** p. 198

4'(1') Saprotrophs on wood or symbiotic with algae and forming lichens; pseudothecia middle-sized to very large, apothecioid . . . . . . . . . . . . . . . . . . . . 5

5(4') Saprotrophs on wood (some genera containing lichenized species) or rarely parasitic on lichens . . . . . . . . . . . . . . . . . . . . . . . . . . . . . . **Patellariaceae** p. 199

5'(4') Symbiotic with algae, forming crustose lichens; pseudothecia dark; ascospores phragmosporous . . . . . . . . . . . . . . . . . . . . . . . . . . . **Lecanactidaceae** p. 200

## A. Hysteriaceae

1. Pseudothecia oblong to cylindrical with thick walls and with lips of opening depressed and forming a longitudinal trough (lips protruding in *Hysterocarina*). Subfamily **Hysterioideae** . . . . . . . . . . . . . . . . . . . . . . . . . . . . . . . . . . . . . . . 2

1'. Pseudothecia shaped like a boat or bivalve mollusk shell or vertically elongated and like the head of an ax; wall more fragile; lips of opening protruding and appressed to form a sharp ridge. Subfamily **Lophioideae** . . . . . . . . . . . . . . . . . . . . . . . 8

    2(1) Ascospores hyaline . . . . . . . . . . . . . . . . . . . . . . . . . . . 3

    2'(1) Ascospores brown . . . . . . . . . . . . . . . . . . . . . . . . . . . 5

3(2) Ascospores 1-septate; pseudothecia seated on the substratum or rarely on a black, felted subiculum . . . . . . . . . . . . . . . . . . . . . . . . . . . . . . . . . **Glonium**
      (Müller and von Arx, 1962: 245, icon.; Zogg, 1962: 64, icon.; Dennis, 1968: 424 Fig. 15 F; Verona and Benedek, 1967, C-254.) Müller and von Arx segregated species lacking a subiculum in *Psiloglonium*. These species were included in *Glonium* by Zogg.

3'(2) Ascospores more than 1-septate . . . . . . . . . . . . . . . . . . . . . . 4

    4(3') Ascospores phragmosporous . . . . . . . . . . . . . . . . . . . **Gloniella**
      (Zogg, 1962: 82, icon.)

    4'(3') Ascospores dictyosporous . . . . . . . . . . . . . . . . . . . . **Gloniopsis**
      (Zogg, 1962: 51, icon.)

5(2') Ascospores nonseptate or with a septum near the lower end cutting off a small hyaline cell . . . . . . . . . . . . . . . . . . . . . . . . . . . . . . . . . . . . . **Farlowiella**
    (Müller and von Arx, 1962: 245, icon.; Zogg, 1962: 64, icon.; Dennis, 1968: 424, Fig. 15 A; Verona and Benedek, 1967, C-257.)

5'(2') Ascospores more than 1-septate . . . . . . . . . . . . . . . . . . . . . . 6

    6(5') Ascospores phragmosporous . . . . . . . . . . . . . . . . . . . **Hysterium**
      (Zogg, 1962: 24, icon.; Arnaud, 1925: 657, icon.)

    6'(5') Ascospores dictyosporous . . . . . . . . . . . . . . . . . . . . . . . 7

7(6') Lips of opening depressed . . . . . . . . . . . . . . . . . . . . **Hysterographium**
    (Zogg, 1962: 35, icon.; Dennis, 1968: 426, Fig. 15 C; Verona and Benedek, 1967, C-255.)

7'(6') Lips of opening protruding . . . . . . . . . . . . . . . . . . . . **Hysterocarina**
    (Zogg, 1962: 88, icon.; Verona and Benedek, 1967, C-258.) Pseudothecium horizontally elongated, cylindrical and heavy-walled as in Hysterioideae, but lips of slit protruding as in Lophioideae.

    8(1') Pseudothecia boat- to shell-shaped . . . . . . . . . . . . . . . . . . . 9

# 7. Loculoascomycetes

8′(1′) Pseudothecia erect, vertically elongated, ax-head-shaped; ascospores scolecosporous, hyaline, becoming brownish .................... **Glyphium**
(Zogg, 1962: 101, icon.; Verona and Benedek, 1967, C-263.)

9(8) Ascospores 1-septate, usually brown ................ **Actidium**
(Zogg, 1962: 101; Dennis, 1968: 426, Fig. 15 I; Verona and Benedek, 1967, C-262.)

9′(8) Ascospores more than 1-septate ....................... 10

    10(9′) Ascospores with cross septa only .................... 11

    10′(9′) Ascospores dictyosporous, brown ................ **Ostreola**
    (Darker, 1963: 1387, icon.; Verona and Benedek, 1965, C-200.)

11(10) Ascospores phragmosporous, ellipsoid to long fusoid, length: width ratio up to 20:1, brown ................................. **Mytilidion**
(Zogg, 1962: 120, icon.; Dennis, 1968: 427, Fig. 15 E; Verona and Benedek, 1961, C-75.)

11′(10) Ascospores scolecosporous, hyaline, becoming brownish ........ **Lophium**
(Zogg, 1962: 93, 94, 96, icon.; Dennis, 1968: 427, Fig. 15 G; Verona and Benedek, 1961, C-74.)

## B. Arthoniaceae

1. Asci grouped in patches in the thallus, separated by loose white hyphal tissue
(Cryptotheciaceae) ................................. 2

1′. Asci grouped in dark-colored pseudothecia, separated by pseudoparaphyses .... 3

    2(1) Ascospores 1- to several-septate .................. **Stirtonia**
    (Santesson, 1952: 61, icon.)

    2′(1) Ascospores dictyosporous .................... **Cryptothecia**
    (Santesson, 1952: 66.)

3(1′) Ascospores 1- to several-septate ................... **Arthonia**
(Müller and von Arx, 1962: 226, icon.; Santesson, 1952: 87, icon.; Redinger, 1938.)

3′(1′) Ascospores dictyosporous ..................... **Arthothelium**
(Santesson, 1952: 92. Redinger, 1938: 166, icon.)

## C. Opegraphaceae

1. Thallus crustose ................................. 2

1′. Thallus fruticose; pseudothecia rounded to linear; ascospores phragmosporous, hyaline (Roccellaceae) ............................. **Roccella**

    2(1) Ascospores phragmosporous, hyaline .................... 3

    2′(1) Ascospores dictyosporous, hyaline ............ **Helminthocarpon**

3(2) Pseudothecia circular, excipuloid tissue slightly radiate in structure ...... **Mazosia**
(Santesson, 1952: 118, icon.)

3′(2) Pseudothecia elongate to irregularly rounded; excipular tissue not radiate ....... 4

    4(3′) Pseudothecia embedded singly in the thallus ............... 5

    4′(3′) Pseudothecia grouped in stromata ............... **Chiodecton**

5(4) Excipular tissue dark, carbonaceous ................ **Opegrapha**
(Santesson, 1952: 103, icon.)

5′(4) Excipular tissue light, hyaline to brownish . . . . . . . . . . . . . . **Enterographa**
    (Santesson, 1952: 107, icon.)

## D. Phillipsiellaceae

1. Ascocarps compound; asci in several disk-shaped groups on or in a flat stroma; ascospores 1-septate . . . . . . . . . . . . . . . . . . . . . . . . . . . . . . . . . . . . . . 2
1′. Ascocarps simple . . . . . . . . . . . . . . . . . . . . . . . . . . . . . . . . . . 4
    2(1) Mycelium internal, forming a hypostroma within the hypertrophied host tissue; asci interspersed with paraphysoidal hyphae, exposed by rupturing of the overlying stromal tissue; ascospores brown . . . . . . . . . . . . . . . . . . . . . . **Englerodothis**
        (Müller and von Arx, 1962: 85, icon.)
    2′(1) Mycelium superficial or lacking; asci embedded in slime . . . . . . . . . . . 3
3(2′) Asci on the surface of the stroma in light-colored disks of slime; basal stroma parenchymatous, radiate at the margin; mycelium lacking; ascospores hyaline . . . . . . . . . . . . . . . . . . . . . . . . . . . . . . . . . . . . . . . **Henningsiella**
    (Müller and von Arx, 1962: 220, icon.)
3′(2′) Asci in rounded to irregular locules in the stroma, exposed by gelatinization or splitting of the stroma; mycelium hyaline or pale, with lateral hyphopodia; ascospores hyaline to pale brown . . . . . . . . . . . . . . . . . . . . . . . . . . . . . . . . . **Asterotexis**
    (Müller and von Arx, 1962: 93, icon.; Verona and Benedek, 1960, C-31.)
    4(1′) Asci on a thin basal weft of hyphae appressed to leaf cuticle; ascospores hyaline . . . . . . . . . . . . . . . . . . . . . . . . . . . . . . . . . . . . 5
    4′(1′) Asci in a disk-shaped to apothecioid pseudothecium . . . . . . . . . . . . 6
5(4) Ascospores phragmosporous, interspersed with a tangle of erect, uncinately branched hyphae . . . . . . . . . . . . . . . . . . . . . . . . . . . . . . . . **Gymnoascopsis**
    (Moreau and Moreau, 1959: 357, icon.)
5′(4) Ascospores 1-septate; interascicular hyphae branched above into short, dark lobes partially covering asci . . . . . . . . . . . . . . . . . . . . . . . . . . . . **Plochmopeltis**
    (Müller and von Arx, 1962: 208, icon.; Verona and Benedek, 1964, C-155; 1961, as *Plectomyriangium* C-38.)
    6(4′) Ascospores 1-septate . . . . . . . . . . . . . . . . . . . . . . . . . . . . 7
    6′(4′) Ascospores more than 1-septate . . . . . . . . . . . . . . . . . . . . . 12
7(6) Pseudothecium depressed globose, setose; ascospores hyaline, finally becoming pale brown . . . . . . . . . . . . . . . . . . . . . . . . . . . . . . . . . **Chaetoscutula**
    (Müller and von Arx, 1962: 213–214, icon.; Verona and Benedek, 1965, C-203.)
7′(6) Pseudothecium disk-shaped . . . . . . . . . . . . . . . . . . . . . . . . . . . 8
    8(7′) Pseudothecia innate in conifer needles, composed of a basal stroma bearing on its surface a layer of asci and paraphysoidal hyphae whose tips form an epithecium, exposed by rupture of the host epidermis; ascospores hyaline . . . . . **Pseudoscypha**
        (Reid and Pirozynski, 1966: 352–353, icon.; Verona and Benedek, 1966, C-235.)
    8′(7′) Pseudothecia superficial . . . . . . . . . . . . . . . . . . . . . . . . . . . 9
9(8′) Pseudothecium a thin, flat disk, seated on a superficial mycelium . . . . . . . . 10
9′(8′) Pseudothecium tapering basally to a narrow foot . . . . . . . . . . . . . . . 11
    10(9) Mycelium setose; pseudothecia setose or bare; ascospores hyaline . . **Johansonia**
        (Müller and von Arx, 1962: 212, icon.; Verona and Benedek, 1969, C-362.)

10′(9) Mycelium and pseudothecia glabrous, ascospores hyaline to pale
brown . . . . . . . . . . . . . . . . . . . . . . . . . . . . . . . . . **Phillipsiella**
(Müller and von Arx, 1962: 217, icon.; Bose and Müller, 1965: 346, icon.; Verona and Benedek, 1968, C-341.)

11(9′) Foot of pseudothecium connected with an innate hypostroma; ascospores septate in the lower third, becoming brown . . . . . . . . . . . . . . . . . . . . . . . **Coccodothis**
(Müller and von Arx, 1962: 83, icon.)

11′(9′) Foot of pseudothecium resting on a superficial mycelium; ascospores septate in the middle, hyaline . . . . . . . . . . . . . . . . . . . . . . . . . . . . **Pseudodiscus**
(Müller and von Arx, 1962: 219, icon.; Verona and Benedek, 1965, C-201.)

    12(6′) Ascospores phragmosporous . . . . . . . . . . . . . . . . . . . . . . 13

    12′(6′) Ascospores dictyosporous, clavate; pseudothecium narrowed at the base to a hypostroma within host tissue, dark-colored . . . . . . . . . . . . **Protoscypha**
(von Arx, 1963: 470.)

13(12) Ascospores ellipsoid-fusoid, rounded at the ends, hyaline, mostly 3-septate; pseudothecium brown, tapering to a basal stalk arising from a hyaline superficial mycelium . . . . . . . . . . . . . . . . . . . . . . . . . . . . . . . **Epibelonium**
(Müller, 1963: 241, icon.; Verona and Benedek, 1965, C-208.) Like *Pseudodiscus* except for 3-septate ascospores.

13′(12) Ascospores fusoid, tapering to both ends and often prolonged into short filamentous appendages; pseudothecium bright-colored; hyperparasitic on *Phyllachora* . . . . . . . . . . . . . . . . . . . . . . . . . . . . . . . **Annajenkinsia**
(von Arx, 1963: 468; Thirumalachar and Narasinham, 1955: 759, icon.; Verona and Benedek, 1965, C-185.)

## E. Patellariaceae

1. Ascospores 1-septate . . . . . . . . . . . . . . . . . . . . . . . . . . . . . 2

1′. Ascospores more than 1-septate . . . . . . . . . . . . . . . . . . . . . . . 7

    2(1) Ascospores remaining hyaline . . . . . . . . . . . . . . . . . . . . . . 3

    2′(1) Ascospores becoming brown . . . . . . . . . . . . . . . . . . . . . . . 4

3(2) Ascospores ciliate at both ends . . . . . . . . . . . . . . . . . . . . **Banhegyia**
(Müller and von Arx, 1962: 260; Kohlmeyer and Kohlmeyer, 1968, Table 71.)

3′(2) Ascospores not ciliate . . . . . . . . . . . . . . . . . . . . . . . . . **Scutula**
(Müller and von Arx, 1962: 260, icon.; Dennis, 1968: 217, Fig. 11 F.)

    4(2′) Ascospores with a germ pore at each end and a thick epispore . **Eutryblidiella**
(Müller and von Arx, 1962: 252, 253, icon.; Pirozynski and Reid, 1966; Verona and Benedek, 1969, C-361.)

    4′(2′) Ascospores lacking germ pores, usually thin-walled . . . . . . . . . . . . 5

5(4′) Exciple well-developed, conspicuous; pseudothecium orbicular; saprotrophs on wood, forming lichens, or lichen parasites . . . . . . . . . . . . . . . . . . . . **Buellia**
(Müller and von Arx, 1962: 255, 256, icon.; Kohlmeyer and Kohlmeyer, 1965, Table 46; Gallϕe, 1932, Pl. 1–4, 8–17.) *Buellia* may be restricted to lichens and *Karschia* retained for nonlichenized species (Dennis, 1968: 219, Fig. 11 C, D).

5′(4′) Exciple poorly developed, inconspicuous . . . . . . . . . . . . . . . . . . 6

    6(5′) Pseudothecia orbicular, parasitic on lichens . . . . . . . . . . . **Abrothallus**
(Nordin, 1964, Pl. I–III; Dennis, 1968: 220, Fig. 11 E.)

6′(5′) Pseudothecia elongated; saprotrophic on wood, lichenized, or parasitic
on lichens .................................. **Melaspilea**
(Müller and von Arx, 1962: 259, icon.; Redinger, 1938.) *Melaspilea* may be restricted to lichen fungi and *Mycomelaspilea* retained for nonlichenized species (Dennis, 1968: 218, Fig. 19 G).

7(1′) Ascospores phragmosporous ........................... 8

7′(1′) Ascospores dictyosporous, hyaline to yellowish brown .......... **Tryblidaria**
(Müller, 1957: 460, icon.; Müller and Ahmad, 1955: 238, icon. as *Hematomyxa*.)

8(7) Ascospores hyaline ........................... **Lecanidion**
(Nannfeldt, 1932: 329, as *Patellaria*, Pl. 19; Dennis, 1968: 221, as *Patellaria*, Fig. 11 I.)

8′(7) Ascospores brown ........................... **Rhytidhysterium**
(Petrak, 1959; Müller and von Arx, 1962: 251.)

### F. Lecanactidaceae

1. Pseudothecia surrounded by a thalline margin or immersed ....... **Schismatomma**
(Santesson, 1952: 110; Galløe, 1936, Pl. 42–43.)

1′. Pseudothecia lacking thalline margin .................. **Lecanactis**
(Santesson, 1952: 110, Galløe, 1930, Pl. 1–4.)

## XII. HEMISPHAERIALES

The Hemisphaeriales (syn. Microthyriales) comprises a large group of Loculoascomycetes with superficial or rarely subcuticular, dimidiate-scutate ascocarps (Figs. 1H, 1I, 9B). They are mostly tropical and subtropical, but a few representatives are common in temperate regions. They occur primarily on living leaves, young stems, and occasionally fruits as hyperparasites on superficial fungi, as ectocommensals apparently growing on exudates from the stomata and cuticle, and as parasites. Some of the parasites cause necrosis and produce mature ascocarps on dead tissue. The mycelium may be entirely superficial and form dark networks of anastomosing hyphae or bands or plates of parallel or radiating hyphae. Less commonly, the mycelium is hyaline and inconspicuous. Parasites may send haustoria into the host epidermal cells, or the mycelium may be partly or entirely internal. The internal mycelium may form hypostromata ranging from knots of cells in substomatal cavities or in the epidermal cells (Fig. 1I) to more extensive layers involving disintegration of host tissue. Single hyphae or stromatic columns emerge through stomata or directly through the epidermal cell walls to initiate the superficial ascostromata.

The ascocarps may be massive and contain several immersed locules. More commonly, they are orbicular to elongated pseudothecia ranging downward in size to barely visible specks on the host surface. These pseudothecia, however, may fuse laterally into extensive crusts. The ascocarp initial is a flat plate of cells which is appressed to the host surface and develops beneath the superficial mycelium or incorporates the mycelium into its structure. The ascostroma thickens on the under surface of the shield (Figs. 1H, 1I, 9B). Internal structure of the ascostroma and the arrangement of asci within it are variable. Asci are globose to cylindrical. Ascospores are mostly one-septate (Fig. 1D), but multiseptate ascospores are characteristic of some families.

The Hemisphaeriales is separated into a large number of families to preserve the classical organization, which stresses characters such as the radiate or nonradiate structure of the shield, and also to indicate the variations in internal structure of the ascocarps which may suggest relationships of these families with other orders. The families have an excessive number of genera which reflect the many variations in substratum relations, presence or absence of hyphopodia, and similar characters of questionable validity at the generic level. The Munkiellaceae, Microthyriaceae (Fig. 9B), Trichopeltinaceae, and Micropeltidaceae have flattened, hemiperithecioid, ostiolate locules with obclavate to cylindrical asci among persistent pseudoparaphyses and in centrum structure closely resemble the Pleosporales.

In the Parmulariaceae the ascostroma is relatively large and contains several linear or ring-shaped locules in which the asci form continuous layers and are exposed by slits in the overlying stroma. The asci are interspersed with vertical hyphae which appear to be pseudoparaphyses, and the Parmulariaceae are generally considered to be hemisphaeriaceous counterparts of the Hysteriaceae. Adequate study of their development, however, might reveal a closer relationship with the Phillipsiellaceae and Patellariaceae.

The severely flattened shields of the ascocarps in the Schizothyriaceae (Fig. 1H) and Stephanothecaceae cover a thin ascostroma in which the globoid asci are individually distributed as in the Myriangiales. At maturity, however, the asci may be crowded into more or less extensive groups or continuous layers. They are exposed by crumbling and splitting of the shield to form broad pores or longitudinal or ring-shaped slits. Ascocarps of the Leptopeltidaceae are similar in structure, but the stromal tissue in which the asci arise is composed of vertically oriented rows of cells (Fig. 1I). The ascocarps are subcuticular or rarely intra- or subepidermal. They may be fused into black crusts, and the host cells are usually filled with dark mycelium. These families might well be placed alongside of the Saccardinulaceae in the Myriangiales.

The Asterinaceae, along with the small families Aulographaceae and Brefeldiellaceae, have relatively small ascocarps in which the erect, short cylindrical asci form palisades in flat locules. The asci apparently arise individually in the centrum tissue and at maturity may be partially separated by more or less persistent strands of this tissue. The species consequently have been variously described as with "paraphysate" or with "aparaphysate" asci. The shield opens over the locules by broad pores, by stellate fissures, or in a few genera with elongated ascocarps by a longitudinal slit. These fungi appear to be hemisphaeriaceous counterparts of the Dothioraceae (Dothideales).

**KEY TO FAMILIES AND IMPORTANT GENERA OF HEMISPHAERIALES**

1. Pseudothecia hemiperithecioid, opening by a small round ostiole in the center of the shield; asci pseudoparaphysate (pseudoparaphyses may be sparse), obclavate to cylindric-obclavate, tending to be semiprostrate with tips converging toward the ostiole . . . . . 2

1′. Pseudothecium apothecioid, opening by crumbling or splitting of the shield to form a broad pore, a longitudinal slit, or irregular clefts; asci ovoid to short cylindric, erect, individually embedded in stromal tissue, standing side by side in a palisade, or separated by paraphysoidal hyphae . . . . . . . . . . . . . . . . . . . . . . . . . . . . 5

    2(1) Shield of pseudothecium radiate, at least at margins (radiate structure sometimes obscured by the irregular contour of the radially arranged cells or by thickening and darkening of the ascocarp) . . . . . . . . . . . . . . . . . . . . . . . . 3

    2′(1) Shield composed of a hyphal reticulum or of inordinately arranged, sinuous, irregularly lobed cells passing into a hyphal reticulum at the margins; mycelium superficial; brown or hyaline and inconspicuous . . . . . . . **Micropeltidaceae** p. 206

3(2) Pseudothecia subcuticular . . . . . . . . . . . . . . . . . **Munkiellaceae** p. 205

3′(2) Pseudothecia superficial . . . . . . . . . . . . . . . . . . . . . . . . . . . 4

    4(3′) Superficial mycelium, if present, forming a loose reticulum
    . . . . . . . . . . . . . . . . . . . . . . . . . . . . **Microthyriaceae** p. 203

    4′(3′) Superficial mycelium forming a branching, band-shaped thallus of parallel hyphae; pseudothecia developing as circular thickenings beneath the thallus . . . . . . . .
    . . . . . . . . . . . . . . . . . . . . . . . . . **Trichopeltinaceae** p. 205

5(1′) Asci standing side by side or interspersed with paraphysoidal hyphae, typically clavate to cylindric . . . . . . . . . . . . . . . . . . . . . . . . . . . . . . . . . 6

5′(1′) Asci distributed individually, separated by remnants of stromal tissue at maturity, typically ovoid to broad clavate . . . . . . . . . . . . . . . . . . . . . . . . 9

    6(5) Mycelium entirely superficial or partially innate; pseudothecia on the superficial mycelium, typically small, uniloculate, opening by a broad pore formed by crumbling of the shield or by radiating clefts or less commonly by a longitudinal slit . . . . . 7

    6′(5) Superficial mycelium lacking; pseudothecia superficial and connected with the internal mycelium or rarely subcuticular, typically larger, often multiloculate, opening by a circular or longitudinal slit in the shield over the ring-shaped or linear locules; shield usually radiate . . . . . . . . . . . . . . . . . **Parmulariaceae** p. 210

7(6) Shield radiate . . . . . . . . . . . . . . . . . . . . . . . . . . . . . . . . 8

7′(6) Shield composed of irregular, inordinately arranged cells; ascospores 1-septate; mycelium lacking hyphopodia . . . . . . . . . . . . . . . . . . **Aulographaceae** p. 209

    8(7) Superficial mycelium forming a loose reticulum . . . . . **Asterinaceae** p. 207

    8′(7) Superficial mycelium forming an orbicular, membranous thallus of radiating cells; pseudothecia developing as circular thickenings beneath the thallus . . . . . . . .
    . . . . . . . . . . . . . . . . . . . . . . . . . **Brefeldiellaceae** p. 210

9(5′) Pseudothecia superficial . . . . . . . . . . . . . . . . . . . . . . . . 10
9′(5′) Pseudothecia subcuticular or rarely intraepidermal, often fused to form black crusts; shield usually radiate . . . . . . . . . . . . . . . . . . . . **Leptopeltidaceae** p. 213
    10(9) Shield of pseudothecium radiate . . . . . . . . . **Stephanothecaceae** p. 212
    10′(9) Shield of pseudothecium of inordinately arranged cells **Schizothyriaceae** p. 213

## A. Microthyriaceae

1. Ascospores nonseptate, ellipsoid, hyaline to brownish; mycelium innate; ascocarps single or fused laterally into dark crusts, on dead stems . . . . . . . . . . . . . . . . . . 2
1′. Ascospores septate . . . . . . . . . . . . . . . . . . . . . . . . . . . . . . 3
    2(1) Pseudothecia connected with internal mycelium only by fine pegs penetrating the outer walls of the epidermal cells . . . . . . . . . . . . . . . . . **Myiocopron**
    (von Arx and Müller, 1954: 92, icon.; Arnaud, 1918, Pl. XLIII; 1931, Pl. IV; Verona and Benedek, 1961, C-39.) *Microdothella* with subcuticular pseudothecia in the Munkiellaceae and the following genus, *Ellisiodothis*, are nearly identical with *Myiocopron*.

    2′(1) Pseudothecia connected by hyphae or hyphal strands to the internal mycelium . . . . . . . . . . . . . . . . . . . . . . . . . . . . . . . . . . . . . . . . . . . **Ellisiodothis**
    (Müller and von Arx, 1962: 96, icon.; Verona and Benedek, 1969, C-360.)
3(1′) Ascospores 1-septate . . . . . . . . . . . . . . . . . . . . . . . . . . . 4
3′(1′) Ascospores more than 1-septate . . . . . . . . . . . . . . . . . . . . . 17
    4(3) Innate hypostroma or mycelium present although it may be only subcuticular and inconspicuous . . . . . . . . . . . . . . . . . . . . . . . . . . . . . . . . 5
    4′(3) Innate mycelium lacking, mycelium entirely superficial . . . . . . . . . . 12
5(4) Superficial mycelium lacking . . . . . . . . . . . . . . . . . . . . . . . 6
5′(4) Superficial mycelium present but often inconspicuous . . . . . . . . . . . 8
    6(5) Innate mycelium forming a subcuticular hypostroma, with hyphae penetrating deeper; ascocarp multiloculate . . . . . . . . . . . . . . . . . . . . **Dothidella**
    (Müller and von Arx, 1962: 534, 535, icon.; Arnaud, 1931, as *Polystomella*, Pl. VI; Verona and Benedek, 1964, C-162.)

    6′(5) Innate mycelium not forming a hypostroma, remaining hyphal, subcuticular and deeper; ascocarps uniloculate but often fused laterally into crusts . . . . . . . 7
7(6′) Asci arranged in a ring around the sterile center of the locule; ascospores septate at or below the middle, hyaline to pale brown . . . . . . . . . . . . . . . **Cyclotheca**
    (Müller and von Arx, 1962: 528, icon.; Arnaud, 1918, as *Hariotula*, Pl. XLIV.)
7′(6′) Asci distributed throughout locule; ascospores septate near the middle, becoming brown . . . . . . . . . . . . . . . . . . . . . . . . . . . . . . . . . . **Seynesiella**
    (Müller and von Arx, 1962: 428, icon.; Arnaud, 1918, Pl. XLV; Dennis, 1968: 430, Fig. 19 H; Verona and Benedek, 1968, C-322.)
    8(5′) Ascospores hyaline; mycelium not setose . . . . . . . . . . . . . . . . 9
    8′(5′) Ascospores brown; mycelium sometimes setose . . . . . . . . . . . . . 10
9(8) Both superficial and internal mycelium inconspicuous; pseudothecia scattered, maturing on dead leaves and stems . . . . . . . . . . . . . . . . . . . . . . . **Microthyrium**

(Müller and von Arx, 1962:510, icon.; Arnaud, 1918, Pl. XXV; Dennis, 1968:429, Pl. XXXVI F; Verona and Benedek, 1962, C-77.)

9'(8) Mycelium forming a hypostromal crust beneath cuticle and in epidermal cells; pseudothecia usually fused laterally into conspicuous crusts, maturing on living leaves .................................... **Polycyclinopsis**
(Müller and von Arx, 1962:533; Batista and Vital, 1958:285 icon.)

10(8') Mycelium setose; pseudothecia multiloculate, maturing on living leaves; hypostroma crustlike, subcuticular ....................... **Seynesiopeltis**
(Müller and von Arx, 1962:532; Arnaud, 1931, Pl. VII.)

10'(8') Mycelium not setose; pseudothecia uniloculate although often fused laterally into crusts, maturing on dead leaves and stems .................. 11

11(10') Pseudothecia usually fused laterally into crusts; internal mycelium forming a compact subcuticular hypostroma ......................... **Palawania**
(Müller and von Arx, 1962:508.)

11'(10') Pseudothecia separate; internal mycelium hyphal or forming small knots in epidermal cells .............................................. **Arnaudiella**
(Müller and von Arx: 1962: 513, icon.)

12(4') Mycelium lacking hyphopodia but sometimes with irregular hyphopodiumlike branches ............................................ 13

12'(4') Mycelium with lateral hyphopodia consistently present .......... 14

13(12) Ascospores remaining hyaline; shield thick, with pore late in appearing ........................................ **Calothyriopsis**
(Müller and von Arx, 1962: 520, 521, icon.)

13'(12) Ascospores ultimately becoming brownish; shield thin, with light spot in center indicating position of pore ........................... **Asterinella**
(Müller and von Arx, 1962: 515, icon.; Arnaud, 1918, Pl. XXIV; Verona and Benedek, 1962, C-110; 1969, C-369.)

14(12') Pseudothecia raised hemisphaerical, fused laterally into stromatic crusts; haustoria large, septate ........................................ **Xenostomella**
(Müller and von Arx, 1962: 527, icon.)

14'(12') Pseudothecia flattened, usually separate; haustoria small, nonseptate . . 15

15(14') Ascospores becoming brown, usually more than 8 $\mu$m wide ..... **Maublancia**
(Müller and von Arx, 1962: 523; Arnaud, 1918, Pl. XXVIII; Verona and Benedek, 1967, C-273.)

15'(14') Ascospores remaining hyaline, usually less than 8 $\mu$m wide ........ 16

16(15') Ascospores drawn out at the lower end into a long slender appendage ............................................ **Caudella**
(Müller and von Arx, 1962: 516, 518, icon.)

16'(15') Ascospores rounded at both ends ............... **Asterinema**
(Müller and von Arx, 1962: 522; Batista and Gayao, 1953: 160, icon.)

17(3') Ascospores phragmosporous ....................... 18

17'(3') Ascospores dictyosporous, 3- to 4-septate with a single longitudinal septum in several cells, reddish; superficial mycelium present ................ **Yatesula**
(Batista, 1959a: 40, icon.) *Petrakina* (Petrak and Ciferri, 1932) is similar to *Yatesula*.

18(17) Superficial mycelium lacking; ascospores hyaline . . . . . . . . . . . 19
18'(17) Superficial mycelium present . . . . . . . . . . . . . . . . . . . . . 22
19(18) Pseudothecia dissolving at the tip, then entirely dissolving into slime
   mass . . . . . . . . . . . . . . . . . . . . . . . . . . . . . . . . . . . **Actinomyxa**
   (Theissen and Sydow, 1917: 417.)
19'(18) Pseudothecia remaining intact . . . . . . . . . . . . . . . . . . . . . 20
   20(19') Pseudothecia setose or hairy . . . . . . . . . . . . . . . . **Caenothyrium**
   (Theissen and Sydow, 1917: 417.)
   20'(19') Pseudothecia glabrous . . . . . . . . . . . . . . . . . . . . . . . 21
21(20') On vascular plants . . . . . . . . . . . . . . . . . . . . . . **Phragmothyrium**
   (Theissen and Sydow, 1917: 416.)
21'(20') On lichens . . . . . . . . . . . . . . . . . . . . . . . . . . . **Micropeltopsis**
   22(18') Hyphopodia lacking; ascospores hyaline . . . . . . . . . . . **Halbaniella**
   (Theissen and Sydow, 1917: 421.)
   22'(18') Hyphopodia present, intercalary; ascospores finally becoming
     brown . . . . . . . . . . . . . . . . . . . . . . . . . . . . . . . . **Platypeltella**
   (Farr and Pollack, 1969: 192–193, icon.)

## B. Trichopeltinaceae

1. Ascospores 1-septate . . . . . . . . . . . . . . . . . . . . . . . . . . . . . . 2
1'. Ascospores phragmosporous, hyaline . . . . . . . . . . . . . . . . **Trichopeltum**
   (Marasas, 1966: 213, icon., Fig. 6.)
   2(1) Ascospores hyaline . . . . . . . . . . . . . . . . . . . . . . . **Trichopeltina**
   (Arnaud, 1930, Pl. XIII; Müller and von Arx, 1962: 836.)
   2'(1) Ascospores brown . . . . . . . . . . . . . . . . . . . . . . . **Trichopeltella**
   (Theissen and Sydow, 1917: 426; Müller and von Arx, 1962: 837.)

## C. Munkiellaceae

1. Ascospores nonseptate . . . . . . . . . . . . . . . . . . . . . . . . . . . . . 2
1'. Ascospores septate . . . . . . . . . . . . . . . . . . . . . . . . . . . . . . 6
   2(1) Ascospores brown with light colored band around middle . . . . . . . . . 3
   2'(1) Ascospores uniformly colored, hyaline to brown . . . . . . . . . . . . . 4
3(2) Intracuticular, with bandlike mycelium; shield of pseudothecium thin, brown, 1-cell
   thick . . . . . . . . . . . . . . . . . . . . . . . . . . . . . . . . . . . **Entopeltis**
   (von Arx and Müller, 1954: 99–100, icon.) *Entopeltis* and the following genus, *Vizella*,
   with hemiperithecioid pseudothecia and *Blasdalea* of the Parmulariaceae with apothecioid pseudothecia were grouped in the family Entopeltaceae by von Arx and
   Müller (1954: 98–105) because of the distinctive ascospores.
3'(2) Subcuticular, usually lacking mycelium; shield of pseudothecium several cells thick,
   black, carbonaceous . . . . . . . . . . . . . . . . . . . . . . . . . . . **Vizella**
   (von Arx and Müller, 1954: 103, icon.)
   4(2') Pseudothecia multiloculate with immersed locules, maturing on living leaves; ascos-
     pores hyaline or becoming brown . . . . . . . . . . . . . . . . . . . **Trabutia**
   (von Arx and Müller, 1954: 86, icon.; Ellis and Everhart, 1892, Pl. 31.)
   4'(2') Pseudothecia uniloculate, single or sometimes fused laterally; ascospores
     hyaline . . . . . . . . . . . . . . . . . . . . . . . . . . . . . . . . . . . . 5

5(4′) Maturing on living leaves ........................ **Parastigmatea**
      (von Arx and Müller, 1954: 82, icon.; Verona and Benedek, 1970, C-479.)
5′(4′) Maturing on dead stems and leaves ................ **Microdothella**
      (von Arx and Müller, 1954: 94–95, icon.; Verona and Benedek, 1970, C-445.)
    6(1′) Ascospores 1-septate ........................... 7
    6′(1′) Ascospores more than 1-septate, phragmosporous ............... 8
7(6) Ascospores septate in the lower fourth, hyaline; pseudothecium
    multiloculate ................................. **Munkiella**
      (Müller and von Arx, 1962: 537–538, icon.; Verona and Benedek, 1964, C-156.)
7′(6) Ascospores septate in the upper or lower third, pale yellowish-green to brown; pseudothecia uniloculate, separate or fused laterally ............... **Hormotheca**
      (Müller and von Arx, 1962: 416, as *Coleroa*, icon.; Dennis, 1968: 378, as *Stigmatea*, Fig. 22 A.)
    8(6′) Ascospores hyaline ....................... **Stigmatodothis**
      (Theissen and Sydow, 1917: 401.)
    8′(6) Ascospores brown ...................... **Melanochlamys**
      (Theissen and Sydow, 1917: 402.)

## D. Micropeltidaceae

1. Superficial mycelium hyaline, inconspicuous ................... 2
1′. Superficial mycelium brown, forming a reticulum over the surface of the host ... 7
    2(1) Ascospores nonseptate, hyaline ................ **Dictyothyrina**
      (Theissen and Sydow, 1917: 429.)
    2′(1) Ascospores septate ........................... 3
3(2′) Ascospores 1-septate ...................... **Dictyothyrium**
      (Müller and von Arx, 1962: 540.)
3′(2′) Ascospores more than 1-septate ..................... 4
    4(3′) Ascospores with several cross-septa .................. 5
    4′(3′) Ascospores dictyosporous ....................... 6
5(4) Ascospores fusoid to cylindrical .................... **Micropeltis**
      (Batista, 1959b: 72, 88–89, 100–101, icon.)
5′(4) Ascospores filiform ....................... **Scolecopeltis**
      (Batista, 1959b: 172–173, icon. as *Scolecopeltidium*.) *Scolecopeltis* intergrades with *Micropeltis* and is doubtfully distinct.
    6(4′) Ascospores hyaline ....................... **Muricopeltis**
      (Batista, 1959b: 312, icon.)
    6′(4′) Ascospores brown ....................... **Neopeltella**
      (Petrak, 1950: 329.)
7(1′) Ascospores 1-septate ........................... 8
7′(1′) Ascospores more than 1-septate ..................... 10

8(7) Posterior cell of ascospore reduced to a filiform appendage to the ellipsoid-clavate anterior cell, ascospores brown; pseudothecia multiloculate . . . . . **Stomiotheca** (Batista, 1959b: 455–456, icon.)

8'(7) Posterior cell of ascospore only slightly smaller than anterior cell, ascospores hyaline . . . . . . . . . . . . . . . . . . . . . . . . . . . . . . . . . . . 9

9(8') Pseudothecia and mycelium glabrous . . . . . . . . . . . . . . . . . **Stomiopeltis** (Müller and von Arx, 1962: 546, icon.; Verona and Benedek, 1965, C-188.)

9'(8') Pseudothecia or mycelium setose . . . . . . . . . . . . . . . . . . **Chaetothyrina** (Müller and von Arx, 1962; 548, icon.)

    10(7') Ascospores phragmosporous, hyaline . . . . . . . . . . . . . . . **Bonaria** (Batista, 1959b: 443–444, icon.)

    10'(7') Ascospores dictyosporous . . . . . . . . . . . . . . . . . . . . . . . 12

11(10') Ascospores hyaline . . . . . . . . . . . . . . . . . . . . . **Dictyostomiopelta** (Batista, 1959b: 435, icon.)

11'(10') Ascospores brown . . . . . . . . . . . . . . . . . . . . . . **Mendoziopeltis** (Batista, 1959: 437, 294, icon.)

## E. Asterinaceae

1. Pseudothecia orbicular, opening by a broad pore or by irregularly stellate fissures; ascospores usually 1-septate, rarely phragmosporous (then so indicated in key) . . . . . . 2

1'. Pseudothecia elongated, opening by a longitudinal slit; ascospores 1-septate . . . 21

    2(1) Mycelium entirely superficial . . . . . . . . . . . . . . . . . . . . . . . . 3

    2'(1) Mycelium partially innate . . . . . . . . . . . . . . . . . . . . . . . . . 11

3(2) Hyphopodia present . . . . . . . . . . . . . . . . . . . . . . . . . . . . . . 4

3'(2) Hyphopodia lacking . . . . . . . . . . . . . . . . . . . . . . . . . . . . . 10

    4(3) Hyphopodia intercalary; ascospores brown . . . . . . . . . . . **Asterolibertia** (Müller and von Arx, 1962: 97; Arnaud, 1918, Pl. XXXI; 1925, Pl. XIII, XIV.)

    4'(3) Hyphopodia lateral . . . . . . . . . . . . . . . . . . . . . . . . . . . . 5

5(4') Pseudothecia fused laterally into irregular crusts; ascospores brown at maturity . . . . . . . . . . . . . . . . . . . . . . . . . . . . . . . . . . **Symphaster** (Müller and von Arx, 1962: 94.)

5'(4') Pseudothecia single . . . . . . . . . . . . . . . . . . . . . . . . . . . . . 6

    6(5') Mycelium and pseudothecia spinose; ascospores brown . . . . . **Trichasterina** (Müller and von Arx, 1962: 94; Arnaud, 1918, Pl. XXXIII.)

    6'(5') Mycelium and pseudothecia glabrous . . . . . . . . . . . . . . . . . . . 7

7(6') Ascospores 1-septate, becoming brown . . . . . . . . . . . . . . . . . . . . 8

7'(6') Ascospores phragmosporous, becoming brown . . . . . . . . . . . . . . . . 9

    8(7) Shield of thick-walled cells, opening by radiating fissures, crumbling or disintegrating into slime; conidia lacking or in hemispherical pycnidia . . . . . . . **Asterina** (Müller and von Arx, 1962: 106, icon.; Kar and Maity, 1970a; Arnaud, 1918, Pl. XXXII, as *Dimerosporium*, Pl. II; Dennis, 1968: 430, Pl. XXXVI G; Verona and Benedek, 1968, C-338.)

8'(7) Shield of bright, thin-walled cells, disintegrating into slime; mycelial conidia consistently present, with 2–3 septa (*Mitteriella*) or irregularly septate and thick-walled (*Sarcinella*) .................................................. **Clypeolella**
(Müller and von Arx, 1962: 103, icon.; Verona and Benedek, 1970, C-442.)

9(7') Ascospores fusoid, 3-septate; shield of several layers of cells; stauriform conidia borne on mycelium ................................................... **Batistinula**
(von Arx, 1960: 9–11, icon.)

9'(7') Ascospores clavate, narrowed below, 2-septate; shield one cell thick; conidia lacking ...................................................... **Patouillardina**
(Arnaud, 1918, Pl. XXXVII.)

    10(3') Ascospores 1-septate, separating at the septum into two cells, finally brown; mycelium with spirally coiled spines .................... **Anariste**
(Müller and von Arx, 1962: 139.)

    10'(3') Ascospores phragmosporous, brown ............... **Kriegeriella**
(Batista, 1959a: 37, icon.)

11(2') Superficial hyphae torulose, forming cell complexes over stromata and hyaline mycelium penetrating leaf; ascospores brown ..................... **Placoasterella**
(Müller and von Arx, 1962: 141, icon.)

11'(2') Superficial hyphae filamentous ........................ 12

    12(11') Hypostroma subcuticular; ascospores brown ....... **Dothidasteromella**
(Müller and von Arx, 1962: 143–144, icon.; Arnaud, 1918, as *Balansina*, Pl. XII.)

    12'(11') Hypostroma intraepidermal or deeper ................. 13

13(12') Lateral hyphopodia present, ascospores becoming brown .......... 14

13'(12') Hyphopodia lacking (or sparse and atypical in *Asterodothis* and *Macowaniella*) ................................................ 15

    14(13) Mycelial conidia present, tretic (porospores), borne on modified hyphopodia ............................................. **Maurodothina**
(Pirozynski and Shoemaker, 1970: 1325, icon., Pl. I.) *Maurodothina* differs from *Viegasia* only in the presence of tretic conidia.

    14'(13) Mycelial conidia lacking ..................... **Viegasia**
(Müller and von Arx, 1962: 91.)

15(13') Marginal hyphae bound together in strands radiating from the periphery of the pseudothecium; hyphopodia sometimes present; ascospores brown ........ **Asterodothis**
(Müller and von Arx, 1962: 89.)

15'(13') Marginal hyphae of the pseudothecium not forming radiating strands .... 16

    16(15') Internal mycelium forming prosenchymatous intercellular complexes; ascospores brown; hyphodia sparse and atypical ............... **Macowaniella**
(Müller and von Arx, 1962: 146.)

    16'(15') Internal mycelium filamentous when intercellular, forming parenchymatous intracellular complexes ........................................ 17

17(16') Mycelium and pseudothecia bearing numerous 3-septate conidia; ascospores brown ................................................. **Eupelte**
(Müller and von Arx, 1962: 137.)

17'(16') Mycelium and pseudothecia lacking conidia ................. 18

18(17') Pseudothecia fused laterally into stromatic crusts; ascospores long hyaline, finally brown . . . . . . . . . . . . . . . . . . . . . . . . . . . . . . . . . . . . . . . . **Neostomella**
(Müller and von Arx, 1962: 145; Batista and Garnier de Souza 1960: 11, icon.; Batista *et al.*, 1963: 20, icon.)
18'(17') Pseudothecia single . . . . . . . . . . . . . . . . . . . . . . . . . . . . . . . . 19
19(18') Ascospores 1-septate . . . . . . . . . . . . . . . . . . . . . . . . . . . . . . . . 20
19'(18') Ascospores phragmosporous, 3-septate, with a middle septum and an additional septum near either end; brown . . . . . . . . . . . . . . . . . . . . . . . . . . **Halbania**
(Theissen and Sydow, 1917: 417; Arnaud, 1918, Pl. XXX; Verona and Benedek, 1962, C-112.)
 20(19) Pseudothecia usually smaller than 100 μm diam.; ascospores hyaline, less than 6 μm wide; mycelium inconspicuous, brownish or hyaline . . . . . **Aphanopeltis**
(Müller and von Arx, 1962: 137.)
 20'(19) Pseudothecia usually larger than 100 μm diam.; ascospores broader and becoming brown; mycelium brown . . . . . . . . . . . . . . . . . . . . . . . **Prillieuxina**
(Müller and von Arx, 1962: 132, icon. from Arnaud, 1918, Pl. XXIX.)
21(1') Mycelium entirely superficial, sending haustoria into epidermal cells; hyphopodia present or lacking . . . . . . . . . . . . . . . . . . . . . . . . . . . . . . . . . . . 22
21'(1') Mycelium partially innate; hyphopodia lacking . . . . . . . . . . . . . . . . 26
 22(21) Mycelium hyphopodiate . . . . . . . . . . . . . . . . . . . . . . . . . . . . . 23
 22'(21) Mycelium lacking hyphopodia . . . . . . . . . . . . . . . . . . . . . . . . . 25
23(22) Hyphopodia intercalary; ascospores hyaline . . . . . . . . . . . . . . . **Cirsosia**
(Müller and von Arx, 1962: 113–115, icon.; Arnaud, 1918, as *Cirsosiella*, Pl. XV.)
23'(22) Hyphopodia lateral . . . . . . . . . . . . . . . . . . . . . . . . . . . . . . . . . . 24
 24(23') Ascospores remaining hyaline . . . . . . . . . . . . . . . . **Lembosiellina**
(Müller and von Arx, 1962: 112.)
 24'(23') Ascospores becoming brown . . . . . . . . . . . . . . . . . . . . . **Lembosia**
(Müller and von Arx, 1962: 111, icon.; Arnaud, 1918, Pl. XVI, XVII, XIX, XX.)
25(22') Mycelium dark, conspicuous; ascospores becoming brown . . . . . **Echidnodella**
(Müller and von Arx, 1962: 118, icon.; Verona and Benedek, 1970, C-480.)
25'(22') Mycelium delicate, pale brown, inconspicuous; ascospores very small, pale brown to hyaline . . . . . . . . . . . . . . . . . . . . . . . . . . . . . . . . . . . . . **Morenoina**
(Müller and von Arx, 1962: 129; Arnaud, 1918, Pl. XXII.)
 26(21') Ascospores brown; internal mycelium forming subcuticular or intraepidermal bands or membranes or deeper-lying layers . . . . . . . . . . . . . **Lembosina**
(Müller and von Arx, 1962: 119, icon.; Arnaud, 1918, as *Maurodothella*, Pl. XIII; Verona and Benedek, 1970, C-449; Arnaud, 1918, Pl. XVI, XVII, XIX, XX.)
 26'(21') Ascospores hyaline; internal mycelium of hyaline hyphae arising from dark cell complexes in depressions over the stomata; superficial mycelium often torulose . . . . . . . . . . . . . . . . . . . . . . . . . . . . . . . . . . . . . **Aulographina**
(Müller and von Arx, 1962: 126, icon.; Verona and Benedek, 1965, C-206.)

## H. Aulographaceae

1. Pseudothecia elongated oblong, opening by a longitudinal slit; mycelium mostly a sub-

cuticular membrane; ascospores hyaline to pale brown . . . . . . . . . **Aulographum**
(Müller and von Arx, 1962: 128, icon.)

1'. Pseudothecia irregularly branched, multiloculate, opening by irregular slits; mycelium superficial, reticulate; ascospores hyaline . . . . . . . . . . . . . . . . **Polyclypeolina**
(Batista, 1959b: 457.)

## I. Brefeldiellaceae

1. Superficial mycelium present, brown, forming appressoriumlike disks or hyphopodia; radiate thallus developing beneath mycelium; ascospores brown . . . . . . . . **Pycnocarpon**
(Müller and von Arx, 1962: 149.)

1'. Superficial mycelium lacking . . . . . . . . . . . . . . . . . . . . . . . . . . . 2

    2(1') Ascospores becoming brown; thallus connected by pegs to internal, intracellular mycelium . . . . . . . . . . . . . . . . . . . . . . . . . . . . . . . **Myriostigmella**
    (Müller and von Arx, 1962: 152, icon.)

    2'(1') Ascospores hyaline, small; internal mycelium lacking or inconspicuous . . . . . . . . . . . . . . . . . . . . . . . . . . . . . . . . . **Brefeldiella**
    (Müller and von Arx, 1962: 149.)

## J. Parmulariaceae

1. Ascospores nonseptate, brown with a pale-colored band around the middle; pseudothecia subcuticular with a ring-shaped locule opening by a circular slit . . . . . . . **Blasdalea**
(von Arx and Müller, 1954: 105, icon.) Compare with *Entopeltis*, Munkiellaceae.

1'. Ascospores 1-septate . . . . . . . . . . . . . . . . . . . . . . . . . . . . . . . 2

    2(1') Internal mycelium lacking; forming a superficial crustlike stroma containing many rounded to elongated locules; penetrating epidermal cells with haustoria; ascospores brown . . . . . . . . . . . . . . . . . . . . . . . . . . . . . . . **Pseudolembosia**
    (Müller and von Arx, 1962; 73 icon.)

    2'(1') Internal mycelium present, subcuticular, intraepidermal, or deeper . . . . . 3

    3(2') Internal mycelium of separate, hyaline hyphae, not forming cellular hypostromatic layers; pseudothecia small, circular, uniloculate; ascospores brown . . . . . . . . . **Ferrarisia**
    (Müller and von Arx, 1962: 75, icon.; Verona and Benedek, 1970, C-481.)

    3'(2') Internal mycelium a membranous or crustlike hypostroma which may be subcuticular, intraepidermal, or deeper; pseudothecia usually larger and multiloculate (but see *Palawaniella* and *Dothidasteroma* which may be uniloculate with rounded locules) . . . . 4

        4(3') Shield not distinctly radiate in structure . . . . . . . . . . . . . . . . . 19

        4'(3') Radiate structure of shield apparent, at least at margins . . . . . . . . . . 5

5(4') Hypostroma subcuticular . . . . . . . . . . . . . . . . . . . . . . . . . . . 6

5'(4') Hypostroma intraepidermal or deeper . . . . . . . . . . . . . . . . . . . . 7

    6(5) Subcuticular mycelium hyphal, forming flat radiate hypostromata; pseudothecia single, fusing laterally into crusts containing elongate to linear locules; ascospores hyaline, becoming brown . . . . . . . . . . . . . . . . . . . . . . . **Aulacostroma**
    (Müller and von Arx, 1962: 72.)

    6'(5) Hypostroma membranous or band-shaped; pseudothecia usually single, rounded

## 7. Loculoascomycetes

to oblong with a rounded, oblong, irregular, or ring-shaped locule; ascospores becoming brown . . . . . . . . . . . . . . . . . . . . . . . . . . **Dothidasteroma**
(Müller and von Arx, 1962: 71 icon.)

7(5') Locules in more or less complete rings surrounding a sterile center . . . . . . . . 8

7'(5') Locules not in rings, linear . . . . . . . . . . . . . . . . . . . . . . . . 12

    8(7) Ascospores separating into 2 cells at the septum, dark brown . . . **Cyclostomella**
    (Müller and von Arx, 1962: 50–51, icon.; Arnaud, 1918, as *Cocconiopsis*, Pl. VIII.)

    8'(7) Ascospores remaining 2-celled . . . . . . . . . . . . . . . . . . . . . . 9

9(8') Ascospores septate near the lower end, brownish . . . . . . . . . . . **Polycyclus**
(Müller and von Arx, 1962: 34, icon.)

9'(8')Ascospores septate near the middle . . . . . . . . . . . . . . . . . . . . . 10

    10(9') Locules in several concentric rings; ascospores small (up to 14 $\mu$m long), hyaline . . . . . . . . . . . . . . . . . . . . . . . . . . . . . **Polycyclina**
    (Müller and von Arx, 1962: 34.)

    10'(9') Locules in single, usually incomplete rings; ascospores larger and usually becoming brown . . . . . . . . . . . . . . . . . . . . . . . . . . . . . . . . . . 11

11(10') Ascocarp connected with hypostroma at many points by single hyphae penetrating cuticle . . . . . . . . . . . . . . . . . . . . . . . . . . . . . . . . . . . . **Cocconia**
(Müller and von Arx, 1962: 58, icon.; Hansford, 1946: 171, icon.; Verona and Benedek, 1970, C-441.)

11'(10') Ascocarp connected with the hypostroma by a columnar foot . . . **Cycloschizon**
(Müller and von Arx, 1962: 52–55, icon.; Arnaud, 1918, Pl. I, VII; Verona and Benedek, 1963, C-133.)

    12(7') Locules linear, extending radially from sterile center or running parallel . . . 13

    12'(7') Locules rounded, oblong or linear, inordinately arranged and often anastomosing, at most radially arranged at margins, rarely single and then rounded . . . . . 17

13(12) Ascocarp elongated; locules forming parallel lines; on ferns . . . **Parmulariopsis**
(Müller and von Arx, 1962: 44.)

13'(12) Ascocarp rounded or fan-shaped, with radiating locules . . . . . . . . . . . 14

    14(13') Ascocarp fan-shaped . . . . . . . . . . . . . . . . . . . . . **Rhipidocarpon**
    (Müller and von Arx, 1962: 46, icon.; Arnaud, 1918, Pl. XI; Verona and Benedek, 1963, C-132.)

    14'(13') Ascocarps rounded; locules radiating in all directions from a sterile center     15

15(14') Ascocarps rather large and thick; ascospores usually broader than 5 $\mu$m
. . . . . . . . . . . . . . . . . . . . . . . . . . . . . . . . . . . . . . **Parmularia**
(Müller and von Arx, 1962: 48; Arnaud, 1918, Pl. IX.) *Parmulariopsella* (Sivanesan, 1970b) is a probable synonym of *Parmularia*.

15'(14') Ascocarps small or thin; ascospores long remaining hyaline, finally brownish, at most 5 $\mu$m wide . . . . . . . . . . . . . . . . . . . . . . . . . . . . . . . 16

    16(15') Ascospores septate at or above the middle; ascocarp flattened conical; on phanerogams . . . . . . . . . . . . . . . . . . . . . . . . . . . . . **Parmulina**
    (Müller and von Arx, 1962: 45; Arnaud, 1918, Pl. X; Verona and Benedek, 1963, C-134.)

16'(15') Ascospores septate near the lower end; ascocarp flat, crustlike; commonly on ferns . . . . . . . . . . . . . . . . . . . . . . . . . . . . . . . . . . . . . . **Inocyclus**
(Müller and von Arx, 1962: 35; Arnaud, 1918, as *Hysterostomella*, Pl. VI.)

17(12') Ascospores at most 6 μm broad, becoming brownish . . . . . . . **Rhagodolobium**
(Müller and von Arx, 1962: 42, icon.)

17'(12') Ascospores usually broader than 6 μm becoming brown . . . . . . . . . . . 18

18(17') Ascocarp multiloculate; locules rounded, oblong, or linear . **Hysterostomella**
(Müller and von Arx, 1962: 61–63, icon.; Verona and Benedek, 1970, C-443.)

18'(17') Ascocarps uniloculate, small but clustered, locules rounded . . **Palawaniella**
(Müller and von Arx, 1962: 69, icon.; Verona and Benedek, 1969, C-363.)

19(4) Ascocarp orbicular, connected by a central foot to an immersed hypostroma, containing a single ring-shaped locule opening by a circular slit; ascospores becoming brown
. . . . . . . . . . . . . . . . . . . . . . . . . . . . . . . . . . . . . . . . **Perischizon**
(Müller and von Arx, 1962: 81.)

19'(4) Ascocarp orbicular connected by a central foot to a subcuticular hypostroma, containing several linear locules radiating toward the periphery but fused toward the center into a ring; ascospores becoming brown . . . . . . . . . . . . . . . **Symphaeophyma**
(Müller and von Arx, 1962: 80.)

## K. Stephanothecaceae

1. Mycelium internal, forming hypostroma; pseudothecia thick; asci clavate, embedded in a hyaline gelatinous tissue; ascospores 1-septate hyaline . . . . . . . . . . . . . . . 2

1'. Mycelium entirely superficial; ascocarps thin, flattened; asci ovoid, embedded in a parenchymatous tissue; ascospores more than 1-septate . . . . . . . . . . . . . 3

2(1) Pseudothecium connected to subcuticular or intraepidermal hypostroma at many points by single hyphae perforating cuticle . . . . . . . . . . . . . **Protothyrium**
(Müller and von Arx, 1962: 77; Arnaud, 1918, Pl. V.)

2'(1) Pseudothecium developed from columns of hyphae erumpent mostly through stomata, often from pseudoparenchymatous cell complexes in substomatal cavities . . . . . . . . . . . . . . . . . . . . . . . . . . . . . . . . . . . . **Campoa**
(Müller and von Arx, 1962: 79; Arnaud, 1925: 651, icon.)

3(1') Ascospores phragmosporous, brown . . . . . . . . . . . . . . . . . **Pycnopeltis**
(Theissen and Sydow, 1917: 418.)

3'(1') Ascospores dictyosporous . . . . . . . . . . . . . . . . . . . . . . . . . . . . 4

4(3') Pseudothecia bright reddish-brown; ascospores hyaline to purplish-red
. . . . . . . . . . . . . . . . . . . . . . . . . . . . . . . . . . . . . . . . **Pycnoderma**
(von Arx, 1963: 458 icon.)

4'(3') Pseudothecia brown; ascospores hyaline . . . . . . . . . . . . . . . . . . . . 5

5(4') Asci embedded in a peripheral fertile zone surrounding a sterile center
. . . . . . . . . . . . . . . . . . . . . . . . . . . . . . . . . . . . . . **Stephanotheca**
(von Arx, 1963: 467.)

5'(4') Asci uniformly distributed throughout pseudothecium . . . . . . . . **Pycnodermina**
(von Arx, 1963: 467.)

## L. Schizothyriaceae

1. Ascospores 1-septate, hyaline ............................................. 1
1'. Ascospores more than 1-septate ........................................ 4
    2(1) Pseudothecia setose ........................................ **Chaetoplaca**
    (Müller and von Arx, 1962: 215.)
    2'(1) Pseudothecia glabrous ............................................. 3
3(2') Pseudothecia dark; shield brown to black, crumbling or splitting at maturity
................................................................ **Schizothyrium**
    (Müller and von Arx, 1962: 198, icon.; Dennis, 1968: 433, as *Microthyriella*, Fig. 18 G; Verona and Benedek, 1967, as *Microthyriella* C-279; as *Schizothyrina* C-282.)
3'(2') Pseudothecia bright, yellowish, grayish, or almost hyaline, gelatinous ... **Leptophyma**
    (Müller and von Arx, 1962: 205; Moreau and Moreau, 1959: 352, as *Plectomyriangium*, icon.; Verona and Benedek, 1961, as *Plectomyriangium* C-38.)
    4(1') Ascospores clavate to cylindrical ........................................ 5
    4'(1') Ascospores filiform; pseudothecium linear, opening by a longitudinal slit
................................................................ **Kerniomyces**
    (Batista, 1959b: 309.)
5(4) Ascospores 2-(rarely 3-) septate ........................................ 6
5'(4) Ascospores many-septate ........................................ **Myriangiella**
    (von Arx, 1963: 471; Batista, 1959b: 400–401, as *Sydowiellina*.) *Sydowiellina*, *Oswaldoa*, and *Spegazziniella* (Batista, 1959b), are probable synonyms of *Myriangiella*.
    6(5) Ascospores hyaline, clavate ........................................ **Metathyriella**
    (Batista, 1959b: 307.) *Ciferrotheca* (Batista, 1959b: 394–396) is probably a synonym of *Metathyriella*.
    6'(5) Ascospores brown, fusoid ........................................ **Amazonotheca**
    (Batista, 1959b: 409, icon.)

## M. Leptopeltidaceae

1. Ascospores nonseptate; shield of pseudothecium radiate ......... **Moeszopeltis**
    (Petrak, 1947; von Arx, 1964.) See *Hypnotheca*, Dothioraceae.
1'. Ascospores septate ............................................. 2
    2(1') Ascospores 1-septate ............................................. 3
    2'(1') Ascospores more than 1-septate ............................................. 5
3(2) Ascospores brown; shield not radiate or only weakly so at the margin ... **Thyriopsis**
    (Müller and von Arx, 1962: 237, icon.: von Arx, 1964.)
3'(2) Ascospores hyaline ............................................. 4
    4(3') Shield radiate; intraepidermal hypostroma of thick-walled dark cells
................................................................ **Leptopeltopsis**
    (Müller and von Arx, 1962: 236, icon.; von Arx, 1964: 187, icon.)
    4'(3') Shield not radiate or only weakly so at the margin; intraepidermal hypostroma lacking or of small hyaline cells ........................................ **Leptopeltis**
    (Müller and von Arx, 1962: 234–235, icon.; von Arx, 1964.)

5(2′) Ascospores phragmosporous, 3- to 4-septate when ripe, hyaline; shield radiate
.................................................... **Pycnothyrium**
(Von Arx, 1964: 185, icon.).

5′(2′) Ascospores dictyosporous, hyaline; pseudothecium intraepidermal; shield
radiate .................................................... **Dothiopeltis**
(Müller, 1956: 198, icon.; von Arx, 1964.)

## REFERENCES

Ammon, H. U. (1963). Über einige Arten aus den Gattung *Pyrenophora* Fries und *Cochliobolus* Drechsler mit *Helminthosporium* als *Nebenfruchtform. Phytopathol. Z.* **47**:244–300.

Arnaud, G. (1918). Les Astérinées. *Ann. Ecole. Nat. Agr. Montpellier* [N. S.] **16**:1–288.

Arnaud, G. (1921). Étude sur les champigons parasites (Parodiellinacées, inclus Erysiphées). *Ann. Epiphyt.* **7**:1–115.

Arnaud, G. (1923). Étude sur les champigons parasites (Parodiellinacées: note complementaire). *Ann. Epiphyt.* **9**:1–40.

Arnaud, G. (1925). Les Astérinées. IV. *Ann. Sci. Natur.: Bot Biol. Veg.* [10] **7**:643–723.

Arnaud, G. (1930). Les Astérinées. V. *Ann. Epiphyt.* **16**:235–302.

Arnaud, G. (1931). Les Astérinées. VII. *Ann. Cryptogam. Exot.* **4**:74–97.

Barr, M. E. (1955). Species of sooty molds from western North America. *Can. J. Bot.* **33**: 497–514.

Barr, M. E. (1956). The development of the ascocarp in *Phaeotrichum hystericinum. Can. J. Bot.* **34**:563–568.

Barr, M. E. (1968). The Venturiaceae in North America. *Can. J. Bot.* **46**:879–864.

Batista, A. C. (1959a). Posição taxonomica de *Fraserula, Kriegeriella* e *Yatesula. An. Soc. Biol. Pernambuco* **16**:35–41.

Batista, A. C. (1959b). Monografia dos fungos Micropeltaceae. *Univ. Recife Inst. Micol. Publ.* **56**:1–519.

Batista, A. C., and R. Ciferri. (1962). The Chaetothyriales. *Sydowia* **3**:1–129.

Batista, A. C., and R. Ciferri. (1963). Capnodiales. *Saccardoa* **2**:1–296.

Batista, A. C., and T. de J. Gayao. (1953). Alguns representantes da flora fungica em Pernambuco. *An. Congr. Nac. Soc. Bot. Brasil.* **4**:75–83.

Batista, A. C., and R. Garnier de Souza. (1960). Duas novas espécies de *Neostomella* Syd. *Univ. Recife Inst. Micol. Publ.* **292**:1–12.

Batista, A. C., and A. F. Vital. (1958). Dois novos gêneros de fungos Polystomellaceae. *Rev. Biol. (Lisbon)* **1**:280–286.

Batista, A. C., H. Da Silva Maia, and J. L. Bezerra. (1963). *Morqueria* n. gen. e alguns outros Asterinaceae. *Univ. Recife Inst. Micol. Publ.* **229**:1–29.

Boedijn, K. B. (1961). Myriangiales from Indonesia. *Persoonia (Leyden)* **2**:63–75.

Booth, C., and W. P. Ting. (1964). *Dolabra nepheliae* gen. nov., sp. nov., associated with canker of *Nephelium lappaceum. Trans. Brit. Mycol. Soc.* **47**:235–237.

Bose, S. K. (1961). Studies on *Massarina* Sacc. and related genera. *Phytopathol. Z.* **41**:151–213.

Bose, S. K., and E. Müller. (1965). Central Himalayan fungi. II. *Indian Phytopathol,* **18**:340–355.

Butin, H. (1963). Über zwei Nebenfruchtformen von *Sydowia polyspora* (Bref. ex. v. Tav.) Müller. *Sydowia* **17**:114–118.

Cain, R. F. (1961). Studies of coprophilous Ascomycetes. VII. *Preussia. Can. J. Bot.* **39**: 1133–1166.

Cain, R. F., and E. R. Luck-Allen. (1969). *Semidelitschia*, a new genus of the Sporormiaceae. *Mycologia* **61**:580–585.

Casagrande, F. (1969). Ricerche biologiche e systematiche su particolari ascomiceti pseudosferiale. *Phytopathol. Z.* **66**:97–136.

Chadefaud, M. (1960). "Les végétaux non vasculaires." *In* "Traité de Botanique Systematique" (M. Chadefaud and L. Emberger), Vol. 1, pp. 1–1018. Masson, Paris.
Chesters, C. G. C., and A. Bell. (1970). Studies in the Lophiostomaceae Sacc. *Mycol. Pap.* **120**: 1–55.
Cooke, W. B., and D. L. Hawksworth. (1970). A preliminary list of the families proposed for fungi (including lichens). *Mycol. Pap.* **121**:1–68.
Corbaz, R. (1957). Recherches sur le genre *Didymella* Sacc. *Phytopathol. Z.* **28**:375–414.
Corlett, M. (1970). Ascocarp development of two species of sooty molds. *Can. J. Bot.* **48**:991–995.
Darker, G. D. (1963). A new genus of the Lophiaceae. *Can. J. Bot.* **41**:1383–1388.
Deighton, F. C., and K. A. Pirozynski. (1966). Microfungi. II. *Brooksia* and *Grallomyces*; *Acrogenotheca ornata* sp. nov; the genus *Xenosporium*. Mycol. Pap. **105**:1–35.
Dennis, R. W. G. (1968). "British Ascomycetes." Cramer, Lehre.
Dodge, B. O. (1939). The ascocarp and ascospore formation in *Stevensea wrightii*. *Mycologia* **31**:96–108.
Doguet, G. (1960). Étude du *Melogramma spiniferum* (Wallr.) de Not. Pyrenomycète ascohymenié, annelascé, bituniqué. *Rev. Mycol. (Paris)* **25**:15–26.
Drechsler, C. (1934). Phytopathological and taxonomic aspects of *Ophiobolus*, *Pyrenophora*, *Helminthosporium*, and a new genus *Cochliobolus*. *Phytopathology* **24**:953–983.
Ellis, J. B., and B. M. Everhart. (1892). "The North American Pyrenomycetes." Ellis & Everhart, Newfield, New Jersey.
Eriksson, O. (1966). On *Eudarluca caricis* (Fr.) O. Eriks. comb. nov., a cosmopolitan uredinicolous pyrenomycete. *Bot. Notis.* **119**: 33–69.
Eriksson, O. (1967a). On graminicolus Pyrenomycetes from Fennoscandia. I. Dictyosporous species. *Ark. Bot.* [2] **6**:339–380.
Eriksson, O. (1967b). On graminicolous Pyrenomycetes from Fennoscandia. II. Phragmosporous and scolecosporous species. *Ark. Bot.* [2] *Ser. 2*, **6**:381–440.
Farr, M. L. (1965). *Dimeriella*, *Wentiomyces*, *Episphaerella*, and *Epipolaeum* (Fungi: Pyrenomycetes). *Taxon* **14**:18–21.
Farr, M. L. (1966). The didymosporous dimeriaceous fungi described from leaves of Gramineae. *Mycologia* **58**:221–248.
Farr, M. L., and F. G. Pollack. (1969). A new species of *Platypeltella* from Mexico. *Mycologia* **61**:191–195.
Gallée, O. (1930). "Natural History of the Danish Lichens," Part III. H. Aschehoug & Co., Copenhagen.
Gallée, O. (1932). "Natural History of the Danish Lichens," Part IV. H. Aschehoug & Co., Copenhagen.
Gallée, O. (1936). "Natural History of the Danish Lichens," Part V. Levin & Munksgaard, Copenhagen.
Graham, J. H., and E. S. Luttrell. (1961). Species of *Leptosphaerulina* on forage plants. *Phytopathology* **51**:680–693.
Hansford, C. G. (1946). The foliicolous Ascomycetes, their parasites and associated fungi. *Mycol. Pap.* **15**:1–240.
Hedjaroude, G. A. (1968). Etudes taxonomiques sur les *Phaeosphaeria* Miyake et leurs formes voisines (Ascomycetes). *Sydowia* **22**:57–107.
Holm, L. (1953). Taxonomical notes on Ascomycetes. III. The herbicolous species of the genus *Didymella* Sacc. *Sv. Bot. Tidsskr.* **47**:520–525.
Holm, L. (1957). Etudes taxonomiques sur les Pleosporacées. *Symb. Bot. Upsal.* **14**(3):1–188.
Holm, L. (1961). Taxonomical notes on Ascomycetes. IV. Notes on *Nodulosphaeria* Rbh. *Sv. Bot. Tidskr.* **55**:63–80.

Holm, L. (1967). Taxonomic notes on Ascomycetes. V. On *Sphaeria parmeliarum* Phill. & Plowr. and the genus *Cucurbidothis* Petr. *Sv. Bot. Tidskr.* **61**:449–456.

Holm, L. (1968). Taxonomic notes in Ascomycetes. VI. On the genus *Gibberidea* Fuck. and some alleged relatives. *Sv. Bot. Tidskr.* **62**:217–241.

Hughes, S. J. (1953). Fungi from the Gold Coast. II. *Mycol. Pap.* **50**:77–97.

Hughes, S. J. (1965). New Zealand Fungi.5. *Trichothallus* and *Plokamidomyces* states of *Trichopeltheca*. *N. Z. J. Bot.* **3**:320–332.

Hughes, S. J. (1967a). New Zealand Fungi. 9. *Ophiocapnocoma* with *Hormiokrypsis* and *Capnophialophora* states. *N. Z. J. Bot.* **5**:117–133.

Hughes, S. J. (1967b). New Zealand Fungi. 10. *Acrogenotheca elegans*. *N. Z. J. Bot.* **5**:504–518.

Kar, A. K., and M. K. Maity. (1970a). New *Asterina* spp. from West Bengal. *Trans. Brit. Mycol. Soc.* **54**:435–444.

Kar, A. K., and M. K. Maity. (1970b). Pyrenomycetes of West Bengal (India). II. *Can. J. Bot.* **48**:1295–1302.

Kern, H. (1959). Über einige Ascomyceten von der Elfenbeinkuste. *Ber. Schweiz. Bot. Ges.* **69**:277–285.

Kohlmeyer, J. (1969). Marine fungi of Hawaii including the new genus *Helicascus*. *Can. J. Bot.* **47**:1469–1487.

Kohlmeyer, J., and E. Kohlmeyer. (1964, 1965, 1967, 1968). "Icones Fungorum Maris," Vol. 1. Cramer, Lehre.

Loeffler, W. (1957). Unterschungen über die Ascomycetengattung *Dothidea* Fr. *Phytopathol. Z.* **30**:349–386.

Luttrell, E. S. (1951). Taxonomy of the Pyrenomycetes. *Univ. Missouri Stud.* **24**(3):1–120.

Luttrell, E. S. (1955). The ascostromatic Ascomycetes. *Mycologia* **47**:511–532.

Luttrell, E. S. (1960). The morphology of an undescribed species of *Dothiora*. *Mycologia* **52**:64–79.

Luttrell, E. S. (1963). A *Trichometasphaeria* perfect stage for a *Helminthosporium* causing leaf blight of *Dactyloctenium*. *Phytopathology* **53**:281–285.

Marasas, W. F. O. (1966). New species of Ascomycetes and a new genus of Sphaeropsidaceae from Transvaal. *Bothalia* **9**:203–215.

Menon, R. (1956). Studies on Venturiaceae on rosaceous plants. *Phytopathol. Z.* **27**:117–146.

Miller, J. H. (1938). Studies in the development of two *Myriangium* species and the systematic position of the order Myriangiales. *Mycologia* **30**:158–181.

Miller, J. H. (1940). The genus *Myriangium* in North America. *Mycologia* **32**:587–600.

Mirza, F. (1968). Taxonomic investigations on the ascomycetous genus *Cucurbitaria* S. F. Gray. *Nova Hedwigia* **16**:161–213.

Moreau, C., and M. Moreau. (1955). Ascomycètes de Côte d' Ivoire. II. Trois Ascomycètes foliicoles du Macaranga. *Rev. Mycol.* **20**: Suppl. Col. 2, 113–122.

Moreau, C., and M. Moreau. (1959). Champigons foliicoles de Guineé. II. Ascomycètes a organization fruste. *Rev. Mycol.* **24**: Suppl. col. 6, 349–359.

Morgan-Jones, G., and T. D. V. Swinscow. (1965). On the genus *Microglaena* Körb. *Lichenologist* **3**:42–54.

Müller, E. (1950). Die schweizerischen Arten der Gattung *Leptosphaeria* und ihrer Verwandten. *Sydowia* **4**:185–319.

Müller, E. (1952). Die schweizerischen Arten der Gattung *Ophiobolus* Riess. *Ber. Schweiz. Bot. Ges.* **62**:307–339.

Müller, E. (1955). *Leptoguignardia*, eine neue Gattung der bitunicaten Ascomyceten. *Sydowia* **9**:216–220.

Müller, E. (1956). Über die neue Ascomycetengattung *Dothiopeltis*. *Sydowia* **10**:197–200.

Müller, E. (1957). Pilze aus dem Himalaya. I. *Sydowia* **11**:455–472.

Müller, E. (1963). *Epibelonium*, eine neue Gattung der Schizothyriaceen (Ascomyceten). *Phytopathol. Z.* **47**:239–243.
Müller, E. (1965). Beobachtungen an Ascomyceten. *Sydowia* **18**:86–105.
Müller, E., and S. Ahmad. (1955). Über einige neue oder bemerkenswerte Ascomyceten aus Pakistan. I. *Sydowia* **9**:233–245.
Müller, E., and S. Ahmad. (1958). Ueber einige neue oder bemerkenswerte Ascomyceten aus Pakistan. III. *Biologia* **4**:25–32.
Müller, E., and S. Ahmad. (1962). Ueber einige neue oder bemerkenswerte Ascomyceten aus Pakistan. V. *Biologia* **8**:151–162.
Müller, E., and R. W. G. Dennis, (1965). Fungi venezuelani. VIII. Plectascales, Sphaeriales, Loculoascomycetes. *Kew Bull.* **19**:357–386.
Müller, E., and J. A. von Arx. (1950). Einige Aspekte zur Systematik pseudosphaerialer Ascomyceten. *Ber. Schweiz. Bot. Ges.* **60**:329–397.
Müller, E., and J. A. von Arx. (1955). Einige Beiträge zur Systematik und Synonymie der Pilze. *Phytopathol. Z.* **24**:353–372.
Müller, E., and J. A. von Arx. (1962). Die Gattungen der didymosporen Pyrenomyceten. *Beitr. Kryptogammenfl. Schweiz* **11**(2):1–922.
Munk, A. (1957). Danish Pyrenomycetes. *Dan. Bot. Ark.* **17**(1):1–491.
Muthappa, B. N. (1969). Morphology of a new loculoascomycete on *Bambusa arundinacea*. *Mycologia* **61**:737–747.
Nannfeldt, J. A. (1932). Studien über die Morphologie und Systematik der nicht-lichenisierten inoperculaten Discomyceten. *Nova Acta Regiae Soc. Sci. Upsal.* [4] **8**(2):1–368.
Nordin, I. (1964). *Abrothallus suecicus*, a common lichenicolous fungus. *Sv. Bot. Tidskr.* **58**: Parts I–III, 225–232.
Nüesch, J. (1960). Beitrag zur Kenntnis der weidenbewohnenden Venturiaceae. *Phytopathol. Z.* **39**:329–360.
Obrist, W. (1959). Untersuchungen über einige "dothideale" Gattungen. *Phytopathol. Z.* **35**:357–388.
Parguey-Leduc, A. (1966–1967). Recherches sur l'ontogénie et l'anatomie comparée des ascocarpes des Pyrenomycètes ascoloculaires. *Ann. Sci. Natur.: Bot. Biol. Veg.* [2] **7**:505–690; **8**:1–110.
Petrak, F. (1929). Mykologische Notizen X.621. *Endodothiora* n. gen. *Ann. Mycol.* **27**:345–346.
Petrak, F. (1947). Über die Leptopeltineen. *Sydowia* **1**:232–247.
Petrak, F. (1948a). *Chaetomelanops* n. gen., eine neue Gattung der Dothiorales. *Sydowia* **2**:68–71.
Petrak, F. (1948b). *Phaeodothiora* n. gen., eine neue Dothiorazeen Gattung aus China. *Sydowia* **2**:80–82.
Petrak, F. (1950). *Neopeltella* n. gen., eine neue Gattung der Dictyopeltineen. *Sydowia* **4**:329–332.
Petrak, F. (1959). Über die Gattungen *Eutryblidiella* v. Höhn, und *Endotryblidium* n. gen. *Sydowia* **13**:239–245.
Petrak, F. (1968). *Endocoleroa*, n. gen., eine neue Gattung der Venturiaceae. *Sydowia* **22**:388–391.
Petrak, F., and R. Ciferri. (1932). Fungi dominicani. *Ann. Mycol.* **30**:149–353.
Pirozynski, K. A., and J. Reid. (1966). Studies on the Patellariaceae. I. *Eutryblidiella sabina* (De Not.) Höhn. *Can. J. Bot.* **44**:655–662.
Pirozynski, K. A., and R. A. Shoemaker. (1970). Some Asterinaceae and Meliolaceae on conifers in Canada. *Can. J. Bot.* **48**:1321–1328.
Punithalingham, E. (1969). Studies on Sphaeropsidales in culture. *Mycol. Pap.* **119**:1–24.

Redinger, K. (1938). Arthoniaceae, Graphidaceae, Chiodectonaceae, Dirinaceae, Rocellaceae, Lecanactidiaceae, Thelotremaceae, Diposchistaceae, Gyalectaceae, Coenogoniaceae. In "Rabenhorst's Kryptogamen-Flora von Deutschland, Österreich und der Schweiz," 2nd ed., Vol. 9, Part 2, No. 1, pp. 1–404. Eduard Kummer, Leipzig.

Reid, J., and K. A. Pirozynski. (1966). A new loculoascomycete on *Abies balsamea* (L.) Mill. *Can. J. Bot.* **44**:351–354.

Reusser, F. A. (1964). Über einige Arten der Gattung *Guignardia* Viala et Ravaz. *Phytopathol. Z.* **51**:205–240.

Reynolds, D. R. (1971). The sooty mold ascomycete genus *Limacinula*. *Mycologia* **63**:1173–1209.

Riedl, H. (1961). Die Arten der Gattung *Mycoporellum* Müll. Arg. sensu A. Zahlbruchner Catol., nebst Bermerkungen zum System dothidealer Flechten. *Sydowia* **15**:257–287.

Riedl, H. (1963). Bemerkungen über *Dermatina*-Arten aus West- und Mitteleuropa. *Sydowia* **17**: 102–113.

Santesson, R. (1952). Foliicolous lichens. I. *Symb. Bot. Upsal.* **12**:1–590.

Scheinpflug, H. (1958). Untersuchungen über die Gattung *Didymosphaeria* Fuck. und einige verwandte Gattungen. *Ber. Schweiz. Bot. Ges.* **68**:325–385.

Shoemaker, R. A. (1965). Revision of some *Dimeriella* and *Dimerosporium* parasites of conifers. *Can. J. Bot.* **43**:631–639.

Sivanesan, A. (1970a). *Plurisperma dalbergiae* gen. et. sp. nov. *Trans. Brit. Mycol. Soc.* **54**:495–496.

Sivanesan, A. (1970b). *Parmulariopsella burseracearum* gen. et sp. nov. and *Microcyclus placodisci* sp. nov. *Trans. Brit. Mycol. Soc.* **55**:509–514.

Theissen, F., and H. Sydow. (1915). Die Dothideales. *Ann. Mycol.* **13**:149–746.

Theissen, F., and H. Sydow. (1917). Synoptische Tafeln. *Ann. Mycol.* **15**:389–491.

Theissen, F., and H. Sydow. (1918). Vorentwürfe zu den Pseudosphaeriales. *Ann. Mycol.* **16**:1–34.

Thirumalachar, M. J., and A. E. Jenkins. (1953). *Bitancourtia cassythae* on *Cassytha filiformis* and proposed nomenclatorial changes among other Myriangiales. *Mycologia* **45**: 781–787.

Thirumalachar, M. J., and M. J. Narasinham. (1955). Notes on Myriangiaceous Fungi, I. *Mycologia* **47**:758–762.

Tilak, S. T. (1963). Ascomycetes on *Celastrus paniculata* Willd. *Mycopathol. Mycol. Appl.* **21**:60–64.

Tilak, S. T., and R. Rao. (1966). Contributions to our knowledge of Ascomycetes of India, V. *Mycopathol. Mycol. Appl.* **28**:90–94.

Tim, S. K. M. (1971). The morphology and development of *Trichothyrium asterophorum* (B. et Br.) v. Höhn. *Bot. Gaz.* **132**:318–326.

Tommerup, I. C. (1970). *Hypnotheca graminis* gen. et sp. nov. perfect state of *Monochaetiella themedae*. *Trans. Brit. Mycol. Soc.* **55**:463–475.

Toro. R. A. (1952). A study of the tropical American black-mildews. *Univ. Puerto Rico J. Agr.* **36**:24–87.

Verona, O., and T. Benedek. (1959–1970). Inconographia mycologica. *Mycopathol. Mycol. Appl., Suppl.* **1–27**.

von Arx, J. A. (1960). *Batistinula*, eine neue Gattung der Asterinaceae. *Univ. Recife Inst. Micol. Publ.* **287**:1–17.

von Arx, J. A. (1963). Die Gattungen der Myriangiales. *Persoonia* (*Leiden*) **2**:421–475.

von Arx, J. A. (1964). Über die Ascomycetenfamilie der Leptopeltaceae. *Acta Bot. Neer.* **13**:182–188.

von Arx, J. A., and E. Müller. (1954). Die Gattungen der amerosporen Pyrenomyceten. *Beitr. Kryptogamenfl. Schweiz* **11**(*1*):1–434.

von Arx, J. A., and E. Müller. (1969). "Keys to the Genera of Amerospored and Didymospored Pyrenomycetes" [translated into English by G. B. Butterfield from *Beitr. Kryptogamenfl. Schweiz* **2**: No. 1 (1954) and **2**: No. 2 (1962)]. Commonwealth Mycological Institute, Kew, Surrey, England.

von Keissler, K. (1938). Pyrenulaceae, Trypetheliaceae, Pyrenidiaceae, Xanthopyreniaceae, Mycoporaceae und Coniocarpineae. In "Rabenhorst's Kryptogamen-Flora von Deutschland, Österreich und der Schweiz," 2nd ed., Vol. 9, Part 1, No. 2, pp. 1–876. Eduard Kummer, Leipzig.

Waterman, A. M. (1945). Tip blight of species of *Abies* caused by a new species of *Rehmiellopsis*. *J. Agr. Res.* **70**:315–337.

Wehmeyer, L. E. (1941). The genus *Thyridaria* (Pyrenomycetes). *Lloydia* **4**:241–261.

Wehmeyer, L. E. (1961). "A World Monograph of the genus *Pleospora* and its Segregates." Univ. of Michigan Press, Ann Arbor.

Wolf, F. A. (1935). Morphology of *Polythrincium* causing sooty blotch of clover. *Mycologia* **27**:58–73.

Zogg, H. (1962). Die Hysteriaceae s. str. und Lophiaceae. *Beitr. Kryptogammenfl. Schweiz* **11**(*3*):1–190.

# Ascomycotina
# Laboulbeniomycetes

# CHAPTER 8

# Laboulbeniomycetes

RICHARD K. BENJAMIN

*Rancho Santa Ana Botanic Garden*
*Claremont, California*

---

## I. INTRODUCTION

Organisms now classified in the Laboulbeniales apparently first were noticed by the French entomologists Alex Laboulbène and Auguste Rouget as early as 1840. In 1850, Rouget published a brief account of a so-called "production parasite" found on several ground-inhabiting beetles. He provided crude drawings of a form later placed in *Laboulbenia*, but he did not present formal descriptions. This first was done by Camille Montagne and Charles Robin (Robin, 1853) when they described *Laboulbenia* with two species, *L. rougetii* and *L. geurinii*, based, in part, on specimens taken from insects provided by Laboulbène and Rouget. Mayr (1852) earlier had figured what probably were specimens of *Laboulbenia* on species of *Nebria* (Coleoptera: Carabidae), but he interpreted these as abnormally developed hairs. During the next two decades few works relating to these organisms appeared. In 1857, Kolenati described *Arthrorhynchus* with two species parasitic on wingless flies (Nycteribiidae) that infest bats, but he regarded these as specialized worms. Diesing (1859) actually proposed a new tribe of Vermes, Arthrorhyngodeae, for them. Knoch (1868) added another species, *Laboulbenia baeri*, parasitic on the European housefly, and the same species was described a year later by Karsten (1869) as *Stigmatomyces muscae*. Karsten depicted several stages of development of this fungus, and, more importantly, he recognized the existence of sexual organs resembling those of the Florideae.

The first definitive works on the Laboulbeniales were those of Peyritsch (1871, 1873, 1875) who formally characterized the family Laboulbeniaceae in 1873 and recognized five genera with twelve species. Berlese, in 1889, again summarized the family as then defined and recognized six genera with fifteen species. The following year the first of Roland Thaxter's many contributions to knowledge of the order appeared (Thaxter, 1890–1896, 1899–1903, 1905, 1908, 1912a,b, 1914–1918a,b, 1920, 1924, 1926, 1931) and a new era in the study of Laboulbeniales had begun.

Thaxter's works on these fungi comprised 22 unillustrated papers and an illustrated monograph published in five parts. He described 103 genera and approximately 1260 species. In addition, he reviewed the work of his predecessors and contemporaries so that his studies encompassed some 128 genera—107 retained as valid—and about 1340 species or varieties. To illustrate his monograph, Thaxter drafted 3427 skillfully executed line drawings. His work now ranks as one of the great classics of taxonomic mycology.

Thaxter's early studies established the ascomycetous nature of the Laboulbeniales. The initial part of his monograph (Thaxter, 1896) provided the first really comprehensive treatment of their morphology, development, and diversity. Thus, many other students of his time were stimulated to investigate these fungi. Among those persons who made significant auxiliary contributions were Carlos Spegazzini (1912, 1914, 1915a,b,c, 1917), François Picard (1908a,b, 1909, 1912, 1913a,b, 1916–1917), and René Maire (1912, 1916a,b,c, 1920).

Subsequent to Thaxter's death in 1932, about 100 publications on the Laboulbeniales have appeared. Many of these are of minor importance. Several are more or less general, nontechnical discussions intended for the nonspecialist or informed layman (Benjamin, 1965; Boelens, 1947; Collart, 1945, 1947; Lepesme, 1946; Shanor, 1955). Some, however, have provided important new data on the taxonomy, morphology, and biology of these fungi. In a recent review, Benjamin (1971) summarized the studies of Thaxter and others on the Laboulbeniales. He also provided a synopsis of the genera together with a key, cited all pertinent literature on the group, and presented a guide for collection and study of these fungi. This work should be consulted for a more complete treatment of the material presented only briefly in the sections below.

## II. OCCURRENCE

### A. Hosts

The Laboulbeniales are predominantly parasites of true insects (class Hexapoda) although several are known on mites (class Arachnida; order Acarina) and millipeds (class Diplopoda; order Juliformia). Most of the species infesting Hexapoda are found on Coleoptera, but species are known also on representatives of ten other orders: Anopleura, Blattaria, Dermaptera, Diptera, Hemiptera, Hymenoptera, Isoptera, Mallophaga, Orthoptera, and Thysanoptera. All species of 90 genera have been found only on species of Coleoptera. Representatives of nine genera have been found not only on members of this order but on those of one or more other orders of arthropods. All members of seventeen genera are known only on orders other than Coleoptera.

## B. Habitat and Distribution

Because Laboulbeniales are obligate parasites, the body of their host is their immediate habitat and they probably are influenced only indirectly by the environment of the latter. In the course of their evolution, these fungi have adapted to insects inhabiting an almost endless variety of habitats—water, soil, decomposing plant and animal remains, foliage and flowers, and even the bodies of living animals, including mammals and birds. Their geographical distribution is dependent on that of their host. Thus, some species have a very restricted range, others are nearly cosmopolitan.

## III. HOST–PARASITE RELATIONSHIPS

### A. Specificity and Transmission

Like other obligate parasites, Laboulbeniales long have been known for their often high degree of host specificity, and with few exceptions they have been observed only on the adult state of the host. Some species, like *Laboulbenia vulgaris* and *L. flagellata*, have a wide host range within the family Carabidae and are found on many species of several genera. But most species appear to be confined to a single host species or to very closely related hosts. Some genera like *Laboulbenia*, *Dimeromyces*, and *Rickia* have extremely wide host ranges and their species are known on representatives of many families of several orders of insects and other arthropods. Species of other genera, like *Amorphomyces*, *Coreomyces*, *Dichomyces*, *Monoicomyces*, *Herpomyces*, and *Ilytheomyces*, are more restricted and may be limited to a single order, family, subfamily, or even genus of hosts.

Not only may species of Laboulbeniales exhibit a narrow host range, but many occur only on very restricted areas of the host body. The position of growth of the fungus on each of the two sexes of insect may differ, and some species are known only on one or the other sex of the host, but not both (Benjamin and Shanor, 1952; Benjamin, 1971). Some examples of position specificity may be correlated with mating phenomena or other behavior patterns resulting in contact transfer of parasites (Whisler, 1968). Several examples, however, require different explanations (Richards and Smith, 1955; Benjamin, 1971) including possibly the existence of physiological variation of different parts of the host integument. Answers to these and many other problems relating to the interaction of Laboulbeniales and their hosts are yet to be resolved experimentally.

In most species of Laboulbeniales it seems likely that transmission is by direct contact of the hosts themselves. However, Arwidsson (1946) and Lindroth (1948) found evidence that soil might play a role in transmission of a few species of *Laboulbenia*. Also, infected insects that swarm in large numbers in a confined area, as in burrows or under stones or other debris, undoubtedly deposit spores onto their surroundings from which these spores subsequently adhere to other individuals. Such random transfer would account for the occurrence of some species of Laboulbeniales on all parts of the body of their hosts. This also would account for the occasional presence of one of these parasites on other than its usual host (Scheloske, 1969; Benjamin, 1971).

## B. Nutrition and Pathogenicity

No member of the Laboulbeniales has been induced to complete its life cycle apart from a living host, although early stages of development have been observed in artificial culture (Richards and Smith, 1954; Whisler, 1968). Whisler placed spores of *Stigmatomyces ceratophorus* on sterile fly wings on brain-heart infusion agar and obtained growth to as much as a twenty-cell stage. A few individuals even formed antheridia and spermatia. However, female structures never were initiated in such cultures. These results and the fact that spores may begin germination prior to discharge from the perithecium indicate that, for some species at least, early stages of development are nutritionally independent of direct contact with living tissues of a host. However, present evidence suggests that maturation and ascospore production in species of Laboulbeniales take place only after the fungus has established haustorial contact with living cells of the epidermis or hemolymph of the body cavity of a susceptible insect or other arthropod.

Few studies of the pathogenic effect, if any, of Laboulbeniales on their hosts have been made. From his many observations, Thaxter believed that these fungi cause little or no harm, although he did note possible minor tissue damage in insects parasitized by species having a penetrating rhizomycelium. Scheloske (1969) suggests that heavy infections on the eyes, legs, or wings might interfere sufficiently with the host's normal movements to make it easier prey for predators. In a recent study of *Chilocorus bipustulatus* (Coleoptera: Coccinellidae) infected with *Hesperomyces virescens*, Kamburov et al. (1966, 1967) implicated this fungus in an observed high rate of premature mortality of infected versus noninfected beetles in a large area of citrus groves in Israel. Bro Larsen (1952), in a field study of three halophilic species of *Bledius* (Coleoptera: Staphylinidae), found a greater tendency for insects living in an area of low salinity to become infected with an unidentified species of Laboulbeniales than those insects living in a neighboring area of high salinity. She noted a much lower mortality rate among noninfected than infected insects, especially those having heavy infections on their antennae and mouthparts. The latter insects became sluggish and ate almost nothing. Whisler (1968) conducted a long-term test, under controlled conditions, of the effect of parasitism of *Stigmatomyces ceratophorus* on the fly

*Fannia canicularis*, but he found the longevity of members of infected and noninfected colonies to be nearly the same. Richards and Smith (1956) studied, also under laboratory conditions, numerous colonies of living cockroaches heavily infected with *Herpomyces* spp. and they could detect no evidence of pathogenicity or debilitation. For the present, it probably is safe to sustain the notion held currently that most Laboulbeniales probably have little or no effect on the well-being of their hosts.

## IV. MORPHOLOGY AND DEVELOPMENT

### A. Asci and Ascospores

The ascus is more or less elongate-clavate and usually contains four ascospores, rarely eight in some species. In most instances, the ascus wall deliquesces prior to spore discharge. Ascospores typically emerge from the perithecium in pairs, and this is true for both monoecious and dioecious species. Ascospore size is relatively uniform for a given species, but in some dioecious forms there may be extreme spore dimorphism. The spore giving rise to the female individual is much larger than that forming the male (Thaxter, 1908; Benjamin, 1970).

The ascospore of all known Laboulbeniales typically is hyaline, elongate, more or less spindle shaped, and once septate (Fig. 10d). With few exceptions, the two segments of the spore are unequal in length with the longer basal cell projecting upwardly and emerging first when spores are discharged from the perithecium (Figs. 6, 7, 29, 30). Only in species of *Amorphomyces* is the two-celled nature of the spore difficult to observe because of the very small upper cell (Fig. 16c), and until recently (Tavares, 1970) members of this genus were thought to be exceptional in having unicellular spores (Thaxter, 1896). Each spore is surrounded by a thin, hyaline envelope that is noticeably more thickened at the lower end (Figs. 10d, 16c). This sheath undoubtedly serves not only to protect the spore following discharge from the perithecium but also to effect its adhesion to the surface of the host.

### B. The Thallus

#### 1. Foot and Receptacle

The first stages of ascospore germination consist of the enlargement of the lower end of the basal spore segment and, in most species, of the concomitant darkening of the adjacent part of the enveloping sheath. In species of some genera such as *Amorphomyces* and *Dioicomyces*, these early modifications may begin while the spores still are inside the perithecium. There results a suckerlike organ, the foot, from the lower surface of which, in most species that have been studied in detail, a small, inconspicuous, simple haustorium arises and penetrates the host's integument, finally reaching the living cells of the epidermis. There are, however, genera such as *Trenomyces*, *Arthrorhynchus*, *Moschomyces*, and *Microsomyces* having species with a simple or branched rhizomycelium that penetrates deeply into host tissues. In such forms there may be little or no blackening of the foot where it contacts the host.

During formation of the foot, the body of the spore undergoes little change. After the haustorium contacts living cells of the host, however, there is a relatively rapid development of the fungus from the original two cells of the spore. This involves an often very precise sequence of cell divisions leading to formation from the basal spore segment of a cellular body, the receptacle, that bears reproductive structures directly or that gives rise to secondary branches or cellular outgrowths bearing sterile or fertile branches. Ontogenesis has been studied in only a few species, but it appears to require one to three weeks.

The receptacle varies greatly in size and complexity from genus to genus as shown in the 29 genera illustrated here, and its structure constitutes a criterion of primary importance for distinguishing genera within the order. In many genera it consists simply of three cells, the lower foot-cell, a subbasal cell bearing one or more perithecia or secondary sterile or fertile outgrowths, and an upper cell subtending an appendage (Figs. 5, 6, 9, 19, 23). In other genera, by secondary development, the receptacle may consist of a small or large number of cells superposed in a single series (Figs. 2, 3, 12, 15, 18) or in tiers of cells variously disposed (Figs. 1, 26, 27, 28). In a few genera such as *Dipodomyces* and *Herpomyces* (Fig. 21) the germinated spore retains, more or less, its original shape and constitutes a simple, primary receptacle. This then gives rise to one or more outgrowths that form a secondary receptacle. The latter assumes the major role of attachment of the fungus to the host and gives rise to reproductive structures.

## 2. Appendages

The term appendage is applied to any branch borne on the receptacle or, in a few genera, the perithecium, and appendage characteristics often are important taxonomic criteria. An appendage may be characterized as being primary or secondary. The primary appendage is distinguished most readily during early stages of development of an individual fungus and is derived directly from the upper spore segment. In some instances it soon loses its identity either by being sloughed away or by being obscured by the formation of secondary appendages. It may remain essentially unchanged as a single, sterile cell, as in the female individual of *Dioicomyces* (Fig. 22a), or it may develop a more or less determinate series of few to many superposed cells one or several of which form male sex organs (Figs. 5–8). In other instances it may produce a more or less indeterminate number of sterile and/or fertile cells and be simple or branched.

Secondary appendages may arise from products of division of the original upper spore segment (Figs. 10, 11, 19) and they also commonly are formed by cells comprising the receptacle (Figs. 1, 13, 17, 30a). Secondary appendages, like the primary, may consist of single cells (Figs. 26–28), be composed of a single series of superposed cells (Figs. 12, 29a, 30a), or become more or less extensively branched (Figs. 2a, 3, 9, 13). They may remain sterile or form male or female sexual organs.

Perithecial appendages usually are outgrowths of outer wall cells; they may be unicellular (Figs. 7, 22) or multicellular (Figs. 3, 4) and always are sterile.

## 3. Male Sexual Organs

Existence of sexuality in the Laboulbeniales first was suggested by Karsten (1869) who observed a trichogyne in *Stigmatomyces baeri* as well as small presumptive male elements, spermatia, formed by cells of the appendage. Subsequent studies have shown that spermatia are produced by some or all species of 91 of the 116 genera now recognized. They usually are formed by cells of the primary or secondary appendages derived from the upper spore segment, but they also may arise from cells of the receptacle or its appendages. The manner of origin of spermatia was utilized by Thaxter (1896) as the main criterion for separating families within the order. Three distinct mechanisms may be distinguished.

*a. The Exogenous Spermatium.* In species of *Drepanomyces*, *Rhynchophoromyces*, *Zodiomyces*, and possibly *Eusynaptomyces* small, rodlike branchlets arising directly on appendages (Fig. 2b,c) are interpreted as spermatia. These cells are small, uninucleate, and clearly distinct from the smaller branchlets that also may develop from branches of the appendages of some of the above genera. In *Zodiomyces*, the branches forming spermatia appear to be specialized for this purpose only. They are very small and are distinguished readily from the large sterile appendages arising from the margin of the fertile upper surface of the receptacle (Fig. 1). In this genus, both perithecia- and spermatia-bearing branchlets arise in profusion from the surface of the fertile, cuplike depression of the receptacle (Thaxter, 1896; Gäumann and Dodge, 1928). The trichogyne of *Zodiomyces* is recurved during early stages of development of the perithecium and extends downward into the region where spermatia are being produced.

*b. The Simple Antheridium.* The type of antheridium most commonly encountered in the Laboulbeniales is one in which spermatia are formed endogenously in a more or less flask-shaped cell having a somewhat attenuated neck through which spermatia pass to the outside (Fig. 10c). In form and position on the thallus, however, such antheridia exhibit much diversity among the 62 genera know definitely to produce them. A single antheridium may terminate a simple primary appendage as in monoecious forms like *Autophagomyces* (Fig. 5) or the male individual as in species of dioecious genera such as *Amorphomyces* (Fig. 16b) or *Dioicomyces* (Fig. 22b). In species of other genera like *Acompsomyces* (Fig. 6), *Stigmatomyces* (Fig. 8), or *Hesperomyces* (Fig. 7), few to many antheridia may be borne laterally on cells comprising the appendage. The antheridium may be free above, united only at its base to the subtending cell (Fig. 10b), or it may be more or less broadly united below not only to its supporting cell but also to adjacent cells or antheridia with only the efferent neck being free (Figs. 6, 8). Finally, one or several cells of the appendage may be converted directly into antheridia with only the divergent discharge tubes being free (Fig. 14).

c. *The Compound Antheridium.* There are 25 genera, 19 monoecious and six dioecious, having male organs in which few to many antheridial cells—structurally like the simple antheridium—are variably united into a compound structure so that their spermatia are discharged into a common chamber before escaping to the outside through a single opening (Fig. 30c). The antheridial cells may be irregularly disposed on the periphery of the antheridium as in *Haplomyces* (Fig. 24), arranged in rows of few to many cells as in *Monoicomyces* (Fig. 23) and *Eucantharomyces* (Fig. 20a,b), or symmetrically grouped in a basal layer as in *Dimeromyces* (Fig. 30b,c), *Dimorphomyces* (Fig. 29b), and *Dichomyces* (Fig. 28).

4. *Female Sexual Organ and the Perithecium*

The perithecium and its included ascogenous system are derived from a single cell of the receptacle or one of its branches. Ontogeny of these structures, like that of the receptacle, usually involves a precise sequence of cell divisions. In no instance has a perithecium been observed to develop from the original upper spore segment.

When mature, the female organ consists of but three superposed cells, a lower carpogenic cell, a median trichophoric cell, and a terminal trichogyne. The first two cells become surrounded by the developing perithecium, whereas the trichogyne is partly or entirely external and may consist of a single cell or a series of superposed cells forming a simple or branched receptive organ. Ultimately, the trichogyne and trichophoric cell disappear and the carpogenic cell gives rise to one, two, four, eight or more ascogenous cells that float freely near the base of the perithecial cavity and give rise to a succession of asci during the functional life of the fungus (Figs. 6, 7, 12, 16a). The perithecium proper consists of a pair of stalk cells, more or less superposed, and three basal cells which give rise to four longitudinal rows of inner and outer wall cells disposed in four or five, sometimes more, tiers of cells. The relationship of the stalk and basal cells often is obscure when the perithecium is sessile or otherwise united with the receptacle.

Three distinct patterns of development of the female organ and perithecium have been described in the Laboulbeniales (Benjamin, 1971). In most genera that have been studied in detail, the developmental sequence originally described by Thaxter (1896) and Gäumann and Dodge (1928) for species of *Stigmatomyces* and *Laboulbenia* has been found. In this type of development, a single cell derived from the receptacle or one of its branches divides into an upper and a lower cell. The upper cell forms the female organ, while the lower cell forms all cells comprising the stalk, base, and walls of the perithecium proper.

In the second type of development, known only in *Coreomyces* (Thaxter, 1908), an intercalary cell of the young fungus proliferates upwardly and forms two cells. One of these gives rise to a single branch that forms one row of true perithecial wall cells and the other cell forms four branches, three producing the other three rows of wall cells and the fourth developing into the female organ. The upward growth of these branches destroys the septa of the cells above so that only their lateral walls remain. These form the apparent walls of the perithecium. All but the upper tiers of true perithecial wall cells formed by the internal upgrowths disappear at maturity. The carpogonial upgrowth eventually forms ascogenic cells in what Thaxter called a pseudoperithecium because of its anomalous origin (Fig. 15).

The third distinctive type of perithecial development is found in *Herpomyces*, and knowledge of this is due almost entirely to the studies of Tavares (1965, 1966) on *H. stylopygae* and *H. paranensis*. In these species, the entire perithecium and the female organ are derived from a single cell of an outgrowth of the subbasal cell of the four-celled primary receptacle (Fig. 21a). Unlike the type of development found in *Stigmatomyces*, *Laboulbenia*, and many other genera, the inner rows of wall cells of the perithecium arise from the third or fourth tiers of outer perithecial wall cells. The carpogonial upgrowth, of which there may be more than one, arises from a cell of the lowermost tier of outer wall cells.

## V. CLASSIFICATION

### A. History

When Montagne and Robin (Robin, 1853) described *Laboulbenia*, the name was underscored, parenthetically, with the statement "e familia *Pyrenomycetum*, novum genus." They obviously considered *L. rougetii* and *L. guerinii* to be ascomycetes; however, they were not cognizant of the true manner of spore formation. Nor, apparently, was Peyritsch (1871, 1873, 1875) although he, too, regarded the fungi he studied as ascomycetes when he characterized the family Laboulbeniaceae. De Bary (1887) and Winter (1887), on the other hand, treated the group as "doubtful" ascomycetes or pyrenomycetes pending better understanding of their presumed asci.

Thaxter's early studies (1890–1895), culminated by Part I of his monograph in 1896, provided the first accurate descriptions of the development of the thallus, perithecium, and ascogenous system of members of the order and clearly established their ascomycetous affinities. He increased the number of genera from six to 28 and recognized 152 species. These were classified in a single family, the Laboulbeniaceae of Peyritsch, with two major categories. Group I, the Endogeneae, included all forms with endogenously formed spermatia. Those species having simple antheridia constituted the "order" Laboulbenieae; those with compound antheridia the "order" Peyritschielleae. Group II, the Exogeneae, included species with exogenous spermatia and comprised the "order" Zodiomyceteae.

In 1897, Lindau accorded the Laboulbeniaceae suprafamilial status as the Laboulbeniinae, and the following year Engler (1898) established the order Laboulbeniales in a separate class, Laboulbeniomycetes, coordinate with the Phycomycetes, Basidiomycetes, and Ascomycetes.

In the second part of his monograph, Thaxter (1908) listed 52 genera and about 400 species and subspecies. He recognized the ordinal status of the group and abandoned the terms Endogeneae and Exogeneae, substituting instead the subordinal names Laboulbeniineae and Ceratomycetineae. In the former, those species having compound antheridia were placed in a new family, Peyritschiellaceae, subdivided into six tribes. Species with simple antheridia were retained in the Laboulbeniaceae with thirteen tribes. Thaxter did not designate a family within the Ceratomycetineae, but did distinguish three tribes. The Ceratomycetaceae, now universally recognized, first was proposed by Maire (1916a). Thaxter's classification, below the ordinal level, is essentially the one accepted by most mycologists ever since (Spegazzini, 1914; Picard, 1913a; Maire, 1916a, 1920; Bánhegyi, 1940; Bessey, 1950; Luttrell, 1951).

## B. Current Status

It generally is agreed that the Laboulbeniales are not closely related to other ascomycetes. Engler's proposal, in 1898, of a separate class for them still is accepted by some (Chadefaud, 1960; Denison and Carroll, 1966), and students who retain the order in the class Ascomycetes readily acknowledge their unique characteristics. Bessey (1950) made no attempt to ally them with other hierarchies within the class. Luttrell (1951) and

Alexopoulos (1962) classified them either as a series or subseries of the subclass Euascomycetidae.

The Laboulbeniales constitute a well-defined natural order. Despite the great diversity in thallus structure exhibited by the numerous genera, the order, on the basis of ascospore morphology, probably is monophyletic. The two-celled ascospore displays only minor structural variation throughout the group. It is unlikely that this condition has arisen more than once in these fungi by convergence from unrelated precursors, but, instead, represents the retention of the presumably primitive spore type (Denison and Carroll, 1966) of a single ancestral form. The nonmycelial, more or less determinate growth pattern of the thallus of all members of the order sets the group apart from other ascomycetes, but, in my opinion, does not warrant its removal from the class. However, the Laboulbeniales probably should be given the status of a separate subclass, Laboulbeniomycetidae.

Thaxter's classification of the Laboulbeniales based on the nature of the presumed male organ is more or less adequate for assigning 91 of the 116 known genera to one of the three families currently recognized, but species of 25 genera lack male organs or these are unconfirmed. Assignment of the latter genera to a family can only be arbitrary, based on morphological resemblance to species having male structures. Most of the tribes distinguished by Thaxter in 1908 circumscribed well-defined and undoubtedly natural groups among the 52 genera then known. As Bessey (1950) suggested, many of these tribes easily could be elevated to family rank, and he proposed this for the Zodiomyceteae. Many of the genera described since 1908 can be accommodated readily in the Thaxterian tribes, others cannot.

Reclassification of the order at the family level will have to incorporate features not only of male and female reproductive structures but also of other aspects of thallus morphology and development. Tavares (1967) has suggested that perithecioid ontogeny may be of primary importance as a basis for classification, but she has not yet published an elaboration of her ideas. Definitive knowledge of the developmental history of many genera still is incomplete or lacking, and, for the present, I prefer not to attempt a realignment of the families and/or tribes until more data are available.

The key presented below and the accompanying illustrations are intended only to provide the reader with a means for recognizing a few representative genera of Laboulbeniales the species of which are likely to be encountered in any collection of common insects. Three of these genera (marked with an *) still have no demonstrated male organs and their inclusion in the family designated for them is arbitrary. A synopsis of the genera and a key to all genera currently recognized by the author is presented elsewhere (Benjamin, 1971).

## C. Subclass Laboulbeniomycetidae[1]

The Laboulbeniomycetidae are minute ectoparasites of living insects or other arthropods. The main body of the thallus (receptacle) is attached to the host by a basal cellular holdfast provided with an internal haustorium which is usually inconspicuous. The receptacle consists of few to many cells that are more or less definitely arranged in uni- or multiseriate rows or else it may become parenchymatous; it bears one or more uni- or multiseriate simple or branched appendages and gives rise to one or more sessile or stalked perithecia. The appendages are sterile or form spermatia externally on one or more cells or internally in flask-shaped antheridia that may be simple, with free exit tubes, or united into a compound structure and discharging spermatia into a common chamber with a single opening to the outside. Simple or compound antheridia are often borne directly on cells of the receptacle. The body of the perithecium usually consists of three basal cells and four longitudinal rows of cells disposed in four or five up to an indeterminate number of tiers. The ascogonium, which becomes surrounded by the perithecium, is initially composed of three superposed cells, the upper forming a simple or branched free trichogyne, the lower eventually giving rise to one, two, four, eight, or more ascogenous cells. Ascogenous cells form asci by budding. The asci are most often four spored, while the ascospores are two celled, elongate, more or less fusiform, surrounded by a thin, hyaline sheath that usually is thicker at one end. The ascus wall usually deliquesces prior to discharge of spores from the ascocarp.

The Laboulbeniomycetidae have a single order, the Laboulbeniales.

### KEY TO THE FAMILIES AND SOME REPRESENTATIVE GENERA OF LABOULBENIALES

1. Spermatia exogenous, formed as tiny branchlike cells on the appendages (Fig. 2b,c) . . . . . . . . . . . . . . . . . . . . . . . . . . . . . **Ceratomycetaceae** 2
1'. Spermatia endogenous, formed inside special cells (antheridia) borne on or within cells of the appendages or receptacle . . . . . . . . . . . . . . . . . . . . . . . . . . . . . 5
    2(1) Receptacle a more or less massive, multicellular, turbinate structure forming a distal, cuplike depression bordered by numerous elongate sterile appendages and bearing stalked perithecia and spermatia-bearing branches
    . . . . . . . . . . . . . . . . . . . . . . . . . . . **Zodiomyces** (Fig. 1)
    Known only on species of Hydrophilidae (Coleoptera).
    2'(1) Receptacle consisting of 3 to many superposed cells subtending a single perithecium and an appendage; perithecium with 7–8 up to 60 or more tiers of wall cells . . 3

---

[1]*Editors' note:* Because in this volume ascomycetes are, for convenience and didactic reasons, treated as a division (Ascomycotina) instead of a class (Ascomycetes), the Laboulbeniales are given class status (Laboulbeniomycetes).

3(2′) Perithecium consisting of an enlarged venter and a slender, elongate, multi-tiered, often recurved neck . . . . . . . . . . . . . . . . **Rhynchophoromyces** (Fig. 2)
   On species of Hydrophilidae. Allied genus: *Thaumasiomyces*.

3′(2′) Perithecium without an elongate neck . . . . . . . . . . . . . . . . . . . . . 4

   4(3′) Perithecium with only 7–8 tiers of wall cells, often with 1–3 cellular appendages variably developed from the upper wall cells . . . . . . . **Autoicomyces*** (Fig. 3)
      On Hydrophilidae, especially species of *Berosus*. Allied genus: *Plectomyces*.

   4′(3′) Perithecium with 9–60 or more tiers of wall cells, often with a single subterminal unicellular or multicellular upgrowth . . . . . . . . . . **Ceratomyces*** (Fig. 4)
      On Hydrophilidae, especially species of *Tropisternus*. Allied genera: *Eusynaptomyces*, *Phurmomyces*, *Synaptomyces*.

5(1′) Antheridia simple, i.e., each with a single opening to the outside, more or less flask-shaped and subtended by cells of the appendages or receptacle (Figs. 10b,c, 16b) or a cell of the appendage converted into an antheridium directly (Fig. 14)
   . . . . . . . . . . . . . . . . . . . . . . . . . . . . . . . **Laboulbeniaceae** 6

5′(1′) Antheridia compound, i.e., 2 or more antheridial cells more or less united and opening into a common chamber which itself has a single opening to the outside (Figs. 20b, 30c) . . . . . . . . . . . . . . . . . . . . . . . . . . . . **Peyritschiellaceae** 20

   6(5) Monoecious; antheridia borne on same individual that bears the perithecium
      . . . . . . . . . . . . . . . . . . . . . . . . . . . . . . . . . . . . . . . . . 7

   6′(5) Dioecious; antheridia and perithecia borne on separate individuals . . . . 18

7(6) Receptacle consisting of (a) 2 superposed cells, the basal forming the foot, the subbasal subtending a single, stalked perithecium, and (b) 1, rarely 2 to 3 smallish cells subtending a simple or branched appendage; or the subbasal cell subtending a cellular appendage and giving rise laterally on 1 or both sides to secondary cellular axes that bear perithecia and appendages . . . . . . . . . . . . . . . . . . . . . . . . . . . . . . . . . . . . 8

7′(6) Receptacle consisting of 3 or more cells superposed in a single series . . . . . . 14

   8(7) Appendage without an elongate sterile cellular termination and lacking sterile branchlets; antheridia borne terminally and/or laterally on 1 or more cells of the appendage . . . . . . . . . . . . . . . . . . . . . . . . . . . . . . . . . . . . 9

   8′(7) Appendage always with a more or less elongate sterile cellular termination or with 1 or more sterile or fertile branchlets; antheridia free or intercalary . . . 12

9(8) Appendage usually simple, consisting of 3 or 4 superposed cells bearing a single terminal antheridium . . . . . . . . . . . . . . . . . . . . . . . . **Autophagomyces** (Fig. 5)
   Common on species of Anthicidae and Pselaphidae (Coleoptera).

9′(8) Appendage usually simple, consisting of a variable number of superposed cells, several of which, including the terminal and sometimes the basal, give rise to 1 or more antheridia externally or internally . . . . . . . . . . . . . . . . . . . . . . . . . . . . . 10

   10(9′) Appendage consisting of only 3–4 cells, the basal cell separating 2–3 small cells distally which develop into simple antheridia; the terminal cell forming a single spinose antheridium and a small accessory cell; the median cell or cells sterile . . . . . .
      . . . . . . . . . . . . . . . . . . . . . . . . . . . **Acompsomyces** (Fig. 6)
      On species of Cryptophagidae and Lathridiidae (Coleoptera).

   10′(9′) Appendage consisting of a variable, sometimes large, number of cells, the basal cell fertile or not and all of the successive cells above bearing one or several divergent antheridia . . . . . . . . . . . . . . . . . . . . . . . . . . . . . . . . . . . 11

11(10′) Apex of the perithecium subtended by a single or double crown of unicellular, slender, more or less erect upgrowths developed from the upper tier of wall cells . . . . . . . . . . . . . . . . . . . . . . . . . . . . . . . . . . . . . . . . . . . . . . **Hesperomyces** (Fig. 7)
    On species of Coccinellidae and Mycetophagidae (Coleoptera) and Anthocoridae (Hemiptera). Allied genera: *Acallomyces, Ilyomyces, Stemmatomyces, Synandromyces*.

11′(10′) Apex of the perithecium without upgrowths developed from the upper tier of wall cells . . . . . . . . . . . . . . . . . . . . . . . . . . . . . . . . . . . . . . **Stigmatomyces** (Fig. 8)
    On species of many families of free-living flies (Diptera) and on a few species of Scydmaenidae and Staphylinidae (Coleoptera). Allied genera: *Arthrorhynchus, Gloeandromyces*.

  12(8′) Subbasal cell of receptacle subtending 2 cells placed side by side or a single, stalked perithecium and an appendage . . . . . . . . . . . . . . . . . . . 13

  12′(8′) Subbasal cell of the receptacle subtending an elongate primary appendage and giving rise laterally on 1 or both sides to single secondary cellular axes; the latter bearing a stalked perithecium from its basal cell and a similar cellular axis from its subbasal cell . . . . . . . . . . . . . . . . . . . . . . . . **Scaphidiomyces** (Fig. 9)
    On species of Scaphidiidae (Coleoptera). Allied genus: *Clonophoromyces*.

13(12) Subbasal cell of the receptacle subtending 2 cells placed more or less side by side, one cell subtending the perithecium, the other cell subtending 2 smallish cells that, in turn, subtend the appendage . . . . . . . . . . . . . . . . . . . **Laboulbenia** (Fig. 10)
    On species of several orders and many families of insects (Class Hexapoda) and a few mites (Class Arachnida; Order Acarina). Allied genera: *Apatomyces, Misgomyces*.

13′(12) Subbasal cell of the receptacle subtending a perithecium on 1 side and a simple or variably branched appendage on the other side . . . . . . . **Corethromyces** (Fig. 11)
    On species of several orders and families of insects, especially Staphylinidae. Allied genera: *Acrogynomyces, Dermapteromyces, Diclonomyces, Distolomyces, Ilytheomyces, Phaulomyces, Rhizomyces*.

  14(7′) Perithecium with at least 8 tiers of wall cells and subtended on both sides by single, elongate, simple appendages . . . . . . . . . . . **Cochliomyces** (Fig. 12)
    On species of Carabidae (Coleoptera).

  14′(7′) Perithecium with only 4–5 tiers of wall cells . . . . . . . . . . . . . . . 15

15(14′) Receptacle of 3 superposed cells subtending numerous smallish cells that bear numerous appendages which more or less surround the bases of 1 or more stalked perithecia . . . . . . . . . . . . . . . . . . . . . . . . . . . . . . . . . . . . . . **Teratomyces** (Fig. 13)
    On species of the subfamily Staphylininae of the Staphylinidae. Allied genera: *Diplomyces, Sandersoniomyces, Symplectromyces*.

15′(14′) Receptacle composed of an indeterminate number of cells superposed in a single series; these cells remaining undivided or separating single appendiculate cells on one side . . . . . . . . . . . . . . . . . . . . . . . . . . . . . . . . . . . . . . . . . . . . . 16

  16(15′) Perithecium terminal, its stalk subtended by several small, flattened cells that give rise laterally to ramiferous branchlets some cells of which may form antheridia directly . . . . . . . . . . . . . . . . . . . . . . . . . . . . . . . . **Coreomyces** (Fig. 15)
    Known only on species of Corixidae (Hemiptera).

  16′(15′) Perithecium not strictly terminal . . . . . . . . . . . . . . . . . . . 17

17(16′) Perithecium borne on a subterminal cell of the axis of the receptacle; successive cells of the axis, below the perithecium, forming single cells on one side, these cells

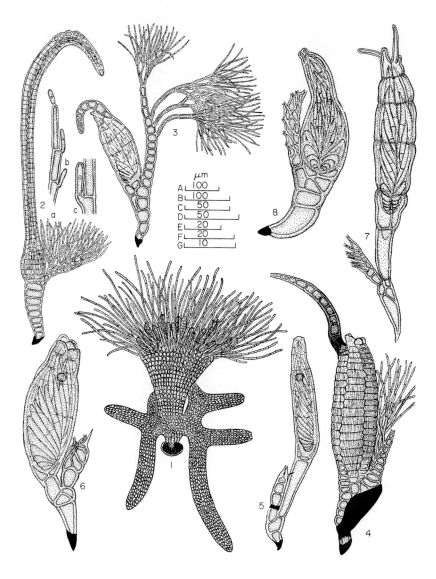

FIGS. 1–8. Laboulbeniales. (1) *Zodiomyces vorticellarius*, scale B. (2) *Rhynchophoromyces sperchopsis*. a, Mature individual, scale A; b, part of an appendage showing origin of exogenous spermatia, scale F; c, portion of appendage showing single spermatium, scale G. (3) *Autoicomyces recurvatus*, scale B. (4) *Ceratomyces mirabilis*, scale B. (5) *Autophagomyces bryaxalis*, scale D. (6) *Acompsomyces brunneolus*, scale E. (7) *Hesperomyces virescens*, scale C. (8) *Stigmatomyces pardyrae*, scale D.

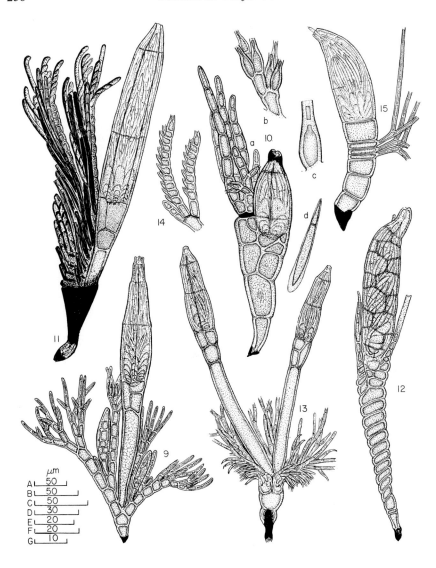

FIGS. 9–15. Laboulbeniales. (9) *Scaphidiomyces baeoceri*, scale F. (10) *Laboulbenia flagellata*. a, Mature individual, scale A; b, fertile appendage bearing several free, simple, flask-shaped antheridia, scale D; c, detail of a simple, flask-shaped antheridium, scale G; d, ascospore, scale F. (11) *Corethromyces cryptobii*, scale B. (12) *Cochliomyces* sp., scale B. (13) *Teratomyces mirificus*, scale A. (14) *Symplectromyces vulgaris*; fertile appendage showing simple antheridia formed directly from intercalary and terminal cells, scale E. (15) *Coreomyces corisae*, scale C.

giving rise to a variable number of simple sterile or fertile appendages which may more or less obscure the axis . . . . . . . . . . . . . . . . . . . **Rhachomyces** (Fig. 17)
On species of Carabidae and Staphylinidae.

17′(16′) One or more perithecia borne laterally on proximal cells of the axis of the receptacle which is prolonged distally and bears sterile branchlets or sessile antheridia
. . . . . . . . . . . . . . . . . . . . . . . . . . . **Ecteinomyces** (Fig. 18)
On species of Ptiliidae (Coleoptera). Allied genus: *Stichomyces*.

18(6′) Perithecium with 5 tiers of wall cells; female individual developed directly from the germinating spore which soon loses its original identity; male individuals minute, with a single terminal antheridium . . . . . . . . . . . . . . . . . . . . . . 19

18′(6′) Perithecium with 9 tiers of wall cells; one or more perithecia arising from a secondary receptacle derived from an outgrowth of a primary receptacle formed by the germinating spore, the latter retaining, more or less, its original identity; male individual forming 2 to many antheridia . . . . . . . . . **Herpomyces** (Fig. 21)
Known only on species of several families of cockroaches (Blattaria).

19(18) Receptacle of female individual without an apparent appendage; spore appearing continuous (actually 2-celled, but the upper cell small and not readily observed)
. . . . . . . . . . . . . . . . . . . . . . . . . . . **Amorphomyces** (Fig. 16)
Known only on species of the subfamily Aleocharinae of the Staphylinidae.

19′(18) Receptacle of female individual with a conspicuous sterile appendage consisting of 1–2 cells subtended by a single small cell . . . . . . . . . . . **Dioicomyces** (Fig. 22)
Common on species of Anthicidae (Coleoptera). Allied genera: *Dicrandromyces*, *Tetrandromyces*, *Triandromyces*.

20(5′) Monoecious; antheridia borne on same individual that bears the perithecium . . . . . . . . . . . . . . . . . . . . . . . . . . . . . . . . . 21

20′(5′) Dioecious; antheridia and perithecia borne on separate individuals . . . . 28

21(20) Receptacle consisting of 2 superposed cells, the subbasal cell subtending a single free perithecium and a small cell bearing a free appendage *or* the subbasal cell giving rise laterally, below the appendage, to 1 or more secondary axes or stalked perithecia and antheridial appendages . . . . . . . . . . . . . . . . . . . . . . . . . 22

21′(20) Receptacle consisting of more than 3 cells . . . . . . . . . . . . . . . . 25

22(21) Receptacle consisting of 2 superposed cells subtending a single perithecium and an appendage . . . . . . . . . . . . . . . . . . . . . . . . . . . . . . . 23

22′(21) Subbasal cell of receptacle subtending a cellular sterile primary appendage and giving rise laterally to 1 or more secondary axes or stalked perithecia and appendages; the latter sterile or forming antheridia; antheridia consisting of 4 tiers of paired cells; the lower tier forming the stalk; the cells of the middle 2 tiers each forming paired antheridial cells from their upper inner angles; the cells of the upper tier giving rise to 2 small cells internally, 1 or all of these forming simple, sterile, elongate branches . . . . . . . . . . . . . . . . . . . . . . . . **Monoicomyces** (Fig. 23)
On species of Staphylinidae, mostly of the subfamily Aleocharinae. Allied genera: *Eumonoicomyces, Kleidiomyces*.

23(22) Appendage with a sterile multicellular termination with or without sterile branchlets in addition to 1 or more compound antheridia derived from 1 or more of the lower cells . . . . . . . . . . . . . . . . . . . . . . . . . . **Cantharomyces** (Fig. 19)
On species of Dryopidae, Limnichidae, and Staphylinidae (Coleoptera).

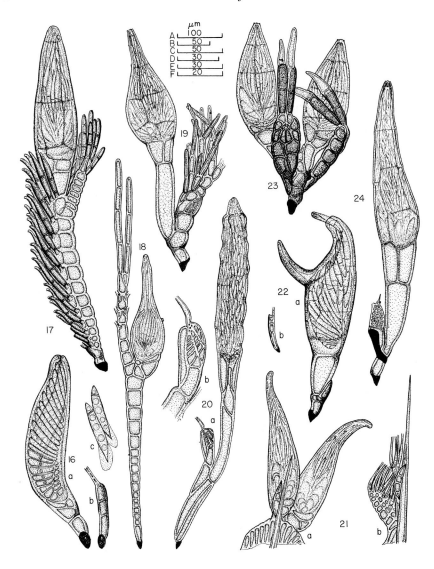

FIGS. 16–24. Laboulbeniales. (16) *Amorphomyces falagriae*. a, Mature female individual, scale E; b, mature male individual, scale E; c, two ascospores showing very small terminal cells characteristic of species of this genus, scale F. (17) *Rhachomyces philonthinus*, scale C. (18) *Ecteinomyces trichopterophilus*, scale D. (19) *Cantharomyces bledii*, scale C. (20) *Eucantharomyces* sp. a, Mature individual, scale A; b, detail of appendage showing compound antheridium in lateral view, scale C. (21) *Herpomyces ectobiae*. a, Mature female individual, scale E; b, mature male individual, scale E. (22) *Dioicomyces spiniger*. a, Mature female individual, scale C; b, mature male individual, scale C. (23) *Monoicomyces* sp., scale E. (24) *Haplomyces texanus*, scale B.

23'(22) Appendage without a sterile multicellular termination or sterile branchlets; the compound antheridium often bearing the spinelike termination of the original spore . . . 24

    24(23') Basal and subbasal cells of the receptacle broadly and obliquely superposed, becoming nearly vertically parallel to one another; antheridial cells disposed in oblique rows; antheridium bordered externally by a single, elongate cell and terminated by a relatively long, slender discharge tube . . . . . . **Eucantharomyces** (Fig. 20) On species of Carabidae.

    24'(23') Basal and subbasal cells of the receptacle transversely superposed; antheridial cells more or less irregularly disposed and completely surrounding the central cavity; antheridium with an inconspicuous, subapical pore . . . . **Haplomyces** (Fig. 24) Known only on species of *Bledius* of the Staphylinidae. Allied genera: *Camptomyces*, *Porophoromyces*.

25(21') Receptacle of 2 superposed cells followed above by 2 more or less elongate cells placed side by side, these subtending a single perithecium on one side and a small number of cells on the other side that are variably united with the perithecium . . . . . . . . . . . . . . . . . . . . . . . . . . . . . . . **Chitonomyces*** (Fig. 25) Common on species of Dytiscidae and Haliplidae (Coleoptera). Allied genus: *Hydraeomyces*.

25'(21') Receptacle elongate, usually flattened, composed of a more or less indeterminate number of superposed cells, these, by secondary divisions, forming tiers containing a definite or indefinite number of cells . . . . . . . . . . . . . . . . . . . . . 26

    26(25') Cells of the receptacle, above the basal cell, disposed in 3, rarely 2, more or less distinct vertical rows that form a variable number of tiers of cells; perithecium sessile, often partly surrounded on either side by the distal cells of the marginal rows; some or all of the marginal cells separating small cells externally and distally, these subtending simple unicellular branches or antheridia . . . . . . . . . . . . . . . . . . . . . . . . . . . . . . . . . **Rickia** (Fig. 27) On species of many families of several orders of insects and on species of a few families of mites. Allied genera: *Diaphoromyces*, *Rhipidiomyces*.

    26'(25') Cells of the receptacle above the basal cell disposed in only 2 or 3 superposed tiers of few to many cells . . . . . . . . . . . . . . . . . . . . . . . . 27

27(26') Receptacle bilaterally symmetrical; the subterminal tier of cells forming a single compound antheridium near the outer margin on each side; the upper tier of cells giving rise above to 1, usually 2, sometimes several parithecia and to more or less numerous simple, unicellular appendages each subtended by a single small cell . . . **Dichomyces** (Fig. 28) On species of Staphylinidae. Allied genus: *Diandromyces*.

27'(26') Receptacle bilaterally asymmetrical; the subterminal tier of cells forming a single compound antheridium near the outer margin on 1 side only; 1 perithecium, rarely more, formed distally and partly surrounded below on each side by cells of the upper tier; marginal cells usually bearing simple, elongate, unicellular appendages subtended by small cells . . . . . . . . . . . . . . . . . . . . . . . . . **Peyritschiella** (Fig. 26) On species of Carabidae and Staphylinidae. Allied genus: *Limnaiomyces*.

    28(20') Subbasal cell of the female receptacle always remaining in contact with the original basal cell below and the lower cell of the primary appendage above; the basal cell growing outwardly and upwardly on 1 side, rarely on both sides, and forming a succession of more or less numerous cells arranged in a lateral axis, the cells paralleling one another and the original subbasal cell; the cells of the secondary axis giving rise above, often alternately, to single perithecia or single sterile appendages;

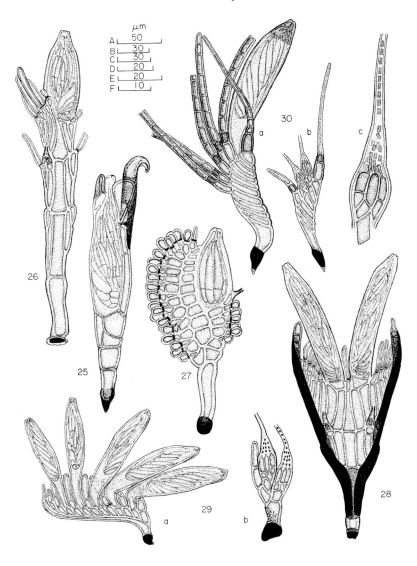

FIGS. 25–30. Laboulbeniales. (25) *Chitonomyces melanurus*, scale D. (26) *Peyritschiella clivinae*, scale A. (27) *Rickia minuta*, scale D. (28) *Dichomyces vulgatus*, scale B. (29) *Dimorphomyces* sp. a, Mature female individual, scale C; b, mature male individual, scale E. (30) *Dimeromyces africanus*. a, Mature female individual, scale A; b, mature male individual, scale A; c, detail of compound antheridium, scale C.

the male individual similar in structure to the female but usually bearing only a very small number of antheridia . . . . . . . . . . . . . **Dimorphomyces** (Fig. 29)
On species of Staphylinidae, especially, but a few species known on Carabidae and Tenebrionidae (Coleoptera) and on mites.

28′(20′) Basal and subbasal cells of the female receptacle becoming more or less distantly separated by the interposition of few to many cells cut off by the basal cell; each intercalary cell of the receptacle remaining sterile or giving rise to a single perithecium or sterile appendage; male individual resembling the female, the receptacle terminated by a sterile primary appendage and bearing 1 or more antheridia laterally, sterile secondary appendages usually lacking . . . . . . . . . . **Dimeromyces** (Fig. 30)
On species of many families of several orders of insects and a few mites. Allied genera: *Eudimeromyces, Nycteromyces, Trenomyces*.

## REFERENCES

Alexopoulos, C. J. (1962). "Introductory Mycology," 2nd ed. Wiley, New York.
Arwidsson, T. (1946). Om svenska laboulbeniacéfynd. *Sv. Bot. Tidskr.* **40**:307–309.
Bánhegyi, J. (1940). Elömunkálatok a magyarországi Laboulbenia-félék monografiájához. (Etudes préliminaires sur les Laboulbéniales de la Hongrie.) *Index Horti Bot. Univ. Budapestinensis* **4**:39–59.
Benjamin, R. K. (1965). Study in specificity. *Natur. Hist., N.Y.* **74**:42–49.
Benjamin, R. K. (1970). Laboulbeniales on semiaquatic Hemiptera. II. *Autophagomyces, Dioicomyces,* and *Prolixandromyces* gen. nov. *Aliso* **7**:165–182.
Benjamin, R. K. (1971). "Introduction and Supplement to Roland Thaxter's Contribution Towards a Monograph of the Laboulbeniaceae" (Bibliotheca Mycol., Vol. 30). Cramer, Lehre.
Benjamin, R. K., and L. Shanor. (1952). Sex of host specificity and position specificity of certain species of *Laboulbenia* on *Bembidion picipes. Amer. J. Bot.* **39**:125–131.
Berlese, A. N. (1889). Revista delle Laboulbeniacee e descrizione d'una nuova specie di questa famiglia. *Malpighia* **3**:44–60.
Bessey, E. A. (1950). "Morphology and Taxonomy of Fungi." McGraw-Hill (Blakiston), New York.
Boelens, W. C. (1947). Insectenschimmels (Laboulbeniaceae). *Tijdschr. Entomol.* **88**:515–520.
Bro Larsen, E. (1952). On subsocial beetles from the salt-marsh, their care of progeny and adaptation to salt and tide. *Trans. Int. Congr. Entomol., 9th, 1951* Vol. 1, pp. 502–506.
Chadefaud, M. (1960). Les Ascomycètes Cladomiens ou Laboulbeniomycètes. *In* "Traité de botanique systématique" (M. Chadefaud and L. Emberger, eds.), Vol. 1, pp. 487–496. Masson, Paris.
Collart, A. (1945). À propos des Laboulbéniacées. *Natur. Belg.* **26**:98–103.
Collart, A. (1947). À la découverte des Laboulbéniales. *Bull. Ann. Soc. Entomol. Belg.* **83**: 21–35.
de Bary A. (1887). "Comparative Morphology and Biology of the Fungi, Mycetozoa and Bacteria." Oxford Univ. Press (Clarendon), London and New York.
Denison, W. C., and G. C. Carroll. (1966). The primitive ascomycete: A new look at an old problem. *Mycologia* **58**:249–269.
Diesing, K. M. (1859). Revision der Rhyngodeen. *Sitzungsber. Kaiserl. Akad. Wiss., Math.-Naturwiss. Kl.* **37**:719–782.
Engler, A. (1898). "Syllabus der Pflanzenfamilien," 2nd ed. Borntraeger, Berlin.
Gäumann, E. A., and C. W. Dodge. (1928). "Comparative Morphology of Fungi." McGraw-Hill, New York.

Kamburov, S. S., D. J. Nadel, and R. Kenneth. (1966). The fungus *Hesperomyces virescens* Thaxter parasitic on *Chilocorus bipustulatus* L. in Israel. *Ktavim* **16**:135–136. (in Hebrew).

Kamburov, S. S., D. J. Nadel, and R. Kenneth. (1967). Observations on *Hesperomyces virescens* Thaxter (Laboulbeniales), a fungus associated with premature mortality of *Chilocorus bipustulatus* L. in Israel. *Isr. J. Agr. Res.* **17**:131–134.

Karsten, H. (1869). "Chemismus der Pflanzenzelle," Wilhelm Braumüller, Wien.

Knoch. (1868). *Laboulbenia baeri* Knoch ein neuer Pilz auf Fliegen. Assemblée des Naturalistes de Russie qui a eu lieu à St. Petersbourg du 28 décembre 1867 an 1$^{er}$ janvier 1868, p. 908.

Kolenati, F. A. (1857). Epizoa der Nycteribien. *Wien. Entomol. Monatsschr.* **1**:66–69.

Lepesme, P. (1946). Quelques mots sur les Laboulbéniales. *Entomologiste* **2**:81–85.

Lindau, G. (1897). Laboulbeniineae. *In* "Die Naturlichen Pflanzenfamilien" (A. Engler and K. Prantl, eds.) Sect. I, pp.491–505. Engelmann, Leipzig.

Lindroth, C. H. (1948). Notes on the ecology of Laboulbeniaceae infesting carabid beetles. *Sv. Bot. Tidskr.* **42**:34–41.

Luttrell, E. S. (1951). Taxonomy of the Pyrenomycetes. *Univ. Mo. Stud.* **24**:1–120.

Maire, R. (1912). Contribution à l'étude des Laboulbéniales de l'Afrique du Nord. *Bul. Soc. Hist. Natur. Afri. Nord* **4**:194–199.

Maire, R. (1916a). Deuxième contribution à l'étude des Laboulbéniales de l'Afrique du Nord. *Bull. Soc. Hist. Natur. Afri. Nord* **7**:6–39.

Maire, R. (1916b). Sur quelques Laboulbéniales. *Bull. Soc. Hist. Natur. Afri. Nord* **7**:100–104.

Maire, R. (1916c). Sur une nouvelle Laboulbéniale parasite des Scaphidiidae. *Bull. Sci. Fr. Belg.* **49**:290–296.

Maire, R. (1920). Troisième contribution à l'étude des Laboulbéniales de l'Afrique du Nord. *Bull. Soc. Hist. Natur. Afri. Nord* **41**:123–138, 143–158, and 159–170.

Mayr, G. (1852). Abnorme Haargebilde an Nebrien und einige Pflanzen Krains. *Verh. Zool.-Bot. Vereins Wien* **2**:75–77.

Peyritsch, J. (1871). Über einige Pilze aus der Familie der Laboulbenien. *Sitzungsber. Kaiserl. Akad. Wiss., Math.-Naturwiss. Kl., Abt. 1* **64**:441–458.

Peyritsch, J. (1873). Beiträge zur Kenntnis der Laboulbenien. *Sitzungsber. Kaiserl. Akad. Wiss., Math-Naturwiss Kl., Abt. 1* **68**:227–254.

Peyritsch, J. (1875). Über Vorkommen und Biologie von Laboulbeniaceen. *Sitzungsber. Kaiserl. Akad. Wiss., Math. Naturwiss. Kl., Abt. 1* **72**:377–385.

Picard, F. (1908a). Sur une Laboulbéniacée marine (*Laboulbenia marina* n. sp.) parasite d'*Aepus robini* Laboulbéne. *Co. Re. Soc. Biol.* **65**:484–486.

Picard, F. (1908b). Les Laboulbéniacées et leur parasitisme chez les insectes. *Feuille Natur.* **39**:29–34.

Picard, F. (1909). Sur une Laboulbéniacée nouvelle (*Hydrophilomyces digitatus* n. sp.) parasite d'*Ochthebius marinus* Paykull. *Bull. Soc. Mycol. Fr.* **25**:245–249.

Picard, F. (1912). Description de deux Laboulbéniacées nouvelles, parasites de Coléoptères. *Bull. Soc. Entomol. Fr.* **1912**:178–181.

Picard, F. (1913a). Contribution à l'étude des Laboulbéniacées d'Europe et du nord de l'Afrique. *Bull. Soc. Mycol. Fr.* **29**:503–571.

Picard, F. (1913b). Sur une Laboulbéniacée nouvelle, parasite de *Stenus aceris* Steph. *Bull. Soc. Entomol. Fr.* **1913**:462–465.

Picard, F. (1916–1917). Sur quelques Laboulbéniales d'Europe. *Bull. Sci. Fr. Belg.* **50**:440–460.

Richards, A. G., and M. N. Smith. (1954). Infection of cockroaches with *Herpomyces* (Laboulbeniales). III. Experimental studies on host specificity. *Bot. Gaz.* **116**:195–198.

Richards, A. G., and M. N. Smith. (1955). Infection of cockroaches with *Herpomyces* (Laboulbeniales). I. Life history studies. *Biol. Bull.* **108**:206–218.

Richards, A. G., and M. N. Smith. (1956). Infection of cockroaches with *Herpomyces* (Laboulbeniales). II. Histology and histopathology. *Ann. Entomol. Soc. Amer.* **49**: 85–93.
Robin, C. (1853). "Histoire naturelle des végétaux parasites qui croissent sur l'homme et sur les animaux vivants." Baillière et Fils, Paris.
Rouget, A. (1850). Notice sur une production parasite observée sur le *Brachinus crepitans. Ann. Soc. Entomol. Fr.* [2] **8**:21–24.
Scheloske, H.-W. (1969). Beiträge zur Biologie, Ökologie und Systematik der Laboulbeniales (Ascomycetes) unter besonderer Berücksichtigung des Parasit-Wirt-Verhältnisses, *Parasitol. Schriftenr.* **19**:1–176.
Shanor, L. (1955). Some observations and comments on the Laboulbeniales. *Mycologia* **47**:1–12.
Spegazzini, C. (1912). Contribución al estudio de las Laboulbeniomicetas argentinas. *An. Mus. Nac. Buenos Aires* **23**:167–244.
Spegazzini, C. (1914). Primo contributo alla conoscenza dell Laboulbeniali italiane. *"Redia"* **10**:21–75.
Spegazzini, C. (1915a). Fungi nonnulli senegalenses et canarienses. *An. Mus. Nac. Buenos Aires* **26**:117–134.
Spegazzini, C. (1915b). Laboulbeniali ritrovate nelle collezioni di alcuni musei italiani. *An. Mus. Nac. Buenos Aires* **26**:451–511.
Spegazzini, C. (1915c). Segunda contribución al conocimiento de las Laboulbeniales italianas. *An. Mus. Nac. Buenos Aires* **27**:37–74.
Spegazzini, C. (1917). Revision de las Laboulbeniales argentinas. *An. Mus. Nac. Buenos Aires* **29**:445–688.
Tavares, I. I. (1965). Thallus development in *Herpomyces paranensis* (Laboulbeniales). *Mycologia* **52**:704–721.
Tavares, I. I. (1966). Structure and development of *Herpomyces stylopygae* (Laboulbeniales). *Amer. J. Bot.* **53**:311–318.
Tavares, I. I. (1967). A new basis for classification of the Laboulbeniales. *Amer. J. Bot.* **54**:648 (abstr.).
Tavares, I. I. (1970). The appendage of *Amorphomyces* (Laboulbeniales). *Mycologia* **62**:741–749.
Thaxter, R. (1890). On some North American species of Laboulbeniaceae. *Proc. Amer. Acad. Arts Sci.* **25**:5–14.
Thaxter, R. (1891). Supplementary note on North American Laboulbeniaceae. *Proc. Amer. Acad. Arts Sci.* **25**:261–270.
Thaxter, R. (1892). Further additions to the North American species of Laboulbeniaceae. *Proc. Amer. Acad. Arts Sci.* **27**:29–45.
Thaxter, R. (1893). New species of Laboulbeniaceae from various localities. *Proc. Amer. Acad. Arts Sci.* **28**:156–188.
Thaxter, R. (1894). New genera and species of Laboulbeniaceae, with a synopsis of the known species. *Proc. Amer. Acad. Arts Sci.* **29**:92–111.
Thaxter, R. (1895). Notes on Laboulbeniaceae, with descriptions of new species. *Proc. Amer. Acad. Arts Sci.* **30**:467–481.
Thaxter, R. (1896). Contributions towards a monograph of the Laboulbeniaceae. I. *Mem. Amer. Acad. Arts Sci.* **12**:187–429.
Thaxter, R. (1899). Preliminary diagnoses of new species of Laboulbeniaceae. I. *Proc. Amer. Acad. Arts Sci.* **35**:153–209.
Thaxter, R. (1900). Preliminary diagnoses of new species of Laboulbeniaceae. II. *Proc. Amer. Acad. Arts Sci.* **35**:407–450.

Thaxter, R. (1901a). Preliminary diagnoses of new species of Laboulbeniaceae. III. *Proc. Amer. Acad. Arts Sci.* **36**:395–414.

Thaxter, R. (1901b). Preliminary diagnoses of new species of Laboulbeniaceae. IV. *Proc. Amer. Acad. Arts Sci.* **37**:19–45.

Thaxter, R. (1902). Preliminary diagnoses of new species of Laboulbeniaceae. V. *Proc. Amer. Acad. Arts Sci.* **38**:7–57.

Thaxter, R. (1903). Notes on the genus *Herpomyces. Science* **17**:463.

Thaxter, R. (1905). Preliminary diagnoses of new species of Laboulbeniaceae. VI. *Proc. Amer. Acad. Arts Sci.* **41**:301–318.

Thaxter, R. (1908). Contribution towards a monograph of the Laboulbeniaceae. II. *Mem. Amer. Acad. Arts Sci.* **13**:217–469.

Thaxter, R. (1912a). New or critical Laboulbeniales from the Argentine. *Proc. Amer. Acad. Arts Sci.* **48**:153–223.

Thaxter, R. (1912b). Preliminary descriptions of new species of *Rickia* and *Trenomyces. Proc. Amer. Acad. Arts Sci.* **48**:363–386.

Thaxter, R. (1914). Laboulbeniales parasitic on Chrysomelidae. *Proc. Amer. Acad. Arts Sci.* **50**:15–50.

Thaxter, R. (1915). New Indo-Malayan Laboulbeniales. *Proc. Amer. Acad. Arts Sci.* **51**:1–51.

Thaxter, R. (1916). New or critical species of *Chitonomyces* and *Rickia. Proc. Amer. Acad. Arts Sci.* **52**:1–54.

Thaxter, R. (1917). New Laboulbeniales, chiefly Dipterophilous American species. *Proc. Amer. Acad. Arts Sci.* **52**:647–721.

Thaxter, R. (1918a). Extra-American Dipterophilous Laboulbeniales. *Proc. Amer. Acad. Arts Sci.* **53**:695–749.

Thaxter, R. (1918b). New Laboulbeniales from Chile and New Zealand. *Proc. Amer. Acad. Arts Sci.* **54**:205–232.

Thaxter, R. (1920). New Dimorphomyceteae. *Proc. Amer. Acad. Arts Sci.* **55**:209–282.

Thaxter, R. (1924). Contribution towards a monograph of the Laboulbeniaceae. III. *Mem. Amer. Acad. Arts Sci.* **14**:309–426.

Thaxter, R. (1926). Contribution towards a monograph of the Laboulbeniaceae. IV. *Mem. Amer. Acad. Arts Sci.* **15**:427–580.

Thaxter, R. (1931). Contribution towards a monograph of the Laboulbeniaceae. V. *Mem. Amer. Acad. Arts Sci.* **16**:1–435.

Whisler, H. (1968). Experimental studies with a new species of *Stigmatomyces* (Laboulbeniales). *Mycologia* **60**:65–75.

Winter, G. (1887). Die Pilze Deutschlands, Oesterreichs und der Schweiz. Ascomyceten: Gymnoasceen und Pyrenomyceten. *In* "Dr. L. Rabenhorst's Kryptogamen-Flora von Deutschland, Oesterreich und der Schweiz," 2nd. ed., Vol. 1, Sect, II, pp. 918–925. Eduard Kummer, Leipzig.

# Ascomycotina
# Discomycetes

CHAPTER 9

# Discomycetes and Tuberales

RICHARD P. KORF[1]

*Plant Pathology Herbarium*
*Cornell University*
*Ithaca, New York*

## I. INTRODUCTION

The ascocarp of the Discomycetes and Tuberales is, in its commonest form, a typical apothecium, taking the shape of a saucer, a cup, or a nearly closed sphere, but varies in complexity from group to group. It is generally characterized by an open hymenium consisting of both asci and paraphyses. In a few forms the hymenium may be devoid of paraphyses, while in others the hymenium may consist solely of paraphyses with the asci scattered in the subhymenial tissues. In the simplest (most reduced) forms, the asci may form a naked fascicle, as in *Ascodesmis*; alternatively, even when sterile ascocarp tissues are present, the number of asci may be reduced to a few, or even to only one, as in some species of *Thelebolus*, and the ascocarp may then best be termed a cleistothecium. A full discussion of the Discomycetes and some of the characters used in recent classifications appears in the recent review article by Kimbrough (137). The Tuberales, or truffles, are for the most part hypogean, and appear to be derived from one order (Pezizales) of the Discomycetes; they represent a group which has evolved following a change or changes in the method of spore dispersal, since they no longer forcibly discharge their ascospores into the air. For convenience, the Elaphomycetaceae, with fairly large, hypogean cleistothecia, are treated here as representatives of the Tuberales, though their true relationships may possibly be with the Eurotiales.

In order to facilitate the use of the keys which follow, a few basic terms and concepts are outlined here. Microanatomy of the ascocarp plays such a significant role in modern taxonomy of this group that a successful identification can seldom be made without a median section through the

---

[1] Supported in part by National Science Foundation Grant GB-8548. The patience and good humor of many students and colleagues who have assisted in the preparation, testing, and revising of these keys is most gratefully acknowledged.

apothecium. For the purposes of the keys, Fig. 1 may serve as a schematic representation of a typical apothecium as viewed in section. The hymenium of asci and paraphyses in a palisade layer is bordered by the apothecial margin. The margin most often ends about even with the apices of the paraphyses and asci, but not infrequently extends upward far beyond the hymenium, sometimes originally enclosing the hymenium before it ruptures to expose the hymenium to the air. In rare instances the margin may end at a point lower down than the position of the ascus and paraphysis apices. The tissues which comprise the margin are, by definition (153), the ectal excipulum, which in turn extends downward to form the flanks of the apothecium. The ectal excipulum is not merely the outermost differentiated tissue, but may consist of one or even several tissues which are morphologically differentiated from one another; this differentiation is frequently less distinct at the growing margin than it is on the flanks. The medullary excipulum usually takes the shape of an inverted cone, and is the inner portion of the apothecial flesh, delimited toward the flanks by the ectal excipulum, and above by the subhymenium which in turn gives rise to the asci and paraphyses. In the stipe,

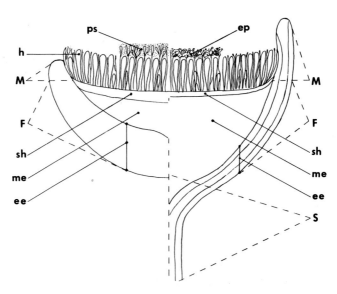

FIG. 1. Schematic drawing of an apothecium in vertical, median section; ee, ectal excipulum, thick and 1-layered on the left, thinner and 3-layered on the right; ep, epithecium (paraphysis apices branching to form a tissue); F, flanks of the apothecium; M, margin of the apothecium, shorter than the hymenium, h, on the left, extending beyond the hymenium on the right; me, medullary excipulum; ps, pseudoepithecium (paraphysis apices merely immersed in an amorphous substance, not forming a tissue); S, stipe, absent on the left, present on the right; sh, subhymenium.

if one is present, the term excipulum is usually avoided, though there may be both inner and outer layers of tissue in the stipe which continue upward into the ectal (or rarely medullary) excipulum, where usually further differentiation takes place in the tissue types. When the paraphyses are longer than the asci, they may be glued together in a matrix to form a pseudoepithecium, or they may branch apically to form a firm, suprahymenial tissue, the epithecium.

An ectal excipulum is wholly lacking in such simplified forms as *Ascodesmis, Karstenella, Ascocorticium,* and most members of the Hemiphacidiaceae. Maas Geesteranus (174) redefined the limits of the Geoglossaceae, separating the family from other stipitate Helotiales on the basis that the hymenium in the Geoglossaceae has no margin where it joins the stipe; by definition, the family lacks an ectal excipulum. Schematic illustrations of these and other major apothecial configurations are presented in Fig. 2.

Within the excipulum, the stipe, or the sclerotium or stroma from which some apothecia arise, a pronounced differentiation of tissue types may be found. These tissue types were named by Starbäck (263), with one addition by Korf (150), and are illustrated here in Fig. 3. Three of the tissue types consist essentially of short-celled tissues: globose cells with intercellular spaces (textura globulosa); tightly packed isodiametric cells without intercellular spaces (textura angularis); and short-celled hyphae with the individual cells more or less brick-shaped (textura prismatica). Long-celled tissues may have interwoven hyphae (textura intricata); more or less parallel, separable, thin-walled hyphae (textura porrecta); more or less parallel, agglutinated, thick-walled hyphae (textura oblita); or hyphae which are agglutinated and interwoven mostly in one plane to form irregularly shaped cells (textura epidermoidea). While intergrades occur, these terms have proved exceptionally useful in describing the tissues of apothecia.

Gel, or mucilaginous material, may be a prominent, and often taxonomically significant, feature in apothecia. Gel may arise either by direct secretion from the hyphae, or by breakdown of thick hyphal walls to form a gelatinous matrix. Sometimes the lumina of the former hyphal cells may give the appearance of being thin-walled hyphae immersed in a gel. Hyphae with gelatinized walls or in a gel usually have a distinctly refractive quality, a "glassy" appearance, especially when the cytoplasm of the hyphae is stained.

Ascus dehiscence remains a major feature distinguishing the orders. An apical or subapical operculum which is thrown back at spore discharge, or more rarely a vertical slit, is found in the Pezizales. A broad pore is present in the Cyttariales, and a definite plug is present in the thickened "inoperculate" ascus apex in the discomycete orders Phacidiales, Ostropales, and Helotiales. These are illustrated in Fig. 4. The exact ascus dehiscence mechanism is unknown in the Medeolariales. Asci no longer rupture forcibly to discharge their ascospores in the Tuberales, though the asci of this group

FIG. 2. Some typical ascocarp configurations, shown in the left half in surface view, in the right half in vertical section (except S, V, W, X, and Y shown only in section). Hymenium in face view is shaded dark, and in section is shown in black or as separate asci and paraphyses. Ascocarp flesh is shown in section lightly shaded. A, Gyrose pileus, e.g., *Gyromitra*, some species of *Discina*; B, pitted, hollow, spongelike pileus, e.g., *Morchella*; C, clavula, typical ascocarp of the Geoglossaceae, with no abrupt notch at the junction of hymenium and stipe; D, saddle-shaped pileus, e.g., some species of *Helvella*; E, campanulate pileus, distinctly notched between the hymenium and the stipe, e.g., *Heyderia*, *Verpatinia*; F, thimblelike, campanulate pileus, e.g., *Verpa*; G, involute pileus, e.g., *Cudonia*, some species of *Leotia*; H, subspherical ascocarp, e.g., *Sphaerosoma*, *Sphaerozone*; J, branching stipe with separate hymenial discs, e.g., *Cordierites*, *Myriodiscus*, *Polydiscidium*; K, cupulate apothecium with a stipe, typical of many genera; L, turbinate apothecium, e.g., *Calycella*, *Pyronema*; M, stipitate-turbinate or ombrophiloid apothecium, e.g., *Cudoniella*, *Neocudoniella*, *Ombrophila*, *Vibrissea*; N, pulvinate apothecium, typical of many genera; P, cupulate apothecium, probably the most common apothecial type; Q, pitted ascocarp, actually many apothecia in a fleshy stroma, e.g., *Cyttaria*; R, deep cupulate or urnulate apothecium, typical of many genera; S, reduced discoid apothecium with a reduced number of asci, e.g., some species of *Ascobolus*, *Thecotheus*, *Thelebolus*; T, semihypogean apothecium, opening by splitting, e.g., *Sarcosphaera*, in the species illustrated the hymenium becoming contorted or saccate as in some species of *Geopora*; U, spathulate or ear-shaped apothecia, arising from a common stalk in the species illustrated

are believed to be evolved from operculate asci. All Discomycetes and Tuberales have unitunicate asci. Some Ascomycetes with bitunicate asci, particularly those assigned to the Patellariaceae (see p. 199), produce ascocarps termed discothecia (156) which mimic apothecia. Such fungi are excluded from the Discomycetes in this treatment (see Chapter 7).

Ascospores of Discomycetes and Tuberales are usually symmetrical in one or two ways. If a spore can be rotated along a longitudinal axis and remain symmetrical, it is said to have radial symmetry. Allantoid spores, or those flattened on one side, are thus not radially symmetrical, and are common in only one large group, the suborder Sarcoscyphineae (Pezizales). Such spores also occur in a few additional genera (*Medeolaria, Selenaspora, Chlorencoelia, Cordierites, Encoelia, Lambertella*, and perhaps others). A spore which can be bisected transversely so as to yield resultant halves which are mirror images is said to have bipolar symmetry. Species of the Medeolariales, Cyttariales, Tuberales and Pezizales, perhaps without exception, have unicellular ascospores with bipolar symmetry. Members of the Ostropales, Phacidiales,

FIG. 3. Tissue types as viewed in section. A, Textura globulosa; B, textura angularis; C, textura prismatica; D, textura intricata; E, textura epidermoidea; F, textura oblita; G, textura porrecta; H1, textura intricata, hyphae widely spaced and immersed in a gel; H2, a gelatinized tissue, appearing to have widely spaced, parallel, thin-walled hyphae immersed in a gel, but probably arising from gelatinization of the thick walls of an earlier textura oblita.

and Helotiales, however, frequently (but not always!) lack bipolar symmetry, usually having the upper portion of the ascospore broader than the basal part. In these three orders, also, the ascospores are frequently transversely septate. Muriform spores are so rare in Discomycetes that one should immediately suspect he has a bitunicate ascomycete in hand if they are found.

---

(*Wynnea*), typically single as in *Otidea, Wynnella*; V, cleistothecium with a single, multispored ascus, e.g., some species of *Thelebolus, Trichobolus*; W, cleistothecium with a few asci (3 in the species illustrated), e.g., some species of *Ascobolus, Thelebolus, Trichobolus*; X, naked fascicle of asci and paraphyses, excipulum lacking, e.g., *Ascodesmis*; Y, palisade of paraphyses and asci arising from a subicular weft, e.g., *Ascocorticium, Ascosorus, Karstenella*; Z, hypogean, closed ascocarp, typical of Tuberales and some Pezizales, in the species illustrated with a greatly folded hymenium as in some species of *Geopora*.

FIG. 4. Ascus apices. A–D, Apical spore discharge mechanisms; E–K, some positive reactions in iodine (Melzer's Reagent), the blue reactions illustrated here as black or shaded. A, Typical apical operculum of most Pezizales; B, subapical operculum, with a thickened pad surrounding the inner portion of the opening, found in many Sarcoscyphineae (Pezizales); C, ascus apex opening by a vertical slit, also showing the subterminal, thickened ring present in some cases, found in a few Pezizales; D, typical inoperculate ascus apex, the thickened apex traversed by a delicate pore, a plug of wall material filling the pore until discharge, as in Helotiales, Ostropales, Phacidiales; E, a very broad pore, as in *Cyttaria* (Cyttariales), in the species illustrated with a distinct ring in the pore which blues in iodine, the plug at times appearing like an operculum; F, diffuse blueing of the ascus wall in iodine, as in some species of Pezizaceae, Ascobolaceae (Pezizales); G, apical blueing of the ascus, here shown restricted to a broad, ringlike zone, sometimes also extending over the apex, as in many Pezizaceae (Pezizales); H, inoperculate apex with the plug blueing in iodine (or with a very broad pore blueing and the plug itself not blue but so small that this can only be determined by viewing the ascus from above rather than from the side); J, inoperculate ascus in which a thin cylinder in the pore blues, seen in optical section as two blue lines; K, inoperculate ascus in which only a small ring surrounding a portion of the plug blues in iodine, seen in optical section as 2 tiny blue dots.

The number of nuclei in ascospores and paraphysis cells, as shown by the studies of Berthet (11), displays a remarkable consistency within the various groups, particularly in the Pezizales. Information on nuclear numbers is provided in the keys to the genera of that order because of the assumed phylogenetic significance, even though these are scarcely "key characters" of use in quick identification. Likewise, Arpin's (1a) studies on carotenoid pigments provide data which, where possible, have also been included in the key to the Pezizales for the sake of completeness and its probable importance in evolution.

The most important mounting medium, other than water, is surely Melzer's Reagent (0.5 gm iodine, 1.5 gm KI, 20 gm chloral hydrate, 20 ml $H_2O$), which is used both as a general differential stain and to test the iodine reaction (under oil immersion!) of the ascus apex or ascus pore. A positive reaction (often abbreviated "J+" from the German word for iodine, *Jod*) is usually some shade of blue, but occasionally violet colors may be produced. The apex of the ascus may blue, or a ring or plug may blue. At times the pore surrounding the plug will blue, and one may see either two thin vertical lines or two dots in the apex, these representing a blueing cylinder or a ring, as seen in optical section. Rarely the reaction may be a diffuse blue over the whole length of the ascus, or even portions of the apothecium may turn blue or violet in Melzer's Reagent. A negative reaction, in which no such blue color is produced, is frequently termed "J–."

Five other mounting media are of particular value with Discomycetes and Tuberales. (1) Cotton blue dye, dissolved in lactic acid or lactophenol, is used in the Pezizales and Tuberales, and to a much lesser extent in the Helotiales, for staining certain cyanophilic ascospore markings; spore markings which fail to turn deep blue when heated in cotton blue solution are termed cyanophobic [Le Gal's (164) classic study on spore markings in Pezizales should be consulted for developmental information.] (2) Phloxine B dye, usually in a 1% aqueous form, is an excellent cytoplasmic stain, a useful general differential reagent, and also will usually serve to distinguish septa in ascospores and hyphae (the cytoplasm stains pink, the septa do not take up the dye); a more critical examination can be made if the phloxine is removed from the mount and replaced with 50% aqueous glycerine, which removes the color from the background. (3) A 2% or 3% aqueous solution of KOH is useful for swelling dried material, and may also be used in conjunction with phloxine, mixing a drop of each at the time the mount is made. Some Discomycetes liberate quantities of a purple dyelike substance into dilute KOH, thus giving an ionomidotic reaction (153). (4) Congo red, as a 1% solution in 10% aqueous $NH_4OH$ or in 2% KOH, is used for differential staining of ascus wall layers in Theleboleae (Pezizales), and also as a general wall stain. (5) An acetocarmine stain (such as Belling's iron–acetocarmine) finds limited use in delimiting the tribe Jafneeae (Pezizales), in which the carminophilic nuclei of the ascospores, paraphyses, and apothecial cells all intensely take up the stain, even in dried material. This is also the case in *Leucoscypha* (61a); so far as tested, the nuclei of other groups of Discomycetes and Tuberales react very faintly, if at all, with acetocarmine, or only react after long exposure to the dye.

## II. A NOTE ON THE REFERENCES

It is most unfortunate that the literature on Discomycetes and Tuberales is exceptionally scattered, and has not been brought together in a form that would encourage even a serious student to study these fungi. Therefore, following most of the genera in the keys, references are provided to at least some significant sources of additional information. Such specific references appear as italic numerals, and refer to the references cited at the end of this chapter. For genera of the Pezizales (only), extensive comments on genera and species, and fuller citations to the literature, are provided elsewhere (158a).

In addition, there are three exceptionally useful guidebooks for which further information is provided by using a special set of symbols. No work on the Discomycetes has even approached in artistic beauty the magnificent plates in Boudier's four-volume "Icones Mycologicae" (17). In the orders

Pezizales, Phacidiales, Ostropales, and Helotiales, if Boudier did not illustrate any species of a genus the symbol "B–" is added. If Boudier provided one or more plates, the plate number(s) is given without mention of the often different generic name which Boudier might have employed. Dennis' "British Ascomycetes" (57), even though a local flora and aimed more at the amateur than the advanced student, is nonetheless probably the most useful general reference work we now have on Discomycetes, providing illustrations of many more genera than appear in the Boudier work. In most cases, Dennis' "British Cup Fungi" (51) will provide similar information if his later book is unavailable. For these same orders, and for the Pezizales also with Seaver's "North American Cup-Fungi (Operculates)" (250, 251), a convention is adopted indicating whether each author treated the genus, and how. Thus "B–/D–/S–" indicates that Boudier, Dennis, and Seaver all failed to treat species of the genus in question. "D+" indicates that Dennis accepted the genus with essentially the same limits as given in this classification. "D+ p.p." means that though Dennis accepted the genus, his concept was broader than that adopted here. As a final example, the conventional code adopted for the genus *Tricharina* in the key reads: "*83, 265*; B 347–51/D as *Tricharia*/S as *Patella* p.p." This conveys the information that more data on species of *Tricharina* may be found in the papers by Gamundi (1966) and by Svrček (1948), that Boudier has five plates illustrating species of the genus (in all three cases not necessarily under that generic name), that Dennis treated the genus under a different name (*Tricharia*), and that Seaver not only treated the species under a different generic name (*Patella*), but his concept of that genus was wider than the concept of *Tricharina* adopted in these keys.

## A. Outline of Classification

Order 1. Medeolariales, p. 257
  Family: Medeolariaceae
Order 2. Cyttariales, p. 258
  Family: Cyttariaceae
Order 3. Tuberales, p. 258
  Family A. Elaphomycetaceae
  Family B. Terfeziaceae
  Family C. Geneaceae
  Family D. Tuberaceae
Order 4. Pezizales, p. 262
  Family A. Sarcosomataceae
  Family B. Sarcoscyphaceae
  Family C. Ascobolaceae
  Family D. Pezizaceae
  Family E. Morchellaceae
  Family F. Helvellaceae
  Family G. Pyronemataceae

Order 5. Phacidiales, p. 277
  Family A. Rhytismataceae
  Family B. Cryptomycetaceae
  Family C. Phacidiaceae
Order 6. Ostropales, p. 28
  Family: Stictidaceae
Order 7. Helotiales, p. 282
  Family A. Ascocorticiaceae
  Family B. Hemiphacidiaceae
  Family C. Geoglossaceae
  Family D. Sclerotiniaceae
  Family E. Orbiliaceae
  Family F. Dermateaceae
  Family G. Hyaloscyphaceae
  Family H. Leotiaceae

## III. ORDERS

### KEY TO ORDERS OF DISCOMYCETES

1. Mycelium endophytic in the stems of *Medeola* (Liliaceae), causing fusiform swellings of the stem, penetrating the epidermis to form an indeterminate feltlike layer of paraphyses among which are dispersed asci . . . . . . . . . . . . . . . . . **Medeolariales** p. 257
1'. Not parasitic on *Medeola* . . . . . . . . . . . . . . . . . . . . . . . . . . . . . . 2
   2(1') Mycelium endophytic in the stems of *Nothofagus* (Fagaceae), causing galls on which are borne compound, more or less spherical ascocarps in which are embedded individual apothecia . . . . . . . . . . . . . . . . . . . . . . **Cyttariales** p. 258
   2'(1') Not forming ascocarps on galls of *Nothofagus* . . . . . . . . . . . . . . . . 3
3(2') Ascospores not liberated by violent discharge into the air, but instead disseminated by animals; ascocarps almost always subterranean, but if above ground then the hymenium covered with a thick, epithecial tissue through which the asci neither penetrate nor elongate to liberate ascospores. Asci spherical to clavate, in a recognizable hymenium or scattered within the tissues . . . . . . . . . . . . . . . . . . . . . . . . . . . . . . **Tuberales** p. 259
3'(2') Ascospores violently discharged into the air; ascocarps almost always produced above ground, but if completely subterranean then the ascocarps becoming epigean either by lifting the soil or by animal action (the genus *Geopora*). Asci clavate to cylindrical, rarely ovoid . . . . . . . . . . . . . . . . . . . . . . . . . . . . . . . . . . . . . . . . . . . 4
   4(3') Ascus apex as thin as, or thinner than, the side walls, opening by throwing back a lid (operculum), or rarely by formation of a longitudinal apical slit . . . . . . . . . . . . . . . . . . . . . . . . . . . . . . . . . . **Pezizales** p. 262
   4'(3') Asci usually distinctly thickened at the apex, with an apical pore through which the ascospores are discharged . . . . . . . . . . . . . . . . . . . . . . . . . . 5
5(4') Apothecia developing within a stroma, such that one or more hymenial areas develop as palisade layers of paraphyses, soon interspersed with asci, the stroma eventually rupturing only after the hymenium is mature (either by one or more longitudinal slits or in a stellate manner) . . . . . . . . . . . . . . . . . . . . . . . . . . . . . . . . . **Phacidiales** p. 278
5'(4') Apothecia not developing within a stroma, either with the hymenium naked from the beginning or with the hymenium developing within a more or less spherical primordium opening by a pore at its apex to expose the hymenium long before the formation of mature asci and ascospores . . . . . . . . . . . . . . . . . . . . . . . . . . . . . . . . . . . . . 6
   6(5') Asci cylindrical, very long and narrow, with a strongly thickened apex (at least in youth) traversed by a delicate pore, resembling the asci of the Clavicipitaceae. Ascospores filiform, nearly as long as the ascus . . . . . . . . . . **Ostropales** p. 282
   6'(5') Asci more or less clavate. Ascospores various, but if filiform then the asci without a greatly thickened apex . . . . . . . . . . . . . . . . . . . . . . **Helotiales** p. 283

## IV. MEDEOLARIALES

In this order, the ascocarps are indefinite and have a palisade layer of paraphyses which have penetrated the epidermis of the swollen stem of the host plant, and that are interspersed with asci. The Medeolariales infect flowering plants which have a greatly shortened internode between the two whorls of leaves, resulting in a rosette symptom, the infected internode and frequently a portion of the stem below the lower whorl of leaves being fusiformly swollen. The asci are large, and are eight-spored; the mechanism of spore discharge is unknown; they are very possibly deliquescent. The ascospores are large, brown, delicately longitudinally ribbed, and subfusiform with one side flattened.

There is a single family, Medeolariaceae, to accommodate one monotypic genus, *Medeolaria* (*M. farlowii*), on the Indian cucumber-root (*Medeola virginiana*) in North America. Ref. *278.*

## V. CYTTARIALES

The ascocarps (sometimes termed stromata) are spherical to pyriform, large, grouped together on galls of host branches, with up to 200 more or less spherical apothecia developing beneath a cortex, each finally having a separate broad opening to the surface. Every apothecium grows independently of the others at the ends of radiating strands of strongly gelatinized tissues within the ascocarp, ascocarps either developing air spaces within or remaining solid. The asci have a broad plug; they blue in iodine in some species but not in others, rupturing irregularly at the apex and at times seeming to form an operculum. All asci are found within each apothecium at precisely the same stage of development; they are eight spored, with a peculiar cushion of epiplasm remaining between the ascospores at maturity. The ascospores are smooth, globose to ellipsoid, hyaline but becoming gray; they are violently discharged and the spore print is (always ?) black. In some species pycnidia (or possibly spermagonia) are found in the same ascocarps in which apothecia develop later.

There is a single family, Cyttariaceae, and only one genus, *Cyttaria*, with about ten species all of which are parasitic on *Nothofagus* in the Southern Hemisphere. The generic name *Cyttariella* has been applied to the pycnidial (or spermagonial) stage, which in some species may be present before the apothecia are delimited in the same ascocarps. Ref. *83a, 149, 181, 208, 229, 246.*

## VI. TUBERALES

In this order, the ascocarps are hypogean, rarely epigean, and large, with fleshy or waxy tissues; at full maturity they are rarely powdery and then sometimes with a corky or woody sterile core. The asci are arranged in a hymenium or scattered in the tissues, and are cylindrical to spherical. They do not discharge ascospores into the air, and have no evident pore or operculum. They have eight, or often fewer, spores, the number of spores often varying among asci of the same ascocarp. The ascospores are hyaline or brown, smooth or variously sculptured, with cyanophilic markings. The spores are always unicellular, with both bipolar and radial symmetry, presumably always disseminated by animals.

### KEY TO FAMILIES AND IMPORTANT GENERA OF TUBERALES

1. Asci scattered, globose, soon disappearing and leaving the ascospores in a powdery mass usually accompanied by capillitial hyphae . . . . . . . . . . . **Elaphomycetaceae** 4
1'. Asci scattered or in a hymenium, persistent; tissues fleshy to waxy . . . . . . . . . 2
    2(1') Asci scattered in spherical, nestlike pockets separated by anastomosing sterile veins, not forming a distinct hymenium; ascocarps neither hollow nor with hypha-filled canals leading to the surface . . . . . . . . . . . . . . . . . **Terfeziaceae** 6
    2'(1') Ascocarps either hollow or with one to several hypha-filled chambers or canals leading to the surface . . . . . . . . . . . . . . . . . . . . . . . . . . . 3
3(2') Paraphyses fused apically to form a secondary cortex over asci which are in a distinct hymenium . . . . . . . . . . . . . . . . . . . . . . . . . . **Geneaceae** 12
3'(2') Paraphyses not fused apically; asci in a distinct hymenium or irregularly arranged . . . . . . . . . . . . . . . . . . . . . . . . . . . . **Tuberaceae** 14
    4(1) Sterile core corky or woody; ascospores ellipsoidal to fusiform, hyaline or light brown . . . . . . . . . . . . . . . . . . . . . . . . . . . . . . **Mesophellia**
        Five Australasian species, probably Basidiomycetes. Ref. *64*.
    4'(1) Sterile core cottony or evanescent; ascospores spherical, dark brown to black . . . . . . . . . . . . . . . . . . . . . . . . . . . . . . . . . . 5
5(4') Ascocarp with a sterile base . . . . . . . . . . . . . . . . . . . **Ascoscleroderma**
    Recognized for 3 species sometimes placed in *Elaphomyces*. Ref. *31*.
5'(4') Ascocarp without a sterile base, fertile throughout . . . . . . . . . **Elaphomyces**
    Dodge recognized 24 species in his monograph. Ref. *64*.
    6(2) Ascospores brown at maturity . . . . . . . . . . . . . . . . . . . . . . 7
    6'(2) Ascospores hyaline, yellowish or pale greenish at maturity . . . . . . . . . . 9
7(6) Spores globose, smooth, or ovoid, delicately warted; asci 8-spored . . . **Carbomyces**
    Trappe has erected a new family for this genus, with 2 species. Ref. *88, 278a*.
7'(6) Spores globose or ovoid, reticulate; asci few-spored . . . . . . . . . . . . . . 8
    8(7') Spores ovoid, meshes of the reticulum very deep; asci (1-) 2–4 (-5)-spored . . . . . . . . . . . . . . . . . . . . . . . . . . . . . **Mukagomyces**
        One species, *M. hiromichii*, in Japan and Australia. Ref. *125, 184, 278a*.
    8'(7') Spores globose, meshes of the reticulum low; asci 1-spored, or rarely 2-spored . . . . . . . . . . . . . . . . . . . . . . . . . . . . . . **Paradoxa**
        A single species, *P. monospora*, in Europe, placed by Fischer in Tuberaceae despite lack of venae externae. Ref. *80, 182*.
9(6') Ascospores smooth, ovoid to limoniform . . . . . . . . . . . . . . . . . . 10
9'(6') Ascospores sculptured, spherical . . . . . . . . . . . . . . . . . . . . . 11
    10(9) Ascocarp with an obvious basal mycelial tuft, peridium whitish, not pseudoparenchymatic . . . . . . . . . . . . . . . . . . . . . . . . . . . . . . **Tirmania**
        Probably only a single species, *T. nivea*. Ref. *80, 278a*.

10′(9) Ascocarp lacking a basal mycelial tuft, peridium dark, verrucose, parenchymatic . . . . . . . . . . . . . . . . . . . . . . . . . . . . . . . . . **Picoa**
There are about 6 species. Ref. *80, 278a.*

11(9′) Asci 5–8-spored . . . . . . . . . . . . . . . . . . . . . . . . . . . . . . **Terfezia**
About 15 species are recognized. Used here to include *Mattirolomyces*, following J. M. Trappe, sometimes recognized for 1 species with ovoid rather than nearly spherical asci. If cells lining the canals are inflated, see *Hydnobolites*, see below. Ref. *80, 88, 278a.*

11′(9′) Asci 2- to 4-spored . . . . . . . . . . . . . . . . . . . . . . . . . . **Delastria**
*Delastria rosea* is the only species. Ref. *80, 88, 278a.*

12(3) Ascospores sculptured . . . . . . . . . . . . . . . . . . . . . . . . . . **Genea**
The genus is taken here in a wide sense to include species placed earlier in *Myrmecocystis* and *Genabea*, about 20 species in all. Ref. *80, 88–90.*

12′(3) Ascospores smooth . . . . . . . . . . . . . . . . . . . . . . . . . . . . . . 13

13(12′) Ascospores ovoid . . . . . . . . . . . . . . . . . . . . . . . . . . **Petchiomyces**
The type species, *P. thwaitesii*, is epigean and looks like an unopened *Peziza*, known from Asia. *Petchiomyces kraspedostoma* from California differs significantly in having hairs around the mouth, and in being hypogeous; it should be compared with *Hydnocystis*. Ref. *15, 80, 88.*

13′(12′) Ascospores spherical . . . . . . . . . . . . . . . . . . . . . . . . . . **Hydnocystis**
The only species is *H. piligera*, in which Burdsall demonstrated the epithecium. Most of the species placed here earlier belong in *Geopora* (Pezizales). See also *Protogenea*, below. Ref. *25.*

14(3′) Hymenium of asci and paraphyses present . . . . . . . . . . . . . . . . 15

14′(3′) Hymenium lacking, or if present consisting of paraphyses only, with the asci scattered below the paraphyses . . . . . . . . . . . . . . . . . . . . . . . . . . 26

15(14) Ascocarp cavities remaining empty . . . . . . . . . . . . . . . . . . . . 16

15′(14) Ascocarp cavities, or at least the smaller ones, filled with hyphae . . . . . . . 22

16(15) Ascocarps with brown, granularly roughened hairs arising from deep within the hyphal (nonpseudoparenchymatic) peridium . . . . . . . . . . **Labyrinthomyces**
The only species, *L. steenisii*, has very short paraphyses and warted spores. *Hydnocystis singeri* differs mainly in having smooth spores and longer paraphyses. Ref. *15, 25.*

16′(15) Ascocarps smooth, or if hairy then the peridium is pseudoparenchymatic . . 17

17(16′) Ascospores strongly sculptured . . . . . . . . . . . . . . . . . . . . **Hydnotrya**
There are about 8 species, including some previously referred to *Geoporella* and *Gyrocratera*. Ref. *80, 88.*

17′(16′) Ascospores smooth or faintly roughened . . . . . . . . . . . . . . . . . . 18

18(17′) Ascospores spherical . . . . . . . . . . . . . . . . . . . . . . . . . . . . 19

18′(17′) Ascospores ovoid to cylindrical . . . . . . . . . . . . . . . . . . . . . . 20

19(18) Ascocarps epigeous, with an open pore, unilocular . . . . . . . . . . **Protogenea**
Unpublished studies of Burdsall and Korf show the only species of the genus, *P. japonica*, to be an epigean form of *Hydnocystis*; an epithecium and hairs are present. Ref. *148.*

19′(18) Ascocarps hypogeous, strongly convoluted . . . . . . . . . . . **Phymatomyces**
One species, *P. yezo-montanensis*, from Japan. Ref. *146.*

## 9. Discomycetes and Tuberales

20(18′) Ascocarps with a distinct stipe; paraphyses about as long as the asci
................................................ **Caulocarpa**
   One species, *C. montana*, in North America. Ref. *87*.

20′(18′) Ascocarps nonstipitate; paraphyses much longer than the asci ..... 21

21(20′) Canals in the ascocarp opening at several points; ascospores delicately roughened .......................................... **Hydnoplicata**
   One Australian species, *H. whitei*. Ref. *89, 90*.

21′(20′) Canals leading to a depressed external cavity; ascospores smooth. ..... **Barssia**
   One species in North America, *B. oregonensis*. Ref. *85*.

22(15′) Canals fertile only at the inner, dilated ends; asci 1- to 4-spored; ascospores globose, yellow to brown, alveolate-reticulate ............. **Piersonia**
   Two North American species. Ref. *80, 86*.

22′(15′) Canals fertile their whole length; asci 1- to 8-spored .......... 23

23(22′) Canals in the ascocarp radiating from a single, external point ........ 24

23′(22′) Canals not radiating from 1 point ....................... 25

24(23) Spores smooth, spherical to elongate, hyaline ........... **Stephensia**
   *Densocarpa* is a synonym. Gilkey recognized four species. Ref. *90*.

24′(23) Spores rough, spherical, brown ................ **Pachyphloeus**
   Fischer recognized 5 species. Ref. *80*.

25(23′) Ascospores globose, hyaline, smooth to minutely pitted ......... **Elderia**
   A single Australian species, *E. arenivaga*. Ref. *184*.

25′(23′) Ascospores globose to ovoid, yellowish, verrucose to reticulate ..... **Choiromyces**
   *Hydnotryopsis* is a synonym. There are about 5 species. Ref. *80, 86*.

26(14′) Ascocarp cavities remaining empty ...................... 27

26′(14′) Ascocarp cavities filled with hyphae ..................... 28

27(26) Ascospores spherical, sculptured; asci 1- to 8-spored ........ **Hydnobolites**
   One species with asci blue in Melzer's Reagent is segregated as *Amylascus* by Trappe; he placed both small genera in Terfeziaceae since the "hymenium" is only an infolding of the peridium. Ref. *80, 278a*.

27′(26) Ascospores ellipsoid, smooth; asci 8-spored ............. **Balsamia**
   There are probably at least 3 species. Ref. *80*.

28(26′) Ascospores ellipsoid, smooth, hyaline ............ **Pseudobalsamia**
   A single North American species is now recognized, *P. magnata*. *P. microspora* has been transferred to the Eurotiales as the type of the genus *Diehliomyces*. Ref. *80*.

28′(26′) Ascospores sculptured ............................. 29

29(28′) Canals in the ascocarp remaining distinct ................. **Tuber**
   A large genus with perhaps 80–100 species. Ref. *80*.

29′(28′) Canals in the ascocarp not distinct ....................... 30

30(29′) Ascospores spherical, yellow to yellow-brown, spinose, the bases of the spines forming a distinct reticulum ................... **Lespiaultinia**
   Two known species. The genus has also been called *Delastriopsis*. Ref. *80, 88*.

30′(29′) Ascospores ovoid, chestnut brown, with irregular crests, becoming sub-reticulate ........................................ **Fischerula**
   One species, *F. macrospora*, in Europe. Ref. *182*.

## VII. PEZIZALES

The ascocarps in this order are epigean, rarely hypogean, forming apothecia of various shapes, and ranging from minute to large. They are composed of tissues which are usually fleshy, sometimes brittle to leathery, rarely gelatinous. The asci are arranged in a distinct hymenium, accompanied by paraphyses (except in *Aparaphysaria*); the ascocarps are rarely reduced to a cleistothecium with a few asci or even a single ascus (some Theleboleae), or to a mycelial tuft (*Ascodesmis*) or a mycelial mat (*Karstenella*). The asci are cylindrical to clavate, rarely ovoid, and at maturity they throw back an apical or subapical operculum, or more rarely open by a vertical, apical slit (some Theleboleae), violently discharging ascospores. For the most part, the asci are eight-spored, although it is not uncommon to find them with two and four spores. Asci with sixteen, 32, 64, 128, and up to more than 7,000 spores are common in the Theleboleae; the number of ascospores is usually regular among asci of the same apothecium, but at times both four- and eight-spored asci occur together. The ascospores are hyaline or brown, rarely purple; they are smooth or variously sculptured, or the wall is sometimes folded or plicate, with both bipolar and radial symmetry (the spores being flattened on one side, or allantoid, rather commonly in the suborder Sarcoscyphineae, but characteristic of only one genus, *Selenaspora*, of the Pezizineae), always unicellular, rarely becoming septate during germination (some Sarcoscyphaceae). Pezizales are found on soil, wood, dung, and plant debris.

**KEY TO SUBORDERS, FAMILIES, SUBFAMILES, TRIBES AND IMPORTANT GENERA OF PEZIZALES**

1. Asci clavate to cylindrical, thick-walled, with a suboperculate apex, operculum terminal or subapical, not blueing in iodine. Ascospores hyaline, guttulate (usually multiguttulate), plurinucleate, frequently blunt at the ends or radially asymmetrical. Paraphyses usually anastomosing. Apothecia leathery, subgelatinous, or even corky, but not fleshy; medullary tissues always long-celled, frequently embedded in a gelatinous matrix ................................ Suborder **Sarcoscyphineae** 2

1'. Asci clavate to cylindrical, thin-walled, with a typical terminal operculum, more rarely broad-clavate to ovoid, and then the walls often thick and in some cases opening by a vertical apical slit (bilabiate), sometimes blueing in iodine. Ascospores hyaline or purple to brown, eguttulate or guttulate, uninucleate or plurinucleate, radially symmetrical (except in *Selenaspora*). Paraphyses usually not anastomosing. Apothecia fleshy to brittle; medullary tissues short-celled, long-celled, or a mixture, usually not embedded in a gelatinous matrix ................................ Suborder **Pezizineae** 3

    2(1) Apothecia dark-colored, with melaninlike pigments, excipulum devoid of carotenoids (bright yellow, orange or red pigments), but such pigments rarely occur in the paraphyses. Asci always maturing successively. Ascospores smooth or with transverse folds, rarely with cyanophilous verruculae, sometimes surrounded by a gelatinous sheath. (Paraphysis cells uninucleate, rarely binucleate) ............ **Sarcosomataceae** 7

2'(1) Apothecia bright-colored, usually with carotenoids in the hymenium (paraphyses) and often also in the excipulum, melaninlike pigments if present confined almost wholly to hairs on the excipulum. Asci maturing successively or simultaneously. Ascospores smooth or with cyanophobic longitudinal ridges or reticulations, rarely with a sheath. (Paraphysis cells multinucleate) .............. **Sarcoscyphaceae** 17

3(1') Ascospores very thick-walled in youth, eguttulate, hyaline and with asci diffusely blue in iodine, or remaining hyaline until just before discharge when they become purple and then brown (by deposition of extrasporal matter from the epiplasm) with asci blue or not in iodine. Carotenoids (yellow, orange or red pigments) present or not. (Ascospores uniformly uninucleate) ............................................. **Ascobolaceae** 3

3'(1') Ascospores usually thin-walled in youth, guttulate or eguttulate, permanently hyaline or if brown then pigments (not deposited from the epiplasm) develop early in cyanophilic spore markings. Asci blue in iodine or not. Carotenoids present or not ........... 4

    4(3') Ascus apex intensely blue or whole wall diffusely blue in iodine. Ascospores hyaline or brown, rarely greenish-yellow. (Ascospores regularly uninucleate)
..................................... **Pezizaceae** 35

    4'(3') Asci not blue in iodine. ........................... 5

5(4') Apothecia large, either discoid, or stalked and spongelike, or stalked and with a pendant, campanulate pileus, always buff or brown, lacking carotenoids. Ascospores eguttulate, hyaline, though the spore-print may be yellow-brown, with a "crown" of tiny guttules in the epiplasm at both ends of the spores in fresh specimens. (Ascospores with 20 to 60 nuclei) ....................................... **Morchellaceae** 38

5'(4') Apothecia large or small, with or without carotenoids. Ascospores guttulate or eguttulate, lacking a "crown" of guttules in the epiplasm. (Ascospores with 1 to 4 nuclei.) ............................................... 6

    6(5') Apothecia large, discoid to cupulate, usually stalked, often saddle-shaped or with a cerebriform pileus (gyromitroid) or with the hymenium covering the upper portion of a chambered stalk. Distinct hairs on the excipulum lacking; rhizoidlike processes present in one genus. Ascospores smooth or warted, hyaline or brown, with 1, 2 or more rarely 3 oil globules (invariably tetranucleate) ............ **Helvellaceae** 40

    6'(5') Apothecia usually small, discoid to cupulate, rarely stalked. Hairs present or absent. Ascospores smooth or variously sculptured, guttulate or eguttulate, hyaline or brown (uninucleate, rarely binucleate) ............... **Pyronemataceae** 45

7(2) Ascospores either completely smooth or plicate (with transverse or very rarely longitudinal folds of the wall) ................... Tribe **Sarcosomateae** 8

7'(2) Ascospores with cyanophilic markings (apiculi, warts, or longitudinal ridges)
.................................... Tribe **Galielleae** 14

    8(7) Apothecium initially a hollow clavula, opening by splitting downward into *Geastrum*-like, recurving rays ............................. **Chorioactis**
        Monotypic, *C. geaster*. Ref. *72*; B−/D−/S as *Urnula* p.p.

    8'(7) Apothecium not splitting into rays ...................... 9

9(8') Apothecium minute, discoid, covered externally with brown setae, hymenium also provided with setalike, brown extensions on some paraphyses .......... **Desmazierella**
    *Desmazierella acicola* on pine needles is the only accepted species, in Europe and North America, with a *Verticicladium* imperfect state. A possible relationship to *Thindia* has been suggested. Ref. *93, 161*; B 363/D+/S−.

9′(8′) Apothecia not setose, usually large. . . . . . . . . . . . . . . . . . . 10

    10(9′) Apothecia exceptionally gelatinous, so much so that when the apothecium is cut the contents flow out; hairs on the excipulum composed of moniliform elements
. . . . . . . . . . . . . . . . . . . . . . . . . . . . . . . . . . **Sarcosoma**

        *Sarcosoma globosum* is a large, northern species. The generic name has been used in a wider sense by Le Gal to include fungi placed here in *Galiella*. Ref. *169, 207*; B−/D−/S as *Bulgaria* p.p.

    10′(9′) Apothecia not as distinctly gelatinous; hairs when present not of distinctly moniliform elements . . . . . . . . . . . . . . . . . . . . . . . . . . . . . . 11

        11(10′) Apothecia deeply cleft down one side, hence appearing like those of *Otidea*, without marginal hairs . . . . . . . . . . . . . . . . . . . . . . . . . . . **Korfiella**

            The single species, *K. karnika*, has 4 or fewer spores in each ascus, orange amorphous granules on the outside of the cup, and is associated with a *Conoplea* imperfect state with germ pores in the conidia; other than in its shape, the genus very closely approaches *Plectania*, and probably should be merged with it. Ref. *123, 209*; B−/D−/S−.

        11′(10′) Apothecia discoid, cupulate, or deep cupulate, not otideoid; marginal hairs present . . . . . . . . . . . . . . . . . . . . . . . . . . . . . . . . . 12

            12(11′) Ascospores globose . . . . . . . . . . . . . . . . . . . . **Pseudoplectania**

                There are two common temperate species, *P. melaena* and *P. nigrella*, but there may be several undescribed species from islands in the Asian and New World tropics. No imperfect states are known. The status of the genus is questionable, and it should very possibly be merged with *Plectania*. Immature specimens of *Plectania* in which only a few asci have delimited ascospores will key here, since their spores remain spherical for a rather long time before elongating. Ref. B 343–4/D +/S + p.p.

            12′(11′) Ascospores ovoid to fusoid . . . . . . . . . . . . . . . . . . . . . . 13

                13(12′) Apothecia deep cupulate. . . . . . . . . . . . . . . . . . . . . . **Urnula**

                    The very common *U. craterium* in North America, rare in Europe, is the causal agent of *Strumella*-canker of oaks, etc. The imperfect state is *Conoplea globosa*, belonging to that group of species in which the conidia have a germ slit. Le Gal and others would merge *Plectania* here. Ref. *123, 169*; B 341/D−/S+ p.p.

                13′(12′) Apothecia shallow cupulate to discoid . . . . . . . . . . . . . . . . **Plectania**

                    The generic name has been used by Seaver and by Le Gal, in particular, for a wholly different genus (*Sarcoscypha* in this classification). The blackish apothecia are somewhat orange or brick-red externally in *P. melastoma* from amorphous granules very like those in *Korfiella*. Some species have a *Conoplea* imperfect state belonging to that group of species which have a germ pore in the conidium. Ref. *123, 152, 207, 239*; B 342/D+/S as *Bulgaria* p.p., *Paxina* p.p.

    14(7′) Apothecia scarcely gelatinous at all, without hairs, with a complex, 3-layered excipulum; ascospores apiculate, markings very like those in *Discina*, distinctly cyanophilic . . . . . . . . . . . . . . . . . . . . . . . . . . . . . . . **Nannfeldtiella**

        *N. aggregata*, the only species, was described with rugose, cyanophobic spore markings, by some error. It was also described as hairy, but this was by misinterpretation of the middle layer of the excipulum after loss in sectioning or in sloughing off of the outermost apothecial layer. It may be closely related instead to *Discina* (Helvellaceae). Ref. *72*; B−/D−/S−.

    14′(7′) Apothecia with or without hairs; ascospores not apiculate, but delicately warted or with longitudinal ridges . . . . . . . . . . . . . . . . . . . . . . . . . 15

## 9. Discomycetes and Tuberales

15(14′) Apothecia distinctly gelatinous, with brown hairs. Ascospores marked with distinct (though often small) warts . . . . . . . . . . . . . . . . . . . . . . . . . **Galiella**
Galiella rufa, sometimes placed in *Sarcosoma*, is the commonest North American species. There are many species, however, mostly in the tropics. Ref. *152, 166, 226*; B−/D−/S as *Bulgaria* p.p.

15′(14′) Apothecia leathery or corky . . . . . . . . . . . . . . . . . . . . . . . 16

    16(15′) Apothecia thick and corky, covered with dark brown hairs; ascospores marked with delicate longitudinal ridges which may anastomose. . . . . . . . . . **Wolfina**
One species, *W. aurantiopsis*, in North America. The orange-red hymenium may indicate the presence of carotenoids. Ref. *72*; B−/D−/S+.

    16′(15′) Apothecia thin-fleshed, leathery, deep-cupulate, with scattered pale brown hairs; ascospores with distinct warts . . . . . . . . . . . . . . . . . . . . . . **Neournula**
One species, *N. pouchetii* (= *N. nordmanensis*), known from both Europe and North America. Ref. *206*; B−/D−/S−.

17(2′) All asci within each apothecium maturing simultaneously
. . . . . . . . . . . . . . . . . . . . . . . . . . . . . Tribe **Boedijnopezizeae** 18

17′(2′) Asci of various ages within a single apothecium, maturing successively.
. . . . . . . . . . . . . . . . . . . . . . . . . . . . Tribe **Sarcoscypheae** 20

    18(17) A distinctly gelatinous layer of tissue present in the apothecium just inside the outermost layer of the ectal excipulum; ascospores remaining smooth
. . . . . . . . . . . . . . . . . . . . . . . . . . . . . . . . . . . . . . . . **Boedijnopeziza**
Recognized for 2 species, *B. institia* in the Asian tropics with a fringe of fasciculate hairs around the mouth and *B. colensoi*, pantropical and devoid of hairs, merged by some authors with *Cookeina* or with *Microstoma*. Ref. *13, 41, 239*; B−/D−/S as *Cookeina* p.p.

    18′(17) No distinct gel layer present in the excipulum. Ascospores smooth, longitudinally ribbed, or reticulate . . . . . . . . . . . . . . . . . . . . . . . . . . . . . . . 19

19(18′) Hairs on the apothecium in fascicles, rarely absent; ascospores with longitudinal ribs or anastomosing bands . . . . . . . . . . . . . . . . . . . . . . . . . . **Cookeina**
There are 3 species, almost wholly restricted to the tropics. Ref. *41, 166, 239*; B−/D−/S+ p.p.

19′(18′) Hairs on the apothecium single, not fascicled; ascospores smooth . . . **Microstoma**
There are a few temperate to arctic or subalpine species. *M. floccosum* is common in North America and Asia, and has been placed in *Anthopeziza* by some authors. *Microstoma protractum* is apparently much commoner in Europe. Both are treated in *Sarcoscypha* by some authors. Ref. *130, 226*; B− /D+ /S as *Plectania* p.p.

    20(17′) Ascospores with longitudinal ribs or with a reticulum, rarely only faintly striolate . . . . . . . . . . . . . . . . . . . . . . . . . . . . . . . . . . . . 21

    20′(17′) Ascospores smooth, rarely with faint transverse folds . . . . . . . . . . 24

21(20) Apothecia spathulate, arising in a dense cluster from a common base and a subterranean "sclerotium" composed of tangled rhizomorphs; ascospores with pronounced longitudinal ribs . . . . . . . . . . . . . . . . . . . . . . . . . . . . . . . . . . . . . . . . . . **Wynnea**
There are only a few species known, worldwide in distribution. These fungi are pink in youth, but soon become nearly black, and have thus been referred in many classifications to the Sarcosomataceae or Urnuleae. Ref. *277*; B−/D−/S+.

21′(20) Apothecia single, cupulate to discoid, occasionally somewhat laterally attached and fan-shaped . . . . . . . . . . . . . . . . . . . . . . . . . . . . . . . . . . . . . 22

22(21′) Ascospores strongly reticulate; apothecia with fascicled hairs . . . . **Geodina**
One species, *G. guanacastensis*, from Central America. Ref. *40*; B−/D−/S−.

22′(21′) Ascospores longitudinally striolate; hairs, when present, single, not fascicled . . . . . . . . . . . . . . . . . . . . . . . . . . . . . . . . . . . . . 23

23(22′) Hyphae of the medullary excipulum embedded in a gelatinous matrix; apothecia fan-shaped, laterally attached . . . . . . . . . . . . . . . . . . . . . . . **Aurophora**
There may be several species; *A. dochmia* is the type. Ref. *42, 239*; B−/D−/S as *Phillipsia* p.p.

23′(22′) Hyphae of the medullary excipulum not embedded in a gelatinous matrix; apothecia discoid to deep cupulate, sessile to distinctly stipitate. . . . . . . . . . . . . **Phillipsia**
This is a fairly large, predominantly tropical genus. Ref. *42, 166*; B−/D−/S+ p.p.

24(20′) Ascospores spherical, more rarely cubical by mutual pressure; apothecia clothed with hyaline, hyphoid hairs . . . . . . . . . . . . . . . . . . . . . . . **Pithya**
This is a small genus, apparently restricted to coniferous foliage. Ref. *43*; B 320/ D+/S+.

24′(20′) Ascospores ovoid to fusoid; hairs present or absent. . . . . . . . . . . 25

25(24′) Asci with a distinct thickened ring below the apex; apothecia glabrous
. . . . . . . . . . . . . . . . . . . . . . . . . . . . . . . . . . **Pseudopithyella**
One species, *P. minuscula*, on *Juniperus* foliage, but possibly at least one more should be recognized. It has been merged with *Sarcoscypha* (as *Plectania*) by Le Gal. Ref. *72, 166*; B−/D−/S+.

25′(24′) Asci without a thickened ring; apothecia glabrous to hairy . . . . . . . . 26

26(25′) Apothecia with a flat disc and a thin stipe, small, with huge ascospores over 50 μm long . . . . . . . . . . . . . . . . . . . . . . . . . . . . . . . . **Pindara**
The asci are clearly suboperculate, though the genus has been referred to the inoperculate discomycetes by Eckblad. One species, *P. terrestris*. Ref. *274, 280*; B−/D−/S−.

26′(25′) Apothecia sessile, or if stipitate, usually large, with smaller ascospores . . . 27

27(26′) Apothecia minute, orange, beset with brown setae . . . . . . . . . . . . **Thindia**
The single species, *T. cupressi*, is on coniferous foliage from India, and looks like a minute *Scutellinia*. It has 4-spored asci. Ref. *161*; B−/D−/S−.

27′(26′) Apothecia larger, without brown setae . . . . . . . . . . . . . . . . . 28

28(27′) Excipulum spongy, with air-filled pockets or cavities; ascospores transversely folded at one stage in development . . . . . . . . . . . . . . . . . **Rickiella**
A single species, *R. transiens*, originally incorrectly placed near *Cyttaria*. The tissues of *Phillipsia* at times become loculate; they may be closely related. Ref. *238*; B−/D−/S−.

28′(27′) Excipulum without obvious cavities; ascospores smooth . . . . . . . . 29

29(28′) Apothecia small, on branches and foliage, with the outer layers of the apothecium of angular isodiametric cells or with the cells fused together to form an epidermoid tissue . . . . . . . . . . . . . . . . . . . . . . . . . . . . . . . . . **Nanoscypha**
Two species are recognized. Close relatives may be either *Sarcoscypha* or *Pithya*. Ref. *43*; B−/D−/S as *Cookeina* p.p.

29′(28′) Apothecia small to large, on woody substrata, rarely on duff, with the outer layers of the apothecium composed of hyphae with long or short cells retaining their hyphal orientation, and with the long axes of the hyphae parallel to the outer surface
. . . . . . . . . . . . . . . . . . . . . . . . . . . . . . . . . . **Sarcoscypha**

This is a fairly large genus, incorrectly termed *Plectania* by some authors. The large *S. coccinea* is common in the early spring as soon as the snow melts. The much smaller *S. occidentalis* is common on oak twigs in summer. Most species have bright red or reddish-orange hymenial pigments, and the hairs, when present, are hyaline and hyphoid. *Phillipsia hartmannii* will key here if the faint longitudinal striations on its ascospores are overlooked. Ref. *43, 130, 166*; B 322–3/D+/S as *Plectania* p.p.

30(3). Ascospores becoming purple and then brown just prior to discharge, by deposition of epiplasmic pigments; asci diffusely blue in iodine, or not reacting
.................................. Tribe **Ascoboleae**   31

30'(3) Ascospores remaining permanently hyaline, or with brown pigments in the spore markings not deposited from the epiplasm; asci diffusely blue in iodine
.................................. Tribe **Iodophaneae**   32

31(30) Ascospores firmly united into a group and discharged as a single projectile, or more rarely at first loosely united and finally free or partially free   ........ **Saccobolus**
There are about 20 species in this genus, almost all on dung. Ref. *20, 21, 187*; B–/D+/S+.

31'(30) Ascospores free from one another, not regularly arranged into a group at any phase of maturation   ............................... **Ascobolus**
Some authors have distinguished *Dasyobolus* for species with hyphoid hairs on the apothecium, and *Sphaeridiobolus* for species with spherical spores. Van Brummelen recognized 48 species in his recent critical world monograph. Though mostly on dung, some species are terrestrial, on wood, leaves, or charred wood or soil. Ref. *20, 187*; B 408–12/D+ also as *Dasyobolus, Sphaeridiobolus*/S+.

32(30') Ascospores spherical, eguttulate or with one inclusion (oil drop? nucleus?)   .................................. 33

32'(30') Ascospores ovoid, eguttulate   ..................... 34

33(32) Apothecia hemispherical to lenticular; ascospores with long spines connected at their bases to form a reticulum, becoming smoky brown at maturity   ....... **Boudiera**
Possibly only one good species, *B. areolata*, but others have been described. Ref. *72*; B 417/D+/S+.

33'(32) Apothecia almost globose; ascospores spinose, remaining hyaline   ... **Sphaerosoma**
The genus has been used in several senses, and also has included species with asci not blueing in iodine which are now referred to *Sphaerozone*. It has been treated in the Tuberales by some authors. *Svrcekia* is a recent synonym. Ref. *162, 242, 256*; B–/D–/S+ p.p.

34(32') Apothecia lenticular to convex, almost always with carotenoid pigments
.................................. **Iodophanus**
About 12 species; some have *Oedocephalum* imperfect states, most are on dung. Ref. *139, 141*; B 413/D+/S as *Ascophanus* p.p., *Humarina* p.p.

34'(32') Apothecia subconical to cylindrical or turbinate, rarely discoid, whitish at first and eventually some shade of purple, devoid of carotenoids   ... **Thecotheus**
The type species, *T. pelletieri*, has 32-spored asci, but the other 3 species are 8-spored. *Ascophanella* (Ref. *78*) (non Korf) is a synonym. The monotypic genus *Ascophanopsis* is said to differ in having the ascospore apiculus surrounded by a collarette; the iodine reaction of its asci is unreported. *Zukalina* (= *Gymnodiscus*) is probably also a synonym. All occur on dung or wood. Ref. *78, 136, 224b, 295*; B 414/D+ /S+ also as *Ascophanus* p.p.

35(4) Apothecia with a delicate, long stipe and a saucer-shaped disc, often tiny, with greenish-brown tones; ascospores with greenish-yellow contents, smooth . . . . . . **Gelatinodiscus**
    One species, *G. flavidus*, on needles of *Chamaecyparis* in western North America. An unusual genus referred to here with great reservation. Despite the generic name, there are no gelatinized hyphae anywhere in the apothecium. Eckblad excluded it from the Pezizales. Ref. *72, 133*; B−/D−/S−.

35'(4) Apothecia discoid to cupulate, rarely convoluted to form a sparassoid mass of fused apothecia . . . . . . . . . . . . . . . . . . . . . . . . . . . . . . . . . . . . . 36

  36(35') Apothecia deep cupulate, wholly immersed in the soil, very large, usually with a violaceous hymenium; ascospores smooth, ovoid, biguttulate . . . . **Sarcosphaera**
    Used here in a restricted sense for *S. crassa* ( = *S. coronaria*). Other immersed, sand-inhabiting species are probably better referred to *Peziza*. Ref. B 302/D+/S+ p.p.

  36'(35') Apothecia usually not deeply immersed in the soil; ascospores smooth or roughened, globose to ovoid to fusoid, eguttulate or with two usually prominent oil guttules, rarely with 3 guttules or only 1 if spores are spherical . . . . . . . . . . . 37

37(36') Apothecia lenticular, in most species broadly attached along the base nearly to the margin, with gel usually found in all or part of the tissues; excipulum giving rise to hyphoid hairs in a gelatinous matrix; asci diffusely blue in iodine; ascospores regularly biguttulate to complete maturity, hyaline . . . . . . . . . . . . . . . . . . . . . . . . **Pachyella**
    Pfister's monograph shows that some of the species usually referred to *Psilopezia* belong here. *Pachyella clypeata* is a common large North American species. The much smaller *P. babingtonii* is worldwide on very wet, even submerged wood. Ref. *224, 224c*; B 310, 312/D as *Psilopezia*/S as *Peziza* p.p., *Psilopezia* p.p.

37'(36') Apothecia discoid to cupulate, usually centrally attached, rarely sparassoid, lacking hyphoid hairs in a gelatinous matrix; asci with a distinct amyloid ring or at least with the apex more strongly blue in iodine than are the lower walls; ascospores eguttulate or guttulate, 1 guttule common in spherical or short-ovoid spores, 2 guttules typical, rarely with 3 guttules in fusoid spores, guttules sometimes present only in young spores; in some species spore walls becoming brown at maturity . . . . . . . . . . . . . . . . . . . . . . . . . . **Peziza**
    Some authors continue to segregate a genus of spherical-spored forms as *Plicaria*. Boudier recognized *Aleuria* (not in the sense used in this classification) for species without oil guttules, and *Galactinia* for those with guttules. Le Gal and others have abandoned *Peziza* as a "nomen confusum," and have unfortunately adopted *Galactinia* as the name for the genus as here delimited; that would not be the correct name for the genus in any case since it is not the oldest available even if *Peziza* were to be discarded. The sparassoid forms are sometimes called *Daleomyces*, and have even been placed in the Tuberales. *Peziza* is a large genus, probably with well over 100 species, in need of careful monographic work. Imperfect states fall in the genera *Oedocephalum* and *Chromelosporium* (earlier called *Ostracoderma* and *Hyphelia*). Ref. *119a, 151, 163, 176, 239, 272*; B 257–73, 275, 301, 303–6/D+ also as *Plicaria*/S+ p.p. also as *Aleurina, Daleomyces, Discina* p.p., *Durandiomyces, Humarina* p.p., *Lamprospora* p.p., *Sarcosphaera* p.p.

  38(5) Ascocarp with a spongelike pileus, stipitate, hollow . . . . . . . . . **Morchella**
    Species concepts vary widely. There are probably 3 good species, but some authors recognize 50 or more. The generic name *Mitrophora* is used by some authors for species with the lower half of the pileus free from the stipe. These fungi are probably the most highly prized edible fungi other than the truffles. Ref. *112, 118, 127*; B 194–217/D+ also as *Mitrophora*/S+.

## 9. Discomycetes and Tuberales

38'(5) Ascocarp without a spongelike pileus, either a large, broad cup or with a campanulate pileus ............................................. 39

39(38') Ascocarp pileus campanulate, like a thimble over the apex of the hollow stipe ................................................................. **Verpa**
   A small genus, from which the species with 2-spored asci is sometimes segregated as *Ptychoverpa*. Edible and good. Ref. *127*; B 218–220/D+ also as *Ptychoverpa*/S+.

39'(38') Ascocarp discoid to cupulate, the large apothecia often with distinct hymenial folds or veinlike roughenings, stipe slight or absent ........................ **Disciotis**
   A single species, *D. venosa*. Seldom eaten, but as good as *Morchella*! Ref. B 254–6 bis/D+/S as *Peziza* p.p.

40(6) Ascospores biguttulate, or apiculate, or spherical ........................ 41

40'(6) Ascospores mostly with a single, large, often yellowish guttule, broadly ovoid, rarely with a large central guttule and 2 smaller ones and then subfusoid, smooth or marked but not apiculate ........................................... Tribe **Helvelleae** 42

41(40) Spores biguttulate, or uniguttulate if spherical, always smooth. Pileus gyrose, brainlike ................................................................. Tribe **Gyromitreae**
   One genus ................................................................. **Gyromitra**
   There are several oval-spored species, notably *G. esculenta*, *G. infula* and their allies, placed in *Physomitra* by Boudier. The genus *Pseudorhizina* (= *Helvellella*) differs only in having spherical spores, and does not deserve recognition. Harmaja has recently merged *Discina* and *Neogyromitra* in this genus. Ref. *115, 127, 225*; B 223–4/D+/S as *Elvela* p.p.

41'(40) Spores apiculate, usually also reticulate, with 3 or more large guttules, or biguttulate and appearing smooth but with a cyanophilous perispore; apothecia discoid to stalked and with a brainlike pileus ................................................. Tribe **Discineae** 44

42(40') Ascocarp clavate, hymenium appressed to the upper part of a vertically chambered stalk; ascospores warted ........................................... **Underwoodia**
   There are 3 good species. *Geomorium* is a synonym. Eckblad has merged the genus into *Helvella*. Ref. *239, 261*; B –/D–/S+

42'(40') Ascocarp not clavate; ascospores smooth or warted ...................... 43

43(42') Ascocarp elongate ear-shaped, with a short stipe; ascospores smooth ..... **Wynnella**
   One species, *W. silvicola*, often incorrectly placed in or near *Otidea* because of its ear-shaped apothecia. Nannfeldt and Le Gal both suggest placing this fungus in a separate tribe. The generic name *Midotis* was probably based on this species, but this cannot now be proved. Ref. *131, 198*; B 250/D–/S as *Scodellina* p.p.

43'(42') Ascocarp cupulate to saddle-shaped, stipe short or long; ascospores smooth or delicately to coarsely warted ....................................... **Helvella**
   This large genus has recently been the subject of several important revisionary papers. In the sense used here it includes the cupulate species sometimes segregated into such genera as *Acetabula* [= *Paxina*], *Macroscyphus* [= *Macropodia*], *Cyathipodia* and *Phaeomacropus* (in which the spores become slightly brown). The slender stalked, saddle-shaped species are sometimes segregated in *Leptopodia*. Species now placed in *Jafnea* might be sought in *Helvella*, but clearly differ in having short, brown hairs. Ref. *60, 61, 62, 63, 135*; B 225–49/D+ also as *Cyathipodia, Leptopodia, Paxina*/S as *Elvela* p.p., *Paxina* p.p.

44(41') Apothecia discoid, attached to the soil by rooting processes; setalike paraphysis elements accompany normal paraphyses ............................. **Rhizina**

Probably only a single species, *R. undulata*, frequently associated with burned areas and implicated in a disease of conifer seedlings. Ref. B 251/D+/S+.

44′(41′) Apothecia short- to long-stalked, discoid or with a contorted, brainlike pileus; hymenial setae absent . . . . . . . . . . . . . . . . . . . . . . . . . . . . . **Discina**

As used here, the genus includes both discoid forms and others with a *Gyromitra*-like cap, often placed in *Neogyromitra* and in *Maublancomyces*. *Fastigiella* and *Paradiscina* are recent, unnecessary splits. Currently a source of controversy in several laboratories. Ref. *10, 115, 183, 227a*; B 221–2, 252–3/D +/S + p.p. also as *Elvela* p.p.

45(6′) Ascospores with brown or brownish walls at maturity, or if hyaline then the apothecium with definite violet to purple pigments; (ascospores uninucleate)
. . . . . . . . . . . . . . . . . . . . . . . . . . . . Subfamily **Ascodesmidoideae** 49

45′(6′) Ascospores hyaline; apothecia lacking violet to purple pigments  . . . . . . . 46

46(45′) Apothecia borne on an obvious subiculum; ascospores smooth or faintly roughened, eguttulate; pale carotenoid pigments frequently present
. . . . . . . . . . . . . . . . . . . . . . . . . . . Subfamily **Pyronematoideae** 52

46′(45′) Apothecia not borne on an obvious subiculum; ascospores smooth to strongly marked, guttulate or eguttulate (uninucleate); carotenoids present or absent  . . . 47

47(46′) Ascospores devoid of guttules (but frequently with gaseous "De Bary bubbles" in certain mountants), nearly or completely lacking carotenoids
. . . . . . . . . . . . . . . . . . . . . . . . . . Subfamily **Ascophanoideae** 53

47′(46′) Ascospores either guttulate, or apothecia provided with large quantities of carotenoids, or both  . . . . . . . . . . . . . . . . . . . . . . . . . . . . . . . . . . . 48

48(47′) Ascospores guttulate; apothecia without carotenoids
. . . . . . . . . . . . . . . . . . . . . . . . . . . Subfamily **Otideoideae** 65

48′(47′) Ascospores guttulate or eguttulate; carotenoids present in the apothecia
. . . . . . . . . . . . . . . . . . . . . . . . . Subfamily **Scutellinioideae** 76

49(45) Ascocarp minute, merely a fascicle of asci and paraphyses without an organized excipulum; ascospores brown at maturity, spherical to ovoid, reticulate  . . . **Ascodesmis**

There are 5 species; the genus is sometimes assigned to the Gymnoascaceae because of the simplicity of the ascocarp, but the asci are clearly operculate. Ref. *200*; B–/D–/S+.

49′(45) Ascocarp larger, either a discoid apothecium or a nearly spherical to irregular ball  . . . . . . . . . . . . . . . . . . . . . . . . . . . . . . . . . . . . . . 50

50(49′) Ascocarp a spherical ball, with the hymenium on the outer surface
. . . . . . . . . . . . . . . . . . . . . . . . . . . . . . . . . . **Sphaerozone**

Used here to accommodate a few, poorly understood species at times included in *Sphaerosoma*, which differs notably in having asci that blue in iodine. All these species have at times been referred to the Tuberales. Ref. *242, 256*; B –/D –/S as *Sphaerosoma* p.p.

50′(49′) Apothecia discoid  . . . . . . . . . . . . . . . . . . . . . . . . . . . . 51

51(50′) Apothecia small, not pustulate; ascospores spherical to ovoid, with one to several guttules, smooth, warted or reticulate  . . . . . . . . . . . . . . . . . . **Pulparia**

A recent synonym is *Marcelleina*; only a few species, most of which have been referred to *Plicaria* at some time. Ref. *20, 158, 165, 239*; B 307?, 308–9/D as *Barlaeina*/S as *Humarina* p.p., *Lamprospora* p.p.

51′(50′) Apothecia small to medium-sized, distinctly pustulate especially towards the margin,

often with a pseudostipe composed of basal hyphae enmeshing sand particles; ascospores ellipsoidal, with two large guttules (at least in youth), and with rounded warts
..................................... **Jafneadelphus**
A small genus, segregated from *Jafnea* from which it differs in pigments, absence of hairs, and noncarminophilous nuclei. *Smardaea* is a synonym. Ref. *21, 239, 271*; B−/D−/S−.

52(46) Apothecia minute, sometimes coalescing to form larger masses, turbinate, naked or with delicate excipular hairs, on burned or steam-sterilized substrata, white, pink, or orange; ascospores very faintly marked (uninucleate) . . . Tribe **Pyronemateae**

One genus ................................. **Pyronema**
There are probably only 2 or 3 good species, but many have been described. Ref. *185*; B 419–21/D+/S+.

52'(46) Apothecia 3–12 mm in diameter, very thin and appressed to the substrate leaves and twigs, brownish-red to greenish-yellow; ascospores smooth (binucleate)
..................................... Tribe **Karstenelleae**

One genus ................................. **Karstenella**
A monotypic genus based on *K. vernalis* from Scandinavia. Consistently binucleate ascospores are unknown elsewhere in the Pezizales. Ref. *116*; B−/D−/S−.

53(47) Apothecia deep-cupulate, lacking hairs, pallid, but pale carotenoids sometimes present ........................... Tribe **Geopyxideae** 55

53'(47) Apothecia shallow-cupulate to discoid, turbinate, or subglobose, with or without distinct hairs ................................. 54

54(53') Apothecia discoid, clothed externally with coarse, hyphoid, brown or brownish hairs; ascospore contents usually yellowish, resinous, somewhat refractive; on soil, dung, and debris ................. Tribe **Pseudombrophileae** 56

54'(53') Apothecia discoid to turbinate or subglobose, naked or fringed at the margin with delicate hyphae or provided with stiff, hyaline setae; on dung
..................................... Tribe **Theleboleae** 59

55(53) Paraphyses accompanying the asci ................. **Geopyxis**
Used here in the restricted sense for *G. carbonaria* and its allies. Some authors also include species with 2 oil guttules in the spores, but these should be referred to *Tarzetta*. Ref. *72, 239*; B−/D+/S+ p.p., also as *Humarina* p.p.

55'(53) Paraphyses absent ................. **Aparaphysaria**
The type species, *A. aparaphysata*, is from South America; a second, undescribed species has been found in India. Ref. *260, 261*; B−/D−/S−.

56(54) Ascospores flattened on one surface, to lunate .......... **Selenaspora**
The only species, *S. guernisacii*, is on dung; the asymmetrical ascospores and anastomosing paraphyses are suggestive of the Sarcoscyphineae. Ref. *119, 165*; B−/D−/S−.

56'(54) Ascospores radially symmetrical ..................... 57

57(56') Hairs acute at the apex, not flexuous ................. 57A

57'(56') Hairs blunt at the apex, more or less flexuous ........... 58

57A(57) Hairs thin-walled, not branched at the base .......... **Tricharina**
Some authors would unite this poorly known genus, once called *Tricharia*, with *Trichophaea*. Mostly on soil, debris, burned areas. Ref. *83, 265*; B 347–51/D as *Tricharia*/S as *Patella* p.p.

57A'(57) Hairs thick-walled, often basally branched . . . . . . . . . . . . . 57B

57B(57A') Ascospores transversely plicate, fusoidal . . . . . . . . . . . **Rhizoblepharia**
   Two species are known on soil. Ref. *76*; B–/D–/S–.

57B'(57A') Ascospores smooth, ellipsoid . . . . . . . . . . . . . . . **Trichophaeopsis**
   *T. bicuspis* is the only known species, on soil, wood, leaves. There is usually one downwardly directed basal prong on each hair. Ref. *132, 160a*; B 366/D as *Trichophaea* p.p./S–.

   58(57') Apothecia without a sterile rim, becoming repand and ombrophiloid; hairs arising in groups . . . . . . . . . . . . . . . . . . . . . **Pseudombrophila**
   Possibly the genus should be restricted to 1 species, *P. deerata*. The 4-spored *P. guldeniae* must surely belong elsewhere. The type species occurs on herbaceous stems and hay, perhaps fruiting only after partial fermentation and heating. Ref. B 390/D+/S+.

   58'(57') Apothecia with a sterile, elevated marginal rim. . . . . . . . . . . . . **Fimaria**
   There are several species on dung. Ref. *19*; B–/D+/S as *Humarina* p.p.

59(54') Asci opening by a typical operculum . . . . . . . . . . . . . . . . . . 60

59'(54') Asci opening by a vertical slit or an irregular tear . . . . . . . . . . . . 61

   60(59) Apothecia naked, discoid; asci 8- to 256-spored . . . . . . . . . . **Coprotus**
   This is a fairly large genus, recently monographed for North America. Many species were formerly referred to *Ascophanus*. *Leporina* is not a synonym according to Kimbrough. Ref. *138, 140, 142*; B 418/D as *Ascophanus* p.p., *Rhyparobius* p.p./S as *Ascophanus* p.p., *Ryparobius* p.p.

   60'(59) Apothecia provided with nonseptate setae; asci 8- to 128-spored . . **Lasiobolus**
   At least 4 species belong here. Ref. *140*; B 415–6/D+/S+.

61(59') Apothecia provided with stiff, 1- to 10-septate setae; only 1 to 3 asci produced per ascocarp; asci with 1500 to 7000+ ascospores . . . . . . . . . . . . . . **Trichobolus**
   Three species are recognized. The genus seems very close to *Lasiobolus*. Ref. *140*; B–/D–/S–.

61'(59') Apothecia without stiff setae . . . . . . . . . . . . . . . . . . . . . 62

   62(61') Asci with an apical plug which stains in ink mounts, cylindric to obclavate with a strongly narrowed base, with 1000 to 1500 ascospores in each of 15 to 25 asci per ascocarp . . . . . . . . . . . . . . . . . . . . . . . . . . . . **Caccobius**
   A single, minute species, *C. minusculus*. Ref. *140*; B–/D–/S–.

   62'(61') No apical plug present in the ascus . . . . . . . . . . . . . . . . . 63

63(62') Ascus apex staining uniformly in congo red, without an apical ring, with a vertical slit, 256- (rarely 64-) spored; ascospores 14 μm long or longer . . . . . . . . **Coprobolus**
   Monotypic, *C. poculiformis*. Ref. *26*; B– /D– /S–.

63'(62') Ascus apex hyaline in congo red, with an apical ring; ascospores 12 μm long or shorter . . . . . . . . . . . . . . . . . . . . . . . . . . . . . . . . . . 64

   64(63') Ascocarp discoid; the apical "nipple" of the ascus hyaline in congo red, but the ring and part of the apex below the nipple staining red; asci with a longitudinal slit at liberation of the spores, 32- to 128-spored . . . . . . . . . . . . **Ascozonus**
   There are many species, poorly known. Ref. *140*; B–/D+/S as *Streptotheca* p.p.

   64'(63') Ascocarps cleistothecial, sometimes opening to discoid; the whole ascus apex above the ring hyaline in congo red; asci opening with an irregular tear at the apex, 8- to 2500-spored . . . . . . . . . . . . . . . . . . . . . . . . . **Thelebolus**

## 9. Discomycetes and Tuberales

The number of asci in 8-spored forms is large, while there may be only a single ascus in some of the multispored species. The uniascal (type) species, *T. stercoreus*, was recently referred to the Erysiphales by Cooke and Barr, but Kimbrough has shown its discomycetous affinities. Ref. *32, 140, 203, 293*; B − /D + also as *Ascophanus* p.p., *Rhyparobius* p.p./S as *Ascophanus* p.p., *Ryparobius* p.p., *Streptotheca* p.p.

65(48) Ascospores spherical, hairs absent. See *Pulvinula* . . . . . . . . . . . . . . . 90
65'(48) Not with the above combination of characters . . . . . . . . . . . . . . . . 66

    66(65') Nuclei of ascospores and paraphyses staining strongly in acetocarmine (carminophilous), even in dried specimens; apothecia deep-cupulate; ascospores biguttulate . . . . . . . . . . . . . . . . . . . . . . . . . . Tribe **Jafneeae** 68

    66'(65') Nuclei of ascospores and paraphyses not carminophilous (if acetocarmine stain is unavailable, take this lead); apothecia discoid to deep-cupulate, rarely sparassoid; ascospores with 1 or more guttules . . . . . . . . . . . . . . . . . . . . . . . 67

67(66') Apothecia lacking hairs, though often distinctly pustulate; ascospores biguttulate; paraphyses often apically hooked, bent, or deformed . . . . . . . Tribe **Otideeae** 69
67'(66') Apothecia with distinct hairs; ascospores with 1 or more guttules; paraphyses usually not apically deformed . . . . . . . . . . . . . . . . . . . Tribe **Mycolachneeae** 71

    68(66) Apothecia with distinct, brown, hyphoid hairs which may be short and appressed, usually with a distinct pseudostipe composed of basal hyphae enmeshing sand or soil particles; ascospores elongate ovoid to fusoid, with irregular warts which are sometimes denser towards the poles so that the ascospores appear apiculate . . . . . . **Jafnea**
    Only 2 species. The pseudostipe led some workers to place these in *Macropodia*. *Jafnea fusicarpa* may easily be confused in the field with *Humaria hemisphaerica*, which also has a worldwide distribution. Ref. *12, 155*; B − /D − /S as *Paxina* p.p.

    68'(66) Apothecia without hairs, but with distinct pustules, often with a pseudostipe; ascospores ellipsoid, smooth (see third lead, 68") . . . . . . . . . . . . **Tarzetta**
    The genus was earlier called *Pustularia* and *Pustulina*, and some species have been placed in *Geopyxis* which differs in having nonguttulate ascospores. The paraphyses may be deformed or hooked apically in some species, recalling *Otidea*. Ref. *12, 72, 158, 196, 210, 241*; B 336–9/D as *Pustularia*/S as *Geopyxis* p.p.

    68"(66) Apothecia covered with hyaline hairs. See *Leucoscypha* . . . . . . . . . 88

69(67) Apothecia broadly attached along their base nearly to the margin, discoid; ascospores over 20 $\mu$m long, smooth, with a perispore which loosens in heated lactic acid; paraphyses often somewhat deformed at the apex . . . . . . . . . . . . . . . . . . . **Psilopezia**
    A small genus, recently monographed by Pfister, who has shown that some species belong instead in *Pachyella*. Ref. *224, 224c*; B 311/D − /S + p.p.

69'(67) Apothecia narrowly attached at the base, or forming a sparassoid mass; ascospores mostly under 20 $\mu$m long, smooth or delicately warted, lacking a perispore which loosens in heated lactic acid . . . . . . . . . . . . . . . . . . . . . . . . . . . . . . 70

    70(69') Ascocarp cupulate to ear-shaped, usually slit down 1 side; paraphysis apices usually straight in youth, but deformed, bent, or hooked at maturity . . . **Otidea**
    This is a fairly large genus, used here inclusive of *Pseudotis* which is sometimes segregated for species with permanently cupulate apothecia. The completely cupulate species of *Tarzetta* will also key here if acetocarmine stain was unavailable at key lead 66. Ref. *131, 198, 202*; B 324–33/D + also as *Pseudotis*/S as *Scodellina* p.p., *Peziza* p.p.

    70'(69') Ascocarp irregular, branched to sparassoid; paraphyses apically hooked

..................................... **Ascosparassis**
One species, *A. shimizuensis*, in Asia. Ref. *147, 157, 198*; B– /D– /S–.

71(67′) Apothecia completely white. See *Leucoscypha* ................ 88

71′(67′) Apothecia not completely white ..................... 72

    72(71′) Hairs hyaline or subhyaline, sometimes with brownish cell sap, thick-walled, blunt; apothecia brown, with a pseudostipe of basal hyphae enmeshing soil particles; ascospores with several oil globules (in youth), delicately warted, ellipsoid
..................................... **Nothojafnea**
Two species are known. Ref. *83b, 239*; B–/D–/S–.

    72′(71′) Hairs brown-walled, or if subhyaline, then neither thick-walled nor blunt . . 73

73(72′) Hairs flexuous, cylindrical, sometimes branched; apothecia partially to wholly immersed in the ground, hymenium lining a cup which may split, or folded and contorted to form a cerebriform mass, sometimes hypogeous for a long time; ascospores smooth, with 1 or 2 guttules ..................................... **Geopora**
    The epigeous and subhypogeous cupulate species have usually been treated under the name *Sepultaria*, while *Geopora* has been considered a member of the Tuberales until recently; all have functional opercula on the asci, and several species show intermediate degrees of folding of the hymenium. A fairly large genus, needing monographic work. Ref. *24, 25, 265*; B 358–62/D as *Sepultaria*/S as *Humarina* p.p., *Sepultaria* p.p.

73′(72′) Hairs stiff, usually tapering to the apex, or if cylindrical at least neither flexuous nor branched; ascospores smooth or rough, biguttulate (or if spherical, then uniguttulate) ..................................... 74

    74(73′) Apothecia usually deep-cupulate, hairs rigid, thick-walled ..... **Humaria**
    This genus is still poorly understood. Some authors choose to avoid the use of the name *Humaria*, since that name has had so many conflicting uses that it constitutes a probable source of confusion, and have adopted *Mycolachnea*. It may grade into *Geopora* and *Trichophaea*. There is no adequate monograph. Species of *Jafnea* will also key here if acetocarmine stain was unavailable at key lead 66; they can easily be distinguished by their very short, blunt hairs and spores with irregularly shaped warts. Ref. *265*; B 352–4, 355?, 356, 357?/D +/S as *Patella* p.p., *Sepultaria* p.p.

    74′(73′) Apothecia discoid, hairs thin- or thick-walled ............. 75

75(74′) Hairs basally forked with a downward directed branch or prong; ectal excipulum a one-cell thick firm tissue of horizontally elongated cells in vertical rows; ascospores smooth. See *Trichophaeopsis* ..................................... 57B′

75′(74′) Hairs simple, not forked at the base; ectal excipulum not as above; ascospores smooth or warted ..................................... **Trichophaea**
    There are still a number of undescribed species in this genus. Some species have *Dichobotrys* imperfect states; a *Rhizoctonia* mycelial state is reported for one species. The spherical-spored species are segregated into *Sphaerosporella* by some authors. Eckblad synonymizes this genus with *Humaria*. Ref. *83, 119a, 132, 239, 265, 292*; B 364–5, 367, 379; D+ also as *Sphaerosporella*/S as *Patella* p.p., *Sphaerospora* p.p.

    76(48′) *Either* with rooting hairs (pointed, brown setae arising from deep within the tissues of the apothecium, the bases usually forked) *or* combining all the following characters: (1) ascospores eguttulate, (2) ascospores with a perispore which loosens in heated lactic acid, (3) medullary and ectal excipula composed of globose cells; (major pigment is $\gamma$ carotene) ............... Tribe **Scutellinieae** 78

## 9. Discomycetes and Tuberales

76′(48′) Rooting hairs lacking; also not combining the 3 characters enumerated
above . . . . . . . . . . . . . . . . . . . . . . . . . . . . . . . . . 77

77(76′) Apothecia large or medium sized, when fresh discoloring greenish or red to brownish on bruising; (major pigment $\beta$ carotene; also present is Arpin's (*1a*) "carotene P. 444," unknown elsewhere in the Pezizales) . . . . . . . . . . . . Tribe **Sowerbyelleae** 81

77′(76′) Apothecia minute to large, not discoloring on bruising; (major pigment $\beta$ carotene, or both $\beta$ and $\gamma$ carotene present) . . . . . . . . . . . . . . Tribe **Aleurieae** 84

    78(76) Hairs absent; paraphyses nearly as broad as the asci, often with globosely swollen tips; ascospores eguttulate . . . . . . . . . . . . . . . . . . . . . . **Coprobia**
        *Coprobia granulata*, worldwide, and some other species are on dung. Ref. *166, 239*; B − /D + /S as *Ascophanus* p.p.

    78′(76) Hairs present; paraphyses gracile . . . . . . . . . . . . . . . . . . . . 79

79(78′) Ascospores eguttulate; hairs rooting or superficial, or both . . . . . . . . . . 80

79′(78′) Ascospores with 1 to many guttules; rooting hairs present . . . . . . . . **Scutellinia**
    Species concepts vary widely among authors, and this may be a very large genus. The commonest species is *S. scutellata*, on wood. *Geneosperma* was recently erected for an Asian species with folliculate ascospores. Ref. *37, 82, 166, 171, 172, 172a, 239, 265*; B 368–71, 373, 375–8/D +/S as *Melastiza* p.p., *Patella* p.p., *Sphaerospora* p.p.

    80(79) Ascospores fusoid, with delicate transverse ridges, outer membrane not loosening in heated lactic acid. See *Rhizoblepharia* . . . . . . . . . . . . . . . . . . 57B

    80′(79) Ascospores ovoid, seemingly smooth but with the outer membrane loosening in heated lactic acid; superficial hairs present, in some species accompanied by rooting hairs . . . . . . . . . . . . . . . . . . . . . . . . . . . . . . . . . . . **Cheilymenia**
        A moderately large genus, mostly on dung. The species with rooting hairs have a strongly differentiated medullary excipulum. Those lacking rooting hairs have both medullary and ectal excipula composed of globose cells, and quite possibly should be transferred to *Coprobia*. Ref. *39, 166, 186, 239, 265*; B 372, 374, 380–5/D +/S as *Patella* p.p.

81(77) Ascospores biguttulate, verrucose, ovoid; apothecia with an olive-beige to clear yellow hymenium, with a distinct stipe, staining brownish on bruising; paraphyses clavate or apically hooked . . . . . . . . . . . . . . . . . . . . . . . . . . . . . . . . . **Sowerbyella**
    In addition to *S. radiculata* and *S. imperialis* (= *S. unicolor*), there appear to be several undescribed species in North America. Some authors adopt *Pseudotis* as the name for this genus. Ref. *117, 196*; B 334–5/D +/S −.

81′(77) Ascospores eguttulate to multiguttulate, smooth, spherical to ovoid or blunt at the ends; apothecia with a bright yellow or orange hymenium; paraphyses quite broad
. . . . . . . . . . . . . . . . . . . . . . . . . . . . . . . . . . . . . . . 82

    82(81′) Ascospores spherical, eguttulate; apothecia deep cupulate, orange, staining bright to dark green on bruising . . . . . . . . . . . . . . . . . . . . . **Caloscypha**
        The single species, *C. fulgens*, is not uncommon under *Picea. Otidella* is a synonym. Ref. B 319/D −/S as *Pseudoplectania* p.p.

    82′(81′) Ascospores ovoid, often blunt at the ends, multiguttulate; apothecia mostly staining reddish on bruising . . . . . . . . . . . . . . . . . . . . . . . 83

83(82′) Apothecia large, clustered, arising from a common base; cells of the ectal excipulum very large (35–80 $\mu$m in diameter) . . . . . . . . . . . . . . . . . . . . **Acervus**
    *Acervus aurantiacus*, originally assigned to the inoperculate discomycetes, is the only

species, from North America. The peculiar asci and anastomosing paraphyses led to it being transferred to the Sarcoscyphineae. It is apparently very closely related to *Phaedropezia* instead. Ref. *157, 254*; B –/D –/S –.

83'(82') Apothecia minute to medium-sized, occurring singly; cells of the ectal excipulum much smaller . . . . . . . . . . . . . . . . . . . . . . . . . . . . . . . . . . . . **Phaedropezia**

A few, primarily tropical, species. Eckblad placed the genus in the Sarcoscyphineae, but cytological evidence that the ascospores are uninucleate confirms Le Gal's placement of the genus in the Pezizineae. Ref. *166, 170, 224a*; B –/D –/S as *Psilopezia* p.p.

    84(77') Easily overlooked, short, appressed, brown hyphoid hairs present, particularly near the margin . . . . . . . . . . . . . . . . . . . . . . . . . . . . . . . . . 85

    84'(77') Hairs either absent, or if present, hyaline . . . . . . . . . . . . . . . . 87

85(84) Ascospores smooth, biguttulate . . . . . . . . . . . . . . . . . . . . **Anthracobia**

A small genus, common around burned areas. Ref. *265*, B 387–9/D +/S as *Patella* p.p.

85'(84) Ascospores warted to reticulate, 1- or 2-guttulate . . . . . . . . . . . . . 86

    86(85') Ascospores broad-ovoid, 1-guttulate, with delicate warts . . . . . . **Hiemsia**

One species, *H. pseudoampezzana*, in Europe. Ref. *271*; B –/D –/S –.

    86'(85') Ascospores ovoid, 2-guttulate, with a definite reticulum or with large warts . . . . . . . . . . . . . . . . . . . . . . . . . . . . . . . . . . . . **Melastiza**

Maas Geesteranus has a key to the several species, mostly on soil. *Melastiza flavorubens* is an older name for *M. greletii*. Ref. *168, 177, 187a, 265*; B 386/D +/S + p.p.

87(84') Apothecia clothed with hyaline hairs; hairs sometimes very short, but then ascospores both ovoid and reticulate . . . . . . . . . . . . . . . . . . . . . . . . . . . 88

87'(84') Apothecia naked or at most with a few hyphoid hairs; ascospores not both ovoid and reticulate . . . . . . . . . . . . . . . . . . . . . . . . . . . . . . . . . . . . . 89

    88(87) Hairs well-developed, thick-walled, blunt or pointed, and sometimes covered with a gelatinous sheath; ascospores smooth, warted, or reticulate, uni- to multi-guttulate; nuclei (in all species?) carminophilic; hymenium white to orange or red; apothecia sessile or subsessile . . . . . . . . . . . . . . . . . . . . **Leucoscypha**

Used here in the wide sense of Rifai. Some species are completely white or develop their carotenoid pigments only very late in development, and would erroneously key to the Mycolachneeae on that basis. Ref. *167, 239, 265*; B 315–6, 340, 345–6/D + also as *Neottiella*/S as *Aleuria* p.p., *Patella* p.p.

    88'(87) Hairs poorly developed, hyphoid, thin-walled, usually blunt; ascospores biguttulate, reticulate or with spinose crests or bands; hymenium red, yellow or orange; apothecia stipitate or sessile . . . . . . . . . . . . . . . . . . . . . . . . **Aleuria**

Used here in a narrower sense than by some authors, since some species formerly placed here are now treated in *Leucoscypha* and *Melastiza*. Denison is surely correct in including here the Central American species he calls "*Aleuria* taxonomic species 4" despite its lack of a reticulum and the presence of peculiar crests. *Aleuria bicucullata* has cuplike markings extending from each end of the spore in addition to a general reticulum. Ref. *38, 187a, 239*; B 313–4, 317–8/D +/S + p.p.

89(87') Apothecia minute, grouped on a rubbery basal stratum ("stroma"), at the edges of pools arising from melting snow . . . . . . . . . . . . . . . . . . . . **Pseudocollema**

*Pseudocollema cartilagineum*, the only species, is known from Washington, California, and Maine. Ref. *133*; B –/D –/S –.

89'(87') Apothecia not grouped on a rubbery stratum . . . . . . . . . . . . . . . 90

    90(89', [65]) Paraphyses very narrow, apically bent to strongly uncinate, ascus usually with a strongly forked base; ascospores smooth, globose or ovoid, multiguttulate in youth, usually with a prominent de Bary bubble in Melzer's Reagent; apothecia discoid, pulvinate, rarely shallow cupulate . . . . . . . . . . . . . . . **Pulvinula**
    This distinctive genus deserves critical monographic work. Most species are on soil, though some are pyrophilous. Some species have only faint traces of carotenoids, and would thus key erroneously to the Otideoideae. Ref. *166, 239*; B 406–7/D + /S as *Humarina* p.p., *Lamprospora* p.p.

    90'(89') Paraphyses slender to stout, straight; ascus without a strongly forked base; ascospores smooth or rough, globose to ovoid or fusoid, guttulate or spumose; apothecia discoid to lenticular or turbinate . . . . . . . . . . . . . . . . . 91

91(90') Ascospores spherical or nearly so . . . . . . . . . . . . . . . . . **Lamprospora**
    A fairly large genus, used here in the restricted sense, excluding species now placed in *Pulvinula* and *Peziza*. Some species, including the common *L. crec'hqueraultii*, have spores at best subspherical, and Le Gal has recently abandoned *Lamprospora*, making it a synonym of *Octospora*, which may be the wisest course in the future. *Ramsbottomia* may be a synonym. Ref. *239*; B 401–5/D+/S+ p.p.

91'(90') Ascospores ellipsoid to fusoid . . . . . . . . . . . . . . . . . . . . . 92

    92(91') Margin of the apothecium composed of long-clavate cells; ascospores smooth or variously sculptured . . . . . . . . . . . . . . . . . . . . . . . . **Octospora**
    Most of the species have at one time or another been treated in *Humaria* or *Humarina*. The genus is accepted here in the restricted sense of Rifai. Other authors would merge into it such genera as *Lamprospora*, *Byssonectria*, and *Kotlabaea*. Ref. *72, 239*; B 391, 395–7, 399–400/D+ p.p./S as *Humarina* p.p.

    92'(91') Margin of the apothecium composed of angular or subglobose cells; ascospores smooth or very faintly marked . . . . . . . . . . . . . . . . . . . . . . . 93

93(92') Ascospores faintly roughened; apothecia broadly attached, on wet wood, at times fruiting under water . . . . . . . . . . . . . . . . . . . "**Peziza**" **lechithina**
    No available generic name is known for this common fungus, which is frequently misidentified as *Psilopezia aquatica*. Ref. *224*; B –/D –/S as *Psilopezia* p.p.

93'(92') Ascospores smooth, apothecia not as above . . . . . . . . . . . . . . . 94

    94(93') Ascospores fusoid-ellipsoid, with two or more guttules . . . . **Byssonectria**
    This genus was recently segregated as *Inermisia*, but *Byssonectria*, originally described as a member of the Hypocreales, provides an older name. *Byssonectria aggregata*, *B. fusispora*, and *B. tetraspora* are all common on soil and mosses. Ref. *158, 239*; B 392–4/D as *Octospora* p.p./S as *Humarina* p.p.

    94'(93') Ascospores ellipsoid, spumose . . . . . . . . . . . . . . . . . . **Kotlabaea**
    *Kotlabaea deformis* is the only species. Ref. *271*; B 398/D as *Octospora* p.p./S –.

## VIII. PHACIDIALES

In this order, the ascocarp begins as a stroma. It is usually externally black, spherical to discoid or frequently hysteriform, and is immersed wholly or partly in host tissues or more rarely is on the surface of the substrate. The ascocarp differentiates to form one or several hymenial areas consisting of asci and paraphyses. The asci mature their ascospores prior to exposure of hymenium brought about by rupture of the upper stromatal layer by one or more slits or in a stellate manner. The asci are thickened apically and the plug or surrounding pore sometimes blues in iodine; they violently discharge ascospores into the air, and are four- or eight-spored. The ascospores are ovoid to filiform, sometimes constricted in the middle, nonseptate to transversely multiseptate, hyaline or rarely brown. Sometimes they have a gelatinous sheath and are bipolar symmetrical or asymmetrical (the upper portion frequently broader than the lower), and radially symmetrical.

**KEY TO FAMILIES, TRIBES AND IMPORTANT GENERA OF PHACIDIALES**

1. Ascospores surrounded by a gelatinous sheath . . . . . . . . . . **Rhytismataceae** 3
1'. Ascospores without a sheath . . . . . . . . . . . . . . . . . . . . . . . . . . . 2
    2(1') Stroma fleshy-gelatinous, composed of hyaline hyphae which are not vertically oriented, with a hyaline to red-brown, olive-brown, or shining black covering layer . . . . . . . . . . . . . . . . . . . . . . . . . . . **Cryptomycetaceae** 23
    2'(1') Stroma externally black, not fleshy-gelatinous, composed of vertically oriented hyphae which may appear to be pseudoparenchymatous . . . . **Phacidiaceae** 25
3(1) Overlying layer of the stroma composed of parallel to interwoven, brown hyphae; ascus pore blue in iodine; ascospores ellipsoid-fusoid, with the gelatinous sheath drawn out at each end into a tapered appendage . . . . . . . . . . . . . . . Tribe **Ceratophacidieae**
    One genus . . . . . . . . . . . . . . . . . . . . . . . . . . . . . . . . . . . **Ceratophacidium**
        *Ceratophacidium aristosporum* on *Sequoia* leaves is the only species. Ref. *236*; D−.
3'(1) Overlying tissue of dark brown, thick-walled cells; ascus pore not blue in iodine; ascospores various, sheath not forming tapered appendages . . . . . . . . . . . . . . . 4
    4(3') Stromata with a single hymenial area, opening by a longitudinal slit or splitting radiately into lobes . . . . . . . . . . . . . . . . . . . . . . . . . . . . . . . . . 5
    4'(3') Stromata with many hymenial areas, opening along several lines of dehiscence . . . . . . . . . . . . . . . . . . . . . . . . . Tribe **Rhytismateae** 6
5(4) Apothecia circular to angular, opening by splitting radiately or irregularly
. . . . . . . . . . . . . . . . . . . . . . . . . . . . . . . . . . . . Tribe **Coccomyceteae** 8
5'(4) Apothecia elliptical to elongate-linear, opening by a single longitudinal slit
. . . . . . . . . . . . . . . . . . . . . . . . . . . . . . . . . . . . . Tribe **Hypodermateae** 9
    6(4') Ascospores brown, ovoid to ellipsoid . . . . . . . . . . . . **Nymanomyces**
        A tropical genus with perhaps only a few species. Ref. *35*; D−.
    6'(4') Ascospores hyaline, filiform to broad clavate . . . . . . . . . . . . . . . . 7
7(6') Ascospores broad clavate, hymenial areas not individualized . . . . . . **Placuntium**
    Recognized for *P. andromedae*, in which Darker reports a pycnidial state. Ref. *35*; D as *Rhytisma* p.p.
7'(6') Ascospores filiform to slightly clavate, hymenial areas individualized . . . . **Rhytisma**
    Though most species have numerous apothecia or fertile areas in the stroma, as in the common tar-spot of maple, *R. acerinum*, the reduced *R. punctatum* also on maple leaves has only a single hymenial area, but is clearly congeneric. Ref. *276*; D+.
    8(5) Ascospores constricted, acuminate below . . . . . . . . . . . . **Duplicaria**

## 9. Discomycetes and Tuberales

*Duplicaria empetri* on leaves of *Empetrum* and *D. acuminata* on *Carex* and *Juncus* are the only species accepted. Ref. *224e, 230*; D—.

8'(5) Ascospores not constricted . . . . . . . . . . . . . . . . . . . **Coccomyces**
There are many species, and no adequate monograph. The monoapothecial *Rhytisma punctatum* on *Acer* leaves will key here, but is not closely related. The plant pathogens on *Prunus* and *Kerria* previously referred to *Coccomyces* belong instead in *Blumeriella* (Helotiales, Dermateaceae). Ref. *81, 231, 276*; D+.

9(5') Stroma erumpent through bark or bare wood, basal layer strongly developed . . . 10

9'(5') Stroma mostly on leaves and stems, basal layer poorly developed . . . . . . . . 11

    10(9) Ascospores constricted in the middle (bifusiform), each half with one or more septa . . . . . . . . . . . . . . . . . . . . . . . . . . . . . . . . . . . . **Bifusepta**
    *Bifusepta tehonii* on *Vaccinium*. Ref. *34*; D—.

    10'(9) Ascospores not constricted, clavate, septate or not . . . . . . . . . . **Colpoma**
    A small genus, of which *C. quercinum* is best known. *Xyloschizon, Sporomega* and *Clithris* appear to be synonymous. Ref. *35, 276*; D+.

11(9') Ascocarps elongate-linear, ascocarps and pycnidia usually arranged in lines along the length of the leaf; all on *Abies* or *Picea* . . . . . . . . . . . . . . . . . . . . 12

11'(9') Ascocarps elliptical, scattered, pycnidia scattered or lacking . . . . . . . . . 14

    12(11) Ascospores constricted in the middle . . . . . . . . . . . . . . **Isthmiella**
    For 3 species formerly referred to *Bifusella*. Ref. *35*; D—.

    12'(11) Ascospores not constricted . . . . . . . . . . . . . . . . . . . . . . . 13

13(12') Ascospores bacillar . . . . . . . . . . . . . . . . . . . . . . . . . . . **Virgella**
*Virgella robusta* is the only species. Ref. *35*; D—.

13'(12') Ascospores clavate . . . . . . . . . . . . . . . . . . . . . . . . . . . . **Lirula**
For 6 species formerly referred to *Hypodermella*. Ref. *35*; D as *Hypodermella* p.p.

    14(11') Ascospores constricted in the middle; all on *Pinus* or *Cunninghamia* . . . 15

    14'(11') Ascospores not constricted . . . . . . . . . . . . . . . . . . . . . . 16

15(14) Apothecia subcuticular . . . . . . . . . . . . . . . . . . . . . . . . . **Bifusella**
In Darker's recent revision, this once larger genus is now reduced to 3 species. A species on *Cunninghamia* occurs in Okinawa. Ref. *35, 200a*; D—.

15'(14) Apothecia subepidermal . . . . . . . . . . . . . . . . . . . . . . . . **Soleella**
Monotypic, *S. striiformis* on *Pinus*. Ref. *35*; D—.

    16(14') Asci clavate to broadly saccate; strong parasites . . . . . . . . . . . . 17

    16'(14') Asci narrowly clavate or cylindrical; weak parasites . . . . . . . . . . 21

17(16) Ascospores elongate-fusoid, 1-septate; ascocarps subepidermal . . . . **Elytroderma**
Two species on pine needles. Ref. *35*; D—.

17'(16) Ascospores clavate to bacillar, nonseptate . . . . . . . . . . . . . . . . . 18

    18(17') Ascocarps subcuticular; on *Larix* needles . . . . . . . . . . **Hypodermella**
    This once large genus is now restricted to *H. laricis*. Ref. *35*; D+ p.p.

    18'(17') Ascocarps subepidermal or subhypodermal; on *Pinus* . . . . . . . . . . 19

19(18') Ascocarps subhypodermal; ascospores clavate . . . . . . . . . . **Lophodermella**
For 6 species previously placed in *Hypodermella*. Ref. *35*; D as *Hypodermella* p.p.

19'(18') Ascocarps subepidermal; ascospores clavate or bacillar . . . . . . . . . . 20

20(19′) Ascospores bacillar . . . . . . . . . . . . . . . . . . . . . . . . **Ploioderma**
    For 3 species previously referred to *Hypoderma*. Ref. *35*; D as *Hypoderma* p.p.

20′(19′) Ascospores clavate . . . . . . . . . . . . . . . . . . . . . . **Davisomycella**
    For 6 species placed earlier in *Hypodermella*. Ref. *35*; D as *Hypodermella* pp.

21(16′) Ascospores bacillar . . . . . . . . . . . . . . . . . . . . . . . . . **Hypoderma**
    Most conifer-inhabiting species have now been removed. Darker recognized *Meloderma* for a species on pine which might perhaps be better placed here. There is no adequate monograph. Ref. *35, 276*; D + p.p.

21′(16′) Ascospores filiform . . . . . . . . . . . . . . . . . . . . . . . . . . . 22

    22(21′) Ascospores multiseptate at maturity . . . . . . . . . . . . **Lophomerum**
        Darker recognized 6 species. Ref. *35, 205*; D–.

    22′(21′) Ascospores nonseptate . . . . . . . . . . . . . . . . . . **Lophodermium**
        Darker lists 21 species, but it is probably a much larger genus. Ref. *35, 275, 276*; D+.

23(2) Stroma an effused crust on living branches, covering layer easily detaching, composed of 3 tissue zones . . . . . . . . . . . . . . . . . . . . . . . . . . **Cryptomyces**
    Monotypic, *C. maximus* on *Salix*. Ref. *5*; D+.

23′(2) Stroma small, subcircular, covering layer not zoned . . . . . . . . . . . . 24

    24(23′) Stroma pseudoparenchymatous; ascus pore blueing in iodine; causing branch cankers . . . . . . . . . . . . . . . . . . . . . . . . . . . **Potebniamyces**
        The genus was previously called *Phacidiella*. *Potebniamyces discolor* on apple and pear has a *Fuckelia* imperfect state. The perfect state of *P. coniferarum* is on pine, with its *Phomopsis* or *Phacidiopycnis* imperfect state on various conifers. *Potebniamyces balsamicola* is on *Abies*. Ref. *258*; D +.

    24′(23′) Stroma plectenchymatous; asci not blue in iodine; on living leaves . . . . . . . . . . . . . . . . . . . . . . . . . . . . **Pseudorhytisma**
        *Pseudorhytisma bistortae* on *Polygonum* is the only species. Ref. *249*; D–.

25(2′) Ascospores multiseptate, fusiform, at least 10 times as long as broad; paraphysis apices swollen, embedded in a pseudoepithecial matrix . . . . . . . . Tribe **Coccophacidieae**

    One genus . . . . . . . . . . . . . . . . . . . . . . . . . . . . . . **Therrya**
        This is the correct name for *Coccophacidium*. Reid and Cain recognize the 8-spored *T. pini* and a 4-spored species, *T. fuckelii*, both previously confused with *Coccomyces*, on pine branches. Ref. *96, 231*; D+.

25′(2′) Ascospores 1-celled, rarely 3-celled, ovoid, ellipsoid, or fusoid, but at most 4 times as long as broad; paraphysis apices swollen or not . . . . . . . . . . . . . . . . . 26

    26(25′) Ascocarps erumpent, cushion-shaped, covering layer not fused with host tissue . . . . . . . . . . . . . . . . . . . . . . . . Tribe **Pseudophacidieae** 27

    26′(25′) Ascocarps not erumpent, but with the host tissue firmly cohering to the upper portion of the stroma . . . . . . . . . . . . . . . . . . . Tribe **Phacidieae** 28

27(26) Occurring on bark . . . . . . . . . . . . . . . . . . . . . . . **Pseudophacidium**
    Four saprobic species on various hosts, the commonest being *P. ledi*. *Myxophacidium* and *Myxophacidiella* are synonyms. Ref. *5, 73, 96, 190*; D+.

27′(26) Occurring on leaves . . . . . . . . . . . . . . . . . . . . . . . . . **Nannfeldtia**
    *Nannfeldtia atra* on *Carex* is the only species. Ref. *214*; D–.

28(26′) Apothecia subcuticular, minute (less than 500 μm in diameter), circular or subcircular . . . . . . . . . . . . . . . . . . . . . . . . . . . . . . . . . . . 29
28′(26′) Apothecia intraepidermal or subepidermal, larger, circular to linear-elongate . . . . . . . . . . . . . . . . . . . . . . . . . . . . . . . . . . . 30
29(28) Ascospores nonseptate . . . . . . . . . . . . . . . . . . . . . . . **Phacidina**
*Phacidina gracile* on leaves of *Lycopodium*, differing from *Phacidium* primarily in position on the host, is the only species. Ref. *5*; D —.
29′(28) Ascospores 3-septate . . . . . . . . . . . . . . . . . . . . . . . **Micraspis**
*Micraspis acicola* on *Picea*, with a very similar appearing imperfect state, *Periperidium*, is the only species. Ref. *33*; D —.
30(28′) Paraphysis apices brownish, greatly swollen with thick walls and united in a mucilaginous pseudoepithecium; ascospores large (over 35 μm long)    . **Neophacidium**
*Neophacidium macrocarpum* on living leaves of *Gynoxis* in South America is the sole species. Ref. *217*; D —.
30′(28′) Paraphysis apices not swollen; ascospores smaller . . . . . . . . . . . 31
31(30′) Apothecia opening by several irregular teeth, black, more or less circular in outline . . . . . . . . . . . . . . . . . . . . . . . . . . . . . . . . . . . **Phacidium**
This is a fairly large genus, with most species on needles of conifers. *Gremmenia*, based on a species earlier referred to *Phragmonaevia*, appears to be synonymous, though originally recognized on the basis that the few-spored asci have multiseptate ascospores and a failure to find the covering stromatic tissue. *Phacidium multivalve* has the stroma extending through the leaf (of *Ilex*) and apothecia on both surfaces, and is sometimes treated separately as *Phacidiostroma*, Ref. *94, 156, 222, 233, 237, 276*; D +.
31′(30′) Apothecia opening by a longitudinal cleft, elongate . . . . . . . . . . . . 32
32(31′) Apothecia obviously black . . . . . . . . . . . . . . . . . . **Cryptomycina**
One species, *C. pteridis*, on fern leaves. Ref. *5*; D+.
32′(31′) Apothecia bright-colored, the blackened outermost layer of the stroma scarcely visible . . . . . . . . . . . . . . . . . . . . . . . . . . . . . . . **Lophophacidium**
*Lophophacidium hyperboreum* on *Picea* needles is the only known species. *Neonaumovia* is a recent synonym. Ref. *233*; D—.

## IX. OSTROPALES

The ascocarps in this order are lignicolous or foliicolous. If superficial, they are pileate-stipitate, clavate and laterally compressed, turbinate, or lenticular. If they are partially to wholly immersed in the substrate, they are discoid to perithecioid. The asci are cylindrical and very long, in youth with a greatly thickened apex traversed by a delicate pore. The apex remains thick or becomes much thinner at maturity. The pore either does or doesn't blue in iodine. The asci are eight-spored. The ascospores are smooth, hyaline, long filiform, and multiseptate, rarely disarticulating while still within the ascus, and are violently discharged.

There is a single family, Stictidaceae.

**KEY TO THE IMPORTANT GENERA OF OSTROPALES**

1. Ascocarps superficial on the substratum ........................ 2
1'. Ascocarps partially to wholly immersed in the substratum ............ 3
  2(1) Ascocarp apothecioid, either stalked and capitate-pileate or sessile, turbinate to discoid ............................................. **Vibrissea**
    Used here to include the sessile forms often placed in a separate genus, *Apostemidium*. Most species are either completely immersed in water or on waterlogged substrata. Sometimes treated in the Geoglossaceae. Ref. *91, 180, 244, 245*; B 432–3/D + also as *Apostemidium*.
  2'(1) Ascocarp clavate, laterally compressed, opening by an apical pore ............................................. **Acrospermum**
    Though placed in this order by Dennis, he claims the asci are bitunicate, as had O. Eriksson. Ref. *77, 283*; B –/D +.
3(1') Ascocarp opening widely at maturity, apothecioid ................ 4
3'(1') Ascocarp with a narrow neck, perithecioid .................... 5
  4(3) Apothecia opening to expose a broad, white margin which frequently splits into lobes ............................................. **Stictis**
    A rather large, though poorly understood genus, mostly on leaves. The genus *Biostictis* has been recognized for a parasitic species on Rubiaceae with a *Fusidium* imperfect state. Compare also *Nanostictis* on the thallus of *Peltigera*. Ref. *29, 219, 230*; B –/D +.
  4'(3) Apothecia with a darker, sterile border ............ **Schizoxylon**
    A poorly understood genus. Ref. *230*; B 568/D +.
5(3') Peritheciumlike cavity vertically oriented with a prominent beak which opens by a slit ............................................. **Ostropa**
    One species, *O. barbata*. Ref. *230*; B –/D +.
5'(3') Peritheciumlike cavity horizontal to the substrate surface, as are the asci; ascocarp beak turning upward, protruding, and opening by a pore ............ **Robergea**
    *Robergea cubicularis* may be the only species. Ref. *230*; B –/D +.

## X. HELOTIALES

In this order, the ascocarps are minute to moderate in size, of varying morphology (clavate, pileate, cupulate, discoid; stipitate or sessile), and are superficial or immersed in host tissues. The asci are clavate to cylindrical, with an apical pore (which will blue in iodine or not) through which the spores are violently discharged. They are two- to eight-spored with the ascospores rarely budding in the ascus to produce conidia (and thus multispored asci). The ascospores are smooth or very rarely marked with cyanophilic ornaments, hyaline to brown, and of various shapes. They are almost always radially symmetrical, although frequently they will have bipolar asymmetry, with the large end of the spore uppermost in the ascus, nonseptate or with one to many transverse septa. Paraphyses accompany the asci (except in *Neolecta*). Helotiales are plant parasites or saprobes, rarely found on soil or dung.

# 9. Discomycetes and Tuberales

## KEY TO FAMILIES, SUBFAMILIES, TRIBES AND IMPORTANT GENERA OF HELOTIALES

1. Asci forming an indefinite, floccose or crustlike layer on bark or leaves; no distinct apothecium .................................. **Ascocorticiaceae** 8

1'. A distinct apothecium produced .................................. 2

    2(1') Excipulum poorly developed, apothecial structure very simple, formed in the leaves of living plants within or beneath the host epidermis which is ruptured and thrown back (usually as a "scale") ............ **Hemiphacidiaceae** 9

    2'(1') Apothecia more complex, not as above .................. 3

3(2') Apothecia clavate, spathulate, or stalked with an irregular pileus, no abrupt edge to the hymenium where it meets the stipe; on soil, rotted wood, or rarely on leaves
.................................. **Geoglossaceae** 15

3'(2') Apothecia cupulate to discoid, rarely clavate or pileate but then with an abrupt edge delimiting the hymenium from the stipe, stalked or sessile, immersed or superficial on plant materials or on a stroma or sclerotium, rarely on soil ............ 4

    4(3') Apothecia arising from a stroma or sclerotium, or from stromatized patches of host tissue, usually brownish and stalked; ascus pore almost always blue in iodine (see also lead 90) .................. **Sclerotiniaceae** 26

    4'(3') Apothecia not arising from a stroma ............ 5

5(4') Apothecia superficial, with a waxy, translucent hymenium of closely adherent asci and paraphyses seemingly glued in a common matrix; excipulum of thin-walled globose to angular cells; asci minute, rarely over 40 μm long; saprobic on wood, rarely on decaying fungi .................................. **Orbiliaceae** 49

5'(4') Not with the above combination of characters .................. 6

    6(5') Apothecia erumpent to sessile, usually brown or black, sometimes yellowish to red, fleshy, rarely cartilaginous to leathery; excipulum of thin-walled to thick-walled, globose to angular cells (textura angularis to textura globulosa), walls of these cells normally dark .................................. **Dermateaceae** 52

    6'(5') Apothecia sessile to stipitate, bright-colored or rarely darker; excipulum chiefly of cells with hyphal characteristics .................. 7

7(6') Apothecia soft-fleshy, almost always provided with distinct and evident hairs at the margin or covering the excipulum; ectal excipulum composed of brick-shaped cells (textura prismatica), rarely of globose cells or elongated hyphae; gelatinized hyphae rare. Raitviir's paper (*227b*) on the family appeared too late to include his concepts in this key; I disagree with many of his conclusions, and reviewed his paper in *Mycologia* **64**:670–671 (1972). He erected 2 new genera, *Albotricha* and *Incrupila*, and took up *Belonidium* in a new sense for many species of *Dasyscyphus* .................. **Hyaloscyphaceae** 97

7'(6') Apothecia fleshy to cartilaginous, without distinctive hairs (except in *Chlorociboria*); excipulum of long-celled hyphae, more rarely of textura prismatica to textura angularis; hyphae immersed in a gel, or with gelatinized walls, present or absent; (some members of the Sclerotiniaceae will key here, but should be distinguishable in having usually larger, buff to dark brown apothecia arising from stromatized or at least blackened host tissues) .................................. **Leotiaceae** 115

    8(1) Ascospores subfusoid, multiseptate .................. **Ascosorus**
        One species, *A. floridanus* on *Quercus* leaves. Ref. *4*; B –/D –.

8′(1) Ascospores unicellular . . . . . . . . . . . . . . . . . . . . . . . . **Ascocorticium**
  The only species, *A. anomalum* on pine bark, has an interesting imperfect state. It has tiny ascospores, (4-) 6–7 × (2-) 3–4 µm, and a distinct pore in the thickened ascus apex. If the spores are much larger, see *Karstenella* (Pezizales) with a flattened apex to the ascus. The genus *Ludwigomyces* has been proposed for a fungus similar to *Ascocorticium*, parasitic on the sporocarps of a myxomycete. Ref. *144, 199*; B−/D+.

9(2) Ascospores filiform, septate . . . . . . . . . . . . . . . . . . . . . . **Naemacyclus**
  *Naemacyclus niveus*, on *Pinus*, is the commonest species. *Lasiostictis* may be a synonym. Ref. *215*; B−/D+.

9′(2) Ascospores 1- to 4-celled, ovoid to clavate . . . . . . . . . . . . . . . . . 10

  10(9′) Ascospores strongly constricted in the middle . . . . . . . . . . . . 11

  10′(9′) Ascospores not constricted . . . . . . . . . . . . . . . . . . . . 12

11(10) Asci 8-spored; ascospores hyaline, eventually becoming 2-celled, at germination 1 cell becoming brown . . . . . . . . . . . . . . . . . . . . . . . . . . . **Rhabdocline**
  Several fungi which have been confused, all on *Pseudotsuga*, have now been sorted out by Parker and Reid. One subspecies of *R. weirii* has a *Rhabdogloeum* imperfect state, others lack this. *Rhabdocline weirii* has asci with an apical plug blue in iodine, while *R. pseudotsugae* has an entirely different ascus apex with no plug. The fungi are so alike in spore characters that convergent evolution seems an unlikely explanation. Ref. *213*; B−/D+.

11′(10) Asci 4-spored; ascospores hyaline, eventually both cells faintly greenish-brown at maturity . . . . . . . . . . . . . . . . . . . . . . . . . . . . . . . . **Fabrella**
  Two subspecies of *F. tsugae* on *Tsuga*. Ref. *156*; B−/D−.

  12(10′) Asci 2- or 4-spored, ascospores brown, 2-celled, with 1 cell very much smaller than the other . . . . . . . . . . . . . . . . . . . . . . . . . **Didymascella**
    This genus was long known as *Keithia*. The 5 species are all on Cupressaceae. Ref. *156, 211, 212*; B−/D+.

  12′(10′) Asci 8-spored, ascospores hyaline or faintly brown at complete maturity, unicellular or 1- to 3-septate in age . . . . . . . . . . . . . . . . . . . . . 13

13(12′) Paraphyses with a gelatinous sheath, longer than the asci and forming an epithecium . . . . . . . . . . . . . . . . . . . . . . . . . . . . . . . **Korfia**
  *Korfia tsugae*, on *Tsuga*, is the only species. Ref. *234*; B−/D−.

13′(12′) Paraphyses without a gelatinous sheath, not forming an epithecium . . . . . 14

  14(13′) Ascus pore not blue in iodine; ascospores ovoid to clavate, remaining hyaline, 1-celled . . . . . . . . . . . . . . . . . . . . . . . . . . . **Hemiphacidium**
    Two species, both on pine. Ref. *156, 233*; B−/D−.

  14′(13′) Ascus pore blue in iodine; ascospores hyaline and nonseptate in youth, becoming brownish and 1- to 3-septate in age . . . . . . . . . . . . . . . . . **Sarcotrochila**
    Three species, on *Larix*, *Abies* and *Picea*. *Stegopezizella* is a synonym. Ref. *16, 156, 233*; B−/D−.

15(3) Ascospores brown at maturity . . . . . . . . . . . . . . . . . . . . . . 16

15′(3) Ascospores hyaline at maturity . . . . . . . . . . . . . . . . . . . . . 17

  16(15) No setae present in the hymenium . . . . . . . . . . . . . . . **Geoglossum**
    A fairly large genus, used here to include also *Gloeoglossum*. Ref. *178, 197*; B 422−4/D+.

9. *Discomycetes and Tuberales*

16'(15) Brown setae present in the hymenium and also clothing the stipe
................................................. **Trichoglossum**
A rather large genus. Ref. *178, 197*; B−/D+.

17(15') Ascospores globose to ovoid; ascocarp often very irregular or deformed or clavate, frequently mistaken for a *Clavaria*; paraphyses absent .............. **Neolecta**
*Neolecta flavovirescens* has globose spores, known from South America. *Neolecta irregularis* and *N. vitellina* have ovoid spores and are widely distributed in the Northern Hemisphere. Other names for the genus are *Spragueola* and *Ascocorynium*. Ref. *9, 126, 158, 179*; B−/D−.

17'(15') Ascospores elongate; paraphyses present ................. 18

18(17') Apothecium spathulate, ascigerous portion flattened and decurrent on opposite sides of the stipe ............... **Spathularia** and **Spathulariopsis**
*Spathularia flavida* may now be the only species of that genus, with several forms, as noted by Maas Geesteranus. In *Spathularia* the stipe tissues are all of thin-walled hyphae. In *Spathulariopsis*, another monotypic genus, based on *S. velutipes*, the stipe has a pseudoparenchymatous outer layer and thick-walled medullary hyphae. Ref. *177a, 179, 197*; B− /D+, *Spathularia*.

18'(17') Apothecium not spathulate ...................... 19

19(18') Apothecia tiny, pileus globose to clavate but separated from the stipe by a distinct groove, on conifer needles. See *Heyderia* ................... 171'

19'(18') Apothecia larger, on various substrata ............... 20

20(19') Apothecia stipitate, with a distinct pileus ............. 21

20'(19') Apothecium a clavula, the apex rarely swollen ........... 23

21(20) Ascospores greater than 2 $\mu$m broad; ascocarps distinctly gelatinous. See *Leotia* ........................................... 140'

21'(20) Ascospores less than 2 $\mu$m broad; ascocarps fleshy ........... 22

22(21') Ascospores acicular to clavate, less than 70 $\mu$m long, often much shorter. See *Cudonia* ........................................ 160

22'(21') Ascospores filiform, more than 85 $\mu$m long, usually much longer; ascocarps usually in or under water. See *Vibrissea* (Ostropales).

23(20') Paraphyses with their apices brown-walled or united into an epithecium by an amorphous, brown matrix ....................... **Thuemenidium**
A small genus, usually called *Corynetes*, possibly merging into *Geoglossum* from which it differs in having permanently hyaline spores. Mains synonymized it with *Microglossum*. Ref. *9, 128, 174, 179, 197*; B−/D as *Corynetes*.

23'(20') Paraphyses with hyaline walls, not united into an epithecium ........ 24

24(23') Hyphae of the stipe axis inseparable and agglutinated; head compressed, hymenium extending down the stipe further on the noncompressed than on the compressed sides; ascospores multicellular at maturity .............. **Microglossum**
Used here in the wide sense of Maas Geesteranus, to include many species previously referred to *Mitrula*. *Ochroglossum* is a synonym. Ref. *9, 128, 174, 179*; B 425−6/D+.

24'(23') Hyphae of the stipe axis easily separable; head not compressed, hymenium separated from the stipe by a horizontal line ............. 25

25(24') Ascospores multicellular at maturity ................. **Nothomitra**
*Nothomitra cinnamomea* is the only species. Maas Geesteranus noted that while

normally only the ascus pore blues in iodine, after sections are boiled in KOH the whole hymenium turns blue in iodine. Ref. *174*; B−/D−.

25′(24′) Ascospores for a long time 1-celled, eventually 2-celled . . . . . . . . . **Mitrula**
*Mitrula paludosa*, the only species now remaining in the genus, is common in the spring on rotting leaves in small pools of water. Ref. *9, 129, 174, 179*; B 427−7 bis/D+ p.p.

   26(4) Apothecia arising from a distinct, free, cylindric or more or less loaf-shaped to globose sclerotium with a dark, differentiated rind . . . . . . . . . . . . . 27

   26′(4) Apothecia arising from stromatized host tissues, often on mummified fruits, an obvious stromatal rind sometimes absent; rarely the stroma a crustlike mantle, associated with microsclerotia . . . . . . . . . . . . . . . . . . . . . . . 33

27(26) Botryose conidial states formed; gel regularly present between the hyphae of the sclerotial medulla . . . . . . . . . . . . . . . . . . . . . . . . . . . . . . 28

27′(26) Conidial states not formed (nearly all members of this family produce spermatia from phialides, whether or not they also produce conidia; spermatia are not considered to be conidia for the purposes of keying these fungi) . . . . . . . . . . . . . . . 29

   28(27) Conidial state with straight conidiophores (*Botrytis, Amphobotrys*)     **Botryotinia**
*Botryotinia fuckeliana* on grapes and many other hosts connects to *Botrytis cinerea*. This is a large genus. Ref. *23, 48, 119a, 120, 286*; B 468? 469?/D as *Sclerotinia* p.p.

   28′(27) Conidial states with twisted conidiophores (*Streptobotrys*)  . . . . **Streptotinia**
*Streptotinia arisaemae* and *S. caulophylli* in North America. Ref. *74, 119a, 286*; B−/D−.

29(27′) Ascospores brown, less than 6 μm long; gel regularly present between the hyphae of the sclerotial medulla . . . . . . . . . . . . . . . . . . . . . . . . . **Martininia**
*Martininia panamaensis*, a worldwide saprobe. Ref. *67, 286*; B−/D as *Martinia*.

29′(27′) Ascospores hyaline, usually larger . . . . . . . . . . . . . . . . . . . . 30

   30(29′) Sclerotia produced free on the mycelium, not digesting host tissues; sclerotial medulla without obvious gel between the hyphae and without remnants of host vascular elements . . . . . . . . . . . . . . . . . . . . . . . . . . . . . . **Whetzelinia**
*Whetzelinia sclerotiorum* (= *Sclerotinia sclerotiorum*) is an important plant pathogen on many hosts. *Whetzelinia tuberosa* is on *Anemone*. This is "*Sclerotinia*" in Whetzel's sense, but excludes all the species developing in culms of sedges and rushes treated here in *Myriosclerotinia*. Ref. *22, 23, 160, 286*; B 470, 477/D as *Sclerotinia* p.p.

   30′(29′) Sclerotia wholly or partially embedded in and digesting host tissues; sclerotial medulla with gel between the hyphae, almost always some remnants of the host vascular tissue detectable in median sections of sclerotia; usually foliicolous or within culms . . . . . . . . . . . . . . . . . . . . . . . . . . . . . . . . 31

31(30′) Sclerotia developed within the culms of sedges and rushes; a *Myrioconium* spermatial state produced; apothecia deep-cupulate to discoid . . . . . . . . . . **Myriosclerotinia**
A small, distinctive genus, monographed by Whetzel and treated in detail by Buchwald. Ref. *22, 23, 48, 208a, 269, 287*; B 472−4/D as *Sclerotinia* p.p.

31′(30′) Sclerotia developing in leaf blades or rarely in twigs . . . . . . . . . . . . 32

   32(31′) Apothecium with a cylindric to conic, campanulate (verpoid) pileus  . **Verpatinia**
A few rare species, sometimes treated in the Geoglossaceae. *Cudoniopsis* may be a synonym. Ref. *48, 179, 286*; B 428?/D+.

   32′(31′) Apothecia deep-cupulate to discoid . . . . . . . . . . . . . **Sclerotinia**

## 9. Discomycetes and Tuberales

*Sclerotinia whetzelii* on eastern poplar, *S. candolleana* on oak and chestnut, *S. erythronii* on *Erythronium*, *S. allii* on *Allium* in the Orient, and many others. Often the genus is incorrectly called *Ciborinia*. Ref. *6, 8, 22, 23, 48, 109, 286*; B 471/D+ p.p.

33(26′) Conidial states formed (nearly all members of this family produce spermatia from phialides, whether or not they also form conidia; spermatia are not considered to be conidial states for the purposes of keying these fungi) . . . . . . . . . . . . . . . 34

33′(26′) Conidal states not formed . . . . . . . . . . . . . . . . . . . . . . . 42

    34(33) Conidial state a *Monilia* . . . . . . . . . . . . . . . . . . . . . . . 35

    34′(33) Conidial state of some other type . . . . . . . . . . . . . . . . . . . 36

35(34) Ascospores hyaline; conidial disjunctors present or absent . . . . . . . **Monilinia**
A large genus including brown-rot of stone fruit pathogens (*M. fructicola, M. laxa, M. fructigena*); *M. ledi* is heteroecious on *Ledum* and *Vaccinium*. Ref. *48, 286*; B−/D+.

35′(34) Ascospores brown; conidial disjunctors absent . . . . . . . . . . **Phaeosclerotinia**
One species, *P. phaeospora* on *Malus* in Japan. Ref. *158*; B−/D−.

    36(34′) Conidia multiseptate, filiform; conidial state pycnidial (*Acarosporium*)
. . . . . . . . . . . . . . . . . . . . . . . . . . . . . . . . . . . **Pycnopeziza**
A few rare, apparently saprobic species. Ref. *288, 291*; B− /D−.

    36′(34′) Conidia neither filiform nor borne in pycnidia . . . . . . . . . . . . 37

37(36′) Conidia multicellular, staurosporus, arms folding down to forcibly discharge the conidium (*Valdensia, = Asterobolus*) . . . . . . . . . . . . . . . . . . . **Valdensinia**
*Valdensinia heterodoxa*, with apothecia on *Vaccinium*, is the only species; it forms conidia on many hosts. Ref. *223*; B−/D−.

37′(36′) Conidia not staurosporous . . . . . . . . . . . . . . . . . . . . . . . 38

    38(37′) Conidia predominantly septate, 1- to 5-celled, in subcuticular sporodochia (*Septotis*) . . . . . . . . . . . . . . . . . . . . . . . . . . . . . . . **Septotinia**
Two species, on *Populus* and *Podophyllum*. Ref. *110, 282a, 286*; B−/D−.

    38′(37′) Conidia one-celled . . . . . . . . . . . . . . . . . . . . . . . . . 39

39(38′) Conidia rough-walled; conidial state *Botrytis*like (*Verrucobotrys*) . . . . **Seaverinia**
*Seaverinia geranii* on *Geranium maculatum* is the only species. Ref. *119a, 253, 286*; B−/D−.

39′(38′) Conidia smooth; conidial state not *Botrytis*like . . . . . . . . . . . . . . 40

    40(39′) Conidia ellipsoid to ovoid, empty disjunctors present (*Ovulitis*) . . **Ovulinia**
Monotypic, *O. azaleae*. Ref. *284*; B−/D−.

    40′(39′) Conidia cylindric to suballantoid, disjunctors absent; (conidial state apparently as yet unnamed) . . . . . . . . . . . . . . . . . . . . . . . . . . . . 41

41(40′) Conidia produced in a copious, pink slime; apothecia discoid, stipitate
. . . . . . . . . . . . . . . . . . . . . . . . . . . . . . . . . . . . **Gloeotinia**
*Gloeotinia temulenta*, causing blind-seed disease on *Secale* and *Lolium*, is the only species. Ref. *22, 48, 294*; B−/D+.

41′(40′) Conidia dry; apothecia with a cylindric head, appearing to be Geoglossaceous
. . . . . . . . . . . . . . . . . . . . . . . . . . . . . . . . . . . **Scleromitrula**
*Scleromitrula shiraiana* on *Morus* in Japan is the only species now in the genus. *Scleroglossum* is a synonym. Ref. *126*; B−/D−.

42(33′) Stroma a thin crust mantelling infected rhizomes, bulbs, or corms; microsclerotia formed, but not giving rise to apothecia . . . . . . . . . . . . . . . **Stromatinia**
*Stromatinia rapulum* (= *S. smilacinae*) on *Polygonatum* is the type species. *Stromatinia narcissi* and *S. gladioli* are of economic importance. Ref. *22, 23, 48, 65, 111, 286*; B 478–9/D +.

42′(33′) Stroma not a mantle; microsclerotia normally absent . . . . . . . . . . 43

43(42′) Apothecia tiny, disc campanulate-reflexed, stipe very long and thin, ascospores less than 6 μm long, on dung . . . . . . . . . . . . . . . . . . . . . . **Coprotinia**
*Coprotinia minutula* is the only species now recognized. Ref. *75, 285, 286*; B –/D + p.p.

43′(42′) Not on dung, or if so then without the above combination of characters . . . 44

44(43′) In section, apothecium with an ectal excipulum of thin-walled, brick-shaped cells (textura prismatica); ascospores hyaline or brown . . . . . . . . . . . 45

44′(43′) In section, apothecium either with a distinct layer of highly gelatinized, long-celled hyphae or with the ectal excipulum of nearly globose cells; ascospores almost always hyaline . . . . . . . . . . . . . . . . . . . . . . . . . . . . . 46

45(44) Ascospores becoming light to dark brown in the ascus, or developing such pigments only after discharge or within asci which malfunction and fail to discharge their spores; ascospores mostly with faintly pitted or otherwise marked walls . . . . . . **Lambertella**
A large, mostly tropical genus. Dumont has shown that *Phaeociboria* and *Phaeodiscus* are synonyms. In addition, *Rutstroemia renispora* on *Nyssa* may belong here. Ref. *7, 66, 288*; B –/D –.

45′(44) Ascospores remaining hyaline, always smooth . . . . . . . . . . . . . **Lanzia**
Following Dumont's suggestion, I am provisionally adopting this generic name to include such species as *Ciboria carunculoides*, causing the popcorn disease of mulberries, and such common species as *Rutstroemia longipes*, *R. luteovirescens*, and *R. pruni-serotinae*. Ref. *66a, 289, 290*; B –/D as *Rutstroemia* p.p.

46(44′) In section, apothecium with a strongly gelatinized layer in the ectal excipulum; outermost layer of the excipulum not composed of globose cells . . . . **Poculum**
Used here tentatively for various species previously placed in *Rutstroemia*, as typified by Dumont with *P. ruborum*. As he has suggested, *Phialea* may be an older name for the genus, but the identity of the type species of that genus (*Peziza phiala*) is still uncertain. Ref. *22, 66a, 290*; B 481, 483/D as *Coprotinia* p.p., *Rutstroemia* p.p.

46′(44′) In section, apothecium with globose cells as the outermost layer of the ectal excipulum (rarely covered with a scurfy layer of hyphae running parallel to the surface); rarely a thin layer of gelatinized hyphae present as an inner layer of the ectal excipulum . . . . . . . . . . . . . . . . . . . . . . . . . . . . . . . . 47

47(46′) Apothecia on leaves and herbaceous stems; ectal excipulum of easily collapsing, thin-walled, nearly hyaline, globose cells; ascospores mostly under 10 μm long
. . . . . . . . . . . . . . . . . . . . . . . . . . . . . . . . . . . **Ciboriopsis**
*Ciboriopsis tenuistipes* on herbs, *C. simulata* on *Acer* leaves; probably a very large, mainly tropical genus, badly in need of monographic work. If the ectal cells are brown, compare *Ciboria*. Ref. *55, 56, 262*; B –/D +.

47′(46′) Apothecia on wood, fruits, cones or catkins, rarely on leaves; ectal excipulum of thin-walled, or rarely thick-walled, hyaline or dark, globose cells . . . . . . . . 48

48(47′) Apothecia greenish, on cones of *Picea*; ascospores less than 10 μm long . . . . . . . . . . . . . . . . . . . . . . . . . . . . . . . . . . . **Rutstroemia**
*Rutstroemia bulgarioides* is now the only species in the genus, which provides a

very different concept for the genus from that adopted by White and most recent authors. The species has also been placed in *Chlorosplenium* and *Chlorociboria*, and is also the type of *Piceomphale*, a recent synonym. Ref. *68, 268*; B 480 bis/D—.

48'(47') Apothecia not greenish, on various substrata, ascospores usually more than 10 µm long . . . . . . . . . . . . . . . . . . . . . . . . . . . . . . . . **Ciboria**

This genus is adopted here, in a very much broader sense than is usual, for a group of anatomically similar forms. Species on fruits have usually been referred to *Ciboria*, but not all belong here. Leaf- and wood-inhabiting species, on the other hand, seem too closely related to the common fruit and ament-attacking species to be removed generically. A fair number of the species referred by White and others to *Rutstroemia* belong here, notably *C. americana* on chestnut burrs and the common wood-inhabiting *C. peckiana* (= *Rutstroemia macrospora*). Ref. *22, 23, 48, 286, 290*; B 482/D + also as *Rutstroemia* p.p.

49(5) Excipular cells and swollen paraphysis apices dark brown . . . . . . . . **Patinella**
*Patinella hyalophaea* in Europe. Ref. *195*; B—/D—.

49'(5) Excipular cells and paraphyses hyaline or brightly colored . . . . . . . . . . 50

50(49') Apothecia with stiff, hyaline, blunt, septate setae . . . . . . . . **Orbiliaster**
Monotypic, *O. pilosus* in the Caribbean. Ref. *46*; B—/D—.

50'(49') Setae not present on the apothecium . . . . . . . . . . . . . . . . . 51

51(50') Margin of the apothecium crenate to distinctly toothed, the teeth made up of cohering hyphae . . . . . . . . . . . . . . . . . . . . . . . . . . . . . . . **Hyalinia**

Accepted here in the sense of Nannfeldt. Boudier used the generic name for *Orbilia*like fungi with filiform paraphyses, restricting *Orbilia* to those with globose apices. Ref. B 465–6/D+.

51'(50') Margin of the apothecium entire . . . . . . . . . . . . . . . . . . **Orbilia**

A large genus, particularly so if *Hyalinia* is considered as a synonym, easily characterized in the field by its waxy-translucent, usually very thin discs, and of course under the microscope by the minute asci. Some species have U-shaped conidia belonging to the genus *Dicranidion*. *Orbiliella* was erected for a species surrounded by a *Trichothecium*, the hyphae of that forming a subiculum, but the connection has not been proved. Ref. *143, 266*; B 460–4, 467/D+.

52(6) Ectal excipulum composed of hyaline to nearly hyaline textura globulosa or textura angularis, rarely with brown textura prismatica towards the margin . . 53

52'(6) Ectal excipulum composed of brown-walled textura globulosa or textura angularis . . . . . . . . . . . . . . . . . . . . . . . . . . . . . . . . . . 54

53(52) Parasitic, rarely saprobic, on woody plant parts; apothecia erumpent, or sunken and then erumpent, sessile or short-stalked; excipulum thick; asci usually large, apical pore usually blue in iodine; *Cryptosporiopsis* or *Discosporiella* imperfect states common . . . . . . . . . . . . . . . . . . . . . . . Subfamily **Peziculoideae** 56

53'(52) Saprobic, on herbaceous or rarely woody plant parts; apothecia immersed, rarely erumpent, sessile; excipulum thin; asci usually small, apical pore blue in iodine; imperfect states not reported . . . . . . . . . . . . . . . . Subfamily **Naevioideae** 60

54(52') Saprobic from the beginning, on woody or herbaceous plant parts; *Phialophora* imperfect states sometimes formed . . . . . . . . . . Subfamily **Mollisioideae** 65

54'(52') Parasitic, apothecia developing either on living tissues or on tissues killed the same or the previous year; conidial states common in the parasitic phase . . . 55

55(54′) Apothecia occurring on woody plant parts, or on the leaves or fruits of Ericaceae . . . . . . . . . . . . . . . . . . . . . . Subfamily **Dermateoideae** 83

55′(54′) Apothecia on herbaceous plant parts or leaves (but not of Ericaceae) . . . . . . . . . . . . . . . . . . . . . . . . . . Subfamily **Pseudopezizoideae** 90

    56(53) Apothecia usually stalked, erumpent early, not surrounded at the margin by torn excipular tissue; ascospores frequently 1- to 3-septate at maturity . . . . . **Pezicula**
        A large genus, needing monographic work. *Cryptosporiopsis* imperfect states common. Ref. *99–101*; B 559–60/D+.

    56′(53) Apothecia immersed, at first closed, erumpent, with torn marginal tissues usually surrounding the disc . . . . . . . . . . . . . . . . . . . . . . . . . 57

57(56′) Hymenium blue in iodine, but the ascus pore itself not blue; ascospores 3- or more septate . . . . . . . . . . . . . . . . . . . . . . . . . . . . . . . . . . **Cryptodiscus**
    *Cryptodiscus pallidus* may be the only species. Ref. B 569/D+.

57′(56′) Hymenium not blued by iodine, but ascus apex with a pore generally blue in iodine; ascospores unicellular . . . . . . . . . . . . . . . . . . . . . . . . . 58

    58(57′) Asci narrow cylindrical; ascospores small, narrow; paraphyses thin, hyaline, not forming an epithecium nor particularly developed in a slime . . . **Habrostictis**
        A small genus, formerly called *Cheilodonta*, characterized by the bright colors and laciniate teeth. Conidial states in the genus *Cryptosporiopsis*. Ref. B 459/D+.

    58′(57′) Asci broad clavate, rather thick-walled; ascospores large, ovoid to ovoid-cylindric; paraphyses formed in an evident slime, apically producing an epithecium . . . . . . . . . . . . . . . . . . . . . . . . . . . . . . . . . . . 59

59(58′) Paraphyses apically swollen and blue-green, strongly branched; excipulum much reduced, basally nearly absent, laterally of a few layers of pale brown cells; apothecia produced beneath the host cuticle which is thrown back, with the angular to globose cells originally produced over the hymenium adhering to the cuticle; ascus pore usually violet in iodine . . . . . . . . . . . . . . . . . . . . . . . . . . . . . . . . **Ploettnera**
    *Ploettnera exigua* on dying canes of *Rubus fruticosus* may be the only species, and apparently has no conidial state. The resemblance to *Laetinaevia* is pronounced. Ref. B–/D+.

59′(58′) Paraphyses hyaline or yellowish; apothecia more highly developed, excipular cells tending to be radially arranged at the margin, but more globular to angular and thick-walled, hyaline below . . . . . . . . . . . . . . . . . . . . . . . . . . . **Ocellaria**
    *Ocellaria ocellata*, and perhaps some other species, with a *Cryptosporiopsis* imperfect state. Ref. B–/D+.

    60(53′) Erumpent from the cones of *Pinus* or from decorticated wood of various kinds, as slitlike areas with a whitish to pinkish hymenium which is decidedly pruinose from the strongly branching paraphysis apices; ascospores large, nonseptate
. . . . . . . . . . . . . . . . . . . . . . . . . . . . . . . . . . . . . Tribe **Propolideae**
    One genus . . . . . . . . . . . . . . . . . . . . . . . . . . . . . . . . . . **Propolis**
        *Propolis versicolor* (= *P. faginea*) is very common on wood. A small genus. Ref. *230*; B–/D+.

    60′(53′) Not on wood, but on herbaceous plant parts, circular or rarely elongated, hymenium not pruinose . . . . . . . . . . . . . . . . . . . . Tribe **Naevieae** 61

61(60′) Apothecia with thick-walled, hyaline, usually unicellular setae at the margin; ascospores brownish, 2-celled . . . . . . . . . . . . . . . . . . . . . . . . . **Chaetonaevia**
    Monotypic, *C. nannfeldtii* on *Arctostaphylos*. Ref. *2*; B–/D–

## 9. Discomycetes and Tuberales

61'(60') Apothecia without marginal setae . . . . . . . . . . . . . . . . . . . . . . 62

    62(61') Basal cells more or less globose, dark-walled, with little or no marginal tissue; developing beneath the epidermis, which splits and rolls back to expose the hymenium; ascospores hyaline, eventually sometimes brownish, unicellular . . . . . . **Trochila**
        This is still a poorly known genus. A *Myxosporium* imperfect state is sometimes present. Ref. *92*; B 570/D+.

    62'(61') Basal cells hyaline-walled . . . . . . . . . . . . . . . . . . . . . . . . 63

63(62') Apothecia erumpent, more or less stipitate, with marginal tissues not extending beyond the hymenial rim; ascospores 1- to 3-septate, hyaline . . . . . . . . . . . **Callorina**
    Recognized for species previously placed in *Calloria*. The type species, *Callorina fusarioides* on *Urtica*, has a *Cylindrocolla* imperfect state. Ref. *158*; B 457/D as *Calloria*.

63'(62') Apothecia remaining immersed, scarcely erumpent, with marginal tissues extending beyond the hymenium as an entire or torn fringe; ascospores unicellular, hyaline, rarely brownish and 2- to 4-celled in age . . . . . . . . . . . . . . . . . . . . . . . 64

    64(63') Basal tissue of a thick-walled textura angularis, at the margin becoming brown-walled and the cells becoming more elongated and hyphalike . . . . . . . **Naevia**
        Adopted here in the restricted sense of Nannfeldt. *Ocellariella* appears to be synonymous. Ref. *195*; B−/D+.

    64'(63') Basal tissue of thin-walled textura angularis, at the margin of hyaline textura prismatica . . . . . . . . . . . . . . . . . . . . . . . . . . . . . . . **Laetinaevia**
        Probably a large genus. *Laetinaevia caulophylli* is very common on the previous year's stems of *Caulophyllum thalyctrioides*. *Europolella* differs only in the clavate paraphysis apices having brown contents, and in that the ascospores may become brown and 1- to 3-septate. Compare also *Ploettnera*. Ref. *36, 189, 195*; B−/D+.

65(54) Apothecia entirely superficial, rarely with a tiny base inserted in the host tissues
    . . . . . . . . . . . . . . . . . . . . . . . . . . . . . . Tribe **Mollisieae** 66

65'(54) Apothecia erumpent from the host tissues . . . . . . . Tribe **Pyrenopezizeae** 79

    66(65) Apothecia on grasses and sedges, originally with an upper tissue composed of radiating hyphae which tears open irregularly to expose the hymenium; asci with a broad pore blue in iodine; ascospores 1- to 3- (to 5-)septate, hyaline or brown
        . . . . . . . . . . . . . . . . . . . . . . . . . . . . . . . . . . . . **Actinoscypha**
        Four species are known. Ref. *191*; B−/D−.

    66'(65) Not with the above combination of characters . . . . . . . . . . . . 67

67(66') Ascospores brown . . . . . . . . . . . . . . . . . . . . . . . . . . . . 68
67'(66') Ascospores hyaline . . . . . . . . . . . . . . . . . . . . . . . . . . . 69

    68(67) Ascospores 1-celled, constricted in the middle . . . . . . . . . . . **Catinella**
        *Catinella olivacea* (= *C. nigro-olivacea*) has a worldwide distribution, on wood, and may be the only species. An ionomidotic reaction is produced in KOH. Ref. B 452/D+.

    68'(67) Ascospores 2-celled, wall faintly longitudinally striate . . . . . . . **Sorokina**
        A small, Southern Hemisphere genus. Ref. *49*; B−/D−.

69(67') Apothecia large, with a laciniate margin, on soil or duff . . . . . **Podophacidium**
    *Podophacidium xanthomelum* is the only known species. *Melachroia* is a synonym. Ref. B 449/D+.

69'(67') On woody or herbaceous plant parts . . . . . . . . . . . . . . . . . . 70

    70(69') Apothecium elongate, hysteriform, rarely branching, margin inrolled; paraphyses

delicate, curved and flexuous, forming a distinct epithecium; ascus pore not blue in iodine; ascospores 0–1-septate .................... **Angelina**
*Angelina rufescens* is the only species. Ref. *69*; B−/D−.

70′(69′) Apothecium not hysteriform ..................... 71

71(70′) Apothecia superficial on an obvious subiculum ............... 72

71′(70′) Apothecia not produced on an obvious subiculum .............. 73

72(71) Ascospores unicellular, rarely 1-septate ............... **Tapesia**
This genus has no real validity, since species with a slight or restricted subiculum bridge the differences between it and *Mollisia*. It is kept apart merely as a convenience, to keep *Mollisia* from being too large a genus to handle. Ref. *1*; B 538–40/D+.

72′(71) Ascospores multicellular ................... **Trichobelonium**
This genus also appears to grade into its nonsubiculate counterpart, *Niptera*. *Neotapesia* may be synonymous. Ref. *1, 193*; B−/D+.

73(71′) Ascospores multiseptate ...................... 74

73′(71′) Ascospores unicellular, rarely 1-septate ................ 75

74(73) Ascospores clavate to subfusoid; paraphyses forming an obvious epithecium ............................. **Patellariopsis**
*Patellariopsis clavispora* and *P. dennisii* are the only known species. Ref. *248*; B−/D+.

74′(73) Ascospores cylindrical; epithecium absent ............. **Niptera**
An often misunderstood genus, grading in part into *Trichobelonium*, in part into *Mollisia*, and adopted here instead of *Belonopsis* following Dennis' recent revision. It is badly in need of monographic work. Ref. *57a, 195*; B 554/D+.

75(73′) Apothecia distinctly hairy ..................... 76

75′(73′) Apothecia essentially hairless .................... 77

76(75) Apothecia large, on wood; ascus pore not blue in iodine ...... **Haglundia**
*Haglundia perelegans* is the type species; one other is known. Ref. *195*; B−/D+.

76′(75) Apothecia small, on grasses; ascus pore blue in iodine ....... **Belonium**
Used here in the restricted sense of Nannfeldt for a few species with short, dark brown hairs. Ref. *195*; B−/D−.

77(75′) Tip cells of the branching paraphyses yellow to olivaceous, enlarged suddenly, forming a distinct epithecium .................... **Cashiella**
*Cashiella atra* and *C. montiicola*, both very distinct from *Mollisia*, have been placed here. Ref. *53, 221*; B−/D−.

77′(75′) Paraphysis apices not forming an epithecium ............... 78

78(77′) Asci large, with a broad apex colored blue-violet in iodine; apothecia tiny, excipular cells nearly hyaline, on *Scirpus, Typha* ........... **Coronellaria**
There are only a few species. Ref. *195*; B−/D−.

78′(77′) Asci usually small, with a small apical pore blue in iodine (or rarely not blue); apothecia small to large, excipular cells decidedly brown at least at the base, on various woody and herbaceous hosts, never yellow-green, nor with an outer, hyphal layer (see third lead, 78″) ...................... **Mollisia**
There is no available monograph for this huge genus. Cultural studies have demonstrated the presence of *Phialophora* states for many species. Some authors recognize a genus *Niptera* for species with 2-celled ascospores, but Dennis has adopted that

name for species formerly placed in *Belonopsis*. Some apothecia have a very slight subiculum at the base, and the genus thus merges imperceptably into *Tapesia*. The genus is in desperate need of critical monographic work. Some authors would merge the genus with *Pyrenopeziza*, which would provide an older name and even more species to consider! *Graddonia* differs in having large ascospores, but is perhaps synonymous. Ref. *1, 45, 47*; B 541–6, 552/D+.

78″(77′) Asci small, with the pore blue in iodine; apothecia small, yellow-green to gray-green, on decorticated wood, brown-walled isodiametric cells of the ectal excipulum covered by an interrupted hyphal layer on the outside, these hyphae running parallel to the outer surface of the apothecium . . . . . . **Chlorosplenium**
This genus is now restricted to *C. chlora* and one other species. Ref. *63a*; B–/D–.

79(65′) Apothecia with distinct, dark brown hairs . . . . . . . . . . . . . . . **Pirottaea**
This is a moderately large genus, used here to include species with unicellular to multicellular ascospores, occurring on herbaceous stems. Ref. *195*; B–/D+.

79′(65′) Distinct, dark hairs not formed . . . . . . . . . . . . . . . . . . . . . . . 80

80(79′) Ectal excipulum of dark brown globose cells, often with pale or hyaline marginal hairs; paraphyses hyaline, filiform; occurring on herbaceous or rarely woody substrata, but not on grasses, sedges or rushes . . . . . . . . . . . . . . . . **Pyrenopeziza**
Hütter recognized 35 species, but it is a much larger genus than that. The genus has at times been confused with *Pseudopeziza* and its allies, but those are parasitic fungi and all species of *Pyrenopeziza* are strictly saprobic. A *Phialophora* state is known for some species. Some authors would merge *Mollisia* in *Pyrenopeziza*. *Dibeloniella* appears to be a synonym. Ref. *36, 124, 192*; B 547, 562/D+.

80′(79′) Ectal excipulum almost wholly of light-brown or hyaline cells forming textura angularis; occurring on grasses, sedges and rushes . . . . . . . . . . . . . 81

81(80′) Excipulum composed of hyphae of different lengths, terminating in free ends at various levels; ascus pore blue in iodine or not; ascospores 0–1-septate, sometimes brown at maturity; paraphyses filiform or fusoid, or clavate-tipped, hyaline or brown
. . . . . . . . . . . . . . . . . . . . . . . . . . . . . . . . . **Hysterostegiella**
Six species. Ref. *36*; B–/D as *Hysteropezizella* p.p.

81′(80′) Excipulum of hyphae all of the same length, terminating at the edge of the hymenial disc or extending beyond it . . . . . . . . . . . . . . . . . . . . . . . 82

82(81′) Marginal hyphae separating to form distinct, hairlike processes which are pale brown to hyaline; ascus pore blue or not in iodine; ascospores unicellular, hyaline, sometimes becoming brown and 1- to 5-septate at maturity; paraphyses hyaline and filiform, lanceolate or apically swollen, or brown-walled and apically swollen
. . . . . . . . . . . . . . . . . . . . . . . . . . . . . . . . . **Hysteropezizella**
The genus is taken here in the restricted sense of Défago, who recognized 10 species. Ref. *36, 195*; B–/D–.

82′(81′) Marginal hyphae agglutinated, not separating, with a delicate fringe of free ends at the margin; ascus pore blue in iodine; ascospores hyaline, 1- to 3- (or 4-)septate; paraphyses filiform, hyaline, sometimes enlarged apically . . . . . . . **Merostictis**
Défago recognized 13 species in her recent monograph for this split from *Hysteropezizella*. Ref. *36*; B–/D as *Hysteropezizella* p.p.

83(55) A distinct epithecium formed above the asci . . . . . . . . . . . . . . . . 84
83′(55) No epithecium formed . . . . . . . . . . . . . . . . . . . . . . . . . . . . 87

84(83) Epithecium in section brownish or yellow . . . . . . . . . . . . . . . . . 85

84′(83) Epithecium in section black, with encrustations on the paraphyses turning KOH solutions blue-green (or in one species, chocolate-brown); ascus pore not blue in iodine . . . . . . . . . . . . . . . . . . . . . . . . . . . . . . . . . . . **Atropellis**
Four North American species on conifer branches. *Fuckelia* imperfect states are formed. Ref. *235*; B−/D−.

85(84) Marginal tissues extending over the hymenium in early stages; ascospores unicellular; ascus pore blue in iodine . . . . . . . . . . . . . . . . . . . . . . . **Dermateopsis**
*Dermateopsis tabacina*, erumpent through the bark of *Quercus*, apparently without an imperfect state, is the only species. Ref. *195*; B−/D−.

85′(84) Marginal tissues not extending beyond the disc . . . . . . . . . . . . . . 86

86(85′) Ascospores transversely 0–7-septate, rarely even muriform, fusoid, with a long, filiform appendage at each end; ascus pore not blue in iodine . . . . . **Waltonia**
One species, *W. pinicola*. Muriform spores are almost unheard of in the true discomycetes, and the possibility exists that the asci are bitunicate. Ref. *243*; B−/D−.

86′(85′) Ascospores unicellular, frequently developing up to 3 septa at maturity; ascus pore usually blue in iodine; ionomidotic reaction in KOH usually prominent
. . . . . . . . . . . . . . . . . . . . . . . . . . . . . . . . . . . . . . . . **Dermea**
This large genus was long investigated by Groves, whose monograph is still the standard work. *Dermatea* is an alternative spelling, and *Bulgariastrum* appears to be synonymous. Imperfect states belong to *Micropera* and *Micula*, Ref. *102*; B−/D+.

87(83′) Ascus pore blue in iodine; ascospores with 1 to many septa . . . . . **Grovesiella**
The genus was erected in 1969 by Morelet for *G. abieticola*; Korf later added two species Groves had placed in *Encoeliopsis*, *G. ericae* and *G. ledi*. The same generic name was independently proposed by Eriksson in 1970, based on *G. ericae*, an odd coincidence. Imperfect states are unknown. Ref. *108*, *188*; B−/D−.

87′(83′) Ascus pore not blue in iodine . . . . . . . . . . . . . . . . . . . . . . . . 88

88(87′) On *Rhododendron*; ascospores 1-septate . . . . . . . . . . . . **Encoeliopsis**
*Encoeliopsis rhododendri* and *E. bresadolae* are the only species recognized when the genus is taken in this restricted sense. *Neogodronia* is a recent synonym. Ref. *108*, *248*; B−/D−.

88′(87′) On conifers; ascospores nonseptate to multiseptate . . . . . . . . . . . 89

89(88′) Ascospores 1- to 7-septate, on various conifers . . . . . . . . . . . . **Ascocalyx**
Adopted here not only to include the 3 species with *Bothrodiscus* pycnidial states recently monographed by Groves, but also to include *A. laricina* and *A. abietina* (= *Scleroderris lagerbergii*), species with *Brunchorstia* states, following Schläpfer-Bernard. *Gremmeniella* (= *Lagerbergia*) is based on the latter species. Ref. *108*, *248*; B−/D−.

89′(88′) Ascospores unicellular, rarely a few spores 1-septate; on *Pinus* . . . **Crumenulopsis**
Groves erected the genus for 2 species, *C. pinicola* and *C. sororia*, usually called *Crumenula*, but that generic name is a synonym of *Godronia*. *Digitosporium* is the imperfect state of *Crumenulopsis sororia*. Ref. *107*; B−/D−.

90(55′) Apothecia arising from a distinct stroma . . . . . Tribe **Pseudopezizeae** 91
90′(55′) Apothecia arising from host tissues . . . . . . . Tribe **Drepanopezizeae** 94

91(90) Stromatized vascular bundles forming rhizomorphlike strands from which the apothecia arise . . . . . . . . . . . . . . . . . . . . . . . . . . . . . . . . . . . **Spilopodia**
A small genus, clearly close to *Leptotrochila*. Ref. B *561*/D+.

91′(90) Stromatized tissues not forming rhizomorphlike strands . . . . . . . . . . . 92

## 9. Discomycetes and Tuberales

92(91′) Excipulum nearly or wholly lacking at the sides of the apothecium, though well-developed below; ascus pore blue in iodine; ascospores 0–1-septate . . **Pseudopeziza**
Taken here in the restricted sense of Schüepp, who has recognized 3 species, and 5 formae speciales of *P. trifolii* (= *P. medicaginis*), an important pathogen on Leguminosae. The covering tissue resembles that in Phacidiales. Imperfect states are not reported. Ref. *249*; B–/D+.

92′(91′) Excipulum well-developed at the sides of the hymenium; ascus pore blue in iodine or not . . . . . . . . . . . . . . . . . . . . . . . . . . . . 93

93(92′) Apothecia on the needles of conifers; ascus pore not blue in iodine; ascospores unicellular . . . . . . . . . . . . . . . . . . . . . . **Nothophacidium**
*Nothophacidium phyllophilum* on *Abies*, causing a snow-blight, is the only species. No conidial state is known. Ref. *232, 259*; B–/D–.

93′(92′) Apothecia on herbaceous plant parts; ascus pore blue in iodine or not; ascospores 0–1-septate . . . . . . . . . . . . . . . . . . . . . . . . **Leptotrochila**
Schüepp recognized 10 species. *Leptotrochila medicaginis* is an important pathogen previously called *Pyrenopeziza medicaginis* and *Pseudopeziza jonesii*. *Fabraea* and *Ephelina* are synonyms, and *Schizothyrioma* may be distinct in having subcuticular ascocarps. *Sporonema* imperfect states are known for some species. Ref. *122a, 249*; B–/D as *Fabraea, Schizothyrioma*.

94(90′) Apothecia remaining immersed-depressed on rupture of the overlying tissues; ascospores and conidia tear-shaped to filiform . . . . . . . . . . . **Blumeriella**
For some important plant pathogens placed by Higgins in *Coccomyces*, by Nannfeldt in *Higginsia*, with *Phloeosporella* (= "*Cylindrosporium*") imperfect states. *Blumeriella jaapii* is adopted by von Arx for all 4 species on *Prunus* previously recognized by Higgins and Nannfeldt. *Blumeriella kerriae* on *Kerria* appears to be a distinct species. Ref. *3, 195*; B–/D–.

94′(90′) Apothecia erumpent, obconic to turbinate; ascospores and conidia ellipsoid to suballantoid . . . . . . . . . . . . . . . . . . . . . . . . . . . . 95

95(94′) Ascospores 1-septate . . . . . . . . . . . . . . . . . . . . . **Diplocarpon**
Three important plant pathogens, *D. rosae* on rose, *D. maculatum* (= *Fabraea maculata*) on apple, quince and *Crataegus*, and *D. earliana* on strawberries, with either *Marssonina* or *Entomosporium* as imperfect states. Ref. *254, 264*; B–/D–.

95′(94′) Ascospores unicellular . . . . . . . . . . . . . . . . . . . . . . . . 96

96(95′) Excipulum poorly developed; ascus pore blue in iodine . . . . **Drepanopeziza**
The genus has recently been monographed by Rimpau; most of the species were previously referred to *Pseudopeziza*. The imperfect states are Melanconiaceous (*Gloeosporidiella, Marssonina*). Ref. *98, 240*; B–/D+.

96′(95′) Excipulum well developed; ascus pore not blue in iodine . . . **Discohainesia**
*Discohainesia oenotherae*, the only species, has a wide host range and was well described by Shear and Dodge (as a *Pezizella*). Two successive imperfect states are formed, a sporodochial *Hainesia* and a pycnidial *Pilidium*. Ref. *195, 257*; B–/D–.

97(7) Apothecia long-lived, reviving after dry periods, leathery from the long-celled excipular tissues . . . . . . . . . . . . . . . . . . . . . Subfamily **Trichoscyphelloideae**
One genus . . . . . . . . . . . . . . . . . . . . . . . . . . . . . **Lachnellula**
Almost all species are on conifers, with apothecia 2 mm or more in diameter, white hairs which are granularly roughened, and yellow, orange or red hymenia. *Trichoscyphella* is the generic name usually applied to the oval- and filiform-spored species,

while *Lachnellula* has traditionally been applied only to the spherical-spored species. Ref. *44, 54, 59*; B 518/D+.

97'(7) Apothecia decaying easily and not reviving after dry periods, fleshy, excipular tissues short-celled . . . . . . . . . . . . . . . . . . . . Subfamily **Hyaloscyphoideae** 98

    98(97') Subiculum present (or absent, and then hairs corkscrew-coiled); excipulum of gelatinized angular cells; paraphyses filiform to subclavate
. . . . . . . . . . . . . . . . . . . . . . . . . . . Tribe **Arachnopezizeae** 102

    98'(97') Subiculum absent, hairs not corkscrew-coiled; excipulum of various types; paraphyses lanceolate or not . . . . . . . . . . . . . . . . . . . . . . . 99

99(98') Excipular cells globose, hairs granulate, brown-walled at least towards the base
. . . . . . . . . . . . . . . . . . . . . . . . . . . . . . . Tribe **Trichodisceae** 105

99'(98') Excipular cells brick-shaped to angular, hairs granulate or smooth, hyaline or brown . . . . . . . . . . . . . . . . . . . . . . . . . . . . . . . . . . . . 100

    100(99') Hairs brown, smooth, thick-walled, usually with glandular tips
. . . . . . . . . . . . . . . . . . . . . . . . . . . . Tribe **Trichopezizelleae** 106

    100'(99') Hairs hyaline to subhyaline, smooth or granularly roughened, or if brown then roughened, walls thick or thin . . . . . . . . . . . . . . . . . . . . . . 101

101(100') Paraphyses lanceolate, or if appearing cylindrical then with definitely pointed tips; hairs septate, smooth or granularly roughened apothecia usually 1 mm in diameter or larger, sessile or stipitate . . . . . . . . . . . . . . . . . . . . . Tribe **Lachneae** 107

101'(100') Paraphyses cylindrical, with obtuse tips; hairs often without septa, smooth; apothecia usually minute, sessile . . . . . . . . . . . . . Tribe **Hyaloscypheae** 110

    102(98) Hairs corkscrew-coiled . . . . . . . . . . . . . . . . . . . . . . . . . 103

    102'(98) Hairs not coiled . . . . . . . . . . . . . . . . . . . . . . . . . . . . 104

103(102) Spores 3-septate, subiculum present . . . . . . . . . . . . . . . . **Velutaria**
*Velutaria griseovitellina* on *Rubus* canes is the only species. The genus was previously called *Tapesina*. Ref. *150*; B−/D−.

103'(102) Spores multiseptate . . . . . . . . . . . . . . . . . . . . . . . . **Lasiobelonium**
Adopted here in the sense of Dennis to include among other species the fairly common *L. miniopsis* on *Acer* bark. Ref. *54*; B−/D−.

    104(102') Ectal excipulum with a layer of dark brown cells; spores unicellular
. . . . . . . . . . . . . . . . . . . . . . . . . . . . . . . . . . . . **Eriopezia**
*Eriopeziza* is a frequent misspelling. The genus is now restricted to one species, *Eriopezia caesia*. Ref. *44, 150*; B−/D−.

    104'(102') Ectal excipulum of hyaline cells; ascospores rarely nonseptate, normally 1- to 7-septate . . . . . . . . . . . . . . . . . . . . . . . . . . . . **Arachnopeziza**
*Arachnoscypha* is a generic split recognized by some authors for species with 0–1-septate spores. Ref. *150, 154*; B 520–1/D+ also as *Arachnoscypha*.

105(99) Hairs thin-walled, tapering to a fine point; on monocot leaves and stems
. . . . . . . . . . . . . . . . . . . . . . . . . . . . . . . . . . . **Trichodiscus**
*Trichodiscus prasinus* in Europe, and some other species. Ref. *44, 54*; B 515/D+.

105'(99) Hairs thick-walled, tapering slightly to a blunt apex; on woody substrata . . **Perrotia**
As now conceived, a fairly large genus with many tropical species; originally said to have operculate asci. Ref. *54*; B 321/D+.

    106(100) Paraphyses lanceolate or with pointed tips . . . . . . . . . **Trichopezizella**

Probably a small genus. Ref. *44, 54, 227*; B 516/D as *Dasyscyphus* p.p.

106'(100) Paraphyses cylindrical, with obtuse tips . . . . . . . . . . . . **Zoellneria**
Only a few species are known. Dennis suggests the affinities may be with Leotiaceae or Sclerotiniaceae. Ref. *50, 54*; B−/D−.

107(101) Apothecium arising beneath the epidermis of the host leaf, which is thrown back as a scale exposing the ascocarps; hairs granularly roughened . . . . . . . **Stegopeziza**
One species, *S. lauri* on *Laurus*. Ref. *156, 264a*; B−/D−.

107'(101) Apothecia not arising beneath a host epidermal scale . . . . . . . . . . 108

108(107') Paraphyses filiform, with a lanceolate, septate tip . . . . . . . **Diplocarpa**
*Diplocarpa bloxamii* is the only species. The generic name should not be confused with *Diplocarpon*. Ref. *44, 54*; B−/D+.

108'(107') Paraphyses lanceolate, or cylindrical with a pointed apex. . . . . . . 109

109(108') Hairs short, cylindrical, thin-walled, obtuse, scarcely more than hyphal outgrowths . . . . . . . . . . . . . . . . . . . . . . . . . . . . . . . . . **Psilachnum**
A small genus. *Psilachum inquilinum* on *Equisetum* is common. Ref. *54*; B− /D+.

109'(108') Hairs well-developed . . . . . . . . . . . . . . . . . . . . . . **Dasyscyphus**
*Dasyscypha* is a more recent spelling. A very large genus. In agreement with Dennis the following names are treated as synonyms: *Arenaea, Dasyscyphella, Discocistella, Erinella, Erinellina, Lachnaster, Lachnella* (the discomycetous, not the agaricaceous, element!), *Lachnum, Pezizellaster*, etc. Also included here are the depauperate, wood-inhabiting species placed by Dennis and by Nannfeldt in *Cistella*, which have the same granularly roughened hairs as do many species of *Dasyscyphus*. *Clavidisculum* is also a synonym, as is *Belonidium*. Ref. *44, 46, 54, 201, 227b, 270*; B 489, 501–14, 517, 523, 526, 536/D+ also as *Cistella*.

110(101') Hairs long, stiff, multiseptate, with fairly thick walls and thinner septa
. . . . . . . . . . . . . . . . . . . . . . . . . . . . . . . . . . . . . **Hyalopeziza**
*Hyalopeziza ciliata* on leaves may be the only species. Ref. *44, 227b*; B−/D+.

110'(101') Hairs unicellular . . . . . . . . . . . . . . . . . . . . . . . . . . 111

111(110') Hairs thin-walled, wider or swollen at the base and drawn out into a fine apex, which in some species is recurved into a hook . . . . . . . . . . . . **Hyaloscypha**
A fairly large genus of poorly understood and rarely collected fungi. As used here the genus includes *Microscypha*, with the hairs less pointed, and also *Unguiculella*, characterized by the hair apices being recurved into a hook. Ref. *44*; B 522, 525, 534/D+ also as *Microscypha, Unguiculella*.

111'(110') Hairs not as above . . . . . . . . . . . . . . . . . . . . . . . . 112

112(111') "Hairs" only noncellular, thin, short, very delicate "processes" at the margin
. . . . . . . . . . . . . . . . . . . . . . . . . . . . . . . . . . . . . . **Mollisina**
A few species of doubtful affinities. Ref. *44*; B−/D+.

112'(111') Apothecium with hairs of definite cellular origin . . . . . . . . . . 113

113(112') Hairs thin-walled, blunt, often with resinous contents, usually short and easily overlooked; a basal brown ring present at the point of attachment to the host, most easily observed when the apothecium is mounted whole and examined from below; usually on leaves . . . . . . . . . . . . . . . . . . . . . . . . . . . **Calycellina**
Some of the species have been placed in "*Helotium*." *Phialina* is a synonym. Ref. *48*; B−/D as *Phialina*.

113'(112') Hairs with thick, glassy walls, cytoplasm visible only at the base or in longhaired species often extending up to the apex as a very delicate lumen . . . . . . . . . . 114

    114(113') Apex of the hairs recurved into a solid hook; ascospores spherical or ovoid . . . . . . . . . . . . . . . . . . . . . . . . . . . . . . . . **Unguiculariopsis**

        *Pithyella*, *Mollisiella* and *Encoeliella* are other names for this small genus. *Unguiculariopsis hysterigena* in the American subtropics has spherical spores and occurs on ascocarps of *Rhytidhysterium* and other bitunicate ascomycetes. *Unguiculariopsis infundibuliformis* has ovoid spores and an unequal sided apothecium, occurring in association with a pycnidial fungus in the Neotropics; the characteristic hairs were overlooked by Durand when he assigned it to *Midotis*. *Unguiculariopsis ilicincola* is spherical spored, on stromatic pyrenomycetes in Europe. *Parencoelia*, in which no hairs are reported, may be related. Ref. *70, 121, 218*; B−/D as *Pithyella*.

    114'(113') Hairs without recurved apices; ascospores not spherical . . . . **Urceolella**

        The genus is in need of critical work. *Hyalotricha* and *Pilatia* have been used for several species with long, often bent hairs. For those short-haired species in which the hairs stain reddish-purple in iodine, the name *Unguicularia* would be available, but a similar iodine reaction can be found in some species of *Unguiculariopsis*. Ref. *44, 280*; B 529–33/D+ also as *Hyalotricha*, *Unguicularia*.

115(7') Apothecia distinctly gelatinous, or with a definite and prominent gel layer in the apothecium . . . . . . . . . . . . . . . . . . . . . . . . . . . . . . . . . 116

115'(7') Apothecia fleshy, not gelatinous, lacking a distinct gel layer . . . . . . . . . 117

    116(115) Margin of the apothecium extending far beyond the hymenium, and there composed of 2 or 3 layers, the outermost scurfy, of nongelatinized hyphae, within which is a broad layer of yellowish, highly gelatinized, long-celled hyphae; apothecia black, erumpent, subsessile; ascospores filiform to narrow-fusoid
. . . . . . . . . . . . . . . . . . . . . . . . . Subfamily **Scleroderroideae** 121

    116'(115) Apothecia not so constructed, hyaline, bright-colored, or dark, even black . . . . . . . . . . . . . . . . . . . . . . . Subfamily **Leotioideae** 122

117(115') Ectal excipulum with the outermost cells globose, loose, forming a mealy or powdery surface, pale brown to black; medullary tissues of thin-walled, hyaline or brownish hyphae . . . . . . . . . . . . . . . . . . . . . . . Subfamily **Encoelioideae** 144

117'(115') Not combining such characters . . . . . . . . . . . . . . . . . . . . 118

    118(117') Ectal excipulum a single or a few layers of angular to nearly globose, dark cells, with thick-walled, hyaline, short-celled hyphae toward the center; apothecia erumpent . . . . . . . . . . . . . . . . . Subfamily **Heterosphaerioideae** 153

    118'(117') Tissue structure not as above . . . . . . . . . . . . . . . . . . . 119

119(118') Apothecia superficial, black or nearly so, uniformly composed of either fairly thin-walled, broad, red hyphae or of thick-walled, dark brown hyphae running more or less parallel to the outer surface . . . . . . . . . . . . . . Subfamily **Durelloideae** 155

119'(118') Apothecia light colored, or if black then erumpent and tissues not as above . . 120

    120(119') Apothecia minute, sessile, completely white; hymenium decidedly pruinose; ascospores large, multiguttulate, unicellular . . . . . Subfamily **Polydesmioideae**

        One genus . . . . . . . . . . . . . . . . . . . . . . . . . . . **Polydesmia**

        One species, *P. pruinosa*, on stromata of old pyrenomycetes. The apically branched paraphyses and often bent ascospores recall *Propolis*. Ref. B 453/D+.

    120'(119') Apothecia various, not as above . . . Subfamily **Hymenoscyphoideae** 156

121(116) Apothecia tiny, goblet-shaped, the gel layer strongly flaring outward toward the

## 9. Discomycetes and Tuberales

margin to form angular, radiating teeth . . . . . . . . . . . . . . . . . . **Asterocalyx**
One species, *A. mirabilis*, on fern leaves in Java. Ref. *122*; B–/D–.

121′(116) Apothecia medium-sized, opening widely at maturity, gel layer not wider at the margin than at the flanks . . . . . . . . . . . . . . . . . . . . . . . . . . . . **Godronia**
*Crumenula* and *Scleroderris* are synonyms. The generic name is used by some authors to include many unrelated fungi. Groves' admirable monograph recognizes 24 species. Conidial states belong in *Topospora* and *Fuckelia*. Ref. *105, 248*; B 563/D+.

    122(116′) Apothecia on living liverworts or mosses, or on algae associated with such bryophytes (probably forming a hemilichen symbiosis), or parasitic on the ascocarps of *Trichoglossum*, never stipitate, usually subglobose . . . Tribe **Mniaecieae** 124

    122′(116′) Apothecia on soil or decaying plant parts, of various shapes, sessile or stipitate . . . . . . . . . . . . . . . . . . . . . . . . . . . . . . . . . . . 123

123(122′) Asci filled with minute conidia which arise from budding of the often difficult-to-detect "primary ascospores" while still within the ascus . . . . Tribe **Tympaneae** 127

123′(122′) Asci not filled with conidia, usually with 8 typical ascospores which may bud conidia after discharge, or even rarely while still within the ascus but then the ascospores remain visible to maturity . . . . . . . . . . . . . . . . . . . . Tribe **Leotieae** 129

    124(122) Parasitic on the ascocarps of *Trichoglossum* . . . . . . . . . . **Micropyxis**
*Micropyxis geoglossi* in North America and the Neotropics is the only known species. Ref. *255*; B–/D–.

    124′(122) On bryophytes or algae . . . . . . . . . . . . . . . . . . . . . . . . . 125

125(124′) Ascospores 2-celled, septum not median, pale brown at maturity; apothecia nearly black . . . . . . . . . . . . . . . . . . . . . . . . . . . . . . . . . . . . . **Paryphydria**
*Paryphydria heimerlii* is the only known species. The generic name has had several different spellings, Rehm's version adopted here being apparently the most appropriate philologically. Ref. *296*; B–/D–.

125′(124′) Ascospores unicellular, hyaline; apothecia bright colored . . . . . . . . . 126

    126(125′) Ascospores multiguttulate; paraphysis apices branched or recurved to form a hook, or apically swollen . . . . . . . . . . . . . . . . . . . . . . . . **Mniaecia**
Used here in a sense including *Epiglia*, for a few, poorly known species. Ref. *18*; B 454–6/D+.

    126′(125′) Ascospores biguttulate; paraphysis apices filiform, unbranched
. . . . . . . . . . . . . . . . . . . . . . . . . . . . . . . . . . . . . . . . **Gloeopeziza**
There are only one or 2 species on liverworts. Quite possibly the genus should be merged with *Mniaecia*. Ref. *296*; B–/D–.

127(123) Apothecia olivaceous, on resin; ascal conidia globose . . . . . . . **Retinocyclus**
Two species, differing from the lichen genus *Biatorella* in having asci which do not blue in iodine. Ref. *113*; B–/D–.

127′(123) Apothecia umber to black, erumpent from bark or on culms of bamboo . . 128

    128(127′) Apothecia fused to form a nearly hemispherical, gelatinous mass, on culms of bamboo; ascal conidia spherical to short ovoid . . . . . . . . . . . **Myriodiscus**
*Ascotremellopsis* also appears to be based on the only species in this genus, *M. sparassoides*. Ref. *14, 204*; B–/D–.

    128′(127′) Apothecia erumpent in groups from bark; ascal conidia bacillar
. . . . . . . . . . . . . . . . . . . . . . . . . . . . . . . . . . . . . . . . **Tympanis**

Groves recognized 36 species in his monograph. If the apothecia are lilac to purple, see *Ascocoryne*, key lead 143′, if green, see *Claussenomyces*, lead 141. Ref. *103*; B−/D+.

129(123′) Ascospores, or at least some in each ascus, dark brown at maturity . . . . 130

129′(123′) Ascospores permanently hyaline . . . . . . . . . . . . . . . . . . . . 133

    130(129) Apothecia large, turbinate; apothecial tissues not homogeneous, but composed of pockets of differentiated tissue; ascus pore blue in iodine; either all 8 ascospores becoming brown, or frequently only 4 spores brown, 4 hyaline; ascospores unicellular . . . . . . . . . . . . . . . . . . . . . . . . . . . . . . . . . . **Bulgaria**
        *Bulgaria inquinans* may be the only species correctly referred to this genus, common on fallen trees. Some authors have chosen a different type species, and have used the generic name for species treated in this classification under *Sarcosoma*, *Galiella* and *Urnula* (Pezizales). They then adopt the name *Phaeobulgaria inquinans* for this species. Ref. *48*; B−/D+.

    130′(129) Apothecia minute, convex-discoid; apothecial tissues not pocketed; ascus pore not blue in iodine; all 8 ascospores brown, unicellular or 1-septate . . . . 131

131(130′). Ascospores unicellular . . . . . . . . . . . . . . . . . . . . . **Bulgariella**
    *Bulgariella pulla*, on wood, is the only species. Ref. *48*; B−/D+.

131′(130′) Ascospores 1-septate at maturity . . . . . . . . . . . . . . . . . . . . 132

    132(131′) Apothecia single, on leaves . . . . . . . . . . . . . . . . **Phaeangellina**
        One species, *P. empetri* on *Empetrum*. The resemblance to *Chloroscypha* has been noted by Dennis. *Diehlia*, with nonseptate, brown spores on leaves of an unknown tropical plant should be compared; if they are synonymous, it would provide an older name. Ref. *47*, *220*; B−/D+.

    132′(131′) Apothecia many on a common, gelatinous, erect tissue . . . **Polydiscidium**
        One species, *P. martynii* in the Neotropics. Ref. *282*; B−/D−.

133(129′) Apothecia black; an epithecium present; ascospores long-filiform to long-fusoid, (0−)3−5(−7)-septate . . . . . . . . . . . . . . . . . . . . . . . **Durandiella**
    Groves recognized 9 species in his monograph. Seaver used the name in different senses, and may have confused a *Tympanis* with the type species. Imperfect states belong in *Micropera*. Ref. *104*; B−/D−.

133′(129′) Apothecia bright-colored, more rarely greenish-black; ascospores various
. . . . . . . . . . . . . . . . . . . . . . . . . . . . . . . . . . . . . . . . 134

    134(133′) Apothecia with gelatinized hyphae only in the outermost layer, always enclosing some nongelatinized tissue . . . . . . . . . . . . . . . . . . . . . . . . 135

    134′(133′) Apothecia with gelatinized hyphae in the medullary tissues; other tissues gelatinous or not . . . . . . . . . . . . . . . . . . . . . . . . . . . . . . 138

135(134) Gel either restricted to the base of the apothecium or extending some distance up the flanks of the apothecium, but not reaching the margin; apothecia pale, sessile, never greenish. . . . . . . . . . . . . . . . . . . . . . . . . . . . . . . . . . **Pezoloma**
    *Sphagnicola* and *Pseudodiscinella* are more recent names. There are about 6 species, mostly on duff. Ref. *58*, *158*; B−/D as *Sphagnicola*.

135′(134) External gel tissue reaching the margin of the apothecium . . . . . . . . . 136

    136(135′) Apothecia yellow-green to blackish-green, on foliage of conifers; ascospores broad fusoid to ellipsoid (mostly at least 15 × 6 μm) . . . . . . . . **Chloroscypha**

## 9. Discomycetes and Tuberales

About 12 species on various conifers. The generic name was abandoned by Seaver, who mistakenly adopted *Kriegeria* instead. *Parksia* is a synonym, based on *C. alutipes*. Ref. *27, 48, 97, 145*; B−/D+.

136'(135') Apothecia without green colors, not on coniferous foliage; ascospores often shorter than 15 μm, usually less than 5 μm broad . . . . . . . . . . . . . . . 137

137(136') Apothecia yellow to orange, on wood or bark. See *Calycella* . . . . . . . . 168

137'(136') Apothecia some other color, mostly on herbaceous stems or on leaves. See *Cyathicula* . . . . . . . . . . . . . . . . . . . . . . . . . . . . . . . . . . . . . . . 168'

    138(134') Ectal excipulum of two layers, outermost layer of thin hyphae immersed in a gel, inner layer of nongelatinized, thin-walled, broad cells . . . . . . . . . 139

    138'(134') Apothecium either wholly gelatinous, or with a distinct ectal excipulum of non-gelatinized cells . . . . . . . . . . . . . . . . . . . . . . . . . . . . . . . 141

139(138) Ascocarp turbinate to irregularly *Tremella*-like . . . . . . . . . . . . **Neobulgaria**
*Neobulgaria pura* (= *Ascotremella turbinata*) is the commonest species. *Evulla* is probably a synonym. Ref. *48, 84, 134, 153*; B−/D+ p.p.

139'(138) Ascocarp distinctly stipitate, with a globose to irregular head, or discoid with a strongly repand pileus . . . . . . . . . . . . . . . . . . . . . . . . . . . . . . . 140

    140(139') Ascospores less than 8 μm long, unicellular; apothecium with a discoid-repand pileus, pallid . . . . . . . . . . . . . . . . . . . . . . **Neocudoniella**
*Neocudoniella albiceps*, which Mains treated in *Leotia*, is probably the only species. Ref. *126, 180*; B−/D−.

    140'(139'[21]) Ascospores longer than 15 μm, unicellular and multiguttulate, septate or pseudoseptate at maturity; apothecia with a subglobose to irregular head, buff to green . . . . . . . . . . . . . . . . . . . . . . . . . . . . . . . . . . . . . **Leotia**
This genus is traditionally treated in the Geoglossaceae, but has very similar microanatomy to that in *Neobulgaria*. Species concepts vary widely among authors. The completely viridus-green species in North America seems best called *L. atrovirens*, while the species with the viridus-green pileus and yellow stem should be *L. viscosa*. *Leotia lubrica* is the commonest species, varying from buff to shades of greenish brown depending on growth conditions. Ref. *9, 180*; B 429/D+.

141(138') Apothecia intensely green to olivaceous green or nearly black; tissues wholly gelatinized . . . . . . . . . . . . . . . . . . . . . . . . . . . . . . . **Claussenomyces**
The genus is usually called *Corynella*, an invalid name. *Claussenomyces prasinulus* is emerald green, and often accompanied by a *Dendrostilbella* imperfect state, and has 3-septate spores which do not bud in the ascus. *Claussenomyces atrovirens* has 5–11-septate spores which bud in the ascus, has a blackish-green apothecium and no associated imperfect state. *Claussenomyces pusilla* differs in having 15-septate spores. *Claussenomyces salviicolor* from North America and the Neotropics differs in having nonseptate spores. The tropical genus *Jacobsonia* of Asia may be related. Ref. *14, 48, 63a, 159*; B 458/D as *Corynella*.

141'(138') Apothecia not green; always some nongelatinized tissue present forming a distinct cortex . . . . . . . . . . . . . . . . . . . . . . . . . . . . . . . . . . . . . . . . 142

    142(141') Ascospores longitudinally ribbed, nonseptate; outermost excipular layer of thin-walled hyphae arranged perpendicular to the surface, constricted at the septa to form

chains of barrel-shaped cells; ascocarps turbinate-gyrose, *Tremella*-
like ................................................. **Ascotremella**
   Following Gamundi and Dennis, the genus is now restricted to *A. faginea*. Seaver's second species is a *Neobulgaria*. Korf and others were wrong in synonymizing the 2 genera. Ref. *48, 84, 153*; B−/D as *Neobulgaria* p.p.

142'(141') Ascospores smooth; outermost excipular layer of thin-walled, angular to brick-shaped cells ........................................... 143

143(142') Apothecia small, discoid, white to pale violet; outermost excipular layer of brick-shaped cells; ascospores unicellular ..................... **Ombrophila**
   In the restricted sense taken here, the genus may contain only a few species ranged around *O. violacea*. *Kubickia* is probably a synonym. Many species have been referred to the genus (on gross shape of the apothecium for the most part), but most have never been subjected to critical microanatomical study. *Ombrophila clavus* is a species of *Cudoniella*, for example, and completely lacks any gelatinous hyphae. Ref. *48, 267*; B−/D+.

143'(142') Apothecia medium-sized, discoid to cupulate or turbinate, lilac to bright purple or reddish brown; outermost excipular layer of angular cells; ascospores usually septate, sometimes budding conidia before or after discharge ............ **Ascocoryne**
   The genus is usually called *Coryne*, but that is technically the correct name for the similar gelatinous imperfect state that accompanies some species. *Ascocoryne sarcoides* with eventually 1- to 3-septate ascospores, and *A. cylichnium* with multi-septate ascospores budding spherical conidia, are quite common. *Ascocoryne turficola* is clearly congeneric, but the spores are nonseptate. Another species with much smaller, nonseptate spores, *A. microspora*, approaches *Ombrophila* in its paler, lilac color. Ref. *28, 48, 114*; B 450−1/D as *Coryne*.

144(117) Apothecial flesh containing many large (ca. 30 μm broad), vesicular cells filled with greenish-brown sap; ascospores hyaline, biguttulate, eventually pale brown (punctate) and 1-septate ........................................ **Velutarina**
   *Velutarina rufo-olivacea*, previously placed in *Velutaria*, may be the only species. Ref. *48, 158*; B 558/D+.

144'(117) Apothecial flesh lacking vesicular cells; ascospores of various types . . 145

145(144') Paraphyses lanceolate ........................... **Cenangiopsis**
   A small genus. Ref. *195*; B−/D−.

145'(144') Paraphyses filiform, sometimes with swollen apices ............ 146

146(145') Ascospores broad-ellipsoidal ............................. 147

146'(145') Ascospores allantoid to fusoid, narrow elliptic or filiform-clavate . . . 148

147(146) On spots on leaves, accompanied by a *Pestalotia* imperfect state; ascus pore blue in iodine ........................................... **Pestalopezia**
   There are 2 species. In *P. rhododendri*, at least, the ascospores turn brown at maturity. Ref. *252*; B−/D−.

147'(146) Saprobic, on conifers, a *Pestalotia* state absent; ascus pore not blue in iodine ............................................... **Cenangium**
   Used here in the very restricted sense for 2 species on pine, *C. ferruginosum* on twigs and branches and *C. acuum* (= *C. acicola*) on needles. Ref. *95, 195, 281*; B−/D+.

## 9. Discomycetes and Tuberales

148(146') Ascospores fusoid, eventually 1-septate; parasitic on *Radulum* on *Betula* .................................................... **Dencoeliopsis**
*Dencoeliopsis johnstonii* is the only species. The excipulum has an outermost, scurfy layer of globose cells and hairlike hyphae arising from a thin, outwardly brown, inwardly yellow layer of hyphae with roughened walls, running parallel to the outer surface. The ascus pore is blue in iodine. Ref. *48*, *108*, *158*; B–/D as *Encoeliopsis*.

148'(146') Ascospores allantoid to cylindric or filiform-clavate, septate or not; not parasitic on *Radulum* ............................................. 149

149(148') Ascospores filiform-clavate; paraphyses forming an obvious epithecium .. 150

149'(148') Ascospores not filiform-clavate; an epithecium lacking .......... 151

150(149) Apothecia opening by irregular tearing of an overlying tissue; ascospores eguttulate, unicellular ........................................... **Discocainia**
One species, *D. treleasei* on conifers. Ref. *235*; B–/D–.

150'(149) Apothecia without an overlying tissue; ascospores multiguttulate, sometimes finally multiseptate ............................................. **Holwaya**
One species, *H. mucida* (= *H. gigantea*), with 2 subspecies, is widely distributed on wood and bark, mostly of *Tilia*. *Crinula caliciiformis* (= *Stilbum giganteum*) is the large, black and gray synnematal state frequently accompanying the apothecia. Ref. *159*; B–/D–.

151(149') Ascospores 1- to 3-septate at maturity; outermost excipular layer of chains of moniliform, brown cells arranged perpendicularly to the surface; apothecia regular, discoid ............................................................. **Nipterella**
Dennis recognized 2 species. Ref. *54*; B–/D–.

151'(149') Ascospores unicellular, often allantoid; apothecia ear-shaped or cupulate, sometimes on a branching common stalk ......................................... 152

152(151') Apothecia giving an ionomidotic reaction in KOH; outermost layer of the excipulum of angular cells usually retaining some hyphal orientation perpendicular to the outer surface ..................................................... **Cordierites**
*Cordierites* is taken here in the sense of Boedijn, and thus becomes the name for the genus Durand and others have called *Ionomidotis*. *Phyllomyces* is also an older name than *Ionomidotis*. *Poloniodiscus* is a recent synonym based on *C. irregularis*, which has fusoid, deciduous apices of the paraphyses. *Cordierites frondosa* from Japan has circinate paraphyses. Ref. *30*, *70*, *170*, *173*, *273*; B–/D–.

152'(151') Apothecia lacking an ionomidotic reaction in KOH; outermost layer of the excipulum of angular cells without definite hyphal orientation, looser and subglobose toward the outer surface; apothecia buff-colored (see third lead, 152″) .. **Encoelia**
This genus, in which the ascospores are mostly suballantoid, now includes most species on wood previously referred to *Cenangium*. *Encoelia heteromera* is the species around which Durand built his concept of the genus *Midotis*, but that generic name still is shrouded in mystery. Some authors recognize *Phaeangella* for species with ascospores brown at maturity. *Phibalis* may be an older generic name. Ref. *48*; B–/D+ also as *Phaeangella*.

152″(151') Apothecia lacking an ionomidotic reaction in KOH; excipulum clothed with clavate or globosely enlarged hyphal outgrowths; apothecia greenish or reddish-brown, never buff ......................................................... **Chlorencoelia**

Recognized for 2 species previously referred to *Chlorosplenium* or *Chlorociboria*. Ref. *63a*; B 486/D as *Chlorosplenium* p.p.

153(118) Hymenium coral red; ascus pore not blue in iodine; ascospores fusoid, nonseptate .................................................. **Grovesia**
One species is known, *G. pulchella*. Ref. *52*; B−/D−.

153'(118) Hymenium gray to black; ascospores either ovoid and unicellular or fusoid to filiform and septate ........................................ 154

    154(153') Apothecial margin fimbriate; ascospores long-ellipsoid to cylindrical, unicellular; ascus pore blue in iodine; paraphyses not forming an
epithecium ............................................... **Heterosphaeria**
There may be several species, on stems, with *Heteropatella* imperfect states. Ref. *48*; B−/D+.

    154'(153') Apothecial margin even; ascospores fusoid to filiform, 1- to many-septate; ascus pore usually blue in iodine; paraphyses forming a definite
epithecium ................................................ **Pragmopora**
Groves recognized 6 species. *Pleurophomella* imperfect states are known. Ref. *106*; B−/D−.

155(119) Apothecial tissues of thin-walled, red-brown cells; ascus pore blue in iodine; ascospores unicellular ...................................... **Patinellaria**
*Patinellaria sanguinea*, with a strikingly red subiculum and nearly black apothecia, may be the only species. The genus is treated as a subgenus of *Durella* by Dennis. Ref. *48*; B−/D−.

155'(119) Apothecial tissues of thick-walled, dark brown hyphae; ascus pore not blue in iodine; ascospores nonseptate or septate ...................... **Xylogramma**
This appears to be the correct name for most of the species previously called *Durella*. Many have *Cystotricha* imperfect states. Species of *Lagerheimia*, differing in having brown, one-celled spores, will also key here. Ref. *48*; B−/D as *Durella*.

    156(120') Ascospores either filiform or regularly 3- or more-septate; apothecia never viridous green ............................................ 157

    156'(120') Ascospores 0–1-septate, not filiform (but if apothecia are viridous green, take this lead) ................................................. 164

157(156) Apothecia goblet-shaped, taller than broad, nearly black, outermost layer very thin and of brown-walled, agglutinated hyphae at the margin and angular cells below; ascus pore not blue in iodine ........................................ **Pocillum**
Several species on leaves. *Pocillum americanum* has somewhat curved, clavate spores, unlike the European *P. cesatii* with filiform spores. Ref. *254*; B−/D+.

157'(156) Apothecia broader than tall, not so constructed .............. 158

    158(157') Apothecia on pine needles; outermost tissues of the excipulum of thinwalled, brown, neither agglutinated nor glassy hyphae; ascospores 3-septate, acicular ................................................. **Pseudohelotium**
*Pseudohelotium pineti* is the only species. Ref. *48*; B−/D+.

    158'(157') Apothecia not on pine needles, not so constructed .......... 159

159(158') Apothecia with a globose to irregular pileus, resembling a *Leotia* ..... 160

159'(158') Apothecia not pileate, but discoid to cupulate, stipitate or sessile ..... 161

    160(159, [22]) Ascospores acicular to clavate, less than 2 μm broad; ascus pore not blue in iodine; paraphyses hyaline ................................ **Cudonia**

## 9. Discomycetes and Tuberales

A small, but common genus usually treated under the Geoglossaceae. *Pachycudonia* has been segregated for 3 species with constricted spores and "long tailed asci." Ref. *128, 180, 197*; B 430–1/D+.

160'(159) Ascospores clavate to fusoid, more than 3 μm broad; ascus pore blue in iodine; paraphyses brown . . . . . . . . . . . . . . . . . . . . . . **Sarcoleotia**
*Sarcoleotia platypus* (= *S. nigra*) has been well-described by Maas Geesteranus. *Sarcoleotia globosa* has been treated by some authors in *Corynetes*. Ref. *71, 126, 175*; B–/D–.

161(159') Apothecia on woody substrata, or if on culms then the ascospores surrounded by a gelatinous sheath . . . . . . . . . . . . . . . . . . . . . . . . . . . . . . 162

161'(159') Apothecia on leaves; no gelatinous sheath around the ascospores. See *Hymenoscyphus* . . . . . . . . . . . . . . . . . . . . . . . . . . . . . . . . . . . . . . 173'

162(161) Apothecia on or among Dematiaceous moulds (*Helminthosporium*like or *Bipolaris*like) on decorticated wood; ectal tissues not glassy-walled; ascus pore not blue in iodine, though some basal tissues of the apothecium sometimes blue strongly, and the ascospores themselves may turn light blue-green in iodine
. . . . . . . . . . . . . . . . . . . . . . . . . . . . . . . . . . . . . . **Strossmayeria**
Cultural studies are needed to determine if the imperfect fungus is a state of this discomycete, or is being parasitized by it. *Strossmayeria basitricha* may represent a very variable fungus in spore characters; whether or not *Gorgoniceps confluens, G. iowensis, G. jamaicensis* and *Belonium introspectum* are synonyms remains to be proved. Ref. *10a, 254*; B–/D+.

162'(161) Apothecia not among molds; ectal tissues of parallel, agglutinated hyphae with yellow, glassy walls; ascus pore blue in iodine . . . . . . . . . . . . . . 163

163(162') Ascospores surrounded by a gelatinous sheath, 3-septate, cylindric-fusoid; apothecia on culms and stems, stipitate . . . . . . . . . . . . . . . . **Belonioscypha**
*Belonioscypha culmicola* is probably the only species; many other that have been placed in the genus have widely different structure. Ref. *48*; B–/D+.

163'(162') Ascospores without a sheath, multiseptate, filiform; apothecia on woody substrata, turbinate . . . . . . . . . . . . . . . . . . . . . . . . . . . . . . . . **Gorgoniceps**
*Gorgoniceps aridula* is common on coniferous wood. Most of the other species treated under this generic name are quite unrelated. Ref. B–/D+.

164(156') Apothecia aeruginous-green; substrate also stained green . . **Chlorociboria**
This genus is now taken in a narrower sense, following Dixon, who placed some species in *Chlorencoelia* and once again separated it from *Chlorosplenium* and *Rutstroemia*. Ref. *63a, 153, 154, 228*; B 485/D as *Chlorosplenium* p.p.

164'(156') Apothecia not aeruginous; substrate not stained green . . . . . . . . 165

165(164') Ascospores brown, 1- to 3-septate at maturity . . . . . . . . . . **Phaeohelotium**
*Phaeohelotium monticola* (= *P. flavum*) can easily be mistaken in the field for the common, yellow "Helotium," *Calycella citrina*. Dennis has expanded the generic concept to include as well species with hyaline unicellular ascospores, and more recently even to include the poorly defined genus *Orbiliopsis*, but then the generic concepts become very confusing. Ref. *56, 154*; B–/D+.

165'(164') Ascospores hyaline, 0–1-septate, rarely 3-septate . . . . . . . . . . . . 166

166(165') Apothecia erumpent, with a tall, sterile collar surrounding the hymenium . . . . . . . . . . . . . . . . . . . . . . . . . . . . . . . . . . . **Stamnaria**

Two species, both on stems of *Equisetum*. *Stamnaria persoonii* is a common saprobe; *S. americana*, in North America and Asia, is parasitic. Ref. *254*; B 498/D+.

166'(165') Apothecia without a distinct collar  . . . . . . . . . . . . . . . . . 167

167(166') Hyphae or cells of the outermost layer with thickened, glassy walls (at fullest development the walls become a refractive gel in which the hyphal lumina are embedded)  . . . . . . . . . . . . . . . . . . . . . . . . . . . . . . . . . . . . 168

167'(166') Outermost layer of the excipulum of thin-walled hyphae or brick-shaped, angular or subglobose cells, not appearing glassy  . . . . . . . . . . . . . . . . . 169

168(167, [137]) Apothecia turbinate to substipitate, rarely stipitate, some shade of yellow to orange; outermost layer of the excipulum of undulating, glassy hyphae, the individual hyphae often difficult to follow; ascospores frequently 1-septate  . . . . **Calycella**
A small genus, but *C. citrina* is worldwide and very common. Ref. *48*; B 441–4/D+.

168'(167, [137']) Apothecia stipitate, of various colors but rarely yellow; outermost layer of the excipulum of glassy, parallel, agglutinated hyphae, behaving as a tough "skin" in squash mounts, the tissue strongly differentiated from the thin-walled hyphae of the medullary excipulum; ascospores usually unicellular, rarely 1- to 3-septate; in some species the excipulum is prolonged at the margin to form long or short teeth  . . . . . . . . . . . . . . . . . . . . . . . . . . . . . . . . . . **Cyathicula**
The genus is taken here in a very much broader sense than is usually understood. Most authors have restricted the genus to those species with distinct marginal teeth, and have assigned here also various unrelated species having only toothed margins in common. The structure of the common *C. coronata* differs not at all, except for the teeth, from the fungi which Dennis treats under *Phialea*.
Most of the species have paraphyses slightly lanceolate or at least pointed at the apex, as in the very common *C. cyathoidea*. *Davincia* was erected for two species with toothed margins and 3-septate spores from Java. Its type species is widely distributed in the Asian tropics, *C. helios*. The type species of *Allophylaria* also has 3-septate spores, lacks teeth, and should apparently be *C. sublicoides*. Ref. *48*, *216*; B 499/D+ also as *Phialea*.

169(167') Apothecia on soil, usually large, substipitate, very soft and fragile, composed throughout of thin-walled, broad hyphae  . . . . . . . . . . . . . . . . . . **Discinella**
Recognized here in the sense of Dennis for terricolous species which on field characters might be taken for operculate discomycetes, doubtfully distinct from *Hymenoscyphus*. Dennis has suggested it might provide an older name for *Pezoloma*, but that has angular excipular cells. Ref. *48*; B 445–8/D+.

169'(167') Apothecia on leaves, wood, or plant debris, not on soil  . . . . . . . . . . 170

170(169') Apothecia with a globose to clavate pileus separated from the stipe by a distinct groove  . . . . . . . . . . . . . . . . . . . . . . . . . . . . . . . . . 171

170'(169') Apothecia with a cupulate to discoid, rarely convex hymenial disc, stipitate or sessile  . . . . . . . . . . . . . . . . . . . . . . . . . . . . . . . . . . . 172

171(170) Stipe very long, threadlike; pileus subglobose; paraphyses apically swollen with thickened, brown and roughened apices; ascus pore not blue in iodine  . . . **Capillipes**
Only one species, *C. cavorum*. Ref. *247*; B –/D–.

171'(170, [19]) Stipe shorter, not threadlike; pileus subglobose to clavate; paraphyses filiform, hyaline, smooth  . . . . . . . . . . . . . . . . . . . . . . . . . . . **Heyderia**

There may be only a single species on conifer needles, *H. abietis*. *Gymnomitrula* is a synonym. The fungus is often placed in *Mitrula*, but Maas Geesteranus has convincingly shown it belongs in this family and not the Geoglossaceae. *Antinoa*, differing in having the ascus pore not blue in iodine, should be compared. Ref. *9, 79, 126, 129, 174, 179, 280*; B− /D as *Mitrula* p.p.

172(170′) Apothecia elongate, erumpent; ectal excipulum of angular cells with yellow, granular deposits on the walls . . . . . . . . . . . . . . . . . **Duebenia**
One species, *D. purpurascens*, on Leguminosae. Ref. *194*; B−/D+.

172′(170′) Apothecia circular, usually superficial; excipulum not as above . . . . 173

173(172′) Apothecia stipitate with a convex hymenium; ascus pore not blue in iodine
. . . . . . . . . . . . . . . . . . . . . . . . . . . . . . . . . . . . . **Cudoniella**
A small number of species, most of which occur in wet places on wood or leaves, not infrequently developing under water. All species are light colored, often with lilaceous tints, though the stipe may be brown. If the medullary hyphae of the apothecia are brown, compare *Chlorencoelia*. Ref. *48*; B 434–4 bis/D+.

173′(172′, [161′]) Apothecia sessile or stipitate, the disc not or scarcely recurved, hymenium concave or plane (only rarely convex); ascus pore almost always blue in iodine
. . . . . . . . . . . . . . . . . . . . . . . . . . . . . . . . . . . . **Hymenoscyphus**
This appears to be the correct name for the large and unwieldy group of species previously referred to *Helotium*. It is doubtfully distinct from *Cudoniella* and remains a "wastebasket genus" to which species without many obvious characters are referred. It must surely be broken into smaller genera, but the limits of those genera are still vague. Dennis has critically studied many species, eliminated some from the genus, and has made several attempts at arranging the species. Artificial characters still form the basis for the subgenera, series and smaller units recognized. As adopted here, the genus would also include *Ciboriella, Pseudociboria,* and *Pezizella* as conceived by Dennis, and *Calycina* as typified by Dumont (*66a*). Aquatic species may have tetraradiate-spored imperfect states. If the hyphae of the flesh are brown, compared *Chlorencoelia*. Ref. *48, 56, 279*; B 435–8, 440, 487–8, 490–2, 494–5, 497/D+ also as *Pezizella* p.p.

REFERENCES

1. Aebi, B. (1972). Untersuchungen über Discomyceten der Gruppe *Tapesia-Trichobelonium*. *Nova Hedwigia* **23**:49–112.
1a. Arpin, N. (1968). "Les Caroténoïdes des Discomycètes: Essai Chimiotaxinomique." Université de Lyon, Villeurbanne; reissued, *Bull. Mens. Soc. Linn. Lyon* **38**: Suppl; 1–169 (1969).
2. Arx, J. A. von. (1951). Eine neue Discomycetengattung aus Skandinavien. *Antonie van Leeuwenhoek; J. Microbiol. Serol.* **17**:85–89.
3. Arx, J. A. von. (1961). Über *Cylindrosporium padi*. *Phytopathol. Z.* **42**:161–165.
4. Arx, J. A. von. (1963). Die Gattungen der Myriangiales. *Persoonia* (*Leyden*) **2**:421–475.
5. Arx, J . A. von, and E. Müller. (1954). Die Gattungen der amerosporen Pyrenomyceten. *Beitr. Kryptogamenfl. Schweiz* **11**(1):1–434.
6. Batra, L. R. (1960). The species of *Ciborinia* pathogenic to *Salix, Magnolia,* and *Quercus*. *Amer. J. Bot.* **47**:819–827.
7. Batra, L. R. (1968). *Phaeodiscus*, a new genus of Sclerotiniaceae (Inoperculatae: Discomycetes). *Amer. J. Bot.* **55**:1205–1209.

8. Batra, L. R., and R. P. Korf. (1959). The species of *Ciborinia* pathogenic to herbaceous angiosperms. *Amer. J. Bot.* **46**:441–450.
9. Benedix, E. H. (1962). Neue über Geoglossaceen: *Coelotiella, Mitrula. Kulturpflanze, Beih.* **3**:389–410.
10. Benedix, E. H. (1969). Art- und Gattungsgrenzen bei höheren Discomyceten, III. *Kulturpflanze* **17**:253–284.
10a. Bertault, R. (1970). Deux espèces du genre *Strossmayeria* Schulzer. *Rev. Mycol.* **35**:130–140.
11. Berthet, P. (1964). "Essai Biotaxinomique sur les Discomycètes." Joanny Lorge, Lyon.
12. Berthet, P., and R. P. Korf. (1969). Sur la position taxonomique du genre *Jafnea* Korf. *Natur. Can.* **96**:247–249.
13. Boedijn, K. B. (1933). The genera *Phillipsia* and *Cookeina* in Netherlands India. *Bull. Jard. Bot. Buitenzorg* III **13**:57–76.
14. Boedijn, K. B. (1935). Two new Malaysian genera of Discomycetes. *Bull. Jard. Bot. Buitenzorg* III **13**:478–483.
15. Boedijn, K. B. (1939). The Tuberales of the Netherlands Indies. *Bull. Jard. Bot. Buitenzorg* III **17**:236–244.
16. Bonar, L. (1962). *Stegopezizella balsameae* and *Gloeosporium balsameae. Mycologia* **54**:395–399.
17. Boudier, E. (1905–1910). "Icones Mycologicae," 4 vols. Klincksieck, Paris.
18. Boudier, E. (1907). "Histoire et Classification des Discomycètes d'Europe." Klincksieck, Paris.
19. Brummelen, J. van. (1962). Studies on Discomycetes–II. On four species of *Fimaria. Persoonia (Leyden)* **2**:321–330.
20. Brummelen, J. van. (1967). A world-monograph of the genera *Ascobolus* and *Saccobolus. Persoonia (Leyden), Suppl.* **1**:1–260.
21. Brummelen, J. van. (1969). Studies on Discomycetes–III. *Persoonia (Leyden)* **5**:225–231.
22. Buchwald, N. F. (1947). Sclerotiniaceae Daniae. *Friesia* **3**:235–330.
23. Buchwald, N. F. (1949). Studies in the Sclerotiniaceae. I. Taxonomy of the Sclerotiniaceae. *Kgl. Vet.- Landbohoejst., Aarsskr.* **1949**:75–191.
24. Burdsall, H. H., Jr. (1965). Operculate asci and puffing of ascospores in *Geopora* (Tuberales). *Mycologia* **57**:485–488.
25. Burdsall, H. H., Jr. (1968). A revision of the genus *Hydnocystis* (Tuberales) and of the hypogeous species of *Geopora* (Pezizales). *Mycologia* **60**:496–525.
26. Cain, R. F., and J. W. Kimbrough. (1969). *Coprobolus*, a new genus of the tribe Theleboleae (Pezizaceae). *Can. J. Bot.* **47**:1911–1914.
27. Cash, E. K. (1945). Some new species of fungi on *Libocedrus. Mycologia* **37**:311–317.
28. Christiansen, M. P. (1963). Danish species of the genus *Coryne. Friesia* **7**:75–85.
29. Christiansen, M. S. (1954). *Nanostictis*, a new genus of scolecosporous Discomycetes. *Bot. Tidsskr.* **51**:59–65.
30. Ciferri, R. (1957). Revision of the genus *Cordierites* Mont. *Atti Ist. Bot. Univ. Pavia* [5] **14**:263–270.
31. Clémencet, M. (1932). Contribution à l'étude du développement et de l'anatomie des Ascomycètes hypogés. Les Elaphomycétacées. *Botaniste* **24**:3–106.
32. Cooke, J. C., and M. E. Barr. (1964). The taxonomic position of the genus *Thelebolus. Mycologia* **56**:763–769.
33. Darker, G. D. (1963). A new genus of Phacidiaceae on *Picea mariana. Can. J. Bot.* **41**:1389–1393.
34. Darker, G. D. (1963). A new phragmosporous genus of the Hypodermataceae. *Mycologia* **55**:812–818.
35. Darker, G. D. (1967). A revision of the genera of the Hypodermataceae. *Can. J. Bot.* **45**:1399–1444.
36. Défago, G. (1968). Les *Hysteropezizella* von Höhnel et leurs formes voisines (Ascomycètes). *Sydowia* **21**:1–76.

37. Denison, W. C. (1961). Some species of the genus *Scutellinia*. *Mycologia* **51**:605–635.
38. Denison, W. C. (1963). A preliminary study of the operculate cup-fungi of Costa Rica. *Rev. Biol. Trop.* **11**:99–129.
39. Denison, W. C. (1964). The genus *Cheilymenia* in North America. *Mycologia* **56**:718–737.
40. Denison, W. C. (1965). Central American Pezizales. I. A new genus of the Sarcoscyphaceae. *Mycologia* **57**:649–656.
41. Denison, W. C. (1967). Central American Pezizales. II. The genus *Cookeina*. *Mycologia* **59**: 306–317.
42. Denison, W. C. (1969). Central American Pezizales. III. The genus *Phillipsia*. *Mycologia* **61**: 289–304.
43. Denison, W. C. (1972). Central American Pezizales. IV. The genera *Sarcoscypha, Pithya*, and *Nanoscypha*. *Mycologia* **64**:609–623.
44. Dennis, R. W. G. (1949). A revision of the British Hyaloscyphaceae with notes on related European species. *Mycol. Pap.* **32**:1–97.
45. Dennis, R. W. G. (1950). Karsten's species of *Mollisia*. *Kew Bull.* **1950**:171–187.
46. Dennis, R. W. G. (1954). Some inoperculate Discomycetes of tropical America. *Kew Bull.* **1954**:289–348.
47. Dennis, R. W. G. (1955). Two proposed new genera of Helotiales. *Kew Bull.* **1955**:359–362.
48. Dennis, R. W. G. (1956). A revision of the British Helotiaceae in the herbarium of the Royal Botanic Gardens, Kew, with notes on related European species. *Mycol. Pap.* **62**:1–216.
49. Dennis, R. W. G. (1958). Critical notes on some Australian Helotiales and Ostropales. *Kew Bull.* **1958**:321–358.
50. Dennis, R. W. G. (1959). The genus *Zoellneria* Velenovsky. *Kew Bull.* **1958**:398–399.
51. Dennis, R. W. G. (1960). "British Cup Fungi and their Allies. An Introduction to the Ascomycetes." Ray Society, London.
52. Dennis, R. W. G. (1960). Fungi Venezuelani: III. *Kew Bull.* **14**:418–458.
53. Dennis, R. W. G. (1961). Some inoperculate Discomycetes from New Zealand. *Kew Bull.* **15**:293–320.
54. Dennis, R. W. G. (1962). A reassessment of *Belonidium* Mont. & Dur. *Persoonia* (*Leyden*) **2**: 171–191.
55. Dennis, R. W. G. (1962). New or interesting British Helotiales. *Kew Bull.* **16**:317–327.
56. Dennis, R. W. G. (1964). Remarks on the genus *Hymenoscyphus* S. F. Gray, with observations on sundry species referred by Saccardo and others to the genera *Helotium, Pezizella* or *Phialea*. *Persoonia* (*Leyden*) **3**:29–80.
57. Dennis, R. W. G. (1968). "British Ascomycetes." Cramer, Lehre.
57a. Dennis, R. W. G. (1972). *Niptera* Fr. versus *Belonopsis* Rehm. *Kew Bull.* **26**:439–443.
58. Dennis, R. W. G., and R. P. Korf. (1958). A Japanese species of *Sphagnicola* Velenovsky. *Kew Bull.* **1958**:181–183.
59. Dharne, C. G. (1965). Taxonomic investigations on the discomycetous genus *Lachnellula* Karst. *Phytopathol. Z.* **53**:101–144.
60. Dissing, H. (1966). The genus *Helvella* in Europe, with special emphasis on the species found in Norden, *Dansk Bot. Ark.* **25**(1):1–172.
61. Dissing, H. (1966). A revision of collections of the genus *Helvella* L. ex St-Amans emend. Nannf. in the Boudier herbarium. *Rev. Mycol.* **31**:189–224.
61a. Dissing, H. (1971). Personal communication.
62. Dissing, H., and M. Lange. (1967). Notes on the genus *Helvella* in North America. *Mycologia* **59**:349–360.
63. Dissing, H., and J. A. Nannfeldt. (1966). *Helvella cupuliformis* sp. nov., *H. villosa* (Hedw. ex O. Kuntze) comb. nov., *H. macropus* (Pers. ex Fr.) Karst., and their allies. *Sv. Bot. Tidskr.* **60**:325–337.

63a. Dixon, J. R. (1972). A monograph of the genera *Chlorosplenium, Chlorociboria,* and *Chlorencoelia* (Helotiales). Ph.D. Thesis, Cornell University, Ithaca, New York.
64. Dodge, C. W. (1929). The higher Plectascales. *Ann. Mycol.* **37**:145–184.
65. Drayton, F. L., and J. W. Groves. (1952). *Stromatinia narcissi,* a new, sexually dimorphic Discomycete. *Mycologia* **44**:119–140.
66. Dumont, K. P. (1971). Sclerotiniaceae II. *Lambertella. Mem. N. Y. Bot. Gard.* **22**:1–178.
66a. Dumont, K. P. (1972). Sclerotiniaceae III. The generic names *Poculum, Calycina* and *Lanzia. Mycologia* **64**:911–915.
67. Dumont, K. P., and R. P. Korf. (1970). Nomenclatural notes. VI. A new name, *Martininia,* to replace *Martinia* (Sclerotiniaceae). *Mycologia* **62**:608–609.
68. Dumont, K. P., and R. P. Korf. (1971). Sclerotiniaceae I. Generic nomenclature. *Mycologia* **63**:157–168.
69. Durand, E. J. (1902). The genus *Angelina* Fr. *J. Mycol.* **8**:108–109.
70. Durand, E. J. (1923). The genera *Midotis, Ionomidotis* and *Cordierites. Proc. Amer. Acad. Arts Sci.* **59**:1–18.
71. Eckblad, F. -E. (1963). Contributions to the Geoglossaceae of Norway. *Nytt Mag. Bot.* **10**: 137–158.
72. Eckblad, F. -E. (1968). The genera of the operculate Discomycetes. A re-evaluation of their taxonomy, phylogeny and nomenclature. *Nytt Mag. Bot.* **15**:1–191.
73. Egger, M. C. (1968). Morphologie und Biologie von *Pseudophacidium*-Arten (Ascomycetes). *Sydowia* **20**:288–328.
74. Elliott, M. E. (1962). *Streptotinia caulophylli* sp. nov. produced in culture. *Can. J. Bot.* **40**: 1197–1201.
75. Elliott, M. E. (1967). *Rutstroemia cuniculi,* a coprophilous species of the Sclerotiniaceae. *Can. J. Bot.* **45**:521–524.
76. Erb, R. W. (1972). A new species of the genus *Rhizoblepharia* from the neotropics, and a redisposition of the genus in the Pyronemataceae, Pseudombrophileae. *Phytologia* **24**:5–14.
77. Eriksson, O. (1967). On graminicolous Pyrenomycetes from Fennoscandia. 2. Phragmosporous and scolecosporous species. *Ark. Bot.* [2] **6**:381–440.
78. Faurel, L., and G. Schotter. (1965). Notes mycologiques VI. —Sur quelques champignons coprophiles d'Afrique equatoriale. *Cah. Maboké* **3**:123–133.
79. Favre, J. (1951). *Antinoa acuum* Vel. *Bull. Soc. Mycol. Fr.* **67**:205–207.
80. Fischer, E. (1938). Tuberineae. *Naturl. Pflanzenfamilien* [2] **5b** (8):1–42.
81. Funk, A. (1967). *Coccomyces heterophyllae* n. sp., a hypodermataceous fungus from the periderm of western hemlock. *Can. J. Bot.* **45**:2263–2266.
82. Gamundi, I. J. (1956). El género *Scutellinia* en la Argentina. *Contr. Cient. Univ. Buenos Aires, Bot.* **1**:69–88.
83. Gamundi, I. J. (1966). Nota sobre Pezizales bonaerenses con comentarios sobre el "status" de algunos géneros. *Rev. Mus. La Plata* [2], *Bot.* **10**:47–68.
83a. Gamundi, I. J. (1971). Las "Cyttariales" Sudamericanas. *Darwiniana* **16**:461–510.
83b. Gamundi, I. J. (1971). Algunos Discomycetes de Chile. *Bol. Soc. Argent. Bot.* **13**:125–128.
84. Gamundi, I. J., and R. W. G. Dennis. (1969). The status of *"Ascotremella"* Seaver. *Darwiniana* **15**:14–21.
85. Gilkey, H. M. (1925). Five new hypogeous fungi. *Mycologia* **17**:250–254.
86. Gilkey, H. M. (1939). Tuberales of North America. *Oreg. State Monogr., Bot.* **1**:1–63.
87. Gilkey, H. M. (1947). New or otherwise noteworthy species of Tuberales. *Mycologia* **39**: 441–452.
88. Gilkey, H. M. (1954). Tuberales. *N. Amer. Flora* [2] **1**:1–29.
89. Gilkey, H. M. (1954). Taxonomic notes on Tuberales. *Mycologia* **46**:783–793.

90. Gilkey, H. M. (1962). New species and revisions in the order Tuberales. *Mycologia* **53**:215–220.
91. Graddon, W. D. (1965). A guide to the lignicolous species of *Apostemidium*. *Trans. Brit. Mycol. Soc.* **48**:639–646.
92. Greenhalgh, G. N., and G. M. Jones. (1964). Some species of *Trochila* and an undescribed Discomycete on leaves of *Prunus laurocerasus*. *Trans. Brit. Mycol. Soc.* **47**:311–320.
93. Gremmen, J. (1949). De imperfecte vorm van *Desmazierella acicola* Lib. *Fungus (Wageningen)* **19**:32–35.
94. Gremmen, J. (1953). Some noteworthy discomycetous fungi on coniferous hosts. *Sydowia* **7**:141–145.
95. Gremmen, J. (1960). A contribution to the mycoflora of the pine forests in the Netherlands. *Nova Hedwigia* **1**:251–288.
96. Gremmen, J. (1960). Conifer inhabiting fungi, I: *Therrya* Penz. et Sacc., *Phacidium lacerum* Fr., and *Pseudophacidium ledi* (A. & S.) Karst. *Nova Hedwigia* **2**:547–553.
97. Gremmen, J. (1963). Conifer inhabiting fungi, II. *Chloroscypha* Seaver and *Fabrella tsugae* (Farl.) Kirschst. in the Netherlands. *Nova Hedwigia* **5**:21–27.
98. Gremmen, J. (1965). Three poplar-inhabiting *Drepanopeziza* species and their life-history. *Nova Hedwigia* **9**:170–176.
99. Groves, J. W. (1939). Some *Pezicula* species and their conidial stages. *Can. J. Res., Sect. C* **17**: 125–143.
100. Groves, J. W. (1940). Three *Pezicula* species occurring on *Alnus*. *Mycologia* **32**:112–123.
101. Groves, J. W. (1941). *Pezicula carnea* and *Pezicula subcarnea*. *Mycologia* **33**:510–522.
102. Groves, J. W. (1946). North American species of *Dermea*. *Mycologia* **38**:351–431.
103. Groves, J. W. (1952). The genus *Tympanis*. *Can. J. Bot.* **30**:571–651.
104. Groves, J. W. (1954). The genus *Durandiella*. *Can. J. Bot.* **32**:116–144.
105. Groves, J. W. (1965). The genus *Godronia*. *Can. J. Bot.* **43**:1195–1276.
106. Groves, J. W. (1967). The genus *Pragmopora*. *Can. J. Bot.* **45**:169–181.
107. Groves, J. W. (1969). *Crumenulopsis*, a new name to replace *Crumenula* Rehm. *Can. J. Bot.* **47**:47–51.
108. Groves, J. W. (1969). Notes on the genus *Encoeliopsis*. *Can. J. Bot.* **47**:1319–1331.
109. Groves, J. W., and C. A. Bowerman. (1955). The species of *Ciborinia* on *Populus*. *Can. J. Bot.* **33**:577–590.
110. Groves, J. W., and M. E. Elliott. (1961). Self-fertility in the Sclerotiniaceae. *Can. J. Bot.* **39**: 215–231.
111. Groves, J. W., and M. E. Elliott. (1961). A new species of *Stromatinia* on *Sanguinaria*. *Bull. Res. Counc. Isr., Sect. D* **10**:150–156.
112. Groves, J. W., and S. C. Hoare. 1953. The Helvellaceae of the Ottawa district. *Can. Field Natur.* **67**:95–102.
113. Groves, J. W., and D. E. Wells. (1956). The genus *Retinocyclus*. *Mycologia* **48**:865–871.
114. Groves, J. W., and D. E. Wilson. (1967). The nomenclatural status of *Coryne*. *Taxon* **16**:35–41.
115. Harmaja, H. (1969). A wider and more natural concept of the genus *Gyromitra* Fr. *Karstenia* **9**:9–12.
116. Harmaja, H. (1969). *Karstenella vernalis* Harmaja, a new genus and species of Discomycetes from Finland. *Karstenia* **9**:20–22.
117. Heim, R. (1962) Quelques Ascomycètes remarquables. IV.–Le *Pseudotis unicolor* (Gill.) nom. nov. et ses sosies. *Bull. Soc. Mycol. Fr.* **77**:299–315.
118. Heim, R. (1966). Quelques Ascomycètes remarquables. V.–Morilles tropicales (Pacifique sud). *Bull. Soc. Mycol. Fr.* **82**:442–449.
119. Heim, R., and M. Le Gal. (1936). Un genre nouveau néerlandais d'Ascobolacés. *Rev. Mycol.* **1**:307–313.
119a. Hennebert, G. L. (1973). *Botrytis* and *Botrytis*-like genera. *Persoonia* (*Leyden*) **7**: (in press).

120. Hennebert, G. L., and J. W. Groves. (1963). Three new species of *Botryotinia* on Ranunculaceae. *Can. J. Bot.* **41**:341–370.
121. Höhnel, F. von. (1910). Fragmente zur Mykologie. 528. *Mollisiella* Phillips und *Unguiculariopsis* Rehm. *Sitzungsber. Akad. Wiss. Wien, Math.-Naturwiss. Kl., Abt. I* **119**:618–622.
122. Höhnel, F. von (1912). Fragmente zur Mykologie. 780. *Asterocalyx* n. G. *Sitzungsber. Akad. Wiss. Wien, Math.-Naturwiss. Kl., Abt. I* **121**:402–404.
122a. Holm, L. (1971). Taxonomic notes on Ascomycetes VII. *Schizothyrioma Ptarmicae* (Desm.) von Höhnel, and its double. *Sv. Bot. Tidskr.* **65**:208–212.
123. Hughes, S. J. (1960). Microfungi V. *Conoplea* Pers. and *Exosporium* Link. *Can. J. Bot.* **38**:659–696.
124. Hütter, R. (1958). Untersuchungen über die Gattung *Pyrenopeziza* Fuck. *Phytopathol. Z.* **33**:1–54.
125. Imai, S. (1940). Second note on the Tuberales of Japan. *Proc. Imp. Acad. (Tokyo)* **16**:153–154.
126. Imai, S. (1941). Geoglossaceae Japoniae. *J. Fac. Agr., Hokkaido Imp. Univ.* **45**:155–264.
127. Imai, S. (1954). Elvellaceae Japoniae. *Sci. Rep. Yokohama Nat. Univ. Sect. 2* **3**:1–35.
128. Imai, S. (1955). Contributiones ad studia monographica Geoglossacearum. II. *Sci. Rep. Yokohama Nat. Univ., Sect. 2* **4**:1–11.
129. Kankainen, E. (1969). On the structure, ecology and distribution of the species of *Mitrula* s. lat. (Ascomycetes, Geoglossaceae). *Karstenia* **9**:23–34.
130. Kanouse, B. B. (1948). The genus *Plectania* and its segregates in North America. *Mycologia* **40**:482–497.
131. Kanouse, B. B. (1950). Studies in the genus *Otidea*. *Mycologia* **41**:660–677.
132. Kanouse, B. B. (1958). Some species of the genus *Trichophaea*. *Mycologia* **50**:121–140.
133. Kanouse, B. B., and A. H. Smith. (1940). Two new genera of Discomycetes from the Olympic National Forest. *Mycologia* **32**:756–759.
134. Kavina, K. (1939). *Evulla carpatica*, un champignon nouveau et admirable. *Stud. Bot. Cech.* **2**:23–27.
135. Kempton, P. E., and V. L. Wells. (1970). Studies on the fleshy fungi of Alaska. IV. A preliminary account of the genus *Helvella*. *Mycologia* **62**:940–959.
136. Kimbrough, J. W. (1969). North American species of *Thecotheus* (Pezizeae, Pezizaceae). *Mycologia* **61**:99–114.
137. Kimbrough, J. W. (1970). Current trends in the classification of Discomycetes. *Bot. Rev.* **36**:91–161.
138. Kimbrough, J. W. (1970). Segregates of *Ascophanus*, *Coprotus* vs. *Leporina* (Thelebolaceae, Pezizales). *Taxon* **19**:779–781.
139. Kimbrough, J. W. (1971). A new species of *Iodophanus* (Pezizaceae) from Ceylon. *Bull. Torrey Bot. Club* **97**:377–379.
140. Kimbrough, J. W., and R. P. Korf. (1967). A synopsis of the genera and species of the tribe Theleboleae (= Pseudoascoboleae). *Amer. J. Bot.* **54**:9–23.
141. Kimbrough, J. W., E. R. Luck-Allen, and R. F. Cain. (1969). *Iodophanus*, the Pezizeae segregate of *Ascophanus* (Pezizales). *Amer. J. Bot.* **56**: 1187–1202.
142. Kimbrough, J. W., E. R. Luck-Allen, and R. F. Cain. (1972). North American species of *Coprotus* (Thelebolaceae, Pezizales). *Can. J. Bot.* **50**: 957–971.
143. Kirschstein, W. (1938). Über neue, seltene und kritische Ascomyceten und Fungi imperfecti. *Ann. Mycol.* **36**:367–400.
144. Kirschstein, W. (1939). Über neue, seltene und kritische Ascomyceten und Fungi imperfecti. II. *Ann. Mycol.* **37**:88–140.
145. Kobayashi, T. (1965). Taxonomic notes on *Chloroscypheae* causing needle blight of Japanese conifers. *Bull. Govt. Forest Exp. Sta. (Tokyo)* **176**:55–74.
146. Kobayasi, Y. (1937). *Phymatomyces*, a new genus of the Tuberaceae. *J. Jap. Bot.* **12**:912–914.

147. Kobayasi, Y. (1960). *Ascosparassis*, a new genus of Discomycetes. *Bull. Nat. Sci. Mus., Tokyo* **5**:44–46.
148. Kobayasi, Y. (1963). On a new genus *Protogenea* of the Tuberales. *Trans. Mycol. Soc. Jap.* **4**:119–120.
149. Kobayasi, Y. (1966). On the genus *Cyttaria*. *Trans. Mycol. Soc. Jap.* **7**:118–132.
150. Korf, R. P. (1952). A monograph of the Arachnopezizeae. *Lloydia* **14**:129–180.
151. Korf, R. P. (1956). *Daleomyces, Durandiomyces*, and other sparassoid forms of operculate Discomycetes. *Mycologia* **48**:711–718.
152. Korf, R. P. (1957). Two bulgarioid genera: *Galiella* and *Plectania*. *Mycologia* **49**:107–111.
153. Korf, R. P. (1958). Japanese Discomycete notes I–VIII. *Sci. Rep. Yokohama Nat. Univ. Sect. 2* **7**:7–35.
154. Korf, R. P. (1959). Japanese Discomycete notes IX–XVI. *Bull. Nat. Sci. Mus., Tokyo* **4**:389–400.
155. Korf, R. P. (1960). *Jafnea*, a new genus of the Pezizaceae. *Nagaoa* **7**:1–8.
156. Korf, R. P. (1962). A synopsis of the Hemiphacidiaceae, a family of the Helotiales (Discomycetes) causing needle blights of conifers. *Mycologia* **54**:12–33.
157. Korf, R. P. (1963). Discomycete flora of Asia, precursor II: A revision of the genera *Acervus* and *Ascosparassis* and their new positions in the Pezizales. *Lloydia* **26**:21–26.
158. Korf, R. P. (1971). Some new Discomycete names. *Phytologia* **21**:201–207.
158a. Korf, R. P. (1972). Synoptic key to the genera of the Pezizales. *Mycologia* **64**:937–994.
159. Korf, R. P., and G. S. Abawi. (1971). On *Holwaya, Crinula, Claussenomyces* and *Corynella*. *Can. J. Bot.* **49**:1879–1883.
160. Korf, R. P., and K. P. Dumont. (1972). *Whetzelinia*, a new generic name for *Sclerotinia sclerotiorum* and *S. tuberosa*. *Mycologia* **64**:248–251.
160a. Korf, R. P., and R. W. Erb. (1972). The genus *Trichophaeopsis*. *Phytologia* **24**:15–19.
161. Korf, R. P., and K. S. Waraitch. (1971). *Thindia*, a new genus of the Sarcoscyphineae. *Mycologia* **63**:98–103.
162. Kubička, J. (1960). *Svrčekia* n. gen. – nový rod terčoplodých hub. *Svrčekia* genus novum Discomycetum. *Česká Myk.* **14**:214–218.
163. Le Gal, M. (1941). Les *Aleuria* et les *Galactinia*. *Rev. Mycol., Suppl.* **6**:56–82.
164. Le Gal, M. (1947). Recherches sur les ornementations sporales des Discomycètes operculés. *Ann. Sci. Natur.; Bot. Biol. Veg.* [11] **8**:73–297.
165. Le Gal, M. (1953). Les Discomycètes de l'herbier Crouan. *Rev. Mycol.* **18**:73–132.
166. Le Gal, M. (1953). Les Discomycètes de Madagascar. *Prodr. Flore Mycol. Madagascar* **4**:1–465.
167. Le Gal, M. (1957). Le genre *Leucoscypha* Boud. *Bull. Jard. Bot. Brux.* **27**:719–728.
168. Le Gal, M. (1958). Le genre *Melastiza* Boudier. *Bull. Soc. Mycol. Fr.* **74**:149–154.
169. Le Gal, M. (1958). Discomycètes du Maroc. I. Un *Urnula* nouveau: *Urnula megalocrater* Malençon et Le Gal sp. nov. Etude de l'espèce, suivie d'une révision des charactères des genres *Urnula* Fr. et *Sarcosoma* Casp. *Bull. Soc. Mycol. Fr.* **74**:155–177.
170. Le Gal, M. (1959). Discomycètes de Congo Belge d'après les récoltes de Madame Goossens-Fontana. *Bull. Jard. Bot. Brux.* **29**:73–132.
171. Le Gal, M. (1966). Contribution à la connaissance du genre *Scutellinia* (Cooke) Lamb. emend. Le Gal. (1[re] étude). *Bull. Soc. Mycol. Fr.* **82**:301–333.
172. Le Gal, M. (1968). Contribution à la connaissance du genre *Scutellinia* (Cooke) Lamb. emend. Le Gal (2[e] étude). *Bull. Soc. Mycol. Fr.* **84**:375–380.
172a. Le Gal, M. (1972). Contribution à la connaissance du genre *Scutellinia* (Cooke) Lamb. emend. Le Gal (3[e] étude). *Bull. Soc. Mycol. Fr.* **87**:433–440.
173. Lloyd, C. G. (1921). *Phyllomyces multiplex* from Otto A. Reinking, Philippines (Fig. 1975). *Mycological Writings of C. G. Lloyd* **6**:1057.
174. Maas Geesteranus, R. A. (1964). On some white-spored Geoglossaceae. *Persoonia* (*Leyden*) **3**:81–96.

175. Maas Geesteranus, R. A. (1966). On *Helvella platypus. Proc. Kon. Ned. Akad. Wetensch.*, Ser. C **69**:191–203.
176. Maas Geesteranus, R. A. (1967). De fungi van Nederland. 2a. Pezizales – deel I. *Kon. Ned. Natuurh. Ver., Wet. Mededel.* **69**:1–72.
177. Maas Geesteranus, R. A. (1967). Studies in cup-fungi–I. *Persoonia (Leyden)* **4**:417–425.
177a. Maas Geesteranus, R. A. (1972). *Spathularia* and *Spathulariopsis. Proc. Kon. Ned. Akad. Wetensch.*, Ser. C **75**:243–255.
178. Mains, E. B. (1954). North American species of *Geoglossum* and *Trichoglossum. Mycologia* **46**:566–631.
179. Mains, E. B. (1955). North American hyaline-spored species of the Geoglosseae. *Mycologia* **47**:846–877.
180. Mains, E. B. (1956). North American species of the Geoglossaceae. Tribe Cudonieae. *Mycologia* **48**:694–710.
181. Marchionatto, J. B. (1940). Las especies de *Cyttaria* y *Cyttariella* en la Argentina. *Darwiniana* **4**:9–32.
182. Mattirolo, O. (1935). Catalogo ragionato dei funghi ipogei raccolti nel Canton Ticino e nelle provincie Italiane confinanti. *Beitr. Kryptogamenfl. Schweiz* **8**(2):1–53.
183. McKnight, K. H. (1969). A note on *Discina. Mycologia* **61**:614–630.
184. McLennan, E. I. (1961). Australian Tuberales. *Proc. Roy. Soc. Victoria* [N.S.] **74**:111–117.
185. Moore, E. J., and R. P. Korf. (1963). The genus *Pyronema. Bull. Torrey Bot. Club* **90**:33–42.
186. Moravec, J. (1968). Příspěvek k poznáni operkulátnich diskomycetů rodu *Cheilymenia* Boud. A study concerning a better recognition of operculate Discomycetes of the genus *Cheilymenia* Boud. *Česká Myk.* **22**:32–41.
187. Moravec, J. (1970). Operkulátní diskomycety čeledi Ascobolaceae Sacc. z okresu Mladá Boleslav v Čechách. Operculate discomycetes of the family Ascobolaceae Sacc. from the Mladá Boleslav district in Central Bohemia. *Česká Myk.* **24**:134–145.
187a. Moravec, J. (1972). Operculate Discomycetes of the genera *Aleuria* Fuck. and *Melastiza* Boud. from the district of Mladá Boleslav (Bohemia). Operkulátní diskomycety rodů *Aleuria* Fuck. a *Melastiza* Boud. z okresu Mladá Boleslav v Čechách. *Česká Myk.* **26**:74–81.
188. Morelet, M. (1969). *Groveseiella* gen. nov. *Bull. Soc. Sci. Natur. Archéol. Toulon. Var* **185**:8.
189. Müller, E. (1957). Die Gattung *Europolella* v. Höhn. *Sydowia* **11**:130–132.
190. Müller, E. (1963). Über zwei neue *Pseudophacidium* Arten. *Phytopathol. Z.* **48**:208–215.
191. Müller, E. (1966). *Actinoscypha* Karsten, eine verkannte Discomyceten-Gattung. *Ber. Schweiz. Bot. Ges.* **76**:230–238.
192. Müller, E., and G. Défago. (1967). *Beloniella* (Sacc.) Boud. und *Dibeloniella* Nannf., zwei wenig bekannte Discomycetengattungen. *Sydowia* **20**:157–168.
193. Müller, E., and R. Hütter. (1963). Eine neue Discomycetengattung aus den Alpen, *Neotapesia* nov. gen. *Ber. Schweiz. Bot. Ges.* **73**:325–331.
194. Nannfeldt, J. A. (1929). *Dübenia* Fr., eine verschollene Discomycetengattung. *Sv. Bot. Tidskr.* **23**:316–322.
195. Nannfeldt, J. A. (1932). Studien über die Morphologie und Systematik der nicht-lichenisierten inoperculaten Discomyceten. *Nova Acta Regiae Soc. Sci. Upsal.* [4] **8**(2):1–368.
196. Nannfeldt, J. A. (1938). Contributions to the mycoflora of Sweden. 5. On *Peziza catinus* Holmskj. ex Fr. and *P. radiculata* Sow. ex Fr. with a discussion of the genera *Pustularia* Fuck. emend. Boud. and *Sowerbyella* Nannf. n. gen. *Sv. Bot. Tidsk.* **32**:108–120.
197. Nannfeldt, J. A. (1942). The Geoglossaceae of Sweden (with regard also to the surrounding countries). *Ark. Bot.* **30A**(4):1–67.
198. Nannfeldt, J. A. (1966). On *Otidea caligata*, *O. indivisa* and *O. platyspora* (Discomycetes Operculatae). *Ann. Bot. Fenn.* **3**:309–318.
199. Oberwinkler, F., F. Casagrande, and E. Müller. (1967). Über *Ascocorticium anomalum* (Ell. et Harkn.) Earle. *Nova Hedwigia* **14**:283–289.

200. Obrist, W. (1961). The genus *Ascodesmis*. *Can. J. Bot.* **39**:943–953.
200a. Ogimi, C., and R. P. Korf. (1972). Discomycete flora of Asia, precursor IV: A new species of *Bifusella* (Rhytismataceae, Hypodermateae) on *Cunninghamia* in Okinawa. *Phytologia* **23**:155–162.
201. Otani, Y. (1967). Notes on some cup fungi of the Hyaloscyphaceae collected in Hokkaido, Japan. *Trans. Mycol. Soc. Jap.* **8**:33–42.
202. Otani, Y. (1969). Some species of the genus *Otidea* collected in Japan. *Trans. Mycol. Soc. Jap.* **9**:101–108.
203. Otani, Y. and S. Kanzawa. (1970). Notes on coprophilous Discomycetes in Japan II. *Trans. Mycol. Soc. Jap.* **11**:43–48.
204. Ou, S. H. (1936). Additional fungi from China IV. *Sinensia* **7**:668–683.
205. Ouellette, G. B., and L. P. Magasi. (1966). *Lophomerum*, a new genus of Hypodermataceae. *Mycologia* **58**:275–280.
206. Paden, J. W., and E. E. Tylutki. (1969). Idaho Discomycetes. I. A new genus of the Sarcoscyphaceae. *Mycologia* **60**:1160–1168.
207. Paden, J. W., and E. E. Tylutki. (1969). Idaho Discomycetes. II. *Mycologia* **61**:683–693.
208. Palm, B. T. (1932). On *Cyttaria* Berk. and *Cyttariella* n. gen. *Ann. Mycol.* **30**:405–420.
208a. Palmer, J. T. (1969). *Myriosclerotinia juncifida* (Nyl.) comb. nov. A little-known parasite of *Juncus*. Investigations into the Sclerotiniaceae. IV. *Friesia* **9**:193–201.
209. Pant, D. C., and V. P. Tewari. (1970). *Korfiella*, a new genus of Sarcoscyphaceae. *Trans. Brit. Mycol. Soc.* **54**: 492–495.
210. Pant, D. C., and V. P. Tewari. (1971). Observations on two species of the genus *Pustulina*. *Mycologia* **62**:1187–1194.
211. Pantidou, M. E., and G. D. Darker. (1963). The species of *Didymascella* on *Juniperus*. *Mycologia* **55**:415–420.
212. Pantidou, M. E., and R. P. Korf. (1954). A revision of the genus *Keithia*. *Mycologia* **46**:386–388.
213. Parker, A. K., and J. Reid. (1969). The genus *Rhabdocline* Syd. *Can. J. Bot.* **47**:1533–1545.
214. Petrak, F. (1947). *Nannfeldtia* n. gen., eine neue Gattung der Diskomyzeten. *Sydowia* **1**:18–20.
215. Petrak, F. (1947). Über die Gattungen *Naemacyclus* Fuck. und *Lasiostictis* Sacc. *Sydowia* **1**:89–93.
216. Petrak, F. (1948). Eine neue Art der Gattung *Davincia* aus Ekuador. *Sydowia* **2**:43–45.
217. Petrak, F. (1950). *Neophacidium* n. gen., eine neue Diskomyzetengattung aus Ekuador. *Sydowia* **4**:333–336.
218. Petrak, F. (1950). *Parencoelia* n. gen., eine neue Gattung der Encoelioideen. *Sydowia* **4**:349–356.
219. Petrak, F. (1950). *Biostictis* n. gen., eine neue Diskomyzetengattung aus Ekuador. *Sydowia* **4**:357–360.
220. Petrak, F. (1951). *Diehlia* n. gen., eine neue Diskomyzetengattung mit gefärbten Sporen. *Sydowia* **5**:311–314.
221. Petrak, F. (1951). *Cashiella* n. gen., eine neue Gattung der Diskomyzeten. *Sydowia* **5**:371–374.
222. Petrak, F. (1963). Über die Gattung *Gremmenia* Korf. *Sydowia* **16**:350–352.
223. Peyronel, B. (1953). La forma ascofora di *Valdensia heterodoxa* Peyronel, tipo di un nuovo genere di Sclerotiniacee: *Valdensinia*. *Nuovo Gi. Bot. Ital.* ˙[N.S.] **59**:181–185.
224. Pfister, D. H. (1971). A monograph of the genera *Psilopezia* and *Pachyella* (Pezizales). Ph.D. Thesis, Cornell University, Ithaca, New York.
224a. Pfister, D. H. (1972). Notes on Caribbean Discomycetes I. Cytological evidence for the exclusion of *Phaedropezia* from the Sarcoscyphaceae. *Caribbean J. Sci.* **12**:39–40.
224b. Pfister, D. H. (1972). The psilopezioid fungi. II. *Thecotheus rivicola* comb. nov. and other Iodophaneae (Pezizales) occurring on water-soaked wood. *Bull. Torrey Bot. Club* **99**: (in press).
224c. Pfister, D. H. (1973). The psilopezioid fungi. I. History, nomenclature, and delimitation of the psilopezioid genera. *Mycologia* **65**:(in press).

224d. Pfister, D. H. (1973). The psilopezioid fungi. III. The genus *Psilopezia* (Pezizales). *Amer. J. Bot.* **60**:(in press).

224e. Powell, P. E., Jr. (1973). The genera *Duplicaria* and *Crandallia* and the taxonomic value of the bifusiform ascospore in the Rhytismataceae. M. Sc. Thesis, Cornell University, Ithaca, New York.

225. Raitviir, A. (1965). Taxonomical notes on the genus *Gyromitra*. *Eesti NSV Tead. Akad. Toim., Ser. Biol.* **14**:320–324.

226. Raitviir, A. (1965). Sarcoscyphaceae in the Far East. *Eesti NSV Tead. Akad. Toim., Ser. Biol.* **14**:529–535.

227. Raitviir, A. (1969). Discomycetes of Middle Asia. I. Descriptions of some new Helotiales. *Eesti NSV Tead. Akad. Toim., Ser. Biol.* **18**:66–69.

227a. Raitviir, A. (1970). Once more on *Neogyromitra caroliniana*. *Trans. Tartu State Univ.* (Pap. Bot. 9) **268**:364–373.

227b. Raitviir, A. (1970). Synopsis of the Hyaloscyphaceae. *Scripta Mycol.* **1**:1–115.

228. Ramamurthi, C. S., R. P. Korf, and L. R. Batra. (1957). A revision of the North American species of *Chlorociboria*. *Mycologia* **49**: 854–863.

229. Rawlings, G. B. (1956). Australasian Cyttariaceae. *Trans. Roy. Soc. N.Z.* **84**:19–28.

230. Rehm, H. (1887–1896). Ascomyceten: Hysteriaceen und Discomyceten. *In* "Dr. L. Rabenhorst's Kryptogamen-Flora von Deutschland, Oesterreich und der Schweiz" (G. Winter and H. Rehm, eds.), 2nd ed., Vol. 1, Part 3, pp. 1–1275. Kummer, Leipzig.

231. Reid. J., and R. F. Cain. (1961). The genus *Therrya*. *Can. J. Bot.* **39**:1117–1129.

232. Reid, J., and R. F. Cain. (1962). Studies on the organisms associated with "snow-blight" of conifers in North America. I. A new genus of the Helotiales. *Mycologia* **54**:194–200.

233. Reid, J., and R. F. Cain. (1962). Studies on the organisms associated with "snow-blight" of conifers in North America. II. Some species of the genera *Phacidium*, *Lophophacidium*, *Sarcotrochila*, and *Hemiphacidium*. *Mycologia* **54**:481–497.

234. Reid, J., and R. F. Cain. (1963). A new genus of the Hemiphacidiaceae. *Mycologia* **55**:781–785.

235. Reid, J., and A. Funk. (1966). The genus *Atropellis*, and a new genus of the Helotiales associated with branch cankers of western hemlock. *Mycologia* **58**:417–439.

236. Reid, J., and K. A. Pirozynski. (1966). Notes on some interesting North American fungi. *Can. J. Bot.* **44**:645–653.

237. Reid, J., and K. A. Pirozynski. (1968). Critical notes on genera of the Hemiphacidiaceae I. *Gremmenia*. *Mycologia* **60**:526–531.

238. Rick, J. (1904). Über einige neue und kritische Pilze Süd-Amerikas. *Ann. Mycol.* **2**:242–247.

239. Rifai, M. A. (1968). The Australasian Pezizales in the Herbarium of the Royal Botanic Gardens Kew. *Verh. Kon. Ned. Akad. Wetensch., Afd. Natuurk., II* **57**(3):1–295.

240. Rimpau, R. H. (1962). Untersuchungen über die Gattung *Drepanopeziza* (Kleb.) v. Höhn. *Phytopathol. Z.* **43**:257–306.

241. Rogers, J. K., K. P. Dumont, and R. P. Korf. (1971). Nomenclatural notes. VIII. *Stromatinia*, an available generic name (not to be replaced by *Tarzetta*, the correct name for *Pustulina*). *Mycologia* **63**:1084–1086.

242. Rouppert, C. (1910). Revision du genre *Sphaerosoma*. *Bull. Int. Acad. Pol. Sci. Lett., Cl. Sci. Math. Natur.* **1909**(2):75–94.

243. Saho, H., and I. Takahashi. (1970). *Waltonia* gen. nov. (Helotiales, Dermateaceae) found on *Pinus* spp. *Trans. Mycol. Soc. Jap.* **11**:3–6.

244. Sánchez, A. (1967). The sections *Apostemium* and *Microstemium* of the genus *Vibrissea* (Fungi). *J. Agr. Univ. P.R.* **51**:79–93.

245. Sánchez, A., and R. P. Korf. (1966). The genus *Vibrissea*, and the generic names *Leptosporium*, *Apostemium*, *Apostemidium*, *Gorgoniceps* and *Ophiogloea*. *Mycologia* **58**:722–737.

246. Santesson, R. (1945). *Cyttaria*, a genus of inoperculate Discomycetes. *Sv. Bot. Tidskr.* **39**:319–345.

247. Santesson, R. (1956). *Capillipes cavorum* g. nov., sp. nov., a new terricolous inoperculate Discomycete from Swedish Lappland. *Friesia* **5**:390–395.
248. Schläpfer-Bernard, E. (1969). Beitrag zur Kenntnis der Discomycetengattungen *Godronia*, *Ascocalyx*, *Neogodronia* und *Encoeliopsis*. *Sydowia* **22**:1–56.
249. Schüepp, H. (1959). Untersuchungen über Pseudopezizoideae sensu Nannfeldt. *Phytopathol. Z.* **36**:213–269.
250. Seaver, F. J. (1928). "The North American Cup-fungi (Operculates)." Seaver, New York.
251. Seaver, F. J. (1942). "The North American Cup-fungi (Operculates)," Supplemented ed. Seaver, New York; reprinted 1961, Hafner, New York.
252. Seaver, F. J. (1942). Photographs and descriptions of cup-fungi–XXXVI. A new species and genus. *Mycologia* **34**:298–301.
253. Seaver, F. J. (1947). Photographs and descriptions of cup-fungi–XLIII. *Seaverinia*. *Mycologia* **39**:113–119.
254. Seaver, F. J. (1951). "The North American Cup-fungi (Inoperculates)." Seaver, New York; reissued as spurious "Supplemented edition," 1961, Hafner, New York.
255. Seeler, E. V., Jr. (1943). Several fungicolous fungi. *Farlowia* **1**:119–133.
256. Setchell, W. A. (1910). The genus *Sphaerosoma*. *Univ. Calif., Berkeley, Publ. Bot.* **4**:107–120.
257. Shear, C. L., and B. O. Dodge. (1921). The life history and identity of "*Patinella fragariae*," "*Leptothyrium macrothecium*," and "*Peziza oenotherae*." *Mycologia* **13**:135–170.
258. Smerlis, E. (1962). Taxonomy and morphology of *Potebniamyces balsamicola* sp. nov. associated with a twig and branch blight of balsam fir in Quebec. *Can. J. Bot.* **40**:351–359.
259. Smerlis, E. (1966). Notes on the nomenclature of *Nothophacidium phyllophilum* comb. nov., its pathogenicity and relationship to *Pezizella minuta*. *Can. J. Bot.* **44**:563–565.
260. Spegazzini, C. (1898). Fungi Argentini novi v. critici. *An. Mus. Nac. Buenos Aires* **6**:81–365.
261. Spegazzini, C. (1922). Cryptogamae nonnullae Fuegianae. *An. Soc. Cient. Argent.* **94**:59–85.
262. Spevak, M. B., and R. P. Korf. (1966). On *Ciboriopsis simulata* and the genus *Ciboriopsis*. *Lloydia* **29**:130–135.
263. Starbäck, K. (1895). Discomyceten-Studien, *Kgl. Sv. Vetenskaps–Akad., Handl., Bihang* [3] **21**(5): 1–42.
264. Stowell, E. A., and M. P. Backus. (1966). Morphology and cytology of *Diplocarpon maculatum* on *Crataegus*. I. The *Entomosporium* stage. *Mycologia* **58**:949–960.
264a. Sutton, B. C., and K. A. Pirozynski. (1963). Notes on British microfungi. I. *Trans. Brit. Mycol. Soc.* **46**:505–522.
265. Svrček, M. (1948). České druhy podčeledi Lachneoideae (čel. Pezizaceae). Bohemian species of Pezizaceae subf. Lachneoideae. *Sb. Narod. Mus. Praze* **4B**(6):1–95.
266. Svrček, M. (1954). Revise Velenovského druhů rodu *Orbilia* (Discomycetes). Revisio critica J. Velenovskýi specierum generis *Orbilia*. *Sb. Narod. Mus. Praze* **10B**(1):1–23.
267. Svrček, M. (1957). *Kubičkia tatrensis* gen. n. et sp. n., a poznámky o rodech *Coryne* a *Ombrophila*. (*Kubičkia tatrensis* Discomycetum genus et species nova čechoslovaca atque adnotationes ad genera *Coryne* et *Ombrophila*.) *Česká Myk.* **11**:32–41.
268. Svrček, M. (1957). *Piceomphale bulgarioides* (Rabenh. in Kalchbr.) Svrček comb. n. a poznámky k problematice diskomycetu *Ombrophila strobilina* v. pojetí Rehmově. Taxonomical and nomenclatorial notes on Discomycete *Ombrophila strobilina* sensu Rehm. *Česká Myk.* **11**:235–240.
269. Svrček, M. (1961). *Sclerotinia dennisii* sp. n. a přehled druhů podrodu *Myriosclerotinia*. *Sclerotinia dennisii* sp. n. cum conspectu specierum subgeneris *Myriosclerotinia*. *Česká Myk.* **15**:35–41.
270. Svrček, M. (1962). O rodu *Discocistella* gen. nov. a některých jeho druzích. De genere *Discocistella* gen. nov. familiae Hyaloscyphacearum. *Česká Myk.* **16**:9–13.

271. Svrček, M. (1969). Nové rody operkulátních diskomycetů (Pezizales). Neue Gattungen operculater Discomyceten. *Česká Myk.* **23**:83–96.
272. Svrček, M. (1970). Über einige Arten der Diskomyzetengattung *Peziza* [Dill.] L. ex St-Amans. O některých druzích terčoplodých hub rodu *Peziza* [Dill.] L. ex St-Amans. *Česká Myk.* **24**:57–77.
273. Svrček, M., and J. Kubička. (1967). *Poloniodiscus fischeri*, nový rod a druh diskomycetů. *Poloniodiscus fischeri*, gen. nov. et sp. nov. Discomycetum. *Česká Myk.* **21**:151–155.
274. Svrček. M., and J. Kubička. (1968). Beitrag zur Kenntnis der operculaten Discomyceten des Gebirges Jeseníky (Hochgesenke) in der Tschechoslowakei. Příspěvek k poznání operkulátních discomycetů Jeseníků. *Česká Myk.* **22**:180–185.
275. Tehon. L. R. (1935). A monographic rearrangement of *Lophodermium*. *Ill. Biol. Monogr.* **13**(4):1–151.
276. Terrier, C.-A. (1942). Essai sur la systématique des Phacidiaceae (Fr.) sensu Nannfeldt (1932). *Beitr. Kryptogamenfl. Schweiz* **9**(2):1–99.
277. Thaxter, R. (1905). Contributions from the Cryptogamic Laboratory of Harvard University. LX. A new American species of *Wynnea*. *Bot. Gaz.* **39**:241–247.
278. Thaxter, R. (1922). Note on two remarkable Ascomycetes. *Proc. Amer. Acad. Arts Sci.* **57**:425–436.
278a. Trappe, J. M. (1971). A synopsis of the Carbomycetaceae and Terfeziaceae (Tuberales). *Trans. Brit. Mycol. Soc.* **57**:85–92.
279. Tubaki, K. (1966). An undescribed species of *Hymenoscyphus*, a perfect stage of *Varicosporium*. *Trans. Brit. Mycol. Soc.* **49**:345–349.
280. Velenovský, J. (1934). "Monographia Discomycetum Bohemiae," Vols. I and II. Velenovský, Prague.
281. Vloten, H. van, and J. Gremmen. (1953). Studies in the Discomycete genera *Crumenula* De Not. and *Cenangium* Fr. *Acta Bot. Neer.* **2**:226–241.
282. Wakefield, E. M. (1934). Contribution to the flora of tropical America: XXI. Fungi collected in British Guiana chiefly by the Oxford University Exploration, 1929. *Kew Bull.* **1934**:238–258.
282a. Waterman, A. M., and E. K. Cash. (1950). Leaf blotch of poplar caused by a new species of *Septotinia*. *Mycologia* **42**:374–384.
283. Webster, J. (1956). Conidia of *Acrospermum compressum* and *A. graminum*. *Trans. Brit. Mycol. Soc.* **39**:361–366.
284. Weiss, F. (1940). *Ovulinia*, a new generic segregate from *Sclerotinia*. *Phytopathology* **30**:236–244.
285. Whetzel, H. H. (1944). A new genus of the Sclerotiniaceae. *Farlowia* **1**:483–488.
286. Whetzel, H. H. (1945). A synopsis of the genera and species of the Sclerotiniaceae, a family of stromatic inoperculate Discomycetes. *Mycologia* **37**:648–714.
287. Whetzel, H. H. (1946). The cypericolous and juncicolous species of *Sclerotinia*. *Farlowia* **2**:385–437.
288. Whetzel, H. H., and W. L. White. (1940). *Mollisia tetrica, Peziza sejournei*, and the genera *Phaeociboria* and *Pycnopeziza*. *Mycologia* **32**:609–620.
289. Whetzel, H. H., and F. A. Wolf. (1945). The cup fungus, *Ciboria carunculoides*, pathogenic on mulberry fruits. *Mycologia* **37**:476–491.
290. White, W. L. (1941). A monograph of the genus *Rutstroemia* (Discomycetes). *Lloydia* **4**:153–240.
291. White, W. L., and H. H. Whetzel. (1938). Pleomorphic life cycles in a new genus of the Helotiaceae. *Mycologia* **30**:187–203.
292. Whitney, H. S., and J. R. Parmeter, Jr. (1964). The perfect stage of *Rhizoctonia hiemalis*. *Mycologia* **56**:114–118.

293. Wicklow, D. T., and D. Malloch. (1971). Studies in the genus *Thelebolus*: Temperature optima for growth and ascocarp development. *Mycologia* **63**:118–131.
294. Wilson, M., M. Noble, and E. Gray. (1954). *Gloeotinia*—a new genus of the Sclerotiniaceae. *Trans. Brit. Mycol. Soc.* **37**:29–32.
295. Zukal, H. (1887). Ueber einige neue Ascomyceten. *Verh. Zool.-Bot. Ges. Wien, Abh.* **38**:39–45.
296. Zukal, H. (1891). Halbflechten. *Flora* **74**:92–107.

# Eumycota
# Deuteromycotina
# Hyphomycetes

# CHAPTER 10

# Hyphomycetes

W. BRYCE KENDRICK

*Department of Biology*
*University of Waterloo*
*Waterloo, Ontario, Canada*

and

J. W. CARMICHAEL

*Mold Herbarium and Culture Collection*
*University of Alberta*
*Edmonton, Alberta, Canada*

## I. INTRODUCTION

The class Hyphomycetes (as part of the Deuteromycotina) is not a part of the main taxonomic classification of the fungi. It is part of an additional, special-purpose cross-classification in which the conidial states of Ascomycotina, Basidiomycotina, and sometimes certain Zygomycotina are grouped together. It also includes conidial states whose perfect state is unknown or lacking. Conidial states are grouped into form-genera for convenience in identification and nomenclature. The species included in a form-genus are related to each other by the form of their conidia and conidiogenous apparatus, but not necessarily by phylogeny. To quote E. W. Mason (1937, p. 75): "In an attempted cross-classification of fungi by their conidial forms, however, phylogeny is out. The loss to applied mycology appears to me to be negligible." The form-genera in the Hyphomycetes are not taxa of the same kind as the genera in the main taxonomic classification. They are demes in the sense of Heslop-Harrison (1962). Failure to appreciate this distinction has led to much confusion regarding nomenclature and generic limits.

### A. *Cross-Reference Names*

It is useful to be able to identify a fungus on the basis of whatever type of spores it is producing when we find or grow it in culture. For this purpose it is necessary to cross-classify some species in one or more form-genera of the

Deuteromycotina. However, it is a disadvantage to have more than one Linnaean binomial for a single species. The purpose of the taxonomic or Linnaean names is to provide a set of unique, all-inclusive and mutually exclusive indexing terms for the storage and retrieval of information about taxa. The cross-indexing terms for additional cross-classifications should not be additional Linnaean names; they should be cross-reference names. Mason (1937) almost proposed such a system, but not quite. Hughes (1953) discussed the problem, and both Hughes (1958) and Ellis (1958) began using a cross-reference naming system of the format "(form-genus name) state of (Linnaean binomial)." However, neither of them explicitly defined the system they were using. This was done by Carmichael (1962) who pointed out that it did not conflict with the International Code of Botanical Nomenclature. The use of cross-reference names instead of additional binomials for the same species is gradually gaining acceptance. In an extended discussion of nomenclatural problems, Hennebert (1971) credits L. R. Tulasne with using the same system in 1851.

We have followed Hughes (1958, 1959) in using 1801 as the starting point for nomenclature of Hyphomycetes.

## B. Arrangement of Form-Genera

Since form-genera are not intended to provide mutually exclusive categories for species, we have abandoned the traditional hierarchical division into orders and families. In addition, we abandoned the emphasis on the aggregation of conidiophores into synnemata, sporodochia, etc. (see also Simmons, 1966). Instead, we have used four independent character sets, all pertaining to the conidia and their ontogeny, to characterize the genera and to order them for identification. The four character sets are (1) Saccardoan spore group; (2) general arrangement of conidia; (3) color of conidia; and (4) type of conidiogenous cell. These character sets are explained further in Sections II,C and III.

## C. Generic Limits and Criteria

Form-genera in the Hyphomycetes are not based on any single concept or uniform set of criteria. Rather, they have been established by various authors, each using his own experience and judgment in an undefined and relatively arbitrary way. This was inevitable when the majority of the Hyphomycetes remained to be described. As a result, the value of the character gap (taxonomic distance) between genera varies greatly. Here, conidium shape or septation segregates genera; there, widely differing shapes and degrees of septation are brought together because of shared ontogenetic mechanisms or other apparently overriding similarity.

E. W. Mason, one of the fathers of modern Hyphomycetology, is reputed to have said, "God made the species, man made the genera," and in our opinion, man has made too many of them. Our approach therefore has been to eliminate names wherever possible (see Section II, A), but we usually deferred to the opinions of others when they were based on careful comparative studies. For example, the recently recognized form-genera *Beltraniella*, *Ellisiopsis* and *Beltraniopsis* are differentiated from the classical *Beltrania* by characters which could easily be regarded as valid only at the species level. Their setae, conidiophores, separating cells and conidia are so similar that they should probably be treated as congeneric in any practical, working classification. Indeed, Pirozynski (1963) has already begun the job of condensation by reducing *Ellisiopsis* to synonymize with *Beltraniella*. These genera, as two of their names imply, were not created in ignorance of *Beltrania*. The case of *Wiesneriomyces* and its synonyms (as we suggest them) is different. Several generic names were apparently published in ignorance of the original one—a problem caused by bad communications among mycologists. It is tempting to suggest that the description of new taxa, at present regulated only by individual taxonomic acumen and conscience, should be monitored by an international screening committee and central registry.

In any event, the time is now past due for some international agreement on a rational basis for the future development of the form-genus system of cross-classification. Since the purpose of form-genera is to facilitate identification, rather than to describe evolution, it might be possible to agree on some of the characters that should be used to delimit form-genera, and then to insure that genera are homogeneous with regard to the chosen characters. Clements and Shear (1931) made the same complaint and the same proposal, but the problem has become worse since then, rather than better. There does seem to be agreement among current workers that form-genera should be homogeneous with regard to the type of conidiogenous cell. Unfortunately, the classification of conidiogenous cells is still somewhat uncertain (see Section III and Kendrick, 1971), and the nature of the conidiogenous cells in many described species is unknown. A section on this topic at the next international mycological congress should be profitable.

### D. The Task of the Compilers

When we began work on this chapter, we intended to illustrate most of the common Hyphomycetes and to list the names of all good genera along with their citations and their common synonyms. It soon became apparent that it was often difficult to decide which genera were "good" enough to include. It then became necessary to keep track of excluded names so that we would know that they had been excluded, and on what basis. Finally, we decided that it was necessary for our own compilation, and would be helpful to others, to have a complete list of the generic names proposed for Hyphomycetes, with an evaluation of every name. The status of many names only became clear when we tried to prepare a diagnostic illustration. Many synonyms came to light in this process, and many names were found to be of questionable application. Our final goal was to be inclusive: to list all generic names proposed for Hyphomycetes and to illustrate all good genera. By the time the enormity of our task had become clear to us, it was too late to back out gracefully.

The task of preparing a key was left until the end, and in the end we did not prepare one in the usual sense. Instead, we sorted the names of good genera into groups or lists according to the four character sets that we used to describe the genera. When a genus varied in one or more of the key characters, or when we thought it might be misinterpreted, we simply included it in as many lists as was appropriate. The resulting "key lists" form a simple and rapid key to a list of generic names. When the list is short, as it is in many cases, it is a simple matter for the would-be identifier to compare his fungus with all the illustrations. When the list is long, as it is in a few cases, the keylists would have been more useful if we had been able to provide additional levels of subdivision.

Our job, as we interpreted it, was to survey and report on the current state of hyphomycete classification. As such, we think of this chapter not as an end product, but as providing a new place to begin.

## E. Sources of Data

This compilation could have been done more quickly and thoroughly by someone with ready access to a major mycological library and herbarium. We were able to complete as much as we did only because of the efforts of previous monographers, indexers, and compilers. Since most of the publications that we used are still available, either in the original edition or as reprints, we will list the most important of the twentieth-century ones along with comments on their particular emphasis or usefulness.

### 1. Descriptions Based Primarily on Literature

Lindau, G. (1907). Hyphomycetes (erste Hälfte). In "Rabenhorst's Kryptogamen-Flora von Deutschland, Oesterreich und der Schweiz," 2nd ed., Vol. 1, Sect. VIII. Eduard Kummer, Leipzig.

Lindau, G. (1910). Hyphomycetes (zweite Hälfte). In "Rabenhorst's Kryptogamen-Flora von Deutschland, Oesterreich und der Schweiz," 2nd ed., Vol. 1, Sect. IX. Eduard Kummer, Leipzig.
 This work is particularly useful because of its broad coverage and many references. The cumulative index at the end of Abt. IX is a useful index also to Saccardo's Sylloge Fungorum, Vol. IV. There are many faithful copies of illustrations and numerous original ones.

Clements, F. E., and Shear, C. L. (1931). "The Genera of Fungi." Hafner, New York.
 This was the last previous attempt to cite and key out all the genera of fungi. It is interesting to note that the authors "... definitely planned to issue a new edition... at intervals

of three to five years...." The keys and illustrations of Hyphomycetes are no longer of much value, but the citations of generic names and types were very helpful for our task.

Dodge, C. W. (1935). "Medical Mycology." St. Louis, Missouri.
This unique work is still the best key to the pre-1935 literature on medical mycology. One rarely finds an error among the thousands of literature citations.

## 2. Descriptions Based Primarily on Specimens

Hughes, S. J. (1953). Conidiophores, conidia and classification. *Can. J. Bot.* **31**:577–659.
This was a proposal for stressing conidiogenesis rather than Saccardoan spore groups as a primary basis for the classification of the Hyphomycetes. The proposal is still being investigated and debated.

Hughes, S. J. (1958). Revisiones Hyphomycetum aliquot cum appendice de nominibus rejiciendis. *Can. J. Bot.* **36**:727–836.
Hughes checked the original descriptions and type or authentic specimens for about 400 generic names of microfungi and provided the first reliable disposition for many nineteenth-century names.

Barnett, H. L. (1960). "Illustrated Genera of Imperfect Fungi," 2nd ed. Burgess, Minneapolis, Minnesota.
This work includes illustrations of 325 Hyphomycetes plus 24 conidial Zygomycotina. For many mycologists, it was the first reference consulted when an unknown mold had to be identified. A third edition appeared in 1972.

Barron, G. L. (1968). "The Genera of Hyphomycetes from Soil." Williams & Wilkins, Baltimore, Maryland.
This was the first major compilation stressing conidiogenous cells. About 200 Hyphomycete genera are keyed out and illustrated with outstanding photomicrographs and line drawings.

von Arx, J. A. (1970). "The Genera of Fungi Sporulating in Pure Culture." Cramer, Lehre.
This gives good keys and citations for 231 Hyphomycete genera and line drawings for 150 of them.

Ellis, M. B. (1971). "Dematiaceous Hyphomycetes." Commonwealth Mycological Institute, Kew, Surrey.
This work raises the taxonomy of the Hyphomycetes to a new level. For each of 295 genera the author has provided a description in standardized format, a diagnostic and reliable drawing (almost all by the author), and a description of the type species. For recently described or questionable genera, type or authentic material was examined. For many genera, there are keys and descriptions for the common species. This work appeared late in our compiling task and necessitated many changes in our lists and drawings.

Kendrick, W. B., ed. (1971). "Taxonomy of Fungi Imperfecti." Univ. of Toronto Press, Toronto.
The proceedings of a workshop conference devoted to the classification and nomenclature of conidia and conidiogenous cell types, and to the implications of these matters for classification of the Deuteromycotina.

## 3. Indexes and Lists

Saccardo, P. A. (1882–1931). "Sylloge Fungorum," 25 vols. Padua.
Has short Latin descriptions of species and genera proposed up to 1920. No illustrations. Volume IV (1886) covers many of the classical Hyphomycete genera.

Petrak, F. (1920–1950). "Index of Fungi" (also known as "Petrak's Lists").
A series of publications giving references for new fungus taxa or names published between 1920 and 1939. The Commonwealth Mycological Institute has published or republished

this material as a series of eight volumes of lists, a cumulative index in two parts, and a supplement in 1969.

"Index of Fungi." (1940–    ). Commonwealth Mycological Institute, Kew.
   Contains citations for all new fungus taxa. Ten years per vol. with a cumulative index.
Bisby, G. R. ed. (1896–1946). "Transactions of the British Mycological Society Fifty Year Index," Vols. 1–30. Cambridge Univ. Press, London and New York.
Webster, J. ed. (1947–1957). "Transactions of the British Mycological Society Ten Year Index," Vols. 31–40. Warren & Son Ltd., Winchester.
Rogerson, C. T. ed., "Mycologia Index," (1968). New York Botanical Garden, New York.
   A cumulative index for Vols. 1–58 (1909–1966) by taxa and subject.
Ainsworth, G. C. (1971). "Ainsworth and Bisby's Dictionary of the Fungi," 6th ed. Commonwealth Mycological Institute, Kew.
   This work lists all generic names with a brief indication of their status and contains many useful notes on authors, higher taxa, etc. The fifth edition (1961) lists the genera of fungi by orders and in some orders by families.

## 4. References Cited in the Introduction

Carmichael, J. W. (1962). *Chrysosporium* and some other aleuriosporic Hyphomycetes. *Can. J. Bot.* **40**:1137–1173.
Clements, F. E., and C. L. Shear. (1931). "The Genera of Fungi." Hafner, New York.
Ellis, M. B. (1958). *Clasterosporium* and some allied Dematiaceae phragmosporae. I. *Mycol. Pap.* **70**:1–89.
Hennebert, G. L. (1971). Pleomorphism in Fungi Imperfecti. *In* "Taxonomy of Fungi Imperfecti" (W. B. Kendrick, ed.), pp. 202–223. Univ. of Toronto Press, Toronto.
Heslop-Harrison, J. (1962). Purposes and procedures in the taxonomic treatment of higher organisms. *In* "Microbial Classification" (G. C. Ainsworth and P. H. A. Sneath, eds.), pp. 14–36. Cambridge Univ. Press, London and New York.
Hughes, S. J. (1953). Conidiophores, conidia and classification. *Can J. Bot.* **31**:577–659.
Hughes, S. J. (1958). Revisiones Hyphomycetum aliquot cum appendice de nominibus rejiciendis. *Can. J. Bot.* **36**:727–836.
Hughes, S. J. (1959). Starting point for nomenclature of Hyphomycetes. *Taxon* **8**:96–103.
Kendrick, W. B., ed. (1971). "Taxonomy of Fungi Imperfecti." Univ. of Toronto Press, Toronto.
Mason, E. W. (1933). Annotated account of fungi received at the Imperial Mycological Institute, List II (Fascicle 2). *Mycol. Pap.* **3**:1–67.
Mason, E. W. (1937). Annotated account of fungi received at the Imperial Mycological Institute, List II (Fascicle 3. General part). *Mycol. Pap.* **4**:1–99.
Pirozynski, K. A. (1963). *Beltrania* and related genera. *Mycol. Pap.* **90**:1–37.
Simmons, E. G. (1966). The theoretical bases for classification of the Fungi Imperfecti. *Quart. Rev. Biol.* **41**:113–123.

*Acknowledgments*

To say the least, this chapter was not produced by the authors alone. The text was compiled by a documentation team at the University of Alberta while the drawings were assembled by an illustration team at the University of Waterloo. The correspondence between the authors fills more than a foot of file space at each institution, since each of us received a constant flow of suggestions and information from the other.

At Edmonton, Mrs. Lynne Sigler typed citations, wrote and ran computer programs for updating and correcting files, and did much of the indexing, cross-checking and proofreading.

At Waterloo, Dr. G. Morgan-Jones skilfully inked many of the pencilled original drawings and lent his taxonomic expertise to the interminable discussions which accompanied the selection and execution of the drawings. Dr. Kris Pirozynski, Dr. T. R. Nag Raj, and Dr. Luella Weresub were also consulted on several occasions. The technical help of Miss Linda Thomasson and Mr. Masood Hashmi was invaluable at certain critical junctures.

Dr. M. B. Ellis at the Commonwealth Mycological Institute in Kew and Dr. S. J. Hughes at the Plant Research Institute in Ottawa each initiated the compilation (at his respective institution) of copies of the original publications of all hyphomycete genera. Their generosity in allowing us to consult these files has made it possible for us to check almost all of our citations against the original description and illustrations. Without this resource, our work would have been much more difficult and much less complete and reliable.

We are also indebted to the hundreds of authors and publications from which we abstracted illustrations and information. There are too many to list them again here, but the references will be found in the list of generic names.

We gratefully acknowledge the financial backing given to us by the University of Alberta, the University of Waterloo, and a National Research Council of Canada grant to Dr. Kendrick.

## II. NOTES ON THE LIST OF GENERIC NAMES

### A. Scope

The list of generic names includes most of the names proposed for Hyphomycetes and sterile mycelia. A few names from the Melanconiales have been included when it was felt that they might be useful. The distinction between the Melanconiales and some of the Hyphomycetes is not sharp, especially when the fungi can be induced to grow and sporulate in culture. Also the line between the Hyphomycetes and the Blastomycetes is sometimes in doubt. We did not include the conidial Zygomycotina or the fossil fungi.

The list includes 1541 generic names. Of these we accepted 595 as applying to recognizable conidial states and rejected 606 as being synonyms, sterile, nomina dubia or confusa, or illegitimate. We felt that 69 of the names newly rejected were established with sufficient reliability to be listed unequivocally. An additional 89 probable synonyms newly established by us are listed as "?=". Most of the remaining 251 questionable names were originally published with descriptions inadequate to indicate the exact nature of the fungus. Often there was no illustration (116 names) or a very poor one (104 names).

At best, the "authors" of these "genera" have published a named reference to an herbarium specimen, leaving it for later workers to unearth the remains and delineate a taxon. All too often there is not even a clear reference to a specimen. In our opinion, every generic name whose type species has never been illustrated and whose publication did not explicitly specify (directly or indirectly) a type specimen should be declared invalid. In addition, all generic names published prior to 1900 and which have not yet been illustrated should be declared invalid whether or not a type specimen was specified. The botanical code should be emended to require illustrations of the conidia and conidiogenous cells as part of the required diagnosis for every new hyphomycete genus or species. The principle of priority in nomenclature becomes a drawback rather than a help when the original publishers of names do not do their job well enough so that subsequent workers can recognize the taxon from the published description.

Perfect state connections are noted for about one quarter of the form-genera of Hyphomycetes. One or more genera of the Ascomycotina are noted as states for 133 form-genera, and clamp connections or genera of Basidiomycotina are noted for 25 form-genera.

## B. Format of Entries

The first part of each entry consists of the genus name, its author(s), date of publication and place of publication, then a diagonal slash followed by the type species, and the author(s) and date of publication of the type species. Where the author and date for the type species are identical to those for the genus, they have not been repeated. When the name was a new combination the author of the epithet is given in brackets, but is followed by the author of the combination only when different from the author of the genus. When the first part of the entry required more than one line, the continuation lines are indented four spaces.

The second part of each entry is indented two spaces and consists of the status of the genus, sometimes followed by comments on the status of perfect state connections, then a diagonal slash, and one or more references to the literature on the genus. The status of the genus is given either as a series of three key-words (see below), or as an "=" sign followed by the name of the genus for which this one has been accepted as a synonym, or followed by a disposition such as "nomen dubium," or else the status is given as "?," which means that the name appears to be validly published, but with insufficient information to determine what kind of fungus the name should refer to. The authority from whom we took synonyms or perfect-state connections is given after the word "fide." Where the authority is followed by a date, it is the date of the publication from which we took the information. When the authority is not followed by a date, it means that the assertion has not been previously published. Questionable synonyms, for which no authority is given, were established by the compilers, usually on the basis of published illustrations. When the second part of the entry is continued, it is also indented four spaces.

The third part of each entry is a list of the synonyms (if any) which we accept for the entry genus. The first synonym is preceded by the flag "SYNO" indented five spaces.

## 10. Hyphomycetes

A few names are listed without complete publication data. These are all invalid or long-established synonyms, mostly taken directly from the "Dictionary of the Fungi" without further bibliographic effort.

### C. Definitions of Key-words

The status of good genera is indicated by a series of three words indicating first the Saccardoan spore group, second the general arrangement of the conidia with respect to each other, and third the color of the conidia. These three aspects were chosen as the most clear-cut and readily determined diagnostic features applicable to all Hyphomycetes. The seven Saccardoan spore groups were used as defined in the "Dictionary of the Fungi" (Ainsworth, 1971 or 1961, Fig. 1.). The names are abbreviated by dropping the suffix -SPORAE to yield the key-words AMERO, DIDYMO, PHRAGMO, DICTYO, SCOLECO, HELICO, and STAURO. Only five general types of conidium arrangement were specified because more specific information is lacking for some genera and because even these types are somewhat intergrading. The five types recognized are as follows:

ARTHRO (arthrocatenate)—conidia are formed in chains by the simultaneous or random fragmentation of a hypha.

BLASTO (blastocatenate)—conidia formed in chains with the youngest spore at the tip or distal end of the chain.

BASO (basocatenate)—conidia formed in chains with the youngest conidium at the basal or proximal end of the chain. Included here are phialoconidia in chains, meristem arthrospores, retrogressive holoblastic conidia, and chains produced from annellated conidiogenous cells.

GLOIO (gloiosporae)—conidia produced in slimy heads or masses. Most commonly these are phialo-conidia.

CETERI (the rest)—conidia not in chains or slime. This includes solitary conidia and conidia produced on inflated or sympodially proliferating conidiogenous cells.

The final pair of key-words, HYALO and PHAEO, refer to the cell-wall color of the conidia only. The color of the hyphae and conidiophores is not taken into account. Where the color is "pale olivaceous" or "hyaline, becoming dark at maturity" we have listed both HYALO and PHAEO as key-words. Similarly, where a genus was considered to be variable or borderline in any of the key characters, then more than one key-word is given.

### III. NOTES ON THE KEY-LISTS

The key-words describing good genera were used to sort the names into sublists. Fungi are listed for 53 of the 70 possible combinations of key-words ($7 \times 5 \times 2 = 70$). Sterile genera are added at the end. The numbers after the names in the key-lists refer to the figure illustrating the genus. Names without numbers are not illustrated.

To further assist the would-be identifier of an unknown hyphomycete, we have arranged the names in each sublist according to the type of conidiogenous cells (or type of conidiogenous axis when several conidiogenous cells form a morphological entity). Unfortunately, for some genera the conidiogenous cells have not been carefully examined and their nature remains in doubt. We have indicated these cases with a question mark. Other genera produce a plurality of conidia by a combination of methods, or by methods which are intermediate or variable with respect to our classification. Nevertheless, the nature of the conidiogenous cells was the most stable and useful character we could find for further subdivision. The types of conidiogenous cells recognized are as follows:

NONSP (nonspecialized)—the conidiogenous cells resemble vegetative hyphae. They are not morphologically specialized for the simultaneous or successive production of many conidia, e.g., *Chrysosporium, Helicoma, Oncopodium, Mammaria, Wardomyces, Pithomyces.*

PHIAL (phialiform)—the conidiogenous cell does not elongate or enlarge during the production of a succession of conidia from a fixed growing point, e.g., *Phialophora, Chalara, Penicillium, Fusarium.*

ANNEL (annelliform)—the conidiogenous axis proliferates percurrently during the production of a succession of conidia so that it becomes annellated or ringed, e.g., *Annellophora, Scopulariopsis, Gyrothrix, Ceratocladium, Spilocaea.*

RACHI (rachiform)—the elongating conidiogenous axis is relatively narrow compared to the width of the conidial attachment and the axis is pushed first to one side and then the other as it grows past a succession of terminal conidia. This sympodial development results in a geniculate or zig-zag rachis, e.g., *Tritirachium, Beauveria, Pyricularia, Drechslera, Spiropes guareicola.* In a few fungi (e.g., *Polythrincium, Costantinella*) the axis usually grows to the same side of each terminal conidium, and is not really rachiform.

RADUL (raduliform)—the elongating conidiogenous axis is relatively wide compared to the width of the conidial attachment and the axis tends to become clavate or somewhat inflated, rather than zig-zag. In some genera, the conidia are initially lateral, as well as (or instead of) terminal, e.g., *Rhinocladiella, Virgaria, Scolecobasidium, Spiropes clavatus, Haplariopsis, Arthrobotrys, Dactylaria, Spadicoides.* The prefix "radul" is derived from Mason (1933, p.9).

AMPUL (ampulliform)—the conidiogenous cell swells to form an ampulliform, clavate, or lobed fertile area which then bears many conidia simultaneously over its surface, e.g., *Oedecephalum, Gonatobotrys, Botrytis, Ostracoderma, Botryosporium.*

MISC (miscellaneous)—a plurality of conidia is achieved in some manner other than the ones above, e.g., *Arthrinium, Basipetospora, Cladobotryum, Trichothecium, Piricauda*. Or else the conidiogenous cells are morphologically specialized, but do not produce a succession of conidia, e.g., *Nigrospora, Spegazzinia, Fusicladiella, Zygosporium*.

## IV. NOTES ON THE ILLUSTRATIONS

### A. Sources

We have drawn illustrations for 566 genera; about twice as many as have been brought together previously. Some few drawings were made from type or authentic material, but in the limited time at our disposal we could not even begin to amass the hundreds of other type specimens from which really authoritative illustrations of most genera could perhaps be derived, so we had to fall back on existing illustrations. In many cases this was not a drawback, since we could consult the excellent compendia by Barron (1968), von Arx (1970), and Ellis (1971), and the papers of many eminent hyphomycete taxonomists and illustrators—Arnaud, Hughes, Linder, Pirozynski, Subramanian, Thaxter, Tubaki, and others. In fact, so far as our copied illustrations go, we stand primarily on the shoulders of giants (and our debt to them is in proportion to their stature); but because we are also, perforce, perched on a few pygmies, it is not surprising that our stance is at times a little uneven. There are drawings in which we have less than perfect faith; for example, our picture of *Articularia* (Fig. 5,D) is hardly believable as it stands, but we reproduced it because it does give some clues to the nature of the fungus. The whorled conidiogenous cells seem to be phialidic and the conidia lie parallel to one another in presumably slimy clusters. Although we have not given the sources of copied drawings directly on the plates, references to illustrations and other important information are cited in the list of generic names (Section V).

### B. Arrangement

Keys, however ingeniously designed, seem to provide stumbling blocks for the uninitiated. Our key, though extremely forgiving, and often providing access to the correct genus despite several incorrect choices by the operator, will doubtless baffle some readers.

With this possibility in mind, we decided to make the plates a kind of visual key in themselves, since "picturebook mycology" is a time-honored and fruitful pastime. We considered several methods of arranging the illustrations—alphabetically by genus; in the same sequence as the key-list; in ontogenetically related groups—but found that no one system, rigidly followed, gave a satisfactory layout. The scheme finally adopted is a compromise, and requires some explanation. There are 67 plates, divided between the Saccardoan spore groups as follows: Amero, Plates 1 to 27; Didymo, Plates 28 to 34; Dictyo, Plates 35 to 38; Helico, Plates 39 to 41; Phragmo, Plates 42 to 54; Scoleco, Plates 55 to 58; Stauro, Plates 59 to 66; and Sterile, Plate 67.

Within each group we have tried to bring together "act-alikes" or "look-alikes." For example, Plates 3 to 12 contain Amero genera which have phialides as their conidiogenous cells. Plate 3 groups those with solitary phialides, Plate 7 those with penicillate heads of phialides, and Plate 10 those producing phialides in sporodochia or synnemata; while Plate 12 illustrates those producing appendaged phialoconidia. As one further example, from the Stauro group Plates 63 and 64 depict mostly aquatic Hyphomycetes with tetraradiate conidia, but while those in plate 63 have three arms growing upward and one downward, those in Plate 64 have one arm growing upward and three downward.

The rationale behind many individual Plates or facing pairs of Plates will often be obvious after even cursory inspection. If it is not, that is probably because several disparate elements have had to be combined to fill the plate.

## V. LIST OF GENERIC NAMES

For an explanation of this list see pp. 329–331

```
ABGLIOPHRAGMA ROY ET GUJARATI 1966. TRANS. BRIT. MYCOL. SOC. 49:363 /A. SETOSUM
    =WIESNERIOMYCES KOORDERS 1907 FIDE KENDRICK

ACAROCYBE SYDOW 1937. ANN. MYCOL. BERLIN 35:285 /A. HANSFORDII
    PHRAGMO,CETERI,PHAEO /ELLIS. MYCOL PAPERS CMI 76:2-5,1960. (ILLUS,KEY)

ACAROCYBELLA ELLIS 1960. MYCOL PAPERS CMI 76:5 (ILLUS) /A. JASMINICOLA
    (HANSFORD)
    PHRAGMO OR SCOLECO,CETERI,PHAEO

ACAULIUM SOPP 1912. SKR. VIDENSK.-SELSK. CHRISTIANIA 11:42-53 /A. ALBONIGRESCENS
    =SCOPULARIOPSIS BAINIER 1907 FIDE ELLIS 1971

ACHITONIUM KUNZE 1819. FLORA II P. 50 /A. ACICOLA
    ?, NOT ILLUSTRATED BY AUTHOR /LINDAU, RABENH. KRYPT. FL. 2 AUFL. 1 BD.
        9:456-457,1910. (REFS)
    SYNO    =PACTILIA FRIES FIDE FRIES (SEE SACCARDO, SYLLOGE FUNG. 4:673,1886)
            =CHROOSTROMA CORDA 1837 FIDE LINDAU 1910
            =LEUCOSPORIUM CORDA 1837 FIDE LINDAU 1910
            =PHYMATOSTROMA CORDA 1837 FIDE LINDAU 1910

ACHORION REMAK 1845. DIAG. PATH. UNTERSUCH. P. 193 /A. SCHOENLEINI (LEBERT)
    =TRICHOPHYTON MALMSTEN 1845

ACICULARIELLA ARNAUD 1954. BULL. SOC. MYCOL. FR. 69:298 AND 301 (ILLUS) /A.
    LASIOSPHAERIAE
    SCOLECO,CETERI,HYALO. BUT SECOND SPECIES INCLUDED BY ARNAUD DOES NOT APPEAR
        TO BE CONGENERIC WITH THE TYPE SPECIES. NO LATIN DIAGNOSIS

ACINULA FRIES 1822. SYSTEMA MYCOL. 2:267 /A. FLAMMEA
    =STERILE
```

ACLADIUM LINK 1809. MAG. GES. NATURF. FREUNDE, BERLIN 3:11 /A. CONSPERSUM
  (LECTOTYPE)
  AMERO,CETERI,HYALO. BASIDIO STATE = BOTRYOBASIDIUM /HUGHES, CAN. J. BOTANY
  36:731-732,1958. (LIST) LINDER, LLOYDIA 5:165-207,1942.(ILLUS) SUB OIDIUM
    SYNO    =SPOROCEPHALIUM CHEVALLIER 1826 FIDE HUGHES 1958

ACMOSPORIUM CORDA 1839. ICONES FUNGORUM 3:12 /A. BOTRYOIDEUM
  =ASPERGILLUS LINK 1809 FIDE HUGHES 1958

ACONTIOPSIS NEGRU 1961. COMUN. ACAD. REPUB. POP. ROM. 11:839 /A. CRATAEGI
  =CYLINDROCLADIUM MORGAN 1892 FIDE KENDRICK

ACONTIUM MORGAN 1902. JOUR. MYCOL. 8:4 /THREE ORIGINAL SPECIES
  ?, NOT ILLUSTRATED BY AUTHOR

ACREMONIELLA SACCARDO 1886. SYLLOGE FUNGORUM 4:302 /A. ATRA (CORDA) (LECTOTYPE)
  AMERO,CETERI,PHAEO. ASCO STATE =MELANOSPORA FIDE MASON 1933 /MASON, MYCOL.
    PAPERS CMI 3:29-39,1933. (REVIS,ILLUS)
    SYNO    =EIDAMIA LINDAU 1904 FIDE ELLIS 1971
            =HARZIA COSTANTIN 1888 FIDE ELLIS 1971
            =MONOPODIUM DELACROIX 1890 FIDE ELLIS 1971

ACREMONIULA CIFERRI 1962. ATTI IST. BOT. UNIV. LAB. CRITTOGAM. PAVIA, SER. 5,
    19:85 /A. SUPRAMELIOLA
  AMERO,CETERI,PHAEO /DEIGHTON, MYCOL PAPERS CMI 118:2-5,1969. (ILLUS)

ACREMONIULA ARNAUD 1954. BULL. SOC. MYCOL. FR. 69:268-269 /A. SARCINELLAE
    (PATOUILLARD ET HAR.)
  =NOMEN ILLEGITIMUM, NO LATIN DIAGNOSIS, NON ACREMONIULA CIFERRI 1962

ACREMONIUM LINK 1809. MAG. GES. NATURF. FREUNDE, BERLIN 3:15 /A. ALTERNATUM
    (LECTOTYPE)
  AMERO,GLOIO,HYALO. ASCO STATE =EMERICELLOPSIS OR NECTRIA FIDE VON ARX 1970
    /VON ARX, GENERA OF FUNGI IN CULTURE P. 177-178,1970. (ILLUS)
    SYNO    =CEPHALOSPORIUM AUCT. NON CORDA FIDE VON ARX 1970

ACROCLADIUM PETRAK 1949. SYDOWIA 3:263 /A. ANDINUM
  ?, NOT ILLUSTRATED BY AUTHOR /SYDOWIA 4:586,1950

ACROCONIDIELLA LINDQUIST ET ALIPPI 1964. DARWINIANA 13:612 /A. TROPAEOLI (BOND)
    LINDQ. ET ALIPPI 1964
  PHRAGMO,CETERI,PHAEO /ELLIS, DEM. HYPH. P. 461-462,1971. (ILLUS)

ACROCONIDIELLINA ELLIS 1971. MYCOL. PAPERS CMI 125:22-27 (ILLUS) /A. LOUDETIAE
  DIDYMO OR PHRAGMO,CETERI,PHAEO. COMPARE WITH DRECHSLERA AND ACROCONIDIELLA

ACROCYLINDRIUM BONORDEN 1851. HANDB. MYK., P.97 /THREE ORIGINAL SPECIES
  =VERTICILLIUM NEES FIDE BARRON 1968

ACRODESMIS SYDOW 1926. ANN. MYCOL. 24:424 /A. CESTRI
  =PERICONIELLA SACCARDO 1885 FIDE ELLIS 1967

ACRODICTYS ELLIS 1961. MYCOL PAPERS CMI 79:5 /A. BAMBUSICOLA
  DICTYO,CETERI,PHAEO /ELLIS, MYCOL PAPERS CMI 79:5-19,1961. (ILLUS,KEY),
    ELLIS, MYCOL PAPERS CMI 103:33-35,1965. (ILLUS)

ACROGENOSPORA ELLIS 1971. DEMATIACEOUS HYPHOMYCETES P. 114-115 (ILLUS) /A.
    SPHAEROCEPHALA (BERKELEY ET BROOME)
  AMERO,CETERI,PHAEO. ASCO STATE = FARLOWIELLA FIDE MASON 1941 (SUB MONOTOSPORA)
    SYNO    =MONOTOSPORA SACCARDO 1880 NON CORDA, SEE MASON 1941

ACROPHIALOPHORA EDWARD 1961. MYCOLOGIA 51:784 (DATED 1959) /A. NAINIANA
  AMERO,BASO,HYALO /VON ARX, GENERA OF FUNGI IN CULTURE P. 179-181,1970.
    (ILLUS) SAMSON AND MAHMOOD, ACTA. BOT. NEERL. 19:804-808,1970. (ILLUS,KEY)

ACROPHRAGMIS KIFFER ET REISINGER 1970. REV. ECOL. BIOL. SOL 7:12 (11-31,ILLUS)
    /A. CORONATA
  PHRAGMO OR STAURO,CETERI,PHAEO

ACROSPEIRA BERKELEY ET BROOME 1857. INTROD. TO CRYPTOGAM. BOT. P. 305 /A.
    MIRABILIS

DICTYO OR HELICO,CETERI,PHAEO /VON ARX, GENERA OF FUNGI IN CULTURE P. 237-238,1970. (ILLUS)
   SYNO    =SPIROSPORA MANGIN ET VINC. 1920 FIDE HUGHES 1958

ACROSPIRA MONTAGNE 1857. ANN. SCI. NAT. IV, 8:300 /A. CROUANII
   ?, NOT ILLUSTRATED BY AUTHOR

ACROSPORIUM NEES 1817. DAS SYSTEM ..., P. 53 /A. MONILIOIDES NEES =ACROSPORIUM STATE OF ERYSIPHE GRAMINIS DE CANDOLLE EX MERAT 1821
   AMERO,BASO,HYALO, ASCO STATE =ERYSIPHE FIDE HUGHES 1958, NON ACROSPORIUM BON. 1851 /HUGHES, CAN. J. BOTANY 36:732-733,1958. (REFS), CLEMENTS AND SHEAR, GENERA OF FUNGI, PLATE 53,1931. (ILLUS)
   SYNO    =OIDIUM AUCT.

ACROSTALAGMUS CORDA 1838. ICONES FUNGORUM 2:15 /A. CINNABARINUS
=VERTICILLIUM NEES FIDE HUGHES 1958

ACROSTAPHYLUS ARNAUD EX SUBRAMANIAN 1956. J. INDIAN BOT. SOC. 35:483 /A. HYPOXYLI ARNAUD 1954
=NODULISPORIUM PREUSS FIDE BARRON 1968

ACROTAMNIUM NEES 1817. DAS SYSTEM P. 75 /A. VIOLACEUM
=STERILE FIDE HUGHES 1958

ACROTHECA FUCKEL 1860. JAHRB. VER. NATURK., NASSAU 15:43 /A. GEI
=RAMULARIA UNGER FIDE HUGHES 1958

ACROTHECIELLA KOORDERS 1907. VERH. AKAD. AMSTERDAM 2:13:250 /A. JAVANICA
?, ORIGINAL ILLUSTRATION NOT DIAGNOSTIC

ACROTHECIUM (CORDA) PREUSS 1851. LINNAEA 24:111 /A. MULTISPORUM PREUSS 1851
=NOMEN DUBIUM FIDE S.J. HUGHES 1958

ACTINICEPS BERKELEY ET BROOME 1876. J. LINN. SOC., LONDON 15:85 /A. THWAITESI
=DIMORPHOCYSTIS, A BASIDIOMYCETE, FIDE BOEDIJN 1959

ACTINOCHAETE FERRO 1907. NUOV. GIORN. BOT. ITAL. 14:232 /A. ARACHNOIDEA
=NOMEN CONFUSUM FIDE M.B. ELLIS

ACTINOCLADIUM EHRENBERG 1819. JAHRB. GEWACHSK. 1:52 /A. RHODOSPORUM
STAURO,CETERI,PHAEO /ELLIS, DEMATIACEOUS HYPHOMYCETES P. 137-138,1971. (ILLUS)

ACTINODOCHIUM SYDOW 1927. ANN. MYCOL. 25:146 (ILLUS. P. 147) /A. CONCINNUM
?, ILLUSTRATION NOT DIAGNOSTIC

ACTINOMMA SACCARDO 1884. ATTI. IST. VENETO, 6 SER., 2:462-463 /A. GASTONIS
?, NOT ILLUS. BY AUTHOR

ACTINONEMA PERSOON 1822. MYCOL. EUROP. 1:52 /A. CAULINICOLA
=STERILE FIDE HUGHES 1958

ACTINOPELTE SACCARDO 1913. ANN. MYCOL. 11:312 /A. JAPONICA
?, PERHAPS A COELOMYCETE, BUT SEE TUBAKI, 1971 /TUBAKI, INST. FERM. OSAKA RES. COM. 5:43-77,1971. (ILLUS,REVIS)

ACTINOSPORA INGOLD 1952. TRANS BRIT MYCOL SOC 35:66 /A. MEGALOSPORA
STAURO,CETERI,HYALO /GOOS, TRANS BRIT MYCOL SOC 55:335-337,1970.(ILLUS)

ACTINOSTILBE PETCH 1925. ANN. BOT. GARD. PERADENIYA 9:327 /A. VANILLAE
?, NOT ILLUSTRATED BY AUTHOR

ADHOGAMINA SUBRAMANIAN ET LODHA 1964. ANTONIE VAN LEEUWENHOEK 30:319 /A. RUCHIRA
=GILMANIELLA BARRON 1964 FIDE HENNEBERT 1968

AEGERITA PERSOON 1801. SYN. METH. FUNG. 684 /A. CANDIDA PERSOON =AEGERITA STATE OF PENIOPHORA CANDIDA LYMAN 1907
=STERILE, BASIDIO STATE =PENIOPHORA FIDE HUGHES 1958 /TUBAKI, J. HATTORI BOTAN. LAB. NO. 20, P. 145-146,1958. (ILLUS)
   SYNO    =CROCYSPORIUM CORDA 1837 FIDE HUGHES 1958
                =DERMOSPORIUM LINK 1815 FIDE LINDAU IN RABENHORST 1910

## 10. Hyphomycetes

AEGERITOPSIS HOHNEL 1903. ANN. MYCOL., BERLIN 1:532-533 /A. NULLIPORIOIDES
  ?. NOT ILLUS. BY AUTHOR

AEROPHYTON ESCHWEILER 1824. SYLL. PLANT. SOC. BOT. RATISB. 1:163-165 /A.
    PRINCIPIS
  ?. ORIG. ILLUS. NOT DIAGNOSTIC

AGARICOSTILBUM WRIGHT 1970. MYCOLOGIA 62:679-682 (ILLUS) /A. PALMICOLA
  AMERO,CETERI,HYALO

AGYRIELLA SACCARDO 1884. MISC. MYCOL. 1:20 /A. NITIDA (LIBERT)
  AMERO,GLOIO,HYALO /ELLIS, DEM. HYPH. P. 525-526, 1971. (ILLUS)

AKANTHOMYCES LEBERT 1858. ZEITSCH. WISSEN. ZOOLOGIE 9:447 /A. ACULEATA
  AMERO,BASO,HYALO, ASCO STATE =CORDYCEPS LINK FIDE MAINS 1958 /MAINS,
    MYCOLOGIA 42:566-589,1950. (ILLUS) MORRIS, WESTERN ILLINOIS UNIV. SER. IN
    THE BIOL. SCI. 3:16,1963

AKENOMYCES ARNAUD 1954. ANN. EC. AGRIC. MONTPELLIER 29:6 (ILLUS) /A. ENIGMATICUS
  =STERILE, PYRIFORM SCLEROTIA

ALATOSPORA INGOLD 1942. TRANS BRIT MYCOL SOC 25:384 /A. ACUMINATA
  STAURO,CETERI,HYALO /TUBAKI, BULL. NAT. SCI. MUSEUM, TOKYO 3:250-252,1957.
    (ILLUS)

ALBOSYNNEMA MORRIS 1967. MYCOPATHOLOGIA 33:179-181 (ILLUS) /A. ELEGANS
  PHRAGMO,GLOIO,PHAEO.

ALEURISMA LINK 1809. MAG. GES. NATURF. FREUNDE, BERLIN 3:19 /A. SPORULOSUM
  =TRICHODERMA PERSOON 1801 FIDE HUGHES 1958

ALEUROSPORIA GRIGORAKIS 1925. COMP. REND. 179:1425 /A. ACUMINATA (BODIN)
  =TRICHOPHYTON MALMSTEN PRO PARTE FIDE DODGE 1935

ALLANTOSPORA WAKKER 1896. MEDED. PROEFST. OOST-JAVA, SER. 2, 28:8-9 /A.
    RADICICOLA WAKKER
  =CYLINDROCARPON WOLLENWEBER 1913 FIDE BOOTH 1966

ALLESCHERIA HARTIG 1899.   SEE HARTIGELLA SYDOW 1900

ALLESCHERIELLA HENNINGS 1897. HEDWIGIA 36:244 /A. UREDINOIDES =A. CROCEA
    (MONT.) HUGHES 1950
  AMERO,CETERI,PHAEO /HUGHES, MYCOL PAPERS CMI 41:1-8,1951. (ILLUS)

ALLIOSPORA PIM 1883. JOUR. BOT., LONDON 21:234-235 /A. SAPUCAYAE
  ?. NOT ILLUS. BY AUTHOR, PROBABLY BASED ON AN ASPERGILLUS

ALLONEMA SYDOW 1934. ANNAL. MYCOL. 32:284 /A. ROSEUM (GROVE) =GEOTRICHUM ROSEUM
    GROVE
  ?. NOT ILLUSTRATED BY SYDOW

ALLOSPHAERIUM LINK = RHIZOCTONIA FIDE SACCARDO FROM AINSWORTH 1971

ALPHITOMYCES REISSEK 1856. S.B. AKAD. WISS. WIEN 21:323-327 AND PLATE /A.
    SCHROTTERI RIESSEK 1856
  ?. LOOKS LIKE A COREMIAL PAECILOMYCES

ALTERNARIA NEES 1817. DAS SYSTEM ... P. 72 /A. TENUIS
  DICTYO,BLASTO,PHAEO, ASCO STATE =CLATHROSPORA OR LEPTOSPHAERIA FIDE SIMMONS
    1952, OR PLEOSPORA FIDE ELLIS 1971 /SIMMONS, MYCOLOGIA 59:67-92,1967.
    (ILLUS) ELLIS, DEM. HYPH. P. 464-497,1971. (ILLUS,KEY)
       SYNO    =MACROSPORIUM FRIES 1832 FIDE HUGHES 1958
               =PRATHODA SUBRAMANIAN 1956 FIDE DEIGHTON 1969
               =RHOPALIDIUM MONTAGNE 1846 FIDE ELLIS 1971
               ?= DICTYOCEPHALA MEDEIROS 1962

ALYSIDIUM KUNZE 1817. IN KUNZE AND SCHMIDT, MYKOL. HEFTE 1:11 /A. FULVUM
  =ALYSIDIUM DUBIUM (PERSOON) ELLIS 1971 BASIONYM =TRICHODERMA DUBIUM PERSOON
    1801, SYN. METH. FUNG. P. 233
  AMERO,BLASTO,HYALO OR PHAEO. THE TYPE SPECIES IS THE FUNGUS THAT LINK (1809)

MIS-IDENTIFIED AS TRICHODERMA AUREUM PERSOON (1801) AND USED TO FOUND THE
GENUS OIDIUM LINK 1809. SINCE A. DUBIUM IS QUITE DISTINCT FROM ACLADIUM
CONSPERSUM LINK, WE HAVE NOT FOLLOWED HUGHES (1958) IN PLACING ALYSIDIUM
INTO SYNONYMY WITH ACLADIUM. /HUGHES, CAN. J. BOTANY 36:731-732,1958.
(REFS) LINDER, LLOYDIA 5:165-207,1942. (ILLUS,KEY,REVIS) SUB OIDIUM
    SYNO    =OIDIUM AUCT. NON LINK 1809

ALYTOSPORIUM (LINK) EHRENBERG 1818. SYLVAE MYC. BEROL. P. 11 /A. BOMBACINUM
    (LECTOTYPE FIDE HUGHES 1958)
=NOMEN DUBIUM FIDE HUGHES 1958

AMALLOSPORA PENZIG 1897. MALPIGHIA 11:461 /A. DACRYDION
    STAURO,GLOIO,HYALO /TUBAKI, TRANS. MYCOL. SOC. JAPAN 6:44-46,1965. (ILLUS)

AMASTIGOSPORIUM BONDARZEVA ET MONTEVERDE 1921. MAT. MYKOL. OBSLED. ROSS. 5:2
    (?1:2) /A. GRAMINICOLA BOND. ET MONT. 1903 (?)
=MASTIGOSPORIUM RIESS 1852 FIDE HUGHES 1951
    SYNO    =AMASTIGIS CLEMENTS AND SHEAR 1931, ORTHOGRAPHIC VARIANT

AMBLYOSPORIOPSIS FAIRMAN 1922. PROC. ROCHESTER ACAD. SCI. 6:132 AND PLATE 21
    /A. PARASPHENOIDES
?, ORIG. ILLUS. NOT DIAGNOSTIC, ASPERGILLUS(?), RHIZOPUS(?)

AMBLYOSPORIUM FRESENIUS 1863. BEITR. ZUR. MYKOL. 3:99 /A. BOTRYTIS
    AMERO,ARTHRO OR CETERI,HYALO /PIROZYNSKI, CAN. J. BOTANY 47:325-334,1969.
    (REVIS,ILLUS)
    SYNO    =GUEGUENIA BAINIER FIDE PIROZYNSKI 1969
              =SEARCHOMYCES MEHROTRA ET MEHROTRA 1965 FIDE PIROZYNSKI

AMBROSIELLA BRADER 1964. MEDED. LANDB-HOOGESCH. WAGENINGEN 64:41 /A. XYLEBORI
    BRADER EX ARX ET HENNEBERT 1965
    AMERO,BLASTO,HYALO /ARX AND HENNEBERT, MYCOPATHOLOGIA 25:312-315,1965.
    (ILLUS) BATRA, MYCOLOGIA 59:976-1017,1967.(REVIS,KEY,ILLUS)

AMEROSPORIELLA HOHNEL 1923. SYST. FUNG. IMP. P. 358-362 /SINE SPECIES NOMINE
=NOMEN ILLEGITIMUM
    SYNO    =AMEROSPORIS CLEMENTS AND SHEAR 1931 FOR AMEROSPORIELLA

AMEROSPORIS CLEMENTS ET SHEAR 1931 = NOM. NOV. FOR AMEROSPORIELLA

AMPHIBLISTRUM CORDA 1837. ICONES FUNGORUM 1:11, TAB. III, FIG. 166 /A.
    HYPOCHNOIDES
?, ORIG. ILLUS. DIFFICULT TO INTERPRET

AMPHICHAETE KLEBAHN 1914. MYCOL. ZBL. 4:16-17 (FIG. 36) /A. ECHINATA
=AMPHICHAETELLA HOHNEL 1916, NON AMPHICHAETA MCALPINE 1904

AMPHICHAETELLA HOHNEL 1916. S.B. AKAD. WISS. WIEN 125:92 (NO ILLUSTRATION) /A.
    ECHINATA (KLEBAHN)
?, NOM. NOV. FOR AMPHICHAETE KLEBAHN 1914

AMPHICHORDA FRIES = 'ISARIA' FIDE FRIES FROM AINSWORTH 1971

AMPHITRICHUM NEES ET NEES 1818. NOVA ACTA ACAD. CAES. LEOP. 9:249,1818 /A.
    EFFUSUM
=CERATOSTOMELLA SACCARDO FIDE AINSWORTH 1961

AMPHOROMORPHA THAXTER 1914. BOT. GAZ. 58:249 (235-253, ILLUSTRATED) /A.
    ENTOMOPHILA
?, PERHAPS NOT A HYPHOMYCETE

AMPULLIFERA DEIGHTON 1960. MYCOL PAPERS CMI 78:36-42 (ILLUS) /A. FOLIICOLA
    AMERO OR DIDYMO OR PHRAGMO,CETERI OR BLASTO,PHAEO /DEIGHTON,MYCOL PAPERS CMI
    101:28-31,1965. (ILLUS,EMEND)

AMPULLIFERELLA BATISTA ET CAVALCANTI 1964. PORT. ACTA BIOL. SER. B, 7:348-350
    AND FIG. 1 /A. AMOEBOIDES
?, PERHAPS THE SAME AS AMPULLIFERA, BUT SEE SUTTON 1969

## 10. Hyphomycetes

AMPULLIFERINA SUTTON 1969. CAN. J. BOTANY 47:609 (609-616, ILLUS) /A. PERSIMPLEX
    DIDYMO,ARTHRO,PHAEO

AMPULLIFEROPSIS BATISTA ET CAVALCANTI 1964. PORT. ACTA BIOL., SER. B, 7:349
    (349-355, ILLUS) /A. HIPPOCRATEACEARUM
    ?, SAME AS AMPULLIFERA?, SEE SUTTON 1969

ANDREAEA PALM ET JOCHEMS 1923, SEE ANDREAEANA

ANDREAEANA PALM ET JOCHEMS 1924. SEE DEPT. PROEF. MEDAN-SUMATRA BULL. 19:19 AND
    PLATE III, 1923 /A. DELIENSIS PALM ET JOCHEMS
    ?, APPEARS TO BE AN ACREMONIUM. ORIG. ILLUS. NOT DIAGNOSTIC.
         SYNO    =ANDREAEA PALM ET JOCHEMS 1923 NON EHRHART
                 =PALMOMYCES MAIRE 1926, NOM. NOV. FOR ANDREAEA

ANEMATIDIUM GRONCHI 1931. BOLL. IST. SIEROTER. MILANO 10:242, FIGS 1-10 /A.
    OXIPHILUM
    ?, ORIG. ILLUS. AND DESCR. NOT DIAGNOSTIC

ANGUILLOSPORA INGOLD 1942. TRANS BRIT MYCOL SOC 25:402 /A. LONGISSIMA (SACC. ET
    SYDOW)
    SCOLECO,CETERI,HYALO /TUBAKI, BULL. NAT. SCI. MUSEUM, TOKYO 3:251-252,1957.
    (ILLUS)

ANGULIMAYA SUBRAMANIAN ET LODHA 1969. ANTONIE VAN LEEUWENHOEK 30:327 (317-330,
    ILLUS) /A. SUNDARA
    AMERO,BASO,HYALO

ANGULOSPORA NILSSON 1962. SVENSK. BOT. TIDSKR. 56:354-355 (ILLUS) /S. AQUATICA
    SCOLECO OR HELICO,CETERI,HYALO

ANKISTROCLADIUM PERROTT 1960. TRANS BRIT MYCOL SOC 43:557 (ILLUS) /A. FUSCUM
    =CASARESIA FRAGOSO 1920 FIDE KENDRICK

ANNELLODOCHIUM DEIGHTON 1969. MYCOL PAPERS CMI 118:28-30 (ILLUS) /A.
    RAMULISPORUM
    AMERO OR DIDYMO,BLASTO OR BASO,PHAEO

ANNELLOPHORA HUGHES 1951. TRANS BRIT MYCOL SOC 34:544 /A. SOLANI (SYDOW)
    PHRAGMO,CETERI OR BLASTO,PHAEO. NOT VERY DIFFERENT FROM SPORIDESMIUM LINK
    /ELLIS, MYCOL PAPERS CMI 70:84-89,1958. (ILLUS,EMEND,KEY)
         SYNO    =CHAETOTRICHUM SYDOW NON RABENHORST FIDE HUGHES 1951

ANNELLOPHORELLA SUBRAMANIAN 1962. PROC. INDIAN ACAD. SCI. SECT B,55:6 /A.
    FAUREAE (HENNINGS) ELLIS 1963
    DICTYO,CETERI,PHAEO /ELLIS, MYCOL PAPERS CMI 87:11-13,1963. (ILLUS)

ANNELLOPHRAGMIA SUBRAMANIAN 1963. PROC. INDIAN ACAD. SCI., SECT. B, 58:349 /A.
    COONOORENSIS (SUBRAMAN.) SUBRAMAN. 1963
    PHRAGMO,CETERI,PHAEO /SUBRAMANIAN, PROC. INDIAN ACAD. SCI. 42B:283-292,1955.
    (ILLUS) SUB ARTHROBOTRYUM

ANODOTRICHUM (CORDA) RABENHORST 1844. KRYPT.-FL. 1:86 /A. OLIGOCARPUM CORDA 1838
    =BLASTOTRICHUM CORDA 1838 FIDE LINDAU 1907

ANOMOMYCES HOHNEL 1919. BERL. DEUTSCHL. BOT. GES. 37:153 /A. ARBUTICOLUS
    (SOWERBY)
    ?, NOT ILLUSTRATED OR DESCRIBED BY HOHNEL

ANSATOSPORA NEWHALL 1944. PHYTOPATHOLOGY 34:98 /A. MACROSPORA (OSTERWALDER)
    =CENTROSPORA NEERGAARD 1942 FIDE ELLIS 1971

ANTENNARIA LINK 1809. LINN. SPEC. PLANT IV 6(1):118 /A. ERICOPHILA (LINK) LINK
    1824
    =NOMEN ILLEGITIMUM FIDE HUGHES 1958

ANTENNATULA FRIES EX STRAUSS 1850. FLORA 33 (BEILAGE):99 /A. PINOPHILA (NEES)
    STRAUSS 1850

PHRAGMO,CETERI,PHAEO /HUGHES, NEW ZEALAND J. BOTANY 8:178-197,1970.
   (ILLUS,EMEND)
   SYNO     =HORMISCIELLA BATISTA 1956 FIDE HUGHES 1970

ANTENNOPSIS HEIM 1952. BULL. SOC. MYCOL. FR. 67:347 /A. GALLICA HEIM ET BUCHLI
   1952 (ON RETICULOTERMES)
   PHRAGMO,CETERI,PHAEO

ANTENNULARIA REICHENBACH 1828. CONSPECT. REGNI VEGET. 1:5 /A. ERICOPHILA (LINK)
   HOHNEL 1909
   =NOMEN DUBIUM FIDE HUGHES 1970 /HUGHES, NEW ZEALAND J. BOTANY 8:156,1970.
   (REVIS)
   SYNO     =ANTENNARIA LINK 1809 FIDE HUGHES 1970
            =ANTENNATARIA REICHENBACH 1841
            =ANTENNINA FRIES 1849
            =GIBBERA FRIES SUBGENUS ANTENNULARIA (REICHENBACH) PETRAK 1947

ANTHINA FRIES 1823. SYST. MYCOL. 2:281 /A. FLAMMEA
   =STERILE

ANTROMYCES FRESENIUS 1850. BEITRAGE ZUR MYKOL. 1:37 (ILLUS) /A. COPRIDIS
   AMERO OR DIDYMO,ARTHRO,HYALO

ANTROMYCOPSIS PAT. ET TRAB. 1897. BULL. SOC. MYC. FR. 13:215 /A. BROUSSONETIAE
   AMERO,ARTHRO,PHAEO /MORRIS, WEST. ILL. UNIV. SER. BIOL. SCI. 3:20-21,1963.
   (ILLUS)

ANULOHYPHA CIFERRI 1962. ATTI IST. BOT. UNIV. PAVIA, SER. 5, 19:88 (ILLUS) /A.
   SERPENS
   =STERILE

ANULOSPORIUM SHERBAKOFF 1933. MYCOLOGIA 25:262 AND PLATE 35 /A. NEMATOGENUM
   =STERILE, BASED ON NEMATODE TRAPPING RINGS

AORATE SYDOW 1929. ANN. MYCOL. 27:84 (ILLUS) /A. COSTARICANA
   =TITAEA SACC. 1876 FIDE AINSWORTH 1961

APHANOCLADIUM GAMS 1970. CEPHALOSPORIUM-ART. SCHIMMELPILZE P. ** /A. ALBUM
   (PREUSS)
   AMERO,GLOIO,HYALO /VON ARX, GENERA OF FUNGI IN CULTURE P. 182-183,1970.
   (ILLUS)

APHOTISTUS HUMB. = RHIZOMORPHA FIDE SACCARDO FROM AINSWORTH 1971

APIOSPORIUM KUNZE 1817. MUK. HEFT. 1:8 /A. SALICIS KUNZE AND SCHMIDT 1817
   =SCLEROTIUM, FIDE HOHNEL FROM AINSWORTH 1971

APOSPORELLA THAXTER 1920. BOT. GAZ. 69:11 AND PLATE III /A. ELEGANS
   ?

APOTEMNOUM CORDA 1833. STURMS DEUTSCHL. FLORA, BD. 3, HEFT 13:77-78 AND TAB. 39
   /A. MACULANS
   ?, SACCARDO TREATED AS CLASTEROSPORIUM, BUT CORDA SAID SLIMY

APPELIA (SACC.) SACCARDO = TRICHOCONIS FIDE AINSWORTH 1971

ARACHNOPHORA HENNEBERT 1963. CAN. J. BOTANY 41:1165-1169 (ILLUS) /A. FAGICOLA
   DICTYO OR STAURO,CETERI,PHAEO

ARANEOMYCES HOHNEL 1909. S.B. AKAD. WISS. WIEN 118:894 /A. ACARIFERUS
   =TITAEA SACCARDO 1876 FIDE DAMON (FROM AINSWORTH 1961)

ARBUSCULA BATISTA ET PERES 1965. MYCOPATHOLOGIA 25:162-165 (ILLUS) /A. EUGENIAE
   =STERILE

ARNAUDINA TROTTER 1931. IN SACCARDO, SYLLOGE FUNGORUM 25:986 /A. MANAOENSIS
   (ARNAUD) TROTTER 1925
   ?, CONIDIAL STATE OF PARODIELLINA MANAOENSIS (HENNINGS) ARNAUD 1919. NEW
   NAME FOR EXOSPORINA ARNAUD NON OUDEMANS
   SYNO     =EXOSPORINELLA BENDER 1932, MYCOLOGIA 24:410-411

ARTHRINIUM KUNZE 1817. MYKOL. HEFTE 1:9 /A. CARICICOLA

  AMERO,CETERI,PHAEO, ASCO STATE =APIOSPORA FIDE HUDSON 1963 /ELLIS, MYCOL
    PAPERS CMI 103:1-33,1965. (MONOGR,KEY,ILLUS)
      SYNO     =PAPULARIA FRIES 1825 FIDE ELLIS 1965
              =CAMPTOUM LINK 1824 FIDE HUGHES 1958
              =RACEMOSPORIUM MOREAU ET MOREAU 1941 FIDE ELLIS 1965
              =GONATOSPORIUM CORDA 1839 FIDE HUGHES 1958
              =GONIOSPORIUM LINK 1824 FIDE HUGHES 1958
              =PHAEOHARZIELLA LOUBIERE 1924 FIDE ELLIS 1965
              =MICROTYPHA SPEGAZZINI 1910 FIDE ELLIS 1971
              =PSEUDOBASIDIUM TENGWALL 1924 FIDE ELLIS 1971
              =SPOROPHLEUM NEES EX LINK 1824 FIDE HUGHES 1958
              =TUREENIA HALL 1915 FIDE ELLIS 1971
              =INNATOSPORA VAN BEYMA 1919 FIDE ELLIS 1971
              ?=RHINOCEPHALUM KAMYSCHKO 1961

ARTHROBOTRYELLA SIBILIA 1928. BOL. STAZ. PAT. ROME, N.S. 8:448, FIGS. 1 AND 2
    /A. HERNICA
  ?=CORDANA PREUSS 1851

ARTHROBOTRYOMYCES BATISTA ET BEZERRA 1961. INST. MYCOL. UNIV. RECIFE PUBL. NO.
    321:9-12 (ILLUS) /A. AMAZONENSIS BAT. ET BEZ. 1961
  ?=ARTHROBOTRYUM CESATI 1854

ARTHROBOTRYS CORDA 1839. PRACHTFLORA P. 43 /A. SUPERBA
  DIDYMO,CETERI,HYALO /HAARD, MYCOLOGIA 60:1140-1159,1968(ILLUS,KEY,REVIS)
      SYNO     =DIDYMOZOOPHAGA SAPRUNOV ET GALIULINA FIDE HAARD 1968
              =GENICULARIA RIFAI ET COOKE 1966 FIDE CARMICHAEL
              ?=GONYELLA SYDOW 1919
              ?=DUDDINGTONIA COOKE 1969

ARTHROBOTRYUM CESATI 1854. HEDWIGIA, 1:TAB. 4,FIG. 1, ERKLARUNG DER TAB. 4 AND
    5 /A. STILBOIDEUM
  PHRAGMO,GLOIO,HYALO OR PHAEO, NON ARTHROBOTRYUM ROSTRUP 1916 = GONYELLA SYDOW
    1919 /HUGHES, THE NATURALIST, OCTOBER-DECEMBER:171-172,1951. (ILLUS)
      SYNO     =LINDAUOMYCES KOORDERS 1907 FIDE CLEMENTS AND SHEAR 1931
              =ARTHROGRAPHIUM CESATI FIDE SACCARDO FROM AINSWORTH 1971
              =WETTSTEINIELLA KUNTZE FIDE AINSWORTH 1971
              ?=ARTHROBOTRYOMYCES BATISTA ET BEZERRA 1961

ARTHROCLADIUM PAPENDORF 1969. TRANS. BRIT. MYCOL. SOC. 52:483-489 (ILLUS) /A.
    CAUDATUM
  PHRAGMO OR SCOLECO,CETERI,PHAEO

ARTHROGRAPHIS COCHET 1939. ANN. PARASIT. HUM. COMP. 17:97-101 AND PL. III /A.
    LANGERONI
  AMERO,ARTHRO,HYALO, 'A. LANGERONI IS PROBABLY THE SAME SPECIES AS OIDIODENDRON
    KALRAI TEWARI ET MACPHERSON 1971 (MYCOLOGIA 63:602-611)

ARTHROGRAPHIUM CESATI =ARTHROBOTRYUM CESATI FIDE SACCARDO FROM AINSWORTH 1971

ARTHROSPORIA GRIGORAKIS 1925. ANN. SCI. NAT. BOT. X,7:414 /SEVEN ORIGINAL
    SPECIES
  =NOMEN ILLEGITIMUM FIDE DODGE 1935

ARTHROSPORIUM SACCARDO 1880. MICHELIA 2:32 /A. ALBICANS
  PHRAGMO,CETERI,HYALO /MORRIS, WESTERN ILLINOIS UNIV. SER. BIOL. SCI.
    3:24-25,1963. (ILLUS)
      SYNO     =PHRAGMOSTILBE SUBRAMANIAN FIDE CARMICHAEL
              ?=STILBOMYCES ELLIS ET EVERHART 1896

ARTICULARIA HOHNEL 1909. S. B. AKAD. WISS. WIEN 118:407 (ILLUS) /A. QUERCINA
    (PECK) =A. STATE OF ASCOMYCETELLA QUERCINA PECK 1881
  AMERO,GLOIO,HYALO, ASCO STATE = ASCOMYCETELLA PECK 1881 FIDE PECK /PECK,
    BULL. TORREY BOT. CLUB 8:49-51 AND PLATE VII, 1881. (ILLUS)

ARTICULARIELLA HOHNEL 1909. S. B. AKAD. WISS. WIEN 118:410 /A. AURANTIACA
    (ELLIS ET MARTIN) HOHNEL
  =HELOSTROMA PAT. FIDE AINSWORTH 1961

ARTICULOSPORA INGOLD 1942. TRANS BRIT MYCOL SOC 25:376 /A. TETRACLADIA
    STAURO,CETERI,HYALO /TUBAKI, BULL. NAT. SCI. MUSEUM, TOKYO 3:251-253,1957.
    (ILLUS)

ARXIELLA PAPENDORF 1967. TRANS. BRIT. MYCOL. SOC. 50:73 /A. TERRESTRIS
    DIDYMO,BLASTO,HYALO /BARRON, HYPHOMYCETES P. 92-93,1968. (ILLUS)

ASCHIZOTRICHUM RIEUF 1962. CAH. RECH. AGRON. RABAT 15:61 /A. GRISEUM
    =WIESNERIOMYCES KOORDERS 1907 FIDE KENDRICK

ASCOCONIDIUM SEAVER 1942. MYCOLOGIA 34:414 /A. CASTANEAE
    PHRAGMO OR DICTYO,CETERI,HYALO, POSSIBLY THE SAME AS EXCIOCONIDIUM PLUNKETT
    1925 /FUNK, CAN. J. BOTANY 44:219-222,1966. (ILLUS)

ASPERGILLOPSIS SPEGAZZINI 1910. ANN. MUS. NAC. BUENOS AIRES SER. 3A, 20:434-436
    (ILLUS) /A. NIGRA (VAN TIEGHEM) SPEG. 1910
    =ASPERGILLUS MICHELI EX LINK 1809 (NIGER SERIES)

ASPERGILLOPSIS SOPP 1912. VID.-SELSK. SKR. I. M.-N. KL. 11:201-204 AND TAFEL 20
    /A. FUMOSIS
    =NOMEN ILLEGITIMUM, NON ASPERGILLOPSIS SPEG. 1910

ASPERGILLUS MICHELI EX LINK  1809. MAG. GES. NATURF. FREUNDE, BERLIN 3:16 /A.
    GLAUCUS LINK 1809 (LECTOTYPE)
    AMERO,BASO,HYALO OR PHAEO, ASCO STATE =EUROTIUM LINK FIDE RAPER AND FENNELL
    1965 OR EMERICELLA FIDE BENJAMIN 1955 /RAPER AND FENNELL, THE GENUS
    ASPERGILLUS 686 PP., 1965 (MONOGR)
        SYNO    =ASPERGILLOPSIS SPEGAZZINI 1910
                =CLADASPERGILLUS RITG. FIDE AINSWORTH 1971
                =EUASPERGILLUS LUDWIG FIDE AINSWORTH 1971
                =RHOPALOCYSTIS GROVE 1911 FIDE RAPER AND FENNELL 1965
                =STERIGMATOCYSTIS CRAMER 1859 FIDE CLEMENTS AND SHEAR 1931
                ?=ALLIOSPORA PIM 1883
                ?=CLADOSARUM YUILL ET YUILL 1938 FIDE AINSWORTH 1971
                ?=GUTTUROMYCES RIVOLTA FIDE AINSWORTH 1971
                ?=REDAELLIA CIFERRI 1930
                ?=SCEPTROMYCES CORDA 1831 SEE LINDAU 1907
                ?=SPERMATOLONCHA SPEGAZZINI 1909
                ?=SPHAEROMYCES MONTAGNE 1845

ASPERISPORIUM MAUBLANC 1913. BULL. SOC. MYCOL. FRANCE 29:357 /A. CARICAE
    (SPEG.) = A. STATE OF SPHAERELLA CARICAE MAUBL. 1913
    DIDYMO,CETERI,PHAEO, ASCO STATE =SPHAERELLA FIDE MAUBLANC 1913 /HUGHES, CAN J
    BOTANY 31:575,1953. (ILLUS)

ASTELECHIA CIFERRI 1962. ATTI IST. BOT. UNIV. PAVIA SER 5, 19:90 AND PLATE V,
    FIGS. 1-3 /A. RADIANS
    ?, ORIGINAL ILLUSTRATION NOT DIAGNOSTIC

ASTEROMYCES MOREAU ET MOREAU EX HENNEBERT 1962. CAN. J. BOTANY 40:1211-1213 /A.
    CRUCIATUS
    AMERO,CETERI,PHAEO /HENNEBERT, CAN J BOTANY 40:1203-1216,1962. (ILLUS)

ASTEROPHORA DITMAR 1809. SCHRAD. J. BOT. 3:56 /A. LYCOPERDOIDES
    AMERO OR STAURO,CETERI,HYALO, BASIDIO STATE =NYCTALIS FRIES FIDE AINSWORTH
    1961 /INGOLD, TRANS BRIT MYCOL SOC 24:29-32,1940. (ILLUS) HUGHES, IN
    KENDRICK 1971 P. 20-23 (ILLUS)
        SYNO    =ASTEROTRICHUM BONORDEN 1851

ASTEROTHECIUM WALLROTH 1836. IN ENDLICHER, GEN. PLANT. P. 25 /A. STRIGOSUM
    WALLROTH ?
    =STEPHANOMA WALLROTH 1833 FIDE LINDAU IN RABENHORST KRYPT.-FL. 2 AUFL., 1
    BD., 8:225, 1907

ASTEROTRICHUM BONORDEN 1851. HANDB. ALLGEM. MYKOL. P. 82-83 AND FIG. 224 /A.
    DITMARI, NOM. NOV. FOR ASTEROPHORA AGARICOIDES DITMAR
    =ASTEROPHORA DITMAR 1809

ASTOMA GRAY = SCLEROTIUM FIDE SACCARDO FROM AINSWORTH 1971

## 10. Hyphomycetes 343

ASTRABOMYCES BATISTA 1961. PUBL. INST. MICOL. UNIV. RECIFE, 320:22-27 (ILLUS)
/A. AMAZONENSIS
?, ORIGINAL ILLUSTRATION DIFFICULT TO INTERPRET

ASTRODOCHIUM ELLIS ET EVERHART 1897. AMER. NAT. 31:430 /A. COLORADENSE
?, NOT ILLUSTRATED BY AUTHORS

ATELEOTHYLAX OTA ET LANGERON 1923. ANN. PARASIT. HUM. COMP. 1:333 /A. CURRII
(CHALMERS ET MARSHALL) =TRICHOPHYTON CURRII CHALM. ET MARSHALL 1914
=NOMEN DUBIUM

ATRACTIELLA SACCARDO 1886. SYLLOGE FUNGORUM 4:578-579 /A. BRUNAUDIANA (SACC.)
SACC. 1886
?, NOT ILLUSTRATED BY AUTHOR

ATRACTILINA DEARNESS ET BARTHOLOMEW 1924. MYCOLOGIA 16:175 /A. CALLICARPAE
DEARN. ET BARTH. 1924
?, NOT ILLUS. BY AUTHORS

ATRACTINA HOHNEL 1904. HEDWIGIA 43:298 /A. BISEPTATA
=STERIGMATOBOTRYS OUDEMANS 1886 FIDE HUGHES 1958

ATRACTIUM LINK 1809. MAG. GES. NATURF. FREUNDE, BERLIN 3:10 /A. STILBASTER
PHRAGMO OR SCOLECO,GLOIO,HYALO /MORRIS, WEST. ILL. UNIV. SER. BIOL. SCI.
3:30-31,1963. (ILLUS)

ATRICHOPHYTON CASTELLANI ET CHALMERS 1919. MANUAL OF TROPICAL MEDICINE, ED. 3,
P. 1008,1919 /A. ALBISCANS (NIEUWENHUIS) CASTEL. ET CHALM. 1919
=CHRYSOSPORIUM CORDA 1833 FIDE CARMICHAEL

AUREOBASIDIUM VIALA ET BOYER 1891. REV. GEN. BOT. 3:369 /A. VITIS =A. PULLULANS
(DE BARY) ARNAUD 1918
AMERO,GLOIO,HYALO OR PHAEO, ASCO STATE =GUIGNARDIA OR DOTHIORA OR SYDOWIA OR
POTEBNIAMYCES FIDE VON ARX 1970,=XENOMERIS FIDE FUNK AND SHOEMAKER 1971
/COOKE, MYCOPATHOLOGIA 12:1-45,1959. (REFS), COOKE, MYCOPATHOLOGIA
21:225-271,1963. (ILLUS,PHYSIOL)
    SYNO    =PULLULARIA BERKHOUT 1923 FIDE COOKE 1962
           =HORMONEMA LAGERB. ET MELIN 1927 FIDE CARMICHAEL
           =SARCINOMYCES LINDNER 1901 FIDE COOKE
           =EXOBASIDIOPSIS KARAKULIN 1922 FIDE CLEMENTS AND SHEAR 1931
           =PROTOCORONOSPORA ATKINSON AND EDGERTON 1907 FIDE VON ARX 1970
           SUB KABATIELLA
           =RHABDOGLOEOPSIS PETRAK 1925 FIDE VON ARX SUB KABATIELLA

AZOSMA CORDA 1831. STURMS DEUTSCHL. FLORA, BD. 3, HEFT 12:35-36 AND TAB. 18 /A.
HELMINTHOSPORIODES
?, ORIG. ILLUS. AND DESCR. NOT DIAGNOSTIC

BACILLISPORA NILSSON 1962. BOT. NOTISER. 115:77-79 (ILLUS) /B. AQUATICA
PHRAGMO OR SCOLECO,CETERI,HYALO

BACTRIDIOPSIS HENNINGS 1904. HEDWIGIA 43:397 /B. ULEI
AMERO,CETERI,PHAEO, NON BACTRIDIOPSIS FRAGOSO ET CIFERRI 1927, BOL. SOC. ESP.
HIST. NAT. 27:330-331 (ILLUS) /DAMON AND DOWNING, MYCOLOGIA
46:209-221,1954. (ILLUS)

BACTRIDIUM KUNZE 1817. MYKOL. HEFTE 1:5 /B. FLAVUM
PHRAGMO,CETERI,HYALO /HUGHES, NEW ZEALAND J. BOTANY 4:522-532,1966.
(REVIS,ILLUS)
    SYNO    =PODOBACTRIDIUM PETCH FIDE HUGHES 1966
           =DAMNOSPORIUM CORDA FIDE AINSWORTH 1971
           =ERICIANELLA BOND. FIDE FRIES FROM AINSWORTH 1971

BACTRODESMIELLA ELLIS 1959. MYCOL. PAPERS CMI 72:14-15 /B. MASONII (HUGHES)
PHRAGMO,BASO OR CETERI,PHAEO

BACTRODESMIUM COOKE 1883. GREVILLEA 12:35 /B. ABRUPTUM (BERK. ET BR.) MASON ET
HUGHES 1953
PHRAGMO,CETERI,PHAEO /ELLIS, MYCOL PAPERS CMI 72:1-15,1959. (REVIS,KEY,ILLUS)

BAHUPAATHRA SUBRAMANIAN ET LODHA 1964. ANTONIE VAN LEEUWENHOEK 30:329 AND FIG. 3, P. 322 /B. SAMALA SUBRAMAN. ET LODHA 1964
AMERO,GLOIO,HYALO. BUT NOT VERY DIFFERENT FROM CLADORRHINUM SACC. ET MARCH.

BAHUSAKALA SUBRAMANIAN 1958. J. INDIAN BOT. SOC. 37:61-63 (ILLUS) /B. OLIVACEO-NIGRA (BERKELEY ET BROOME)
DIDYMO OR PHRAGMO,ARTHRO,PHAEO, ASCO STATE = AULOGRAPHINA ARX ET MULLER FIDE MULLER, HARR AND SULMONT 1969 /VON ARX, GENERA OF FUNGI IN CULTURE P. 168-169,1970. (ILLUS)

BAHUSANDHIKA SUBRAMANIAN 1956. J. INDIAN BOT. SOC. 35:469 /B. INDICA (SUBRAMAN.) SUBRAMAN. 1956
=TORULA (PERSOON) LINK 1809 FIDE KENDRICK /BARNETT, IMPERFECT FUNGI NO 237, 1960. (ILLUS)

BAINIERIA ARNAUD 1952. BULL. SOC. MYCOL. FR. 68:188-189 (ILLUS) /B. HYALINA
AMERO,CETERI,HYALO, NO LATIN DIAGNOSIS

BALANIUM WALLROTH 1833. FLORA CRYPT. GERMAN. 2:159 /B. STYGIUM
DIDYMO,CETERI,PHAEO /ELLIS, MYCOL PAPERS CMI 79:19-20,1961. (ILLUS) HUGHES AND HENNEBERT, CAN. J. BOTANY 39:1505-1508,1961. (ILLUS)

BARGELLINIA BORZI 1888. MALPIGHIA 2:469 /B. MONOSPORA
=WALLEMIA JOHAN-OLSON 1887 FIDE VON ARX 1970

BARNETTELLA RAO ET RAO 1964. MYCOPATHOLOGIA 22:56 /B. SPECIOSA
DICTYO,PASO,PHAEO
   SYNO   =VINCULUM ROY, DWIVEDI ET KHANNA 1965 FIDE VERONA 1967

BARTHELETIA ARNAUD 1954. BULL. SOC. MYCOL. FRANCE 69:300 /B. PARADOXA
=STERILE

BARYEIDAMIA KARSTEN 1888. HEDWIGIA 27:132-137 (ILLUS) /B. PARASITICA
=PAPULASPORA PREUSS 1851 FIDE SACCARDO

BASIASCUM CAVARA 1888. ATTI IST. BOT. PAVIA II, 1:433 /B. ERIOBOTRYAE
=SPILOCAEA FRIES 1825 FIDE HUGHES 1958

BASIDIDYMA CIFERRI 1962. ATTI IST. BOT. UNIV. PAVIA, SER 5 19:94-95 AND PLATE VIII, FIGS. 1-3 /B. PEREXIGUA
?, ORIGINAL ILLUSTRATION NOT DIAGNOSTIC

BASIDIELLA COOKE 1878. GREVILLEA 6:118 /B. SPHAEROCARPA
? /COOKE, J. QUE. MICRO. CLUB, SER 2, 2:PL IX, FIG. 1A TO E

BASIDIOBOTRYS HOHNEL 1909. S.B. AKAD. WISS. WIEN 118:420 /B. CLAUTRIAVII (PATOUILLARD)
AMERO,CETERI,HYALO, ASCO STATE =HYPOXYLON FIDE BARNETT 1957 /BARNETT, MYCOLOGIA 49:588-595,1957. (ILLUS,ECOL) MARTIN, SOUTH AFRICAN J. BOTANY 33:205-240,1967. (ILLUS,ECOL)

BASIFIMBRIA SUBRAMANIAN ET LODHA 1968. CURR. SCI. 37:247 /B. AUREA SUBRAMAN. ET LODHA 1968
AMERO,CETERI,HYALO, CLOSE TO GILMANIELLA

BASIPETOSPORA COLE ET KENDRICK 1968. CAN. J. BOTANY 46:991-992 (987-992,ILLUS) /B. RUBRA =BASIPETOSPORA STATE OF MONASCUS RUBER VAN TIEGHEM 1884
AMERO,BASO,HYALO, ASCO STATE =MONASCUS VAN TIEGHEM FIDE COLE AND KENDRICK 1968

BASISPORIUM MOLLIARD 1902. BULL. SOC. MYCOL. FR. 18:170 /B. GALLARUM MOLLIARD = NIGROSPORA ORYZAE
=NIGROSPORA ZIMM. 1902 FIDE MASON 1933

BASITORULA ARNAUD 1954. BULL. SOC. MYCOL. FR. 69:276 /B. CINGULATA
=GLIOMASTIX GUEGUEN FIDE DICKINSON 1968, NO LATIN DIAGNOSIS

BEAUVERIA VUILLEMIN 1912. BULL. BOT. SOC. FR. 59:40 /B. BASSIANA (BALS.) VUILL. 1912

## 10. Hyphomycetes

AMERO,CETERI,HYALO /MACLEOD, CAN J. BOTANY 32:818-890,1954. (REVIS,ILLUS)
   MACLEOD, ANN. N.Y. ACAD. SCI. 60:58-70,1954 (ECOL,DESCR)

BELTRANIA PENZIG 1882. NUOV. GIORN. BOT. ITAL. 14:72 /B. RHOMBICA
   AMERO,CETERI,PHAEO /BARRON, HYPHOMYCETES P. 98-99,1968. (ILLUS) PIROZYNSKI,
   MYCOL PAPERS CMI 90:1-16,1963. (REVIS,KEY,ILLUS)

BELTRANIELLA SUBRAMANIAN 1952. PROC. INDIAN ACAD. SCI. B,36:227 /B. ODINAE
   AMERO,CETERI,PHAEO. ASCO STATE= PSEUDOMASSARIA FIDE HODGES AND BARR 1971
      (MYCOLOGIA 58:562) /PIROZYNSKI, MYCOL PAPERS CMI 90:26-28,1963. (ILLUS)
      PIROZYNSKI, CAN. J. BOTANY 48:571-575,1970. (ILLUS)
      SYNO   =ELLISIOPSIS BATISTA ET NASCIMENTO 1956 FIDE PIROZYNSKI 1970

BELTRANIOPSIS BATISTA ET BEZERRA 1960. PUBL. INST. MIC. UNIV. RECIFE 296:7 /B.
   ESENBECKIAE
   AMERO,CETERI,PHAEO /PIROZYNSKI, MYCOL PAPERS CMI 90:24-26,1963. (ILLUS)

BENIOWSKIA RACIBORSKI 1900. PARAS. ALGEN PILZE JAVAS, 2:37 /B. GRAMINIS
   RACIBORSKI =B. SPHAEROIDEA (KALCHBRENNER ET COOKE) MASON
   AMERO,CETERI,HYALO /MASON, MYCOL PAPERS CMI 2:26-27,1928. (REFS) BARRON,
   HYPHOMYCETES P. 100, 1968. (ILLUS)

BERKELEYNA KUNZE 1898. REVIS. GEN. PL. 3:477 /B. CURTA (BERKELEY)
   =PERICONIA TODE EX PERSOON 1801 FIDE ELLIS 1971

BERKLEASMIUM ZOBEL 1854. IN CORDA, ICONES FUNGORUM 6:4 /B. CORDAEANUM =B.
   CONCINNUM (BERK.) HUGHES 1958
   DICTYO,CETERI,PHAEO /GOOS, CAN. J. BOTANY 47:503-504 AND PL. 1, 1969.(ILLUS)

BERTEROMYCES CIFERRI 1954. SYDOWIA 8:267 /B. AENEUS
   =CERCOSPORIDIUM EARLE FIDE DEIGHTON 1967

BIHARIA THIRUMALACHAR ET MISHRA 1953. SYDOWIA 7:79 /B. VANGUERIAE
   =STENELLA SYDOW 1930 FIDE ELLIS 1970

BILBOQUE VIEGAS 1960. BRAGANTIA 19:903 AND FIG 1, P. 898 /B. MAGNIFICUM
   ?, ORIGINAL ILLUSTRATION AND DESCRIPTION NOT DIAGNOSTIC

BIOCONIOSPORIUM BATISTA ET BEZERRA 1964. PUBL. INST. MICOL. UNIV. RECIFE 417:4
   AND FIG. 2, P.14 /B. BACCHARIDIS
   =PITHOMYCES BERKELEY ET BROOME 1873 FIDE VON ARX 1970

BIPOLARIS SHOEMAKER 1959. CAN. J. BOTANY 37:882 /B. MAYDIS (NISIKADO)
   =DRECHSLERA ITO 1930 FIDE ELLIS 1971

BISPORA CORDA 1837. ICONES FUNGORUM 1:9 /B. MONILIOIDES CORDA =B. ANTENNATA
   (PERSOON) MASON 1953
   DIDYMO,BLASTO,PHAEO /HUGHES, CAN. J. BOTANY 36:740,1958 (LIST) BARRON,
   HYPHOMYCETES P. 102-103,1968. (ILLUS)

BISPOROMYCES VAN BEYMA 1940. A. VAN LEEUWENHOEK J. MICROBIOL. SEROL. 6:277 /B.
   CHLAMYDOSPORIS
   =CHLORIDIUM LINK FIDE HUGHES 1958

BIZZOZERIELLA SPEGAZZINI 1888. AN. SOC. ARG. 26:2:73 /B. PHYLLOGENA
   ?, NOT ILLUSTRATED BY AUTHOR

BLASTOBOTRYS KLOPETEK 1967. ARCH. MIKROBIOL. 58:92-95 (ILLUS) /B. NIVEA
   AMERO,CETERI,HYALO

BLASTOCONIUM CIFERRI 1931. J. DEP. AGRIC. P. R. 15:233 /B. TROPICUM
   ?, NOT ILLUS. BY AUTHOR. BASED ON A DARK, YEAST-LIKE FUNGUS

BLASTOMYCES COSTANTIN ET ROLLAND 1888. BULL. SOC. MYCOL. FRANCE 4:153 /B.
   LUTEUS COST. ET ROLL. 1888
   =CHRYSOSPORIUM CORDA 1833 FIDE CARMICHAEL 1962

BLASTOMYCOIDES CASTELLANI 1928. AMER. J. TROP. MED. 8:381 /B. IMMITIS (STILES)
   CASTEL. = COCCIDIOIDES IMMITIS STILES
   =NOMEN ILLEGITIMUM FIDE CARMICHAEL 1962

BLASTOPHORELLA BOEDIJN 1937. BLUMEA, SUPPL. 1:140 /B. SMITHII
DIDYMO,CETERI,HYALO /ELLIS. DEM. HYPH. P. 397-398,1971. (ILLUS)

BLASTOTRICHUM CORDA 1838. ICONES FUNGORUM 2:10 AND TAB. IX, FIG. 50 /B.
CONFERVOIDES
=NOMEN DUBIUM, SEE GAMS AND HOOZEMANS, PERSOONIA 6:99,1970

BLODGETTIA WRIGHT 1881. TRANS. IRISH ACAD. 28:25 /B. BORNETI
=STERILE

BLOXAMIA BERKELEY ET BROOME 1854. ANN. MAG. NAT. HIST. 13:468 /B. TRUNCATA
BERK. ET BR. 1854
AMERO,BASO,HYALO /PIROZYNSKI AND MORGAN-JONES, TRANS BRIT MYCOL SOC
51:185-187,1968. (ILLUS)
SYNO    =ENDOSPOROSTILBE SUBRAMANIAN 1958 FIDE NAG RAJ
        ?=EXOSPORINA OUDEMANS 1904

BODINIA OTA ET LANGERON 1923. ANN. PARAS. HUM. COMP. 1:330 /B. VIOLACEA
(SABOURAUD) BODIN 1923
=TRICHOPHYTON MALMSTEN 1845

BOMPLANDIELLA SPEGAZZINI 1886. AN. SOC. ARG. 22:222 /B. GUARANITICA
?. NOT ILLUSTRATED BY AUTHOR, SOMETIMES GIVEN AS BONPLANDIELLA

BONORDENIELLA PENZIG ET SACCARDO 1901. MALPIGHIA 15:259 /B. MEMORANDA PENZ. ET
SACC. 1901
=CONIOSPORIUM LINK 1809 FIDE ELLIS, MYCOL. PAPERS CMI 125:2.1971

BONPLANDIELLA, SEE BOMPLANDIELLA

BOSTRYCHONEMA CESATI 1859. ERBAR. CRITT. ITAL. NO. 149 /B. ALPESTRE
? /LINDAU IN RABENHORST KRYPT.-FLORA 2 ED, 1 BAND, 8:381-382,1907., (ILLUS)

BOTRYDIELLA BADURA 1963. ALLIONIA 9:182 /B. BICOLOR =STAPHYLOTRICHUM COCCOSPORUM
=STAPHYLOTRICHUM MEYER ET NICOT 1956 FIDE VON ARX 1970

BOTRYOCLADIUM PREUSS 1851. LINNAEA 24:134 /B. DELECTATUM
=NEMATOGONIUM DESMAZIERES 1834 FIDE SACCARDO FROM AINSWORTH 1971

BOTRYODERMA PAPENDORF ET UPADHYAY 1969. TRANS. BR. MYCOL. SOC. 52:257 /B.
LATERITIUM
AMERO,CETERI,PHAEO

BOTRYONIPHA PREUSS 1852. LINNAEA 25:79 /B. ALBA
=STERILE, BASED ON THE ACANTHOPHYSES OF A HYMENOMYCETE /HUGHES, FRIESIA
9:61-63,1969. (ILLUS)

BOTRYOPHIALOPHORA LINDER 1944. FARLOWIA 1:404 /B. MARINA
=MYRIOCONIUM SYDOW FIDE VON ARX 1970

BOTRYOSPORIUM CORDA 1831. STURMS DEUTSCHL. FL. III:9 AND TAB. 5 /B. DIFFUSUM
(ALBERTINI ET SCHWEINITZ)
AMERO,CETERI,HYALO /MASON, MYCOL PAPERS CMI 2:27-29,1928. (LIST) BARRON,
HYPHOMYCETES P. 103-104, 1968. (ILLUS)
SYNO    =PEYLIA OPIZ FIDE AINSWORTH 1971

BOTRYOTRICHUM SACCARDO ET MARCHAL 1885. BULL. SOC. ROY. BOT. BELG. 24:66 /B.
PILULIFERUM =BOTRYOTRICHUM STATE OF CHAETOMIUM PILULIFERUM DANIELS 1961
AMERO,CETERI,HYALO, ASCO STATE =CHAETOMIUM FIDE DANIELS 1961 /DANIELS, TRANS
BRIT MYCOL SOC 44:79-86,1961. (ILLUS)

BOTRYOXYLON CIFERRI 1962. ATTI IST. BOT. UNIV. PAVIA, SER. 5, 19:97 /B.
GENICULATUM (CORDA) =CONOPLEA GENICULATA (CORDA) HUGHES 1958
=CONOPLEA PERSOON 1801 FIDE HUGHES 1958

BOTRYPES PREUSS 1852. LINNAEA 25:740 /B. ROSEA
=CILICOPODIUM CORDA FIDE LINDAU IN RABENHORST 1910

BOTRYTIS MICHELI EX PERSOON 1801. SYNOPSIS METH. FUNGORUM P. 690 /B. CINEREA
PERSOON 1801 =B. STATE OF SCLEROTINIA FUCKELIANA (DEBARY) FUCKEL

    AMERO,CETERI,PHAEO, ASCO STATE =SCLEROTINIA FIDE BARRON 1968 /ELLIS, DEM.
        HYPH. P. 178-184,1971. (ILLUS,KEY)
            SYNO    =CEPHALOCLADIUM REICHARDT FIDE AINSWORTH 1971
                    ?=PTERODINIA CHEVALLIER FIDE AINSWORTH 1971

BOTRYTOIDES MOORE ET ALMEIDA 1936. ANN. MISSOURI BOT. GARD. 23:545-547 AND PL.
    26, FIG. 3 /B. MONOPHORA, NOMEN ILLEGITIMUM
    =NOMEN CONFUSUM, BASED ON CHROMOBLASTOMYCOSIS FUNGI

BRACHYCLADIUM CORDA 1838. ICONES FUNGORUM 2:14 /B. PENICILLATUM
    =DENDRYPHION WALLROTH 1833 FIDE HUGHES 1958

BRACHYDESMIELLA ARNAUD EX HUGHES 1961. CAN. J. BOTANY 39:1095-1097 (ILLUS) /B.
    BISEPTATA
    PHRAGMO OR AMERO,CETERI,PHAEO

BRACHYDESMIUM (SACCARDO) COSTANTIN 1888. LES MUCEDINEES SIMPLES P. 167-168
    (ILLUS) /B. OVOIDEUM (CORDA) COSTANTIN 1888
    ?

BRACHYHELICOON ARNAUD 1952. BULL. SOC. MYCOL. FR. 68:208-209, FIG. 7U /B.
    XYLOGENUM
    HELICO,GLOIO,HYALO, NO LATIN DIAGNOSIS

BRACHYSPORIELLA BATISTA 1952. BOL. SOC. AGRIC. PERNAMBUCO 19:108 /B. GAYANA
    PHRAGMO,CETERI,PHAEO /ELLIS, MYCOL PAPERS CMI 72:15-19,1959. (ILLUS,KEY)
        SYNO    =EDMUNDMASONIA SUBRAMANIAN 1958 FIDE KENDRICK
                =MONOSPORELLA HUGHES NON KEILIN FIDE ELLIS 1959
                =MONOTOSPORELLA HUGHES 1958 FIDE ELLIS 1959

BRACHYSPORIUM SACCARDO 1880. SYLLOGE FUNG. 4:427 /B. OBOVATUM (BERK.)
    PHRAGMO,CETERI,PHAEO /ELLIS, MYCOL PAPERS CMI 106:43-54,1966. (KEY,ILLUS)

BRIAREA CORDA 1831. STURMS DEUTSCHL. FL. III (PILZE),BD 3,HEFT 11:11 /B.
    ELEGANS CORDA =ASPERGILLUS GLAUCUS (GROUP)
    =ASPERGILLUS LINK 1809 FIDE THOM AND RAPER  1945 (FROM HUGHES 1958)

BRIOSIA CAVARA 1888. ATTI. IST. BOT. PAVIA 2:321 /B. AMPELOPHAGA
    AMERO,BASO,PHAEO /MORRIS, WEST ILLINOIS UNIV SER BIOL SCI 3:32-33,1963.
    (ILLUS)

BROMICOLLA EICHW. = SCLEROTIUM FIDE SACCARDO FROM AINSWORTH 1971

BROOMEOLA KUNZE 1891. REV. GEN. PLANT. 2:845 /B. GLAUCA (BERKELEY ET BROOME)
    ?, NOM. NOV. FOR ENDODESMIA BERK. ET BROOME 1874 NON BENTH. 1862

BRYOCHYSIUM LINK = RHIZOCTONIA FIDE RABENHORST FROM AINSWORTH 1971

BURGOA GOIDANICH 1937. BOLL. STAZ. PAT. VEG., ROMA 17, N.S., 354 /SIX ORIGINAL
    SPECIES
    =STERILE

BYSSOCLADIUM LINK 1815. MAG. GES. NATURF. FREUNDE, BERLIN 7:35 /B. CANDIDUM
    =STERILE FIDE HUGHES 1958

CACUMISPORIUM PREUSS 1851. LINNAEA 24:30 /C. TENEBROSUM =C. CAPITULATUM (CORDA)
    HUGHES 1958
    PHRAGMO,CETERI OR GLOIO,PHAEO, COMPARE WITH PLEUROTHECIUM /GOOS, MYCOLOGIA
    61:52-56,1969. (ILLUS)

CADOPHORA LAGERBERG ET MELIN 1928. SVENSKA SKOGSVFOREN. TIDSKR. 25:263 /C.
    FASTIGIATA =PHIALOPHORA FASTIGIATA (L. ET M.) CONANT
    =PHIALOPHORA MEDLAR 1915 FIDE CONANT 1937

CALCARISPORA MARVANOVA ET MARVAN 1963. CAS. SLEZSK. MUS. OPAVE, SER. A,
    12:106-109 (ILLUS) /C. HIEMALIS
    ?=LUNULOSPORA INGOLD 1942

CALCARISPORIUM PREUSS 1851. LINNAEA 24:124 /C. ARBUSCULA
    AMERO,CETERI,HYALO /HUGHES, MYCOL PAPERS CMI 43:1-6,1951. (ILLUS)

CALDARIOMYCES WORONICHIN 1926. ANN. MYCOL. 24:261-264 /C. FUMAGO WORON. 1926 = 'FUMAGO' OF ZOPF 1878
?. ILLUSTRATED BY ZOPF, NOVA ACTA ACAD. CAES. LEOPOLD-CAROL. 40:TAF. 5 & 6, 1878

CALOSTILBELLA HOHNEL 1919. BERL. DEUTSCHL. BOT. GES. 37:160 /C. CALOSTILBE PHRAGMO,GLOIO,HYALO OR PHAEO, ASCO STATE =CALOSTILBE FIDE MORRIS 1963 /MORRIS, WEST. ILLINOIS SER BIOL SCI 3:34-35,1963. (ILLUS)

CAMPOSPORIUM HARKNESS 1884. BULL. CALIF. ACAD. SCI. 1:37 /C. ANTENNATUM PHRAGMO,CETERI,PHAEO /HUGHES, MYCOL PAPERS CMI 36:3-16,1951. (ILLUS)
    SYNO    =CAMAROSPORIUM CLEMENTS ET SHEAR (ORTHOGRAPHIC VARIANT),NON CAMAROSPORIUM SCHULZER

CAMPSOTRICHUM EHRENBERG 1819. IN LINK, JAHRB. F. GEWACHSK. 1(2):55 /C. BICOLOR
?. HUGHES (1958) EXAMINED THE TYPE SPECIMEN BUT DID NOT ILLUSTRATE

CAMPTOMERIS SYDOW 1927. ANN. MYCOL. BERLIN 25:142 /C. CALLIANDRAE DIDYMO OR PHRAGMO,CETERI,PHAEO /HUGHES, MYCOL PAPERS CMI 49:14-19,1952. (ILLUS)

CAMPTOSPORIUM LINK EX DUBY 1830. BOT. GALLICUM 2:928 /C. GLAUCUM (PERSOON) DUBY AS "EHRENB."
=NOMEN ILLEGITIMUM FIDE HUGHES 1958

CAMPTOUM LINK 1824. LINN. SPEC. PLANT. IV,6(1):44 /C. CURVATUM (KUNZE)
=ARTHRINIUM CURVATUM KUNZE
=ARTHRINIUM KUNZE 1817 FIDE HUGHES 1958

CAMPYLOSPORA RANZONI 1953. FARLOWIA 4:373 AND FIG. 11 /C. CHAETOCLADIA STAURO,CETERI,HYALO

CANDELABRELLA RIFAI ET COOKE 1966. TRANS. BRIT. MYCOL. SOC. 49:160 /C. JAVANICA
=ARTHROBOTRYS CORDA 1839 FIDE BARRON 1968

CANDELABRUM VAN BEVERWIJK 1951. ANTONIE VAN LEEUWENHOEK J. MICROBIOL. SEROL. 17:283-285 (ILLUS) /C. SPINULOSUM
STAURO,CETERI,HYALO
    SYNO    =DIONYSIA ARNAUD 1952 FIDE KENDRICK

CANDELOSPORA REA ET HAWLEY 1912. PROC. ROYAL IRISH ACAD. 31:11 /C. ILICICOLA
=CYLINDROCLADIUM  MORGAN 1892 FIDE BOOTH 1960

CAPILLARIA PERSOON 1822. MYCOL. EUROP. 1:50 /?. SIX ORIGINAL SPECIES
=NOMEN ILLEGITIMUM (NON CAPILLARIA STACKH. 1809)

CAPNOBOTRYS HUGHES 1970. NEW ZEALAND J. BOTANY 8:205 /C. NEESII
?. CONIDIA ONLY ILLUSTRATED BY AUTHOR

CAPNOCYBE HUGHES 1966. NEW ZEALAND J. BOTANY 4:335 (333-353, ILLUS) /C. FRASERAE =CAPNOCYBE STATE OF LIMACINIA FRASERAE HUGHES 1966
PHRAGMO,GLOIO,PHAEO, ASCO STATE =LIMACINIA NEGER FIDE HUGHES 1966

CAPNOPHIALOPHORA HUGHES 1966. NEW ZEALAND J. BOTANY 4:352 (333-353, ILLUS) /C. FRASERAE =C. STATE OF LIMACINIA FRASERAE HUGHES 1966
AMERO,GLOIO,HYALO, ASCO STATE =LIMACINIA NEGER FIDE HUGHES 1966

CAPNOSTYSANUS SPEGAZZINI 1918. PHYSIS 4(17):295 /C. STYSANOPHORUS (PENZIG ET SACCARDO)
?. NOT ILLUSTRATED BY SPEGAZZINI

CARLOSIA ARNAUD 1954. BULL. SOC. MYCOL. FR. 69:294 /C. MELIOLAE (ZIMMERMAN)
=ISTHMOSPORA STEVENS 1918 FIDE KENDRICK

CAPRIONIA BRICENO-IRAGORRY 1938. REV. CLIN. LUIZ RAZETTI, CARACAS 1:121 /C. PEDROSOI (BRUMPT) BRIC.-IRAG. = HORMODENDRUM PEDROSOI BRUMPT
=FONSECAEA NEGRONI 1936

CASARESIA FRAGOSO 1920. BOL. ESPAN. HIST. NAT. 20:112-114 (ILLUS) /C. SPHAGNORUM STAURO,CETERI,PHAEO

## 10. Hyphomycetes 349

    SYNO    =ANKISTROCLADIUM PERROTT 1960 FIDE KENDRICK

CATENULARIA GROVE 1886. IN SACCARDO, SYLLOGE FUNGORUM 4:303 /C. SIMPLEX GROVE
    =CATENULARIA STATE OF CHAETOSPHAERIA CUPULIFERA (BERK. ET BR.) SACCARDO 1883
    AMERO,BASO,PHAEO. ASCO STATE =CHAETOSPHAERIA FIDE HUGHES 1965 /HUGHES, NEW
    ZEALAND J. BOTANY 3:136-150,1965. (REVIS,ILLUS)
        SYNO    =PSILOÑIELLA COSTANTIN 1888 FIDE ELLIS 1971
                    =HAPLOCHALARA LINDER 1933 FIDE MASON 1941

CATTANEA GAROVAGLIO 1875. RC. IST. LOMB. SCI. LETT., 2 SER., 8:125 /C.
    HEPTASPORA GAROV. = D. HEPTOSPORUM (GAROV.) DAMON 1952
  =DICTYOSPORIUM CORDA 1836 FIDE ELLIS 1971

CENOCOCCUM FRIES 1825. SYST. ORB. VEG. P. 364 /C. GEOPHILUM
  =STERILE

CENTROSPORA NEERGAARD 1942. ZENTRALBL. BAKT. 2,54:411 /C. OHLSENII =C. ACERINA
    (HARTIG) NEWHALL 1946
    SCOLECO OR PHRAGMO,CETERI,HYALO OR PHAEO /HUGHES, MYCOL PAPERS CMI
    36:39,1951. (DESCR,REFS)
        SYNO    =ANSATOSPORA NEWHALL 1944 FIDE ELLIS 1971

CEPHALIOPHORA THAXTER 1903. BOT. GAZ. 37:157 /C. TROPICA (LECTOTYPE)
    PHRAGMO,CETERI,PHAEO /BARRON, HYPHOMYCETES P. 112-113,1968 (ILLUS)
        SYNO    =CEPHALOMYCES BAINIER 1907 FIDE CARMICHAEL

CEPHALOCLADIUM REICHARDT = BOTRYTIS FIDE AINSWORTH 1971

CEPHALODIPLOSPORIUM KAMYSCHKO 1961. BOT. MATER. (NOT. SYST. SECT. CRYPT. INST.
    BOT. ACAD. SCI. USSR 14:221 /C. ELEGANS
  =CEPHALOSPORIOPSIS PEYRONEL FIDE BARRON 1968

CEPHALODOCHIUM BONORDEN 1851. HANDB. MYKOL. P. 135 AND TAF. XI, FIG 227 /C.
    ALBUM
    ?

CEPHALOMYCES BAINIER 1907. BULL. SOC. MYCOL. FRANCE 23:109 /C. NIGRICANS
  =CEPHALIOPHORA THAXTER 1903 FIDE CARMICHAEL (FROM THE DESCRIPTIONS, C.
  NIGRICANS APPEARS TO BE THE SAME SPECIES AS C. IRREGULARIS THAXT.)

CEPHALOPHORUM NEES = STILBUM FIDE SACCARDO FROM AINSWORTH 1971

CEPHALOSPORIOPSIS PEYRONEL    . MEM. R. ACC. SCI. TORINO, SER 2, 66:52 (ILLUS)
    /C. ALPINA PEYRONEL
    DIDYMO,GLOIO,HYALO /BARRON, GENERA OF HYPHOMYCETES P. 113-114,1968. (ILLUS)
        SYNO    =CEPHALODIPLOSPORIUM KAMYSCHKO 1961 FIDE BARRON 1968

CEPHALOSPORIUM CORDA 1839. ICONES FUNGORUM 3:11-12 AND TAF. II, FIG. 29 /C.
    ACREMONIUM CORDA 1842
    ?, SEE VON ARX 1970 /BARRON, HYPHOMYCETES P. 114-116,1968. (ILLUS) SUKAPURE
    AND THIRUMALACHAR, MYCOLOGIA 58:351-361,1966. (KEY)

CEPHALOTHECIUM CORDA 1838. ICONES FUNGORUM 2:14 /C. ROSEUM ?=TRICHOTHECIUM
    ROSEUM (PERSOON) LINK 1809
  =TRICHOTHECIUM LINK 1809 (PROBABLY, BUT TYPE OF C. ROSEUM NOT SEEN) FIDE
  HUGHES 1958

CEPHALOTRICHUM LINK 1809. MAG. GES. NATURF. FREUNDE, BERLIN 3:20 AND TAB. 1,
    FIG. 34 /C. STEMONITIS (PERSOON) (LECTOTYPE FIDE HUGHES 1958)
    AMERO,EASO,PHAEO /HUGHES, CAN. J. BOTANY 36:744,1958. (REFS) MORTON AND
    SMITH, MYCOL PAPERS CMI 86:1-96,1963. (REVIS,KEY, ILLUS) SUB DORATOMYCES
        SYNO    =DORATOMYCES CORDA 1829 FIDE HUGHES 1958
                    =STYSANUS CORDA 1837 FIDE HUGHES 1958
                    =STYSANOPSIS FERRARIS 1909 IN HUGHES 1958
                    ?=STILBELLULA BOEDIJN 1951
                    =SYNPENICILLIUM COSTANTIN 1888 FIDE MORTON AND SMITH 1963 SUB
                    DORATOMYCES
                    ?=TRICHURIS CLEMENTS ET SHEAR 1896
                    ?=SAROPHORUM SYDOW ET SYDOW 1916

CERACEA CRAGIN 1885. BULL. WASHBURN COLL. LAB. NAT. HIST. P. 82 /C. VERNICOSA
    ?. TREMELLALES? /MARTIN, MYCOLOGIA 41:77-86,1949.

CERATOCLADIUM CORDA 1839. PRACHT-FLORA ... , P. 41 /C. MICROSPERMUM
    AMERO,CETERI,HYALO, NON CERATOCLADIUM PATOUILLARD 1898, BULL. SOC. MYCOL. FR.
    14:196-197,. (NO ILLUS) /HUGHES, MYCOL. PAPERS CMI 47:5-8,1951. (ILLUS)

CERATOPHORUM SACCARDO 1880. MICHELIA 2:22 /C. HELICOSPORUM (SACCARDO)
    PHRAGMO OR HELICO,CETERI,PHAEO /ELLIS, MYCOL PAPERS CMI 70:13-16,1958. (ILLUS)

CERATOPODIUM CORDA 1837. ICONES FUNGORUM 1:20, TAV. V, FIG. 264 /C. ALBUM
    ?=GRAPHIUM (FIDE SACCARDO 1886)

CERATOSPORELLA HOHNEL 1919. BERL. DEUTSCHL. BOT. GES. 37:155 /C. BICORNIS
    (MORGAN) HOHNEL 1923
    STAURO,CETERI,PHAEO /HUGHES, MYCOL PAPERS CMI 46:22-25,1951. (ILLUS)

CERATOSPORIUM SCHWEINITZ 1832. TRANS. AMER. PHIL. SOC. II, 4:300 /C. FUSCESCENS
    STAURO,CETERI,PHAEO /HUGHES, MYCOL PAPERS CMI 39:1-11,1951. (ILLUS) HUGHES,
    NEW ZEALAND J. BOTANY 2:305-309,1964. (ILLUS)

CERCODEUTEROSPORA CURZI 1932. BOL. STAZ. PAT. VEG. ROMA 12:3-5 (ILLUS) /C.
    TRICHOPHILA
    =MYCOVELLOSIELLA RANGEL 1917 FIDE CASTELLANI ET CIFERRI

CERCOSEPTORIA PETRAK 1925. ANN. MYCOL. 23:69 /C. CHAMAESYCEAE (STEVENS ET
    DALBEY)
    ?. NOM. NOV. FOR SEPTORIOPSIS STEVENS ET DALBEY, NOT VERY DIFFERENT FROM
    CERCOSPORA FRESENIUS 1863

CERCOSPERMA ARNAUD 1954. BULL. SOC. MYCOL. FR. 69:289-290 (ILLUS) /C. SUBSESSILE
    SCOLECO,CETERI,PHAEO, NO LATIN DIAGNOSIS

CERCOSPORA FRESENIUS 1863. BEITR. ZUR. MYKOL. P. 90 /C. APII (LECTOTYPE)
    SCOLECO OR PHRAGMO,CETERI,PHAEO OR HYALO, ASCO STATE = MYCOSPHAERELLA FIDE
    MULLER IN KENDRICK 1971 /CHUPP, MONOGR. OF THE GENUS CERCOSPORA 1954.
    (MONOGR) BARRON, HYPHOMYCETES P. 116-117,1968. (ILLUS)
    SYNO    =VIRGASPORIUM COOKE 1875 FIDE ELLIS 1971
            =PSEUDOCERCOSPORA SPEGAZZINI 1911 FIDE HODGES AND HAASIS

CERCOSPORELLA SACCARDO 1880. MICHELIA 2:20 /C. PERSICA (LECTOTYPE)
    SCOLECO OR PHRAGMO,CETERI,HYALO, ASCO STATE = MYCOSPHAERELLA FIDE MULLER IN
    KENDRICK 1971 /BARRON, HYPHOMYCETES P. 116-117,1968. (REFS)
    SYNO    ?=STENOSPORA DEIGHTON 1969

CERCOSPORIDIUM EARLE 1901. MUHLENBERGIA 1:16 /C. EUPHORBIAE (TRACY ET EARLE)
    INED. =C. CHAETOMIUM (COOKE) DEIGHTON 1967
    DIDYMO OR PHRAGMO,CETERI,HYALO OR PHAEO /DEIGHTON, MYCOL PAPERS CMI
    112:27-77,1967. (REVIS,ILLUS)
    SYNO    =BERTEROMYCES CIFERRI 1954 FIDE DEIGHTON 1967

CERCOSPORINA SPEGAZZINI 1910. AN. MUS. NAC. BUENOS AIRES 20(SER 3A):424-429
    /THIRTEEN ORIGINAL SPECIES
    ?. PROBABLY CERCOSPORA

CERCOSPORIOPSIS MIURA 1928. FLORA OF MANCHURIA AND EAST MONGOLIA, 3,
    CRYPTOGAMS, P.527 /EIGHT ORIGINAL SPECIES
    ?

CERCOSPORULA ARNAUD 1954. BULL. SOC. MYCOL. FR. 69:289-290 (ILLUS) /C.
    CRASSIUSCULA
    SCOLECO,CETERI,HYALO, NO LATIN DIAGNOSIS

CEREBELLA CESATI 1851. APUD RABENHORST, BOT. ZTG. 9:669 /C. ANDROPOGONIS CESATI
    = EPICOCCUM ANDROPOGONIS (CESATI) SCHOL-SCHWARZ 1959
    =EPICOCCUM LINK 1815 FIDE SCHOL-SCHWARZ 1959 /ELLIS, DEM. HYPH. P.
    73-74,1971. (ILLUS)

CHAETOBASIDIELLA HOHNEL 1918. BERL. DEUTSCHL. BOT. GES. 35:317 /C.

VERMICULARIOIDEA, DESCRIBED BUT NOT ILLUSTRATED IN MITT. BOT. LAB. TECH.
  HOCHSCH. WIEN. BD. II, HEFT 2:35-37
?

CHAETOCHALARA SUTTON ET PIROZYNSKI 1965. TRANS. BRIT. MYCOL. SOC. 48:556-560
  (ILLUS) /C. AFRICANA
  AMERO,BASO OR GLOIO,PHAEO OR HYALO, BUT DISTINGUISHED FROM CHALARA ONLY BY
    PRESENCE OF SETAE

CHAETOCONIDIUM ZUKAL 1887. VERH. ZOOL. BOT. GES. WIEN 37:45 AND TAF. I, FIG. 7
    /C. ARACHNOIDEUM
  PHRAGMO OR AMERO,CETERI,HYALO /LINDAU, RABENHORST   KRYPT. FL. 2 ED.
    8:185-186,1907. (ILLUS)

CHAETODOCHIS CLEMENTS ?='CHAETOSTROMA' FIDE AINSWORTH 1971

CHAETODOCHIUM HOHNEL 1932. MITT. BOT. INST. TECHN. HOCHSCH. WIEN 9:44 /C. BUXI
    (DE CANDOLLE EX FRIES) HOHNEL
  =VOLUTELLA TODE EX FRIES 1832 FIDE VON ARX 1970

CHAETOPSELLA HOHNEL 1930. MITT. BOT. LAB. TECH. HOCHSCH. WEIN 7:44 /C. GRISEA
    (EHRENB.)
  =CHAETOPSIS GREVILLE 1885 FIDE HUGHES 1951

CHAETOPSINA RAMBELLI 1956. R. C. ACCAD. BOLOGNA (FIS.) 3:5 /C. FULVA
  AMERO,GLOIO,HYALO /BARRON, HYPHOMYCETES P. 117-120,1968. (ILLUS)

CHAETOPSIS GREVILLE 1825. EDINB. PHIL. J. 13:63 /C. WAUCHII =C. GRISEA
    (EHRENBERG) SACCARDO 1880 FIDE HUGHES 1951
  AMERO OR DIDYMO,GLOIO,HYALO /HUGHES, TRANS. BRIT. MYCOL. SOC.
    34:569-573,1951. (ILLUS)
    SYNO    =CHAETOPSELLA HOHNEL 1930 FIDE HUGHES 1951

CHAETOSIRA CLEMENTS 1931 FOR WIESNERIOMYCES

CHAETOSPERMUM SACCARDO 1892. SYLLOGE FUNGORUM 10:706 /C. TUBERCULARIS
  AMERO,GLOIO,HYALO, A COELOMYCETE /MYCOLOGIA 38:187,1946, MYCOPATHOLOGIA 16:114
    SYNO    =MASTIGONEMA SPEGAZZINI 1926

CHAETOSTROMA CORDA 1829. IN STURM. DEUTSCHL. CRYPT. FLORA 2:122 /?
  ?, SEE SACCARDO, SYLLOGE FUNGORUM 4:749,1886 FOR REFERENCES

CHAETOTRICHUM SYDOW 1927. ANN. MYCOL. 25:150 /C. SOLANI SYDOW = ANNELOPHORA
    SOLANI (SYDOW) HUGHES 1951
  =ANNELLOPHORA HUGHES 1951, NON CHAETOTRICHUM RABENHORST 1844

CHALARA (CORDA) RABENHORST 1844. DEUTSCHL. KRYPT.-FLORA 1:38 /C. FUSIDIOIDES
  AMERO OR DIDYMO OR PHRAGMO,BASO OR GLOIO,HYALO OR PHAEO, ASCO STATE
    =CERATOCYSTIS FIDE ELLIS 1971, SEE ALSO CHAETOCHALARA /BARRON, HYPHOMYCETES
    P. 120-122,1968. (ILLUS) NAG RAJ, THE GENUS CHALARA, THESIS, UNIV. OF
    WATERLOO,1971. (MONOGR.)
    SYNO    =CHALAROPSIS PEYRONEL 1916, PRO PARTE, FIDE BARRON 1968
            =THIELAVIOPSIS WENT 1893, PRO PARTE, FIDE BARRON 1968
            =CYLINDROCEPHALUM BONORDEN 1851 FIDE HUGHES 1958

CHALAROPSIS PEYRONEL 1916. STAZ. SPER. AGRIC. ITAL. 49:595 /C. THIELAVIOIDES
  AMERO,CETERI,PHAEO, ASCO STATE =CERATOCYSTIS FIDE HENNEBERT 1967,
    PHIALOCONIDIAL STATE =CHALARA /HENNEBERT, ANTONIE VAN LEEUWENHOEK
    33:333-340,1967. (ILLUS)

CHANTRANSIOPSIS THAXTER 1914. BOT. GAZ. 58:246 (235-253, ILLUS) /C. DECUMBENS
    (LECTOTYPE)
  AMERO,CETERI,HYALO

CHARDONIA CIFERRI 1930. IN CHARDON AND TORO, HOUR. DEPT. AGRIC. PORTO RICO
    14(4):295 /C. ROSEA
  ?, NOT ILLUSTRATED BY AUTHOR

CHEIROCONIUM HOHNEL 1910. S. B. AKAD. WISS. WIEN 119:665 /C. BEAUMONTII (BERK.

```
      ET CURT. EX HOHNEL) HOHNEL =S. TINCTUM (PECK) HUGHES 1958
   =SIROTHECIUM KARSTEN 1887 FIDE HUGHES 1958

CHEIROMYCELLA HOHNEL 1910. S.B. AKAD. WISS. WIEN 119:664 /C. SPEIROIDES
   (HOHNEL) =C. MICROSCOPICA (KARSTEN) HUGHES 1958
   STAURO,GLOIO,PHAEO /ELLIS, DEM. HYPH. P. 325-326,1971. (ILLUS)

CHEIROMYCES BERKELEY ET CURTIS 1857. INTROD. CRYPT. BOT. P. 313 AND FIG. 70C
   /C. STELLATUS
   STAURO,CETERI,PHAEO /MOORE. MYCOLOGIA 50:682,1958. (REFS)
      SYNO    =STIGMOPSIS BUBAK 1914 FIDE MOORE 1958

CHEIROPODIUM SYDOW 1915. ANN. MYCOL. 13:42 /C. FLAGELLATUM SYDOW
      =CLASTEROSPORIUM FLAGELLATUM (SYDOW) ELLIS 1958
   =CLASTEROSPORIUM SCHWEINITZ 1832 FIDE ELLIS 1958

CHEIROSPORA MOUGEOT ET FRIES 1825. FRIES, SYSTEMA MYCOL. 3:484 /C. BOTRYOSPORA
   (MONTAGNE) HUGHES 1958
   DICTYO,CETERI,PHAEO /HUGHES, CAN. J. BOTANY 36:748,1958. (REFS)
      SYNO    =THYRSIDIUM MONTAGNE 1846 FIDE HUGHES 1958
              =RHABDOSPORIUM CHEVALLIER 1826 FIDE HUGHES 1958 (AS ?=)
              =HYPEROMYXA CORDA 1839 FIDE HUGHES 1958
              =MYRIOCEPHALUM DE NOTARIS 1842 FIDE HUGHES 1958

CHELISPORIUM SPEGAZZINI 1911. AN. MUS. NAC. BUENOS AIRES 3(13):463-464 (ILLUS)
   /C. HYSTERIOIDES
   ?

CHLAMYDOALEUROSPORIA GRIGORAKIS 1924. COMP. REND. 179:1425 /C. GRANULOSA
   (SABOURAUD) GRIG. =TRICHOPHYTON MENTAGROPHYTES (LECTOTYPE)
   =TRICHOPHYTON MALMSTEN 1845

CHLAMYDOMYCES BAINIER 1907. BULL. SOC. MYC. FR. 23:240 /C. DIFFUSUS BAINIER =C.
   PALMARUM (COOKE) MASON 1928
   DIDYMO OR AMERO,CETERI,HYALO /MASON, MYCOL PAPERS CMI 2:37,1928 (REFS)
   BARRON, HYPHOMYCETES P. 122-123,1968. (ILLUS)

CHLAMYDORUBRA DESHPANDE ET DESHPANDE 1966. MYCOPATHOLOGIA 29:272 /C. VERRUCOSA
   DESHP. ET DESHP. 1966
   ?, ORIGINAL DESCRIPTION AND ILLUSTRATION NOT DIAGNOSTIC

CHLAMYDOSPORIUM PEYRONEL 1913 =DUBIUS FIDE CLEMENTS AND SHEAR 1931

CHLORIDIELLA ARNAUD 1954. BULL. SOC. MYCOL. FR. 69:270 AND FIG. 1F AND G, P.
   269 /C. LEUCOPODA (BONORDEN)
   AMERO,CETERI,HYALO. NO LATIN DIAGNOSIS
      SYNO    =IDRIELLA NELSON ET WILHELM 1956 FIDE CARMICHAEL

CHLORIDIUM LINK 1809. MAG. GES. NATURF. FREUNDE, BERLIN 3:13 /C. VIRIDE
   AMERO,GLOIO,HYALO OR PHAEO, ASCO STATE = CHAETOSPHAERIA FIDE MULLER IN
   KENDRICK 1971 /BARRON, HYPHOMYCETES P. 123-125,1968. (ILLUS)
      SYNO    =BISPOROMYCES VAN BEYMA 1921 FIDE HUGHES 1958
              =CIRRHOMYCES HOHNEL 1903 FIDE HUGHES 1958
              =PSILOBOTRYS SACCARDO 1879 FIDE HUGHES 1958
              ?=PIMINELLA ARNAUD 1954
              ?=SPHAEROMYCETELLA ARNAUD 1954

CHMELIA SVOBODOVA 1966. BIOLOGIA, BRATISLAVA 21:82 /C. SLOVACA
   ?, FOR A BLACK YEAST-LIKE AGENT OF CHROMOMYCOSIS. ORIGINAL DESCRIPTION AND
   ILLUSTRATION DIFFICULT TO INTERPRET /EMMONS ET AL., MEDICAL MYCOLOGY P.
   363,1970. (DESCR)

CHRISTIASTER KUNTZE =GONATOBOTRYUM SACCARDO FIDE AINSWORTH 1971

CHROMATIUM LINK ='DEMATIUM' FIDE AINSWORTH 1971

CHROMELOSPORIUM CORDA 1833. STURMS DEUTSCHL. FLORA III (PILZE),BD. 3, HEFT
   13:81 /C. OCHRACEUM CORDA =OSTRACODERMA OCHRACEUM (CORDA) HUGHES 1958
   =OSTRACODERMA FRIES 1825 FIDE HUGHES 1958
```

## 10. Hyphomycetes

CHROMOSPORIUM CORDA 1825. STURMS DEUTSCHL. FLORA III (PILZE) BD. 2 HEFT 9:119
/C. ROSEUM
=NOMEN DUBIUM SEE HUGHES 1958

CHROMOSTYLUM GIARD =METARRHIZIUM FIDE AINSWORTH 1971

CHROOSTROMA CORDA 1837. ICONES FUNGORUM 1:5, FIG. 48 /C. PINI
=ACHITONIUM KUNZE FIDE LINDAU IN RABENHORST 1910

CHRYSACHNE CIFERRI 1938. ANN. MYCOL. 36:239 /C. HYPHOCHORIA
?

CHRYSOSPORIUM CORDA 1833. STURMS DEUTSCHL. FLORA III, BD 2, HEFT 9:119 /C.
    CORII CORDA = C. MERDARIUM (LINK) CARMICHAEL 1962
    AMERO,CETERI,HYALO, ASCO STATE =ARTHRODERMA FIDE CARMICHAEL 1962 OR
    AJELLOMYCES FIDE MCDONOUGH 1968 SUB BLASTOMYCES (MYCOLOGIA 60:76-83)
    /CARMICHAEL, CAN. J. BOTANY 40:1137-1173,1962. (REVIS,KEY,ILLUS)
        SYNO    =ALEURISMA AUCT. NON LINK FIDE CARMICHAEL 1962
                  =ATRICHOPHYTON CASTEL. ET CHALM. 1919 FIDE CARMICHAEL
                  =BLASTOMYCES COSTANTIN ET ROLLAND FIDE CARMICHAEL 1962
                  =BLASTOMYCES GILCHRIST ET STOKES FIDE CARMICHAEL 1962
                  =EMMONSIA CIFERRI ET MONTEMARTINI FIDE CARMICHAEL 1962
                  =GEOMYCES TRAAEN FIDE CARMICHAEL 1962
                  =GILCHRISTIA CIFERRI ET REDAELLI FIDE CARMICHAEL 1962
                  =GLENOSPORELLA NANNIZZI FIDE CARMICHAEL 1962
                  =MYCELIOPHTHORA COSTANTIN FIDE CARMICHAEL 1962
                  =RHINOCLADIOPSIS KAMYSCHKO 1961 FIDE CARMICHAEL
                  =ZYMONEMA DE BEURMANN ET GOUGEROT FIDE CARMICHAEL 1962

CHUPPIA DEIGHTON 1965. MYCOL PAPERS CMI 101:32-34 (ILLUS) /C. SARCINIFERA
    DICTYO,CETERI,PHAEO

CILICIOPODIUM CORDA 1833. STURM DEUTSCHL. FLORA, PILZE 3:57 /C. VIOLACEUM
    AMERO,CETERI,HYALO /BARRON, HYPHOMYCETES P. 127-128,1968. (ILLUS)
        SYNO    =BOTRYPES PREUSS 1852 FIDE LINDAU IN RABENHORST 1910
                  =CLAVULARIA KARSTEN 1883 FIDE CLEMENTS AND SHEAR 1931

CILIOFUSARIUM ROSTRUP 1892. BOT. TIDSSKR. 18:77 /C. UMBROSUM ROSTRUP =MENISPORA
    GLAUCA PERSOON 1822 FIDE HUGHES 1958
    =MENISPORA PERSOON 1822 FIDE HUGHES 1958

CIRCINOCONIS BOEDIJN 1942. 150TH ANNIV. VOL. R. BOT. GARD. CALCUTTA P. 209-211
    (ILLUS) /C. PARADOXA
    HELICO,CETERI,PHAEO /BARNETT, IMPERFECT FUNGI NO. 133,1960. (ILLUS)

CIRCINOTRICHUM NEES 1817. DAS SYSTEM ... P. 19 /C. MACULIFORME
    SCOLECO,CETERI,HYALO. DIFFERS FROM GYROTHRIX ONLY IN HAVING UNBRANCHED SETAE
    /PIROZYNSKI, MYCOL PAPERS CMI 84:3-8,1962. (ILLUS)
        SYNO    =GYROTRICHUM SPRENGEL 1827 FIDE HUGHES 1958
                  =DEPHILIPPIA RAMBELLI 1959 FIDE PIROZYNSKI 1962

CIRRENALIA MEYERS ET MOORE 1960. AMER. J. BOTANY 47:346 /C. MACROCEPHALA
    (KOHLMEYER)
    HELICO,CETERI,PHAEO

CIRRHOMYCES HOHNEL 1903. ANN. MYCOL. 1:529 /C. CAUDIGERUS
    =CHLORIDIUM LINK FIDE HUGHES 1958

CLADASPERGILLUS RITG. =ASPERGILLUS FIDE AINSWORTH 1971

CLADOBOTRYUM NEES 1817. DAS SYSTEM ... P. 56 /C. VARIUM NEES = C. VARIOSPERMUM
    (LINK) HUGHES 1958
    DIDYMO,BASO OR CETERI,HYALO. ASCO STATE =HYPOMYCES TULASNE 1865 FIDE BARRON
    1968 /GAMS AND HOOZEMANS, PERSOONIA 6:95-110,1970 (ILLUS,KEY) COLE AND
    KENDRICK, CAN. J. BOTANY 49:595-599,1971. (ILLUS)
        SYNO    =DIDYMOCLADIUM SACCARDO 1886 FIDE HUGHES 1958
                  =DIPLOCLADIUM BONORDEN 1851 FIDE HUGHES 1958
                  ?=CYLINDROPHORA BONORDEN 1851. SEE GAMS AND HOOZEMANS P. 99

CLADOBYSSUS RITG. =HYPHA PERSOON FIDE SACCARDO FROM AINSWORTH 1971

CLADOGRAPHIUM PEYRONEL 1918. NUOV. GIORN. BOT. ITAL. 25:439-440 (ILLUS) /C.
    RIVULORUM
    ?

CLADORRHINUM SACCARDO ET MARCHAL 1885. BULL. SOC. BOT. BELG. 24:64 /C.
    FOECUNDISSIMUM SACC. ET MARCHAL 1885.
    AMERO,GLOIO,HYALO /ARX AND GAMS, NOVA HEDWIGIA 13:199-208,1966. (ILLUS,EMEND)
        SYNO    ?=BAHUPAATHRA SUBRAMANIAN ET LODHA 1964

CLADOSARUM YUILL ET YUILL 1938. TRANS. BRIT. MYCOL. SOC. 22:199 /C. OLIVACEUM
    ?=ASPERGILLUS FIDE RAPER FROM AINSWORTH 1971

CLADOSPORIELLA DEIGHTON 1965. MYCOL PAPERS CMI 101:34-36 (ILLUS) /C.
    CERCOSPORICOLA
    SCOLECO,BLASTO,PHAEO /DEIGHTON, MYCOL PAPERS CMI 118:33-39,1969. (ILLUS)

CLADOSPORIUM LINK 1815. MAG. GES. NATURF. FREUNDE, BERLIN 7:37 /C. HERBARUM
    (PERSOON) =CLADOSPORIUM STATE OF MYCOSPHAERELLA TASSIANA (DE NOTARIS)
    JOHANSON 1884 FIDE BARR 1958
    AMERO OR DIDYMO OR PHRAGMO,BLASTO,PHAEO, ASCO STATE =MYCOSPHAERELLA OR
    AMORPHOTHECA OR VENTURIA FIDE ELLIS 1971 /ELLIS, DEM. HYPH. P.
    308-319,1971. (ILLUS,KEY)
        SYNO    =HETEROSPORIUM KLOTZSCH 1877 FIDE HUGHES 1958
                =HORMODENDRUM BONORDEN 1853 FIDE BARRON 1968
                =DIDYMOTRICHUM BONORDEN 1851 FIDE HUGHES 1958
                ?=FULVIA CIFERRI 1954
                =MYDONOSPORIUM CORDA 1833 FIDE LINDAU IN RABENHORST 1907
                =MYXOCLADIUM CORDA 1837 FIDE HUGHES 1958
                =SPOROCLADIUM CHEVALLIER 1826 FIDE HUGHES 1958
                ?=TANDONELLA PRASAD ET VERMA 1970

CLADOSTERIGMA PATOUILLARD 1892. BULL. SOC. MYC. FRANCE 8:138 AND PL. XII, FIG.
    3A TO F /C. FUSISPORA
    ?, ORIGINAL ILLUSTRATION DIFFICULT TO INTERPRET

CLADOTRICHUM CORDA 1832. STURM DEUTSCHL. FLORA III (PILZE), BD 3, 12:39 /C.
    POLYSPORUM
    =OEDEMIUM LINK 1824 FIDE HUGHES 1958
        SYNO    =SPHINCTROSPORIUM KUNZE EX FRIES FIDE SACCARDO FROM AINSWORTH 1971

CLASTEROSPORIUM SCHWEINITZ 1832. TRANS. AMER. PHIL. SOC. II, 4:300 /C. CARICINUM
    PHRAGMO,CETERI,PHAEO, ASCO STATE =HEMISPHAERIALES FIDE ELLIS 1958 /ELLIS,
    MYCOL PAPERS CMI 70:1-13,1958. (KEY,ILLUS)
        SYNO    =CLASTERISPORIUM (ORTHOGRAPHIC VARIANT)
                =HYMENOPODIUM CORDA 1837 FIDE ELLIS 1958
                =COMETELLA SCHWEINITZ 1835 FIDE ELLIS 1958
                =CHEIROPODIUM SYDOW 1915 FIDE ELLIS 1958
                =SPORHELMINTHIUM SPEGAZZINI 1918 FIDE HUGHES 1958

CLATHROCOCCUM HOHNEL 1911. S.B. AKAD. WISS. WIEN 120:473 /C. COMPACTUM
    (BERKELEY ET CURTIS)
    ?=EPICOCCUM FIDE AINSWORTH 1971

CLATHROSPHAERA ZALEWSKI 1888. ROZPRAWY I SPRAWY. Z. POSIEDZ.-WYDZIALU MAT.
    PRZYROD, KRAKOW 18:151-191 /C. SPIRIFERA
    =NOMEN CONFUSUM FIDE VAN BEVERWIJK 1951

CLATHROSPHAERINA VAN BEVERWIJK 1951. TRANS. BRIT. MYCOL. SOC. 34:289 /C.
    ZALEWSKII
    DICTYO,CETERI,HYALO /TUBAKI, J. HATTORI BOT. LAB. NO. 20, P.152-154,1958.
    (ILLUS)

CLATHROTRICHUM PATOUILLARD 1921. BULL. SOC. MYCOL. FR. 37:33-35 (ILLUS) /C.
    SUBCARNEUM PAT. 1921
    =MARTINDALIA SACCARDO ET ELLIS 1884 FIDE CARMICHAEL, ?=HEYDENIA

CLAVARIOPSIS DE WILDEMAN 1895. ANN SOC. BELG. MICROSC. 19:200 /C. AQUATICA DE
    WILD. 1895
    STAURO,CETERI,HYALO, ASCO STATE= COROLLOSPORA FIDE SHEARER ET CRANE 1971
    /ANASTASIOU, MYCOLOGIA 53:11-I6,1961. (ILLUS)

## 10. Hyphomycetes

CLAVATOSPORA NILSSON 1964. SYMB. BOT. UPSALIENS. 18(2):88 /(?)C. LONGIBRACHIATA (INGOLD)
 STAURO,CETERI,HYALO. IS A SEGREGATE FROM HELISCUS

CLAVULARIA KARSTEN 1883. SYMB. MYCOL. 9:67 /C. FUSISPORA
 =CILICIOPODIUM CORDA FIDE CLEMENTS AND SHEAR 1931

CLINOTRICHUM COOKE 1871. POP. SCI. REV. 10:30 /C. LANOSUM COOKE =RHINOTRICHUM LANOSUM, POP. SCI. REV. 10:29-30.
 =NOMEN ILLEGITIMUM FIDE HUGHES 1958. THE FUNGUS BECOMES ACLADIUM LANOSUM (COOKE) CARMICHAEL COMB. NOV., BASIONYM: RHINOTRICHUM LANOSUM COOKE

CLONOSTACHYOPSIS HOHNEL 1907. S.B. AKAD. WISS. WIEN 116:149 /C. POPULI (HARZ)
 =CLONOSTACHYS CORDA 1839 FIDE CLEMENTS AND SHEAR 1931

CLONOSTACHYS CORDA 1839. PRACHTFLORA P. 31 /C. ARAUCARIA
 AMERO,BASO OR GLOIO,HYALO, BUT IS INCLUDED IN GLIOCLADIUM BY SOME AUTHORS /TUBAKI, TRANS. MYCOL. SOC. JAPAN 4:83-90,1963. (ILLUS) BARRON, HYPHOMYCETES P. 177-179,1968. (ILLUS) SUB GLIOCLADIUM
  SYNO    =CLONOSTACHYOPSIS HOHNEL 1907 FIDE CLEMENTS AND SHEAR 1931

CLOSTEROALEUROSPORIA GRIGORAKIS 1924. COMP. REND. 179:1424 /C. AUDOUINI (GRUBY) GRIG. (LECTOTYPE)
 =NOMEN ILLEGITIMUM, AN UNNECESSARY NEW NAME FOR MICROSPORUM

\* CLOSTEROSPORIA GRIGORARIS 1925. ANN. SCI. NAT., SER. 10 7:410-411 /C. LANOSA
 =MICROSPORUM CANIS (LECTOTYPE)
 =MICROSPORUM GRUBY 1834

COCCIDIOIDES STILES 1896. IN RIXFORD AND GILCHRIST, JOHNS HOPKINS HOSP. REP. 1:243 /C. IMMITIS
 AMERO,ARTHRO,HYALO. A HUMAN PATHOGEN, CONIDIAL STATE CLOSE TO SPORENDONEMA /HUPPERT, SUN AND BAILEY, IN PROC. SECOND COCCI. SYMPOS., ARIZONA, 323-328,1967. (ILLUS)
  SYNO    =BLASTOMYCOIDES CASTELLANI 1928 FIDE CARMICHAEL 1962

COCCOBOTRYS BOUDIER ET PATOUILLARD 1900. BULL. SOC. MYCOL. FR. 16:141 /C. XYLOPHILUS (FRIES) B. ET P.
 =STERILE

COCCOPLEUM EHRENBERG 1818 =SCLEROTIUM FIDE LINDAU IN RABENHORST 1910

COCCOSPORA WALLROTH 1833. FLORA CRYPT. GERMAN. 2:176 /C. AURANTIACA
 =STERILE (SCLEROTIUM) FIDE DAMON AND DOWNING 1954

COCCOSPORELLA KARSTEN 1893. ACTA SOC. F. F. FENNICA 9:11 /C. CALCSPORA
 =MYCOGONE LINK 1809 FIDE HUGHES 1958

COCCOSPORIUM CORDA 1832. STURMS DEUTSCHL. FL. III (PILZE) BD 3 HEFT 12:49 /C. MACULIFORME
 =NOMEN DUBIUM SEE HUGHES 1958
  SYNO    =FREYNELLA KUNTZE FIDE AINSWORTH 1971

COCCOTRICHUM LINK 1824. LINN. SPEC. PLANT. IV 6(1):26 /C. MARTII =ALEURISMA GRANULOSUM MARTIUS 1817
 =NOMEN DUBIUM SEE HUGHES 1958

CODINAEA MAIRE 1937. PUBL. INST. BOT. BARCELONA 3:15 /C. ARISTATA
 AMERO OR DIDYMO OR PHRAGMO,GLOIO,HYALO. ASCO STATE =CHAETOSPHAERIA FIDE HUGHES AND KENDRICK 1968 /HUGHES AND KENDRICK, NEW ZEALAND J. BOTANY 6:323-375,1968., (REVIS,ILLUS)
  SYNO    =MENISPORELLA AGNIHOTHRUDU 1962 FIDE HUGHES AND KENDRICK 1968

COELOGRAPHIUM (SACCARDO) GAUMANN 1920. BULL. JARD. BOT. BUITENZ. 3(2):13 AND TAB. 2 AND 3 /C. CAVICEPS (OUDEMANS) GAUMANN 1920
 ?, ORIGINAL ILLUSTRATION DIFFICULT TO INTERPRET

COELOSPORIUM LINK 1824. LINN. SPEC. PLANT. IV, 6(1):127 /C. FRUTICULOSUM LINK (=DEMATIUM ARTICULATUM PERSOON)
 =NOMEN DUBIUM FIDE HUGHES 1958

\*_Closterosporia_ Grigorakis 1925.

COLEODICTYOSPORA CHARLES 1929. PHYTOPATHOLOGY 19:1051-1053 (ILLUS) /C. CUBENSIS
    DICTYO,CETERI,PHAEO

COLEOMYCES MOREAU ET MOREAU 1937. BULL. SOC. MYCOL. FR. 53:33-38 (ILLUS) /C.
    RUFUS
    =CYLINDROCARPON WOLLENWEBER 1913 FIDE CARMICHAEL

COLLETOSPORIUM LINK 1824. SPEC. PLANT. 1:25 /C. UMBRINUM (LECTOTYPE FIDE HUGHES
    1958)
    =STERILE FIDE HUGHES 1958

COLLETOTRICHUM CORDA 1831. STURMS DEUTSCHL. KRYPT. FLORA III;3:41 /C. LINEOLA
    AMERO,GLOIO,HYALO, (MELANCONIALES) /VON ARX, PHYTOPATH. ZEITSCHR.
    29:413-508,1957. (REVIS) BARRON, HYPHOMYCETES P.131-132,1968. (ILLUS)
        SYNO    =ELLISIELLA SACCARDO 1880 FIDE VON ARX 1970
                =DICLADIUM CESATI 1852 FIDE VON ARX 1970
                ?=DICTYOCHAETA SPEGAZZINI 1923

COLLODOCHIUM HOHNEL.1902. S.B. AKAD. WISS. WIEN 111:1029 /C. ATROVIOLACEUM
    ?. NOT ILLUSTRATED BY AUTHOR

COLUMNOPHORA BUBAK ET VLEUGEL 1916. ANN. MYCOL. 14:349 /C. RHYTISMATIS BUBAK
    1916
    =STIGMINA SACCARDO 1880 FIDE NAG RAJ AND KENDRICK

COMETELLA SCHWEINITZ 1835. IN FRIES, ? /C. CARICINA (FRIES) FRIES 1849 (SUMMA
    VEG. SCAN. 2:506)
    =CLASTEROSPORIUM SCHWEINITZ 1832 FIDE HUGHES 1958

CONIODICTYUM HARIOT ET PATOUILLARD 1909. BULL. SOC. MYCOL. FR. 25:13-14 (ILLUS)
    /C. CHEVALIERI HAR. ET PAT. 1909
    ?. ORIGINAL ILLUSTRATION SUGGESTS THAT THIS MAY BE AN EARLIER NAME FOR
    DIHETEROSPORA AND DICTYOARTHRINOPSIS, BUT SEE AINSWORTH, DICTIONARY OF THE
    FUNGI, 1971

CONIOSCYPHA HOHNEL 1904. ANN. MYCOL. 2:58 (NO ILLUS) /C. LIGNICOLA
    ?=MAMMARIA CESATI 1854

CONIOSPORIOPSIS SPEGAZZINI 1918. PHYSIS 4:291 /C. FUMAGO (SCHWEINITZ)
    =NOMEN CONFUSUM FIDE HUGHES 1958

CONIOSPORIUM LINK 1809. MAG. GES. NATURF. FREUNDE, BERLIN 3:8 /C. OLIVACEUM
    DICTYO,BASO,PHAEO, ASCO STATE =HYSTERIUM FIDE HUGHES 1958 /HUGHES, MYCOL
    PAPERS CMI 37:10-17,1950. (ILLUS) HUGHES, THE NATURALIST 93-98,1952.
    (ILLUS) SUB SIRODESMIUM
        SYNO    =SIRODESMIUM DE NOTARIS 1849 FIDE HUGHES 1958
                =BONORDENIELLA PENZ. ET SACC. 1901 FIDE HUGHES 1958
                ?=DISCOSPORIUM SACCARDO ET SYDOW (NON HOHNEL) FIDE AINSWORTH 1971

CONIOTHECIELLA SPEGAZZINI 1919. PHYSIS 4:295 /C. PHYLLOGENA (DESMAZIERES)
    ?. NOT ILLUSTRATED BY SPEGAZZINI

CONIOTHECIUM CORDA 1833. STURMS DEUTSCHL. FLORA III (PILZE), BD. 3,HEFT 13:71
    /CONIOTHECIUM ATRUM
    =NOMEN DUBIUM, HUGHES 1958 COULD NOT FIND TYPE

CONOPLEA PERSOON 1801. SYN METH. FUNG. P. 234 /C. SPHAERICA
    AMERO,CETERI,PHAEO /HUGHES, CAN J. BOTANY 38:659-696,1960. (REVIS,ILLUS,KEY)
        SYNO    =STREPTOTHRIX CORDA 1839 FIDE HUGHES 1960
                ?=STREBLOCAULIUM CHEVALLIER 1837

COPROTRICHUM BONORDEN 1851. HANDB. ALLGEM. MYKOL. P. 76 AND TAF. VI, FIG. 132
    /C. PURPURASCENS
    =GEOTRICHUM LINK 1809 FIDE CARMICHAEL 1957

CORALLINOPSIS LAGARDE 1917. ARCH. ZOOL. EXPER. GEN. 56:286-288 AND PL. X /C.
    PILULIFERA
    ?. A SYNNEMATOUS CEPHALOSPORIUM ON INSECTS. SEE ALSO SYNNEMATIUM.

CORALLODENDRON JUNGHUHN 1838. PRAEM. FL. CRYPT. JAVAE P. 7 /C. LEUCOCEPHALUM

## 10. Hyphomycetes

? /MORRIS, WESTERN ILLINOIS SER BIOL SCI 3:36-37,1963. (ILLUS)
SYNO =STILBODENDRON SYDOW 1916 FIDE MORRIS 1963
=HASSKARLINDA KUNTZE 1891 BASED ON SAME TYPE

CORDALIA GOBI 1885. MEM. ACAD. IMP. SCI. ST. PETERSBOURG 32(14):13 /C. PERSICINA (DITMAR)
=TUBERCULINA SACCARDO 1880 FIDE LINDAU IN RABENHORST 1910

CORDANA PREUSS 1851. LINNAEA 24:129 /C. PAUCISEPTATA
DIDYMO,CETERI,PHAEO /HUGHES, CAN. J. BOTANY 33:259-263,1955. (ILLUS)
SYNO =PREUSSIASTER KUNZE 1891 FIDE HUGHES 1958
?=ARTHROBOTRYELLA SIBILIA 1928

CORDELLA SPEGAZZINI 1886. AN. SOC. CIENT. ARGENTINA 22:210 /C. CONIOSPORIOIDES
AMERO,CETERI,PHAEO /ELLIS, MYCOL PAPERS CMI 103:30-33,1965. (ILLUS)

COREMIELLA BUBAK ET KRIEGER 1912. ANN. MYCOL. 10:52 /C. CYSTOPOIDES =C. CUBISPORA (BERK. ET CURTIS) ELLIS 1971
AMERO,ARTHRO,PHAEO /ELLIS, DEMATIACEOUS HYPHOMYCETES P. 32-34,1971. (ILLUS)
SYNO ?=GEOTRICHELLA ARNAUD 1954

COREMIOPSIS SIZOVA ET SUPRUN 1957. VESTN. MOSKOV. UNIV. 2:55 /C. ROSEA
?, ORIGINAL ILLUSTRATION INADEQUATE TO CHARACTERIZE, PROBABLY PAECILOMYCES

COREMIUM LINK 1809. MAG. GES. NATURF. FREUNDE, BERLIN 3:19 /C. GLAUCUM
=PENICILLIUM LINK 1809 FIDE AINSWORTH 1960

CORETHROPSIS CORDA 1839. PRACHTFLORA P. 1 /C. PARADOXA
?, NO ONE SINCE CORDA HAS REPORTED SEEING THE TYPE OR A FUNGUS THAT MATCHES THE ORIGINAL ILLUSTRATION. THE NAME SHOULD BE CONSIDERED A NOMEN DUBIUM.

COROLLIUM SOPP 1912. MONOGR. DER PILZGR. PENICILLIUM, P. 98-100 AND TAF. X, FIG. 108 /C. DERMATOPHAGUM
?, PROBABLY BASED ON PAECILOMYCES VARIOTI

CORONOSPORA ELLIS 1971. MYCOL. PAPERS CMI 125:16-17 (ILLUS) /C. DENDROCALAMI
STAURO,CETERI,PHAEO

CORYMBOMYCES APPEL ET STRUNK 1904. CENT. BAKT. 2(11):632-634 (ILLUS) /C. ALBUS
=GLIOCLADIUM CORDA 1840 FIDE CARMICHAEL

CORYNESPORA GUSSOW 1906. ZEITSCHR. PFLANZENKR. 16:13 /C. MAZEI =C. CASSIICOLA (BERK. ET CURT.) WEI 1950
PHRAGMO,BLASTO OR CETERI,PHAEO /ELLIS, MYCOL PAPERS CMI 65:1-15,1957. (KEY,ILLUS) ELLIS, MYCOL PAPERS CMI 76:19-36,1960. (ILLUS)

CORYNESPORELLA MUNJAL ET GILL 1961. INDIAN PHYTOPATHOL. 14:7 /C. URTICAE
PHRAGMO,CETERI OR BLASTO,PHAEO /ELLIS, DEM. HYPH. P.379-380,1971. (ILLUS)

CORYNEUM NEES 1817. DAS SYSTEM ... P.34 /C. UMBONATUM NEES EX LINK 1825
=CORYNEUM STATE OF PSEUDOVALSA LONGIPES (TULASNE) SACCARDO 1879 FIDE HUGHES 1958
PHRAGMO,CETERI,PHAEO, (MELANCONIALES) ASCO STATE =PSEUDOVALSA FIDE HUGHES 1958 /BARNETT, IMPERFECT FUNGI NO. 456,1960. (ILLUS) HUGHES, CAN. J. BOTANY 33:343-349,1955. (ILLUS)

CORYNODESMIUM WALLROTH 1828. IN REICHENBACH, /?
=NOMEN NUDUM, SEE HUGHES, CAN. J. BOTANY 36:757-758,1958

COSMARIOSPORA SACCARDO 1880. MICHELIA 2:34, ILLUS. IN FUNGI ITALICI, FIG. 769,1881 /C. BIZZOZERIANA SACC. 1881
?

COSTANTINELLA MATRUCHOT 1892. RECHERCHES SUR DEVEL. DE QUELQUES MUCEDIN. P. 97 /C. CRISTATA =C. TERRESTRIS (LINK) HUGHES 1958
AMERO,CETERI,HYALO, BASIDIO STATE =BOTRYOBASIDIUM SEE HUGHES IN KENDRICK, P.20-21, 1971 /BARRON, HYPHOMYCETES P. 135-136,1968. (ILLUS)

CREMASTERIA MEYERS ET MOORE 1960. AMER. J. BOTANY 47:348 /C. CYMATILIS
?, ?=STERILE, ?=HUMICOLA

CRINULA FRIES 1821. SYSTEMA MYCOLOGICUM 1:493 /C. CALICIIFORMIS =CRINULA STATE
OF HOLWAYA GIGANTEA FIDE MORRIS 1963
AMERO,GLOIO,HYALO, ASCO STATE =HOLWAYA SACCARDO FIDE MORRIS 1963 /MORRIS,
WESTERN ILLINOIS SER BIOL SCI 3:38-39,1963. (ILLUS)

CRISTULA CHENANTAIS 1919. BULL. SOC. MYCOL. FR. 36:208-209, PL. 18, FIGS. 11 &
12 /C. INTEGRA
?, ORIGINAL ILLUSTRATION DIFFICULT TO INTERPRET

CRISTULARIA (SACCARDO) COSTANTIN 1888. LES MUCEDINEES SIMPLES P. 134-135
(ILLUS) /C. GRANULIFORMIS (SACC.) COST. 1888
?

CRISTULARIELLA HOHNEL 1930. S.B. AKAD. WISS. WIEN 125:124 /C. DEPRAEDANS (COOKE)
?, NOT ILLUSTRATED BY HOHNEL

CROCYSPORIUM CORDA 1837. ICONES FUNGORUM 1:5 /C. AEGERITA CORDA =AEGERITA STATE
OF PENIOPHORA CANDIDA LYMAN
=AEGERITA PERSOON 1801 FIDE HUGHES 1958

CRYPTOCORYNEUM FUCKEL 1869. SYMBOL. MYCOL. 372 /C. FASCICULATUM = C.
CONDENSATUM (WALLROTH) MASON ET HUGHES 1953
STAURO,CETERI,PHAEO /ELLIS. DEM. HYPH. P. 108-109,1971 (ILLUS)

CRYPTOPHIALE PIROZYNSKI 1968. CAN. J. BOTANY 46:1123-1127 (ILLUS) /C.
KAKOMBENSIS
SCOLECO OR DIDYMO,GLOIO,HYALO

CRYPTOSTROMA GREGORY ET WALLER 1951. TRANS. BRIT. MYCOL. SOC. 34:593 /C.
CORTICALE (ELLIS ET EVERHART)
AMERO,BASO,PHAEO

CULCITALNA MEYERS ET MOORE 1960. AMER. J. BOTANY 47:349 /C. ACHRASPORA
=TRICHOCLADIUM HARZ 1871 FIDE HUGHES 1969

CULICIDOSPORA PETERSEN 1960. BULL. TORREY BOT. CLUB 87:342-347 (ILLUS) /C.
AQUATICA
STAURO,CETERI,HYALO

CURCULIOSPORA ARNAUD 1954. BULL. SOC. MYCOL. FR. 69:287 AND FIG. 3 R, P. 273
/C. SYDOWII
HELICO,CETERI,PHAEO, NO LATIN DIAGNOSIS

CURVIDIGITUS SAWADA 1943. REP. GOVT. RES. INST. FORMOSA 86:176 /C. DAPHNIPHYLLI
?, JAPANESE DIAGNOSIS, APPEARS TO BE DREPANOCONIS

CURVISPORIUM CORBETTA 1963. RISO 12:3 /C. RIBALDII
=CURVULARIA BOEDIJN 1933 FIDE CARMICHAEL
  SYNO    =CURVUSPORIUM AND CURVOSPORIUM (ORTHOGRAPHIC VARIANTS IN ORIG. PUBL.)

CURVULARIA BOEDIJN 1933. BULL. JARD. BOT. BUITENZ. III, 13:127 /C. LUNATA
(WAKKER) =CURVULARIA STATE OF COCHLIOBOLUS LUNATA NELSON ET HAASIS 1964
PHRAGMO,CETERI,PHAEO, ASCO STATE =COCHLIOBOLUS FIDE ELLIS 1966 /ELLIS, MYCOL
PAPERS CMI 106:1-43,1966. (KEY,ILLUS)
  SYNO    =CURVISPORIUM CORBETTA 1963 FIDE CARMICHAEL
          =MALUSTELLA BATISTA ET LIMA 1960 FIDE ELLIS 1971

CURVULARIOPSIS ELLIS 1961. MYCOL PAPERS CMI 82:39-41 (ILLUS) /C. CYMBISPERMA
(PATOUILLARD)
PHRAGMO OR HELICO,CETERI,PHAEO

CUSPIDOSPORIUM CIFERRI 1955. SYDOWIA 9:303 /C. CUSPIDATUM (SACCARDO)
=CORYNESPORA PULVINIFORMIS
=EXOSPORIUM LINK 1809 FIDE ELLIS 1971

CUSTINGOPHORA STOLK, HENNEBERT ET KLOPOTEK 1968. PERSOONIA 5:195 /C. OLIVACEA
AMERO,GLOIO,PHAEO

CUTICULARIA DUCOMET 1907. ANN. AGR. RENNES 1:235 /C. ILICIS
=STERILE

## 10. Hyphomycetes

CYCLOCONIUM CASTAGNE 1845. CATAL. PLANTES . . . MARSEILLES P. 220 /C. OLEAGINEUM =SPILOCAEA OLEAGINEA (CAST.) HUGHES 1953
=SPILOCAEA FRIES 1825 FIDE HUGHES 1958

CYLINDRIUM BONORDEN 1851. HANDB. MYKOL. P. 34 /C. FLAVOVIRENS (DITMAR) (LECTOTYPE)
=FUSIDIUM LINK 1809 FIDE HUGHES 1958

CYLINDROCARPON WOLLENWEBER 1913. PHYTOPATHOLOGY 3:225 /C. CYLINDROIDES
   PHRAGMO,GLOIO,HYALO, ASCO STATE =NECTRIA FRIES 1825 FIDE BOOTH 1966 /BOOTH,
   MYCOL PAPERS CMI 104:1-56,1966. (REVIS,KEY,ILLUS)
      SYNO    =EURICOA BATISTA ET MAIA 1955 FIDE CARMICHAEL
                =HYALOFLOREA BATISTA ET MAIA 1955 FIDE CARMICHAEL
                =MOESZIA BUBAK 1914 FIDE VON ARX 1970
                =COLEOSPORIUM MOREAU ET MOREAU 1937 FIDE CARMICHAEL
                =ALLANTOSPORA WAKKER 1896 FIDE BOOTH 1966

CYLINDROCEPHALUM BONORDEN 1851. HANDB. MYK. P. 103 /C. AUREUM (CORDA)
=CHALARA (CORDA) RABENHORST 1844 FIDE HUGHES 1958

CYLINDROCLADIUM MORGAN 1892. BOTANICAL GAZETTE 46:191 /C. SCOPARIUM
   DIDYMO OR PHRAGMO,GLOIO,HYALO, ASCO STATE =CALONECTRIA DE NOTARIS 1864 FIDE
   BARRON 1968 /BARRON, HYPHOMYCETES P. 139-140,1968. (ILLUS,REFS)
      SYNO    =CANDELOSPORA REA ET HAWLEY 1912 FIDE BOOTH 1960
                =TETRACYTUM VANDERWALLE 1945 FIDE WORMALD (SUB CANDELOSPORA) FROM AINSWORTH 1971

CYLINDROCOLLA BONORDEN 1851. HANDB. ALLGEM. MYKOL. P. 149 /C. URTICAE (PERSOON)
=CYLINDROCOLLA STATE OF CALLORIA FUSARIOIDES (BERK.) FRIES 1849 FIDE LINDAU 1907
AMERO,ARTHRO,HYALO, ASCO STATE =CALLORIA FRIES 1849 FIDE LINDAU 1907 /LINDAU, RABENHORST KRYPT.-FL. 2 ED. 1 BD. 9 ABT. P. 477-479,1907 (ILLUS,REFS)

CYLINDRODENDRUM BONORDEN 1851. HANDB. ALLGEM. MYKOL. P. 98 AND FIG. 127 /C. ALBUM (LECTOTYPE)
DIDYMO,GLOIO,HYALO /TRANS. BRIT. MYCOL. SOC. 27:91,1944

CYLINDROPHORA BONORDEN 1851. HANDB. ALLGEM. MYKOL. P. 92 AND FIGS. 110 AND 131 /THREE ORIGINAL SPECIES
?=CLADOBOTRYUM FIDE GAMS AND HOOZEMANS 1970

CYLINDROSPORA KUHN 1877. HEDWIGIA 16:120-121 /TWO SPECIES MENTIONED
=NOMEN ILLEGITIMUM, PUBLICATION OF NEW GENUS APPARENTLY NOT INTENDED

CYLINDROTRICHUM BONORDEN 1851. HANDB. ALLGEM. MYKOL. P.88 /C. OLIGOSPERMUM (CORDA) (LECTOTYPE)
DIDYMO,GLOIO,HYALO /BARRON, HYPHOMYCETES P. 141-142,1968. (ILLUS)

CYPHELLOPHORA DE VRIES 1962. MYCOPATHOLOGIA 16:47-50 (ILLUS) /C. LACINIATA
DIDYMO OR PHRAGMO,GLOIO,PHAEO

CYSTODENDRON BUBAK 1914. ANN. MYCOL. 12:212 /C. DRYOPHILUM (PASSERINI)
AMERO,GLOIO,HYALO, NOT VERY DIFFERENT FROM PHIALOPHORA /ELLIS, DEM. HYPH. P. 511,1971 (ILLUS)

CYSTOPHORA RABENHORST 1844 NON AGARDH 1841, SEE VOGLINOANA

DACRINA FRIES 1832 (NON FRIES 1825) =STRUMELLA SACCARDO FIDE LINDAU IN RABENHORST 1910

DACRYMYCELLA BIZZOZERO 1885. ATTI IST. VENET. SCI. LETT. ARTI 6 SER. 3:308 AND TAB. III, FIG. 8 /D. FERTILISSIMA
?, ORIGINAL ILLUSTRATION DIFFICULT TO INTERPRET

DACRYODOCHIUM KARSTEN 1896. HEDWIGIA 35:47 /D. FLUXILE
?, NOT ILLUSTRATED BY AUTHOR

DACTULIOPHORA LEAKY 1964. TRANS. BRIT. MYCOL. SOC. 47:341 /D. TARII
=STERILE, HAS BULBILS WITH SETAE

DACTYLARIA SACCARDO 1880. MICHELIA 2:20 /D. PURPURELLA (SACC.)
   DIDYMO OR PHRAGMO,CETERI,HYALO OR PHAEO /BHATT AND KENDRICK, CAN. J. BOTANY
      46:1253-1257,1968. (REVIS, ILLUS)
      SYNO     =DIPLORHINOTRICHUM HOHNEL 1902 FIDE BHATT AND KENDRICK 1968

DACTYLARIOPSIS MECHTIEVA 1967. MIKOL. I FITOPATOL. 1:278 /D. BROCHOPAGA
      (DRECHSLER)
   ?, A SEGREGATE FROM DACTYLELLA GROVE 1884

DACTYLELLA GROVE 1884. JOUR. BOTANY 22:199 /D. MINUTA
   PHRAGMO,CETERI,HYALO /COOKE AND DICKINSON, TRANS. BRIT. MYCOL. SOC.
      48:621-629,1965. (ILLUS)
      SYNO     =MONACROSPORIUM OUDEMANS 1884 FIDE BARRON 1968

DACTYLINA ARNAUD EX SUBRAMANIAN 1964. J. INDIAN. BOT. SOC. 42:297 (DATED 1963)
      /D. TYLOPAGA (DRECHSLER) SUBRAMANIAN 1964
   =NOMEN ILLEGITIMUM, A SEGREGATE FROM DACTYLELLA WITH A TERMINAL PROLONGATION
      OF THE CONIDIA. NON DACTYLINA NYLANDER 1860 (LICHEN). SEE ALSO
      DRECHSLERELLA SUBRAMANIAN 1964.

DACTYLIUM NEES 1817. DAS SYSTEM ... P. 58 /D. CANDIDUM
   =NOMEN DUBIUM FIDE BARRON 1968, BUT SEE AINSWORTH, DICTIONARY OF THE FUNGI,
      1971.

DACTYLOSPORIUM HARZ 1871. BULL. SOC. IMPER. NAT. MOSCOW 44:131 /D. MACROPUS
      (CORDA)
   DICTYO,CETERI,PHAEO, NON DACTYLOSPORIUM MECHTIEVA 1967 /HUGHES, THE
      NATURALIST 1952(APRIL-JUNE):63-64,1952. (ILLUS)

DACTYLOSPORIUM MECHTIEVA 1967. MIKOL. I FITOPATOL. 1:278 /D. LEPTOSPORA
      (DRECHSLER) =DACTYLELLA LEPTOSPORA DRECHSLER
   =NOMEN ILLEGITIMUM, NON DACTYLOSPORIUM HARZ 1871

DAMNOSPORIUM CORDA =BACTRIDIUM FIDE AINSWORTH 1971

DEIGHTONIELLA HUGHES 1952. MYCOL PAPERS CMI 48:27 /D. AFRICANA
   DIDYMO OR PHRAGMO,CETERI,PHAEO /ELLIS, MYCOL PAPERS CMI 66:1-12,1957.
      (KEY,ILLUS)

DELORTIA PATOUILLARD ET GAILL. 1888. BULL. SOC. MYC. FRANCE 4:43-44 AND PL.
      XIII, FIG. 5A TOP /D. PALMICOLA PATOUILLARD 1888
   HELICO,GLOIO,HYALO
      SYNO     ?=HYALOTROCHOPHORA FINLEY ET MORRIS 1967

DEMATIUM PERSOON 1801. SYN. METH. FUNG. /?, SEE HUGHES, CAN. J. BOTANY
      36:760,1958
   =NOMEN CONFUSUM SEE HUGHES 1958

DEMATOIDIUM STAUTZ 1931. PHYTOPATH. Z. 3:204-208 (ILLUS) /D. NIGRESCENS
   ?, FOR A DARK YEAST-LIKE FUNGUS

DEMATOPHORA HARTIG 1883. UNTERSUCH FORSTBOT. INST. MUNCH. 3:95-140 AND TAF. 7
      /D. NECATRIX
   =NODULISPORIUM PREUSS 1849 FIDE CARMICHAEL /MARTIN, S. AFRICAN J. BOT.
      34:187-192,1968. (ILLUS)

DENDRINA FRIES 1832. SYSTEMA MYCOLOGICUM 3:454 /?D. FLAVUM
   ?=NOMEN DUBIUM, SEE LINDAU IN RABENHORST KRYPT.-FL. 2 AUFL. 1 BD. 8:203 AND
      659,1907

DENDRODOCHIUM BONORDEN 1851. HANDB. MYKOL. P. 135 AND FIG. 228 /D. AURANTIACUM
      (LECTOTYPE)
   AMERO,CETERI,HYALO /BARRON, HYPHOMYCETES P. 148-149,1968. (ILLUS)
      SYNO     =DICHITONIUM BERKELEY ET CURTIS FIDE AINSWORTH 1971
               =PATOUILLARDIA ROUMEGUERE 1885 FIDE HOHNEL FROM CLEMENTS AND SHEAR 19

DENDROGRAPHIUM MASSEE 1892. GREVILLEA 21:5 /D. ATRUM
   PHRAGMO,CETERI,PHAEO /ELLIS, DEM. HYPH. P. 398-400,1971. (ILLUS)

DENDROSPORA INGOLD 1944. TRANS. BRIT. MYCOL. SOC. 26:104-107, DATED 1943

## 10. Hyphomycetes 361

    (ILLUS) /D. ERECTA
   STAURO,CETERI,HYALO

DENDROSPORIUM PLAKIDAS ET EDGERTON 1936. MYCOLOGIA 28:83-84 (ILLUS) /D. LOBATUM
   DIDYMO OR STAURO,CETERI,HYALO

DENDROSTILBELLA HOHNEL 1905. OEST. BOT. ZEITS. 55:22 /D. PRASINULA
   ?, NOT ILLUSTRATED BY AUTHOR /BARRON, HYPHOMYCETES P. 149-150,1968. (ILLUS)

DENDRYPHIELLA BUBAK ET RANOJ. 1914. ANN. MYCOL. 12:417 /D. INTERSEMINATA (BERK.
   ET RAV.)
   =DENDRYPHION WALLROTH 1833 FIDE HUGHES 1958

DENDRYPHION WALLROTH 1833. FLORA CRYPT. GERMAN. 2:300 /D. COMOSUM
   PHRAGMO OR DIDYMO,BLASTO OR CETERI,PHAEO, ASCO STATE = PLEOSPORA FIDE ELLIS
   1971 /HUGHES, CAN. J. BOTANY 31:636-638,1953. (ILLUS) ELLIS, DEM. HYPH., P.
   501-505,1971. (ILLUS,KEY)
      SYNO     =DENDRYPHIELLA BUBAK ET RANOJ. 1914 FIDE HUGHES 1958
            =ENTOMYCLIUM WALLROTH 1833 FIDE HUGHES 1958
            =BRACHYCLADIUM CORDA 1838 FIDE HUGHES 1958
            ?=DWAYAMALA SUBRAMANIAN 1956
            ?=PSEUDOTORULA SUBRAMANIAN 1958

DENDRYPHIOPSIS HUGHES 1953. CAN. J. BOTANY 31:655 /D. ATRA (CORDA)
   =DENDRYPHIOPSIS STATE OF AMPHISPHAERIA INCRUSTANS ELLIS ET EVERHART 1892
   FIDE HUGHES 1958
   PHRAGMO,CETERI,PHAEO, ASCO STATE =AMPHISPHAERIA FIDE HUGHES 1958 /HUGHES,
   CAN. J. BOTANY 31:638-639,1953. (ILLUS) HUGHES, CAN. J. BOTANY 36:762,1958.
   (LIST)

DEPHILIPPIA RAMBELLI 1959. MYCOPATHOLOGIA 11:137 /D. HELICOTRICHOIDES
   =CIRCINOTRICHUM NEES 1817 FIDE PIROZYNSKI 1962

DERMOSPORIUM LINK 1815. MAG. GES. NATURF. FREUNDE, BERLIN 7:32, TAB. 1, FIG. 4A
   /D. FLAVICANS
   =AEGERITA PERSOON FIDE LINDAU IN RABENHORST 1910

DESMIDIOSPORA THAXTER 1891. BOT. GAZ. 16:203 AND PL. XX, FIGS. 1-9 /D.
   MYRMECOPHILA
   STAURO,CETERI,PHAEO

DESMOTRICHUM LEVILLE 1843. ANN. SCI. NAT. 2 SER., 19:217, TAB. VII, FIG. 8 /D.
   SIMPLEX
   =GONATOBOTRYS CORDA FIDE LINDAU IN RABENHORST 1907

DEXHOWARDIA TAYLOR 1970. MYCOPATHOLOGIA 40:306 (305-308,ILLUS) /D. TETRASPORA
   AMERO,CETERI,HYALO, APPEARS TO BE CONIDIAL STATE OF TRECHISPORA, SEE HUGHES
   IN KENDRICK 1971, P.20-21.

DIAPHANIUM FRIES 1835. FL. SCAND. 307 /D. MAXIMUM
   ?

DICELLISPORA SAWADA 1944. REP. GOVT. RES. INST. FORMOSA 87:74 /D. LELEBAE
   ?, JAPANESE DIAGNOSIS, PROBABLY GONATOBOTRYUM

DICHITONIUM BERKELEY ET CURTIS ?=DENDRODOCHIUM FIDE AINSWORTH 1971

DICHOTOMELLA SACCARDO 1914. ANN. MYCOL. 12:312 /D. AREOLATA SACCARDO
   =NIGROSPORA ORYZAE FIDE HUGHES 1958
   =NIGROSPORA ZIMMERMAN 1902 FIDE HUGHES 1958

DICHOTOMOPHTHORA MEHRLICH ET FITZPATRICK EX ELLIS 1971. DEMATIACEOUS
   HYPHOMYCETES P. 388 /D. PORTULACAE MEHRLICH ET FITZPATRICK 1935
   PHRAGMO,CETERI,PHAEO /BARRON, HYPHOMYCETES P. 151-152,1968. (ILLUS)

DICHOTOMOPHTHOROPSIS ELLIS 1971. MYCOL. PAPERS CMI 125:20-22 (ILLUS) /D.
   NYMPHAEARUM (RAND) =HELICOSPORIUM NYMPHAEARUM RAND 1917
   HELICO,CETERI,PHAEO

DICLADIUM CESATI 1852 =COLLETOTRICHUM CORDA FIDE VON ARX 1970

DICOCCUM CORDA 1829. STURMS DEUTSCHL. FLORA, III, BD 2, HEFT 9:117 /D.
    MINUTISSIMUM
    =NOMEN DUBIUM FIDE HUGHES 1958

DICRANIDION HARKNESS 1885. BULL. CALIFORNIA ACAD. SCI. 1:163 /D. FRAGILE
    STAURO,CETERI,HYALO /PEEK AND SOLHEIM, MYCOLOGIA 50:857-860,1958(ILLUS,EMEND)
        SYNO    =PEDILOSPORA HOHNEL 1902 FIDE HUGHES 1951

DICTYOARTHRINIUM HUGHES 1952. MYCOL PAPERS CMI 48:29 /D. QUADRATUM = D.
    SACCHARI (STEVENSON) DAMON 1953
    DICTYO,CETERI,PHAEO /HUGHES, MYCOL PAPERS CMI 48:29-33,1952. (ILLUS) ELLIS,
    DEM. HYPH. P. 580-582,1971. (ILLUS)

DICTYOARTHRINOPSIS BATISTA ET CIFERRI 1957. PUBL. INST. MIC. UNIV. RECIFE 94:3
    /D. COSTARICENSIS
    ?, PROBABLY AN EARLIER NAME FOR DIHETEROSPORA KAMYSCHKO 1962

DICTYOCATENULATA FINLEY ET MORRIS 1967. AMER. MIDL. NAT. 77:200-202 (ILLUS) /D.
    ALBA
    DICTYO,BASO,HYALO

DICTYOCEPHALA MEDEIROS 1962. PUBL. INST. MICOL. UNIV. RECIFE 372:13 /D.
    ULMIFOLIAE (OBREGON-BOTERO)
    ?, PROBABLY AN ALTERNARIA

DICTYOCHAETA SPEGAZZINI 1923. PHYSIS 7:18-19 (ILLUS) /D. FUEGIANA
    ?=COLLETOTRICHUM CORDA 1831. AINSWORTH 1971 ACCIDENTALLY LISTS AS AN
    HYMENOMYCETE

DICTYODESMIUM HUGHES 1951. MYCOL PAPERS CMI 36:27-29 (ILLUS) /D. ULMICOLA
    (ELLIS ET KELLERMAN)
    DICTYO,CETERI,PHAEO

DICTYOPHRYNELLA BATISTA ET CAVALCANTI 1964. PORT. ACTA BIOL. SER. B, 7:356 /D.
    BIGNONIACEARUM
    ?, ORIGINAL ILLUSTRATION NOT DIAGNOSTIC

DICTYOSPORIUM CORDA 1836. WEITENWEBERS BEITRAGE ZUR NAT. P. 1836 /D. ELEGANS
    DICTYO OR STAURO,CETERI,PHAEO /HUGHES, CAN. J. BOTANY 36:762-763,1958. (LIST)
    BARRON, HYPHOMYCETES P.152-153,1968. (ILLUS)
        SYNO    =SPEIRA CORDA 1837 FIDE HUGHES 1958
                =CATTANEA GAROVAGLIO 1875 FIDE ELLIS 1971

DICYMA BOULANGER 1897. REV. GEN. BOT. 9:17-20 AND PL. 1 AND 2 /D. AMPULLIFERA
    BOUL. 1897 = D. STATE OF ASCOTRICHA CHARTARUM BERKELEY FIDE ELLIS 1971
    AMERO,CETERI,PHAEO, ASCO STATE =ASCOTRICHA FIDE AINSWORTH 1961 /ELLIS, DEM.
    HYPH. P.212-213,1971. (ILLUS) HAWKSWORTH, MYCOL. PAPERS CMI 126:1-28,1971.
    (ILLUS,KEY)
        SYNO    ?=GONYTRICHELLA EMOTO ET TUBAKI 1970

DIDYMARIA CORDA 1842. ANLEIT. STUD. MYKOL. PP. 32 AND 199 /D. UNGERI
    =RAMULARIA UNGER FIDE VON ARX 1970

DIDYMARIOPSIS SPEGAZZINI 1910. ANN. MUS. NAC. BUENOS AIRES T. 20, SER 3A:424
    (ILLUS) /D. CUPHAEICOLA SPEG. 1910
    ?

DIDYMOBOTRYOPSIS HENNINGS 1902. HEDWIGIA 41:149 /D. PARASITICA
    ?, NOT ILLUSTRATED BY AUTHOR

DIDYMOBOTRYUM SACCARDO 1886. SYLLOGE FUNGORUM 4:626 /D. RIGIDUM (BERKELEY ET
    BROOME) SACCARDO (LECTOTYPE FIDE ELLIS 1971)
    DIDYMO,BLASTO,PHAEO /ELLIS, DEM. HYPH. P. 381-384,1971. (ILLUS)

DIDYMOCLADIUM SACCARDO 1886. SYLLOGE FUNGORUM 4:186 /D. TERNATUM (BONORDEN)
    =CLADOBOTRYUM NEES 1817 FIDE HUGHES 1958

DIDYMOPSIS SACCARDO ET MARCHAL 1885. BULL. SOC. BOT. BELG. 24:61-62 AND PL IV,
    FIGS 10 AND 11 /D. PEREXIGUA SACC. ET MARCHAL 1885

## 10. Hyphomycetes 363

? /LINDAU, IN RABENHORST KRYPT.-FL. 2 AUFL. 1 BD. VIII:363-364,1907. (ILLUS)

DIDYMOSTILBE HENNINGS 1902. HEDWIGIA 41:148 /D. COFFEAE
   DIDYMO,GLOIO,HYALO, NON DIDYMOSTILBE BRESADOLA ET SACCARDO 1902 (3 WEEKS
     AFTER HENNINGS) /MORRIS, WEST. ILLINOIS SER BIOL SCI 3:48-49,1963. (ILLUS)
     SYNO    ?=KUTILAKESOPSIS AGNIHOTHRUDU ET BARUA 1957

DIDYMOTHOZETIA RANGEL 1915. BOL. AGR. S. P. 16:325-327 AND TAB V FIGS. 1-14 /D.
   MIMOSENSIS
? 
     SYNO    =DITHOZETIA CLEMENTS ET SHEAR 1931 (ORTHOGRAPHIC VARIANT)

DIDYMOTRICHELLA ARNAUD 1954. BULL. SOC. MYCOL. FR. 69:284 AND FIG. 16B, P. 301
   /D. QUERCINA
   DIDYMO,CETERI,PHAEC, NO LATIN DIAGNOSIS

DIDYMOTRICHUM BONORDEN 1851. HANDB. MYK. P. 89 /D. NODULOSUM (CORDA) (LECTOTYPE
   FIDE HUGHES 1958)
   =CLADOSPORIUM LINK 1815 FIDE HUGHES 1958

DIDYMOTRICHUM HOHNEL 1914. S. B. AKAD. WISS. WIEN. 123:140 /D. CHRYSOSPERMUM
   (SACCARDO) HOHNEL =DACTYLARIA CHRYSOSPERMA (SACC.) BHATT ET KENDRICK 1968
   =NOMEN ILLEGITIMUM, NON DIDYMOTRICHUM BONORDEN 1851 (FIDE HUGHES 1958)

DIDYMOZOOPHAGA SOPRUNOV ET GALIULINA 1951. MIKROBIOLOGIYA 20:489-499 (ILLUS)
   /D. ARTHROBOTRYOIDES (BERLESE) SOP. ET GAL. 1951
   =ARTHROBOTRYS CORDA 1839 FIDE HAARD 1968

DIHETEROSPORA KAMYSCHKO 1962. BOTAN. MATER. (NOTUL. SYST. SECT. CRYPT. INST.
   BOTAN. ACAD. SCI. USSR) 15:138 /D. HETEROSPORA KAMYSCHKO =D. CHLAMYDOSPORIA
   (GODDARD) BARRON ET ONIONS 1966
   DICTYO,CETERI,HYALO, MAY BE THE SAME AS CONIODICTYUM HAR. ET FAT. 1909 OR
   DICTYOARTHRINOPSIS BATISTA ET CIFERRI 1957 /BARRON AND ONIONS, CAN. J.
   BOTANY 44:861-869,1966. (ILLUS)

DIMERA FRIES 1825. SYST. ORB. VEG. P. 183 /BOTRYTIS DIDYMA SCHM.
   =OEDEMIUM LINK 1824 FIDE HUGHES 1958

DIMORPHOSPORA TUBAKI 1958. J. HATTORI BOTAN. LAB. NO. 20:156-158 (ILLUS) /D.
   FOLIICOLA
   AMERO,GLOIO OR BLASTO,HYALO, BUT GENUS FOUNDED ON TWO KINDS OF CONIDIA
     SYNO    =FLUMINISPORA INGOLD 1958 FIDE KENDRICK

DIONYSIA ARNAUD 1952. BULL. SOC. MYCOL. FR. 68:210-213 (ILLUS) /D. CAPITATA
   =CANDELABRUM VAN BEVERWIJK 1952 FIDE KENDRICK
     SYNO    =DYONISIA ARNAUD 1952 FIDE AINSWORTH 1960 (ORTHOGR. VAR.)

DIPLOCLADIELLA ARNAUD 1954. BULL. SOC. MYCOL. FR. 69:303 AND FIG. 13, I TO P
   /D. SCALAROIDES
   STAURO,CETERI,PHAEO, NO LATIN DIAGNOSIS, COMPARE WITH IYENGARINA SUBRAMANIAN
   /TUBAKI, J. HATTORI BOT. LAB. NO. 20:158-159,1958. (ILLUS)

DIPLOCLADIUM BONORDEN 1851. HANDB. ALLGEM. MYCOL. P. 98 /TWO ORIGINAL SPECIES
   =CLADOBOTRYUM NEES 1817 FIDE HUGHES 1958

DIPLOCOCCIUM GROVE 1885. JOUR. BOTANY, LONDON 23:167 /D. SPICATUM
   DIDYMO,CETERI OR BLASTO,PHAEO, ASCO STATE =HELMINTHOSPHAERIA FIDE ELLIS 1963
   /BARRON, HYPHOMYCETES P. 155-156,1968. (ILLUS)

DIPLOIDIUM ARNAUD 1923. ANN. EPIPHYT. 9:33 /D. SWEETIAE
   =SEPTOIDIUM ARNAUD 1921 FIDE ELLIS 1971

DIPLOOSPORA GROVE 1916. JOUR. BOTANY 54:220 AND TAB. 542, FIG. 8 /D. ROSEA
   DIDYMO,BLASTO,HYALO

DIPLORHINOTRICHUM HOHNEL 1902. S. B. AKAD. WISS. WIEN 111:1040 /D. CANDIDULUM
   =DACTYLARIA SACCARDO 1880 FIDE BHATT AND KENDRICK 1968

DIPLORHYNCHUS ARNAUD 1952. BULL. SOC. MYCOL. FR. 68:207-211 (ILLUS) /D. BILOBA
   HELICO,CETERI,HYALO, NO LATIN DIAGNOSIS

DIPLOSPORIUM LINK 1824. LINNAEUS SPEC. PLANT. IV,6(1):64 /D. NIGRESCENS
    =OEDEMIUM LINK 1824 FIDE HUGHES 1958

DISCOCOLLA PRILLIEUX ET DELACROIX 1894. BULL. SOC. MYC. FRANCE 10:86 /D. PIRINA
    PRILL. ET DELACR. 1894
    ?, ORIGINAL ILLUSTRATION NOT DIAGNOSTIC /LINDAU, RABENH. KRYPT. FL. 2 AUFL. 1
    BD. 9:506. (ILLUS)

DISCOFUSAPIUM PETCH 1921. TRANS. BRIT. MYCOL. SOC. 7:164 /D. TASMANIENSE
    (MCALPINE)
    =FUSARIUM LINK FIDE CLEMENTS AND SHEAR 1931

DISCOSPORIUM SACCARDO ET SYDOW (NON HOHNEL 1914) ?=CONIOSPORIUM FIDE AINSWORTH
    1971

DITHOZETIA CLEMENTS ET SHEAR 1931 (ORTHOGRAPHIC VARIANT), SEE DIDYMOTHOZETIA
    RANGEL 1915

DIVINIA CIFERRI 1955. SYDOWIA 9:326 /D. DIATRICHA
    ?, ORIGINAL ILLUSTRATION INADEQUATE TO CHARACTERIZE

DOMINGOELLA PETRAK ET CIFERRI 1932. ANN. MYCOL. 30:339 /D. ASTERINARUM
    AMERO,CETERI,PHAEO /ELLIS, DEM. HYPH. P. 113-114,1971. (ILLUS)

DORATOMYCES CORDA 1829. STURMS DEUTSCHL FLORA III, BD 2, HEFT 7:65 /D. NEESII
    =CEPHALOTRICHUM LINK 1809 FIDE HUGHES 1958

DRECHSLERA ITO 1930. PROC. IMP. ACAD. TOKYO 6:355 /D. TRITICI-VULGARIS
    (NISIKADO) = D. STATE OF PYRENOPHORA TRITICI-REPENTIS (DIED.) DRECHSL.
    PHRAGMO,CETERI,PHAEO, ASCO STATE = PYRENOPHORA OR TRICHOMETASPHAERIA OR
    COCHLIOBOLUS FIDE ELLIS 1971 /ELLIS, DEM. HYPH. P. 403-452,1971.
    (ILLUS,KEY,REVIS)
        SYNO    =BIPOLARIS SHOEMAKER 1959 FIDE ELLIS 1971

DRECHSLERELLA SUBRAMANIAN 1964. J. INDIAN BOT. SOC. 42:299 (DATED 1963) /D.
    ACROCHAETA (DRECHSLER)
    PHRAGMO,CETERI,HYALO, SEE ALSO DACTYLINA ARNAUD EX SUBRAMANIAN

DREPANOCONIS SCHROETER ET HENNINGS 1896. HEDWIGIA 35:211-212 (NO ILLUS) /D.
    BRASILIENSIS
    HELICO,CETERI,HYALO, BUT SEE AINSWORTH, DICTIONARY OF FUNGI, 1971
        SYNO    ?=CURVIDIGITUS SAWADA 1943

DREPANOSPORA BERKELEY ET CURTIS 1875. GREVILLEA 3:105 /D. PANNOSA BERK. ET
    CURTIS 1875
    =HELICOSPORIUM NEES 1817 FIDE MOORE 1957

DRUMOPAMA SUBRAMANIAN 1957. PROC. INDIAN ACAD. SCI. 46:333 /D. GIRISA
    AMEPO,CETERI,HYALO /MORRIS, WEST. ILLINOIS SER. BIOL SCI 3:52-53,1963. (ILLUS)

DUBIOMYCES LLOYD 1921. MYCOLOGICAL NOTES 65:1034 /D. VIRIDIS
    =USTILAGINOIDEA BREFELD FIDE DIEHL FROM AINSWORTH 1971

DUDDINGTONIA COOKE 1969. TRANS. BRIT. MYCOL. SOC. 53:315-318 (ILLUS) /D.
    FLAGRANS (DUDDINGTON) =TRICHOTHECIUM FLAGRANS
    ?=ARTHROBOTRYS CORDA 1839

DUOSPORIUM THIND ET RAWLA 1961. AMER. J. BOTANY 48:859-862 (ILLUS) /D. CYPERI
    PHRAGMO,CETERI,PHAEO, FOUNDED ON TWO KINDS OF CONIDIA, BUT MACROCONIDIA ARE
    DISTINCTIVE

DWAYABEEJA SUBRAMANIAN 1958. J. INDIAN BOT. SOC. 37:53-57 (ILLUS) /D. SUNDARA
    PHRAGMO OR SCOLECO,CETERI,PHAEO

DWAYALOMA SUBRAMANIAN 1957. J. INDIAN BOT. SOC. 36:61-64 (ILLUS) /D. TRINA
    DIDYMO,CETERI,HYALO

DWAYAMALA SUBRAMANIAN 1956. J. INDIAN BOT. SOC. 35:473-476 (ILLUS) /D
    PRATHILOMAKA

## 10. Hyphomycetes 365

?=DENDRYPHION WALLROTH 1833, D. PRATHILOMAKA APPEARS TO BE DENDRYPHIELLA INFUSCANS (THUMEN) ELLIS 1971

DYONISIA ARNAUD 1952, SEE DIONYSIA

ECHINOBOTRYUM CORDA 1832. STURMS DEUTSCHL. FLORA III(PILZE), BD. 3, HEFT 12:51 /E. ATRUM
   AMERO,CETERI,PHAEO /BARRON, HYPHOMYCETES 1968, P. 161. (ILLUS)

ECHINOIDIA PATOUILLARD 1918. BULL. SOC. MYCOL. FR. 24:199-201 (ILLUS) /E. THEOBROMAE PAT. 1918
   ?, A STATE OF POLYPORUS HYDNOPHORUS FIDE PETCH FROM AINSWORTH 1971

ECHINOSPORIUM WORONICHIN 1913. MONIT. JARD. BOT. TIFLIS 28:25 AND FIGS. 12-14 /E. ACERIS WORON. 1913
   =PETRAKIA SYDOW 1913 FIDE CARMICHAEL

ECTOSTROMA FRIES 1823. SYSTEMA MYCOL. 2:601 /E. LIRIODENDRI (KUNAE) (LECTOTYPE)
   =STERILE

ECTOTRICHOPHYTON CASTELLANI ET CHALMERS 1919. MAN. TROP. MED. ED. 3 P. 1003-1008 /E. MENTAGROPHYTES (ROBIN) CAST. ET CHALM. 1919
   =TRICHOPHYTON MALMSTEN 1845

EDMUNDMASONIA SUBRAMANIAN 1958. J. INDIAN BOT. SOC. 37:403 /E. PULCHRA
   =BRACHYSPORIELLA BATISTA 1952 FIDE KENDRICK

EIDAMIA LINDAU 1904. RABENHORST KRYPT.-FL. I, 8:123 /E. ACREMONIOIDES (HARZ) LINDAU
   =ACREMONIELLA SACCARDO 1886 FIDE ELLIS 1971

ELADIA SMITH 1961. TRANS. BRIT. MYCOL. SOC. 44:47 /E. SACCULA (DALE)
   AMERO,BASO,PHAEO /BARRON, HYPHOMYCETES 1968, P. 162-163. (ILLUS)

ELAEODEMA SYDOW 1922. ANN. MYCOL., BERLIN 20:64 /E. CINNAMOMI
   ?, NOT ILLUSTRATED BY AUTHOR

ELLETEVERA DEIGHTON 1969. MYCOL. PAPERS CMI 118:17-21 (ILLUS) /E. PARASITICA (ELL. ET EV.)
   PHRAGMO,CETERI,PHAEO

ELLISIA BATISTA ET PERES 1965. MYCOPATHOLOGIA 25:166 /E. INGAE
   ?, IS ANTEDATED BY THE HIGHER PLANT GENUS ELLISIA. E. INGAE PROBABLY BELONGS IN ORMATHODIUM SYDOW

ELLISIELLA SACCARDO 1880. MICHELIA 2:26, ILLUS. IN FUNGI ITALICI FIG. 781 /E. CAUDATA (PECK) SACC. 1880
   =COLLETOTRICHUM CORDA 1831 FIDE VON ARX 1970

ELLISIELLINA DA CAMARA 1949. AGRON. LUSIT. 11:72 /ONE OF THE TWO ORIGINAL SPECIES IS THE TYPE OF ELLISIELLA SACC. 1880
   =NOMEN ILLEGITIMUM FIDE CARMICHAEL

ELLISIOPSIS BATISTA ET NASCIMENTO 1956. AN. SOC. BIOL. PERNAMBUCO 14:21 /E. GALLESIAE
   =BELTRANIELLA SUBRAMANIAN 1952 FIDE PIROZYNSKI 1970

EMBELLISIA SIMMONS 1971. MYCOLOGIA 63:380-386 (ILLUS) /E. ALLII (CAMPANILE)
   PHRAGMO OR DICTYO,CETERI,PHAEO, BUT NOT VERY DIFFERENT FROM CURVULARIA

EMMONSIA CIFERRI ET MONTEMARTINI 1959. MYCOPATHOLOGIA 10:314 /E. PARVA (EMMONS ET ASHBURN) CIF. ET MONT. 1959
   =CHRYSOSPORIUM CORDA FIDE CARMICHAEL 1962

ENDOCALYX BERKELEY ET BROOME 1875. J. LINN. SOC. LONDON 15:84 /E. THWAITESII BERK. ET BR. 1875 (LECTOTYPE)
   AMERO,CETERI,PHAEO /ELLIS, DEM. HYPH. P.579-580,1971. (ILLUS)
      SYNO   ?=NIGROCUPULA SAWADA 1944

ENDOCONIDIUM PRILLIEUX ET DELACROIX 1891. BULL. SOC. MYCOL. FR. 7:116 (ILLUS)
/E. TEMULENTUM PRILL. ET DELACR. 1891, =E. STATE OF PHIALEA TEMULENTA
PRILL. ET DELACR.
AMERO,BASO,HYALO, ASCO STATE =PHIALEA (FRIES) GILLET FIDE PRILL. ET DEL.
/LINDAU, RABENHORSTS KRYPT.-FL. 2 ED. 1BD. 9:479-481,1910. (ILLUS)
AINSWORTH, DICTIONARY OF THE FUNGI 5 ED. P. 167-168,1961. (REFS)

ENDODERMOPHYTON CASTELLANI 1910. MAN. TROP. MED., ED. 1, P. 610 /THE AGENT OF
TINEA IMBRICATA (TRICHOPHYTON CONCENTRICUM)
=TRICHOPHYTON MALMSTEN 1845

ENDODESMIA BERKELEY ET BROOME 1874. ANN. MAG. NAT. HIST. 4, 7:432 /E. GLAUCA
BERK. ET BR. 1874
=BROOMEOLA KUNZE 1891, NON ENDODESMIA BENTH. 1862

ENDOPHRAGMIA DUVERNOY ET MAIRE 1920. BULL. SOC. MYCOL. FRANCE 36:86 /E.
MIRABILIS
PHRAGMO,CETERI,PHAEO /ELLIS, MYCOL PAPERS CMI 72:19-36,1959. (ILLUS, KEY)
SYNO     =PHRAGMOCEPHALA MASON ET HUGHES FIDE ELLIS 1959

ENDOPHRAGMIOPSIS ELLIS 1966. MYCOL. PAPERS CMI 106:55-57 (ILLUS) /E. PIROZYNSKII
PHRAGMO,CETERI,PHAEO

ENDOSPOROSTILBE SUBRAMANIAN 1958. JOUR. INDIAN BOT. SOC. 37:49 /E. NILAGIRICA
=BLOXAMIA BERKELEY ET BROOME 1854 FIDE NAG RAJ

ENDOSTILBUM MALENCON 1964. BULL. SOC. MYCOL. FR. 80:103-112 (ILLUS) /E. CERASI
(BOURD. ET GALZ.)
AMERO,GLOIO,HYALO. COMPARE WITH MICROSTROMA

ENTOMYCLIUM WALLROTH 1833. FLORA CRYPT. GERMAN. 2:189 /E. FOLLICULATUM WALLR.
=DENDRYPHION NANUM (NEES) HUGHES 1958
=DENDRYPHION WALLROTH 1833 FIDE HUGHES 1958

ENTYLOMELLA HOHNEL 1916. OESTRR. BOT. ZEITS. 66:105 /E. FICARIAE (BERKELEY)
(LECTOTYPE FIDE WEESE 1932)
=NOMEN NUDUM, FOR CONIDIAL ENTYLOMA, ?=ITERSONILIA DERX FIDE SOWELL AND KORF
1960. SEE ALSO CIFERRI, OMAGIU LUI TRAIAN SAVULESCU, P. 176, 1959.

EPICLINIUM FRIES 1849. SUMMA VEG. SCAND. P. 475 /E. PEZIZOIDEUM (SCHWEINITZ)
?, NOT ILLUSTRATED BY FRIES

EPICOCCUM LINK 1815. MAG. GES. NATURF. FREUNDE, BERLIN 7:32 /E. NIGRUM
DICTYO,CETERI,PHAEO /SCHOL-SCHWARZ, TRANS. BRIT. MYCOL. SOC.
42:149-173,1959.(REVIS,ILLUS) ELLIS, DEM. HYPH. P. 72-74,1971. (ILLUS)
SYNO     =CEREBELLA CESATI 1851 FIDE SCHOL-SCHWARZ 1959
=PARATRICHAEGUM FAUREL ET SCHOTTER 1966 FIDE KENDRICK
?=CLATHROCOCCUM HOHNEL 1911 FIDE AINSWORTH 1971

EPIDERMOPHYTON SABOURAUD 1907. ARCH. MED. EXP. ANAT. PATHOL. 19(5):565 /E.
INGUINALE = E. FLOCCOSUM (HARZ) LANGERON ET MILOCHEVITCH 1930
PHRAGMO,CETERI,HYALO /AJELLO, SABOURAUDIA 6:147-159,1968. (REVIS) REBEL AND
TAPLIN, THE DERMATOPHYTES 124 PP.,1970. (ILLUS)

EPIDOCHIOPSIS KARSTEN 1892. HEDWIGIA 31:294 (NO ILLUS) /E. ATROVIRENS (KARSTEN)
?

EPIDOCHIUM FRIES 1849. SUMMA VEG. SCAND. P. 471 /E. ATROVIRENS
=TREMELLA DILLENIUS EX FRIES FIDE AINSWORTH 1961

EPILITHIA NYLANDER 1855. BOT. ZEIT. 1855:66 /E. CRISTATA
=HARPOGRAPHIUM SACCARDO FIDE SACCARDO 1886

EPOCHNIELLA SACCARDO 1880. MICHELIA 2:176 /?
=STEMPHYLIUM WALLROTH 1833 FIDE LINDAU IN RABENHORST 1910

EPOCHNIUM LINK 1809. MAG. GES. NATURF. FREUNDE, BERLIN 3:18 /E. MONILIOIDES
LINK =MONILIA FRUCTIGENA PERSOON FIDE HUGHES 1958

## 10. Hyphomycetes

=MONILIA PERSOON 1801 FIDE HUGHES 1958
ERICIANELLA BOND. =BACTRIDIUM FIDE FRIES FROM AINSWORTH 1971

ERIOCERCOSPORA DEIGHTON 1969. MYCOL. PAPERS CMI 118:5 (5-17,ILLUS) /E. BALLADYNAE (HANSFORD)
   PHRAGMO OR SCOLECO,CETERI,PHAEO

ERIOMENE (SACCARDO) MAIRE EX PEYRONEL 1919. BULL. SOC. MYCOL. FR. 35:179 /E. CILIATA (CORDA) MAIRE EX PEYRONEL
   =MENISPORA PERSOON 1822 FIDE HUGHES 1958 ALSO ERIOMENE (SACCARDO) CLEMENTS ET SHEAR 1931 AND ERIONEMA MAIRE 1906 (? LAPSUS PRO ERIOMENE) AND ERIOMENELLA PEYRONEL 1919. (SEE HUGHES 1958)

ERIOMYCOPSIS SPEGAZZINI 1910. AN. MUS. NAC. 3(13):429 (NO ILLUS) /E. BOMPLANDI
   ?, SEE HANSFORD, BOTHALIA VOL. 4,1942

EUASPERGILLUS LUDWIG =ASPERGILLUS FIDE AINSWORTH 1971

EURASINA ARNOLD 1970. Z. PILZK. 35:305-310,(DATED 1969) /E. BONDARZEWIAE
   PHRAGMO,CETERI,HYALO

EURICOA BATISTA ET MAIA 1955. ANN. SOC. BIOL. PERNAMBUCO 13:151-153 (ILLUS) /E. DOMINGUESII
   =CYLINDROCARPON WOLLENWEBER 1913 FIDE CARMICHAEL

EVERHARTIA SACCARDO ET ELLIS 1882. MICHELIA 2:580 /E. HYMENULOIDES SACC. ET ELLIS 1882
   HELICO,GLOIO,HYALO, SEE ALSO HYALOTROCHOPHORA AND VANBEVERWIJKIA

EXCIOCONIDIUM PLUNKETT 1925. IN STEVENS, BERNICE P. BISHOP MUSEUM BULL. 19:156 & FIG. 34, P. 157 /E. CIBOTII
   ?, POSSIBLY SPOROSCHISMA OR ASCOCONIDIUM

EXCIPULARIA SACCARDO 1884. SYLLOGE FUNGORUM 3:689 /E. FUSISPORA (BERK. ET BROOME)
   PHRAGMO,CETERI,PHAEO, EMENDED HOHNEL 1904, ANN. MYCOL. 2:52, COMPARE WITH LISTEROMYCES

EXOBASIDIOPSIS KARAKULIN 1922. NOT. SYST. INST. CRYPT., PETROGRAD 1:83 /E. VICIAE
   =AUREOBASIDIUM VIALA ET BOYER FIDE CLEMENTS AND SHEAR 1931

EXOPHIALA CARMICHAEL 1966. SABOURAUDIA 5:120-123 (ILLUS) /E. SALMONIS
   AMERO OR DIDYMO,GLOIO,PHAEO

EXOSPORELLA HOHNEL 1912. S.B. AKAD. WISS. WIEN 121:414 /E. SYMPLOCI
   ?, NOT ILLUSTRATED BY AUTHOR

EXOSPORIELLA KARSTEN 1892. FINL. MOGELSV. 160-161 /E. FUNGORUM (FRIES)
   PHRAGMO,CETERI,PHAEO /ELLIS, DEM. HYPH. P.148-149,1971. (ILLUS)
     SYNO    =PHANEROCORYNELLA HOHNEL 1919, BASED ON SAME TYPE

EXOSPORINA OUDEMANS 1904. KON. AKAD. AMSTERDAM 6:498-501 AND FIGS. A TO D /E. LARICIS
   ?=BLOXAMIA. EXOSPORINA ARNAUD 1921 = ARNAUDINA TROTTER 1931 AND EXOSPORINELLA BENDER 1932

EXOSPORIUM LINK 1809. MAG. GES. NATURF. FREUNDE, BERLIN 3:10 /E. TILIAE
   PHRAGMO,CETERI,PHAEO /ELLIS, MYCOL PAPERS CMI 82:21-39,1961. (ILLUS,KEY,REVIS)
     SYNO    =CUSPIDOSPORIUM CIFERRI 1955 FIDE ELLIS 1971

EXOTRICHUM SYDOW 1914. ANN. MYCOL. 12:571 (NO ILLUS) /E. LEUCOMELAS
=MYROTHECIUM TODE EX FRIES 1821 FIDE HOHNEL FROM AINSWORTH 1971

FAVOTRICHOPHYTON (CASTELL. ET CHALM.) NEVEU-LEMAIRE 1921. IN PRECIS DE PARASITOL. HUMAINE, ED. 5, P. 55 /F. OCHRACEUM (SABOURAUD) NEVEU-LEMAIRE = TRICHOPHYTON VERRUCOSUM
=TRICHOPHYTON MALMSTEN 1848

FLABELLOSPORA ALASOADURA 1968. NOVA HEDWIGIA 15:415-418 AND TAB. 44 & 45 /F. CRASSA
STAURO,CETERI,HYALO

FLAGELLOSPORA INGOLD 1942. TRANS. BRIT. MYCOL. SOC. 25:404 /F. CURVULA
SCOLECO,CETERI,HYALO

FLAHAULTIA ARNAUD 1951. BULL. SOC. MYCOL. FRANCE 67:195 AND FIG. 4D & E /F. HYALINA
?, LOOKS LIKE GLIOCLADIUM

FLOCCARIA GREVILLE 1828. SCOTTISH CRYPT. FL. P. 6 AND PL. 301 /F. GLAUCA GREV. 1828
?, PROBABLY A COREMIAL PENICILLIUM

FLUMINISPORA INGOLD 1958. TRANS. BRIT. MYCOL. SOC. 41:369 /F. OVALIS
=DIMORPHOSPORA TUBAKI 1958 FIDE KENDRICK

FONSECAEA NEGRONI 1936. REV. INST. BAC. DEP. NAC. HIG., BUENOS AIRES 7:424 /F. PEDROSOI (BRUMPT) NEGRONI = HORMODENDRUM PEDROSOI BRUMPT
?, USED FOR HUMAN PATHOGENS WITH PHIALOPHORA AND CLADOSPORIUM STATES
    SYNO    =CARRIONIA BRICENO-IRAGORRY 1938
              =HORMODENDROIDES MOORE ET ALMEIDA 1936
              =PHIALOCONIDIOPHORA MOORE ET ALMEIDA 1936 FIDE CARMICHAEL

FRASERIELLA CIFERRI ET CORTE 1957. ATTI IST. BOT. UNIV. PAVIA, SER. 5, 14:107-109 (ILLUS) /F. BISPORA = F. STATE OF XEROMYCES BISPORA FRASER
?, ASCO STATE = XEROMYCES FIDE CIF. ET CORTE 1957. THE CONIDIAL STATE IS CLOSE TO EPIDERMOPHYTON

FREYNELLA KUNTZE =COCCOSPORIUM FIDE AINSWORTH 1971

FUCKELINA SACCARDO 1875. NUOV. GIORN. BOT. ITAL. 7:326 /F. SOCIA SACC
=STACHYBOTRYS CORDA 1837 FIDE BOOTH 1957

FULVIA CIFERRI 1954. ATTI IST. BOT. UNIV. PAVIA, SER. 5, 10:246 /F. FULVA (COOKE) CIFERRI =CLADOSPORIUM FULVUM COOKE
?=CLADOSPORIUM LINK 1815

FUMAGO PERSOON 1822. MYCOL. EUROP. 1:9 /F. VAGANS (LECTOTYPE FIDE HUGHES 1958)
=NOMEN CONFUSUM FIDE FRIEND FROM AINSWORTH 1971
    SYNO    =SOREDOSPORA CORDA 1837 FIDE HUGHES 1958

FUMAGOPSIS SPEGAZZINI 1911. AN. MUS. NAC. BUENOS AIRES 3(13):464-465 (ILLUS) /F. TRIGLIFIOIDES
=TRIDENTARIA PREUSS 1852 FIDE KENDRICK

FUSARIELLA SACCARDO 1884. ATTI R. ISTIT. VEN. SCI. LETT. ARTI VI, 2:463 /F. ATROVIRENS
PHRAGMO,BASO,PHAEO /HUGHES, MYCOL. PAPERS CMI 28:1-11,1949. (ILLUS,REVIS)

FUSARIUM LINK 1809. MAG. GES. NATURF. FREUNDE, BERLIN 3:10 /F. ROSEUM
PHRAGMO,GLOIO,HYALO. ASCO STATE =GIBBERELLA OR NECTRIA OR CALONECTRIA /TOUSSOUN AND NELSON, IDENTIFICATION OF FUSARIUM SPECIES, PENN. STATE UNIV. PRESS 51 PP., 1968. (ILLUS) BOOTH, THE GENUS FUSARIUM, CMI, 236PP., 1971 (MONOGR.)
    SYNO    =SPOROTRICHELLA KARSTEN 1887 FIDE HUGHES 1958
              =DISCOFUSARIUM PETCH 1921 FIDE CLEMENTS AND SHEAR 1931
              ?=FUSIDOMUS GROVE 1929 FIDE GROVE 1934 FROM AINSWORTH 1971
              =LACHNIDIUM GIARD FIDE SACCARDO FROM AINSWORTH
              =PIONNOTES FRIES 1849 FIDE WOLLENWEBER FROM AINSWORTH 1971
              =PSEUDOMICROCERA PETCH 1921 FIDE WOLLENWEBER FROM CLEMENTS AND SHEAR 1931
              =RACHISIA LINDNER 1913 FIDE CLEMENTS AND SHEAR 1931
              =SELENOSPORIUM CORDA 1837 FIDE LINDAU IN RABENHORST 1910

FUSELLA SACCARDO 1886. SYLLOGE FUNGORUM 4:246 (NO ILLUS) /THREE ORIGINAL SPECIES
?

## 10. Hyphomycetes

FUSICLADIELLA HOHNEL 1919. BERLIN DEUTSCH. BOT. GES. 37:155 /F. MELAENA
(FUCKEL) HUGHES 1952
  DIDYMO,CETERI,HYALO, ASCO STATE =MYCOSPHAERELLA FIDE HOHNEL 1919 /HUGHES,
MYCOL PAPERS CMI 49:20-24,1952. (ILLUS) DEIGHTON. MYCOL PAPERS CMI
101:23-28,1965. (ILLUS)
    SYNO     =KURSSANOVIA PIDOPLICHKO 1948 FIDE KENDRICK

FUSICLADINA ARNAUD 1952. BULL. SOC. MYCOL. FR. 68:200 AND 215, FIG. 8C /F.
LASIOSPHAERIAE
?, ORIGINAL ILLUSTRATION NOT DIAGNOSTIC

FUSICLADIOPSIS MAIRE 1906. BULL. SOC. BOT. FR. 53:CLXXXVII /F. CONVIVA MAIRE
=STEMPHYLIUM BOTRYOSUM WALLROTH
=STEMPHYLIUM WALLROTH 1833 FIDE HUGHES 1958, FOR FUSICLADIOPSIS KARAK. ET
VASSIL. SEE KARAKULINIA.

FUSICLADIUM BONORDEN 1851. HANDB. ALLGEM. MYKOL. P. 80 /F. VIRESCENS BONORDEN
=F. STATE OF VENTURIA PIRINA ADERH. FIDE HUGHES 1958
  AMERO OR DIDYMO,CETERI,PHAEO, ASCO STATE =VENTURIA FIDE HUGHES 1958 /BARRON,
HYPHOMYCETES P.166-167,1968. (ILLUS) DEIGHTON, MYCOL PAPERS CMI
112:16-27,1967. (ILLUS)
    SYNO     =MEGACLADOSPORIUM VIENNOT-BOURGIN 1949 FIDE HUGHES 1958

FUSICOLLA BONORDEN 1851. HANDB. ALLGEM. MYKOL. P. 150 AND FIG. 248 /F. BETAE
?, ORIGINAL ILLUSTRATION NOT DIAGNOSTIC /LINDAU, RABENHORST KRYPT-FL. 2 AUFL.
1 BD. 9:455,1910. (ILLUS)

FUSIDIUM LINK 1809. MAG. GES. NATURF. FREUNDE, BERLIN 3:8 /F. GRISEUM
(LECTOTYPE)
  AMERO,BLASTO,HYALO /BARRON, HYPHOMYCETES P. 167-169,1968. (ILLUS)
    SYNO     =CYLINDRIUM BONORDEN 1851 FIDE HUGHES 1958

FUSIDIOMUS GROVE 1929 ?=FUSARIUM FIDE GROVE 1934 FROM AINSWORTH 1971

FUSISPORELLA SPEGAZZINI 1911. AN. MUS. NAC. BUENOS AIRES 3(13):454 (ILLUS) /F.
BUFONIS
?, ORIGINAL ILLUSTRATION NOT DIAGNOSTIC

FUSISPORIUM LINK 1809. MAG. GES. NATURF. FREUNDE, BERLIN 3:19 /F. AURANTIACUM
?, HUGHES (1958) DID NOT SEE THE TYPE

FUSOMA CORDA 1837. ICONES FUNGORUM 1:7 /TWO ORIGINAL SPECIES
=NOMEN DUBIUM FIDE HUGHES 1958

GEMMOPHORA SCHKROBATOW 1912. BERL. DEUTSCHL. BOT. GES. 30:478-482 (ILLUS) /G.
PURPURASCENS SCHKR. 1912
?, =STERILE?

GENICULARIA RIFAI ET COOKE 1966. TRANS. BRIT. MYCOL. SOC. 49:153 /G.
CYSTOPORIA (DUDDINGTON)
=ARTHROBOTRYS CORDA FIDE CARMICHAEL

GENICULOSPORA NILSSON 1964. SYMB. BOT. UPSALIENS. 18(2):95 AND FIG. 13 /(?) G.
GRANDIS (GREATHEAD)
?, A SEGREGATE FROM ARTICULOSPORA INGOLD

GENICULOSPORIUM CHESTERS ET GREENHALGH 1964. TRANS. BRIT. MYCOL. SOC. 47:400
/G. SERPENS CHEST. ET GREENHALGH 1964 =NODULISPORIUM CORTICIOIDES (FERR. ET
SACC) HUGHES 1958 =NODULISPORIUM STATE OF HYPOXYLON SERPENS (PERS. EX
FRIES) KICKX 1835 FIDE CARMICHAEL
=NODULISPORIUM PREUSS 1849 FIDE CARMICHAEL

GEOMYCES TRAAEN 1914. NYT. MAG. NATURV. 52:28-31 AND TAF. IV, FIGS. 1 & 2 /FOUR
ORIGINAL SPECIES = CHRYSOSPORIUM PANNORUM FIDE CARMICHAEL 1962
=CHRYSOSPORIUM CORDA 1833 FIDE CARMICHAEL 1962

GEOTRICHELLA ARNAUD 1954. BULL. SOC. MYCOL. FR. 69:273 AND FIG. 3A AND B /G.
ALTERNATA

?=COREMIELLA BUBAK ET KRIEGER 1912 /KOBAYASI ET AL., ANN. REP. INST. FOR
    FERMENTATION, OSAKA  NO. 3:116-117 (K. TUBAKI),1967.  (ILLUS)

GEOTRICHUM LINK 1809. MAG. GES. NATURF. FREUNDE, BERLIN 3:17 /G. CANDIDUM
    AMERO,ARTHRO OR GLOIO,HYALO. ASCO STATE =ENDOMYCES FIDE BUTLER 1970 (SCIENCE
    169:481-482) /BARRON, HYPHOMYCETES 172-173,1968. (ILLUS,REFS) CARMICHAEL,
    MYCOLOGIA 49:820-830,1957. (ILLUS)
        SYNO    =OOSPOROIDEA SUMSTINE 1913 FIDE CARMICHAEL 1957
                =COPROTRICHUM BONORDEN 1851 FIDE CARMICHAEL 1957
                =POLYMORPHOMYCES COUPIN 1914 FIDE CARMICHAEL

GIBELLULA CAVARA 1894. ATTI IST. BOT. PAVIA 2:347 AND TAB. XXVI FIGS. 13-19 /G.
    PULCHRA (SACCARDO)
    AMERO,BASO,HYALO. ASCO STATE =TORRUBIELLA FIDE MORRIS 1963 /MORRIS, WESTERN
    ILLINOIS UNIV SER BIOL SCI 3:58-59,1963. (ILLUS)
        SYNO    =SYNSTERIGMATOCYSTIS COSTANTIN FIDE PETCH FROM AINSWORTH 1971

GIBELLULOPSIS BATISTA ET MAIA 1959. ANN. SOC. BIOL. PERNAMBUCO 16:153 /G. PISCIS
    =VERTICILLIUM NEES 1817 FIDE CARMICHAEL

GILCHRISTIA CIFERRI ET REDAELLI 1934. J. TROP. MED. HYG. 37:281 /G.
    DERMATITIDIS (GILCHRIST ET STOKES) CIF. ET RED. 1934
    =CHRYSOSPORIUM CORDA 1833 FIDE CARMICHAEL 1962

GILLETIA TORREND 1914. BULL. JARD. BOT. BRUX. 4(1):12-13 (NO ILLUS) /G.
    CINNABARINA
    =TELLIGIA HENDRICKX 1948, NON GILLETIA SACCARDO ET PENZIG 1882
    (PERONOSPORALES)

GILMANIELLA BARRON 1964. MYCOLOGIA 56:514 /G. HUMICOLA
    AMERO,CETERI,PHAEO /BARRON, HYPHOMYCETES 173-174,1968. (ILLUS)
        SYNO    =ADHOGAMINA SUBRAMANIAN ET LODHA 1964 FIDE BARRON 1968

GLENOSPORA BERKELEY AND DESMAZIERES 1849 =SEPTOBASIDIUM PATOUILLARD 1892 NOMEN
    CONSERVANDUM FIDE CARMICHAEL 1962

GLENOSPORELLA NANNIZZI 1931. ATTI IST. BOT. UNIV. PAVIA, SER IV, 2:98 /G.
    ALBISCANS (NIEUW) (LECTOTYPE)
    =CHRYSOSPORIUM CORDA 1833 FIDE CARMICHAEL 1962

GLENOSPOROPSIS FONSECA 1943 =NOMEN NUDUM FIDE CARMICHAEL 1962

GLIOBOTRYS HOHNEL 1902. S.B. AKAD. WISS. WIEN, ABT 1, 111:1048-1049 /G.
    ALBOVIRIDIS
    =STACHYBOTRYS CORDA 1837 FIDE ELLIS 1971

GLIOCEPHALIS MATRUCHOT 1899. BULL. SOC. MYCOL. FR. 15:254 /G. HYALINA
    AMERO,GLOIO,HYALO /BARRON, HYPHOMYCETES P. 174-176,1968. (ILLUS)

GLIOCEPHALOTRICHUM ELLIS ET HESSELTINE 1962. BULL. TORREY BOT. CLUB 89:21 /G.
    BULBILIUM
    AMERO,GLOIO,HYALO /BARRON, HYPHOMYCETES P. 175-177,1968. (ILLUS) WILEY AND
    SIMMONS, MYCOLOGIA 58:575-585, 1971.(ILLUS,KEY)

GLIOCLADIOPSIS SAKSENA 1954. MYCOLOGIA 46:662 /G. SAGARIENSIS
    =CYLINDROCLADIUM MORGAN 1892 FIDE BARRON 1968

GLIOCLADIUM CORDA 1840. ICONES FUNGORUM 4:30 /G. PENICILLIOIDES
    AMERO,GLOIO,HYALO OR PHAEO. ?=CLONOSTACHYS CORDA 1839. ASCO STATE =NECTRIA
    FRIES FIDE BARRON 1968 /BARRON, HYPHOMYCETES P. 177-179,1968. (ILLUS)
        SYNO    =CORYMBOMYCES APPEL ET STRUNK 1903 FIDE CARMICHAEL
                ?=FLAHAULTIA ARNAUD 1951

GLIOCLADOCHIUM HOHNEL 1926. MITT. BOT. LAB. TECH. HOCHSCH. WIEN, BD. 3, HEFT
    1:4 /G. TOMENTOSUM (FRIES) = PERIOLA TOMENTOSA 1823
    ?

GLIODENDRON SALONEN ET RUOKOLA 1969. MYCOPATHOLOGIA 38:332 /G. BALNICOLA (AS
    BALNICOLUM)
    =STERIGMATOBOTRYS OUDEMANS 1886 FIDE KENDRICK

## 10. Hyphomycetes

GLIOMASTIX GUEGUEN 1905. BULL. SOC. MYCOL. FR. 21:240 /G. MURORUM (CORDA) HUGHES 1958
 AMERO,BASO OR GLOIO,PHAEO. ASCO STATE =WALLROTHIELLA FIDE HUGHES AND DICKINSON 1968 /DICKINSON, MYCOL PAPERS CMI 115:1-24,1968. (ILLUS,KEY,REVIS)
  SYNO =BASITORULA ARNAUD 1954 FIDE DICKINSON 1968
   =HAPLOTRICHELLA ARNAUD 1954 FIDE CARMICHAEL

GLIOPHRAGMA SUBRAMANIAN ET LODHA 1964. CANAD. J. BOTANY 42:1059 /G. SETOSUM SUBRAM. ET LODHA 1964
 =WIESNERIOMYCES KOORDERS 1907 FIDE KENDRICK

GLIOSTROMA CORDA 1837. ICONES FUNGORUM 1:5 AND TAB. 1, FIG. 46 /G. PEZIZOIDES
 ?, =AEGERITA FIDE SACCARDO FROM AINSWORTH 1971

GLIOTRICHUM ESCHWEILER 1824. SYLL. PLANT. A SOC. BOT. R. RATISB. EDITA 1:166 /G. CASSELIAE
 =NOMEN DUBIUM FIDE HUGHES 1958

GLOEOCERCOSPORA BAIN ET EDGERTON 1943. PHYTOPATHOLOGY 33:225 /G. SORGHI
 ?

GLOIOSPHAERA HOHNEL 1902. S.B. AKAD. WISS. WIEN. 111:1038 /G. GLOBULIGERA
 AMERO,GLOIO,HYALO /WANG, MYCOLOGIA 63:890-893,1971. (ILLUS)
  SYNO =SCOPULA ARNAUD 1952 FIDE WANG 1971

GLOMERULARIA PECK 1879. REP. N. Y. MUSEUM 32:43 /G. CORNI
 =GLOMOPSIS, NON GLOMERULARIA KARSTEN 1849

GLOMOPSIS HENDERSON 1961. NOTES ROY. BOT. GARDEN EDINBURGH 23:500 /G. CORNI (PECK)
 AMERO,CETERI,HYALO. NOM. NOV. FOR GLOMERULARIA PECK 1879 NON KARSTEN 1849
  SYNO =GLOMERULARIA PECK 1879

GLOMOSPORA HENDERSON 1961. NOTES ROY. BOT. GARDEN EDINBURGH 23:497 /G. EMPETRI
 AMERO,CETERI,HYALO

GLYCYPHILA MONTAGNE 1851. COMP. REND. 33:393-397 (NO ILLUS) /TWO ORIGINAL SPECIES
 ?, THE NAME HAS ALSO BEEN CITED AS GLYCYOPHILA AND GLYCOPHILA

GOIDANICHIELLA ARNAUD EX BARRON 1968. THE GENERA OF HYPHOMYCETES FROM SOIL, P. 180-182 (ILLUS) /G. SCOPULA (PREUSS)
 AMERO,GLOIO,HYALO, NO LATIN DIAGNOSIS
  SYNO =GOIDANICHIA ARNAUD 1954 NON TOM. ET CIFERRI 1952 (A LICHEN)

GOLOVINIA MECHTIEVA 1967. MIKOL. I FITOPATOL. 1:275 /G. BEMBICOIDES (DRECHSLER)
 =DACTYLELLA BEMBICOIDES
 ?, A SEGREGATE FROM DACTYLELLA GROVE 1884

GOMPHINARIA PREUSS 1851. LINNAEA 24:130, ILLUS. IN STURMS DEUTSCHL. FL., BD. 6, HEFT 35-36, TAB. 60, 1862 /G. AMOENA
 ?, =ACROTHECA FIDE SACCARDO FROM AINSWORTH 1971

GONATOBOTRYS CORDA 1839. PRACHT-FLORA .... P. 9 /G. SIMPLEX
 AMERO,CETERI,HYALO /BARRON, HYPHOMYCETES, P. 182-183,1968. (ILLUS)
  SYNO =DESMOTRICHUM LEVILLE 1843 FIDE LINDAU 1907

GONATOBOTRYUM SACCARDO 1880. MICHELIA 2:24 /G. FUSCUM (SACC.)
 AMERO,BLASTO,PHAEO. PERHAPS THE SAME AS GONATORRHODUM CORDA /BARRON, HYPHOMYCETES P. 183-184,1968. (ILLUS) KENDRICK ET AL., CAN. J. BOTANY 46:591-596,1968. (ILLUS)
  SYNO =THOMIELLA DODGE 1935 FIDE CARMICHAEL
   =CHRISTIASTER KUNTZE FIDE AINSWORTH 1971
   ?=DICELLISPORA SAWADA 1944

GONATOPHRAGMIUM DEIGHTON 1969. IN CEJP AND DEIGHTON, MYCOL. PAPERS CMI 117:13-20 (ILLUS) /G. MORI (SAWADA)
 PHRAGMO,CETERI,PHAEO

GONATORRHODIELLA THAXTER 1891. BOTAN. GAZ. 16:501 /G. PARASITICA

=NEMATOGONIUM DESMAZIERES FIDE HUGHES 1953

GONATORRHODUM CORDA 1839. PRACHTFLORA P. 5 /G. SPECIOSUM
  ?, PERHAPS AN EARLIER NAME FOR GONATOBOTRYUM, BUT TYPE NOT EXAMINED BY RECENT
  AUTHOR /LINDAU, IN RABENHORST KRYPT. FL. 2 ED. 1 BD. 8:686-687,1907. (ILLUS)

GONATOSPORIUM CORDA 1839. ICONES FUNGORUM 3:8 /G. PUCCINIOIDES (DE CANDOLLE)
  CORDA
  =ARTHRINIUM KUNZE 1817 FIDE HUGHES 1958

GONGROMERIZA PREUSS 1851. LINNAEA 24:106 /G. CLAVAEFORMIS
  =CHLORIDIUM LINK FIDE HUGHES 1969

GONGYLOCLADIUM WALLROTH 1833. FL. CRYPT. GERMAN. 2:160 /G. ATRUM (LINK)
  WALLROTH =OEDEMIUM DIDYMIUM
  =OEDEMIUM LINK 1824 FIDE HUGHES 1958

GONIOSPORIUM LINK 1824. LINN. SPEC. PLANT., IV,6(1):45 /G. PUCCINIOIDES (DE
  CANDOLLE)
  =ARTHRINIUM KUNZE 1817 FIDE HUGHES 1958

GONYELLA SYDOW 1919. ANN. MYCOL. 17:44 /G. TYPICA (ROSTRUP)
  ?, FOR ARTHROBOTRYUM ROSTRUP 1916 NON CESATI 1854, APPEARS TO BE ARTHROBOTRYS
  CORDA

GONYTRICHELLA EMOTO ET TUBAKI 1970. TRANS. MYCOL. SOC. JAPAN 11:95-97 (ILLUS)
  /G. OLIVACEA
  ?=DICYMA BOULANGER 1897

GONYTRICHUM NEES ET NEES 1818. NOVA ACTA ACAD CAES. LEOP. 9:244 /G. CAESIUM
  NEES ET NEES =G. STATE OF MELANOPSAMELLA INAEQUALIS (GROVE) HOHNEL 1919
  AMERO,GLOIO,PHAEO OR HYALO, SIMILAR TO UNCIGERA. ASCO STATE =MELANOPSAMELLA
  FIDE HUGHES 1958 /BARRON, HYPHOMYCETES P. 184-186,1968. (ILLUS)
    SYNO    =MESOBOTRYS SACCARDO 1880 FIDE HUGHES 1958

GRALLOMYCES STEVENS 1918. BOTANICAL GAZ. 45:245 /G. PORTORICENSIS
  STAURO,CETERI,PHAEO /DEIGHTON AND PIROZYNSKI, MYCOL PAPERS CMI
  105:10-17,1966. (ILLUS)
    SYNO    =PHIALETEA BATISTA ET NASCIMENTO 1960 FIDE DEIGHTON AND PIROZYNSKI
            1966
            =OPHIOPODIUM ARNAUD 1954 FIDE DEIGHTON AND PIROZYNSKI 1966

GRANULARIA SACCARDO 1882. MICHELIA 2:648 (NO ILLUS) /G. EUROTIOIDES SACCARDO ET
  ELLIS 1882
  ?

GRAPHIDIUM LINDAU 1910. RABENH. KRYPT. FL. 2 AUFL. 1 BD. 9:748 /G. COFRENSI
  ?, THE ILLUSTRATION SUGGESTS PAECILOMYCES, BUT THE CONIDIA ARE REPORTED AS
  NOT IN CHAINS OR SLIME.

GRAPHIOLA POIT. 1824. ANN. SCI. NAT. 1824:473 /G. PHOENICIS (MOUG.)
  AMERO,CETERI,HYALO. ALSO CONSIDERED TO BE IN USTILAGINALES /HUGHES, CAN. J.
  BOTANY 31:644-645,1953. (ILLUS) TUBAKI, MYCOPATHOLOGIA 43:49-60,1971.
  (ILLUS)

GRAPHIOPSIS BAINIER 1907. BULL. SOC. MYCOL. FR. 23:19 /G. CORNUI BAINIER
  =PHAEOISARIA CORNUI (BAINIER) MASON 1937
  =PHAEOISARIA HOHNEL 1909 FIDE MORRIS 1963, NON GRAPHIOPSIS TRAIL 1889

GRAPHIOPSIS TRAIL 1899. SCOTTISH NATURALIST 10:75 /G. CHLOROCEPHALA (FRESENIUS)
  =CLADOSPORIUM LINK FIDE MASON AND ELLIS 1953

GRAPHIOTHECIUM FUCKEL 1869. SYMBOLAE MYCOL. P. 366 AND TAB. 1, FIG 30 /G.
  FRESENII
  AMERO,BASO,PHAEO /MORRIS, WESTERN ILLINOIS UNIV SER BIOL SCI 3:60-61,1963.
  (ILLUS)
    SYNO    =STROMATOSTYSANUS HOHNEL 1919 FIDE CLEMENTS AND SHEAR 1931

GRAPHIUM CORDA 1837. ICONES FUNGORUM 1:18 /G. PENICILLIOIDES (LECTOTYPE)
  AMERO,GLOIO,PHAEO. ASCO STATE =PETRIELLA CURZI FIDE BARRON 1968 OR

## 10. Hyphomycetes 373

ALLESCHERIA FIDE CARMICHAEL /BARRON, HYPHOMYCETES P. 186-188,1968. (ILLUS)
   SYNO   ?=CERATOPODIUM CORDA 1837 FIDE SACCARDO 1886

GRUBYELLA OTA ET LANGERON 1923. ANN. PARASIT. HUM. COMP. 1:330 /G. SCHOENLEINII
   (LEBERT) OTA ET LANG. = TRICHOPHYTON SCHOENLEINII (LECTOTYPE)
   =TRICHOPHYTON MALMSTEN 1848

GUEGUENIA BAINIER 1907. BULL. SOC. MYCOL. FRANCE 23:107 /G. CESPITOSA
   =AMBLYOSPORIUM FRESENIUS FIDE PIROZYNSKI 1969

GUELICHIA SPEGAZZINI 1886. AN. SOC. ARG. 22:220-221 /G. PARADOXA
   ?, NOT ILLUSTRATED BY AUTHOR

GUTTUROMYCES RIVOLTA ?=ASPERGILLUS FIDE AINSWORTH 1971

GYMNODOCHIUM MASSEE ET SALMON 1902. ANN. BOT. 16:89 AND PL. IV, FIGS 49-51 /G.
   FIMICOLUM
   DIDYMO,BASO,HYALO

GYMNOSPORIUM CORDA 1833. STURMS DEUTSCHL. FLORA III (PILZE), BD. 3, HEFT 13:69
   /G. OLIVACEUM
   =NOMEN DUBIUM SEE HUGHES 1958

GYOERFFIELLA KOL 1928. FOLIA CRYPTOG. 1:618 /G. TATRICA =G. ROTULA (HOHNEL)
   MARVANOVA 1967
   STAURO,CETERI,HYALO /VON ARX, GENERA OF FUNGI IN CULTURE P. 215,1970. (REFS)
      SYNO   =INGOLDIA PETERSEN FIDE MARVANOVA 1967, PERSOONIA 5:29-44

GYROCERUS CORDA 1837. ICONES FUNGORUM 1:9 /G. AMMONIS
   =NOMEN DUBIUM SEE HUGHES 1958, ALSO AS GYROCERAS SEE AINSWORTH 1961

GYROTHRIX (CORDA) CORDA 1842. ICONES FUNGORUM 5:13 /G. PODOSPERMA (CORDA)
   RABENHORST 1844.
   SCOLECO,CETERI,HYALO, DIFFERS FROM CIRCINOTRICHUM ONLY IN HAVING BRANCHED
   SETAE /PIROZYNSKI, MYCOL PAPERS CMI 84:8-27,1962. (ILLUS,REVIS,KEY)
      SYNO   =PEGLIONIA GOIDANICH 1934 FIDE HUGHES AND PIROZYNSKI 1971

GYROTRICHUM SPRENGEL 1827. LINN. SYST. VEG. XVI,4(1):554 /G. MACULIFORME (NEES)
   =CIRCINOTRICHUM NEES 1917 FIDE HUGHES 1958

HADRONEMA SYDOW 1909. ANN. MYCOL. 7:172 /H. ORBICULARE
   DIDYMO,CETERI,PHAEO, COMPARE WITH SPILODOCHIUM /ELLIS, DEM. HYPH. P.
   198,1971. (ILLUS)

HADROSPORIUM SYDOW 1938. ANN. MYCOL. 36:311-312 /H. FRASERIANUM
   ?, NOT ILLUSTRATED BY AUTHOR

HADROTRICHUM FUCKEL 1865. FUNGI RHENANI NO. 1522 /H. PHRAGMITIS
   AMERO,CETERI,PHAEO /HUGHES, CAN J. BOTANY 31:574-575,1953. (ILLUS)
      SYNO   =MICROBASIDIUM BUBAK ET RANOJEVIC FIDE HOHNEL FROM AINSWORTH 1971

HAINESIA ELLIS ET SACCARDO 1884. SYLL. FUNG. 3:699 /H. RHOINA (SACC.) ELLIS ET
   SACC. =H. LYTHRI (DESM.) HOHNEL
   AMERO,GLOIO,HYALO, (COELOMYCETE) ASCO STATE =DISCOHAINESIA FIDE VON ARX 1970
   /VON ARX, GENERA OF FUNGI IN CULTURE P. 181 AND 185,1970. (ILLUS)

HALOBYSSUS ZUKAL 1893. OST. BOT. Z. 43:279-281 AND TAF. XII, FIG. 12 /H.
   MONILIFORMIS
   ?, =MONILIA FIDE CLEMENTS AND SHEAR 1931

HALYSIUM CORDA 1837. ICONES FUNGORUM 1:17 /H. ATRUM CORDA
   =NOMEN CONFUSUM, FIDE HUGHES 1958

HANSFORDIA HUGHES 1951. MYCOL. PAPERS CMI 43:15-25 (ILLUS) /H. OVALISPORA
   AMERO,CETERI,PHAEO

HANSFORDIELLA HUGHES 1951. MYCOL. PAPERS CMI 47:10-15 (ILLUS) /H. ASTERINEARUM
   PHRAGMO OR DICTYO,CETERI,PHAEO

HANSFORDIELLOPSIS DEIGHTON 1960. MYCOL PAPERS CMI 78:33-35 (ILLUS) /H.
   ABURIENSIS

PHRAGMO OR DICTYO,CETERI,PHAEO

HANSFORDIULA MORRIS 1963. AMER. MIDL. NATURALIST 69:103 /H. FASCICULATA
=PHAEOISARIA HOHNEL 1909 FIDE MORRIS 1963

HANTZSCHIA AUERSWALD 1862. HEDWIGIA 10:60 AND TAB. XI /H. PHYCOMYCES AUERSW. 1862
=NOM. REJ. FIDE KENDRICK 1964 (=PHIALOCEPHALA KENDRICK) /KENDRICK, CAN. J. BOTANY 42:1291-1295,1964

HAPLARIA LINK 1809. MAG. GES. NATURF. FREUNDE, BERLIN 3:11 /H. GRISEA
=BOTRYTIS PERSOON 1801 FIDE HUGHES 1958

HAPLARIELLA SYDOW 1908. ANN. MYCOL. 6:497 /H. CORDIAE (HENNINGS)
?. FOR HAPLARIOPSIS HENNINGS 1908 NON OUDEMANS 1903, NOT ILLUSTRATED BY HENNINGS OR SYDOW

HAPLARIOPSIS OUDEMANS 1903. NED. ARCH. 3(2):902 /H. FAGICOLA
DIDYMO,CETERI,HYALO /ELLIS, DEM. HYPH. P. 293,1971 (ILLUS)

HAPLOBASIDION ERIKSSON 1889. BOT. CENTRALBL. 38:786-787 /H. THALICTRI
AMERO,BLASTO,PHAEO, CLOSE TO LACELLINOPSIS /ELLIS, MYCOL. PAPERS CMI 67:1-6,1957. (ILLUS, KEY)

HAPLOCHALARA LINDER 1933. MYCOLOGIA 25:347 AND FIGS 7-8 /H. ANGULOSPORA
=CATENULARIA GROVE 1886 FIDE MASON 1941

HAPLOGRAPHIUM BERKELEY ET BROOME 1859. ANN. MAG. NAT. HIST. III 3:360 /H. DELICATUM BERK. ET BR. 1859 =H. STATE OF HYALOSCYPHA DEMATIICOLA (BERK. ET BR.) NANNFELDT 1936
AMERO,GLOIO,HYALO /HUGHES, CAN. J. BOTANY 31:588-589,1953. (ILLUS) BARRON, GENERA OF HYPHOMYCETES, P.192-193,1968. (ILLUS)

HAPLOTRICHELLA ARNAUD 1954. BULL. SOC. MYCOL. FR. 69:280 AND 282 FIG. 6R /H. LINKII ?=GLIOMASTIX MURORUM VAR. FELINA FIDE CARMICHAEL
=GLIOMASTIX GUEGUEN FIDE CARMICHAEL

HAPLOTRICHUM ESCHWEILER 1824. SYLL. PLANT. A SOC. BOT. R. RATISB. EDITA 1:167 /H. CANDIDUM
=NOMEN DUBIUM, SEE HUGHES 1958, NON HAPLOTRICHUM LINK 1824 (=ACLADIUM)

HARPAGOMYCES WILCZ. 1911. KOSMOS 36:314-316 (ILLUS) /H. LOMNICKI
PHRAGMO OR STAURO,CETERI,HYALO

HARPOCEPHALUM ATKINSON 1897. BULL. CORNELL UNIV. 3:41 /H. DEMATIOIDES
=PERICONIA TODE EX PERS. 1801. FIDE ELLIS 1971

HARPOGRAPHIUM SACCARDO 1880. MICHELIA 2:33 /H. FASCICULATUM
AMERO OR SCOLECO,GLOIO,HYALO /ELLIS, DEM. HYPH. P. 562-563,1971. (ILLUS)
SYNO    =EPILITHIA NYLANDER 1855 FIDE SACCARDO 1886

HARPOSPORIUM LOHDE 1874. TAGEBLATT DER 47 VERSAMMLUNG DEUTSCHER NATURF. UND ARZTE IN BRESLAU P. 206 /H. ANGUILLULAE
SCOLECO,GLOIO,HYALO, NOT ILLUSTRATED BY AUTHOR, TYPE NOT EXAMINED BY RECENT AUTHOR(?) /BARRON, GENERA OF HYPHOMYCETES, P. 193-195,1968. (ILLUS)
SYNO    =POLYRRHINA SORCKIN 1876 FIDE AINSWORTH 1971

HARTIGIELLA SYDOW 1900. IN LINDAU IN ENGLER UND PRANTL, 1(1):558 /H. LARICIS (HARTIG)
=MERIA VUILLEMIN 1896 FIDE AINSWORTH 1971, FOR ALLESCHERIA HARTIG NON SACCARDO ET SYDOW

HARZIA COSTANTIN 1888. LES MUCEDINEES SIMPLES, P. 42 /H. ACREMONIOIDES (HARZ)
=ACREMONIELLA SACCARDO 1886 FIDE ELLIS 1971

HARZIELLA COSTANTIN ET MATRUCHOT 1899. BULL. SOC. MYCOL. FR. 15:104-107 AND PL. VII, FIGS. 1-9 /H. CAPITATA COST. ET MATR. 1899
AMERO,GLOIO,HYALO, BUT IS A LATER HOMONYM OF HARZIELLA KUNZE 1891 (=TRICHOCLADIUM FIDE AINSWORTH 1960) /ARNAUD, BULL. SOC. MYCOL. FR. 68:197-199,1952. (ILLUS)

## 10. Hyphomycetes

HASSKARLINDA KUNTZE 1891 =CORALLODENDRON (BASED ON SAME TYPE)

HEIMIODORA NICOT ET PARGUEY 1960. ANN. SCI. NAT., BOTAN., SER 12, 1:384 /H. VERTICILLATA
 AMERO,GLOIO,HYALO /BENJAMIN, ALISO 5:278-281,1963. (ILLUS)

HELICOCEPHALUM THAXTER 1891. BOT. GAZ. 16:201-204 AND PL. XIX, FIG. 1-6 /H. SARCOPHILUM
 AMERO OR HELICO,ARTHRO OR GLOIO,PHAEO

HELICOCERAS LINDER 1931. ANN. MISSOURI BOT. GARDEN 18:3 /H. CELTIDIS (BIV.-BERNH.)
 =SIROSPORIUM BUBAK ET SEREB. FIDE ELLIS 1963

HELICOCORYNE CORDA 1854. ICONES FUNGORUM 6:9 /H. VIRIDE CORDA = HELICOMA VIRIDIS (CORDA) HUGHES
 =HELICOMA CORDA 1837 FIDE HUGHES 1958

HELICODENDRON PEYRONEL 1918. NUOV. GIORN. BOT. ITAL. N.S. 25:461 AND FIGS. 71-76 /H. PARADOXUM
 HELICO,BLASTO,HYALO OR PHAEO, EMENDED LINDER 1929 /BARRON, GENERA OF HYPHOMYCETES, P. 195-196,1968. (ILLUS)
  SYNO    =HELICODESMUS LINDER 1925 FIDE ELLIS 1971

HELICODESMUS LINDER 1925. AMER. J. BOT. 12:267 /H. ALBUS
 =HELICODENDRON PEYRONEL 1918 FIDE ELLIS 1971

HELICOMA CORDA 1837. ICONES FUNGORUM 1:15 /H. MUELLERI CORDA =H. STATE OF LASIOSPHAERIA PEZIZULA (BERK. ET CURT.) SACC. FIDE HUGHES 1958
 HELICO,CETERI,HYALO OR PHAEO. ASCO STATE =LASIOSPHAERIA FIDE HUGHES 1958 NOW KNOWN AS THAXTERIELLA FIDE MULLER IN KENDRICK 1971 /ELLIS, DEM. HYPH., P. 190-191,1971. (ILLUS)
  SYNO    =HELICOCORYNE CORDA 1854 FIDE HUGHES 1958
          =LITUARIA RIESS 1853 FIDE LINDER FROM AINSWORTH 1971
          =MOORELLA RAO ET RAO 1964 FIDE KENDRICK
          ?=HÉLICOMINOPSIS DEIGHTON 1960
          ?=HELICOSPORELLA ARNAUD 1954
          ?=HELICOSPORINA ARNAUD 1954
          ?=HELICOPSIS KARSTEN 1889 FIDE LINDER FROM AINSWORTH 1971

HELICOMINA OLIVE 1948. MYCOLOGIA 40:16 /H. CAPERONIAE
 HELICO OR PHRAGMO,CETERI,PHAEO /ELLIS, DEM. HYPH. P.209-210,1971. (ILLUS)

HELICOMINOPSIS DEIGHTON 1960. MYCOL PAPERS CMI 78:20 /H. FICI
 ?=HELICOMA CORDA 1837

HELICOMYCES LINK 1809. MAG. GES. NATURF. FREUNDE, BERLIN 3:21 /H. ROSEUS
 HELICO,CETERI,HYALO /BARNETT, INPERFECT FUNGI NO. 141,1960. (ILLUS)
  SYNO    =XYPHASMA REBENTISCH FIDE RABENHORST FROM AINSWORTH 1971

HELICOON MORGAN 1892. J. CINCINNATI SOC. NAT. HIST. 14:49 /H. SESSILE
 HELICO,CETERI,HYALO OR PHAEO /BARRON, GENERA OF HYPHOMYCETES, P. 196-197,1968. (ILLUS)

HELICOPSIS KARSTEN 1889. REV. MYCOL. 11:96 /H. OLIVACEUS KARST. 1889
 ?, NOT ILLUSTRATED BY AUTHOR, = HELICOMA CORDA 1837 FIDE LINDER FROM AINSWORTH 1971

HELICORHOIDION HUGHES 1958. CAN. J. BOTANY 36:773 /H. BOTRYOIDEUM (COOK)
 HELICO,CETERI,PHAEO /HUGHES, CAN. J. BOTANY 36:773,1958. (ILLUS)

HELICOSPORANGIUM KARSTEN 1865. BOT. UNTERS. LAB. LANDW. 1:76 /H. PARASITICUM
 ?=PAPULASPORA PREUSS 1851 /KARSTEN, HEDWIGIA 27:132-137,1888. (ILLUS)

HELICOSPORELLA ARNAUD 1954. BULL. SOC. MYCOL. FR. 69:292-293, FIG. 12 M /H. LUTEO-FUSCA
 ?=HELICOMA CORDA

HELICOSPORINA ARNAUD 1954. BULL. SOC. MYCOL. FR. 69:292 AND 293 (ILLUS) /H. GLOBULIFERA

?=HELICOMA CORDA 1837, RAMBELLI (1960) EMENDED HELICOSPORINA BASED ON A
    SPECIES OTHER THAN THE TYPE

HELICOSPORIUM NEES 1817. DAS SYSTEM ... P. 68 /H. VEGETUM NEES =H. STATE OF
    OPHIONECTRIA CEREA (BERK. ET CURT.) ELLIS ET EVERHART FIDE HUGHES 1958
    HELICO,CETERI,HYALO, ASCO STATE =OPHIONECTRIA FIDE HUGHES 1958 OR TUBEUFIA
    FIDE ELLIS 1971 /BARRON, GENERA OF HYPHOMYCETES 188-189,1968. (ILLUS)
        SYNO    =HELICOTRICHUM NEES ET NEES 1818 FIDE HUGHES 1958

HELICOSTILBE HOHNEL 1902. S. B. AKAD. WISS. WIEN, ABT. 1:1028-1029 (NO ILLUS)
    /H. HELICINA
    ?, LINDER (1929) EMENDED THE GENUS TO INCLUDE ONLY H. SIMPLEX AND EXCLUDE
    HOHNELS TYPE = TROCHOPHORA MOORE 1955.

HELICOTRICHUM NEES ET NEES 1818. NOVA ACTA ACAD. CAES. LEOP. 9:246 /H.
    PULVINATUM NEES ET NEES =HELICOSPORIUM STATE OF OPHIONECTRIA CEREA
    =HELICOSPORIUM NEES 1817 FIDE HUGHES 1958

HELISCUS SACCARDO 1880. MICHELIA 2:1 /H. LUGDUNENSIS SACC. ET THERRY 1880
    STAURO OR PHRAGMO,CETERI,HYALO, ASCO STATE =NECTRIA FIDE VON ARX 1970 /VON
    ARX, GENERA OF FUNGI 188-189,1970. (ILLUS)

HELMINTHOPHORA BONORDEN 1851. HANDB. ALLGEM. MYKOL. P. 93 AND TAF. VI, FIG. 137
    /H. TENERA
    ?, =DACTYLIUM NEES 1817 FIDE SACC. FROM AINSWORTH 1971

HELMINTHOSPORIOPSIS SPEGAZZINI 1880. AN. SOC. CIENT. ARGENTINA 10:166 (NO
    ILLUS) /H. TYPICA
    =PODOSPORIUM SCHWEINITZ 1832 FIDE AINSWORTH 1961

HELMINTHOSPORIUM LINK 1809. MAG. GES. NATURF. FREUNDE, BERLIN 3:10 /H.
    VELUTINUM =H. CILIARE (PERSOON) HUGHES 1958
    PHRAGMO,CETERI,PHAEO, ORTHOGRAPHIC VARIANT OF HELMISPORIUM LINK /ELLIS, MYCOL
    PAPERS CMI 82:2-21,1961. (ILLUS,REVIS,KEY)
        SYNO    =HELMISPORIUM LINK 1809 FIDE HUGHES 1958

HELOSTROMA PATOUILLARD 1902. BULL. SOC. MYCOL. FRANCE 18:47 /H. ALBA
    (DESMAZIERES)
    =MICROSTROMA NIESSL 1861 FIDE VON ARX 1970
        SYNO    =ARTICULARIELLA HOHNEL 1909 FIDE AINSWORTH 1961

HEMIBELTRANIA PIROZYNSKI 1963. MYCOL. PAPERS CMI 90:30-34 (ILLUS,KEY) /H.
    CINNAMOMI (DEIGHTON)
    AMERO,CETERI,HYALO OR PHAEO

HEMISPORA VUILLEMIN 1906. BULL. SOC. MYCOL. FR. 22:125 /H. STELLATA
    =WALLEMIA JOHAN-OLSEN FIDE CARMICHAEL 1969

HEPTASTER CIFERRI, BATISTA ET NASCIMENTO 1956. PUBL. INST. MICOL. UNIV. RECIFE
    46:3 /H. HUGHESII CIFERRI ET AL. 1956
    STAURO,CETERI,PHAEO

HERMATOMYCES SPEGAZZINI 1911. AN. MUS. NAC. BUENOS AIRES 3, 13:446 /H.
    TUCUMANENSIS
    DICTYO,CETERI,PHAEO /ELLIS, DEM. HYPH. P. 110-111,1971. (ILLUS)

HERPOSIRA SYDOW 1938. ANN. MYCOL. 36:312 /H. VELUTINA
    AMERO,CETERI,PHAEO /ELLIS, DEM. HYPH. P. 170-171,1971. (ILLUS)

HETEROBOTRYS SACCARDO 1880. MICHELIA 2:21 AND FUNGI ITALICI FIG. 807, 1881 /TWO
    ORIGINAL SPECIES
    ?=ATICHIA (MYRIANGIALES) FIDE HOHNEL FROM AINSWORTH 1971

HETEROCEPHALUM THAXTER 1903. BOT. GAZ. 35:157, PL.IV, FIGS. 1-7 /H. AURANTIACUM
    AMERO,GLOIO,HYALO /BARRON, GENERA OF HYPHOMYCETES P. 202-203,1968.
    (ILLUS,REFS) MORRIS,WESTERN ILLINOIS UNIV SER BIOL SCI 3:68-69,1963. (ILLUS)

HETEROCONIDIUM SAWADA 1944. REP. GOVT. RES. INST. FORMOSA 87:92, FIGS. 52-54
    /H. CASSIAE
    ?, NO LATIN DIAGNOSIS

## 10. Hyphomycetes

HETEROCONIUM PETRAK 1949. SYDOWIA 3:246 /H. CITHAREXYLI
   PHRAGMO,BLASTO,PHAEO, BUT DOUBTFULLY DISTINCT FROM TAENIOLELLA /ELLIS, DEM.
   HYPH. P. 96-97,1971. (ILLUS)

HETEROSPOPIOPSIS PETRAK 1950. SYDOWIA 4:521 /H. MONILIFERA (PATOUILLARD)
   ?, NOT ILLUSTRATED BY PETRAK

HETEROSPORIUM KLOTZSCH 1877. IN COOKE, GREVILLEA 5:123 /H. ORNITHOGALI KLOTZSCH
   =CLADOSPORIUM ORNITHOGALI (KLOTZSCH) DE VRIES
   =CLADOSPORIUM LINK 1815 FIDE HUGHES 1958

HEYDENIA FRESENIUS 1852. BEITR. MYKOL. 2:47, PL. V, FIGS. 37-45 /H. ALPINA
   AMERO,CETERI OR BLASTO,HYALO /LINDAU, RABENH. KRYPT. FL. 2 ED. 1 BD.
   9:373-374,1910. (ILLUS)
      SYNO    =RICCOA CAVARA 1903 FIDE CLEMENTS AND SHEAR 1931
              =RUPINIA ROUM. ET SPEGAZ. 1880 FIDE LINDAU 1910
              ?=MARTINDALIA SACCARDO ET ELLIS 1884
              ?=CLATHROTRICHUM PATOUILLARD 1921

HEYDENIOPSIS NAUMOV 1915. MAT. MIKOL. FITOPAT. ROSSTOK. 1:23-25 AND FIGS. 28-29
   /H. INGRICA
   ?

HIMANTIA PERSOON 1801. SYN. METH. FUNG. P. 703 /THREE ORIGINAL SPECIES
   =STERILE, MYCELIAL STAGES OF CORTICIUM ETC. FIDE LINDAU 1910

HIOSPIRA MOORE 1962. TRANS. BRIT. MYCOL. SOC. 45:145 /H. HENDRICKXII (HANSFORD)
   HELICO,CETERI OR BLASTO,PHAEO, ASCO STATE =BROOKSIA HANSFORD FIDE DEIGHTON
   AND PIROZYNSKI 1966 /DEIGHTON AND PIROZYNSKI, MYCOL PAPERS CMI
   105:2-10,1966. (ILLUS)

HIPPOCREPIDUM SACCARDO 1875. IN THUMEN, MYCOTH. UNIV. NO. 85 /H. MESPILI
   SACCARDO APUD THUMEN 1875
   =HIRUDINARIA CESATI 1856 FIDE HUGHES 1951

HIRSUTELLA PATOUILLARD 1892. REV. MYCOL. 14:69 /H. ENTOMOPHILA
   AMERO,GLOIO OR CETERI,HYALO, ASCO STATE =CORDYCEPS OR OPHIOCORDYCEPS OR
   TORRUBIELLA FIDE MORRIS 1963 /MORRIS, WESTERN ILLINOIS UNIV SER BIOL SCI
   3:70-71,1963. (ILLUS)
      SYNO    =MAHEVIA LAGARDE 1917 FIDE CARMICHAEL
              =TRICHOSTERIGMA PETCH 1923 FIDE TRANS. BRIT. MYCOL. SOC. VOL. 9
              (FROM AINSWORTH 1971)

HIRUDINARIA CESATI 1856. KLOTZSCH HERB. MYC. EDIT. NOVA. NO. 269 /H. MACROSPORA
   (LECTOTYPE)
   STAURO,CETERI,PHAEO /HUGHES, MYCOL. PAPERS CMI 39:11-24,1951. (ILLUS, REVIS)
      SYNO    =HIPPOCREPIDUM SACCARDO 1875 FIDE HUGHES 1951

HISTOPLASMA DARLING 1906. J. AM. MED. ASSOC. 46:1283-1285 /H. CAPSULATUM
   AMERO,CETERI,HYALO /HOWELL, MYCOLOGIA 31:191-216,1939. (ILLUS)

HOBSONIA BERKELEY 1891. ANN. BOT. 5:509 /H. GIGASPORA
   HELICO,CETERI,HYALO

HOEHNELIELLA BRESADOLA ET SACCARDO 1902. VERH. Z-B. GES. WIEN 52:437 (NO ILLUS)
   /H. PERPLEXA BRES. ET SACC. 1902
   =COELOMYCETES

HORMIACTELLA SACCARDO 1886. SYLLOGE FUNGORUM 4:311 /H. FUSCA (PREUSS) =
   HORMIACTIS FUSCA PREUSS
   AMERO,BLASTO,HYALO, CLOSE TO FUSIDIUM LINK

HORMIACTINA BUBAK 1916. HEDWIGIA 57:336 (ILLUS) /H. WROBLEWSKII
   ?, DOES NOT APPEAR TO DIFFER FROM HORMIACTIS

HORMIACTIS PREUSS 1851. FUNGI HOYERSWERDA P. 128 /H. ALBA PREUSS AND H. FUSCA
   PREUSS ORIGINALLY DESCRIBED
   DIDYMO,BLASTO,HYALO /BARRON, GENERA OF HYPHOMYCETES P. 206-207,1968. (ILLUS)
      SYNO    ?=HORMIACTINA BUBAK 1916

HORMIOKRYPSIS BATISTA ET NASCIMENTO 1957. ANN. SOC. BIOL. PERNAMBUCO 15:345 /H.
    LIBOCEDRI =H. STATE OF OPHIOCAPNOCOMA PHLOIOPHILIA (FISHER) HUGHES FIDE
    HUGHES 1967
  STAURO,CETERI,PHAEO, ASCO STATE =OPHIOCAPNOCOMA FIDE HUGHES 1967 /HUGHES, NEW
    ZEALAND J. BOTANY 117-133,1967. (ILLUS)

HORMISCIELLA BATISTA 1956. ANN. SOC. BIOL. PERNAMBUCO 14:100 /H. ATRA
    =ANTENNATULA ATRA (BATISTA) HUGHES 1970
    =ANTENNATULA FRIES 1850 FIDE HUGHES 1970

HORMISCIOMYCES BATISTA ET NASCIMENTO 1957. ANN. SOC.BIOL. PERNAMBUCO 15:349 /H.
    PREPUSUM
  AMERO,GLOIO,HYALO, MUCH LIKE PLOKAMIDOMYCES /HUGHES, NEW ZEALAND J. BOTANY
    8:180,181 AND 187,1970. (ILLUS)

HORMISCIOPSIS SUMSTINE 1914. MYCOLOGIA 6:32 /H. GELATINOSA
  ?, ORIGINAL ILLUSTRATION INADEQUATE TO CHARACTERIZE

HORMISCIUM KUNZE 1817. MYKOL. HEFTE 1:13 /H. EXPANSUM KUNZE =TORULA HERBARUM
    (PERSOON) LINK
    =TORULA (PERSOON) LINK 1809 FIDE HUGHES 1958

HORMOCEPHALUM SYDOW 1939. ANN. MYCOL. 37:424 /H. ECUADORENSE
  PHRAGMO,CETERI,PHAEO /ELLIS, DEM. HYPH. P. 99-100,1971. (ILLUS)

HORMOCLADIUM HOHNEL 1923. ZENTRALBL. BAKT., ABT. 2, BD. 60:5 /H. KAKI (HORI ET
    YOSHIMO)
  ?, NOT ILLUSTRATED BY HOHNEL

HORMODENDROIDES MOORE ET ALMEIDA 1936. ANN. MISSOURI BOT. GARDEN 23:547 /H.
    PEDROSOI (BRUMPT)
    =FONSECAEA NEGRONI 1936

HORMODENDRUM BONORDEN 1853. BOT. ZEIT. 11:286 /H. OLIVACEUM (CORDA)
    =CLADOSPORIUM LINK FIDE BARRON 1968

HORMODOCHIS CLEMENTS 1909. GENERA OF FUNGI P. 163 AND 176 (NO ILLUS) /H.
    MELANOCHLORA (DESMAZIERES)
  ?
    SYNO    =HORMODOCHIUM (SACCARDO) TROTTER (ORTHOGRAPHIC VARIANT)

HORMODOCHIUM (SACCARDO) TROTTER   SEE HORMODOCHIS

HORMONEMA LAGERBERG ET MELIN 1927. SVENSKA SKOGSV. TIDSKR. HEFT II, OCH IV:233,
    FIGS. 38-39 /H. DEMATIOIDES LAGERB. ET MELIN 1927
    =AUREOBASIDIUM VIALA ET BOYER 1891 FIDE CARMICHAEL

HUGHESIELLA BATISTA ET VITAL 1956. ANN. SOC. BIOL. PERNAMBUCO 14:141 /H.
    EURICOI = THIELAVIOPSIS STATE OF CERATOCYSTIS PARADOXA (DADE) MOREAU FIDE
    ELLIS 1971
    =THIELAVIOPSIS WENT 1893 FIDE ELLIS 1971

HUGHESINIA LINDQUIST ET GAMUNDE 1970. BOL. SOC. ARGENT. BOT. 13:53-57 (ILLUS)
    /H. CHUSQUEAE
  STAURO,CETERI,PHAEO /ELLIS, MYCOL. PAPERS CMI 125:19-20,1971. (ILLUS)

HUMICOLA TRAAEN 1914. NYT. MAG. NATURVIDENSK. 52:33 /H. FUSCOATRA
  AMERO,CETERI,PHAEO, PERHAPS NOT DIFFERENT FROM THERMOMYCES TSIKLINSKY 1899
    /BARRON, GENERA OF HYPHOMYCETES P. 207-208,1968. (ILLUS,REFS) FASSATIOVA,
    CASOPIS CESKOSLOV. VED. SPOLEC. MYKOL. 21:78-89, 1967. (ILLUS,KEY)
    SYNO    =MELANOGONE WOLLENWEBER ET RICHTER 1934 FIDE MASON 1941
            ?=CREMASTERIA MEYERS ET MOORE 1960

HYALOBOTRYS PIDOPLICHKO 1948. MIKROBIOL. ZH. 9:2-3 /H. ELEGANS
    =STACHYBOTRYS CORDA 1837 FIDE CARMICHAEL

HYALODENDRON DIDDENS 1934. ZENTRALBL. BAKT. PARASIT. INFEKTIONSKR. 90:316 /H.
    LIGNICOLA
  AMERO,BLASTO,HYALO /BARRON, GENERA OF HYPHOMYCETES, P. 208-210,1968. (ILLUS)

## 10. Hyphomycetes

HYALODICTYS SUBRAMANIAN 1962. PROC. INDIAN ACAD. SCI., SECT. B 55:8-10 (ILLUS)
 /H. DEGENERANS (SYDOW)
 DICTYO,CETERI,HYALO

HYALOFLOREA BATISTA ET MAIA 1955. ANN. SOC. BIOL. PERNAMBUCO 13:154 /H. RAMOSA
 =CYLINDROCARPON WOLLENWEBER 1913 FIDE CARMICHAEL

HYALOPUS CORDA 1838. ICONES FUNGORUM II:16 /?, 6 ORIGINAL SPECIES TRANSFERRED
 FROM STILBUM
 ?, PROBABLY A SYNONYM OF ACREMONIUM LINK FIDE LINDAU 1907

HYALOSTACHYBOTRYS SRINIVASAN 1958. J. INDIAN BOT. SOC. 37:340 /H. BISBYI
 =STACHYBOTRYS CORDA FIDE BARRON 1968

HYALOTROCHOPHORA FINLEY ET MORRIS 1967. AMER. MIDL. NAT. 77:202-204, FIGS. 4-7
 /H. LIGNITALIS (THAXTER) = EVERHARTIA LIGNITALIS THAXTER 1891
 ?=DELORTIA PATOUILLARD ET GAILL. 1888

HYMENELLA FRIES 1823. SYSTEMA MYCOL. 2:234 (NO ILLUS) /THREE ORIGINAL SPECIES
 ? /LINDAU, RABENH. KRYPT. FL. 2 AUFL. 1 BD. 9:629-630,1910. (ILLUS)
 SYNO    =HYMENULA FRIES 1828 (ORTHOGRAPHIC VARIANT)

HYMENOBACTRON (SACCARDO) HOHNEL 1923. SYST. FUNG. IMP. P. 342 /H. DESMAZIERI
 (CASTAGNE) HOHNEL 1923
 ?, NOT ILLUSTRATED BY HOHNEL

HYMENOPODIUM CORDA 1837. ICONES FUNGORUM 1:7 /H. SARCOPODIOIDES CORDA
 =CLASTEROSPORIUM CARICINUM SCHWEINITZ
 =CLASTEROSPORIUM SCHWEINITZ 1832 FIDE HUGHES 1958

HYMENOPSIS SACCARDO 1886. SYLLOGE FUNGORUM 4:744-747 /SIXTEEN ORIGINAL SPECIES
 ?, =MYXORMIA BERK. ET BR. (COELOMYCETES) FIDE CLEMENTS AND SHEAR 1931

HYMENOSTILBE PETCH 1931. THE NATURALIST 1931:101 /H. MUSCARIUM
 ?, ASCO STATE =CORDYCEPS FIDE MORRIS 1963 /MORRIS, WESTERN ILLINOIS UNIV SER
    BIOL SCI 3:74-75,1963. (ILLUS)

HYMENULA FRIES, SEE HYMENELLA

HYPEROMYXA CORDA 1839. ICONES FUNGORUM 3:34 /H. STILBOSPOROIDES =CHEIROSPORA
    BOTRYOSPORA (MONT.) HUGHES
 =CHEIROSPORA MOUGEOT 1825 FIDE HUGHES 1958

HYPHA PERSOON 1822. MYC. EUR. 1:63 /H. BOMBYCINA
 =STERILE
    SYNO    =CLADOBYSSUS RITG. FIDE SACCARDO FROM AINSWORTH 1971
            =HYPHASMA FIDE LINDAU IN RABENHORST 1910

HYPHASMA REBENTISCH 1805 =HYPHA FIDE LINDAU IN RABENHORST 1910

HYPHELIA FRIES 1825. SYST. ORB. VEG. P. 149 /H. ROSEA (PERSOON) FRIES
 =TRICHOTHECIUM ROSEUM (PERSOON) LINK
 =TRICHOTHECIUM LINK 1809 FIDE HUGHES 1958

HYPHOCHLAENA CIFERRI 1962. ATTI IST. BOT. UNIV. PAVIA SER. 5, 19:110 /H.
    CAESPITOSA
 =STERILE

HYPHOSOMA SYDOW 1924. ANN. MYCOL. 22:315 /H. HYPOXYLOIDES
 =NOMEN CONFUSUM FIDE HUGHES 1970

IDRIELLA NELSON ET WILHELM 1956. MYCOLOGIA 48:550 /I. LUNATA
 =CHLORIDIELLA ARNAUD 1954 FIDE CARMICHAEL

ILLOSPORIUM MARTIUS 1817. FL. CRYPT. ERLANG. P. 325 /I. ROSEUM
 ?, PROBABLY A NOMEN DUBIUM SEE HUGHES 1958.  SEE STRUMELLA SACCARDO

INDIELLA BRUMPT 1906. ARCH. PARASIT. 10:60-63 /I. MANSONI
 =STERILE, FOR ISOLATES FROM HUMAN MYCETOMAS

INFRAFUNGUS CIFERRI 1951. MYCOPATHOLOGIA 6:26 /I. MICROPUS (SACCARDO) =FUSARIUM MICROPUS SACC.
?, NOT ILLUSTRATED BY CIFERRI

INGOLDIA PETERSEN 1962. MYCOLOGIA 54:147 /I. CRAGINIFORMIS
=GYOERFFIELLA KOL 1928 FIDE VON ARX 1970

INNATOSPORA VAN BEYMA 1929. VERH. K. AKAD. WET., SECT. 2, 26:5 /I. ROSEA
=ARTHRINIUM KUNZE 1817 FIDE ELLIS 1971

INSECTICOLA MAINS 1950. MYCOLOGIA 42:577-578 AND FIG. 20 ON P. 584 /I. CLAVATA
=AKANTHOMYCES LEBERT 1858 FIDE KENDRICK /MORRIS, WESTERN ILLINOIS SER BIOL SCI 3:76-77,1963. (ILLUS)

IRPICOMYCES DEIGHTON 1969. MYCOL PAPERS CMI 118:24-26 (ILLUS) /I. SCHIFFNERULAE
SCOLECO,CETERI,PHAEO

ISARIA PERSOON 1801. SYN. METH. FUNG. P. 687 /? - SEE HUGHES CAN. J. BOTANY 36:781,1958
=NOMEN DUBIUM - SEE BARRON, GENERA OF HYPHOMYCETES, P. 212-213, 1968. MANY FUNGI PLACED IN ISARIA MAY GO IN BEAUVERIA

ISARIELLA HENNINGS 1909. HEDWIGIA 48:19 /I. AUERSWALDIAE
?, NOT ILLUSTRATED BY AUTHOR

ISARIOPSELLA HOHNEL 1929. MITTEII. BOT. INST. TECHN. HOCHSCH. WIEN VI, 2 HEFT, P. 68 /I. VOSSIANA (THUMEN) =RAMULARIA VOSSIANA THUMEN
?, NOT ILLUSTRATED BY HOHNEL

ISARIOPSIS FRESENIUS 1865. BEITRAGE ZUR MYKOLOGIE 3:87-88 AND FIGS. 18-28 /I. PUSILIS
?

ISTHMOSPORA STEVENS 1918. BOT. GAZETTE 45:244 /I. SPINOSA =I. STATE OF TRICHOTHYRIUM ASTEROPHORUM FIDE HUGHES 1953
DICTYO OR STAURO,CETERI,PHAEO, ASCO STATE =TRICHOTHYRIUM FIDE HUGHES 1953 /HUGHES, MYCOL PAPERS CMI 50:77-97,1953. (ILLUS)
   SYNO    =CARLOSIA ARNAUD 1954 FIDE KENDRICK

ITERSONILIA DERX 1948. BULL. JARD. BOT. BUITENZORG 3,17:465-466 (ILLUS) /I. PERPLEXANS
AMERO,CETERI,HYALO, HAS CLAMP CONNECTIONS ON HYPHAE /BARNETT, IMPERFECT FUNGI NO. 33,1960. (ILLUS) SOWELL AND KORF, MYCOLOGIA 52:934-945,1960. (EMEND,REFS)
   SYNO    =ENTYLOMELLA HOHNEL 1916 (NOMEN NUDUM) FIDE SOWELL AND KORF 1960

IYENGARINA SUBRAMANIAN 1958. J. INDIAN BOT. SOC. 37:407 /I. ELEGANS SUBRAMAN. 1958
STAURO,CETERI,PHAEO, COMPARE WITH DIPLOCLADIELLA ARNAUD 1954

JACOBASCHELLA KUNTZE ='DIPLOSPORIUM BON.' FIDE AINSWORTH 1971

JACOBIA ARNAUD 1951. BULL. SOC. MYCOL. FRANCE 67:195 AND FIG. 4 A-C /J. CONSPICUA
AMERO,GLOIO,HYALO, HAS CLAMP CONNECTIONS ON HYPHAE

JACULISPORA HUDSON ET INGOLD 1960. TRANS. BRIT. MYCOL. SOC. 43:475-478 (ILLUS) /J. SUBMERSA
AMERO OR STAURO,CETERI,HYALO

JACZEWSKIELLA MURASH 1926. MAT. MIK. FIT. 5:5-6 (ILLUS) /J. ALTAJENSIS
?, ORIGINAL ILLUSTRATION DIFFICULT TO INTERPRET

JAINESIA FRAGOSO ET CIFERRI 1925. BOL. ESPAN. HIST. NAT. 25:514 (NO ILLUS) /J. MELIOLICOLA FRAG. ET CIFERRI 1925
PHRAGMO,CETERI,PHAEO

JOHNSTONIA ELLIS 1971. MYCOL. PAPERS CMI 125:13-15 (ILLUS) /J. COLOCASIAE
DIDYMO,CETERI,PHAEO

KABATIELLA BUBAK 1907. HEDWIGIA 46:297 /K. MICROSTICTA
=AUREOBASIDIUM VIALA ET BOYER FIDE COOKE 1959

KARAKULINIA GOLOVINA 1964. NOV. SIST. NIZ. RAST. P. 21'2 /FUSICLADIOPSIS CERASI
(RABENHORST) KARAKULIN ET VASSILJEOSKY 1937 (LECTOTYPE)
=FUSICLADIUM FIDE DEIGHTON FROM AINSWORTH 1971. PROPOSED AS A NOM. NOV. FOR
FUSICLADIOPSIS KARAKULIN ET VASSILJEOSKY 1937 NON MAIRE 1906.

KAUFMANNWOLFIA GALGOCZY ET NOVAK 1962. ON THE WORK OF THE NATIONAL INST. FOR
PUBLIC HEALTH FOR 1961, P. 225 /K. INTERDIGITALIS (PRIESTLY) GAL. ET NOVAK
=TRICHOPHYTON MENTAGROPHYTES (ROBIN) BLANCHARD 1896
=TRICHOPHYTON MALMSTEN 1845

KERATINOMYCES VANBREUSEGHEM 1952. BULL. ACAD. BELG. CL. SCI. 38:1075 /K.
AJELLOI VANBREUS. 1952 =K. STATE OF ARTHRODERMA UNCINATUM FIDE DAWSON AND
GENTLES 1961
PHRAGMO,CETERI,HYALO. ASCO STATE =ARTHRODERMA FIDE DAWSON AND GENTLES 1961
/TUBAKI, J. HATTORI BOT. LAB. 20:184,1958. (ILLUS)

KMETIA BRESADOLA ET SACCARDO 1902. SYLLOGE FUNGORUM 16:1158 /K. EXIGUA BRES. ET
SACC. 1902
SCOLECO,GLOIO,HYALO /COOKE, MYCOLOGIA 44:808-809,1952. (DESCR)

KMETIOPSIS BATISTA ET PERES 1960. PUBL. INST. MIC. UNIV. RECIFE 245:4
(1-9,ILLUS) /K. HYMENAEAE
AMERO OR SCOLECO,GLOIO,HYALO

KNYARIA KUNTZE =TUBERCULARIA FIDE AINSWORTH 1971

KOORCHALOMA SUBRAMANIAN 1953. J. INDIAN BOT. SOC. 32:124 /K. MADREEYA SUBRAM.
1953
AMERO,GLOIO,HYALO
SYNO    =KOORCHALOMELLA CHONA MUNJAL ET KAPOOR 1958 FIDE CARMICHAEL
        =STARKEYOMYCES AGNIHOTHRUDU 1956 FIDE CARMICHAEL
        =LOMACHASHAKA SUBRAMANIAN 1956 FIDE CARMICHAEL

KOORCHALOMELLA CHONA, MUNJAL ET KAPOOR 1958. INDIAN PHYTOPATH. 11:130 /K. ORYZAE
=KOORCHALOMA SUBRAMANIAN FIDE CARMICHAEL

KORDYANELLA HOHNEL 1904. ANN. MYCOL. 2:274 /K. AUSTRIACA
?, NOT ILLUSTRATED BY AUTHOR

KOSTERMANSINDA RIFAI 1968. REINWARDTIA 7:4 /K. MAGNA (BOEDIJN) =SCLEROGRAPHIUM
MAGNUM BOEDIJN
DICTYO,CETERI,PHAEO /ELLIS, DEM. HYPH. P. 112-113,1971. (ILLUS)

KUMANASAMUHA RAO ET RAO 1964. MYCOPATHOLOGIA 22:330-334 (ILLUS) /K. SUNDARA
AMERO,CETERI,PHAEO

KURSSANOVIA PIDOPLICHKO 1948. MIKROBIOL. ZH. 9(2-3):57 /K. TRISEPTATA
=FUSICLADIELLA HOHNEL 1919 FIDE KENDRICK

KUTILAKESA SUBRAMANIAN 1956. J. INDIAN BOT. SOC. 35:478-481 (ILLUS) /K.
MADREEYA SUBRAMAN. 1956
=MYROTHECIUM TODE EX FRIES 1821 FIDE NAG RAJ

KUTILAKESOPSIS AGNIHOTHRUDU ET BARUA 1957. J. INDIAN BOT SOC. 36:306-311
(ILLUS) /K. MACALPINEI AGNIHOT. ET BARUA 1957
?=DIDYMOSTILBE HENNINGS 1902

LACELLINA SACCARDO 1913. ANN. MYCOL. BERL. 11:418 /L. LIBYCA SACC. ET TROTTER
1913 =L. GRAMINICOLA (BERKELEY ET BROOME) PETCH 1924
AMERO,BLASTO,PHAEO /ELLIS, MYCOL PAPERS CMI 67:10-15,1957 (ILLUS,KEY)

LACELLINOPSIS SUBRAMANIAN 1953. PROC. INDIAN ACAD. SCI. 37:100-105 /L. SACCHARI
SUBRAMAN. 1953
AMERO,BLASTO,PHAEO /ELLIS, MYCOL PAPERS CMI 67:6-10,1957. (ILLUS,KEY)

LACHNIDIUM GIARD =FUSARIUM FIDE SACCARDO FROM AINSWORTH 1971

LACHNODOCHIUM MARCHAL 1895. BULL. SOC. ROY. BOT. BELG. 34:144-145 AND PL. 1,
   FIG. 2 /L. CANDIDUM
   ? /LINDAU, RABENHORST KRYPT-FL. 2 AUFL. 1 BD. 9:472-473,1910. (ILLUS)

LACINIOCLADIUM PETRI 1917. ANN. IST. SUP. AGR. FOR. NAZ. 2:31-32 /L. CASTANEAE
   ?, ORIGINAL ILLUSTRATION NOT DIAGNOSTIC

LANGERONIA VANBREUSEGHEM 1950. ANN. SOC. BELGE MED. TROP. 30:888 /L.
     SOUDANENSIS (JOYEUX) VANBREUS. 1950
   =TRICHOPHYTON MALMSTEN 1845 FIDE AJELLO 1968

LANGLOISULA ELLIS ET EVERHART 1889. JOUR. MYCOL. 5:68-69 /L. SPINOSA ELL. ET
     EVERH. 1889
   =NOMEN CONFUSUM FIDE AINSWORTH 1971

LEANDRIA RANGEL 1915. BOL. AGR. SAO PAULO 16:324-325 AND FIGS. 4-15 /L.
     MOMORDICAE
   ?, ORIGINAL ILLUSTRATION DIFFICULT TO INTERPRET

LECYTHOPHORA NANNFELDT 1934. SVENSKA SKOGSV. TIDSKR. 3-4:433-436 (ILLUS) /L.
     LIGNICOLA NANNF. 1934
   =PHIALOPHORA MEDLAR FIDE GOIDANICH FROM AINSWORTH 1971

LEIOSEPIUM SACCARDO 1900. BULL. SOC. MYCOL. FR. 16:24 /L. AUREUM SACC. ET
     FAUTREY 1900
   =SEPEDONIUM LINK 1809 FIDE DAMON 1952

LEMONNIERA DE WILDEMAN 1894. ANN. SOC. BELGE MICR. 18:143 /L. AQUATICA
   STAURO,CETERI,HYALO /TUBAKI, BULL. NAT. SCI. MUS. TOKYO 3:259-260,1957.
     (ILLUS) VON ARX, GENERA OF FUNGI IN CULTURE P. 188-189,1970. (ILLUS)

LEPIDOPHYTON TRIBONDEAU 1899. ARCH. MED. NAVALE 72:5-52 /NO SPECIES DESCRIBED
   =NOMEN ILLEGITIMUM

LEPTODISCELLA PAPENDORF 1969. TRANS. BRIT. MYCOL. SOC. 53:145 /L. AFRICANA
     (PAPENDORF) =LEPTODISCUS AFRICANA PAPENDORF 1967 (TRANS. BRIT. MYCOL. SOC.
     50:687)
   DIDYMO,GLOIO,HYALO

LEPTODISCUS GERDEMANN 1953. MYCOLOGIA 45:552 /L. TERRESTRIS
   =MYCOLEPTODISCUS OSTAZESKI, NON LEPTODISCUS HERTWIG

LEPTOGRAPHIUM LAGERBERG ET MELIN 1928. SVENSKA SKOGSV. TIDSKR. 25:257 (DATED
     1927) /L. LUNDBERGII LAGERB. ET MELIN 1928
   AMERO,GLOIO,HYALO, ASCO STATE =CERATOCYSTIS OR EUROPHIUM FIDE BARRON 1968
     /BARRON, GENERA OF HYPHOMYCETES, P. 215-217,1968. (ILLUS)

LEPTOSPORIUM (SACCARDO) HOHNEL 1923. SYST. FUNGI IMP. 436 /FUSARIUM
     SALMONICOLOR BERKELEY ET CURTIS
   =NOMEN ILLEGITIMUM, NON LEPTOSPORIUM BONORDEN 1851

LEPTOTRICHUM CORDA 1842. ICONES FUNGORUM 5:10 AND TAF. II, FIG. 16 /L. GLAUCUM
   ? /LINDAU, RABENHORST KRYPT. FL. 2 AUFL., 1 BD., 9:502,1910. (ILLUS)

LEUCODOCHIUM SYDOW 1917. ANN. MYCOL. 15:266 /L. PIPTURI
   ?, NOT ILLUSTRATED BY AUTHOR

LEUCOSPORIUM CORDA 1836 =ACHITONIUM KUNZE FIDE LINDAU IN RABENHORST 1910 SUB
     PACTILIA

LIBERTELLA DESMAZIERES 1830. ANN. SCI. NAT. BOT. SER. 1, 19:277 /L. BETULINA
     DESM. 1930
   SCOLECO,GLOIO,HYALO, (MELANCONIALES) ASCO STATE =DIATRYPE, DIATRYPELLA,
     QUATERNARIA OR DIAPORTHE FIDE VON ARX 1970 /BARRON, GENERA OF HYPHOMYCETES
     P. 216-218,1968. (ILLUS)

LINDAUOMYCES KOORDERS 1907. BOTANISCHE UNTERSUCHUNGEN, AMSTERDAM, P. 240-241
     (ILLUS) /L. JAVANICUS
   =ARTHROBOTRYUM FIDE CLEMENTS AND SHEAR 1931

## 10. Hyphomycetes

LINDAUOPSIS ZAHLBRUCKNER 1906. BERL. DEUTSCHL. BOT. GES. 24:145-146 AND TAF. 10 /L. CALOPLACAE
    ?, ORIGINAL ILLUSTRATION DIFFICULT TO INTERPRET

LINDAVIA NIEUWLAND 1916 =SCOPULARIA PREUSS

LINODOCHIUM HOHNEL 1909. S.B. AKAD. WISS. WIEN 118:1239 (NO ILLUS) /L. HYALINUM (LIBERT)
    SCOLECO,GLOIO,HYALO

LISTEROMYCES PENZIG ET SACCARDO 1901. MALPIGHIA 15:259 /L. INSIGNIS PENZ. ET SACC. 1901
    PHRAGMO,CETERI,PHAEO, COMPARE WITH EXCIPULARIA /GOOS, MYCOLOGIA 63:213-218,1971(ILLUS)
        SYNO    =CYLOMYCES CLEMENTS 1931 FIDE AINSWORTH 1960

LITUARIA RIESS 1853. BOT. ZEITUNG 11:136 AND TAF. III, FIGS. 8-10 /L. STIGMATEA
    =HELICOMA CORDA 1837 FIDE LINDER FROM AINSWORTH 1971

LOLIOMYCES MAIRE 1937. MEM. SOC. SCI. NAT. MAROC 45:140 (NO ILLUS) /L. TEMULENTIS
    =STERILE

LOMAANTHA SUBRAMANIAN 1954. J. INDIAN BOT. SOC. 33:31 /L. POOGA
    PHRAGMO,CETERI,PHAEO, BUT COMPARE WITH SPORIDESMIUM BRACHYPUS, SEE ELLIS, DEM. HYPH. P.117, FIG. 77B,1971.

LOMACHASHAKA SUBRAMANIAN 1956. J. INDIAN BOT. SOC. 35:66 /L. KERA SUBRAMAN. 1956
    =KOORCHALOMA SUBRAMANIAN FIDE CARMICHAEL

LOPHOPHYTON MATRUCHOT ET DASSONVILLE 1889. REV. GEN. BOT. 11:432-444 AND PL. 19 /L. GALLINAE (MEGNIN) MATR. ET DASS. = TRICHOPHYTON GALLINAE
    =TRICHOPHYTON MALMSTEN 1845

LUNULOSPORA INGOLD 1942. TRANS. BRIT. MYCOL. SOC. 25:408 /L. CURVULA
    SCOLECO,CETERI,HYALO /BARNETT, IMPERFECT FUNGI NO. 111,1960. (ILLUS)
        SYNO    ?=CALCARISPORA MARVANOVA ET MARVAN 1963

LUTZIOMYCES FONSECA 1939. ANN. BRAS. DERM. SIF. 14:107 /L. HISTOSPOROCELLULARIS (HABERFELD) FONSECA = PARACOCCIDIOIDES BRASILIENSIS
    =PARACOCCIDIOIDES ALMEIDA 1930 SEE DODGE 1935, P.179

MACRAEA SUBRAMANIAN 1952 NON LINDL. 1930 =PRATHIGADA

MACROSPORIUM FRIES 1832. SYST. MYCOL. 3:374 /M. CHEIRANTHI (LIB.) FRIES
    =ALTERNARIA CHEIRANTHI (LIB.) BOLLE
    =ALTERNARIA NEES 1817 FIDE HUGHES 1958

MACROSTILBUM PATOUILLARD 1898. BULL. SOC. MYCOL. FR. 14:197 /M. RADICOSUM
    ?, NOT ILLUSTRATED BY AUTHOR

MACROTRICHUM GREVILLE 1825. EDINB. PHIL. J. 25:64-65 AND PL. 1 /TWO ORIGINAL SPECIES
    ?, ORIGINAL ILLUSTRATIONS AND DESCRIPTION DIFFICULT TO INTERPRET

MADURELLA BRUMPT 1905. COMP. REND. SOC. BIOL. 57:999 /M. MYCETOMI (LAVERAN)
    =STERILE, MADURELLA MYCETOMI HAS A PHIALOPHORA STATE /BORELLI, BOL. VENEZ. LAB. CLIN., CARACAS 2:1-15,1957. (ILLUS)

MAGDALAENAEA ARNAUD 1952. BULL. SOC. MYCOL. FR. 68:209 AND 211 (ILLUS) /M. MONOGRAMMA
    STAURO,CETERI,HYALO, NO LATIN DIAGNOSIS

MAHABALELLA SUTTON ET PATIL 1966. NOVA HEDWIGIA 11:203-207 AND TAB. 21 /M. ACUTISETOSA
    AMERO,GLOIO,HYALO

MAHEVIA LAGARDE 1917. ARCH. ZOOL. EXPER. GEN. 56:292-294 AND FIGS. 9-11 /M.

```
     GUIGNARDII (MAHEU)
=HIRSUTELLA PATOUILLARD 1892 FIDE CARMICHAEL

MALBRANCHEA SACCARDO 1882. MICHELIA 2:639 /M. PULCHELLA SACCARDO ET PENZIG 1882
   AMERO,ARTHRO,HYALO /COONEY AND EMERSON, THERMOPHILIC FUNGI P.93-102,1964.
   (ILLUS)
      SYNO    =THERMOIDIUM MIEHE 1907 FIDE SACCARDO IN AINSWORTH 1961

MALLEOMYCES RIVOLTA ?=HYPHOMYCETE FROM AINSWORTH 1971

MALUSTELLA BATISTA ET LIMA 1960. PUBL. INST. MICOL. UNIV. RECIFE 263:7 /M.
   AERIA BATISTA ET AL. 1960 =CURVULARIA LUNATA FIDE VON ARX 1970
=CURVULARIA BOEDIJN 1933 FIDE VON ARX 1970

MAMMARIA CESATI 1854. BOT. ZEITUNG 12:190 /M. ECHINOBOTRYOIDES
   AMERO,CETERI,PHAEO /BARRON, GENERA OF HYPHOMYCETES P. 218-219,1968. (ILLUS)
      SYNO    ?=CONIOSCYPHA HOHNEL 1904

MARCOSIA SYDOW 1916. ANN. MYCOL. 14:96 (NO ILLUS) /M. ULEI
   ?, SEE MORGAN-JONES, CAN. J. BOT. 49:1007,1971 FOR ILLUS OF TYPE

MARGARINOMYCES LAXA 1930. ZENTRLB. BAKT. PARASITKDE, ABT. 2,81:392 /M. BUBAKII
    LAXA =P. BUBAKII (LAXA) SCHOL-SCHWARZ
=PHIALOPHORA MEDLAR 1915 FIDE VON ARX 1970

MARGARITISPORA INGOLD 1942. TRANS. BRIT. MYCOL. SOC. 25:352 /M. AQUATICA
   AMERO OR PHRAGMO,CETERI,HYALO /TUBAKI, BULL. NAT. SCI. MUS., TOKYO
   3:259-261,1957. (ILLUS)

MARIANNAEA ARNAUD 1952. BULL. SOC. MYCOL. FR. 68:196 AND FIG. 2A /M. ELEGANS
=PAECILOMYCES BAINIER FIDE BROWN AND SMITH 1957

MARTINDALIA SACCARDO ET ELLIS 1884. MISCEL. MYCOL. 2:16 /M. SPIRONEMA
   ?=HEYDENIA FRESENIUS 1852 /MORRIS, WEST. ILLINOIS SER BIOL SCI 3:82-83,1963.
   (ILLUS)
      SYNO    =CLATHROTRICHUM PATOUILLARD 1921 FIDE CARMICHAEL

MASONIA SMITH 1952. TRANS. BRIT. MYCOL. SOC. 35:149 /M. GRISEA
   =SCOPULARIOPSIS BAINIER 1907 FIDE SMITH 1963 (NON MASONIA HANSFORD 1944)

MASONIELLA SMITH 1952. TRANS. BRIT. MYCOL. SOC. 35:237 /M. GRISEA (SMITH) SMITH
     =SCOPULARIOPSIS BRUMPTII
=SCOPULARIOPSIS BAINIER 1907 FIDE SMITH 1963

MASTIGONEMA SPEGAZZINI 1926. BOL. ACAD. CORDOBA 29:177 (NO ILLUS) /M.
     BRUCHIANUM SPEG. 1926
=CHAETOSPERMUM FIDE AINSWORTH 1971

MASTIGOSPORIUM RIESS 1852. IN FRESENIUS, BEITR. MYKOL. P. 56 /M. ALBUM
   PHRAGMO,GLOIO OR CETERI,HYALO /HUGHES, MYCOL PAPERS CMI 36:1-3,AND
   40-41,1951. (ILLUS)
      SYNO    =AMASTIGOSPORIUM BONDARZEVA-MONTEVERDE 1903 FIDE HUGHES 1951

MATRUCHOTIELLA GRIGORAKIS 1924. C.R. ACAD. SCI., PARIS 179:1424 /M. CURRII
     (CHALMERS ET MARSHALL)
=NOMEN DUBIUM, (PROBABLY, SEE DODGE, MEDICAL MYCOLOGY, P.431,1935)

MAUGINIELLA CAVARA 1925. ATTI R. ACCAD. NAZ. LINCEI, ROMA, SER 6 1:65-67 /M.
     SCAETTAE CAVARA
   ?, ILLUSTRATIONS IN BOL. ORTO. BOT. R. UNIV. NAPOLI 8:TAV. V,1925, ARE
     DIFFICULT TO INTERPRET

MAXILLOSPORA HOHNEL 1914. S. B. AKAD. WISS. WIEN. ABT. 1, 123:138 (NO ILLUS)
   /M. MAXILLIFORMIS (ROSTRUP)
=TETRACLADIUM DE WILDEMAN 1894 FIDE INGOLD FROM AINSWORTH 1971

MEDUSULA CORDA 1837. ICONES FUNGORUM 1:18, FIG. 241 /M. OCHRACEA
=VOLUTELLA FIDE LINDAU IN RABENHORST 1910
```

## 10. Hyphomycetes

MEGACLADOSPORIUM VIENNOT-BOURGIN 1949. CHAMPIGNONS PARASITES 1:489 /F. PYRORUM (LIB.) VIEN.-BOURGIN 1949
 =FUSICLADIUM BONORDEN 1851 FIDE HUGHES 1958

MEGALODOCHIUM DEIGHTON 1960. MYCOL. PAPERS CMI 78:17-20 (ILLUS) /M. PALMICOLA =M. ELAEIDIS (BEELI) DEIGHTON 1969
 AMERO,BASO,PHAEO /DEIGHTON, MYCOL. PAPERS CMI 117:8,1969. (REVIS)

MEGASTER CIFERRI ET AL. 1956. PUBL. INST. MIC. UNIV. RECIFE 48:2 /M. LONGICORNIS STAURO,CETERI,PHAEO

MEGATRICHOPHYTON NEVEU-LEMAIRE 1921. PRECIS PARASITOL. HUM. P. 54 /M. ROSEUM (BODIN)
 =TRICHOPHYTON MALMSTEN 1845

MELANCONIUM LINK 1809. MAG. GES. NATURF. FREUNDE, BERLIN 3:9 /M. ATRUM
 AMERO,GLOIO,PHAEO, (MELANCONIALES) ASCO STATE = MELANCONIS TULASNE /VON ARX, GENERA OF FUNGI IN CULTURE P. 155 AND 157,1970. (ILLUS)
  SYNO    =TRIMMATOSTROMA CORDA 1837 FIDE HUGHES 1958

MELANODISCUS HOHNEL 1918. BERL. DEUTSCHL. BOT. GES. 36:309 /M. NERVISEQUIUS
 ?, NOT ILLUSTRATED BY AUTHOR

MELANODOCHIUM SYDOW 1938. ANN. MYCOL. 36:310-311 /M. AUSTRALIENSE
 ?, NOT ILLUSTRATED BY AUTHOR

MELANOGONE WOLLENWEBER ET RICHTER 1934. ZENTRALBL. BAKT., ABT. 2, 90:74-76 (ILLUS) /M. PUCCINIOIDES WOLL. ET RICHTER 1934
 =HUMICOLA TRAAEN FIDE MASON 1941

MELANOGRAPHIUM SACCARDO 1913. ANN. MYCOL. 11:558 /M. SPLENISPORIUM SACCARDO =M. SPINULOSUM (SPEG.) HUGHES 1958
 AMERO,CETERI,PHAEO /ELLIS, MYCOL PAPERS CMI 93:14-23,1963. (KEY,ILLUS)
  SYNO    =SPOROSTACHYS SACCARDO 1917 FIDE HUGHES 1958
          =PSEUDOCAMPTOUM FRAGOSO ET CIFERRI 1925 FIDE HUGHES 1958

MELANOTRICHUM CORDA 1833. IN STURMS DEUTSCHL. FL., III(PILZE) BD. 3, HEFT 13:89-90 AND TAB. 45 /M. MICROSPERMUM
 ?, ORIGINAL ILLUSTRATION NOT DIAGNOSTIC

MEMNONIELLA HOHNEL 1923. ZENTRALBL. BAKT. PARASITENKUNDE ABT. 2:60 /M. ATERRIMA HOHNEL =M. ECHINATA (RIVOLTA) GALLOWAY
 =STACHYBOTRYS CORDA 1837 FIDE SMITH 1962

MEMNONIUM CORDA 1833. IN STURMS DEUTSCHL. FL. III (PILZE), BD 3, HEFT 13:91 /M. EFFUSUM
 =NOMEN DUBIUM SEE HUGHES 1958 AND LINDAU 1907

MENISPORA PERSOON 1822. MYCOLOGIA EUROPAEA 1:32 /M. GLAUCA
 AMERO OR DIDYMO OR PHRAGMO,GLOIO,HYALO. ASCO STATE = CHAETOSPHAERIA FIDE MULLER IN KENDRICK 1971 /HUGHES AND KENDRICK, NEW ZEALAND J. BOT. 6:323-375,1968. (ILLUS,KEY) HUGHES AND KENDRICK, CAN. J. BOTANY 41:693-718,1963. (ILLUS,KEY)
  SYNO    =ERIOMENE (SACC.) MAIRE EX PEYRONEL 1919 FIDE HUGHES 1958
          =ERIONEMA PENZIG 1898 FIDE HUGHES 1958
          =CILIOFUSARIUM ROSTRUP 1892 FIDE HUGHES 1958
          =ERIOMENELLA PEYRONEL 1919 FIDE HUGHES 1958

MENISPORELLA AGNIHOTHRUDU 1962. PROC. INDIAN ACAD. SCI. 56B:98 /M. ASSAMICA
 =CODINAEA MAIRE 1937 FIDE ELLIS IN KENDRICK AND HUGHES 1968

MENISPOROPSIS HUGHES 1952. MYCOL PAPERS CMI 48:59 /M. THEOBROMAE
 AMERO OR DIDYMO,GLOIO,HYALO /HUGHES AND KENDRICK, NEW ZEALAND J. BOT. 6:323-375,1968. (ILLUS)

MENOIDEA MANGIN ET HARIOT 1907. BULL. SOC. MYC. FRANCE 23:65-68 (ILLUS) /M. ABIETIS
 ?, ORIGINAL ILLUSTRATION NOT DIAGNOSTIC

MERIA VUILLEMIN 1896. BULL. SOC. SCI. NANCY 2,14:13 /M. LARICIS
    AMERO,CETERI,HYALO. ORIGINAL ILLUSTRATION DIFFICULT TO INTERPRET
        SYNO    =HARTIGIELLA SYDOW 1900 FIDE AINSWORTH 1971

MEROSPORIUM CORDA 1832. IN STURMS DEUTSCHL. FL. III (PILZE), BD. 3, HEFT 12:45
    /M. MINUTUM
    =NOMEN CONFUSUM FIDE HUGHES 1958

MESOBOTRYS SACCARDO 1880. MICHELIA 2:27 /M. FUSCA (CORDA) =GONYTRICHUM
    MACROCLADUM
    =GONYTRICHUM NEES ET NEES 1818 FIDE HUGHES 1958

METARRHIZIUM SOROKIN 1883. LES MALADIES DES PLANTES, ETC. 2:268 /M. ANISOPLIAE
    (METSCHNIKOV)
    AMERO,BASO,HYALO /BARRON, HYPHOMYCETES P. 223-225,1968. (ILLUS)
        SYNO    =MYROTHECIELLA SPEGAZZINI 1910 FIDE CARMICHAEL
                =CHROMOSTYLUM GIARD FIDE AINSWORTH 1971

MIAINOMYCES CORDA 1833. IN STURMS DEUTSCHL. FL. III (PILZE), BD. 3, HEFT
    13:83-84 AND TAB. 42 /M. FUNGICOLUS
    ?, ORIGINAL ILLUSTRATION NOT DIAGNOSTIC

MICROBASIDIUM BUBAK ET RANOJEVIC 1914. ANN. MYCOL. 12:415-416 (ILLUS) /M.
    SORGHI (PASSER.) BUB. ET RAN. 1914
    =HADROTRICHUM FIDE HOHNEL FROM AINSWORTH 1971

MICROCERA DESMAZIERES 1848. ANN. SCI. NAT., SER. 3, 10:359-360 /M. COCCOPHILA
    DESM. 1848
    ?, NOT ILLUSTRATED BY AUTHOR

MICROCLAVA STEVENS 1917. TRANS. ILLINOIS ACAD. SCI. 10:205 /M. MICONIAE
    AMERO OR DICTYO,CETERI,PHAEO /BONAR, MYCOLOGIA 59:596-597,1967. (ILLUS)
        SYNO    =OMMATOSPORA BATISTA ET CAVALCANTI 1964 FIDE KENDRICK
                ?=OMMATOSPORELLA BAT., BEZ. ET POROCA 1967

MICRODOCHIUM SYDOW 1924. ANN. MYCOL. 22:267-268 (ILLUS) /M. PHRAGMITIS
    SCOLECO,CETERI,HYALO

MICROSPATHA KARSTEN 1889. REV. MYC. 11:207 /M. GLAUCA
    ?, NOT ILLUSTRATED BY AUTHOR

MICROSPORUM GRUBY 1843. C. R. SEANC. HEBD. ACAD. SCI. PARIS 17:301 /M. AUDOUINI
    PHRAGMO,CETERI,HYALO. ASCO STATE =NANNIZZIA FIDE STOCKDALE 1961 /BARRON,
        HYPHOMYCETES P. 225-227,1968. (ILLUS,REFS) REBELL AND TAPLIN,
        DERMATOPHYTES. 124PP.,1970. (ILLUS,KEY)
        SYNO    =MICROSPORON GRUBY (ORTHOGRAPHIC VARIANT)
                =SABOURAUDITES OTA ET LANGERON 1923
                =THALLOMICROSPORON BENEDEK 1964

MICROSTROMA NIESSL 1861. OST. BOT. Z. 11:250-253 (NO ILLUS) /M. QUERCIGENUM
    NIESSL =M. ALBUM (DESMAZIERES) SACCARDO
    AMERO,GLOIO,HYALO. GROWS AS A YEAST IN CULTURE, COMPARE WITH ENDOSTILBUM
        SYNO    =HELOSTROMA PATOUILLARD 1902 FIDE VON ARX 1970
                =ARTICULARIELLA HOHNEL 1909 FIDE VON ARX 1970

MICROTRICHOPHYTON (CAST. ET CHALM.) NEVEU-LEMAIRE 1921. PRECIS PARASITOL. HUM.,
    ED. 5, P. 53-54 /?
    =TRICHOPHYTON MALMSTEN 1848

MICROTYPHA SPEGAZZINI 1910. ANN. MUS. NAC. HIST. NAT. BUENOS AIRES 20:432 /M.
    SACCHARICOLA
    =ARTHRINIUM KUNZE 1817 FIDE ELLIS 1971

MICROXYPHIUM (HARV.) SPEGAZZINI 1918 =MYCELIUM, SEE HUGHES 1958

MILOWIA MASSEE 1884. J. R. MICROSCOP. SOC., SER. 2, 4:841-845 AND PL. XII /M.
    NIVEA
    PHRAGMO,CETERI OR ARTHRO,PHAEO. TYPE SPECIES APPEARS FROM THE ILLUSTRATION TO
        BE THIELAVIOPSIS BASICOLA (BERK. ET BROOME) FERRARIS 1912

## 10. Hyphomycetes

MINIMIDOCHIUM SUTTON 1969. CAN. J. BOTANY 47:2095-2100 /M. SETOSUM
  AMERO,GLOIO,HYALO

MIRANDINA ARNAUD 1952. BULL. SOC. MYCOL. FR. 68:205-206 AND FIG. 8D AND E /M. CORTICOLA
  SCOLECO,CETERI,HYALO

MITTERIELLA SYDOW 1933. ANN. MYCOL. 31:95 /M. ZIZYPHINA
  PHRAGMO,CETERI,PHAEO /ELLIS, DEM. HYPH. P. 228-229,1971. (ILLUS)

MIURAEA HARA 1948. BYOGAICHU-HOTEN (MANUAL OF PESTS AND DISEASES) P. 779 /M. DEGENERANS (SYDOW) = CLASTEROSPORIUM DEGENERANS SYDOW
  ?, FOR HYALINE-SPORED CLASTEROSPORIUM

MOESZIA BUBAK 1914. BOT. KOEZLEM. 13:94 /M. CYLINDROIDES BUBAK =C. MAGNUSIANUM WOLLENWEBER FIDE VON ARX
  =CYLINDROCARPON WOLLENWEBER 1913 FIDE VON ARX 1970

MONACROSPORIUM OUDEMANS 1885. NED. KRUIDK. ARCH. 2,4:250 /M. ELEGANS
  =DACTYLELLA GROVE 1884 FIDE BARRON 1968

MONILIA PERSOON 1801. SYN. METH. FUNG. P. 693 /M. FRUCTIGENA
  AMERO,BLASTO,HYALO OR PHAEO, ASCO STATE =MONILINIA OR NEUROSPORA FIDE VON ARX 1970 /BARRON, HYPHOMYCETES P. 227,1968. (ILLUS)
    SYNO    =EPOCHNIUM LINK 1809 FIDE HUGHES 1958
            ?=HALOBYSSUS ZUKAL 1893 FIDE CLEMENTS AND SHEAR 1931

MONILIELLA STOLK ET DAKIN 1966. ANTONIE VAN LEEUWENHOEK 32:399 /M. ACETOBUTANS
  AMERO,ARTHRO AND BLASTO,HYALO, DOUBTFULLY DISTINCT FROM HYALODENDRON DIDDENS 1934 /VON ARX, GENERA OF FUNGI IN CULTURE P. 166-167,1970. (ILLUS)

MONILIGER LETEL. ?=PENICILLIUM FIDE AINSWORTH 1971

MONILIOPSIS RUHLAND 1908. ARB. BIOL. ANST., BERLIN 6:71-76 (ILLUS) /M. ADERHOLDII
  =STERILE(?), ORIGINAL ILLUSTRATION DIFFICULT TO INTERPRET

MONILOCHAETES HALSTEAD EX HARTER 1916. J. AGRIC. RES. 5:791 AND PL. LVII /M. INFUSCANS
  ?, ORIGINAL ILLUSTRATION NOT DIAGNOSTIC /BARNETT, IMPERFECT FUNGI NO. 164, 1960. (ILLUS)

MONOCHAETOPSIS PATOUILLARD, LISTED IN AINSWORTH 1971

MONOCILLIUM SAKSENA 1955. INDIAN PHYTOPATH. 8:8 /M. INDICUM
  =TORULOMYCES DELITSCH 1943 FIDE KENDRICK /BARRON, HYPHOMYCETES P. 228-230,1968. (ILLUS)

MONOCONIDIA ROZE 1897. BULL. SOC. MYC. FR. 13:82-83 AND PL. VI, FIG. 27 /M. MINUTISSIMA
  ?, ORIGINAL ILLUSTRATION DIFFICULT TO INTERPRET

MONODICTYS HUGHES 1958. CAN. J. BOTANY 36:785 (NO ILLUS) /M. PUTREDINIS (WALLROTH)
  DICTYO,CETERI,PHAEO /ELLIS, DEM. HYPH. P. 68-70,1971. (ILLUS,KEY)

MONOGRAMMIA STEVENS 1917. TRANS. ILLINOIS ACAD. SCI. 10:202 /M. MINCONIAE
  =TITAEA SACCARDO 1876 FIDE DAMON FROM AINSWORTH 1971

MONOPODIUM DELACROIX 1890. BULL. SOC. MYCOL. FR. 6:99 /M. UREDOPSIS DEL. 1890
  =ACREMONIELLA SACCARDO 1886 FIDE ELLIS 1971

MONOSPORELLA HUGHES 1953. CAN. J. BOTANY 31:654 /M. SETOSA (BERK. ET CURTIS)
  =MONOTOSPORELLA HUGHES 1958 =BRACHYSPORIELLA BATISTA 1952, NON MONOSPORELLA KEILIN 1920

MONOSPORIELLA SPEGAZZINI 1918. PHYSIS 4:293 /M. MELIOLICOLA
  ?

MONOSPORIUM BONORDEN 1851. HANDB. MYKOLOGIE P. 95 /?
=NOMEN ILLEGITIMUM FIDE HUGHES 1958. M. APIOSPERMUM IS THE GRAPHIUM STATE OF ALLESCHERIA BOYDII FIDE CARMICHAEL

MONOSTACHYS ARNAUD 1954. BULL. SOC. MYCOL. FR. 69:277-278 AND FIG. 5G,E AND F /M. TRANSVERSALIS ARNAUD
AMERO,BASO,HYALO. COULD BE PAECILOMYCES OR GLIOCLADIUM

MONOTOSPORA CORDA 1837. ICONES FUNGORUM 1:11 /M. TORULOIDES
=NOMEN DUBIUM FIDE CARMICHAEL 1962

MONOTOSPORA SACCARDO 1880. MICHELIA 2:25 AS 'MONOTOSPORA CORDA EM.' /M. SPHAEROCEPHALA BERKELEY ET BROOME 1854 (TYPICAL SPECIES FIDE MASON 1941)
=NOMEN ILLEGITIMUM (NON MONOTOSPORA CORDA). SEE ACROGENOSPORA.

MONOTOSPORELLA HUGHES 1958. CAN. J. BOTANY 36:786 /M. SETOSA (BERKELEY ET CURTIS)
=BRACHYSPORIELLA BATISTA 1952 FIDE ELLIS 1959

MONOTRICHUM GAUMANN 1922. ANN. MYCOL. 20:261 (ILLUS) /M. COMMELINAE
DIDYMO,CETERI,HYALO. ORIGINAL ILLUSTRATION SUGGESTS CONIDIAL BROOMELLA VITALBAE. SEE MULLER, IN KENDRICK, TAXONOMY OF FUNGI IMPERFECTI, FIG. 13.2, P.194,1971. COMPARE WITH TRICHOCONIS

MOORELLA RAO ET RAO 1964. MYCOPATHOLOGIA 22:51 /M. SPECIOSA
=HELICOMA CORDA 1837 FIDE KENDRICK

MORRISOGRAPHIUM MORELET 1968. TAXON 17:528 /M. PILOSUM (EARLE) = ISARIOPSIS PILOSA EARLE 1897
SCOLECO OR PHRAGMO,GLOIO,HYALO
    SYNO    =PHRAGMOGRAPHIUM MORRIS 1966 NON HENNINGS 1905

MUCHMORIA SACCARDO 1906. ANN. MYCOL. 4:277 /M. PORTORICENSIS
=NOMEN CONFUSUM FIDE HUGHES 1958

MUCROSPORIUM PREUSS 1851. LINNAEA 24:128 /TWO ORIGINAL SPECIES
?, NOT ILLUSTRATED BY AUTHOR

MUIARIA THAXTER 1914. BOT. GAZ. 58:241 (235-253,ILLUS) /M. GRACILIS (LECTOTYPE)
DICTYO,CETERI,PHAEO

MUIOGONE THAXTER 1914. BOT. GAZ. 58:239 (235-253,ILLUS) /M. CHROMOPTERI
DICTYO,CETERI,PHAEO

MUIRELLA SPRAGUE 1958. MYCOLOGIA 50:827 /M. ALASCENSIS (SPRAGUE)
=GLOEOCERCOSPORA ALASCENSIS
?, NO ILLUSTRATION PUBLISHED EITHER 1954 OR 1958 /SPRAGUE, MYCOLOGIA 46:77-78,1954. (DESCR) SUB GLOEOCERCOSPORA

MULTICLADIUM DESHPANDE ET DESHPANDE 1966. MYCOPATHOLOGIA 30:184-186 (ILLUS) /M. DIGITATUM DESHP. ET DESHP. 1966
?,PROBABLY THE SAME AS SPEIROPSIS TUBAKI 1958

MULTIPATINA SAWADA 1928. REP. AGR. RES. INST., FORMOSA 35:121-122 AND FIGS. 24-25 /M. CITRICOLA
?, ORIGINAL ILLUSTRATION NOT DIAGNOSTIC

MUROGENELLA GOOS ET MORRIS 1965. MYCOLOGIA 57:776 /M. TERROPHILA
PHRAGMO,CETERI,PHAEO /BARRON, HYPHOMYCETES P. 231-232,1968. (ILLUS)

MYCELIOPHTHORA COSTANTIN 1894. REV. GEN. BOT. 6:289 /M. LUTEA COSTANTIN =C. LUTEUM (COSTANTIN) CARMICHAEL
=CHRYSOSPORIUM CORDA 1833 FIDE CARMICHAEL 1962

MYCELODERMA DUCOMET 1907. RECH. DEV. CHAMP., THESE, RENNES, P. 119-120 AND PL. 111-112 /M. CUTICULARE
?, ORIGINAL ILLUSTRATION DIFFICULT TO INTERPRET

## 10. Hyphomycetes

MYCOBACILLARIA NAUMOV 1915. MAT. MIK. FIT. 1:26 AND FIG. 30 /M. SIMPLEX
 ?, LOOKS LIKE OSCILLATORIA IN ILLUSTRATION

MYCOBANCHE PERSOON 1818. TRAITE SUR CHAMPIGN. P. 133 /M. CHRYSOSPERMA
 =SEPEDONIUM LINK 1809 FIDE HUGHES 1958

MYCOCHAETOPHORA HARA ET OGAWA 1931. IN HARA, FUNGI, THE NIPPON FUNGOLOG. SOC.
   NR. 3-4, P.112 /M. JAPONICA
 ?=HYPHOMYCETE

MYCOENTEROLOBIUM GOOS 1970. MYCOLOGIA 62:171-175 (ILLUS) /M. PLATYSPORUM
   DICTYO,CETERI,PHAEO

MYCOGONE LINK 1809. MAG. GES. NATURF. FREUNDE, BERLIN 3:18 /M. ROSEA
   DIDYMO,CETERI,HYALO, ASCO STATE =HYPOMYCES FIDE BARRON 1968 /BARRON,
   HYPHOMYCETES P. 232-233,1968. (ILLUS)
      SYNO    =COCCOSPORELLA KARSTEN 1893 FIDE HUGHES 1958

MYCOMYCES WYSS-CHODAT 1928. BULL. SOC. BOT. GENEVE 19(1927):209-210 (ILLUS) /M.
   FUNGOIDES
 ?, ORIGINAL ILLUSTRATION DIFFICULT TO INTERPRET, PERHAPS NOT A FUNGUS

MYCOVELLOSIELLA RANGEL 1917. ARCH. JARD. BOT. RIO JAN. 2:71 /M. CAJANI
   (HENNINGS)
   AMERO OR DIDYMO OR PHRAGMO,BLASTO,PHAEO, NOT VERY DIFFERENT FROM CLADOSPORIUM
   /ELLIS, DEM. HYPH. P. 303-305,1971. (ILLUS)
      SYNO    =VELLOSIELLA RANGEL 1915 NON VELLOZIELLA BAILL. 1886
              =CERCODEUTEROSPORA CURZI 1932 FIDE CASTELLANI AND CIFERRI

MYDONOSPORIUM CORDA 1833. IN STURM DEUTSCHL. FLORA III (PILZE), P. 95, TAB.48
   /M. OLIVACEUM
 =CLADOSPORIUM LINK 1815 FIDE LINDAU IN RABENHORST 1907

MYDONOTRICHUM CORDA 1833. IN STURM DEUTSCHL. FLORA III (PILZE), P. 37, TAB. 19
   /M. ATRUM
 ?, SEE LINDAU IN RABENHORST, KRYPT.-FL. 2 AUFL., 1 BD., 9:49,1910

MYRIOCONIUM SYDOW 1912. ANNALES MYCOLOGICI 10:449 /M. SCIRPI SYDOW =M.
   SCIRPICOLA (FERD. ET WINGE) FERD. ET WINGE
   AMERO,GLOIO OR BASO,HYALO, ASCO STATE =SCLEROTINIA OR CIBORIA OR RUTSTROEMIA.
   MAY BE ASSOCIATED WITH BOTRYTIS OR MONILIA CONIDIAL STATES. /VON ARX,
   GENERA OF FUNGI IN CULTURE P. 180,1970. (REFS) BARNETT, IMPERFECT FUNGI NO.
   42,1960. (ILLUS) SUB BOTRYOPHIALOPHORA
      SYNO    =CRISTULARIELLA HOHNEL 1916 FIDE VON ARX 1970
              =BOTRYOPHIALOPHORA LINDER 1944 FIDE VON ARX 1970
              =RHACODIELLA PEYRONEL 1919 FIDE CARMICHAEL

MYRIOPHYSA FRIES 1849. SUMMA VEG. SCAND. 481 /M. ATRA
 ?=ATICHIA FIDE AINSWORTH 1971

MYROPYXIS CESATI 1851. FLORA 34:73 /M. CARICICOLA
 ?, PROBABLY A NOMEN DUBIUM

MYROTHECIELLA SPEGAZZINI 1911. AN. MUS. NAC. BUENOS AIRES 3(13):460-461 (ILLUS)
   /M. CATENULIGERA
 =METARRHIZIUM SOROKIN 1883 FIDE CARMICHAEL (FROM ILLUS)

MYROTHECIUM TODE EX FRIES 1821. FUNGI MECKLENBERGENSIS 1:25 /M. INUNDATUM TODE
   EX GRAY 1821
   AMERO,GLOIO,HYALO OR PHAEO, ASCO STATE =NECTRIA FIDE BARRON 1968 /ELLIS, DEM.
   HYPH. P. 552-556,1971. (ILLUS,KEY)
      SYNO    ?=PIROBASIDIUM HOHNEL 1902
              =KUTILAKESA SUBRAMANIAN 1956 FIDE NAG RAJ
              =EXOTRICHUM SYDOW 1914 FIDE HOHNEL FROM AINSWORTH 1971

MYSTROSPORIELLA MUNJAL ET KULSHRESTHA 1969. MYCOPATHOLOGIA 39:356 /M. LITSEAE
   DICTYO OR PHRAGMO,CETERI,PHAEO /ELLIS, DEM. HYPH. P. 301-302,1971. (ILLUS)

MYSTROSPORIUM CORDA 1837. ICONES FUNGORUM 1:12 /M. DUBIUM
=NOMEN DUBIUM SEE HUGHES 1958

MYXOCLADIUM CORDA 1837. ICONES FUNGORUM 1:12 /M. ARUNDINIS CORDA =CLADOSPORIUM
    HERBARUM (PERSOON) LINK 1815
=CLADOSPORIUM LINK 1815 FIDE HUGHES 1958

MYXODOCHIUM ARNAUD 1951. BULL. SOC. MYCOL. FRANCE 67:196 AND FIG. 4H & I /M.
    HYALINUM
    ?. HAS CLAMP CONNECTIONS ON HYPHAE

MYXOTRICHELLA (SACCARDO) SACCARDO 1892. SYLLOGE FUNGORUM 10:593 (NO ILLUS) /M.
    SPELAEA (SACC.) SACC. 1892
=STERILE. FOR EMPTY ASCOCARPS OF MYXOTRICHUM

NAKATAEA HARA 1939. THE DISEASES OF THE RICE PLANT, 2 ED., P. 185 /N. SIGMOIDEA
    (CAVARA) HARA = HELMINTHOSPORIUM SIGMOIDEUM CAVARA 1880
    PHRAGMO,CETERI,PHAEO. ASCO STATE = LEPTOSPHAERIA FIDE SHOEMAKER 1959 /ELLIS,
    DEM. HYPH. P. 219-221,1971. (ILLUS)
        SYNO     =VAKRABEEJA SUBRAMANIAN 1956 FIDE CARMICHAEL

NALANTHAMALA SUBRAMANIAN 1956. J. INDIAN BOT. SOC. 35:476 /N. MADREEYA
    AMERO,BASO,HYALO

NAPICLADIUM THUMEN 1875. MYCOTH. UNIV. NO 91, ANNO 1875 /N. SORAUERI THUMEN
    =SPILOCAEA POMI FRIES
    =SPILOCAEA FRIES 1825 FIDE HUGHES 1958

NASCIMENTOA CIFERRI ET BATISTA 1956. PUBLCOES INST. MICOL. RECIFE 44:4 /N.
    PSEUDOENDOGENA =S. GUAREICOLA
    =SPIROPES CIFERRI 1955 FIDE ELLIS 1968

NECATOR MASSEE 1898. KEW BULLETIN 1898:119 (NO ILLUS) /N. DECRETUS
    ?. FOR CONIDIAL STATE OF CORTICIUM SALMONICOLOR FIDE AINSWORTH 1961 /BROOKS
    AND SHARPLES, ANN. APPL. BIOL. 2:58 (FROM AINSWORTH 1961)

NECRAPHIDIUM CIFERRI 1951. MYCOPATHOLOGIA 6:25 /N. ZOOPHILUM (SACC. ET TROTTER)
    CIFERRI =RAMULARIA ZOOPHILA
    ?. NOT ILLUSTRATED BY CIFERRI

NEGERIELLA HENNINGS 1897. HEDWIGIA 36:244 (NO ILLUS) /N. CHILENSIS
    PHRAGMO,CETERI,PHAEO. MAY BE PODOSPORIUM /MORRIS, MYCOPATHOLOGIA
    33:181-184,1967. (ILLUS) HOHNEL, S. B. AKAD. WISS. WIEN, ABT. 1,
    120:401-402,1911 (EMEND, NO ILLUS)

NEMATOCTONUS DRECHSLER 1941. PHYTOPATHOLOGY 31:777-778 AND FIGS. ON P. 774 AND
    776 /TWO ORIGINAL SPECIES
    AMERO,CETERI,HYALO. HAS CLAMP CONNECTIONS /VON ARX, GENERA OF FUNGI IN
    CULTURE P. 163-164,1970. (ILLUS)

NEMATOGONIUM DESMAZIERES 1834. ANN. SCI. NAT. II,2:70 /N. AURANTIACUM DESM. =N.
    FERRUGINEUM (PERSOON) HUGHES 1958
    AMERO,BLASTO,HYALO /BARRON, HYPHOMYCETES P. 235-236,1968. (ILLUS)
        SYNO     =BOTRYOCLADIUM PREUSS 1851 FROM AINSWORTH 1971

NEMATOGRAPHIUM GOIDANICH 1935. ANNALI DE BOT. 21:46 (NO ILLUS) /FIVE ORIGINAL
    NEW COMBINATIONS
    ?. FOR GRAPHIUM WITH SIMPLE SYNNEMATA

NEMATOMYCES FAUREL ET SCHOTTER 1965. REV. MYCOL. 30:330 /N. COPROPHILA
    =GLIOMASTIX GUEGUEN 1905 FIDE VON ARX 1970

NEOMICHELIA PENZIG ET SACCARDO 1901. MALPIGHIA 15:246 /N. MELAXANTHA
    =PITHOMYCES BERKELEY ET BROOME 1873 FIDE CLEMENTS AND SHEAR 1931

NEOTRICHOPHYTON CASTELLANI ET CHALMERS 1919. MAN. TROP. MED. ED 3 P. 1001 /N.
    FLAVUM (BODIN) CAST. ET CHALM. 1919
    =TRICHOPHYTON MALMSTEN 1845

## 10. Hyphomycetes 391

NECTTIOSPORELLA HOHNEL EX GRANITI 1951. NUOVO G. BOT. ITAL. 58:148 /N. TRISETI
    GRANITI 1951
    AMERO,GLOIO,HYALO

NETA SHEARER ET CRANE 1971. MYCOLOGIA 63:239-242 (ILLUS) /N. PATUXENTICA
    DIDYMO,CETERI,HYALO

NIA MOORE ET MEYERS 1959. MYCOLOGIA 51:874 (871-876,ILLUS) /N. VIBRISSA
    AMERO OR STAURO,CETERI,HYALO

NIGROCUPULA SAWADA 1944. REP. GOVT. RES. INST. FORMOSA 87:91 AND PL. 5, FIGS.
    46-49 /N. FORMOSANA
    ?, PERHAPS ENDOCALYX

NIGROSPORA ZIMMERMAN 1902. ZENTRALBL. BAKT. ABT. 2,8:220 /N. PANICI
    AMERO,CETERI,PHAEO. ASCO STATE =KHUSKIA HUDSON (=SCIRRHIELLA SPEG. =APIOSPORA
    SACC.) FIDE VON ARX 1970 /HUDSON, TRANS BRIT MYCOL SOC 46:355-360,1963.
    (ILLUS,REFS)
        SYNO    =PHAEOCONIS CLEMENTS 1909 FIDE HUGHES 1958
                =BASISPORIUM MOLLIARD 1902 FIDE MASON 1933
                =DICHOTOMELLA SACCARDO 1914 FIDE HUGHES 1958

NODULISPORIUM PREUSS 1849. KLOTZSCHII HERB. VIV. MYCOL. NO 1272 /N. OCHRACEUM
    AMERO,CETERI,HYALO OR PHAEO, ASCO STATE =HYPOXYLON AND RELATED GENERA FIDE
    MARTIN 1967 /MARTIN, J. S. AFRICAN BOTANY 33:205-240,1967. (ILLUS)
    GREENHALGH AND CHESTERS, TRANS. BRIT. MYCOL. SOC. 51:57-82,1968. (ILLUS)
        SYNO    =GENICULISPORIUM CHESTERS ET GREENHALGH 1964 FIDE CARMICHAEL
                =ACROSTAPHYLUS ARNAUD 1954 FIDE VON ARX 1970
                =DEMATOPHORA HARTIG 1883 FIDE CARMICHAEL
                =PLEUROGRAPHIUM GOIDANICH 1935 FIDE AINSWORTH 1971 SUB DEMATOPHORA
                =RHINOTRICHELLA ARNAUD 1954 FIDE CARMICHAEL
                =(?)VRIKSHOPAMA RAO ET RAO 1964

NOMURAEA MAUBLANC 1903. BULL. SOC. MYCOL. FR. 19:295-296 AND PL. 14, FIG. 8 /N.
    PRASINA MAUBL. 1903
    AMERO,BLASTO,HYALO

NOTHOSPORA PEYRONEL 1913 =DUBIUS FIDE CLEMENTS AND SHEAR 1931

NYCTALINA ARNAUD 1952. BULL. SOC. MYCOL. FRANCE 68:189-190 /N. LIGNICOLA
    ?, HAS CLAMP CONNECTIONS /ARNAUD, BULL. SOC. MYCOL. FRANCE 67:197, FIG.
    10,1951. (ILLUS) SUB ?

OEDEMIUM LINK 1824. LINN. SPEC. PLANT. IV,6(1):42 /O. ATRUM LINK = O. STATE OF
    THAXTERIA FUSCA (FUCKEL) BOOTH
    DIDYMO,BLASTO,PHAEO, ASCO STATE =THAXTERIA FIDE ELLIS 1971 OR
    CHAETOSPHAERELLA FIDE MULLER IN KENDRICK 1971 /HUGHES AND HENNEBERT, CAN.
    J. BOTANY 41:773-809,1963. (ILLUS,REVIS)
        SYNO    =DIPLOSPORIUM LINK 1824 FIDE HUGHES 1958
                =CLADOTRICHUM CORDA 1831/32 FIDE HUGHES 1958
                =GONGYLOCLADIUM WALLROTH 1833 FIDE HUGHES 1958
                =DIMERA FRIES 1825 FIDE HUGHES 1958
                ?=SPHINCTROSPORIUM KUNZE EX FRIES. SEE AINSWORTH 1971

OEDOCEPHALUM PREUSS 1851. LINNAEA 24:131 /O. GLOMERULOSUM (BULLIARD) SACCARDO
    (LECTOTYPE)
    AMERO,CETERI,HYALO, ASCO STATE =PEZIZA OR PYRONEMA, BASIDIO STATE =PENIOPHORA
    OR HETEROBASIDIUM OR FOMES FIDE BARRON 1968 AND VON ARX 1970 /VON ARX,
    GENERA OF FUNGI IN CULTURE P. 208-209,1970. (ILLUS,REFS)

OEDOTHEA SYDOW 1930. ANN. MYCOL. 28:202 /O. VISMIAE
    DIDYMO,CETERI,PHAEO /ELLIS, DEM. HYPH. P. 143-144,1971. (ILLUS)

OIDIODENDRON ROBAK 1932. NYT. MAG. NATURVIDENSK. 71:251 /O. FUSCUM = O.
    TENUISSIMUM (PECK) HUGHES 1958 (LECTOTYPE)
    AMERO,ARTHRO,HYALO OR PHAEO, ASCO STATE =ARACHNIOTUS FIDE BARRON AND BOOTH
    1966 /BARRON, CAN. J. BOTANY 40:589-607,1962. (ILLUS)

OIDIOPSIS SCALIA 1902. AGRICOLT. CALABRO-SICULO 27:393-397 (NO ILLUS) /O. SICULA
    ?, ASCO STATE =LEVEILLULA FIDE AINSWORTH 1961 /MYCOL. PAPERS CMI 40:14,
    TRANS. BRIT. MYCOL. SOC. 30:114 AND 32: 289

OIDIUM LINK 1809. MAG. GES. NATURF. FREUNDE, BERLIN 3:18 /O. AUREUM (PERSOON)
    LINK =SPOROTRICHUM AUREUM
    =SPOROTRICHUM LINK FIDE HUGHES 1958, OIDIUM AUCT. NON LINK USUALLY WAS USED
    FOR ACROSPORIUM STATES OF ERYSIPHE SPECIES. OIDIUM SENSU LINDER 1942 WAS
    USED FOR ACLADIUM, ALYSIDIUM AND OLPITRICHUM.

OLPITRICHUM ATKINSON 1849. BOT. GAZ. 19:244 /O. CARPOPHILUM ATKINSON =O.
    MACROSPORUM (FARLOW) SUMSTINE 1937
    AMERO,CETERI,HYALO /BARRON, HYPHOMYCETES P. 242-243,1968. (ILLUS)
        SYNO    =OIDIUM LINDER 1942 PRO PARTE, NON OIDIUM LINK 1809

OMMATOSPORA BATISTA ET CAVALCANTI 1964. RIV. PATOL. VEG. PAVIA SER. 3, 4:565
    /O. RORAIMENSIS
    =MICROCLAVA STEVENS 1917 FIDE KENDRICK

OMMATOSPORELLA BATISTA, BEZERRA ET POROCA 1967. ATAS INST. MICOL. UNIV. RECIFE
    5:424 AND FIG. 1 /O. BURSERACEARUM BATISTA ET AL. 1967
    ?, PROBABLY MICROCLAVA

ONCOCLADIUM WALLROTH 1833. FLOR. CRYPT. GERMAN. P. 289 (NO ILLUS) /O. FLAVUM
    WALLROTH
    AMERO,ARTHRO,HYALO /HUGHES, CAN. J. BOTANY 46:941-942 AND PL. II,1968.
    (ILLUS) ORR AND KUEHN, MYCOLOGIA 63:198-199,1971.(REVIS)
        SYNO    =ACTINODENDRON ORR AND KUEHN 1963 PRO PARTE

ONCOPODIELLA ARNAUD EX RIFAI 1965. PERSOONIA 3:407-411 (ILLUS) /O. TETRAEDRICA
    ARNAUD 1954 =O. TRIGONELLA (SACCARDO) RIFAI 1965
    STAURO,CETERI,PHAEO /ARNAUD, BULL. SOC. MYCOL. FR. 69:296 AND FIG. 130 AND
    R,1954. (ILLUS)

ONCOPODIUM SACCARDO 1904. ANN. MYCOL., BERLIN 2:19 /O. ANTONIAE SACCARDO ET
    SACCARDO
    DICTYO,CETERI,PHAEO /ELLIS, MYCOL PAPERS CMI 87:17-20,1963. (ILLUS)

OOSPORA WALLROTH 1833. FL. CRYPT. GERM. 2:182 /?, MANY ORIGINAL SPECIES
    INCLUDING THE TYPES OF FOUR OTHER GENERA
    =NOMEN ILLEGITIMUM FIDE HUGHES 1958, HAS BEEN USED FOR SPECIES OF ACROSPORIUM
    AND GEOTRICHUM AMONG OTHERS

OOSPOROIDEA SUMSTINE 1913. MYCOLOGIA 5:53 /O. LACTIS (FRESENIUS) =GEOTRICHUM
    CANDIDUM LINK
    =GEOTRICHUM LINK 1809 FIDE CARMICHAEL 1957

OPHIOCLADIUM CAVARA 1893. SEITS. PFLANZENKR. 3:26 /L. HORDEI
    =OVULARIA SACCARDO 1880 FIDE SPRAGUE FROM AINSWORTH 1971

OPHIODENDRON ARNAUD 1952. BULL. SOC. MYCOL. FR. 69:208 AND 211 (ILLUS) /O.
    LAOCOONI
    =NOMEN DUBIUM FIDE HENNEBERT 1968. SEE ALSO SPIROSPHAERA

OPHIOPODIUM ARNAUD 1954. BULL. SOC. MYCOL. FR. 69:300 /O. HUMIRIAE
    =GRALLOMYCES STEVENS 1918 FIDE DEIGHTON AND PIROZYNSKI 1966

OPHIOTRICHUM KUNZE APUD FRIES 1849. SUMMA VEG. SCAND. P. 503 (NO ILLUS) /O.
    PHLOMIDIS FRIES 1849
    =NOMEN DUBIUM, NO TYPE SPECIMEN FIDE NANNFELDT 1960 (IN LIT.)

ORBIMYCES LINDER 1944. FARLOWIA 1:404 AND PL. I, FIG. 5-6 /O. SPECTABILIS
    STAURO,CETERI,PHAEO /BARNETT, IMPERFECT FUNGI NO. 272,1960. (ILLUS)

OREOPHYLLA CIFERRI 1954. SYDOWIA 8:253 /O. ANGELAEMARIAE
    ?, ORIGINAL ILLUSTRATION NOT DIAGNOSTIC

ORMATHODIUM SYDOW 1928. ANN. MYCOL. 26:138-139 (NO ILLUS) /O. STYRACIS
    PHRAGMO OR SCOLECO,BLASTO,PHAEO
        SYNO    ?=ELLISIA BATISTA ET PERES 1965

## 10. Hyphomycetes

OSTEOMORPHA ARNAUD 1952. BULL. SOC. MYCOL. FR. 68:192 /O. FRAGILIS
 ?, GEOTRICHUM WITH CLAMP CONNECTIONS /ARNAUD, BULL SOC. MYCOL. FR. 67:197 AND FIG. 1 P.1951. (ILLUS)

OSTRACODERMA FRIES 1825. SYSTEMA MYCOL. 3:214 /O. PULVINATUM
 AMERO,CETERI,HYALO, ASCO STATE =PEZIZA FIDE KORF 1960 /BARRON, HYPHOMYCETES P. 243-244,1958. (ILLUS,REFS)
   SYNO  =CHROMELOSPORIUM CORDA 1833 FIDE HUGHES 1958
         ?=PHYMATOTRICHUM BONORDEN 1851

OVEREEMIA ARNAUD 1954. BULL. SOC. MYCOL. FR. 69:300 AND FIG. 3N AND O /O. GLOMERELAE (OVEREEM)
 =STERILE MYCELIUM (OF BROOKSIA TROPICALIS) /DEIGHTON AND PIROZYNSKI, MYCOL. PAPERS CMI 105:3,1966

OVULARIA SACCARDO 1880. MICHELIA 2:17 /O. OBOVATA (FUCKEL) SACCARDO =O. OBLIQUA (COOKE) SACCARDO
 AMERO,CETERI,HYALO, ASCO STATE =MYCOSPHAERELLA FIDE VON ARX 1970 /VON ARX, GENERA OF FUNGI IN CULTURE P. 202-203,1970. (ILLUS)
   SYNO  =OPHIOCLADIUM CAVARA 1893 FIDE SPRAGUE FROM AINSWORTH 1971
         =PSEUDOVULARIA SPEGAZZINI 1911 FIDE CLEMENTS AND SHEAR 1931

OVULARIOPSIS PATOUILLARD ET HARIOT 1900. JOUR. DE BOTANIQUE 14:245 /O. ERYSIPHOIDES PAT. ET HARIOT 1900
 AMERO,BASO,HYALO, ASCO STATE =PHYLLACTINIA FIDE LINDAU 1910. OVULARIOPSIS IS SCARCELY DISTINCT FROM ACROSPORIUM NEES

OXYSPORIUM LEVILLE ='HELMINTHOSPORIUM' FIDE SACCARDO FROM AINSWORTH 1971

OZONIUM LINK 1809. MAG. GES. NATURF. FREUNDE, BERLIN 3:21 /O. AURICOMUM
 =NOMEN DUBIUM, FOR STERILE MYCELIUM, HUGHES 1958 DID NOT FIND TYPE

PAATHRAMAYA SUBRAMANIAN 1956. J. INDIAN BOT. SOC. 35:70 /P. SUNDARA
 AMERO,CETERI,PHAEO /ELLIS, DEM. HYPH. P. 321-322,1971. (ILLUS)

PACHNOCYBE BERKELEY 1836. SMITHS ENGLISH FLORA 5(2):333 /P. FERRUGINEA (SOWERBY)
 AMERO,CETERI,HYALO /ELLIS, DEM. HYPH. P. 184-185,1971. (ILLUS)

PACHYBASIDIELLA BUBAK ET SYDOW 1915. ANN. MYCOL. 13:9 /P. POLYSPORA BUBAK ET SYDOW =K. POLYSPORA (BUBAK ET SYDOW) KARAKULIN
 =KABATIELLA BUBAK 1907 FIDE VON ARX 1970

PACHYBASIUM SACCARDO 1885. REV. MYCOLOGIQUE 7:160 /P. HAMATUM (BONORDEN) SACCARDO = TRICHODERMA SPORULOSUM (LINK) HUGHES
 =TRICHODERMA PERSOON 1801 FIDE HUGHES 1958

PACHYMA FRIES =SCLEROTIA OF PORIA AND LENTINUS FIDE AINSWORTH 1971

PACHYTRICHUM SYDOW 1925. ANN. MYCOL. 23:420 (ILLUS) /P. GAUZUMAE
 =PERICONIA PERSOON 1801 FIDE LINDER FROM AINSWORTH 1971

PACTILIA FRIES 1835. FL. SCAND. P. 365 (NO ILLUS) /P. MYCOPHILA MONTAGNE ET FRIES 1849
 =ACHITONIUM KUNZE 1819 FIDE FRIES FROM SACCARDO SYLLOGE FUNGORUM 4:673,1886

PAECILOMYCES BAINIER 1907. BULL. SOC. MYCOL. FR. 23:26 /P. VARIOTI
 AMERO,BASO,HYALO OR PHAEO, ASCO STATE =BYSSOCHLAMYS /BARRON, HYPHOMYCETES P. 244-246,1968. (ILLUS) BROWN AND SMITH, TRANS. BRIT. MYCOL. SOC. 40:17-89,1957. (MONOGR)
   SYNO  =SPICARIA AUCT. NON HARTING
         =MARIANNAEA ARNAUD 1952 FIDE BROWN AND SMITH 1957
         ?=GRAPHIDIUM LINDAU 1910
         ?=PSEUDOFUSIDIUM DEIGHTON 1969
         ?=SPICARIOPSIS HEIM 1939

PAEPALOPSIS KUHN, SEE PAIPALOPSIS, ORTHOGRAPHIC VARIANT

PAGIDOSPORA DRECHSLER 1960. SYDOWIA 14:246-247 AND PL. 34 AND 35 /P. AMOEBOPHILA
 AMERO,CETERI,HYALO, HAS CLAMP CONNECTIONS

PAIPALOPSIS KUHN 1883. HEDWIGIA 22:11 AND 28-31 (NO ILLUS) /P. IRMISCHIAE
?. BASIDIO STATE =UROCYSTIS FIDE KUHN IN LINDAU 1907

PALMOMYCES MAIRE 1926, SEE ANDREAEANA

PANCHANANIA SUBRAMANIAN ET NAIR 1966. ANTONIE VAN LEEUWENHOEK 32:381-383
    (ILLUS) /P. JAIPURENSIS SUBR. ET NAIR 1966
    AMERO OP DIDYMO,CETERI,PHAEO

PANTOSPORA CIFERRI 1938. ANN. MYCOL. 36:240-242 /P. GUAZUMAE
?. NOT ILLUSTRATED BY AUTHOR

PAPULARIA FRIES 1825. SYST. ORB. VEG. P. 195 /P. FAGI
    =ARTHRINIUM KUNZE 1817 FIDE ELLIS 1965

PAPULASPORA PREUSS 1851. LINNAEA 24:112 /P. SEPEDONIOIDES
    DICTYO,CETERI,HYALO OR PHAEO. GENUS IS OFTEN REFERRED TO MYCELIA STERILIA
    /BARRON, HYPHOMYCETES  P. 246-247,1968. (ILLUS)
        SYNO    =BARYEIDAMIA KARSTEN 1888 FIDE SACCARDO
                ?=HELICOSPORANGIUM KARSTEN 1865

PARACOCCIDIOIDES ALMEIDA 1930. C. R. SOC. BIOL. PARIS 105:315 /P. BRASILIENSIS
    (SPLENDORE)
    AMERO,CETERI,HYALO. CAUSE OF SOUTH AMERICAN BLASTOMYCOSIS.  CONIDIAL STATE
    NOT DIFFERENT FROM CHRYSOSPORIUM, SEE BORELLI, CONGR. VENEZ. DE CIENCEAS
    MED. IV:2241-2253,1955 FOR ILLUS. OF CONIDIAL STATE.
        SYNO    =LUTZIOMYCES FONSECA 1939 SEE DODGE 1935, P.179

PARASPORA GROVE 1884. JOUR. BOT. 22:196 AND TAB. 246, FIG. 9 /P. SEPTATA
?. ORIGINAL ILLUSTRATION DIFFICULT TO INTERPRET

PARATRICHAEGUM FAUREL ET SCHOTTER 1966. REVU MYCOL. 30:348-349 AND FIG. 9
    (DATED 1965) /P. STERCORARIUM
    =EPICOCCUM LINK 1815 FIDE KENDRICK

PASPALOMYCES LINDER 1933. MYCOLOGIA 25:345 AND PLATE 42, FIGS. 4-6 /P. AUREUS
    DIDYMO,BLASTO,PHAEO
        SYNO    ?=SCOLECOBASIDIELLA ELLIS 1971

PASSALOPA FRIES 1849. SUMMA VEG. SCAND. 2:500 /P. BACILLIGERA (MONTAGNE ET
    FRIES) MONT. ET FRIES 1856
    DIDYMO OR PHRAGMO,CETERI,PHAEO. ASCO STATE =MYCOSPHAERELLA FIDE VON ARX 1970
    /HUGHES, CAN. J. BOTANY 31:570-572,1953. (ILLUS)

PATOUILLARDIA ROUMEGUERE 1885. REV. MYK. 7:177 /P. LICHENOIDES
    =DENDRODOCHIUM FIDE HOHNEL FROM CLEMENTS AND SHEAR 1931

PATOUILLARDIELLA SPEGAZZINI 1889. BOL. ACAD. CIENC. CORDOBA 11:620 /P.
    GUARANITICA
?. NOT ILLUSTRATED BY AUTHOR

PEDILOSPORA HOHNEL 1902. S.B. AKAD. WISS. WIEN 111:1047 (NO ILLUS) /P.
    PARASITANS
    =DICRANIDION HARKNESS 1885 FIDE PEEK AND SOLHEIM 1958

PEGLIONIA GOIDANICH 1934. MALPIGHIA 33:4 /P. VERTICICLADA
    =GYROTHRIX (CORDA) CORDA 1842 FIDE HUGHES AND PIROZYNSKI 1971

PENDULISPORA ELLIS 1961. MYCOL. PAPERS CMI 82:41-43 (ILLUS) /P. VENEZUELANICA
    DICTYO,CETERI,PHAEO

PENICILLIFER VAN EMDEN 1968. ACTA BOT. NEERL. 17:54 /P. PULCHER
    DIDYMO OR PHRAGMO,BASO,HYALO /VON ARX, GENERA OF FUNGI IN CULTURE P.187,1970.
    (ILLUS)

PENICILLIUM LINK 1809. MAG. GES. NATURF. FREUNDE BERLIN 3:16 /P. EXPANSUM
    AMERO,BASO,HYALO OR PHAEO. ASCO STATE =EUPENICILLIUM OR TALAROMYCES FIDE VON
    ARX 1970 /RAPER AND THOM, A MANUAL OF THE PENICILLIA, 1949. (MONOGR,ILLUS)
        SYNO    ?=FLOCCARIA GREVILLE 1828
                =PRITZELIELLA HENNINGS 1903 FIDE CLEMENTS AND SHEAR 1931

## 10. Hyphomycetes

?=MONILIGER LETEL FIDE AINSWORTH 1971
=RHODOCEPHALUS CORDA 1837 FIDE LINDAU 1907

PENOMYCES GIARD 1891. COMP. REND. 112:1518-1520 /P. TELARIUS GIARD
   ?, NOT ILLUSTRATED BY AUTHOR

PENTAPOSPORIUM BATISTA 1957. REV. BIOL. LISBOA 1:106 AND FIGS. 4-5 /P.
   FOURCROYAE
   STAURO,CETERI,PHAEO

PERIBOTRYON FRIES 1832. SYSTEMA MYCOL. 3:287 (NO ILLUS) /P. PAVONI
   =LICHEN FIDE AINSWORTH 1971

PERICONIA TODE EX PERSOON 1801. SYN. METH. FUNGORUM P. 686 /P. BYSSOIDES
   PERSOON 1801
   AMERO,BLASTO,PHAEO, ASCO STATE =DIDYMOSPHAERIA FIDE ELLIS 1971 /MASON AND
      ELLIS, MYCOL PAPERS CMI 56:1-127,1953. (MONOGR,ILLUS) ELLIS, DEM. HYPH. P.
      344-353,1971. (ILLUS,KEY)
      SYNO    =SPOROCYBE FRIES 1825 FIDE HUGHES 1958
              =SPORODUM CORDA 1837 FIDE HUGHES 1958
              =TRICHOCEPHALUM COSTANTIN 1887 FIDE ELLIS 1971
              =HARPOCEPHALUM ATKINSON 1897 FIDE ELLIS 1971
              =BERKLEYNA KUNZE 1898 FIDE ELLIS 1971
              =PACHYTRICHUM SYDOW 1925 FIDE LINDER FROM AINSWORTH 1971
              ?=SUBRAMANIA RAO ET RAO 1964

PERICONIELLA SACCARDO 1884. MISC. MYCOL. 2:17 /P. VELUTINA (WINTER)
   AMERO OR DIDYMO OR PHRAGMO,CETERI OR BLASTO,PHAEO /BARRON, HYPHOMYCETES, P.
      250,1968. (ILLUS) ELLIS, MYCOL. PAPERS CMI 111:2-35,1967. (ILLUS,KEY)
      SYNO    =ACRODESMIS SYDOW 1926 FIDE ELLIS 1967
              =RAMICHLORIDIUM STAHEL 1937 FIDE ELLIS 1967

PERIDIOMYCES KARSTEN ?=HYPHOMYCETE, FROM AINSWORTH 1971

PERIOLA FRIES 1822. SYSTEMA MYCOL. 2:266-267 (NO ILLUS) /P. TOMENTOSA
   (LECTOTYPE FIDE AINSWORTH 1971)
   ?, SEE ALSO GLIOCLADOCHIUM

PERIOLOPSIS MAIRE 1913. ANN. MYCOL. 11:357 /P. HELICOCHAETA
   =SARCOPODIUM EHRENBERG 1818 FIDE PIROZYNSKI

PESTALOTIA DE NOTARIS 1839. MEM. R. ACCAD. SCI. TORINO 2,3:80 /P. PEZIZOIDES
   PHRAGMO,GLOIO,PHAEO, (MELANCONIALES) ASCO STATE =BROOMELLA SACCARDO FIDE VON
      ARX 1970 /VON ARX, GENERA OF FUNGI IN CULTURE P. 156-159,1970. (ILLUS,REFS)
      SYNO    =PESTALOTIOPSIS STEYAERT 1949 FIDE VON ARX 1970
              =TRUNCATELLA STEYAERT 1949 FIDE VON ARX 1970

PESTALOTIOPSIS STEYAERT 1949. BULL. JARD. BOT. ETAT, BRUXELLES 19:300 /P.
   GUEPINI (DESMAZIERES)
   =PESTALOTIA DE NOTARIS 1839 FIDE VON ARX 1970

PESTALOZZIELLA SACCARDO ET ELLIS 1882. MICHELIA 2:575 /P. SUBSESSILIS
   AMERO,CETERI,HYALO, (MELANCONIALES) /NAG RAJ AND KENDRICK, CAN. J. BOTANY 50:
      ,1972

PESTALOZZINA (SACCARDO) SACCARDO 1894. IN SORAUER, Z. PFLKRANKH. 4:213-215 /?
   ?, SEE HUGHES, MYCOL. PAPERS CMI 36:2-3,1951

PETRAKIA SYDOW 1913. ANN. MYCOL. 11:406-407 (ILLUS) /P. ECHINATA (PEGLION)
   STAURO,CETERI,PHAEO /VON ARX, GENERA OF FUNGI IN CULTURE P. 159-160,1970.
      (ILLUS) VAN DER AA, ACTA BOT. NEERL. 17:221-225,1968. (ILLUS)
      SYNO    =ECHINOSPORIUM WORONICHIN 1913 FIDE CARMICHAEL

PETRAKIOPSIS SUBRAMANIAN ET REDDY 1968. SYDOWIA 20:340 (DATED 1966) /P. ELEGANS
   SCOLECO OR STAURO,CETERI,HYALO. CLOSE TO OR THE SAME AS TETRACRIUM

PEYLIA OPIZ =BOTRYOSPORIUM CORDA 1831 FIDE AINSWORTH 1971

PEYRONELIA FRAGOSO ET CIFERRI 1927. BOL. REAL SOC. ESPAN. HIST. NAT. 27:333-334

(ILLUS) /P. SIRODESMIOIDES CIF. ET FRAG. 1927
DICTYO,BLASTO,PHAEO /HUGHES, CAN. J. BOTANY 36:795,1958. (REFS)

PEYRONELINA ARNAUD 1952. BULL. SOC. MYCOL. FR. 68:213 AND 215 (ILLUS) /P.
    GLOMERULATA
    DICTYO,CETERI,HYALO

FEZIOTRICHUM (SACCARDO) LINDAU 1900. NAT. PFLANZENF. 1,1:467 /P. LACHNELLA
    =NOMEN DUBIUM SEE DOWNING, MYCOLOGIA 45:938,1953

PHACELLULA SYDOW 1927. ANN. MYCOL. 25:139-140 /P. GOUANIAE
    ?, NOT ILLUSTRATED BY AUTHOR

PHAEOCONIS CLEMENTS 1909. GENERA OF FUNGI P. 148 /P. PANICI (ZIMMERMANN) CLEM.
    =NIGROSPORA ZIMMERMANN 1902 FIDE CLEMENTS AND SHEAR 1931

PHAEODACTYLIUM AGNIHOTHRUDU 1968. PROC. INDIAN ACAD. SCI. B 68:206 /P.
    VENKATESANUM AGNIHOT. 1968
    PHRAGMO,CETERI,HYALO /ELLIS, DEM. HYPH. P. 214-216,1971. (ILLUS)

PHAEODOCHIUM FARR 1968. NOVA HEDWIGIA 15:268-269 AND FIGS. 52,70,87 (IN FARR
    AND HORNER) /P. MYROTHECIOIDES
    ?

PHAEOHARZIELLA LOUBIERE 1924. RECHERCHES SUR QUELQUE MUCEDINEES CASEICOLES,
    THESE, PARIS, P. 52 /P. HETEROSPORA
    =ARTHRINIUM KUNZE 1817 FIDE ELLIS 1965

PHAEOHYMENULA PETRAK 1954. SYDOWIA 8:77-79 /P. FUSISPORA
    ?, NOT ILLUSTRATED BY AUTHOR

PHAEOISARIA HOHNEL 1909. S. B. AKAD. WISS. WIEN 118:329 /P. BAMBUSAE = P.
    CLEMATIDIS (FUCKEL) HUGHES 1958
    AMERO,CETERI,HYALO /ELLIS, DEM. HYPH. P. 213-214,1971. (ILLUS)
        SYNO    =GRAPHIOPSIS BAINIER 1907 FIDE MORRIS 1963
                =HANSFORDIULA MORRIS 1963 FIDE MORRIS 1963
                ?=SPICULOSTILBELLA MORRIS 1963 FIDE CARMICHAEL
                ?=THAROOPAMA SUBRAMANIAN 1956

PHAEOISARIOPSIS FERRARIS 1909. ANN. MYCOL. 7:280 /P. GRISEOLA (SACCARDO)
    (LECTOTYPE FIDE ELLIS 1971)
    PHRAGMO,CETERI,PHAEO /ELLIS, DEM. HYPH. P. 268-269,1971. (ILLUS)

PHAEORAMULARIA MUNTANOLA 1960. LILLOA 30:182-183 AND 209-213 (ILLUS) /P.
    GOMPHRENICOLA (SPEGAZZINI) = CERCOSPORA GOMPHRENICOLA SPEGAZZINI 1882
    DIDYMO OR PHRAGMO,BLASTO,PHAEO, CLADOSPORIUM-LIKE

PHAEOSCOPULARIOPSIS OTA 1928. JAPANESE J. DERM. UROL. 28:405 /THREE ORIGINAL
    SPECIES
    =SCOPULARIOPSIS BAINIER 1907 FIDE MORTON AND SMITH 1963

PHAEOSTILBELLA HOHNEL 1925. MITT. BOT. TECHN. HOCHSCH. WIEN 2:72 /P. ATRA
    (DESMAZIERES) =SACCARDAEA ATRA (DESM.) MASON ET ELLIS
    =SACCARDAEA CAVARA 1894 FIDE HUGHES 1958

PHAEOTRICHOCONIS SUBRAMANIAN 1956. PROC. INDIAN ACAD. SCI., SECT. B,44:2 /P.
    CROTALARIAE (SALAM ET RAO)
    PHRAGMO,CETERI,PHAEO /BARRON, HYPHOMYCETES P. 251-254,1968. (ILLUS) SUB
    PHAEOTRICHONIS

PHANEROCORYNELLA HOHNEL 1919. BERL. DEUTSCHL. BOT. GES. 37:157 /P. FUNGORUM
    (FRIES)
    =EXOSPORIELLA KARSTEN 1892, BASED ON SAME TYPE

PHIALETEA BATISTA ET NASCIMENTO 1960. ATAS INST. MICOL. UNIV. RECIFE 1:264 /P.
    AEROSPORA
    =GRALLOMYCES STEVENS 1918 FIDE DEIGHTON AND PIROZYNSKI 1966

PHIALOCEPHALA KENDRICK 1961. CAN. J. BOTANY 39:1079 /P. DIMORPHOSPORA

## 10. Hyphomycetes

AMERO,GLOIO,HYALO, ASCO STATE = LILLIPUTIA FIDE CARMICHAEL OR
  CHAETOSPHAERELLA FIDE MULLER IN KENDRICK 1971 /KENDRICK, CAN. J. BOTANY
  41:1015-1023,1963. (ILLUS,KEY)

PHIALOCONIDIOPHORA MOORE ET ALMEIDA 1936. ANN. MISSOURI BOT. GARD. 23:547-550
  AND PL. 26 /TWO ORIGINAL SPECIES
  =FONSECAEA NEGRONI 1936 FIDE CARMICHAEL

PHIALOMYCES MISRA ET TALBOT 1964. CAN. J. BOTANY 42:1287 /P. MACROSPORUS
  AMERO,BASO,PHAEO /BARRON, HYPHOMYCETES P. 255,1968. (ILLUS)

PHIALOPHORA MEDLAR 1915. MYCOLOGIA 7:201-203 (ILLUS) /P. VERRUCOSA
  AMERO,GLOIO,HYALO, ASCO STATES =CONIOCHAETA, GAEUMANNOMYCES, PYRENOPEZIZA,
  MOLLISIA, CORYNE, ETC. FIDE VON ARX 1970 /SCHOL-SCHWARZ, PERSOONIA
  6:59-94,1970. (ILLUS,KEY,REVIS)
    SYNO    =MARGARINOMYCES LAXA 1930 FIDE VON ARX 1970
            =CADOPHORA LAGERB. ET MELIN 1928 FIDE HUGHES 1958
            =LECYTHOPHORA NANNFELDT 1934 FIDE GOIDANICH FROM AINSWORTH 1971

PHIALOPHOROPSIS BATRA 1968. MYCOLOGIA 59:1008 (DATED 1967), ILLUS. P. 995 /P.
  TRYPODENDRI
  AMERO,BASO,HYALO

PHIALOSTELE DEIGHTON 1969. MYCOL. PAPERS CMI 117:11-13 (ILLUS) /P. SCYTOPETALI
  AMERO,BASO,HYALO

PHIALOTUBUS ROY ET LEELAVATHY 1966. TRANS. BRIT. MYCOL. SOC. 49:495 /P.
  MICROSPORUS
  AMERO,BASO,PHAEO /BARRON, HYPHOMYCETES P.256-258,1968.(REFS)

PHLOEOCONIS FRIES 1849. SUMMA VEG. SCAND. 2:520 /P. VIOLACEA (CESATI)
  =STERILE
    SYNO    =PSEUDOPROTOMYCES GIBELLI FIDE SACCARDO FROM AINSWORTH 1971
            ?=RHIZOSPORIUM RABENHORST FIDE AINSWORTH 1971

PHRAGMOCEPHALA MASON ET HUGHES 1951. NATURALIST, LONDON 1951:97 /P. ATRA
  (BERKELEY ET BROOME)
  =ENDOPHRAGMIA DUVERNOY ET MAIRE 1920 FIDE ELLIS 1959

PHRAGMODOCHIUM HOHNEL 1924. BULL. JARD. BOT. BUITENZ. 3(6):6 (NO ILLUS) /P.
  MODESTUM
  =FUSARIUM LINK 1809 FIDE CLEMENTS AND SHEAR 1931

PHRAGMOGRAPHIUM MORRIS 1966. MYCOPATHOLOGIA 28:99-100 (ILLUS) /P. ULMI
  =NOMEN ILLEGITIMUM, NON PHRAGMOGRAPHIUM HENNINGS 1905, NOW REPLACED BY
  MORRISOGRAPHIUM MORELET 1968

PHRAGMOSPATHULA SUBRAMANIAN ET NAIR 1966. ANTONIE VAN LEEUWENHOEK 32:384 /P.
  PHOENICIS
  DIDYMO OR PHRAGMO,CETERI,PHAEO /ELLIS, DEM. HYPH. P. 162-163,1971. (ILLUS)

PHRAGMOSTACHYS COSTANTIN 1888. LES MUCEDINEES SIMPLES P. 97 /P. ELATA (SACCARDO)
  =STERIGMATOBOTRYS OUDEMANS 1886 FIDE HUGHES 1958

PHRAGMOSTILBE SUBRAMANIAN 1959. MYCOPATHOLOGIA 10:350-354 (ILLUS) /P. LINDERI
  =A. CANDIDUM (SCHWEINITZ) HUGHES 1958 FIDE CARMICHAEL
  =ARTHROSPORIUM SACCARDO 1880 FIDE CARMICHAEL

PHYMATOSTROMA CORDA 1837. ICONES FUNGORUM 1:5 /?
  =ACHITONIUM KUNZE FIDE LINDAU IN RABENHORST 1910 SUB PACTILIA

PHYMATOTRICHUM BONORDEN 1851. HANDB. ALLGEM. MYKOL. P. 116 AND TAF. 8, FIGS.
  138,181,197 /? ONE OF THE THREE ORIGINAL SPECIES APPEARS TO BE A
  BOTRYOSPORIUM, ONE A BOTRYTIS AND ONE AN OSTRACODERMA.
  ?, HAS BEEN USED MOSTLY FOR OSTRACODERMA SPP.

PHYSALIDIUM MOSCA 1965. ALLIONIA 11:78 /P. ELEGANS
  AMERO OR STAURO,CETERI,PHAEO /ELLIS, DEM. HYPH. P. 161-162,1971. (ILLUS)

FHYSOSPORA FRIES 1849. SUMMA VEG. SCAND. P.495 /P. RUBIGINOSA
?

PILULINA ARNAUD 1954. BULL. SOC. MYCOL. FR. 69:268 AND 269 (ILLUS) /P.
NIGRISPORA
AMERO,CETERI,PHAEO, NO LATIN DIAGNOSIS

PIMINA GROVE 1888. J. BOTANY. LONDON 26:206 /P. PARASITICA GROVE =Z. GIBBUM
(SACC., ROUSS. ET BOMM.) HUGHES 1958
=ZYGOSPORIUM MONTAGNE 1842 FIDE HUGHES 1958

PIMINELLA ARNAUD 1954. BULL. SOC. MYCOL. FR. 69:280 AND 281 (ILLUS) /P.
CASTANEAE
?, DOES NOT APPEAR TO DIFFER FROM CHLORIDIUM LINK 1809. SEE ALSO
SPHAEROMYCETELLA.

PINOYELLA CASTELLANI ET CHALMERS 1919. MAN. TROP. MED., ED. 3 P.1023 /P. SIMII
(PINOY)
=TRICHOPHYTON MALMSTEN FIDE STOCKDALE, MACKENZIE AND AUSTWICK 1965

PIONNOTES FRIES 1849. SUMMA VEG. SCAND. 481 /P. CAPITATA (SCHWEINITZ)
=FUSARIUM LINK FIDE WOLLENWEBER FROM AINSWORTH 1971

PIRICAUDA BUBAK 1914. ANN. MYCOL. 12:218 /P. ULEANA (SACC. ET SYDOW) BUBAK =P.
PARAGUAYENSE (SPEG.) MOORE 1959
DICTYO,CETERI,PHAEO /HUGHES, CAN. J. BOTANY 38:921-924,1960. (ILLUS,REVIS)

PIRICULARIA SACCARDO 1880. MICHELIA 2:20 /P. GRISEA (COOKE)
=PYRICULARIA SACCARDO 1880 (ORTHOGRAPHIC VARIANT)

PIROBASIDIUM HOHNEL 1902. S. B. AKAD WISS. WIEN 111:1001 /P. SARCOIDES (JACQ)
HOHNEL =P. STATE OF CORYNE SARCOIDES
?=MYROTHECIUM, ASCO STATE = CORYNE TULASNE /MORRIS, WESTERN ILLINOIS UNIV SER
BIOL SCI 3:94-95,1963. (ILLUS)

PIROSTOMELLA SACCARDO 1914. ANN. MYCOL. 12:308 /P. RAIMUNDI
?, NO ILLUSTRATION WITH ORIGINAL DIAGNOSIS

PITHOMYCES BERKELEY ET BROOME 1873. J. LINN. SOC. LONDON 14:100 /P. FLAVUS
BERK. ET BROOME 1873
PHRAGMO OR DICTYO,CETERI,PHAEO. COMPARE WITH MONODICTYS /ELLIS, MYCOL PAPERS
CMI 76:7-19,1960. (ILLUS,KEY)
    SYNO    =BIOCONIOSPORIUM BATISTA ET BEZERRA 1964 FIDE VON ARX 1970
            =SCHELEOBRACHEA HUGHES 1958 FIDE VON ARX 1970
            =NEOMICHELIA PENZIG ET SACCARDO 1901 FIDE CLEMENTS AND SHEAR 1931
            ?=POLYSCHEMA UPADHYAY 1966

PITHOSIRA PETRAK 1949. SYDOWIA 3:259-261 /P. SYDOWII
?, NOT ILLUSTRATED BY AUTHOR

PLACENTARIA AUERSWALD ET RABENHORST ?='PERIOLA' FIDE AINSWORTH 1971

PLECOTRICHUM CORDA 1833. STURMS DEUTSCHL. FL. III (PILZE) BD 3, HEFT 13:87 /P.
FUSCUM
=STERILE FIDE HUGHES 1958

PLECTOTHRIX SHEAR 1902. BULL. TORREY BOT. CLUB 29:457 /P. GLOBOSA
?, NOT ILLUSTRATED BY AUTHOR

PLEIOCHAETA (SACCARDO) HUGHES 1951. MYCOL. PAPERS CMI 36:32-39 (ILLUS) /P.
SETOSA (KIRCHNER) HUGHES 1951
PHRAGMO,CETERI,HYALO OR PHAEO

PLEUROCATENA ARNAUD 1952. BULL. SOC. MYCOL. FR. 68:193 AND 195 /P. ACICULARIS
AMERO,BASO,HYALO

PLEUROCOLLA PETRAK 1924. ANN. MYCOL. 22:15 /P. TILIAE
?, NO ILLUSTRATION WITH ORIGINAL DIAGNOSIS

## 10. Hyphomycetes

PLEUROGRAPHIUM GOIDANICH 1935. ANNALI DI BOT. 19(1):48 /P. DESMAZIERI (SACCARDO)
 =NODULISPORIUM FIDE AINSWORTH 1971 SUB DEMATOPHORA

PLEUROPHRAGMIUM COSTANTIN 1888. LES MUCEDINEES SIMPLES P. 100 /P. BICOLOR
 COSTANTIN =P. SIMPLEX (BERKELEY ET BROOME) HUGHES 1958
 PHRAGMO,CETERI,PHAEO /ELLIS, MYCOL PAPERS CMI 114:42-44,1968. (ILLUS)
  SYNO   ?=SCOLECOBASIDIUM ABBOTT 1927

PLEUROPYXIS CORDA 1837. ICONES FUNGORUM 1:23 /P. MICROSPERMA
 =NOMEN CONFUSUM FIDE HUGHES 1958

PLEUROTHECIUM HOHNEL 1919. BERL. DEUTSCHL. BOT. GES. 37:154 (NO ILLUS) /P.
 RECURVATUM (MORGAN) = ACROTHECIUM RECURVATUM MORG. 1895
 PHRAGMO,CETERI OR GLOIO,HYALO, COMPARE WITH CACUMISPORIUM /GOOS, MYCOLOGIA
 1048-1053,1969. (ILLUS)

PLOKAMIDOMYCES BATISTA, COSTA ET CIFERRI 1957. PUBL. INST. MICOL. RECIFE 90:15
 /P. COLENSOI BAT. COST. ET CIF. =P. STATE OF TRICHOPELTHECA ASIATICA BAT.
 COST ET CIF 1957 FIDE HUGHES 1965
 AMERO,GLOIO,HYALO, ASCO STATE =TRICHOPELTHECA FIDE HUGHES 1965 /HUGHES, NEW
 ZEALAND J. BOTANY 3:320-332,1965. (ILLUS,REVIS)

POCHONIA BATISTA ET FONSECA 1965. PUBL. INST. MICOL. RECIFE NO. 462 /P.
 HUMICOLA BATISTA ET FONSECA =D. HETEROSPORA KAMYSCHKO 1962
 =DIHETEROSPORA KAMYSCHKO 1962 FIDE BARRON AND ONIONS 1966

PODOBACTRIDIUM PETCH 1916. ANN. R. BOT. GARD. PERADENIYA 6:180 /P. CLAVATUM
 (BERKELEY ET BROOME) PETCH =B. CLAVATUM BERK. ET BR. 1873
 =BACTRIDIUM KUNZE 1817 FIDE HUGHES 1966

PODOCONIS BOEDIJN 1933. BULL. JARD. BOT. BUITENZ. 3,13:133 /P. THEAE (BERN.)
 =S. THEAE (BERN.) HUGHES
 =SPORIDESMIUM LINK 1809 FIDE ELLIS 1958

PODOSPORIELLA ELLIS ET EVERHART 1895. PROC. ACAD. NAT. SCI. PHILA. 1894:385 /P.
 HUMILIS ELLIS ET EVERHART = P. GLOMERATA (HARKNESS) BONAR 1965
 PHRAGMO,CETERI,PHAEO /ELLIS, DEM. HYPH. P. 291-292,1971. (ILLUS)

PODOSPORIUM SCHWEINITZ 1832. TRANS. AMER. PHIL. SOC. II,4:278 /P. RIGIDUM
 (LECTOTYPE)
 PHRAGMO,CETERI,PHAEO /ELLIS, DEM. HYPH. P. 378-380,1971. (ILLUS)
  SYNO   =HELMINTHOSPORIOPSIS SPEGAZZINI 1880 FIDE MORRIS 1963
         =(?)NEGERIELLA HENNINGS 1897 FIDE KENDRICK

POLLACCIA BALDACCI ET CIFERRI 1937. ATTI IST. BOT. UNIV. PAVIA IV,10:61 (NO
 ILLUS) /P. RADIOSA (LIBERT) BALD. ET CIFERRI 1937 = OIDIUM RADIOSUM LIBERT
 ?=SPILOCAEA, FOR CONIDIAL VENTURIA TREMULAE FIDE AINSWORTH 1971 /HUGHES, CAN.
 J. BOTANY 31:572-573,1953. (ILLUS)

POLYACTIS LINK 1809. MAG. GES. NATURF. FREUNDE, BERLIN 3:16 /P. VULGARIS
 =BOTRYTIS CINEREA PERSOON 1801
 =BOTRYTIS PERSOON 1801 FIDE HUGHES 1958

POLYCEPHALOMYCES KOBAYASI 1941. SCI. REP. TOKYO BUN. DAIG. 5:245-246 (ILLUS)
 /P. FORMOSUS
 ?,=STILBELLA LINDAU 1900

POLYCLADIUM INGOLD 1959. TRANS. BRIT. MYCOL. SOC. 42:114 (ILLUS) /P. EQUISETI
 STAURO,CETERI,HYALO

POLYDESMUS DURIEU ET MONTAGNE 1845. ANN. SCI. NAT. III,4:365 /P. ELEGANS DURIEU
 ET MONTAGNE IN MONTAGNE 1845 =S. DENSUM (SACC. ET ROUM.) MASON ET HUGHES IN
 HUGHES 1953
 =SPORIDESMIUM LINK 1809, SEE HUGHES 1958 AND ELLIS 1958

POLYMORPHOMYCES COUPIN 1914. REV. GEN. BOT. 26:245-248 AND PL. 5 /P. BONNIERI
 =GEOTRICHUM LINK 1809 FIDE CARMICHAEL

POLYPAECILUM SMITH 1961. TRANS. BRIT. MYCOL. SOC. 44:437 (ILLUS) /P. INSOLITUM
 AMERO,BASO,HYALO, ASCO STATE =DICHOTOMOMYCES OR THERMOASCUS FIDE VON ARX 1970

POLYRRHINA SOROKIN 1876. ANN. SCI. NAT. 6:4:65 (ILLUS) /P. MULTIFORMIS
=HARPOSPORIUM LOHDE 1874 FIDE 'MYCOLOGIA 1938' FROM AINSWORTH 1971

POLYSCHEMA UPADHYAY 1966. MYCOPATHOLOGIA 30:276-282 (ILLUS) /P. TERRICOLA
?,PROBABLY THE SAME AS PITHOMYCES BERK. ET BROOME 1873

POLYSCYTALINA ARNAUD 1954. BULL. SOC. MYCOL. FR. 69:283 AND 284 (ILLUS) /P.
   GRISEA
   DIDYMO,BLASTO,HYALO, NO LATIN DIAGNOSIS, SAME AS CURRENT CONCEPT OF
   POLYSCYTALUM

POLYSCYTALUM RIESS 1853. BOT. ZEIT. 11:138-139 (ILLUS) /P. FECUNDISSIMUM
   AMERO OR DIDYMO,BLASTO,HYALO, BUT TYPE NOT EXAMINED BY MODERN AUTHOR /BARRON,
   HYPHOMYCETES  P. 261-263,1968. (ILLUS)

POLYTHRINCIELLA BATISTA ET MAIA 1960. PUBL INST. MICOL. UNIV. RECIFE 283:20 /P.
   BOMBACISFOLIA
   ?, ORIGINAL ILLUSTRATION NOT DIAGNOSTIC

POLYTHRINCIOPSIS WALKER 1966. AUSTRALIAN J. BOTANY 14:195-200 AND PL. 1 /P.
   PHRAGMITIS
   =POLYTHRINCIUM KUNZE ET SCHMIDT 1817 FIDE KENDRICK

POLYTHRINCIUM KUNZE ET SCHMIDT 1817. MYK. HEFT 1:13 /P. TRIFOLII =P. STATE OF
   MYCOSPHAERELLA KILLANII PETRAK (=CYMADOTHEA TRIFOLII WOLF) FIDE VON ARX 1970
   DIDYMO,CETERI,PHAEO. ASCO STATE =MYCOSPHAERELLA FIDE VON ARX 1970 /HUGHES,
   MYCOL PAPERS CMI 49:19-20,1952. (ILLUS)
      SYNO    =POLYTHRINCIOPSIS WALKER 1966 FIDE KENDRICK

PRATHIGADA SUBRAMANIAN 1956. J. MADRAS UNIV. 26:366 /P. CRATAEVEA (SYDOW)
   PHRAGMO,CETERI,PHAEO /ELLIS, DEM. HYPH. P. 278-279,1971. (ILLUS)
      SYNO    =MACRAEA SUBRAMANIAN 1952, NON LINDL. 1830

PRATHODA SUBRAMANIAN 1956. J. INDIAN BOT. SOC. 35:73 /P. SAPARVA SUBRAMANIAN
   =A. SAPARVA (SUBR.) DEIGHTON 1969
   =ALTERNARIA NEES 1817 FIDE DEIGHTON 1969

PREUSSIASTER KUNTZE 1891. REV. GEN. PLANT. 2:850 /P. PAUCISEPTATUS (PREUSS)
   =CORDANA PREUSS 1851 FIDE HUGHES 1958

PRISMARIA PREUSS 1851. LINNAEA 24:125 AND FUNGI HOYERSW. NO. 86 /P. ALBA
   ?, THE ILLUSTRATION (IN STURMS DEUTSCHL. FL., VI, 35-6:TAB. 55) IS DIFFICULT
   TO INTERPRET /LINDAU, RABENHORST KRYPT. FL. 2 AUFL. 1 BD. 8:537-539,1907.
   (ILLUS)

PRITZELIELLA HENNINGS 1903. HEDWIGIA 43:88 /P. CAERULEA
   =PENICILLIUM LINK 1809 FIDE CLEMENTS AND SHEAR 1931 (AS COREMIUM LINK)

PROPHYTROMA SOROKIN 1877. HEDWIGIA 16:87-88 AND FIGS. 1-12 /P. TUBULARIS
   ?, ORIGINAL ILLUSTRATION DIFFICULT TO INTERPRET

PROTEOPHIALA CIFERRI 1957. SYDOWIA 11:284-289 /P. MATTIROLIANA
   ?, NOT ILLUSTRATED BY AUTHOR

PROTOCORONIS CLEMENTS AND SHEAR 1931 FOR PROTOCORONOSPORA

PROTOCORONOSPORA ATKINSON AND EDGERTON 1907. JOUR. MYCOL. 13:186 /P. NIGRICANS
   =AUREOBASIDIUM FIDE VON ARX 1970 SUB. KABATIELLA

PSEUDALLESCHERIA NEGRONI ET FISCHER 1944. REV. INST. BACT. BUENOS AIRES 12:201
   /P. SHEARI =ALLESCHERIA BOYDII FIDE CARMICHAEL
   =ALLESCHERIA SACCARDO ET SYDOW FIDE CARMICHAEL.  THIS ASCOMYCETOUS GENUS IS
   INCLUDED HERE BECAUSE AINSWORTH 1961 ASSIGNED IT TO THE MONILIALES.

PSEUDOBASIDIUM TENGWALL 1924. MEDED. PHYTOPATH. LAB. WILLIE COMMELIN SCHOLTEN
   6:38 /P. BICOLOR
   =ARTHRINIUM KUNZE 1817 FIDE ELLIS 1971

PSEUDOBELTRANIA HENNINGS 1902. HEDWIGIA 41:310 /P. CEDRELAE
   AMERO,CETERI,PHAEO /PIROZYNSKI, MYCOL PAPERS CMI 90:28-30,1963. (ILLUS)

PSEUDOBOTRYTIS KRZEMIENIEWSKA ET BADURA 1954. ACTA SOC. BOT. POLON. 23:727 /P.
    FUSCA KRZEM. ET BADURA =P. TERRESTRIS (TIMONIN) SUBRAMANIAN 1956
    AMERO OR DIDYMO,CETERI,PHAEO /BARRON, HYPHOMYCETES P. 263-264,1968. (ILLUS)
        SYNO    =UMBELLULA MORRIS 1955 FIDE SUBRAMANIAN 1956

PSEUDOCAMPTOUM FRAGOSO ET CIFERRI 1925. BOT. SOC. ESPANA. HIST. NAT. 25:454 /P.
    CITRI FRAGOSO ET CIFERRI =M. CITRI (FRAG. ET CIF.) ELLIS
    =MELANOGRAPHIUM SACCARDO 1913 FIDE ELLIS 1963

PSEUDOCERCOSPORA SPEGAZZINI 1911. AN. MUS. NAC. BUENOS AIRES 3:13:437 /P.
    SPORA-VITIS (LEVEILLE)
    =CERCOSPORA FRESENIUS 1863 FIDE HODGES AND HAASIS, MYCOLOGIA 54:452

PSEUDOCOCCIDIOIDES FONSECA 1928. BOL. INST. CLIN. QUIR., BUENOS AIRES 4:495 /P.
    MAZZAE FONSECA =COCCIDIOIDES IMMITIS
    =COCCIDIOIDES STILES 1896 FIDE DODGE 1935

PSEUDOCORDYCEPS HAUMAN 1936. BULL. SOC. BOT. BELG. 69:116-117 AND FIG. 5 /P.
    SEMINICOLA
    ?, ORIGINAL ILLUSTRATION DIFFICULT TO INTERPRET

PSEUDOEPICOCCUM ELLIS 1971. DEMATIACEOUS HYPHOMYCETES P. 270 /P. COCOS (STEVENS)
    AMERO,CETERI,PHAEO, CLOSE TO HADROTRICHUM

PSEUDOFUMAGO BRIOSI ET FARNETI 1906. ATT. INST. PAVIA 2:10:31 AND TAV.X, FIGS.
    18-24 /P. CITRI
    ?, A DARK YEAST-LIKE FUNGUS

PSEUDOFUSIDIUM DEIGHTON 1969. MYCOL. PAPERS CMI 118:26-28 (ILLUS) /P. HANSFORDII
    AMERO,BASO,HYALO, BUT PERHAPS NOT DISTINCT FROM PAECILOMYCES

PSEUDOGASTER HOHNEL 1907. DENK. AKAD. WIEN 83:38 /P. SINGULARIS
    ?, NOT ILLUSTRATED BY AUTHOR

PSEUDOGRAPHIELLA MORRIS 1966. MYCOPATHOLOGIA 28:97-98 (ILLUS) /P. VARIISEPTATA
    DIDYMO OR PHRAGMO,GLOIO,HYALO

PSEUDOHANSFORDIA ARNOLD 1970. Z. PILZK. 35:305-310 (DATED 1969) (ILLUS) /P.
    IRREGULARIS
    ?, NATURE OF CONIDIOGENOUS CELLS NOT CLEAR--POLYPHIALIDES?  SEE ALSO EURASINA

PSEUDOMICROCERA PETCH 1921. TRANS. BRIT. MYCOL. SOC. 7:164 /P. HENNINGSI
    =FUSARIUM FIDE WOLLENWEBER FROM CLEMENTS AND SHEAR 1931

PSEUDOPETRAKIA ELLIS 1971. MYCOL. PAPERS CMI 125:3-4 (ILLUS) /P. KAMBAKKAMENSIS
    (SUBRAMANIAN)
    STAURO,CETERI,PHAEO

PSEUDOPOLYSTIGMINA MURASH. 1928. TRANS. SIBER. INST. 9:235 /P. SPIRAEICOLA
    ?

PSEUDOPROTOMYCES GIBELLI =PHLOEOCONIS FIDE SACCARDO FROM AINSWORTH 1971

PSEUDOSPIROPES ELLIS 1971. DEMATIACEOUS HYPHOMYCETES P. 258-260 (ILLUS,KEY) /P.
    NODOSUS (WALLROTH)
    PHRAGMO,CETERI,PHAEO

PSEUDOSTEMPHYLIUM (WILTSHIRE) SUBRAMANIAN 1961. CURR. SCI. 30:423 /P.
    CONSORTIALE (THUMEN) SUBRAMANIAN 1961 =U. CONSORTIALE (THUM.) SIMMONS
    =ULOCLADIUM PREUSS 1851 FIDE SIMMONS 1967

PSEUDOTORULA SUBRAMANIAN 1958. J. INDIAN BOT. SOC. 37:57-61 (ILLUS) /P.
    HETEROSPORA
    =(?)DENDRYPHION WALLROTH 1833

PSEUDOVULARIA SPEGAZZINI 1911. ANN. MUS. NAC. BUENOS AIRES 3:13:418 /P. TRIFOLII
    =OVULARIA SACCARDO 1880 FIDE CLEMENTS AND SHEAR 1931

PSILOBOTRYS SACCARDO 1879. MICHELIA 1:538. /P. MINUTA
    =CHLORIDIUM LINK 1809 FIDE HUGHES 1958

PSILONIA FRIES 1825. SYST. ORB. VEG. 1:187 /P. GILVA
  ?=VOLUTELLA FIDE AINSWORTH 1971

PSILONIELLA COSTANTIN 1888. LES MUCEDINEES SIMPLES. P. 190-191 (ILLUS) /P.
  CUNEIFORMIS (RICHON) COST. 1888
  =CATENULARIA GROVE 1886 FIDE ELLIS 1971

PTEROCONIUM SACCARDO EX GROVE 1914. HEDWIGIA 55:146 /P. PTEROSPERMUM (COOKE ET
  MASEE) GROVE 1914
  AMERO,CETERI,PHAEO. ASCO STATE = APIOSPORA FIDE ELLIS 1971 /ELLIS, DEM. HYPH.
  P. 577-579,1971. (ILLUS,KEY)

PTERODINIA CHEVALLIER ?=BOTRYTIS FIDE AINSWORTH 1971

PTERULOPSIS WAKEFIELD ET HANSFORD 1943. PROC. LINN. SOC. LONDON 1942-3, P. 64
  AND FIG. 15, R. 63 /P. DUMMERI
  SCOLECO,CETERI,HYALO /MORRIS, WEST. ILLINOIS UNIV SER BIOL SCI
  3:104-105,1963. (ILLUS)

PUCCINIOPSIS SPEGAZZINI 1888. AN. SOC. ARG. 26:2:74 /P. GUARANITICA
  ?, NOT ILLUSTRATED BY AUTHOR /BARNETT, IMPERFECT FUNGI NO. 310,1960. (ILLUS)

PULLULARIA BERKHOUT 1923. DIE SCHIMMELGESCHLACTEN..., THESIS, UTRECHT, P. 54-55
  AND FIG. 32 /P. PULLULANS (DE BARY) BERKH. 1923
  =AUREOBASIDIUM VIALA ET ROYER 1891 FIDE COOKE 1962

PYCNOSTYSANUS LINDAU 1904. VERH. BOT. VER. BRANDENB. 45:160 /P. RESINAE
  =SOROCYBE RESINAE (FRIES) FRIES 1849
  =SOROCYBE FRIES 1849 (SEE HUGHES 1958 AND LINDAU 1910)

PYRAMIDOSPORA NILSSON 1962. SVENSK BOT. TIDSKR. 56:358-359 (ILLUS) /P.
  CASUARINAE
  STAURO,CETERI,HYALO

PYRENIOPSIS KUNTZE 1898. REVIS. GEN. PL. 3(3):508 /NEW NAME FOR TRICHODERMA
  =TRICHODERMA PERSOON 1801 FIDE RIFAI 1969

PYRENIUM PERSOON 1801. SYN. METH. FUNG. P. 236 /P. TERRESTRE
  =NOMEN DUBIUM SEE HUGHES 1958

PYRICULARIA SACCARDO 1880. MICHELIA 2:20 /P. GRISEA (COOKE)
  PHRAGMO OR DIDYMO,CETERI,HYALO OR PHAEO, ASCO STATE =MASSARINA SACCARDO FIDE
  VON ARX 1970 /BARRON, HYPHOMYCETES P. 265-266,1968. (ILLUS)
  SYNO    =PIRICULARIA SACCARDO 1880 (ORTHOGRAPHIC VARIANT)

PYRICULARIOPSIS ELLIS 1971. DEMATIACEOUS HYPHOMYCETES P. 206 /P. PARASITICA
  (SACCARDO ET BERLESE)
  PHRAGMO,CETERI,PHAEO

QUESTERIELLA ARNAUD 1954. BULL. SOC. MYCOL. FR. 69:284 AND 287, FIG. 8A /Q.
  CORNI
  ?, CONIDIAL STATE OF QUESTIERIA (=SCHIFFNERULA FIDE PETRAK FROM AINSWORTH
  1971)

RACEMOSPORIUM MOREAU ET MOREAU 1941. REV. MYCOLOGIE N.S. 6:80 /R. SATURNUS
  =ARTHRINIUM KUNZE 1817 FIDE ELLIS 1965

RACHISIA LINDNER 1913. DEUT. ESSIGIND. 17:467 (ILLUS) /R. SPIRALIS
  =FUSARIUM LINK FIDE CLEMENTS AND SHEAR 1931

RACODIUM PERSOON 1801. SYN. METH. FUNG. P. 701 /R. CELLARE (LECTOTYPE FIDE DE
  VRIES)
  =STERILE (LICHEN) FIDE ELLIS. (FOR RACODIUM AUCT. NON PERSOON, SEE ZASMIDIUM)
  SYNO    =RHACODIUM (ORTHOGRAPHIC VARIANT)

RADICISETA SAWADA ET KATSUKI 1959. PUBL. COLL. AGRIC. NAT. TAIWAN UNIV. 8:205
  /R. BLECHNI
  ?, NOT ILLUSTRATED BY AUTHORS

## 10. Hyphomycetes 403

RAFFAELEA VON ARX ET HENNEBERT 1965. MYCOPATHOLOGIA 25:309-312 (ILLUS) /R. AMBROSIAE
  AMERO,CETERI,HYALO /BATRA, MYCOLOGIA 59:976-1017,1967. (REVIS,KEY,ILLUS)

RAGNHILDIANA SOLHEIM 1931. MYCOLOGIA 23:402 /R. AGERATI (STEVENS) STEVENS ET SOLHEIM
  ?. A SEGREGATE FROM CERCOSPORA

RAMALIA BATISTA 1957. REV. BIOL. LISBOA 1:111 /R. VERONICAE
  ?. ORIGINAL ILLUSTRATION NOT DIAGNOSTIC

RAMICHLORIDIUM STAHEL 1937. TROP. AGRIC. TRIN. 14:44 /R. MUSAE
  =PERICONIELLA SACCARDO 1885 FIDE ELLIS 1967

RAMULARIA UNGER 1833. DIE EXANTH. DER PFL. P. 169 /R. DIDYMA (LECTOTYPE FIDE VON ARX 1970)
  DIDYMO,BLASTO,HYALO, BUT TYPE SPECIMEN IS NOT KNOWN. ASCO STATE = MYCOSPHAERELLA FIDE MULLER IN KENDRICK 1971 /VON ARX, GENERA OF FUNGI IN CULTURE P. 202-203,1970. (ILLUS)
    SYNO   =ACROTHECA FUCKEL 1860 FIDE HUGHES 1958
           =DIDYMARIA CORDA 1842 FIDE VON ARX 1970

RAMULARIOPSIS SPEGAZZINI 1910. AN. MUS. NAC. BUENOS AIRES 20:421 /R. CNIDOSCOLI
  ?. ORIGINAL ILLUSTRATION DIFFICULT TO INTERPRET

RAMULASPERA LINDROTH 1902. ACT. SOC. FENN. 22:5-6 /R. SALICINA (VEST.)
  ?. NOT ILLUSTRATED BY LINDROTH, VESTGREN'S ILLUS. NOT DIAGNOSTIC

RAMULISPORA MIURA 1920. S. MANCH. AGR. BULL. 11:43 /R. ANDROPOGONIS
  SCOLECO,GLOIO,HYALO

RANOJEVICIA BUBAK 1910. ANN. MYCOL. 8:400-401 (ILLUS) /R. VAGANS RANOJEVIC ET BUBAK
  ?. ORIGINAL ILLUSTRATION AND DESCRIPTION NOT DIAGNOSTIC

RECTICHARELLA SCHEER 1944. Z. PARASITENK. 13:275-282 (ILLUS) /R. ASELII
  ?. NO LATIN DIAGNOSIS, ORIGINAL ILLUSTRATION DIFFICULT TO INTERPRET

REDAELLIA CIFERRI 1930. ARCH. PROTISTENK. 71:424-428 (ILLUS) /R. ELEGANS
  ?. POSSIBLY AN OSMOPHILIC ASPERGILLUS. TYPE CULTURE IN CBS

RHABDOGLOEOPSIS PETRAK 1925. ANN. MYCOL. 23:52 /R. BALSAMEAE (DAV.)
  =AUREOBASIDIUM FIDE VON ARX 1970 SUB KABATIELLA

RHABDOSPORIUM CHEVALLIER 1826. FLORE GEN. ENV. PARIS 1:428 /R. DIFFUSUM CHEV. 1826
  =CHEIROSPORA MOUGEOT ET FRIES 1825 FIDE HUGHES (AS ?=)

RHACODIELLA PEYRONEL 1919. STAZ. SPER. AGR. ITAL. 52:39-41 (ILLUS) /R. CASTANEAE (BAINIER)
  =MYRIOCONIUM SYDOW 1912 FIDE CARMICHAEL

RHINOCEPHALUM KAMYSCHKO 1961.. BOT. MATER. (NOT. SYST. SECT. CRYPT. INST. BOT. ACAD. SCI. USSR) 14:224 (ILLUS) /R. CHOCHRAJAKOVII
  ?=ARTHRINIUM KUNZE 1817

RHINOCLADIELLA KAMYSCHKO 1960. NOT. SYST. SECT. CRYPT. INST. BOT. ACAD. SCI. USSR 13:165 AND FIG. 5 /R. SPOROTRICHOIDES
  =NOMEN ILLEGITIMUM, NON RHINOCLADIELLA NANNF. 1934 REPLACED BY RHINOCLADIOPSIS KAMYSCHKO 1961

RHINOCLADIELLA NANNFELDT 1934. SVENSKA SKOGSVFOREN. TIDSKR. 1934:462 /R. ATROVIRENS
  AMERO,CETERI,HYALO OR PHAEO. SCHOL-SCHWARZ GIVES DOUBTFUL SYNONYMS /BARRON, HYPHOMYCETES P. 267-268,1968. (ILLUS) SCHOL-SCHWARZ, ANT. V. LEEUWENHOEK 34:119-152,1968. (ILLUS,REVIS) ELLIS, DEM. HYPH. P. 246-248,1971. (ILLUS)

RHINOCLADIOPSIS KAMYSCHKO 1961. BOT. MATER. (NOT. SYST. SECT. CRYPT. INST. BOT. ACAD. SCI. USSR) 14:243 /NOM. NOV. FOR RHINOCLADIELLA
  =CHRYSOSPORIUM CORDA 1833 FIDE CARMICHAEL

RHINOCLADIUM SACCARDO ET MARCHAL 1885. BULL. SOC. BOT. BELG. 24:65-66 AND FIGS. 10-12 /R. COPROGENUM SACC. ET MARCHAL 1885
AMERO,CETERI,PHAEO /BARRON, HYPHOMYCETES P. 268-269,1968. (ILLUS)

RHINOTRICHELLA ARNAUD 1954. BULL. SOC. MYCOL. FR. 69:272 AND 269 /R. GRISEA (SACCARDO)
=NODULISPORIUM PREUSS 1849 FIDE CARMICHAEL

RHINOTRICHUM CORDA 1837. ICONES FUNGORUM 1:17 /R. SIMPLEX
=NOMEN DUBIUM FIDE HUGHES 1958. FUNGI PLACED IN RHINOTRICHUM MAY BELONG IN ACLADIUM, ALYSIDIUM OR OLPITRICHUM

RHIPIDOCEPHALUM TRAIL 1886. REPORT FOR 1886 ON THE FUNGI OF THE EAST OF SCOTLAND /R. ABIETIS
=NOMEN NUDUM FIDE KENDRICK 1951 (CAN. J. BOTANY 39:818-820)

RHIZOCTONIA DE CANDOLLE 1815. FLOR. FR. 6:111 /R. (?)
=STERILE, BASIDIO STATE =THANATEPHORUS OR CERATOBASIDIUM OR OTHERS FIDE BARRON 1968 /BARRON, HYPHOMYCETES P. 269-270,1968. (ILLUS,REFS) VON ARX, GENERA OF FUNGI IN CULTURE P. 245-246,1970. (ILLUS,REFS)
    SYNO    =ALLOSPHAERIUM LINK FIDE SACCARDO FROM AINSWORTH 1971
                 =BRYOCHYSIUM FIDE RABENHORST FROM AINSWORTH 1971
                 =THANATOPHYTUM NEES 1817 FIDE LINDAU IN RABENHORST 1910

RHIZOHYPHA CHODAT ET SIGRIANSKY 1911. BULL. SOC. BOT. GENEVE 2(3):350 /R. RADICIS CHOD. ET SIGR. 1911
=STERILE

RHIZOMORPHA ROTH EX PERSOON 1801. SYN. METH. FUNGORUM 704 /R. SUBCORTICALIS PERSOON 1801
=STERILE, FOR RHIZOMORPHS
    SYNO    =APHOTISTUS HUMB. FIDE SACCARDO FROM AINSWORTH 1971

RHIZOSPORIUM RABENHORST ?=PHLOEOCONIS FIDE AINSWORTH 1971

RHIZOSTILBELLA VAN DER WOLK 1914. MYCOL. CENTRALB. 4:236-241 AND PLATE /R. RUBRA
? ORIGINAL ILLUSTRATION NOT DIAGNOSTIC

RHODOCEPHALUS CORDA 1837. ICONES FUNGORUM 1:21, FIG. 282 /R. CANDIDUS
=PENICILLIUM LINK 1809 FIDE LINDAU 1907

RHOMBOSTILBELLA ZIMMERMANN 1902. ZENTRALBL. BAKT. 8,ABT. 2:221 /R. ROSEA
AMERO,CETERI,HYALO /PIROZYNSKI, MYCOL PAPERS CMI 90:34-37,1963. (ILLUS,KEY)

RHOPALIDIUM MONTAGNE ET FRIES 1836. ANN. SCI. NAT. 2:30 /R. BRASSICAE (BERK.) MONT. ET FRIES 1836
=ALTERNARIA NEES 1817 FIDE AINSWORTH 1971

RHOPALOCONIDIUM PETRAK 1952. SYDOWIA 6:299-301 /R. ASIMINAE (ELLIS ET MORGAN) PETRAK =PHLEOSPORA ASIMINAE
?, NOT ILLUSTRATED BY PETRAK

RHOPALOCYSTSIS GROVE 1911. J. ECONOM. BOTANY 6:40 /R. NIGRA (VAN TIEGHAM)
=ASPERGILLUS NIGER VAN TIEGHAM
=ASPERGILLUS LINK 1809 FIDE RAPER AND FENNELL 1965

RHYNCHOMYCES WILLK. 1866. MIKR. FEIND. WALD. P. 87 /R. VIOLACEUS
=NOMEN ILLEGITIMUM, NON RHYNCHOMYCES SACCARDO 1885

RHYNCHOSPORINA ARX 1957. VERH. AKAD. WET. AMST. 51:19 /R. MEINERSII (SPRAGUE)
=GLOEOSPORIUM MINERSII SPRAGUE
?, NOT ILLUSTRATED BY VON ARX

RHYNCHOSPORIUM HEINSEN 1901. JB. HAMB. WISSEN. VEREIN. 1843 /R. GRAMINICOLA
DIDYMO,CETERI,HYALO. SCARCELY DIFFERENT FROM MARSSONINA MAGNUS 1906, SEE VON ARX 1970

RICCOA CAVARA 1903. ANN. MYCOL. 1:44 (ILLUS) /R. AETENSIS
=HEYDENIA FIDE HOHNEL FROM CLEMENTS AND SHEAR 1931

## 10. Hyphomycetes

RICLARETIA PEYRONEL 1916. MEM. R. ACC. SCI. TORINO, II, 66:39-40 (ILLUS) /R. URTICAE
SCOLECO,BASO,HYALO
    SYNO    ?=STACHYBOTRYNA TUBAKI ET YOKOYAMA 1971

RIESSIA FRESENIUS 1852. BEITR. Z. MYCOL. 2:74 /R. SEMIOPHORA
  DICTYO OR STAURO,CETERI,HYALO. HYPHAE HAVE CLAMP CONNECTIONS /BARRON, HYPHOMYCETES P. 272-273,1968. ILLUS

ROTAEA CESATI 1851. BOT. ZEIT. 9:180 OR FLORA, P.75 /R. FLAVA
  ?, NOT ILLUSTRATED IN FLORA 1851 /LINDAU, RABENH. KRYPT. FL. 2 AUFL. 1 BD. 8:401,1907. (ILLUS)

RUPINA ROUMEGUERE ET SPEGAZZINI 1880. REV. MYCOL. 2:2 /R. BAYLACII
  =HEYDENIA FRESENIUS 1852 FIDE SACCARDO 1886

SABOURAUDIELLA BOEDIJN 1951. MYCOPATHOLOGIA 6:123 /S. PURPUREA (BANG) BOED. 1951 =T. RUBRUM (CASTELLANI) SABOURAUD 1911
  =TRICHOPHYTON MALMSTEN 1845

SABOURAUDITES OTA ET LANGERON 1923. ANN. PARASIT. HUM. COMP. 1:326 /THREE SUBGENERA, EACH WITH ITS OWN TYPE
  =MICROSPORUM PRO PARTE AND TRICHOPHYTON PRO PARTE

SACCARDAEA CAVARA 1894. ATTI. IST. BOT. PAVIA 3:346 /S. ECHINOCEPHALA
  AMERO,GLOIO,PHAEO. DIFFERS FROM GRAPHIUM IN POSSESSION OF SETAE /MASON AND ELLIS, MYCOL PAPERS CMI 56:40-41,1953. (ILLUS) MORRIS, WESTERN ILLINOIS SER BIOL SCI 3:112-113,1963. (ILLUS)
    SYNO    =PHAEOSTILBELLA HOHNEL 1925 FIDE HUGHES 1958

SADASIVANIA SUBRAMANIAN 1957. J. INDIAN BOT. SOC. 36:66 /S. GIRISA
  AMERO,CETERI,PHAEO /ELLIS, DEM. HYPH. P. 343-344,1971. (ILLUS)

SAPROCHAETE WAGNER ET DAWES 1970. MYCOLOGIA 62:792-796 (ILLUS) /S. SACCHAROPHILA (COKER ET SHANOR)
  =STERILE. ORIGINALLY DESCRIBED AS AN ALGA

SAPROPHRAGMA DESHPANDE ET DESHPANDE 1966. MYCOPATHOLOGIA 30:200-202 (ILLUS) /S. ACEROSUM DESHP. ET DESHP. 1966
  ?, ORIGINAL ILLUSTRATION NOT DIAGNOSTIC

SARCINELLA SACCARDO 1877. FUNGI ITALICI TAF. 126 /S. HETEROSPORA =S. STATE OF SCHIFFNERULA PULCHRA (SACC.) PETRAK 1928 FIDE ELLIS 1971
  DICTYO,CETERI,PHAEO. ASCO STATE =SCHIFFNERULA FIDE ELLIS 1971 /ELLIS, DEM. HYPH. P. 49-50,1971. (ILLUS)

SARCINODOCHIUM HOHNEL 1905. OEST. BOT. ZEITS. 55:15 /S. HETEROSPORIUM
  ?, NOT ILLUSTRATED BY AUTHOR

SARCINOMYCES LINDNER 1901. MKR. BETRIEBS. ED. 3:300 /S. CRUSTACEUS
  =AUREOBASIDIUM VIALA ET BOYER 1891 FIDE COOKE FROM AINSWORTH 1971

SARCOPODIUM EHRENBERG 1818. SYLVAE MYC. BEROL. P. 12 AND 23 /S. CIRCINATUM
  ?, NO RECENT ILLUSTRATION /HUGHES, CAN. J. BOTANY 36:801-802,1958. (REFS)
    SYNO    =TRICHOLECCNIUM CORDA 1837 FIDE HUGHES 1958
            =PERIOLOPSIS MAIRE 1913 FIDE PIROZYNSKI

SAROPHORUM SYDOW ET SYDOW 1916. BOT JB. 54:260-261 (ILLUS) /S. LEDERMANNII SYD. (SIC) 1916
  ?=CEPHALOTRICHUM LINK 1809

SATWALEKERA RAO,RAO ET RAO 1969. NOVA HEDWIGIA 18:640 /S. SUNDARA RAO,RAO ET RAO ?=TORULA HEREARUM (PERSOON) LINK 1815
  =TORULA (PERSOON) LINK 1809 FIDE CARMICHAEL

SAVULESCUELLA CIFERRI 1959. OMAGIU LUI TRAIAN SAVULESCU P. 179 (NO ILLUS) /S. ALISMACEARUM (SACCARDO) CIFERRI =CYLINDROSPORIUM ALISMACEARUM SACC.
  ?, BASIDIO STATE =DOASSANSIA AND TRACYA (TILLETIACEAE) FIDE INDEX OF FUNGI 2:561,1960

SCEDOSPORIUM SACCARDO 1911. MENTIONED BUT NOT PUBLISHED IN ANN. MYCOL. 9:254 /A
    SUGGESTED NAME FOR THE CONIDIAL STATE OF ALLESCHERIA BOYDII
    =MISCITATION, SACCARDO DID NOT FORMALLY PROPOSE SCEDOSPORIUM

SCENOMYCES STEVENS 1927. ILL. BIOL. MONOGR. 11:60-61 AND FIGS. 89-91 AND
    123-125 /S. PERPLEXANS
    =STERILE

SCEPTRIFERA DEIGHTON 1965. MYCOL. PAPERS CMI 101:37-39 (ILLUS) /S. PULCHRA
    AMERO OR DIDYMO,CETERI,PHAEO

SCEPTROMYCES CORDA 1831. STURMS, DEUTSCHL. FL. III (PILZE) 3:7-8 AND TAB. 4 /S.
    OPIZII
    ?, LINDAU 1907 LISTS THIS AS A MONSTROUS FORM OF ASPERGILLUS NIGER, BASED ON
    A REPORT OF ENGELKE, IN HEDWIGIA 41:219,1902

SCHELEOBRACHEA HUGHES 1958. CAN. J. BOTANY 36:802 /S. ECHINULATA (SPEGAZZINI)=
    PITHOMYCES CHARTARUM (BERK. ET CURT.) ELLIS 1960
    =PITHOMYCES BERKELEY ET BROOME FIDE ELLIS 1960

SCHIZOCEPHALUM PREUSS 1852. LINNAEA 25:77 /S. ATROFUSCUM
    ?, SEE LINDAU IN RABENHORST 1907, P.696, FOR DESCRIPTION

SCHIZOTRICHELLA MORRIS 1956. MYCOLOGIA 48:733 /S. LUNATA
    =VOLUTELLA TODE EX FRIES 1823 FIDE CARMICHAEL

SCHIZOTRICHUM MCALPINE 1903. PROC. LINN. SOC. N.S. WALES 28:562 (NO ILLUS) /S.
    LOBELIAE
    SCOLECO,CETERI,HYALO

SCHOENLEINIUM JOHAN-OLSEN 1897. ZENTRALBL. BAKT. ABT 2, 3:276-284 AND TAF. IV
    /TWO ORIGINAL SPECIES
    =TRICHOPHYTON MALMSTEN 1845 FIDE CARMICHAEL

SCINIATOSPORIUM "(REICHB.) CORDA" 1844. IN RABENHORST, DEUTSCHL. KRYPT.-FL.
    1:47,1844 /SEIMATOSPORIUM ROSAE CORDA 1833, IN STURM'S DEUTSCHL. FL.
    III,3,PL. 40
    =SEIMATOSPORIUM CORDA (COELOMYCETES). SCINIATOSPORIUM IS AN ORTHOGRAPHIC
    VARIANT BASED ON IMPERFECTIONS IN THE PRINT FOR THE NAME UNDER CORDA'S
    PLATE 40. BUT SEE MORGAN-JONES, CAN. J. BOTANY 49:993-1009,1971

SCLEROCOCCUM FRIES 1819. NOV. FLOR. SUECICAE, LUNDAE, P. 79 (NO ILLUS) /S.
    SPHAERALE FRIES 1825
    ?

SCLERODISCUS PATOUILLARD 1890. JOUR. DE BOT. 4:66-67 (ILLUS) /S. NITENS
    ?, ORIGINAL ILLUSTRATION NOT DIAGNOSTIC

SCLEROGRAPHIUM BERKELEY 1854. HOOKER J. BOT. 6:209 /S. ATERRIMUM
    DICTYO,CETERI,PHAEO /HUGHES, INDIAN PHYTOPATH. 4:5-6,1951. (ILLUS)

SCLEROSTILBUM POVAH 1932. MYCOLOGIA 24:242 /S. SEPTENTRIONALE
    ?, ORIGINAL ILLUSTRATION NOT DIAGNOSTIC /MORRIS, WESTERN ILLINOIS UNIV. SER.
    BIOL. SCI. 3:116-117,1963. (ILLUS)

SCLEROTIOMYCES WORONICHIN 1926 ?=SCLEROTIUM FIDE AINSWORTH 1971

SCLEROTIUM TODE 1790 EX FRIES 1823. FUNGI MECKLENB. 1:36,1790 /S. COMPLANATUM
    TODE 1790
    =STERILE,BASIDIO STATE =PELLICULARIA OR TYPHULA, ASCO STATE =SCLEROTINA COELO
    STATE =MACROPHOMINA OR COLLETOTRICHUM FIDE VON ARX 1970 AND BARRON 1968
    /BARRON, HYPHOMYCETES P. 272-274,1968. ILLUS
        SYNO    =ASTOMA GRAY FIDE SACCARDO FROM AINSWORTH 1971
                =BROMICOLLA EICHW. FIDE SACCARDO FROM AINSWORTH 1971
                =COCCOPLEUM EHRENBERG FIDE LINDAU IN RABENHORST 1910
                =PACHYMA FRIES, SEE AINSWORTH 1971
                ?=SCLEROTIOMYCES WORONICHIN FIDE AINSWORTH 1971
                =SPERMOEDIA FRIES 1823, SEE AINSWORTH 1971
                =XYLOCHAERAS FRIES FIDE SACCARDO FROM AINSWORTH 1971

## 10. Hyphomycetes 407

SCOLECOBASIDIELLA ELLIS 1971. MYCOL. PAPERS CMI 125:12-13 (ILLUS) /S. AVELLANEA
 (SAPPA ET MOSCA)
 ?=PASPALOMYCES LINDER 1933

SCOLECOBASIDIUM ABBOTT 1927. MYCOLOGIA 19:29 /S. TERREUM
 DIDYMO OR PHRAGMO,CETERI,PHAEO, BUT DIFFERS FROM PLEUROPHRAGMIUM ONLY IN
  SHORTER CONIDIOPHORE /BARRON AND BUSCH, CAN. J. BOTANY 40:77-84,1962. ILLUS

SCOLICOTRICHUM KUNZE 1817. IN KUNZE AND SCHMIDT. MYK. HEFTE 1:10 /S. VIRESCENS
 =NOMEN CONFUSUM FIDE HUGHES 1958
  SYNO   =SCOLECOTRICHUM KUNZE EX AUCT. (ORTHOGRAPHIC VARIANT)

SCOPULA ARNAUD 1952. BULL. SOC. MYCOL. FR. 68:197 AND 198 /S. HYALINA
 =GLOIOSPHAERA HOHNEL 1902 FIDE WANG 1971

SCOPULARIA PREUSS 1851. LINNAEA 24:134 /S. VENUSTA (=NOMEN DUBIUM FIDE KENDRICK
  1964)
 =NOMEN ILLEGITIMUM, NON SCOPULARIA LINDLEY 1835, SEE KENDRICK, CAN. J. BOT.
  42:1119-1122,1964.  LINDAVIA NIEUWLAND 1916 (AMER. MIDL. NAT. 4:385) WAS
  PROPOSED TO REPLACE SCOPULARIA PREUSS

SCOPULARIOPSIS BAINIER 1907. BULL. SOC. MYCOL. FR. 23:98 /S. BREVICAULIS
  (SACCARDO)
 AMERO,BASO,HYALO OR PHAEO, ASCO STATE =MICROASCUS /MORTON AND SMITH, MYCOL
  PAPERS CMI 86:1-96,1963. (ILLUS,MONOGR)
  SYNO   =ACAULIUM SOPP 1912 FIDE ELLIS 1971
         =MASONIA SMITH 1952 FIDE SMITH 1963
         =MASONIELLA SMITH 1952 FIDE SMITH 1963
         =PHAEOSCOPULARIOPSIS OTA 1928 FIDE MORTON AND SMITH 1963
         ?=TORULOIDEA SUMSTINE 1913

SCORIOMYCES ELLIS ET SACCARDO 1884. MISC. MYC. 2:18 /S. CRAGINI ELLIS 1884
 ?, NOT ILLUSTRATED BY AUTHORS

SCUTISPORIUM PREUSS 1851. LINNAEA 24:112 /TWO ORIGINAL SPECIES
 =STEMPHYLIUM WALLROTH FIDE LINDAU IN RABENHORST 1910

SCYTALIDIUM PESANTE 1957. ANN. SPER. AGR. N. S. 11:249 /S. LIGNICOLA
 AMERO OR DIDYMO OR PHRAGMO,ARTHRO,PHAEO /VON ARX, GENERA OF FUNGI IN CULTURE
  P. 168,1970. ILLUS

SEARCHOMYCES MEHROTRA ET MEHROTRA 1963. SYDOWIA 16:213-214 (ILLUS,DATED 1962)
  /M. COPROPHILOIDES MEHR. ET MEHR. 1963
 =AMBLYOSPORIUM FRESENIUS 1863 FIDE PIROZYNSKI 1969

SEIMATOSPORIUM CORDA 1833. STURMS DEUTSCHL. FLORA III (PILZE), BD. 3, HEFT
  13:79 & PL.40 /S. ROSAE
 PHRAGMO,GLOIO,PHAEO, (MELANCONIALES) ASCO STATE =GRIPHOSPHAERIA OR
  CLATHRIDIUM /VON ARX, GENERA OF FUNGI IN CULTURE P. 156-158,1970. ILLUS
  SYNO   =BASIPILUS SUBRAMANIAN 1961 FIDE VON ARX 1970
         =MONOCERAS GUBA 1961 FIDE VON ARX 1970
         =SCINIATOSPORIUM (ORTHOGRAPHIC VARIANT)

SELENOSPORELLA ARNAUD 1954. BULL. SOC. MYCOL. FR. 69:292 AND 293 /S. CURVISPORA
 AMERO OR SCOLECO,GLOIO,HYALO /MACGARVIE, SCIENT. PROC. R. DUBL. SOC. SER. B
  2:153,1968.(VALIDATED)

SELENOSPORIUM CORDA 1837. ICONES FUNGORUM 1:7 /?
 =FUSARIUM FIDE LINDAU IN RABENHORST 1910

SELENOTILA LAGERHEIM 1892. BERL. DEUTSCHL. BOT. GES. 10:53 AND FIGS. 24-28 /S.
  NIVALIS
 ?, PERHAPS AN ALGA OR A SMUT

SEPEDONIUM LINK 1809. MAG. GES. NATURF. FREUNDE, BERLIN 3:18 /U. MYCOPHILA
  PERSOON =S. MYCOPHILUM (PERSOON) NEES 1816/17
 AMERO,CETERI,HYALO. ASCO STATE =HYPOMYCES OR APIOCREA OR THIELAVIA /BARRON,
  HYPHOMYCETES P. 278-279,1968. ILLUS
  SYNO   =MYCOBANCHE PERSOON 1818 FIDE HUGHES 1958
         =LEIOSEPIUM SACCARDO 1900 FIDE DAMON 1952

SEPTOCYLINDRIUM BONORDEN EX SACCARDO 1877. MICHELIA 1:89 /S. SEPTATUM
    (BONORDEN) LINDAU 1907 =S. BONORDENII SACCARDO 1877
    PHRAGMO,BLASTO,HYALO. S. SEPTATUM WAS DESCRIBED AS A HYALINE COUNTERPART OF
    SEPTONEMA SECEDENS /BONORDEN, HANDB. MYKOL. P. 35 AND FIG 16,1851. (ILLUS)
        SYNO    =TAPEINOSPORIUM BONORDEN 1853 FIDE LINDAU IN RABENHORST 1907

SEPTOIDIUM ARNAUD 1921. ANN. EPIPHYT. 7:106 (NO ILLUS) /S. CLUSIACEAE
    PHRAGMO,CETERI,HYALO. ASCO STATE= PARODIOPSIS FIDE ARNAUD 1921 /ELLIS, DEM.
    HYPH. P. 115-116,1971. (ILLUS)
        SYNO    =DIPLOIDIUM ARNAUD 1923 FIDE ELLIS 1971

SEPTONEMA CORDA 1837. ICONES FUNGORUM 1:9 /S. SECEDENS
    PHRAGMO,BLASTO,PHAEO. SEE ALSO TAENIOLELLA /HUGHES, THE NATURALIST  P.
    173-176,1951. ILLUS HUGHES, THE NATURALIST P. 7-12,1952. ILLUS

SEPTORIOPSIS STEVENS ET DALBY NON ALIORUM =CERCOSEPTORIA

SEPTOSPORIUM CORDA 1832. STURMS DEUTSCHL. FLORA III, BD. 3, HEFT 12:33 /S. ATRUM
    DICTYO,CETERI,PHAEO /ELLIS, MYCOL PAPERS CMI 79:1-5,1961. (ILLUS)

SEPTOTIS BUCHWALD 1949. K. VET. HOJSK. AARS. SKR. P. 104-105 /GLOEOSPORIUM
    PODOPHYLLINUM ELLIS ET EVERHART 1888.  ASCO STATE =SEPTOTINIA PODOPHYLLINA
    FIDE BUCHWALD 1949
    ?. NO LATIN DIAGNOSIS. NO NEW COMBINATION FOR THE TYPE SPECIES, NO
    ILLUSTRATION.

SEPTOTRICHUM CORDA 1840. ICONES FUNGORUM 4:4-6 /?
    =LEAF HAIRS FIDE HUGHES 1958

SEPTOTRULLULA HOHNEL 1902. S.B. AKAD. WISS. WIEN, ABT. 1, 111:1025-1027 /S.
    BACILLIGERA (LECTOTYPE)
    PHRAGMO,GLOIO OR ARTHRO,HYALO OR PHAEO /ELLIS, DEMAT. HYPHO. P. 98-99,1971.
    (ILLUS)

SESQUICILLIUM GAMS 1968. ACTA BOT. NEERL. 17:455-460 (ILLUS) /S. BUXI (SCHMIDT)
    AMERO,GLOIO,HYALO /VON ARX, GENERA OF FUNGI IN CULTURE P. 179-181,1970.
    (ILLUS)

SETODOCHIUM BATISTA ET CIFERRI 1957. PUBL. INST. MIC. UNIV. RECIFE 94:5 /S.
    CASEARIAE
    ?=VOLUTINA PENZIG ET SACCARDO 1904

SIBIRINA ARNOLD 1970. NOVA HEDWIGIA 19:299-301 (ILLUS) /S. FUNGICOLA
    DIDYMO,GLOIO,HYALO

SIGMATOMYCES SACCARDO ET SYDOW 1913. ANN. MYCOL. 11:319 /S. BAKERI SACC. ET
    SYDOW 1913
    ?. NOT ILLUS. BY AUTHORS

SIGMOIDEA CRANE 1968. AMER. J. BOTANY 55:997-998 (ILLUS) /S. PROLIFERA
    (PETERSEN) =FLAGELLOSPORA PROLIFERA PETERSON
    SCOLECO,CETERI,HYALO

SIGMOIDEOMYCES THAXTER 1891. BOT. GAZ. 45:22-23 AND PL. IV, FIGS. 15-18 /S.
    DISPIROIDES
    AMERO,CETERI,HYALO

SIRODESMIUM DE NOTARIS 1849. MEM. ACCAD. TORINO, II,10:347 /S. GRANULOSUM =C.
    STATE OF HYSTERIUM INSIDENS SCHWEINITZ FIDE HUGHES 1958
    =CONIOSPORIUM LINK 1809 FIDE HUGHES 1958

SIRODOCHIELLA HOHNEL 1925. MITT. BOT. TECHN. HOCHSCH. WIEN 2:67-68 (NO ILLUS)
    /S. RHODELLA
    ?. NON SIRODOCHIELLA HOHN. 1919, BER. DTSCH. BOT. GES. 37:153. NO ILLUS

SIROSPORIUM BUBAK ET SEREBRIANIKOW 1912. HEDWIGIA 52:272 /S. ANTENNIFORME
    (BERK. ET CURTIS) BUBAK ET SEREBR. 1912
    DICTYO OR PHRAGMO,CETERI,PHAEO /ELLIS, MYCOL PAPERS CMI 87:2-11,1961. (ILLUS)
        SYNO    =HELICOCERAS LINDER 1931 FIDE ELLIS 1963

## 10. Hyphomycetes

SOLHEIMIA MORRIS 1967. MYCOPATHOLOGIA 33:181-182 (ILLUS) /S. COSTASPORA MORRIS
AMERO,GLOIO,PHAEO /MORRIS, MYCOPATHOLOGIA 33:181-182,1967. (ILLUS)

SOREDOSPORA CORDA 1837. ICONES FUNGORUM 1:12 /S. GRAMINIS =FUMAGO GRAMINIS
(CORDA) HUGHES 1958
= FUMAGO PERSOON 1822 FIDE HUGHES 1958

SOREYMATOSPORIUM CAMARA 1930 =STEMPHYLIUM FIDE AINSWORTH 1971

SOROCYBE FRIES 1849. SUMMA VEGET. SCAND. 2:468 /S. RESINAE (FRIES)
AMERO,BLASTO,PHAEO, ASCO STATE =AMORPHOTHECA PARBERY 1969 /BARRON,
HYPHOMYCETES P. 264-265,1968. ILLUS SUB PYCNOSTYSANUS PARBERY, AUSTRAL. J.
BOTANY 17:331-357,1969. (ILLUS)
    SYNO    =PYCNOSTYSANUS LINDAU 1904 SEE HUGHES 1958 AND PARBERY 1969

SOROSPORELLA SOROKIN 1888. ZENTRALBL. BAKT. ABT. 2, 4:644-647 AND TAF. 4, FIGS.
6-18 /S. AGROTIDIS
?, YEAST-LIKE(?) INSECT PARASITE

SPADICOIDES HUGHES 1958. CAN. J. BOTANY 36:805 /S. BINA CORDA
AMERO OR DIDYMO OR PHRAGMO,CETERI,PHAEO /ELLIS, MYCOL PAPERS CMI
93:6-14,1963. (ILLUS,KEY)

SPEGAZZINIA SACCARDO 1880. MICHELIA 2:37 /S. TESSARTHRA (BERKELEY ET CURTIS)
STAURO OR DICTYO,CETERI,PHAEO /HUGHES, MYCOL PAPERS CMI 50:62-66,1953. (ILLUS)
    SYNO    =TETRACHIA SACCARDO 1921 FIDE BOEDIJN FROM AINSWORTH 1961

SPEIRA CORDA 1837. ICONES FUNGORUM 1:9 /S. TORULOIDES
=DICTYOSPORIUM CORDA 1836 FIDE HUGHES 1958

SPEIROPSIS TUBAKI 1958. J. HATTORI BOT. LAB. NO. 20, P. 171 /S. PEDATOSPORA
STAURO,CETERI,PHAEO /TUBAKI, J. HATTORI BOT. LAB. 20:171-173,1958. (ILLUS)
    SYNO    ?=MULTICLADIUM DESHP. ET DESHP. 1966

SPELAEOMYCES FRESENIUS ?=XYLOSTROMA FIDE AINSWORTH 1971

SPERMATOLONCHA SPEGAZZINI 1909. AN. MUS. NAC., SER. 3, 10:139 (ILLUS) /S.
MATICOLA
?, PROBABLY AN ASPERGILLUS

SPERMODERMIA TODE 1790. FUNGI MECKLENB. 1:1 /S. CLANDESTINA
=A YOUNG HYPOXYLON FIDE HOHNEL FROM AINSWORTH 1971

SPERMOEDIA FRIES 1823 =SCLEROTIUM SEE AINSWORTH 1971

SPERMOSPORA SPRAGUE 1948. MYCOLOGIA 40:177 /S. SUBULATA (SPRAGUE)
PHRAGMO,GLOIO OR CETERI,HYALO /DEIGHTON, TRANS. BRIT. MYCOL. SOC.
51:41-49,1968. (ILLUS,REVIS)
    SYNO    ?=SPERMOSPORELLA DEIGHTON 1969

SPERMOSPORELLA DEIGHTON 1969. MYCOL PAPERS CMI 118:21-22 (ILLUS) /S. AGGREGATA
?=SPERMOSPORA SPRAGUE 1948

SPHACELIA LEVEILLE 1827. MEM. SOC. LINN., PARIS 5:578 /S. SEGETUM
?, NOT ILLUSTRATED BY AUTHOR, ASCO STATE =CLAVICEPS FIDE AINSWORTH

SPHAERIDIUM FRESENIUS 1852. BEITR. ZUR MYKOL. P. 46 AND TAF. V, FIGS. 31-36 /S.
VITELLINUM
AMERO,BLASTO,HYALO, BUT NOT VERY DIFFERENT FROM FUSIDIUM LINK /BARRON,
HYPHOMYCETES P. 282-283,1968. (ILLUS)

SPHAEROCOLLA KARSTEN 1892. HEDWIGIA 31:294 /S. AURANTIACA
?, NOT ILLUSTRATED BY AUTHOR

SPHAEROCYBE MAGROU 1945. C. R. ACAD. SCI. PARIS 220:220-222 /S. CONCENTRICA
?, NOT ILLUSTRATED BY AUTHOR, NO LATIN DAIGNOSIS

SPHAEROMYCES ARNAUD 1954. BULL. SOC. MYCOL. FR. 69:279 AND 280 (ILLUS) /S.
CLAVISPORUS

AMERO,GLOIO,HYALO. NO LATIN DIAGNOSIS, AN ACREMONIUM (=CEPHALOSPORIUM) WITH
LONG BROWN CONIDIOPHORES. NON SPHAEROMYCES MONTAGNE 1845 (?= ASPERGILLUS)

SPHAEROMYCETELLA ARNAUD 1954. BULL. SOC. MYCOL. FR. 69:279 AND 280 (ILLUS) /S.
LEUCOCEPHALA
?. DOES NOT APPEAR TO DIFFER FROM CHLORIDIUM LINK 1809. SEE ALSO PIMINELLA.

SPHAEROSPORIUM SCHWEINITZ 1832. TRANS. AMER. PHIL. SOC. II,4:303 /S. LIGNATILE
AMERO,BLASTO,HYALO /DAMON AND DOWNING, MYCOLOGIA 46:214-216,1954. (ILLUS)

SPHINCTROSPORIUM KUNZE EX FRIES =CLADOTRICHUM FIDE SACCARDO FROM AINSWORTH 1971

SPICARIA HARTING 1846. NIEWE VERH. KON. INST. WETENSCH. AMSTERDAM 12:203-297
/S. SOLANI
=NOMEN CONFUSUM SEE BROWN AND SMITH, TRANS. BRIT. MYCOL. SOC. 40:22-24,1957.
SPICARIA HARZ 1871 WAS AN ILLEGITIMATE REVISION WHICH EXCLUDED HARTINGS
TYPE SPECIES.  MOST FUNGI IDENTIFIED AS SPICARIA ARE SPECIES OF
PAECILOMYCES.

SPICARIOPSIS HEIM 1939. IN HEIM AND BOURIQUET, REV. PATH. VEG. 26:25-29 (ILLUS)
/S. TROPICALE
?=PAECILOMYCES BAINIER 1907

SPICULARIA PERSOON 1822. MYC. EUR. 1:39 /?. SEE HUGHES, CAN. J. BOTANY
36:806,1958
=NOMEN DUBIUM SEE HUGHES 1958

SPICULOSTILBELLA MORRIS 1963. AMER. MIDL. NATURALIST 69:101 /S. DENDRITICA
MORRIS
?=PHAEOISARIA HOHNEL 1909

SPILOCAEA FRIES 1825. SYST. ORB. VEG. P. 198 /S. POMI = S. STATE OF VENTURIA
INAEQUALIS (COOKE) WINTER APUD THUMEN 1875
AMERO OR DIDYMO,CETERI,PHAEO, ASCO STATE =VENTURIA /HUGHES, CAN. J. BOTANY
31:560-565,1953. (ILLUS)
SYNO     =CYCLOCONIUM CASTAGNE 1845 FIDE HUGHES 1958
         =NAPICLADIUM THUMEN 1875 FIDE HUGHES 1958
         =BASIASCUM CAVARA 1888 FIDE HUGHES 1958
         ?=POLLACCIA BALDACCI ET CIFERRI 1937

SPILODOCHIUM SYDOW 1927. ANN. MYCOL. 25:158 /S. VERNONIAE
AMERO OR DIDYMO,BLASTO,PHAEO, COMPARE WITH HADRONEMA /ELLIS, DEMAT. HYPHO. P.
94-95,1971. (ILLUS)

SPILOMIUM NYLANDER 1856. PROD. LICH. GALL. P. 91 /S. SILICEUM (FEE)
?. NOT ILLUSTRATED BY NYLANDER

SPIRALIA GRIGORAKIS 1924. COMP. REND. 179:1424 /NO SPECIES MENTIONED
=NOMEN ILLEGITIMUM (FOR TRICHOPHYTON PRO PARTE)

SPIRALOTRICHUM YATES 1918. PHILIPP. J. SCI. (BOT.) 13:383-384 /S. PIPERIS
?. NOT ILLUSTRATED BY AUTHOR

SPIROPES CIFERRI 1955. SYDOWIA 9:303 /S. GUAREICOLA (STEVENS)
PHRAGMO,CETERI,PHAEO /ELLIS, MYCOL PAPERS CMI 114:1-42,1968. (ILLUS,MONOGR)
SYNO     =NASCIMENTOA CIFERRI ET BATISTA 1956 FIDE ELLIS 1968

SPIROSPHAERA VAN BEVERWIJK 1953. TRANS. BRIT. MYCOL. SOC. 36:120 /S. FLORIFORMIS
HELICO,CETERI,HYALO, COMPARE WITH STROMELLA SACCARDO /HENNEBERT, TRANS. BRIT.
MYCOL. SOC. 51:13-24,1968. (ILLUS,REVIS)

SPIROSPORA MANGIN ET VINC. 1920. BULL. SOC. MYCOL. FR. 36:96 /S. CASTANEAE
MANG. ET VINC. =A. MIRABILIS BERK. ET BROOME
=ACROSPEIRA BERKELEY ET BROOME 1857 FIDE HUGHES 1958

SPIROTRICHUM SAITO =TRITIRACHIUM FIDE 'REV. MYCOL., 14:134' FROM AINSWORTH 1971

SPONDYLOCLADIELLA LINDER 1934. MYCOLOGIA 26:437 /S. BOTRYTIOIDES
PHRAGMO,CETERI,PHAEO /ELLIS, DEMAT. HYPHO. P. 386-387,1971. (ILLUS)

## 10. Hyphomycetes

SPONDYLOCLADIOPSIS ELLIS 1963. MYCOL. PAPERS CMI 87:15-17 (ILLUS) /S. CUPULICOLA
  PHRAGMO,CETERI,HYALO

SPONDYLOCLADIUM MARTIUS 1817. FLORA CRYPT. ERLANG. P. 355 /S. FUMOSUM MARTIUS
    =STACHYLIDIUM VERTICILLATUM (PERS.) HUGHES
    =STACHYLIDIUM LINK 1809 FIDE HUGHES 1958

SPORENDOCLADIA ARNAUD 1954. BULL. SOC. MYCOL. FR. 69:277 AND 279 (ILLUS) /S.
    CASTANEAE
  AMERO,BASO,HYALO. NO LATIN DIAGNOSIS

SPORENDONEMA DESMAZIERES 1827. ANN. SCI. NAT. 11:246-249 AND PL. XXI /S. CASEI
  AMERO,ARTHRO,HYALO /VON ARX, GENERA OF FUNGI IN CULTURE P. 166-167,1970.
  (ILLUS)

SPORHELMINTHIUM SPEGAZZINI 1918. PHYSIS 4:292 /S. ANOMALUM SPEG.
    =CLASTEROSPORIUM ANOMALUM (SPEG.) HUGHES 1958
  =CLASTEROSPORIUM SCHWEINITZ 1832 FIDE HUGHES 1958

SPORIDESMIUM LINK 1809. MAG. GES. NATURF. FREUNDE, BERLIN 3:41 /S. ATRUM
  PHRAGMO OR SCOLECO,CETERI,PHAEO, ASCO STATE =EUPELTE FIDE ELLIS 1958 /ELLIS,
    MYCOL PAPERS CMI 70:16-84,1958. (ILLUS,MONOGR)
      SYNO    =PODOCONIS BOEDIJN 1933 FIDE HUGHES 1958
               =POLYDESMUS DURIEU ET MONTAGNE 1845 SEE ELLIS 1958
               ?=LOMAANTHA SUBRAMANIAN 1954

SPORIDIOBOLUS NYLAND 1949. MYCOLOGIA 41:686 /S. JOHNSONII
  ?, HAS CLAMP CONNECTIONS, NOT VERY DIFFERENT FROM TILLETIOPSIS AND
    ITERSONILIA /SOWELL AND KORF, MYCOLOGIA 52:934-945,1960. (REVIS)

SPOROCEPHALIUM CHEVALLIER 1826. FLORE GEN. ENV. PARIS 1:60 /S. CAPITATUM (LINK)
  CHEV. =ACLADIUM STATE OF BOTRYOBASIDIUM CANDICANS
  =ACLADIUM LINK 1809 FIDE HUGHES 1958

SPOROCEPHALUM ARNAUD 1952. BULL. SOC. MYCOL. FR. 68:187-189 (ILLUS) /S.
    PENIOPHORAE
  AMERO,CETERI,HYALO. NO LATIN DIAGNOSIS, COULD BE ACCOMMODATED IN OEDOCEPHALUM

SPOROCLADIUM CHEVALLIER 1826. FLORE GEN. ENV. PARIS 1:36 AND P. 647
    /CLADOSPORIUM HERBARUM (PERSOON) LINK (LECTOTYPE FIDE HUGHES 1958)
  =CLADOSPORIUM LINK 1815 FIDE HUGHES 1958

SPOROCLEMA TIESENH. 1912. ARCH. HYDR. PLANKT. 7:302-303 (ILLUS) /S. PIRIFORME
  ?, PERHAPS AN IMMATURE OOMYCETE

SPOROCYBE FRIES 1825. SYST. ORB. VEG. 1:170 /PERICONIA BYSSOIDES PERSOON 1801
  =PERICONIA PERSOON 1801 FIDE HUGHES 1958

SPOROCYSTIS MORGAN 1902. JOUR. MYCOL. 8:169 /S. CONDITA
  ?, NOT ILLUSTRATED BY AUTHOR

SPORODERMA MONTAGNE 1856. SYLLOGE CRYPT. NO. 1069, 1856 AND IN SACCARDO,
    SYLLOGE FUNG. 4:676,1886 /S. CHLOROGENUM
  =TRICHODERMA PERSOON FIDE HOHNEL 1910 FROM RIFAI 1969

SPORODINIOPSIS HOHNEL 1903. ANN. MYCOL. 1:528 /S. DICHOTOMUS
  =NOMEN DUBIUM FIDE HAWKSWORTH 1971 (MYCOL. PAPERS CMI 126:26)

SPORODUM CORDA 1837. ICONES FUNGORUM 1:18 /S. CONOPLEOIDES CORDA =PERICONIA
    HISPIDULA (PERSOON) MASON ET ELLIS
  =PERICONIA PERSOON 1801 FIDE HUGHES 1958

SPOROGLENA SACCARDO 1894. ENGLERS BOT. JB. 18 BAND, 3:40 /S. VELUTINA
  ?, NOT ILLUSTRATED BY AUTHOR

SPOROPHIALA RAO 1970. MYCOPATHOLOGIA 41:315-320 (ILLUS) /S. PROLIFICA
  PHRAGMO,BLASTO,PHAEO

SPOROPHLEUM NEES EX LINK 1824. LINN. SPEC. PLANT. IV,6(1):45 /S. GRAMINEUM

=ARTHRINIUM SPOROPHLEUM KUNZE 1817
=ARTHRINIUM KUNZE 1817 FIDE HUGHES 1958

SPOROPHORA LUTERAAN 1952. REV. BIOL. GEN. THEOR. APPL. 2:1-12 AND PLATE /S.
  TORULIFORMIS
  ?. NO LATIN DIAGNOSIS, PERHAPS YEAST STAGE OF USTILAGO

SPOROPHRAGMA DESHPANDE ET DESHPANDE 1966. MYCOPATHOLOGIA 30:200-202 /S.
  ACEROSIAE DESHP. ET DESHP. 1966
  ?

SPOROSCHISMA BERKELEY ET BROOME 1847. GARD. CHRON. P. 540 /S. MIRABILE BERK. ET
  BR. 1847
  PHRAGMO,BASO,PHAEO OR HYALO, ASCO STATE = MELANOCHAETA FIDE MULLER IN
    KENDRICK 1971 /HUGHES, MYCOL PAPERS CMI 31:1-22,1949. ILLUS,KEY
    SYNO    ?=EXCIOCONIDIUM PLUNKETT 1925

SPOROSTACHYS SACCARDO 1917. ATTI ACCAD. SCI. VEN.-TRENT.-ISTR. III,10:92 /S.
  MAXIMA SACCARDO =MELANOGRAPHIUM SPINULOSUM (SPEGAZZINI) HUGHES
=MELANOGRAPHIUM SACCARDO 1913 FIDE HUGHES 1958

SPOROTHRIX HEKTOEN ET PERKINS 1900. J. EXPER. MEDICINE 5:77 /S. SCHENCKII
  AMERO,CETERI,HYALO, ASCO STATE =CERATOCYSTIS /BARRON, HYPHOMYCETES P.
    284-285,1968. (ILLUS)
    SYNO    =SPOROTRICHUM AUCT. NON LINK FIDE CARMICHAEL 1962
            =SPOROTRICHOPSIS DE BEURMANN ET GOUGEROT 1911 FIDE CARMICHAEL

SPOROTRICHELLA KARSTEN 1887. MEDDEL. SOC. F. F. FENNICA 14:96 /S. ROSEA KARST.
  1887
=FUSARIUM LINK 1809 FIDE HUGHES 1958

SPOROTRICHOPSIS DE BEURMANN ET GOUGEROT 1911. ARCH. PARASIT. 15:103-105 (ILLUS)
  /"SPOROTRICHOPSIS? BEURMANNI"
=SPOROTHRIX HEKTOEN ET PERKINS 1900 FIDE CARMICHAEL

SPOROTRICHUM LINK 1809. MAG. GES. NATURF. FREUNDE, BERLIN 3:13 /S. AUREUM
  AMERO,CETERI,HYALO OR PHAEO, S. AUREUM HAS CLAMP CONNECTIONS FIDE VON ARX
    1970 /VON ARX, GENERA OF FUNGI IN CULTURE P. 163-164,1970. (ILLUS) VON ARX,
    PERSOONIA 6:179-184,1971. (ILLUS,REVIS)
    SYNO    =OIDIUM LINK 1809 FIDE HUGHES 1958

STACHYBOTRYELLA ELLIS ET BARTHOLOMEW 1902. JOUR. MYCOL. 8:177 /S. REPENS ELLIS
  ET BARTH. 1902
  ?. NOT ILLUSTRATED BY AUTHORS

STACHYBOTRYNA TUBAKI ET YOKOYAMA 1971. TRANS. MYCOL. SOC. JAPAN 12:18-20
  (ILLUS) /S. COLUMARE
  ?. LOOKS LIKE RICLARETIA PEYRONEL 1916

STACHYBOTRYS CORDA 1837. ICONES FUNGORUM 1:21 /S. ATRA CORDA =S. CHARTARUM
  (EHRENBERG) HUGHES 1958
  AMERO,BASO OR GLOIO,HYALO OR PHAEO, ASCO STATE = MELANOPSAMMA FIDE HUGHES
    1958 OR CHAETOSPHAERIA FIDE MULLER IN KENDRICK 1971 (SUB GLIOBOTRYS)
    /ELLIS, DEMAT. HYPHO. P. 540-546,1971. (ILLUS,KEY)
    SYNO    =MEMNONIELLA HOHNEL 1923 FIDE SMITH 1962
            =GLIOBOTRYS HOHNEL 1902 FIDE ELLIS 1971
            =HYALOBOTRYS PIDOPLICHKA 1948 FIDE CARMICHAEL
            =HYALOSTACHYBOTRYS SRINIVASAN 1958 FIDE BARRON 1964
            =FUCKELINA SACCARDO 1875 FIDE BOOTH 1957
            =SYNSPORIUM PREUSS 1849 FIDE HUGHES 1958

STACHYLIDIUM LINK 1809. MAG. GES. NATURF. FREUNDE, BERLIN 3:15 /S. BICOLOR LINK
  =S. VERTICILLATUM (PERSOON) HUGHES 1958
  AMERO,GLOIO,PHAEO OR HYALO /BARRON, HYPHOMYCETES P. 286-289,1968. (ILLUS)
    SYNO    =SPONDYLOCLADIUM MARTIUS 1817 FIDE HUGHES 1958

STAGONOSTROMA DIEDICKE 1914. KRYPT. FL. MARK BRANDENB. 9:561-562 /S. DULCAMARAE
  (PASSER.) DIED. 1914 =STAGONOSPORA DULCAMARAE PASSER.
  ?. NOT ILLUSTRATED BY DIEDICKE. ASCO STATE =GIBBERELLA FIDE AINSWORTH 1971

## 10. Hyphomycetes

STAPHYLOTRICHUM MEYER ET NICOT 1956. BULL. SOC. MYCOL. FR. 72:322 /S. COCCOSPORUM
  AMERO,CETERI,HYALO, BUT NOT VERY DIFFERENT FROM BOTRYOTRICHUM SACC. ET MARCHAL /BARRON, HYPHOMYCETES P. 289-291,1968. (ILLUS)

STARKEYOMYCES AGNIHOTHRUDU 1956. J. INDIAN BOT. SOC. 35:40 /S. KOORCHALOMOIDES
  =KOORCHALOMA SUBRAMANIAN 1953 FIDE CARMICHAEL

STEMMARIA PREUSS 1851. LINNAEA 24:137, ILLUS IN STURM'S DEUTSCHL. FL. 6, 35-6:TAB. 67,1862 /S. GLOBOSA
  ?, ORIGINAL ILLUSTRATION NOT DIAGNOSTIC

STEMPHYLIOMMA SACCARDO ET TRAVERSO 1911. SYLLOGE FUNG. 20:886 /S. VALPARADISIACUM (SPEGAZZINI) SACC ET TRAV. 1911
  PHRAGMO,CETERI,PHAEO /ELLIS, DEMAT. HYPHO. P. 82-83,1971. (ILLUS)
    SYNO     =STEMPHYLIOPSIS SPEGAZZINI 1910, NON SMITH 1901

STEMPHYLIOPSIS SPEGAZZINI 1910. REV. FAC. AGRON. 6:193 /S. VALPARADISIACUM SPEG. 1910
  =STEMPHYLIOMMA SACC. ET TRAV. 1911, NON STEMPHYLIOPSIS SMITH 1901.

STEMPHYLIUM WALLROTH 1833. FL. CRYPT. GERMANIAE P. 300 /S. BOTRYOSUM =S. STATE OF P. HERBARUM (PERSOON) RABENHORST FIDE SIMMONS 1952
  DICTYO,CETERI,PHAEO, ASCO STATE =PLEOSPORA FIDE SIMMONS 1952 /SIMMONS, MYCOLOGIA 59:67-92,1967. (ILLUS)
    SYNO     =FUSICLADIOPSIS MAIRE 1906 FIDE HUGHES 1958
             =EPOCHNIELLA SACCARDO 1880 FIDE LINDAU IN RABENHORST 1910
             =SCUTISPORIUM PREUSS 1851 FIDE LINDAU IN RABENHORST 1910
             =SOREYMATOSPORIUM CAMARA 1930 FIDE AINSWORTH 1971
             =THYRODOCHIUM WERDERMANN 1924 FIDE WILTSHIRE FROM AINSWORTH 1971
             =THYROSPORA TEHON ET DANIELS 1925 FIDE SMITH FROM AINSWORTH 1971

STENELLA SYDOW 1930. ANN. MYCOL. 28:205 /S. ARAGUATA
  PHRAGMO,BLASTO,PHAEO /DEIGHTON, TRANS. BRIT. MYCOL. SOC. 56:411-418,1971 (ILLUS).

STENELLOPSIS HUGUENIN 1966. BULL. SOC. MYCOL. FR. 81:695 (DATED 1965) /S. FAGRAEAE
  PHRAGMO,CETERI,PHAEO /ELLIS, DEMAT. HYPHO. P. 267-268,1971. (ILLUS)

STENOSPORA DEIGHTON 1969. MYCOL PAPERS CMI 118:22-24 (ILLUS) /S. UREDINICOLA
  ?, NOT VERY DIFFERENT FROM CERCOSPORELLA

STEPHANOMA WALLROTH 1833. FLORA CRYPT. GERMAN. 2:269 /S. STRIGOSUM
  AMERO OR STAURO,CETERI,HYALO /BARRON, HYPHOMYCETES P. 292-293,1968. (ILLUS)
    SYNO     =ASTEROTHECIUM WALLROTH 1836 FIDE LINDAU 1907
             =SYNTHETOSPORA MORGAN 1892 FIDE CLEMENTS AND SHEAR 1931

STEPHANOSPORIUM DAL VESCO 1961. ALLIONIA 7:182 /S. ATRUM =S. CEREALE (THUMEN) SWART 1965
  AMERO,ARTHRO,PHAEO /ELLIS, DEMAT. HYPHO. P. 35-36,1971. (ILLUS)

STERIGMATOBOTRYS OUDEMANS 1886. NEDERL. KRUIDK. ARCH. II,4:548 /S. ELATA (SACC.) OUDEMANS =S. MACROCARPA (CORDA) HUGHES 1958
  PHRAGMO,GLOIO,PHAEO OR HYALO /HUGHES, CAN. J. BOTANY 36:814,1958. (REFS)
    SYNO     =ATRACTINA HOHNEL 1904 FIDE HUGHES 1958
             =PHRAGMOSTACHYS COSTANTIN 1888 FIDE HUGHES 1958
             =GLIODENDRON SALONEN ET RUOKOLA 1969 FIDE KENDRICK

STERIGMATOCYSTIS CRAMER 1859. VIERT. NAT. GES., ZURICH 4:323 /S. NIGRA (VAN TIEGHEM) =A. NIGER VAN TIEGHEM
  =ASPERGILLUS LINK 1809 FIDE CLEMENTS AND SHEAR 1931

STEVENSOMYCES MORRIS ET FINLEY 1965. MYCOLOGIA 57:483-485 (ILLUS) /S. PALMAE (STEVENS ET KING)
  AMERO,CETERI,HYALO

STIGMINA SACCARDO 1880. MICHELIA 2:22 /S. PLATANI (FUCKEL) (LECTOTYPE, FIDE HUGHES 1958)

```
PHRAGMO OR DICTYO,CETERI,PHAEO /ELLIS, MYCOL PAPERS CMI 72:36-71,1959.
    (ILLUS,KEY)
    SYNO    =THYROSTROMA HOHNEL 1911 FIDE ELLIS 1971
            =THYROSTROMELLA SYDOW 1924 FIDE ELLIS 1971
            =COLUMNOPHORA BUEAK ET VLEUGEL 1916 FIDE NAG RAJ AND KENDRICK

STIGMOPSIS BUBAK 1914. ANN. MYCOL. 12:218 /S. MONTELLICA (SACCARDO) (LECTOTYPE)
    =CHEIROMYCES MONTELLICA (SACC.) MOORE 1958
    =CHEIROMYCES BERKELEY ET BROOME 1857 FIDE MOORE 1958

STILBELLA LINDAU 1900. NAT. PFLANZENFAM. TEIL I, ABT. 1**, P. 489 (NO ILLUS)
    /S. ERYTHROCEPHALA (DITMAR) (LECTOTYPE)
    =STILBUM PERSOON 1801

STILBELLULA BOEDIJN 1951. SYDOWIA 5:227-228 (ILLUS) /S. PALLIDA
    =(?)CEPHALOTRICHUM LINK (SEE SUB STYSANUS PHILLIPSII IN MYCOL. PAPERS CMI
        56:40,1953)

STILBOCHALARA FERDINANDSEN ET WINGE 1910. BOT. TIDSSKR. 30:220-222 (ILLUS) /S.
    DIMORPHOSPORA FERD. ET WINGE 1910
    =THIELAVIOPSIS WENT 1893 FIDE ELLIS 1971

STILBODENDRON SYDOW 1916. ANN. MYCOL. 14:260-262 (ILLUS) /S. CAMERUNENSE
    ?

STILBOMYCES ELLIS ET EVERHART 1896. PROC. ACAD. PHIL. 1895:441 /S. BERENICE
    ?, NOT ILLUS. BY AUTHORS.  DESCR. SUGGESTS ARTHROSPORIUM SACC.

STILBOTHAMNIUM HENNINGS 1897. ENGLER'S BOT. JHB. 23, 4:542 AND TAF 14, FIG. 69A
    TO C /S. TOGOENSE
    ?, ORIGINAL ILLUSTRATION NOT DIAGNOSTIC

STILBUM PERSOON 1801. SYN. METH. FUNG. P. 680 /S. RIGIDUM (LECTOTYPE, FIDE
        HUGHES 1958)
    AMERO,GLOIO,HYALO /BARRON, HYPHOMYCETES P. 294-295,1968. (ILLUS)
        SYNO    =STILBELLA LINDAU 1900
                =CEPHALOPHORUM NEES FIDE SACCARDO FROM AINSWORTH 1971
                ?=POLYCEPHALOMYCES KOBAYASI 1941

STREBLOCAULIUM CHEVALLIER 1837. FUNG. ET BYSS. ILLUS., FASC. 1, NO. 19 /S.
    ATRO-VIRENS
    ?, LOOKS LIKE CONOPLEA PERSOON 1801

STREPTOTHRIX CORDA 1839. PRACHTFLORA, P 27 /S. FUSCA =CONOPLEA FUSCA PERSOON
        1822
    =CONOPLEA PERSOON 1801 FIDE HUGHES 1958

STROMATERIA CORDA 1837. ICONES FUNGORUM 1:5, FIG. 81 /S. CARNEA
    =TUBERCULARIA FIDE SACCARDO FORM AINSWORTH 1971

STROMATOCREA COOKE 1952. MYCOLOGIA 44:248-250 (ILLUS) /S. CEREBRIFORME
    AMERO,BASO,PHAEO, ASCO STATE =HYPOCREOPSIS FIDE AINSWORTH 1971

STROMATOGRAPHIUM HOHNEL 1907. DENK. AKAD. WIEN 83:37 /S. STROMATICUM (BERK.)
        =STILBUM STROMATICUM BERK.
    ?, NOT ILLUSTRATED BY HOHNEL

STROMATOSTYSANUS HOHNEL 1919. BERL. DEUTSCHL. BOT. GES. 37:153 /S.
        CAPRIFOLIORUM (DESMAZIERES)
    =GRAPHIOTHECIUM FUCKEL FIDE CLEMENTS AND SHEAR 1931

STRUMELLA FRIES 1825. SUMMA VEG. SCAND. 2:482 /S. CUCURBITACEARUM (LECTOTYPE
        FIDE HUGHES 1958)
    =NOMEN DUBIUM FIDE HUGHES 1958

STRUMELLA SACCARDO 1880. MICHELIA 2:36 (NO ILLUS) /S. OLIVATRA (SACC.) SACC
    1880 =ILLOSPORIUM OLIVATRUM (LECTOTYPE)
    HELICO,CETERI,PHAEO, ILLEGITIMATE BUT IN USE (NON STRUMELLA FRIES 1825).
        ASCO STATE =URNULA FIDE WOLF 1958
        SYNO    =DACRINA FRIES 1832 (NON FRIES 1825) FIDE LINDAU IN RABENHORST 1910
```

## 10. Hyphomycetes

STRUMELLOPSIS HOHNEL 1909. S.B. AKAD. WISS. WIEN 118:896-897 /S. ANNULARIS (RACIBORSKI)
?, NOT ILLUSTRATED BY HOHNEL

STYSANOPSIS FERRARIS 1909. ANN. MYCOL. 7:281 /S. MEDIA (SACCARDO) FERRARIS
=CEPHALOTRICHUM MEDIUM (SACC.) HUGHES 1958 (LECTOTYPE FIDE CARMICHAEL)
=CEPHALOTRICHUM LINK 1809

STYSANUS CORDA 1837. ICONES FUNGORUM 1:22 /S. STEMONITIS (PERSOON) =C. STEMONITIS (PERS.) LINK 1809
=CEPHALOTRICHUM LINK 1809 FIDE HUGHES 1958

SUBRAMANIA RAO ET RAO 1964. TRANS. AM. MICROSC. SOC. 83:399-406 (ILLUS) /S. SUNDARA
?=PERICONIA TODE EX PERSOON 1801

SUBULISPORA TUBAKI 1971. TRANS. MYCOL. SOC. JAPAN 12:20-22 (ILLUS) /S. PROCURVATA
SCOLECO,CETERI,HYALO

SYMPHYOSIRA PREUSS 1852. LINNAEA 25:274 /S. LUTEA
?, NOT ILLUS. BY AUTHOR, ASCO STATE =SYMPHYOSIRINIA FIDE AINSWORTH

SYMPODIELLA KENDRICK 1958. TRANS. BR. MYCOL. SOC. 41:519-521 (ILLUS) /S. ACICOLA
AMERO,ARTHRO,HYALO /BARRON, HYPHOMYCETES P. 296-297,1968. (ILLUS)

SYMPODINA SUBRAMANIAN ET LODHA 1964. ANTONIE VAN LEEUWENHOEK 30:317 /S. COPROPHILA SUBR. ET LODHA 1964
=VERONAEA CIFERRI ET MONTEMARTINI 1957 FIDE VON ARX 1970

SYMPODIOPHORA ARNOLD 1970. NOVA HEDWIGIA 19:301-304 (ILLUS) /S. STEREICOLA
DIDYMO,CETERI,HYALO

SYNCOLLESIA AGARDH 1824. SYST. ALG., P.32 /S. FOLIORUM
?, PROBABLY A NOMEN DUBIUM

SYNGLIOCLADIUM PETCH 1932. TRANS. BRIT. MYCOL. SOC. 17:177 /S. ARANEARUM
?, NOT ILLUSTRATED BY AUTHOR /MORRIS, WESTERN ILLINOIS UNIV SER BIOL SCI 3:126-127,1963. ILLUS

SYNNEMATIUM SPEARE 1920. MYCOLOGIA 12:74-75 AND PL. 4, FIGS. 1-15 /S. JONESII
AMERO,GLOIO,HYALO, A SYNNEMATOUS ACREMONIUM (CEPHALOSPORIUM AUCT.). SEE ALSO CORALLINOPSIS /MORRIS, WESTERN ILLINOIS UNIV SER BIOL SCI 3:128-129,1963. (ILLUS)

SYNPENICILLIUM COSTANTIN 1888. BULL. SOC. MYCOL. FR. 4:67 /S. ALBUM
=DORATOMYCES PUTREDINIS (CORDA) MORTON ET SMITH 1963
=CEPHALOTRICHUM LINK 1809 FIDE MORTON AND SMITH 1963 SUB DORATOMYCES

SYNPHRAGMIDIUM STRAUSS 1853. STURMS DEUTSCHL. FLORA III (PILZE) BD. 7, HEFT 33/34:41 /S. KUMMERI
?, HUGHES 1958 COULD NOT FIND THE TYPE, ?NOMEN DUBIUM

SYNSPORIUM PREUSS 1849. KLOTZSCHII HERB. VIV. MYCOL. NO 1285, ANNO 1849 /S. BIGUTTATUM PREUSS =STACHYBOTRYS CHARTARUM (EHRENBERG) HUGHES 1958
=STACHYBOTRYS CORDA 1837 FIDE HUGHES 1958

SYNSTERIGMATOCYSTIS COSTANTIN =GIBELLULA FIDE PETCH FROM AINSWORTH 1971

SYNTHETOSPORA MORGAN 1892. BOT. GAZ. 46:192 /S. ELECTA
=STEPHANOMA WALLROTH 1833 FIDE CLEMENTS AND SHEAR 1931

TAENIOLA BONORDEN 1851. HANDB. ALLGEM. MYKOL. P.36 /SIX ORIGINAL SPECIES
=NOMEN ILLEGITIMUM FIDE HUGHES 1958

TAENIOLELLA HUGHES 1958. CAN. J. BOTANY 36:816-817 (REFS) /T. EXILIS (KARSTEN)
PHRAGMO,BLASTO,PHAEO. DIFFERENCE FROM HETEROCONIUM NOT CLEAR /ELLIS, DEM. HYPH. P. 91-94,1971. (ILLUS,KEY)

TANDONELLA PRASAD ET VERMA 1970. INDIAN PHYTOPATH. 23:111-113 (ILLUS) /T. ZIZYPHI
?, SIMILAR TO CLADOSPORIUM

TAPEINOSPORIUM BONORDEN 1853. BOT. ZEIT. 11:285, TAB. VII, FIG. 6 /T. VIRIDE
=SEPTOCYLINDRIUM BONORDEN 1851 FIDE LINDAU IN RABENHORST 1907

TAWDIELLA DESHPANDE ET DESHPANDE 1966. MYCOPATHOLOGIA 28:206-208 (ILLUS) /T. DOLIIFORMIS DESHP. ET DESHP. 1966
?, ORIGINAL ILLUSTRATION NOT DIAGNOSTIC

TELLIGIA HENDRICKX 1948. PUBL. INST. NAT. ETUDE AGRON. CONGO BELGE, SER. SCI. 35:8 /T. CINNABARINA (TORREND) HENDR. =GILLETIA CINNABARINA TORR. 1914
?, NOT ILLUS. BY HENDRICKX. FOR GILLETIA TORREND NON SACCARDO

TERATOSPERMA SYDOW 1909. ANN. MYCOL. 7:172-173 (ILLUS) /T. SINGULARE
PHRAGMO OR STAURO,CETERI,PHAEO /ELLIS, MYCOL PAPERS CMI 69:1-7,1957. (ILLUS,KEY)

TERMITOSPHAERA CIFERRI 1935. ATTI IST. BOT. UNIV. PAVIA, SER. 4 6:242 /T. DUTHIEI (BERKELEY) =AEGERITA DUTHEI BERK.
?, ORIGINAL ILLUSTRATION NOT DIAGNOSTIC. BASIDIO STATE =TERMITOMYCES HEIM 1942 FIDE AINSWORTH 1971

TETRACHAETUM INGOLD 1942. TRANS. BR. MYCOL. SOC. 25:377-379 AND FIGS. 27-29 /T. ELEGANS
STAURO,CETERI,HYALO /TUBAKI, BULL. NAT. SCI. MUSEUM, TOKYO 41:261-262,1957. (ILLUS)

TETRACHIA SACCARDO 1921. BULL. ORT. BOT. NAPOLI 6:65 (NO ILLUS) /T. SINGULARIS SACCARDO
=SPEGAZZINIA SACC. 1880 FIDE BOEDIJN FROM AINSWORTH 1961

TETRACLADIUM DE WILDEMAN 1899. ANN. SOC. BELG. MICR. 17:35 /T. MARCHALIANUM
STAURO,CETERI,HYALO /TUBAKI, BULL. NAT. SCI. MUSEUM, TOKYO 41:261-262,1957. (ILLUS)
SYNO    =MAXILLOSPORA HOHNEL 1914 FIDE INGOLD FROM AINSWORTH 1971

TETRACOCCOSPORIUM SZABO 1905. HEDWIGIA 44:77 /T. PAXIANUM
DICTYO,CETERI,PHAEO /ELLIS DEMAT. HYPHO. P. 81-83,1971. (ILLUS)

TETRACOLIUM KUNZE EX LINK 1824. IN LINN. SPEC. PLANT. IV, 61(1):125 /T. TUBERCULARIAE (NEES ET NEES) KUNZE IN LINK 1824
=NOMEN DUBIUM FIDE HUGHES 1958

TETRACRIUM HENNINGS 1902. HEDWIGIA 41:116 (NO ILLUS) /T. AURANTII
STAURO,CETERI,HYALO, CLOSE TO OR THE SAME AS PETRAKIOPSIS. ASCO STATE =PUTTEMANSIA FIDE HOHNEL 1911

TETRACYTUM VANDERWALLE 1945. PARASITICA 1:149 /T. LAURI
=CYLINDROCLADIUM MORGAN 1892 (SUB CANDELOSPORA) FIDE WORMALD FROM AINSWORTH 1971

TETRAPLOA BERKELEY ET BROOME 1850. ANN. NOT. HIST. 2:5:459 /T. ARISTATA
STAURO,CETERI,PHAEO /ELLIS, DEMAT. HYPHO. P. 51-52,1971 (ILLUS)

TETRAPOSPORIUM HUGHES 1951. MYCOL. PAPERS CMI 46:25-28 (ILLUS) /T. ASTERINEARUM
STAURO,CETERI,PHAEO

THALLOMICROSPORON BENEDEK 1964. MYCOPATHOLOGIA 23:96 /T. KUEHNII BENEDEK ?=M. GYPSEUM FIDE WEITZMAN IN AJELLO 1968
=MICROSPORUM GRUBY 1843

THALLOSPORA OLIVE 1948. MYCOLOGIA 40:11-12 AND FIG. 3B-E ON P. 18 /T. ASPERA
STAURO,CETERI,HYALO /BARNETT, IMPERFECT FUNGI NO. 127,1960. (ILLUS)

THANATOPHYTUM NEES 1817 =RHIZOCTONIA FIDE LINDAU IN RABENHORST 1910

THAROOPAMA SUBRAMANIAN 1956. J. INDIAN BOT. SOC. 35:85 /T. TRINA

?=PHAEOISARIA HOHNEL 1909 /MORRIS, WESTERN ILLINOIS UNIV SER BIOL SCI
   3:130-131,1963. (ILLUS)

THECLOSPORA HARKNESS 1884. BULL. CALIF. ACAD. SCI. 1:41 /T. BIFIDA
   =EMERICELLA BERKELEY ET BROOME (ASCO) FIDE PEEK AND SOLHEIM 1958

THERMOIDIUM MIEHE 1907. BER. DEUTSCH. BOT. GES. 25:515 /T. SULFUREUM
   =MALBRANCHEA SACCARDO 1892 FIDE SACCARDO FROM AINSWORTH 1961

THERMOMYCES TSIKLINSKY 1899. ANN. INST. PASTEUR 13:500 /T. LANUGINOSUS
   AMERO,CETERI,PHAEO, BUT NOT VERY DIFFERENT FROM HUMICOLA /BARRON,
   HYPHOMYCETES P. 297-298,1968. (ILLUS)

THIELAVIOPSIS WENT 1893. MEDED. PROEFST. W. JAVA 7:4 /T. ETHACETICA =T. STATE
   OF CERATOCYSTIS PARADOXA FIDE ELLIS 1971
   AMERO,BASO,PHAEO, ASCO STATE =CERATOCYSTIS, PHIALOCONIDIAL STATE =CHALARA
   /ELLIS, DEMATIACEOUS HYPHOMYCETES P. 31-32,1971. (ILLUS)
      SYNO    =HUGHESIELLA BATISTA ET VITAL 1956 FIDE ELLIS 1971
              =STILBOCHALARA FERD. ET WINGE 1910 FIDE ELLIS 1971

THOMIELLA DODGE 1935. MEDICAL MYCOLOGY P. 834 (NO ILLUS) /T. DESSYI (SPEGAZZINI)
   =GONATOBOTRYUM SACCARDO 1886 FIDE CARMICHAEL

THOZETELLA KUNTZE 18  . REVISIO GEN. PLANT. P. 873 (NO ILLUS) /T. NIVEA
   (BERKELEY) KUNTZE =THOZETIA NIVEA BERKELEY
   =NOMEN DUBIUM FIDE AGNIHOTHRUDU 1958 /AGNIHOTHRUDU, MYCOLOGIA
   50:575-576,1958. (REFS)
      SYNO    =THOZETIA BERKELEY ET MUELLER 1881 NON THOZETIA MUELLER

THOZETELLOPSIS AGNIHOTHRUDU 1958. MYCOLOGIA 50:570-579 (ILLUS) /T. TOCKLAIENSIS
   AMERO,GLOIO,HYALO /BARRON, HYPHOMYCETES P. 299-300,1968. (ILLUS)

THOZETIA BERKELEY ET MUELLER 1881. JOUR. LINN. SOC. 18:388 /T. NIVEA
   =THOZETELLA, NON THOZETIA MUELLER

THYRODOCHIUM WERDERMANN 1924. ANN. MYCOL. 22:188 (ILLUS) /T. DRACAENAE
   =STEMPHYLIUM FIDE WILTSHIRE FROM AINSWORTH 1971

THYROSPORA TEHON ET DANIELS 1925. PHYTOPATH. 15:718 (ILLUS) /T. SARCINIFORME
   =STEMPHYLIUM FIDE SMITH FROM AINSWORTH 1971

THYROSTROMA HOHNEL 1911. S. B. AKAD. WISS. WIEN., ABT. 1, 120:472 /T. COMPACTUM
   (SACCARDO)
   =STIGMINA SACCARDO 1880 FIDE ELLIS 1971

THYROSTROMELLA HOHNEL 1919. BERL. DEUTSCHL. BOT. GES. 37:157 /T. MYRIANA
   (DESMAZIERES)
   DICTYO,CETERI,PHAEO, NON THYROSTROMELLA SYDOW 1924 /HUGHES, CAN. J. BOTANY
   33:341-343,1955. (ILLUS)

THYROSTROMELLA SYDOW 1924. ANN. MYCOL. 22:406-407 /T. TRIMERA (SACCARDO)
   =STIGMINA SACCARDO 1880 FIDE ELLIS 1971

THYRSIDIUM MONTAGNE 1849. IN DURIEU, EXPL. SCI. ALGERIE BOT. 1:325 /T.
   BOTRYOSPORUM (MONT.) MONT. 1849
   =CHEIROSPORA MOUGEOT ET FRIES 1825 FIDE HUGHES 1958

THYSANOPHORA KENDRICK 1961. CAN. J. BOTANY 39:817-832 (ILLUS) /T.
   PENICILLIOIDES (ROUMEGUERE)
   AMERO,BASO,HYALO

THYSANOPYXIS RABENHORST 1864. ABH. NAT. GES. HALLE 8:136 /T. PULCHELLA (CESATI)
   =VOLUTELLA PULCHELLA RABENH. 1850
   =VOLUTELLA TODE EX FRIES 1832 FIDE LINDAU 1910

TILACHLIDIOPSIS KEISSLER 1924. ANN. NATURH. MUS. WIEN 37:215-216 (ILLUS) /T.
   RACEMOSA
   ? /MORRIS, WESTERN ILLINOIS UNIV SER BIOL SCI 3:132-133,1963. (ILLUS)

TILACHLIDIUM PREUSS 1851. LINNAEA 24:126-127 (NO ILLUS) /T. PINNATUM =T.
    BRACHIATUM (BATSCH EX FRIES) PETCH 1937
    ?, NOT VERY DIFFERENT FROM ACREMONIUM /MORRIS, WESTERN ILLINOIS UNIV SER BIOL
    SCI 3:134-135,1963. (ILLUS)

TITAEA SACCARDO 1876. NUOV. GIORN. ITAL. 8:193 (NO ILLUS) /T. CALLISPORA
    STAURO,CETERI,HYALO
        SYNO    =AORATE SYDOW 1929 FIDE AINSWORTH 1961
                =ARANEOMYCES HOHNEL 1909 FIDE DAMON (FROM AINSWORTH 1961)
                =MONOGRAMMIA STEVENS 1917 FIDE DAMON (FROM AINSWORTH 1961)

TITAEELLA ARNAUD 1951. BULL. SOC. MYCOL. FRANCE 67:196 AND FIG. 5I /T.
    CAPNOPHILA
    ?, NO LATIN DIAGNOSIS, HAS CLAMP CONNECTIONS

TOLYPOCLADIUM GAMS 1971. PERSOONIA 6:185-191 (ILLUS) /T. INFLATUM
    AMERO,GLOIO,HYALO, COMPARE WITH TRICHODERMA AND HARPOSPORIUM

TOLYPOMYRIA PREUSS 1852. LINNAEA 25:726 /T. PRASINA
    =TRICHODERMA PERSOON 1801 FIDE HUGHES 1969

TOMENTICOLA DEIGHTON 1969. MYCOL. PAPERS CMI 117:20-25 (ILLUS) /T. TREMATIS
    PHRAGMO,CETERI,PHAEO

TORULA (PERSOON) LINK 1809. MAG. GES. NATURF. FREUNDE, BERLIN 3:21 /MONILIA
    HERBARUM PERSOON =TORULA HERBARUM (PERSOON) LINK 1815
    PHRAGMO,BLASTO,PHAEO /ELLIS. DEMAT. HYPHO. P. 336-339,1971. (ILLUS,KEY)
        SYNO    =HORMISCIUM KUNZE 1817 FIDE HUGHES 1958
                =SATWALEKERA RAO,RAO ET RAO 1969 FIDE CARMICHAEL
                =BAHUSANDIHIKA SUBRAMANIAN 1966 FIDE KENDRICK

TORULELLA GYELNIK 1938. LILLOA 4:64 /T. ASPERELLAE
    ?, ORIGINAL ILLUSTRATION NOT DIAGNOSTIC

TORULINA SACCARDO ET SACCARDO 1906. SYLLOGE FUNGORUM 18:566 /T. SEROTINAE
    (OUDEMANS) =G. MURORUM
    =GLIOMASTIX GUEGUEN 1905 FIDE DICKINSON 1968

TORULOIDEA SUMSTINE 1913. MYCOLOGIA 5:53 AND PL. 34, FIG. 1 /T. EFFUSA
    ?, ILLUS. AND DESCR. SUGGEST SCOPULARIOPSIS

TORULOMYCES DELITSCH 1943. IN LEMBKE UND DELITSCH, SYSTEMATIK DER
    SCHIMMELPILZE, P. 91-19 AND TAF. 30, FIGS. 232-235 /T. LAGENA
    AMERO,BASO,PHAEO /BARRON, HYPHOMYCETES P. 302-304,1968. (ILLUS)
        SYNO    =MONOCILLIUM SAKSENA 1955 FIDE KENDRICK

TORULOPSIELLA BENDER 1932. MYCOLOGIA 24:411 (NO ILLUS) /TORULOPSIS FUMAGINEA
    SPEGAZZINI
    ?, FOR TORULOPSIS SPEGAZZINI NON BERLESE

TORULOPSIS OUDEMANS NON BERLESE, SEE TORULINA

TORULOPSIS SPEGAZZINI NON BERLESE, SEE TORULOPSIELLA

TRETOPILEUS DODGE 1946. BULL. TORREY BOT. CLUB 73:223 AND FIGS. 1 & 2 /T.
    OPUNTIAE =T. SPHAEROPHORUS (BERKELEY ET CURTIS) HUGHES ET DEIGHTON 1960
    DICTYO,CETERI,PHAEO /DEIGHTON, MYCOL PAPERS CMI 78:1-4,1960. (ILLUS)

TRIADELPHIA SHEARER ET CRANE 1971. MYCOLOGIA 63:247-249 (ILLUS) /T. HETEROSPORA
    PHRAGMO,CETERI,PHAEO

TRICELLULA VAN BEVERWIJK 1954. ANTONIE VAN LEEUWENHOEK 20:11-15 (ILLUS) /T.
    INAEQUALIS
    STAURO,CETERI,HYALO /VON ARX, GENERA OF FUNGI IN CULTURE P. 216,1970. (ILLUS)
        SYNO    =VOLUCRISPORA HASKINS 1958 FIDE VON ARX 1970

TRICHAEGUM CORDA 1837. ICONES FUNGORUM 1:15 (ILLUS) /T. CLADOSPORIS
    ?, ORIGINAL ILLUSTRATION NOT DIAGNOSTIC

## 10. Hyphomycetes

TRICHOBOTRYS PENZIG ET SACCARDO 1901. MALPIGHIA 15:245 /T. PANNOSA
    AMERO,BLASTO,PHAEO /ELLIS, DEMAT. HYPHO. P. 340-341,1971. (ILLUS)

TRICHOCEPHALUM COSTANTIN 1887. LES MUCEDINEES SIMPLES P. 106 /T. CURTUM
    (BERKELEY)
    =PERICONIA TODE EX PERSOON 1801 FIDE ELLIS 1971

TRICHOCLADIUM HARZ 1871. BULL. SOC. IMP. MOSCOW 44:125 /T. ASPERUM
    DIDYMO OR PHRAGMO,CETERI,PHAEO /BARRON, HYPHOMYCETES P. 305-306,1968. (ILLUS)
    ELLIS, DEMAT. HYPHO. P. 66-68,1971. (ILLUS,KEY)

TRICHOCONIS CLEMENTS 1909. GEN. FUNG. 145,176 (NO ILLUS) /T. CAUDATA (APPEL ET
    STRUNK) =PIRICULARIA CAUDATA APPEL ET STRUNK (ILLUS IN ZENTRALB. BAKT.,
    ABT. 2,11:556-557,1903)
  PHRAGMO,CETERI,HYALO. COMPARE WITH MONOTRICHUM
        SYNO    =APPELIA (SACC.) SACCARDO FIDE AINSWORTH 1971

TRICHODERMA PERSOON 1801. SYN. METH. FUNG. P. 23 /T. VIRIDE (LECTOTYPE FIDE
    HUGHES 1958)
    AMERO,GLOIO,HYALO, ASCO STATE =HYPOCREA FIDE RIFAI 1969 /RIFAI, MYCOL PAPERS
    CMI 116:1-56,1969. (ILLUS,KEY,REVIS)
        SYNO    =ALEURISMA LINK 1809 FIDE HUGHES 1958
                =PACHYBASIUM SACCARDO 1885 FIDE HUGHES 1958
                =PYRENIOPSIS KUNTZE 1898 FIDE RIFAI 1969
                =SPORODERMA MONTAGNE 1856 SEE RIFAI 1969
                =TOLYPOMYRIA PREUSS FIDE HUGHES 1969
                ?=TOLYPOCLADIUM GAMS 1971

TRICHODERMIA HOFFMAN =TRICHOTHECIUM FIDE SACCARDO FROM AINSWORTH 1971

TRICHODOCHIUM SYDOW 1927. ANNALES MYCOLOGICI 25:159 /T. DISSEMINATUM
    DIDYMO,BASO,PHAEO /ELLIS, MYCOL PAPERS CMI 111:36-39,1967. (ILLUS)

TRICHOFUSARIUM BUBAK 1906. BULL. HERB. BOISS. 2(6):488 (NO ILLUS) /FUSARIUM
    ROSEUM VAR. RUSCI SACCARDO
    =FUSARIUM LINK 1809 FIDE AINSWORTH 1961

TRICHOLECONIUM CORDA 1837. ICONES FUNGORUM 1:17 /T. ROSEUM CORDA =SARCOPODIUM
    CIRCINATUM EHRENBERG 1818 FIDE HUGHES
    =SARCOPODIUM EHRENBERG 1818 FIDE HUGHES 1958

TRICHOMYCES MALMSTEN 1845. TRANSLATION IN ARCH. ANAT. PHYSIOL. WISS. MED.
    1848:1-19 /T. TONSURANS
    =TRICHOPHYTON MALMSTEN 1845 FIDE DODGE 1935

TRICHOPHYTON MALMSTEN 1845. ARCH. ANAT. PHYS. WISS. MED. 1848:1 (THIS IS A
    TRANSLATION INTO GERMAN OF THE RARE SWEDISH ORIGINAL PUBLICATION OF 1845)
    /T. TONSURANS
  PHRAGMO,CETERI,HYALO, ASCO STATE =ARTHRODERMA /REBELL AND TAPLIN,
    DERMATOPHYTES 124 PP., 1970. (ILLUS,KEY) PADHYE AND CARMICHAEL, CAN. J.
    BOTANY 49:1525-1540,1971(ILLUS,KEY)
        SYNO    =FAVOTRICHOPHYTON NEVEU-LEMAIRE 1921
                =ACHORION REMAK 1845
                =CHLAMYDOALEUROSPORIA GRIGORAKIS 1924
                =ECTOTRICHOPHYTON CASTELLANI ET CHALMERS 1919
                =ENDODERMOPHYTON CASTELLANI 1910
                =GRUBYELLA OTA ET LANGERON 1923
                =KAUFMANNWOLFIA GALGOCZY ET NOVAK 1962
                =LANGERONIA VANBREUSEGHEM 1950 FIDE AJELLO 1968
                =LOPHOPHYTON MATRUCHOT ET DASSONVILLE 1889
                =MEGATRICHOPHYTON NEVEU-LEMAIRE 1921
                =MICROTRICHOPHYTON NEVEU-LEMAIRE 1921
                =NEOTRICHOPHYTON CASTELLANI ET CHALMERS 1919
                =PINOYELLA CASTELLANI ET CHALMERS 1919 FIDE STOCKDALE, MACKENZIE
                AUSTWICK 1965
                =SABOURAUDIELLA BOEDIJN 1951
                =SCHOENLEINIUM JOHAN-OLSEN 1897 FIDE CARMICHAEL
                =TRICHOMYCES MALMSTEN 1845 FIDE DODGE 1935

TRICHOSPORIELLA KAMYSCHKO EX GAMS ET DOMSCH 1969. NOVA HEDWIGIA 18:19 /T. HYALINA
  ? /VON ARX, GENERA OF FUNGI IN CULTURE P. 232-234,1970. (ILLUS)

TRICHOSPORUM FRIES 1825. SUMMA VEG. SCAND. P. 492,1849 (WITHOUT ANY SPECIES IN 1825) /?
  =NOMEN ILLEGITIMUM, NON TRICHOSPORIUM DON 1822 FIDE HUGHES 1958

TRICHOSTEPIGMA PETCH 1923. TRANS. BRIT. MYCOL. SOC. 8:215 /T. CLAVISPORUM
  =HIRSUTELLA FIDE TRANS. BRIT. MYCOL. SOC. 9, FROM AINSWORTH 1971

TRICHOSTROMA CORDA 1829. STURMS DEUTSCHL. FLORA III, (PILZE), BD. 2, HEFT 9:131 /T. PURPURASCENS
  =NOMEN DUBIUM, SEE HUGHES 1958, NON TRICHOSTROMA LINK 1826 SINE SPECIEI NOMINE

TRICHOTHECA KARSTEN 1887. SYMB. MYC. 20:101 /T. ALBA
  ?, NOT ILLUSTRATED BY AUTHOR

TRICHOTHECIUM LINK 1809. MAG. GES. NATURF. FREUNDE, BERLIN 3:18 /T. ROSEUM (PERSOON)
  DIDYMO,BASO,HYALO, ASCO STATE =HYPOMYCES FIDE TUBAKI 1960 /RIFAI AND COOKE, TRANS. BRIT. MYCOL. SOC. 49:147-168,1966. (ILLUS, REVIS)
    SYNO    =HYPHELIA FRIES 1825 FIDE HUGHES 1958
            =CEPHALOTHECIUM CORDA 1838 (PROBABLY), SEE HUGHES 1958
            =TRICHODERMIA HOFFMAN FIDE SACCARDO FROM AINSWORTH 1971

TRICHURUS CLEMENTS ET SHEAR 1896. BOT. SURV. NEBRASKA 4:7 /T. CYLINDRICUS
  AMERO,BASO,PHAEO, BUT NOT VERY DIFFERENT FROM CEPHALOTRICHUM /BARRON, HYPHOMYCETES P. 310-311,1968. (ILLUS)

TRICLADIUM INGOLD 1942. TRANS. BRIT. MYCOL. SOC. 25:388 /T. SPLENDENS
  STAURO,CETERI,HYALO /TUBAKI, NAGAOA 1960:18-19,1960. (ILLUS)

TRICORNISPORA BONAR 1967. MYCOLOGIA 59:597 /T. BAMBUSAE
  =TRIDENTARIA PREUSS 1852 FIDE KENDRICK

TRIDENTARIA PREUSS 1852. LINNAEA 25:74 (NO ILLUS) /T. ALBA
  STAURO,CETERI,HYALO
    SYNO    =FUMAGOPSIS SPEGAZZINI 1910 FIDE KENDRICK
            =TRICORNISPORA BONAR 1967 FIDE KENDRICK

TRIGLYPHIUM FRESENIUS 1852. BEITR. ZUR. MYKOL. P. 44 AND PL. 9, FIG. 6 /T. ALBUM
  ? /LINDAU, RABENH. KRYPT. FL. 2 AUFL. 1 BD. 9:590-591,1910. (ILLUS)

TRIMMATOSTROMA CORDA 1837. ICONES FUNGORUM 1:9 /T. SALICIS= M. SALICIS (CORDA) HUGHES 1958
  =MELANCONIUM LINK 1809 FIDE HUGHES 1958

TRINACRIUM RIESS 1852. IN FRESENIUS, BEITR. MYKOLOGIE P. 42 AND TAF. V, FIGS. 14-17 /T. SUBTILE
  STAURO,CETERI,HYALO /BARNETT, IMPERFECT FUNGI NO. 112,1960. (ILLUS)

TRIPLICARIA KARSTEN 1889. HEDWIGIA 28:195 (NO ILLUS) /T. HYPOXYLOIDES
  ?, ASCO STATE =HYPOXYLON FIDE AINSWORTH 1961

TRIPOSPERMUM SPEGAZZINI 1918. PHYSIS 4:295 /T. ACERINUM (SYDOW)
  STAURO,CETERI,PHAEO /HUGHES, MYCOL PAPERS CMI 46:10-22,1951. (ILLUS,REVIS)

TRIPOSPORINA SPEGAZZINI 1918. S.B. AKAD. WISS. WIEN 121:410-411 (ILLUS) /T. UREDINICOLA HOHNEL 1912
  STAURO,CETERI,HYALO /DRECHSLER, MYCOLOGIA 29:531-535,1937. (ILLUS)

TRIPOSPORIUM CORDA 1837. ICONES FUNGORUM 1:16 AND TAB. IV, FIG. 220 /T. ELEGANS
  STAURO,CETERI,PHAEO /HUGHES, MYCOL PAPERS CMI 46:1-10,1951. (ILLUS)

TRISCELOPHORUS INGOLD 1943. TRANS. BRIT. MYCOL. SOC. 26:148-152 AND PL. VIII /T. MONOSPORUS
  STAURO,CETERI,HYALO /TUBAKI, BULL. NAT. SCI. MUS. TOKYO 3:263-264,1957. (ILLUS)

## 10. Hyphomycetes

TRISULCOSPORIUM HUDSON ET SUTTON 1964. TRANS. BRIT. MYCOL. SOC. 47:200 AND FIG. 1 /T. ACERINUM
   STAURO,CETERI,HYALO

TRITIRACHIUM LIMBER 1940. MYCOLOGIA 32:23-26 (ILLUS) /T. DEPENDENS
   AMERO,CETERI,HYALO /BARRON, HYPHOMYCETES P. 313-314,1968. (ILLUS)
      SYNO     =SPIROTRICHUM SAITO FIDE AINSWORTH 1971

TROCHOPHORA MOORE 1955. MYCOLOGIA 47:90 (NO ILLUS) /T. SIMPLEX (PETCH)
   HELICO,CETERI,PHAEO, NOM. NOV. FOR HELICOSTILBE HOHNEL EMEND. LINDER 1929
   /ELLIS, DEMAT. HYPHO. P. 185-186,1971 (ILLUS)

TROPOSPORELLA KARSTEN 1892. HEDWIGIA 31:299 /T. FUMOSA
   HELICO,CETERI,PHAEO

TROPOSPORIUM HARKNESS 1884. BULL. CALIFORNIA ACAD. SCI. 1:39-40 (NO ILLUS) /T. ALBUM
   HELICO,CETERI,HYALO /PEEK AND SOLHEIM, MYCOLOGIA 50:847-851,1958. (ILLUS)

TUBERCULARIA TODE EX PERSOON 1801. SYN. METH. FUNG. P. 111 /T. VULGARIS TODE 1790
   AMERO,GLOIO,HYALO, ASCO STATE =NECTRIA FIDE BARRON 1968 /BARRON, HYPHOMYCETES P. 314-315,1968. (ILLUS)
      SYNO     =KNYARIA KUNTZE FIDE AINSWORTH 1971
                  =STROMATERIA CORDA 1837 FIDE SACCARDO FROM AINSWORTH 1971

TUBERCULARIOPSIS HOHNEL 1909. S.B. AKAD. WISS. WIEN 118:421-422 (ILLUS) /T. ANOMALA
   ?

TUBERCULARIS CLEMENTS AND SHEAR 1931 FOR TUBERCULARIOPSIS

TUBERCULINA SACCARDO 1880. MICHELIA 2:34 /T. PERSICINA (DITMAR)
   AMERO,CETERI,PHAEO /VON ARX, GENERA OF FUNGI IN CULTURE P. 156-157,1970. (ILLUS)
      SYNO     =CORDALIA GOBI 1885 FIDE LINDAU IN RABENHORST 1910
                  =UREDINULA FIDE SPEGAZZINI FROM AINSWORTH 1971

TUREENIA HALL 1915. PHYTOPATHOLOGY 5:57 /T. JUNCOIDEA
   =ARTHRINIUM KUNZE 1817 FIDE ELLIS 1971

TYLOMYCES CORTINI 1921. RENDIC. R. ACCAD. LINCEI ROMA 30 SER. 5 FASC. 1-2:63 /T. GUMMIPARUS
   ?, ORIGINAL PUBLICATION NOT SEEN

ULOCLADIUM PREUSS 1851. IN STURM, DEUTSCHL. FLORA III (PILZE), HEFT 30:83 /U. BOTRYTIS
   DICTYO,CETERI,PHAEO /SIMMONS, MYCOLOGIA 59:67-92,1967. (ILLUS,KEY)
      SYNO     =PSEUDOSTEMPHYLIUM (WILTSHIRE) SUBRAMANIAN 1961 FIDE SIMMONS 1967

UMBELLULA MORRIS 1955. MYCOLOGIA 47:602-605 (ILLUS) /U. TERRESTRIS (TIMONIN)
   =PSEUDOBOTRYTIS KRZEM. ET BADURA 1954 FIDE SUBRAMANIAN 1956

UMBELOPSIS AMOS ET BARNETT 1966. MYCOLOGIA 58:805-808 (ILLUS) /U. VERSIFORMIS
   AMERO,CETERI,HYALO /BARRON, HYPHOMYCETES P. 316-318,1968. (ILLUS)

UNCIGERA SACCARDO 1885. IN SACCARDO AND BERLESE, ATTI R. ISTIT. VEN SCI LETT. ARTI VI,3:741 /U. CORDAE SACCARDO ET BERLESE 1885, TAB IX, FIG. 10
   AMERO OR DIDYMO,GLOIO,HYALO, SIMILAR TO GONYTRICHUM /LINDAU, RABENH. KRYPT. FL., 2 AUFL., 1 BD.,8:338,1907. (ILLUS)

UREDINULA SPEGAZZINI 1880. ANAL. SOC. CIENTIF. ARGENT. 213 /?
   =TUBERCULINA FIDE SPEGAZZINI FROM AINSWORTH 1971

UROBASIDIUM GIESENHAGEN 1892. FLORA, JENA, 76:139-141 /U. ROSTRATUM GIESENH. 1892
   =ZYGOSPORIUM MONTAGNE 1842 FIDE ELLIS 1971

UROPHIALA VUILLEMIN 1910. C. R. ACAD. SCI. PARIS 150:883 /U. MYCOPHILA
   =ZYGOSPORIUM MONTAGNE 1842 FIDE HUGHES 1958

UROSPORIUM FINGERHUTH 1836. LINNAEA 10:231-232 /U. CURVATUM FINGERH. 1836
  ?, NOT ILLUSTRATED BY AUTHOR

USTILAGINOIDEA BREFELD 1895. UNTERS. MYKOL. 12:195 AND TAF. XII, FIGS. 22-30
  /U. ORYZAE (PATOUILLARD) BREFELD =U. VIRENS (COOKE) TAKAHASHI 1896 FIDE
  ELLIS 1971 (LECTOTYPE)
    AMERO,CETERI,PHAEO /ELLIS, DEMAT. HYPHO. P. 42-43,1971. (ILLUS)
      SYNO    =DUBIOMYCES LLOYD 1921 FIDE DIEHL FROM AINSWORTH 1971

VAKRABEEJA SUBRAMANIAN 1956. J. INDIAN BOT. SOC. 35:466 /V. SIGMOIDEA (CAVARA)
  =NAKATAEA HARA 1939 FIDE CARMICHAEL

VANBEVERWIJKIA AGNIHOTHRUDU 1961. TRANS. BRIT. MYCOL. SOC. 44:51-54 (ILLUS) /V.
  SPIROSPORA
    HELICO,GLOIO,HYALO, EMENDED BY SHEARER AND CRANE 1971, PERHAPS THE SAME AS
    EVERHARTIA /SHEARER AND CRANE, MYCOLOGIA 63:249-251,1971. (ILLUS,EMEND)

VANBREUSEGHEMIA BALABANOFF 1965. MYCOPATHOLOGIA 25:345 /?
  =KERATINOMYCES VANBREUSEGHEM 1952 FIDE INDEX OF FUNGI 3:448,1968

VARICOSPORINA MEYERS ET KOHLMEYER 1965. CAN. J. BOTANY 43:915-921 (ILLUS) /V.
  RAMULOSA
    STAURO,CETERI,HYALO

VARICOSPORIUM KEGEL 1906. BERL. DEUTSCHL. BOT. GES. 24:213-216 (ILLUS) /V.
  ELODEAE
    STAURO,CETERI,HYALO /BARRON, HYPHOMYCETES P. 318-319,1968. (ILLUS)

VASCULOMYCES ASHBY 1913. BULL. DEPT. AGR. JAMAICA 2:151 AND PL. 28, FIGS. 22-23
  /V. XANTHOSOMAE
    =STERILE

VELLOSIELLA RANGEL 1915. BOL. AGRIC. SAO PAULO 16A:144-151 (ILLUS) /V. CAJANI
  (HENNINGS)
    =MYCOVELLOSIELLA RANGEL 1917, NON VELLOZIELLA BAILL. 1886

VERMICULARIOPSIELLA BENDER 1932. MYCOLOGIA 24:410 /FOR VERMICULARIOPSIS HOHNEL
  1929 NON TORREND
  ?, NOT ILLUSTRATED BY BENDER OR HOHNEL

VERONAEA CIFERRI ET MONTEMARTINI 1957. ATTI IST. BOT. LAB. CRITT. PAVIA
  5,15:68-70 (ILLUS) /V. BOTRYOSA CIF. ET MONTEM. 1957
    DIDYMO,CETERI,PHAEO OR HYALO, CLOSE TO SCOLECOBASIDIUM AND DACTYLARIA /VON
    ARX, GENERA OF FUNGI IN CULTURE P. 217-218,1970. (ILLUS)
      SYNO    =SYMPODINA SUBRAMANIAN ET LODHA 1964 FIDE VON ARX 1970

VERONAIA BENEDEK 1961. MYCOPATHOLOGIA 14:115 /V. CASTELLANII
  =NOMEN CONFUSUM SEE AJELLO 1968

VERRUCISPORA SHAW ET ALCORN 1967. PROC. LINN. SOC. N. S. WALES 92:171-173 AND
  PL. VI /V. PROTEACEARUM
    PHRAGMO,CETERI,PHAEO /ELLIS, DEMAT. HYPHO. P. 274-275,1971. (ILLUS)

VERTICICLADIELLA HUGHES 1953. CAN. J. BOTANY 31:653 /V. ABIETINA (PECK)
    AMERO,GLOIO,HYALO, ASCO STATE =CERATOCYSTIS FIDE KENDRICK 1962 /KENDRICK,
    CAN. J. BOTANY 40:771-797,1962. (ILLUS)

VERTICICLADIUM PREUSS 1851. LINNAEA 24:127 /V. TRIFIDUM
    AMERO,CETERI,HYALO, ASCO STATE =DESMAZIERELLA FIDE HUGHES 1951 /HUGHES, MYCOL
    PAPERS CMI 43:6-15,1951. (ILLUS)

VERTICILLIASTRUM DASZEWSKA 1912. BULL SOC. BOT. GENEV, SER. 2, 4:302-303 AND
  313 (ILLUS) /V. GLAUCUM DASZ. 1912
    ?, ORIGINAL ILLUS. AND DESCR. NOT DIAGNOSTIC

VERTICILLIODOCHIUM BUBAK 1914. ANN. MYCOL. 12:219-220 AND TAB. VIII, FIGS. 1-6
  /V. TUBERCULARIOIDES (SPEGAZZINI)
    ?, ORIGINAL ILLUSTRATION AND DESCRIPTION NOT DIAGNOSTIC
      SYNO    =VERTICILLIS CLEMENTS ET SHEAR 1931 (ORTHOGRAPHIC VARIANT)

## 10. Hyphomycetes

VERTICILLIOPSIS COSTANTIN 1892. COMPT. REND. 114:850 /V. INFESTANS
?, NOT ILLUSTRATED BY AUTHOR

VERTICILLIS CLEMENTS ET SHEAR 1931 FOR VERTICILLIODOCHIUM

VERTICILLIUM NEES 1817. DAS SYSTEM .... P. 57 AND TAB. IV, FIG. 55 /V. TENERUM
 = V. STATE OF NECTRIA INVENTA FIDE HUGHES 1958
 AMERO,GLOIO,HYALO, ASCO STATE =NECTRIA /VON ARX, GENERA OF FUNGI IN CULTURE
 P. 184-185,1970. (ILLUS,REFS)
  SYNO  =ACROSTALAGMUS CORDA 1838 FIDE HUGHES 1958
        =GIBELLULOPSIS BATISTA ET MAIA 1959 FIDE CARMICHAEL

VINCULUM ROY, DWIVEDI ET KHANNA 1965. TRANS. BRIT. MYCOL. SOC. 48:113-115 AND
 PL. 9 /V. INDICUM ROY, DWIVIDI ET KHANNA 1965
 =BARNETTELLA RAO ET RAO 1964 FIDE VERONA 1967

VIRGARIA NEES 1816. DAS SYSTEM .....P. 54 /V. NIGRA (LINK)
 AMERO,CETERI,PHAEO, NOT VERY DIFFERENT FROM NODULISPORIUM /BARRON,
 HYPHOMYCETES P. 323-324,1968. (ILLUS)

VIRGARIELLA HUGHES 1953. CAN. J. BOTANY 31:654 /V. GLOBIGERA (SACCARDO ET ELLIS)
 AMERO,CETERI,PHAEO, LIKE NODULISPORIUM WITH UNBRANCHED CONDIDIOPHORES /ELLIS,
 DEMAT. HYPHO. P. 195-196,1971. (ILLUS)

VIRGASPORIUM COOKE 1875. GREVILLEA 3:182 AND PL. 48, FIG. 5 /V. MACULATUM
 =CERCOSPORA FRESENIUS 1863 FIDE ELLIS 1971

VIRGATOSPORA FINLEY 1967. MYCOLOGIA 59:538-541 (ILLUS) /V. ECHINOFIBROSA
 PHRAGMO,GLOIO,PHAEO /ELLIS, DEMAT. HYPHO. P. 557-558,1971. (ILLUS)

VISCOMACULA SPRAGUE 1950. MYCOLOGIA 42:758-759 /V. AENEA
 ?, NOT ILLUSTRATED BY AUTHOR

VOGLINOANA KUNTZE 1891. REV. GEN. PLANT. 2:874 (NO ILLUS) /V. CRATERIOIDES
 (RABENHORST) KUNTZE =CYSTOPHORA CRATERIOIDES RABENH.
 =NOMEN DUBIUM FIDE LINDAU 1907 (SUB CYSTOPHORA)

VOLUCRISPORA HASKINS 1958. CAN. J. MICROBIOL. 4:278 /V. AURANTIACA
 =TRICELLULA VAN BEVERWIJK 1954 FIDE VON ARX 1970

VOLUTELLA TODE EX FRIES 1832. SYSTEMA MYCOL. III:467 /V. CILIATA (ALBERTINI ET
  SCHWEINITZ) FRIES (NEOTYPE)
 AMERO,GLOIO,HYALO, ASCO STATE =PSEUDONECTRIA FIDE VON ARX 1970, VOLUTELLA IS
 A HOMONYM OF AN EARLIER NON-FUNGUS NAME /VON ARX, GENERA OF FUNGI IN
 CULTURE P. 182-183,1970. (ILLUS)
  SYNO  =CHAETODOCHIUM HOHNEL 1932 FIDE VON ARX 1970
        =MEDUSULA CORDA 1837 FIDE LINDAU IN RABENHORST 1910
        =THYSANOPYXIS RABENHORST 1864 FIDE LINDAU 1910
        =SCHIZOTRICHELLA MORRIS 1956 FIDE CARMICHAEL
        ?=PSILONIA FRIES 1825 FIDE AINSWORTH 1971

VOLUTELLARIA (SACC.) SACCARDO 1886. SYLLOGE FUNGORUM 4:682 /V. ACAROIDES
 SACCARDO
 ?, NOT ILLUSTRATED BY AUTHOR

VOLUTELLIS CLEMENTS ET SHEAR 1931. GENERA OF FUNGI P. 389 /V. SULPHUREA TORREND
 1914, BULL. JARD. BOT. BRUX 4:12
 ?, FOR VOLUTELLOPSIS TORREND NON SPEGAZZINI 1910

VOLUTELLOPSIS SPEGAZZINI 1910. TEV. FAC. AGRON. 6:197-198 (ILLUS) /V. CHILENSIS
 ?, ORIGINAL ILLUSTRATION AND DESCRIPTION NOT DIAGNOSTIC

VOLUTINA PENZIG ET SACCARDO 1904. ICONES FUNGORUM JAVAN. P. 113 AND TAB. 79,
 FIG. 2 /V. CONCENTRICA
 AMERO,BASO,HYALO /BARNETT, IMPERFECT FUNGI NO. 294,1960. (ILLUS)
  SYNO  ?=SETODOCHIUM BATISTA ET CIFERRI 1957

VRIKSHOPAMA RAO ET RAO 1964. MYCOPATHOLOGIA 23:287-290 (ILLUS) /V. SWETASAKHA
 ?,PROBABLY BASED ON A NODULISPORIUM

WALLEMIA JOHAN-OLSON 1887. FORH. CHRISTIANIA VID. SELSK. 12:6 AND TAB I-IV /W.
ICHTHYOPHAGA JOHAN-OLSON =W. SEBI (FRIES) VON ARX 1970
AMERO,BASO OR ARTHRO,HYALO /BARRON, HYPHOMYCETES P. 49, 134, AND
326-327,1968. (ILLUS)
 SYNO  =HEMISPORA VUILLEMIN 1906 FIDE CARMICHAEL 1969
     =BARGELLINIA BORZI 1888 FIDE VON ARX 1970
     =MANGINIELLA CABARA 1925 FIDE VON ARX 1970

WARDOMYCES BROOKS ET HANSFORD 1922. TRANS. BRIT. MYCOL. SOC. 8:137 /W. ANOMALA
 BROOKS ET HANSFORD
AMERO OR DIDYMO,CETERI,PHAEO /BARRON, GENERA OF HYPHOMYCETES P. 327-329,1968.
(ILLUS,REFS)

WETTSTEINIELLA KUNTZE =ARTHROBOTRYUM FIDE AINSWORTH 1971

WIESNERIOMYCES KOORDERS 1907. VERH. AKAD. AMSTERDAM 2,13:246-247 (ILLUS) /W.
 JAVANICUS
AMERO OR PHRAGMO OR SCOLECO,BLASTO OR GLOIO OR ARTHRO,HYALO /BARRON,
HYPHOMYCETES P. 83,1968. (ILLUS) SUB ABGLIOPHRAGMA ELLIS, DEMAT. HYPHO. P.
362-364,1971. (ILLUS)
 SYNO  =CHAETOSIRA CLEMENTS 1931 FIDE CLEMENTS AND SHEAR 1931
     =ABGLIOPHRAGMA ROY ET GUJARATI 1966 FIDE KENDRICK
     =ASCHIZOTRICHUM RIEUF 1962 FIDE KENDRICK
     =GLIOPHRAGMA SUBRAMANIAN ET LODHA 1964 FIDE KENDRICK

XENODIELLA SYDOW 1935. ANN. MYCOL. 33:98 /X. PETRAKII =X. STATE OF XENODIUM
 PETRAKII SYDOW 1935
?, NOT ILLUSTRATED BY AUTHOR. ASCO STATE =XENODIUM SYDOW 1935

XENOPLACA PETRAK 1949. SYDOWIA 3:261 /X. AEQUATORIENSIS
?, NOT ILLUSTRATED BY AUTHOR

XENOPUS PENZIG ET SACCARDO 1901. MALPIGHIA 15:240 /X. FARINOSUS
=BASIDIOMYCETE FIDE HUGHES 1958

XENOSPORELLA HOHNEL 1923. ZENTRALBL. BAKT. ABT. 2, 60:17 /X. PLEUROCOCCA HOHNEL
 =X. PLEUROCOCCUM (HOHNEL) PIROZYNSKI 1966
=XENOSPORIUM PENZIG ET SACCARDO 1902 FIDE PIROZYNSKI 1966

XENOSPORIUM PENZIG ET SACCARDO 1901. MALPIGHIA 15:248 /X. MIRABILE PENZ. ET
 SACC. 1901
DICTYO OR HELICO,CETERI,PHAEO /ELLIS, MYCOL PAPERS CMI 87:13-15,1963. (ILLUS)
PIROZYNSKI, MYCOL PAPERS CMI 105:24-35,1966. (ILLUS,KEY)
 SYNO  =XENOSPORELLA HOHNEL 1923 FIDE PIROZYNSKI 1966

XENOSTILBUM PETRAK 1959. SYDOWIA 13:105-108 /X. SYDOWII
?, NOT ILLUSTRATED BY AUTHOR

XIPHOMYCES SYDOW 1916. ANN. MYCOL. 14:374 /X. SACCHARI
?, NOT ILLUSTRATED BY AUTHOR

XYLOCHAERAS FRIES =SCLEROTIUM FIDE SACCARDO FROM AINSWORTH 1971

XYLOCLADIUM SYDOW 1900. IN LINDAU, NAT. PFLANZENF. 1:1:494 /?
=NOMEN CONFUSUM FIDE HOHNEL FROM AINSWORTH 1971. SEE ALSO CLEMENTS AND
SHEAR, GENERA OF FUNGI, P.386 AND 409, 1931.

XYLOHYPHA (FRIES) MASON 1960. IN DEIGHTON, MYCOL. PAPERS CMI 78:43 (REFS) /X.
 NIGRESCENS (PERSOON) MASON 1960
AMERO,BLASTO,PHAEO

XYLOSTROMA TODE 1790 =SHEETS OF MYCELIUM FIDE AINSWORTH 1971

XYPHASMA REBENTISCH =HELICOMYCES FIDE RABENHORST FROM AINSWORTH 1971

ZALERION MOORE ET MEYERS 1962. CANAD. J. MICROBIOL. 8:407-416 (ILLUS) /Z.
 EISTLA MOORE ET MEYERS =Z. MARITIMA (LINDER) ANASTASIOU 1963
HELICO,CETERI,PHAEO /ANASTASIOU, CANAD. J. BOTANY 41:1135-1139,1963.
(ILLUS,REVIS)

## 10. Hyphomycetes

ZANCLOSPORA HUGHES ET KENDRICK 1965. NEW ZEALAND J. BOTANY 3:151-158 (ILLUS)
  /Z. NOVA-ZELANDIAE
  AMERO,GLOIO,HYALO, ASCO STATE = CHAETOSPHAERIA FIDE MULLER IN KENDRICK 1971

ZASMIDIUM FRIES 1849. SUMMA VEG. SCAND. 2:407 (NO ILLUS) /Z. CELLARE, NON
  RACODIUM CELLARE PERSOON 1801 (LECTOTYPE, SEE HUGHES 1958)
  AMERO,CETERI,PHAEO /BARRON, HYPHOMYCETES P. 266-267,1968. (ILLUS) SUB
  RACODIUM ELLIS, DEMAT. HYPHO. P. 246-248,1971. (ILLUS) SUB RHINOCLADIELLA

ZYGODESMELLA FRAGOSO 1917. BOLL. ESPAN. HIST. NAT. 17:260 (ILLUS) /Z. CASARESI
  ='RHINOTRICHUM' FIDE NANNFELDT FROM AINSWORTH 1971

ZYGOPHIALA MASON 1945. IN MARTYN IN MYCOL. PAPERS CMI 13:5 /Z. JAMAICENSIS
  DIDYMO,CETERI,HYALO /HUGHES, MYCOL. PAPERS CMI 48:82,1952. (ILLUS)

ZYGOSPORIUM MONTAGNE 1842. ANN. SCI. NAT. BOT. SER. 2, 17:121 /Z. OSCHEOIDES
  AMERO,CETERI,HYALO OR PHAEO /HUGHES, MYCOL PAPERS CMI 44:1-18,1951.
  (ILLUS,REVIS) ELLIS, DEMAT. HYPHO. P. 322-326,1971. (ILLUS,KEY) WANG AND
  BAKER, CAN. J. BOTANY 45:1945-1951,1967. (ILLUS)
    SYNO    =PIMINA GROVE 1888 FIDE HUGHES 1958
            =UROPHIALA VUILLEMIN 1910 FIDE HUGHES 1958
            =UROBASIDIUM GIESENHAGEN 1892 FIDE ELLIS 1971

## VI. KEY-LISTS

For an explanation of these lists see pp. 331–333

### KEY-LISTS FOR GENERA OF HYPHOMYCETES

| AMEROSPOROUS GENERA | | Plate | AMEROSPOROUS GENERA | | Plate |
|---|---|---|---|---|---|
| ARTHRO | HYALO | | BASO | HYALO | |
| MISC | *Wiesneriomyces* | 55 E | ANNEL | *Scopulariopsis* | 13 B |
| | (not true arthro) | | MISC | *Basipetospora* | 27 A |
| NONSP | *Amblyosporium* | 1 A | NONSP | *Acrosporium* | 27 D |
| NONSP | *Antromyces* | 33 A | NONSP | *Ovulariopsis* | 27 C |
| NONSP | *Arthrographis* | 2 I | PHIAL | *Acrophialophora* | 5 C |
| NONSP | *Coccidioides* | | PHIAL | *Akanthomyces* | 10 A |
| | (see *Sporendonema*) | | PHIAL | *Angulimaya* | 4 E |
| NONSP | *Cylindrocolla* | 2 A | PHIAL | *Aspergillus* | 8 B |
| NONSP | *Geotrichum* | 2 G | PHIAL | *Bloxamia* | 6 D |
| NONSP | *Malbranchea* | 1 C | PHIAL | *Chaetochalara* | 6 A |
| NONSP | *Moniliella* | 2 F | PHIAL | *Chalara* | 6 B |
| NONSP | *Oidiodendron* | 2 B | PHIAL | *Clonostachys* | |
| NONSP | *Oncocladium* | 2 E | | (see *Gliocladium*) | |
| NONSP | *Sporendonema* | 1 B | PHIAL? | *Endoconidium* | |
| PHIAL | *Wallemia* | 2 H | | (no illus. available) | |
| RACHI | *Sympodiella* | 1 F | PHIAL | *Gibellula* | 8 C |
| | | | PHIAL | *Metarrhizium* | 9 E |
| ARTHRO | PHAEO | | PHIAL | *Monostachys* | 3 B |
| MISC | *Helicocephalum* | 41 H | PHIAL | *Myrioconium* | 9 I |
| NONSP | *Antromycopsis* | 1 E | PHIAL | *Nalanthamala* | 10 D |
| NONSP | *Coremiella* | 1 D | PHIAL | *Paecilomyces* | 3 D, 7 H |
| NONSP | *Oidiodendron* | 2 B | PHIAL | *Penicillium* | 7 G |
| NONSP | *Scytalidium* | 2 D | PHIAL? | *Phialophoropsis* | 4 F |
| NONSP | *Stephanosporium* | 2 C | PHIAL | *Phialostele* | 11 F |

| AMEROSPOROUS GENERA | | Plate | AMEROSPOROUS GENERA | | Plate |
|---|---|---|---|---|---|
| PHIAL | *Pleurocatena* | 3 G | NONSP | *Monilia* | 17 B |
| PHIAL | *Polypaecilum* | 13 C | NONSP | *Moniliella* | 2 F |
| PHIAL? | *Pseudofusidium* | | | | |
| | (see *Paecilomyces*) | | BLASTO | PHAEO | |
| PHIAL | *Sporendocladia* | 6 C | AMPUL | *Gonatobotryum* | 17 G |
| PHIAL | *Stachybotrys* | 7 C | AMPUL | *Haplobasidion* | 18 F |
| PHIAL | *Thysanophora* | 7 B | AMPUL | *Lacellinopsis* | 18 I |
| PHIAL? | *Volutina* | | ANNEL | *Annellodochium* | 31 G |
| | (no illustration) | | NONSP | *Alysidium* | 18 E |
| PHIAL | *Wallemia* | 2 H | NONSP | *Ampullifera* | 43 G |
| | | | NONSP | *Cladosporium* | 29 B |
| BASO | PHAEO | | NONSP | *Lacellina* | 18 H |
| ANNEL | *Annellodochium* | 31 G | NONSP | *Monilia* | 17 B |
| ANNEL | *Cephalotrichum* | 13 G | NONSP | *Periconia* | 18 B |
| ANNEL | *Scopulariopsis* | 13 B | NONSP | *Sorocybe* | 18 A |
| ANNEL | *Torulomyces* | 3 F | NONSP | *Spilodochium* | 29 C |
| ANNEL | *Trichurus* | 13 A | NONSP | *Trichobotrys* | 18 G |
| MISC? | *Briosia* | 27 G | NONSP | *Xylohypha* | 18 D |
| MISC? | *Thielaviopsis* | | RACHI | *Cladosporium* | 29 B |
| | (see *Chalara*) | | RADUL | *Mycovellosiella* | |
| PHIAL | *Aspergillus* | 8 B | | (see *Cladosporium*) | |
| PHIAL | *Catenularia* | 3 H | RADUL | *Periconiella* | 46 F |
| PHIAL | *Chaetochalara* | 6 A | | | |
| PHIAL | *Chalara* | 6 B | CETERI | HYALO | |
| PHIAL | *Cryptostroma* | 1 G | AMPUL | *Basidiobotrys* | 14 B |
| PHIAL | *Eladia* | 7 A | AMPUL | *Blastobotrys* | 16 D |
| PHIAL | *Gliomastix* | 3 A | AMPUL | *Botryosporium* | 16 H |
| PHIAL | *Graphiothecium* | 11 C | AMPUL? | *Glomospora* | 20 F |
| PHIAL? | *Megalodochium* | 27 F | AMPUL | *Gonatobotrys* | 16 F |
| PHIAL | *Paecilomyces* | 3 D, 7 H | AMPUL | *Graphiola* | 15 B |
| PHIAL | *Penicillium* | 7 G | AMPUL | *Oedocephalum* | 16 E |
| PHIAL | *Phialomyces* | 7 E | AMPUL | *Ostracoderma* | 16 A |
| PHIAL | *Phialotubus* | 7 F | AMPUL | *Sigmoideomyces* | 17 F |
| PHIAL | *Stachybotrys* | 7 C | AMPUL? | *Sporocephalum* | 16 C |
| PHIAL | *Stromatocrea* | 1 H | ANNEL | *Bainieria* | 13 E |
| | | | ANNEL | *Ceratocladium* | 14 A |
| BLASTO | HYALO | | ANNEL? | *Stevensomyces* | 27 J |
| AMPUL | *Nematogonium* | 16 G | ANNEL | *Umbelopsis* | 19 C |
| MISC | *Nomuraea* | 17 H | MISC | *Itersonilia* | |
| MISC | *Polyscytalum* | 28 E | | (Ballistospores) | |
| MISC | *Sphaeridium* | 17 D | MISC | *Zygosporium* | 22 F |
| MISC | *Sphaerosporium* | 27 E | NONSP | *Amblyosporium* | 1 A |
| MISC | *Wiesneriomyces* | 55 E | NONSP | *Asterophora* | 60 B |
| NONSP | *Alysidium* | 18 E | NONSP | *Beniowskia* | 15 F |
| NONSP | *Ambrosiella* | 19 D | NONSP | *Botryotrichum* | 19 A |
| NONSP | *Dimorphospora* | 22 H | NONSP | *Chaetoconidium* | 19 I |
| NONSP | *Fusidium* | 17 A | NONSP | *Chantransiopsis* | 22 C |
| NONSP | *Heydenia* | 19 B | NONSP | *Chlamydomyces* | 32 F |
| NONSP | *Hormiactella* | 17 E | NONSP | *Chrysosporium* | 19 G |
| NONSP | *Hyalodendron* | 17 C | NONSP | *Dexhowardia* | 15 H |

## 10. Hyphomycetes

| AMEROSPOROUS GENERA | | Plate | | AMEROSPOROUS GENERA | | Plate | |
|---|---|---|---|---|---|---|---|
| NONSP | *Glomopsis* | 20 | F | ANNEL | *Spilocaea* | 13 | D |
| NONSP | *Heydenia* | 19 | B | MISC | *Arthrinium* | 20 | B |
| NONSP | *Histoplasma* | 20 | G | MISC | *Cordella* | 20 | C |
| NONSP | *Jaculispora* | 63 | D | MISC | *Nigrospora* | 21 | D |
| NONSP | *Nematoctonus* | 15 | A | MISC | *Pteroconium* | 20 | A |
| NONSP | *Nia* | 64 | C | MISC | *Sadasivania* | 18 | C |
| NONSP | *Pagidospora* | 19 | H | MISC | *Zygosporium* | 22 | F |
| NONSP | *Paracoccidioides* (see *Chrysosporium*) | | | NONSP | *Acremoniella* | 21 | I |
| | | | | NONSP | *Acremoniula* | 21 | K |
| NONSP | *Pestalozziella* (Melanconiales) | | | NONSP | *Allescheriella* | 19 | E |
| | | | | NONSP | *Ampullifera* | 43 | G |
| NONSP | *Sepedonium* | 20 | E | NONSP | *Bactridiopsis* | 21 | H |
| NONSP | *Sporotrichum* | 22 | E | NONSP | *Botryoderma* | 21 | E |
| NONSP | *Staphylotrichum* | 19 | F | NONSP | *Chalaropsis* | 22 | I |
| NONSP | *Stephanoma* | 60 | A | NONSP | *Echinobotryum* | 23 | C |
| PHIAL | *Agaricostilbum* | 10 | C | NONSP | *Endocalyx* | 20 | D |
| PHIAL | *Ciliciopodium* | 11 | H | NONSP | *Gilmaniella* | 21 | G |
| PHIAL | *Dendrodochium* | 6 | H | NONSP? | *Herposira* | 22 | B |
| PHIAL? | *Hirsutella* | 8 | G | NONSP | *Humicola* | 21 | A |
| PHIAL? | *Margaritispora* | 45 | B | NONSP | *Mammaria* | 23 | B |
| PHIAL | *Meria* | 4 | G | NONSP | *Microclava* | 22 | D |
| RACHI | *Beauveria* | 24 | E | NONSP | *Nigrospora* | 21 | D |
| RACHI | *Calcarisporium* | 26 | B | NONSP | *Physalidium* | 60 | G |
| RACHI | *Costantinella* | 25 | B | NONSP? | *Pilulina* | 21 | C |
| RACHI | *Drumopama* | 25 | E | NONSP | *Rhinocladium* | 22 | G |
| RACHI | *Nodulisporium* | 25 | G | NONSP | *Sporotrichum* | 22 | E |
| RACHI | *Ovularia* | 24 | I | NONSP | *Thermomyces* | 21 | B |
| RACHI | *Raffaelea* | 24 | B | NONSP | *Ustilaginoidea* | 22 | A |
| RACHI | *Rhombostilbella* | 15 | I | NONSP | *Wardomyces* | 32 | D |
| RACHI | *Sporothrix* | 24 | F | PHIAL | *Tuberculina* | 1 | I |
| RACHI | *Tritirachium* | 26 | A | RADUL | *Asteromyces* | 23 | F |
| RADUL | *Acladium* | 15 | G | RADUL | *Beltrania* | 23 | D |
| RADUL | *Basifimbria* | 24 | H | RADUL | *Beltraniella* | 23 | A |
| RADUL | *Chloridiella* | 24 | K | RADUL | *Beltraniopsis* | 23 | E |
| RADUL | *Hemibeltrania* | 15 | E | RADUL | *Brachydesmiella* | 47 | D |
| RADUL | *Nodulisporium* | 25 | G | RADUL | *Conoplea* | 25 | D |
| RADUL | *Olpitrichum* | 15 | C | RADUL | *Dicyma* | 25 | A |
| RADUL? | *Pachnocybe* | 14 | F | RADUL | *Fusicladium* | 24 | D |
| RADUL | *Phaeoisaria* | 11 | E | RADUL | *Hadrotrichum* | 27 | H |
| RADUL | *Rhinocladiella* | 24 | A | RADUL | *Hansfordia* | 26 | E |
| RADUL? | *Sporocephalum* | 16 | C | RADUL | *Hemibeltrania* | 15 | E |
| RADUL | *Verticicladium* | 26 | D | RADUL | *Kumanasamuha* | 25 | C |
| | | | | RADUL | *Melanographium* | 27 | I |
| | | | | RADUL | *Nodulisporium* | 25 | G |
| CETERI | PHAEO | | | RADUL | *Paathramaya* | 11 | D |
| AMPUL | *Botrytis* | 16 | B | RADUL | *Panchanania* | 32 | B |
| ANNEL | *Acrogenospora* | 21 | J | RADUL | *Periconiella* | 46 | F |
| ANNEL | *Domingoella* | 21 | F | RADUL | *Pseudobeltrania* | 23 | G |
| ANNEL? | *Physalidium* | 60 | G | RADUL | *Pseudobotrytis* | 32 | J |

| AMEROSPOROUS GENERA | | Plate | AMEROSPOROUS GENERA | | Plate |
|---|---|---|---|---|---|
| RADUL | *Pseudoepicoccum* | | PHIAL | *Hormisciomyces* | |
| | (see *Hadrotrichum*) | | | (see *Plokamidomyces*) | |
| RADUL | *Rhinocladiella* | 24 A | PHIAL | *Kmetiopsis* | 57 B |
| RADUL | *Sceptrifera* | 30 D | PHIAL | *Koorchaloma* | 12 G |
| RADUL | *Spadicoides* | 31 E | PHIAL | *Mahabalella* | 12 C |
| RADUL | *Virgaria* | 25 F | PHIAL | *Menispora* | 12 D |
| RADUL | *Virgariella* | 24 G | PHIAL | *Menisporopsis* | 34 C |
| RADUL | *Zasmidium* | 24 J | PHIAL | *Minimidochium* | 12 B |
| | | | PHIAL | *Myrioconium* | 9 I |
| GLOIO | HYALO | | PHIAL | *Myrothecium* | 10 G |
| ANNEL | *Leptographium* | 13 H | PHIAL | *Neottiosporella* | 12 A |
| MISC? | *Hirsutella* | 8 G | PHIAL | *Phialocephala* | 6 G |
| MISC | *Wiesneriomyces* | 55 E | PHIAL | *Phialophora* | 4 C |
| NONSP | *Dimorphospora* | 22 H | PHIAL | *Plokamidomyces* | 9 H |
| NONSP | *Geotrichum* | 2 G | PHIAL | *Selenosporella* | 5 G |
| PHIAL | *Acremonium* | 3 C | PHIAL | *Sesquicillium* | 9 B |
| PHIAL | *Agyriella* | 9 D | PHIAL | *Sphaeromyces* | 3 I |
| PHIAL? | *Aphanocladium* | 15 D | PHIAL | *Stachybotrys* | 7 C |
| PHIAL? | *Articularia* | 5 D | PHIAL | *Stachylidium* | 5 A |
| PHIAL? | *Aureobasidium* | 4 B | PHIAL? | *Stilbum* | 11 I |
| PHIAL | *Bahupaathra* | 4 A | PHIAL | *Synnematium* | |
| PHIAL | *Capnophialophora* | 4 H | | (see *Acremonium*) | |
| PHIAL | *Chaetochalara* | 6 A | PHIAL | *Thozetellopsis* | 12 F |
| PHIAL | *Chaetopsina* | 5 H | PHIAL | *Tolypocladium* | |
| PHIAL | *Chaetopsis* | 34 B | | (see *Trichoderma*) | |
| PHIAL | *Chaetospermum* | | PHIAL | *Trichoderma* | 4 I |
| | (Melanconiales) | | PHIAL | *Tubercularia* | 10 H |
| PHIAL | *Chalara* | 6 B | PHIAL | *Uncigera* | 5 B |
| PHIAL | *Chloridium* | 3 E | PHIAL | *Verticillium* | 9 F |
| PHIAL | *Cladorrhinum* | 4 D | PHIAL | *Volutella* | 10 F |
| PHIAL | *Clonostachys* | | PHIAL | *Zanclospora* | 5 E |
| | (see *Gliocladium*) | | RADUL | *Haplographium* | 6 F |
| PHIAL | *Codinaea* | 12 E | RADUL | *Harpographium* | 57 H |
| PHIAL | *Colletotrichum* | | RADUL | *Heimiodora* | 14 E |
| | (Melanconiales) | | RADUL? | *Heterocephalum* | 10 E |
| PHIAL | *Crinula* | 11 G | RADUL? | *Jacobia* | 14 C |
| PHIAL | *Cystodendron* | 6 E | RADUL? | *Microstroma* | 10 B |
| PHIAL | *Dimorphospora* | 22 H | RADUL | *Selenosporella* | 5 G |
| PHIAL? | *Endostilbum* | 9 G | RADUL | *Verticicladiella* | 26 C |
| PHIAL | *Gliocephalis* | 8 A | | | |
| PHIAL | *Gliocephalotrichum* | 8 E | GLOIO | PHAEO | |
| PHIAL | *Gliocladium* | 6 I | ANNEL | *Graphium* | 13 F |
| PHIAL | *Gloiosphaera* | 8 F | ANNEL | *Melanconium* | |
| PHIAL | *Goidanichiella* | 8 D | | (Melanconiales) | |
| PHIAL | *Gonytrichum* | 5 F | NONSP | *Aureobasidium* | 4 B |
| PHIAL | *Hainesia* | | NONSP | *Helicocephalum* | 41 H |
| | (Coelomycete?) | | PHIAL | *Aureobasidium* | 4 B |
| PHIAL | *Harpographium* | 57 H | PHIAL | *Chaetochalara* | 6 A |
| PHIAL | *Harziella* | 9 A | | | |

## 10. Hyphomycetes

| AMEROSPOROUS GENERA | | Plate | AMEROSPOROUS GENERA | | Plate |
|---|---|---|---|---|---|
| PHIAL | *Chalara* | 6 B | PHIAL | *Gonytrichum* | 5 F |
| PHIAL | *Chloridium* | 3 E | PHIAL | *Myrothecium* | 10 G |
| PHIAL | *Custingophora* | 7 D | PHIAL | *Saccardaea* | 11 A |
| PHIAL? | *Exophiala* | 33 G | PHIAL | *Solheimia* | 11 B |
| PHIAL | *Gliocladium* | 6 I | PHIAL | *Stachybotrys* | 7 C |
| PHIAL | *Gliomastix* | 3 A | PHIAL | *Stachylidium* | 5 A |

| DICTYOSPOROUS GENERA | | Plate | DICTYOSPOROUS GENERA | | Plate |
|---|---|---|---|---|---|
| BASO | HYALO | | NONSP | *Cheirospora* | 41 E |
| MISC | *Dictyocatenulata* | 35 C | NONSP | *Chuppia* | 36 I |
| BASO | PHAEO | | NONSP | *Coleodictyospora* | 38 I |
| MISC? | *Barnettella* | 35 G | NONSP | *Dictyodesmium* | 38 E |
| MISC | *Coniosporium* | 35 D | NONSP | *Dictyosporium* | 66 E |
| BLASTO | PHAEO | | NONSP | *Epicoccum* | 35 I, 35 J |
| NONSP | *Alternaria* | 36 H | NONSP | *Hansfordiella* | 37 C |
| NONSP | *Peyronelia* | 35 B | NONSP | *Hansfordiellopsis* | 37 A |
| RADUL | *Alternaria* | 36 H | NONSP | *Hermatomyces* | 38 J |
| CETERI | HYALO | | NONSP | *Isthmospora* | 59 H |
| PHIAL | *Ascoconidium* | 38 G | NONSP | *Kostermansinda* | 38 H |
| PHIAL | *Clathrosphaerina* | 38 D | NONSP | *Microclava* | 22 D |
| PHIAL | *Diheterospora* | 9 C | NONSP | *Monodictys* | 37 I |
| PHIAL | *Hyalodictys* | 35 A | NONSP | *Muiaria* | 38 B |
| PHIAL | *Papulaspora* | 67 G | NONSP | *Muiogone* | 38 C |
| PHIAL | *Peyronelina* | 35 F | NONSP | *Mycoenterolobium* | 67 A |
| PHIAL | *Riessia* | 59 A | NONSP | *Oncopodium* | 35 H |
| CETERI | PHAEO | | NONSP | *Papulaspora* | 67 G |
| ANNEL | *Acrodictys* | 37 G | NONSP | *Pithomyces* | 47 A |
| ANNEL | *Annellophorella* | 37 J | NONSP | *Sarcinella* | 37 E |
| ANNEL | *Arachnophora* | 60 F | NONSP | *Septosporium* | 35 E |
| ANNEL | *Stemphylium* | 36 A | NONSP | *Spegazzinia* | 59 I |
| ANNEL | *Stigmina* | 36 G | NONSP | *Tetracoccosporium* | 37 D |
| MISC | *Dictyoarthrinium* | 37 F | NONSP | *Thyrostromella* | 36 J |
| MISC | *Piricauda* | 37 B | NONSP | *Xenosporium* | 40 F |
| MISC | *Spegazzinia* | 59 I | RACHI | *Embellisia* | 48 B |
| MISC | *Tretopileus* | 67 F | RACHI | *Pendulispora* | 38 A |
| NONSP | *Acrospeira* | 38 F | RACHI | *Thyrostromella* | 36 J |
| NONSP | *Arbuscula* | 67 D | RACHI | *Ulocladium* | 36 E |
| NONSP | *Berkleasmium* | 37 H | RADUL | *Dactylosporium* | 36 D |
| NONSP | *Burgoa* | 67 E | RADUL | *Mystrosporiella* | 36 C |
| | | | RADUL | *Sclerographium* | 36 B |
| | | | RADUL | *Sirosporium* | 36 F |

| DIDYMOSPOROUS GENERA | | Plate | DIDYMOSPOROUS GENERA | | Plate |
|---|---|---|---|---|---|
| ARTHRO | HYALO | | NONSP | *Scytalidium* | 2 D |
| NONSP | *Antromyces* | 33 A | BASO | HYALO | |
| ARTHRO | PHAEO | | ANNEL | *Gymnodochium* | 28 C |
| NONSP | *Ampulliferina* | 29 D | MISC | *Cladobotryum* | 27 B, 28 B |
| NONSP | *Bahusakala* | 42 B | MISC | *Trichothecium* | 28 D |

| DIDYMOSPOROUS GENERA | | Plate | | DIDYMOSPOROUS GENERA | | Plate | |
|---|---|---|---|---|---|---|---|
| PHIAL | *Chalara* | 6 | B | RADUL | *Veronaea* | 30 | B |
| PHIAL | *Penicillifer* | 34 | E | RADUL | *Zygophiala* | 30 | C |
| BASO | PHAEO | | | CETERI | PHAEO | | |
| ANNEL | *Annellodochium* | 31 | G | AMPUL | *Camptomeris* | 46 | G |
| ANNEL | *Trichodochium* | 29 | F | ANNEL | *Deightoniella* | 31 | D |
| PHIAL | *Chalara* | 6 | B | ANNEL | *Oedothea* | 31 | F |
| | | | | ANNEL | *Phragmospathula* | 42 | F |
| BLASTO | HYALO | | | ANNEL | *Spilocaea* | 13 | D |
| NONSP | *Arxiella* | 28 | J | MISC | *Dendryphion* | 29 E, 43 | D |
| NONSP | *Diploospora* | 28 | F | MISC | *Johnstonia* | | |
| NONSP | *Hormiactis* | 28 | H | | (no illustration) | | |
| NONSP | *Polyscytalina* | 28 | I | NONSP | *Ampullifera* | 43 | G |
| NONSP | *Polyscytalum* | 28 | E | NONSP | *Balanium* | 31 | A |
| RADUL | *Ramularia* | 28 | G | NONSP | *Diplococcium* | 29 | I |
| BLASTO | PHAEO | | | NONSP | *Hadronema* | 32 | A |
| AMPUL | *Oedemium* | 29 | G | NONSP | *Trichocladium* | 32 | H |
| ANNEL | *Annellodochium* | 31 | G | NONSP | *Wardomyces* | 32 | D |
| NONSP | *Ampullifera* | 43 | G | RACHI | *Cercosporidium* | 45 | F |
| NONSP | *Bispora* | 29 | A | RACHI | *Hadronema* | 32 | A |
| NONSP | *Didymobotryum* | 29 | H | RACHI | *Polythrincium* | 32 | E |
| NONSP | *Diplococcium* | 29 | I | RACHI | *Pyricularia* | 50 | G |
| NONSP | *Spilodochium* | 29 | C | RADUL | *Acroconidiellina* | | |
| RACHI? | *Paspalomyces* | 28 | A | | (see *Acroconidiella*) | | |
| RADUL | *Cladosporium* | 29 | B | RADUL | *Asperisporium* | 31 | B |
| RADUL | *Dendryphion* | 29 E, 43 | D | RADUL | *Camptomeris* | 46 | G |
| RADUL | *Diplococcium* | 29 | I | RADUL | *Cordana* | 31 | C |
| RADUL | *Mycovellosiella* | | | RADUL | *Dactylaria* | 30 | F |
| | (see *Cladosporium*) | | | RADUL | *Dendryphion* | 29 E, 43 | D |
| RADUL | *Periconiella* | 46 | F | RADUL | *Didymotrichella* | 31 | H |
| RADUL | *Phaeoramularia* | | | RADUL | *Diplococcium* | 29 | I |
| | (see *Cladosporium*) | | | RADUL | *Fusicladium* | 24 | D |
| | | | | RADUL | *Panchanania* | 32 | B |
| CETERI | HYALO | | | RADUL | *Passalora* | 32 | C |
| AMPUL | *Blastophorella* | 30 | A | RADUL | *Periconiella* | 46 | F |
| MISC | *Cladobotryum* | 27 B, 28 | B | RADUL | *Pseudobotrytis* | 32 | J |
| MISC | *Fusicladiella* | 30 | G | RADUL | *Sceptrifera* | 30 | D |
| NONSP | *Chlamydomyces* | 32 | F | RADUL | *Scolecobasidium* | 32 | G |
| NONSP? | *Dwayaloma* | 34 | F | RADUL | *Spadicoides* | 31 | E |
| NONSP | *Monotrichum* | | | RADUL | *Veronaea* | 30 | B |
| | (see *Spermospora*) | | | | | | |
| NONSP | *Mycogone* | 32 | I | GLOIO | HYALO | | |
| PHIAL? | *Rhynchosporium* | 33 | D | PHIAL | *Cephalosporiopsis* | 33 | H |
| PHIAL? | *Sympodiophora* | 33 | B | PHIAL | *Chaetopsis* | 43 | B |
| RACHI | *Cercosporidium* | 45 | F | PHIAL | *Chalara* | 6 | B |
| RACHI | *Pyricularia* | 50 | G | PHIAL | *Codinaea* | 12 | E |
| RADUL | *Arthrobotrys* | 30 | I | PHIAL | *Cryptophiale* | 57 | G |
| RADUL | *Dactylaria* | 30 | F | PHIAL | *Cylindrocladium* | 54 | G |
| RADUL | *Dendrosporium* | 60 | D | PHIAL | *Cylindrodendrum* | 33 | E |
| RADUL | *Haplariopsis* | 30 | J | PHIAL | *Cylindrotrichum* | 34 | A |
| RADUL | *Neta* | 30 | H | PHIAL | *Didymostilbe* | 33 | C |

## 10. Hyphomycetes

| DIDYMOSPOROUS GENERA | | Plate |
|---|---|---|
| PHIAL | *Leptodiscella* | 34 D |
| PHIAL | *Menispora* | 12 D |
| PHIAL | *Menisporopsis* | 34 C |
| PHIAL? | *Pseudographiella* | 54 D |
| PHIAL | *Sibirina* | 34 G |
| PHIAL | *Uncigera* | 5 B |

| DIDYMOSPOROUS GENERA | | Plate |
|---|---|---|
| RACHI | *Leptodiscella* | 34 D |
| GLOIO | PHAEO | |
| MISC? | *Exophiala* | 33 G |
| PHIAL | *Chalara* | 6 B |
| PHIAL | *Cyphellophora* | 33 F |
| PHIAL? | *Exophiala* | 33 G |

| HELICOSPOROUS GENERA | | Plate |
|---|---|---|
| ARTHRO | PHAEO | |
| MISC | *Helicocephalum* | 41 H |
| BLASTO | HYALO | |
| NONSP | *Helicodendron* | 41 D |
| BLASTO | PHAEO | |
| NONSP | *Helicodendron* | 41 D |
| NONSP | *Hiospira* | 41 G |
| CETERI | HYALO | |
| NONSP | *Angulospora* | 58 E |
| NONSP | *Diplorhynchus* | 39 E |
| NONSP | *Helicoma* | 39 G |
| NONSP? | *Helicomyces* | 39 H |
| NONSP | *Helicoon* | 41 C |
| NONSP | *Helicosporium* | 39 C |
| NONSP | *Hobsonia* | 41 F |
| NONSP | *Spirosphaera* | 41 J |
| NONSP | *Troposporium* | 41 B |
| PHIAL | *Helicomyces* | 39 H |
| RADUL? | *Drepanoconis* | 39 B |
| CETERI | PHAEO | |
| NONSP | *Acrospeira* | 38 F |
| NONSP | *Ceratophorum* | 52 E |
| NONSP | *Circinoconis* | 40 B |

| HELICOSPOROUS GENERA | | Plate |
|---|---|---|
| NONSP | *Cirrenalia* | 40 H |
| NONSP | *Curculiospora* | 40 A |
| NONSP | *Helicoma* | 39 G |
| NONSP | *Helicoon* | 41 C |
| NONSP | *Helicorhoidion* | 41 A |
| NONSP | *Hiospira* | 41 G |
| NONSP | *Strumella* | 40 E |
| NONSP | *Troposporella* | 40 J |
| NONSP | *Xenosporium* | 40 F |
| NONSP | *Zalerion* | 40 C |
| RACHI | *Curvulariopsis* | 40 D |
| RACHI | *Helicomina* | 40 I |
| RACHI | *Helicorhoidion* | 41 A |
| RADUL | *Dichotomophthoropsis* (no illustration) | |
| RADUL | *Trochophora* | 40 G |
| GLOIO | HYALO | |
| PHIAL? | *Brachyhelicoon* | 41 I |
| PHIAL? | *Delortia* | 39 D |
| PHIAL? | *Everhartia* | 39 A |
| PHIAL | *Vanbeverwijkia* | 39 F |
| GLOIO | PHAEO | |
| MISC | *Helicocephalum* | 41 H |

| PHRAGMOSPOROUS GENERA | | Plate |
|---|---|---|
| ARTHRO | HYALO | |
| MISC | *Wiesneriomyces* (not true phragmo) | 55 E |
| NONSP | *Septotrullula* | 42 E |
| ARTHRO | PHAEO | |
| NONSP | *Bahusakala* | 42 B |
| NONSP | *Milowia* | 42 C |
| NONSP | *Scytalidium* | 2 D |
| NONSP | *Septotrullula* | 42 E |
| BASO | HYALO | |
| PHIAL | *Chalara* | 6 B |
| PHIAL | *Penicillifer* | 34 E |
| PHIAL | *Sporoschisma* | 41 I |

| PHRAGMOSPOROUS GENERA | | Plate |
|---|---|---|
| BASO | PHAEO | |
| ANNEL | *Bactrodesmiella* | 42 A |
| PHIAL | *Chalara* | 6 B |
| PHIAL | *Fusariella* | 42 H |
| PHIAL | *Sporoschisma* | 42 I |
| BLASTO | HYALO | |
| MISC | *Wiesneriomyces* | 55 E |
| NONSP | *Septocylindrium* (see *Septonema*) | |
| BLASTO | PHAEO | |
| ANNEL | *Annellophora* | 49 A |
| ANNEL | *Corynespora* | 53 H |
| MISC | *Torula* | 43 C |

| PHRAGMOSPOROUS GENERA | | Plate | PHRAGMOSPOROUS GENERA | | Plate |
|---|---|---|---|---|---|
| NONSP | *Ampullifera* | 43 G | AMPUL | *Dichotomophthora* | 48 E |
| NONSP | *Corynesporella* | 53 A | ANNEL | *Acrophragmis* | 44 F |
| NONSP | *Heteroconium* | 43 H | ANNEL | *Annellophora* | 49 A |
| NONSP | *Ormathodium* | 43 I | ANNEL | *Annellophragmia* | 53 F |
| NONSP | *Septonema* | 43 F | ANNEL | *Antennopsis* | 44 B |
| NONSP | *Sporophiala* | 42 D | ANNEL | *Bactrodesmiella* | 42 A |
| NONSP | *Taeniolella* | 43 E | ANNEL | *Brachysporiella* | 44 A |
| RADUL | *Cladosporium* | 29 B | ANNEL | *Clasterosporium* | 53 G |
| RADUL | *Dendryphion* | 29 E, 43 D | ANNEL | *Corynespora* | 53 H |
| RADUL | *Mycovellosiella* (see *Cladosporium*) | | ANNEL | *Coryneum* (Melanconiales) | |
| RADUL | *Periconiella* | 46 F | ANNEL | *Deightoniella* | 31 D |
| RADUL | *Phaeoramularia* (see *Cladosporium*) | | ANNEL | *Endophragmia* | 47 I |
| | | | ANNEL | *Endophragmiopsis* | 47 F |
| RADUL | *Stenella* | 48 H | ANNEL | *Exosporiella* | 49 C |
| | | | ANNEL | *Phragmospathula* | 42 F |
| CETERI | HYALO | | ANNEL | *Podosporium* | 49 B |
| ANNEL | *Septoidium* | 44 G | ANNEL | *Spiropes* | 50 A |
| NONSP | *Bacillispora* | 54 H | ANNEL | *Sporidesmium* | 53 D |
| NONSP | *Bactridium* | 51 G | ANNEL | *Stigmina* | 36 G |
| NONSP | *Chaetoconidium* | 19 I | ANNEL | *Teratosperma* | 52 F |
| NONSP | *Dactylella* | 51 A | MISC | *Acarocybe* | 44 H |
| NONSP | *Drechslerella* | 51 F | MISC | *Cacumisporium* | 45 G |
| NONSP | *Epidermophyton* | 51 J | MISC | *Dwayabeeja* | 43 B |
| NONSP | *Harpagomyces* | 60 H | MISC | *Hormocephalum* | 49 E |
| NONSP | *Keratinomyces* | 51 D | MISC | *Spondylocladiella* | 44 D |
| NONSP | *Mastigosporium* | 51 C | NONSP | *Acarocybella* | 52 D |
| NONSP | *Microsporum* | 51 I | NONSP | *Ampullifera* | 43 G |
| NONSP? | *Spermospora* | 51 B | NONSP | *Antennatula* | 49 F |
| NONSP? | *Trichoconis* | 51 E | NONSP | *Arthrocladium* | 55 F |
| NONSP | *Trichophyton* | 51 H | NONSP | *Bactrodesmium* | 47 B |
| PHIAL | *Ascoconidium* | 38 G | NONSP | *Ceratophorum* | 52 E |
| PHIAL | *Heliscus* | 63 A | NONSP | *Clasterosporium* | 53 G |
| PHIAL | *Margaritispora* | 45 B | NONSP | *Corynesporella* | 53 A |
| RACHI | *Arthrosporium* | 54 C | NONSP | *Dendryphiopsis* | 48 D |
| RACHI | *Centrospora* | 55 C | NONSP | *Excipularia* | 47 C |
| RACHI | *Cercospora* | 56 C | NONSP | *Hansfordiella* | 37 C |
| RACHI | *Cercosporella* | 56 E | NONSP | *Hansfordiellopsis* | 37 A |
| RACHI | *Cercosporidium* | 45 F | NONSP | *Jainesia* | 47 H |
| RACHI | *Pleiochaeta* | 52 A | NONSP | *Listeromyces* | 47 E |
| RACHI | *Pleurothecium* | 45 E | NONSP | *Lomaantha* | 52 B |
| RACHI | *Pyricularia* | 50 G | NONSP | *Milowia* | 42 C |
| RADUL | *Dactylaria* | 30 F | NONSP | *Murogenella* | 50 C |
| RADUL? | *Eurasina* | 45 A | NONSP? | *Negeriella* (see *Podosporium*) | |
| RADUL | *Phaeodactylium* | 49 G | | | |
| RADUL | *Spondylocladiopsis* | 45 C | NONSP | *Pithomyces* | 47 A |
| | | | NONSP | *Sporidesmium* | 53 D |
| CETERI | PHAEO | | NONSP | *Stemphyliomma* | 44 I |
| AMPUL | *Camptomeris* | 46 G | NONSP | *Triadelphia* | 52 H |
| AMPUL | *Cephaliophora* | 30 E | | | |

# Hyphomycetes

| PHRAGMOSPOROUS GENERA | | Plate | |
|---|---|---|---|
| NONSP | *Trichocladium* | 32 | H |
| RACHI | *Acroconidiella* | 44 | E |
| RACHI | *Centrospora* | 55 | C |
| RACHI | *Cercospora* | 56 | C |
| RACHI | *Cercosporidium* | 45 | F |
| RACHI | *Curvularia* | 49 | D |
| RACHI | *Curvulariopsis* | 40 | D |
| RACHI | *Duosporium* | 44 | J |
| RACHI | *Embellisia* | 48 | B |
| RACHI | *Exosporium* | 53 | B |
| RACHI | *Helicomina* | 40 | I |
| RACHI | *Nakataea* | 50 | E |
| RACHI | *Phaeotrichoconis* | 52 | G |
| RACHI | *Pleiochaeta* | 52 | A |
| RACHI | *Pyricularia* | 50 | G |
| RACHI | *Spiropes* | 50 | A |
| RACHI | *Stenellopsis* | 48 | F |
| RACHI | *Tomenticola* | 50 | F |
| RADUL | *Acroconidiellina* | | |
| | (see *Acroconidiella*) | | |
| RADUL | *Annellophragmia* | 53 | F |
| RADUL | *Brachydesmiella* | 47 | D |
| RADUL | *Brachysporium* | 44 | C |
| RADUL | *Camposporium* | 52 | C |
| RADUL | *Camptomeris* | 46 | G |
| RADUL | *Cercospora* | 56 | C |
| RADUL | *Curvularia* | 49 | D |
| RADUL | *Dactylaria* | 30 | F |
| RADUL | *Dendrographium* | 43 | A |
| RADUL | *Dendryphion* | 29 E, 43 | D |
| RADUL | *Dichotomophthora* | 48 | E |
| RADUL | *Drechslera* | 50 | D |
| RADUL | *Elletevera* | 56 | F |
| RADUL | *Eriocercospora* | 48 | G |
| RADUL | *Gonatophragmium* | 45 | D |
| RADUL | *Helminthosporium* | 53 | C |
| RADUL | *Mitteriella* | 47 | G |
| RADUL | *Mystrosporiella* | 36 | C |
| RADUL | *Nakataea* | 50 | E |
| RADUL | *Passalora* | 32 | C |
| RADUL | *Periconiella* | 46 | F |
| RADUL | *Phaeoisariopsis* | 53 | I |
| RADUL | *Pleurophragmium* | 42 | G |

| SCOLECOSPOROUS GENERA | | Plate | |
|---|---|---|---|
| ARTHRO | HYALO | | |
| MISC | *Wiesneriomyces* | 55 | E |
| BASO | HYALO | | |

| PHRAGMOSPOROUS GENERA | | Plate | |
|---|---|---|---|
| RADUL | *Podosporiella* | 48 | A |
| RADUL | *Prathigada* | 50 | H |
| RADUL | *Pseudospiropes* | 50 | B |
| RADUL | *Pyriculariopsis* | 50 | I |
| RADUL | *Scolecobasidium* | 32 | G |
| RADUL | *Sirosporium* | 36 | F |
| RADUL | *Spadicoides* | 31 | E |
| RADUL | *Spiropes* | 50 | A |
| RADUL | *Verrucispora* | 48 | C |
| GLOIO | HYALO | | |
| ANNEL | *Arthrobotryum* | 46 | B |
| MISC | *Wiesneriomyces* | 55 | E |
| NONSP? | *Mastigosporium* | 51 | C |
| NONSP | *Septotrullula* | 42 | E |
| NONSP? | *Spermospora* | 51 | B |
| PHIAL? | *Atractium* | 54 | E |
| PHIAL? | *Calostilbella* | 46 | A |
| PHIAL | *Chalara* | 6 | B |
| PHIAL | *Codinaea* | 12 | E |
| PHIAL | *Cylindrocarpon* | 54 | F |
| PHIAL | *Cylindrocladium* | 54 | G |
| PHIAL | *Fusarium* | 54 | B |
| PHIAL | *Menispora* | 12 | D |
| PHIAL? | *Morrisographium* | 54 | A |
| PHIAL? | *Pseudographiella* | 54 | D |
| RACHI | *Pleurothecium* | 45 | E |
| RACHI | *Sterigmatobotrys* | 45 | H |
| GLOIO | PHAEO | | |
| ANNEL | *Arthrobotryum* | 46 | B |
| ANNEL | *Seimatosporium* | | |
| | (Melanconiales) | | |
| MISC | *Cacumisporium* | 45 | G |
| NONSP | *Septotrullula* | 42 | E |
| PHIAL? | *Albosynnema* | 46 | D |
| PHIAL? | *Calostilbella* | 46 | A |
| PHIAL | *Chalara* | 6 | B |
| PHIAL | *Cyphellophora* | 33 | F |
| PHIAL | *Pestalotia* | | |
| | (Melanconiales) | | |
| PHIAL | *Virgatospora* | 46 | C |
| RACHI | *Sterigmatobotrys* | 45 | H |
| RADUL | *Capnocybe* | 46 | E |

| SCOLECOSPOROUS GENERA | | Plate | |
|---|---|---|---|
| PHIAL | *Riclaretia* | 57 | E |
| BLASTO | HYALO | | |
| MISC | *Wiesneriomyces* | 55 | E |

| SCOLECOSPOROUS GENERA | | Plate | | SCOLECOSPOROUS GENERA | | Plate | |
|---|---|---|---|---|---|---|---|
| BLASTO | PHAEO | | | CETERI | PHAEO | | |
| NONSP | *Oramathodium* | 43 | I | ANNEL | *Sporidesmium* | 53 | D |
| RADUL | *Cladosporiella* | 56 | A | MISC | *Dwayabeeja* | 43 | D |
| | | | | NONSP | *Acarocybella* | 52 | D |
| | | | | NONSP | *Arthrocladium* | 55 | F |
| CETERI | HYALO | | | NONSP? | *Cercosperma* | 53 | E |
| ANNEL | *Circinotrichum* | 14 | D | NONSP | *Irpicomyces* | 56 | D |
| ANNEL | *Gyrothrix* | | | NONSP | *Sporidesmium* | 53 | D |
| | (see *Circinotrichum*) | | | RACHI | *Centrospora* | 55 | C |
| MISC | *Circinotrichum* | 14 | D | RACHI | *Cercospora* | 56 | C |
| NONSP? | *Aciculariella* | 57 | I | RADUL | *Eriocercospora* | 48 | G |
| NONSP | *Anguillospora* | 58 | G | | | | |
| NONSP | *Angulospora* | 58 | E | GLOIO | HYALO | | |
| NONSP? | *Bacillispora* | 54 | H | MISC | *Wiesneriomyces* | 55 | E |
| NONSP? | *Lunulospora* | 57 | F | PHIAL? | *Atractium* | 54 | E |
| NONSP | *Petrakiopsis* | 55 | A | PHIAL | *Cryptophiale* | 57 | G |
| NONSP? | *Sigmoidea* | 58 | B | PHIAL | *Harpographium* | 57 | H |
| PHIAL? | *Cercosporula* | 57 | J | PHIAL | *Harposporium* | 57 | A |
| PHIAL? | *Flagellospora* | 58 | D | PHIAL | *Kmetia* | 57 | D |
| PHIAL? | *Microdochium* | 57 | C | PHIAL | *Kmetiopsis* | 57 | B |
| RACHI | *Centrospora* | 55 | C | PHIAL | *Libertella* | | |
| RACHI | *Cercospora* | 56 | C | | (Melanconiales) | | |
| RACHI | *Cercosporella* | 56 | E | PHIAL? | *Morrisographium* | 54 | A |
| RACHI | *Pterulopsis* | 58 | A | PHIAL | *Selenosporella* | 5 | G |
| RACHI | *Subulispora* | 24 | C | RADUL | *Harpographium* | 57 | H |
| RADUL | *Cercosporula* | 57 | J | RADUL? | *Linodochium* | 58 | C |
| RADUL | *Mirandina* | 55 | D | RADUL? | *Ramulispora* | 58 | F |
| RADUL | *Schizotrichum* | 56 | B | RADUL | *Selenosporella* | 5 | G |

| STAUROSPOROUS GENERA | | Plate | | STAUROSPOROUS GENERA | | Plate | |
|---|---|---|---|---|---|---|---|
| CETERI | HYALO | | | | | | |
| NONSP | *Actinospora* | 64 | E | NONSP | *Pyramidospora* | 59 | E |
| NONSP | *Alatospora* | 63 | F | NONSP | *Riessia* | 59 | A |
| NONSP | *Articulospora* | 63 | E | NONSP | *Stephanoma* | 60 | A |
| NONSP | *Asterophora* | 60 | B | NONSP | *Tetrachaetum* | 63 | G |
| NONSP | *Campylospora* | 61 | F | NONSP | *Tetracladium* | 66 | B |
| NONSP | *Candelabrum* | 59 | G | NONSP | *Tetracrium* | 55 | B |
| NONSP | *Clavariopsis* | 63 | C | NONSP | *Thallospora* | 61 | I |
| NONSP | *Clavatospora* | 63 | B | NONSP | *Titaea* | 59 | F |
| NONSP | *Culicidospora* | 61 | D | NONSP | *Tricellula* | 59 | B |
| NONSP | *Dendrospora* | 61 | E | NONSP | *Tricladium* | 61 | B |
| NONSP | *Flabellospora* | 64 | F | NONSP | *Tridentaria* | 63 | H |
| NONSP | *Gyoerffiella* | 61 | H | NONSP | *Trinacrium* | 63 | J |
| NONSP | *Harpagomyces* | 60 | B | NONSP | *Triposporina* | 59 | D |
| NONSP? | *Jaculispora* | 63 | D | NONSP | *Triscelophorus* | 64 | G |
| NONSP | *Magdalaenaea* | 63 | I | NONSP | *Trisulcosporium* | 64 | H |
| NONSP | *Nia* | 64 | C | NONSP | *Varicosporina* | 61 | G |
| NONSP? | *Petrakiopsis* | 55 | A | NONSP | *Varicosporium* | 61 | A |
| NONSP | *Polycladium* | 61 | C | PHIAL | *Heliscus* | 63 | A |

## 10. Hyphomycetes

| STAUROSPOROUS GENERA | | Plate | STAUROSPOROUS GENERA | | Plate |
|---|---|---|---|---|---|
| PHIAL | *Lemonniera* | 64 A | NONSP | *Hughesinia* | |
| RACHI | *Dicranidion* | 66 D | | (no illustration) | |
| RADUL | *Dendrosporium* | 60 D | NONSP | *Isthmospora* | 59 H |
| | | | NONSP? | *Megaster* | 62 E |
| CETERI | PHAEO | | NONSP | *Oncopodiella* | 60 C |
| ANNEL | *Acrophragmis* | 44 F | NONSP? | *Orbimyces* | 65 B |
| ANNEL | *Actinocladium* | 62 F | NONSP? | *Pentaposporium* | 62 D |
| ANNEL | *Arachnophora* | 60 F | NONSP | *Petrakia* | 60 E |
| ANNEL | *Ceratosporella* | 65 D | NONSP | *Physalidium* | 60 G |
| ANNEL | *Grallomyces* | 62 I | NONSP | *Pseudopetrakia* | |
| ANNEL | *Iyengarina* | 65 C | | (see *Petrakia*) | |
| ANNEL | *Teratosperma* | 52 F | NONSP | *Spegazzinia* | 59 I |
| ANNEL | *Triposporium* | 62 C | NONSP | *Speiropsis* | 66 A |
| MISC | *Spegazzinia* | 59 I | NONSP | *Tetraploa* | 65 A |
| NONSP | *Actinocladium* | 62 F | NONSP | *Tetraposporium* | 64 B |
| NONSP | *Casaresia* | 62 H | NONSP | *Tripospermum* | 62 B |
| NONSP | *Ceratosporium* | 62 A | RACHI | *Coronospora* | |
| NONSP? | *Cheiromyces* | 66 H | | (no illustration) | |
| NONSP | *Cryptocoryneum* | 66 F | RACHI | *Diplocladiella* | 65 E |
| NONSP | *Desmidiospora* | 59 C | GLOIO | HYALO | |
| NONSP | *Dictyosporium* | 66 E | NONSP? | *Amallospora* | 66 C |
| NONSP | *Heptaster* | 62 G | | | |
| NONSP | *Hirudinaria* | 65 F | GLOIO | PHAEO | |
| NONSP | *Hormiokrypsis* | 64 D | RADUL | *Cheiromycella* | 66 G |

| STERILE GENERA | Plate | STERILE GENERA | Plate |
|---|---|---|---|
| *Acinula* | | *Hypha* | |
| *Acrotamnium* | | *Hyphochlaena* | |
| *Actinonema* | | *Indiella* | |
| *Aegerita* | 67 B | *Loliomyces* | |
| *Akenomyces* | | *Madurella* | |
| *Anthina* | | *Moniliopsis* | |
| *Anulohypha* | | *Mycoenterolobium* | 67 A |
| *Anulosporium* | | *Myxotrichella* | |
| *Arbuscula* | 67 D | *Overeemia* | |
| *Bartheletia* | | *Papulaspora* | 67 G |
| *Blodgettia* | | *Phloeoconis* | |
| *Botryonipha* | | *Plecotrichum* | |
| *Burgoa* | 67 E | *Racodium* | |
| *Byssocladium* | | *Rhizoctonia* | |
| *Cenococcum* | | *Rhizohypha* | |
| *Coccobotrys* | | *Rhizomorpha* | |
| *Coccospora* | | *Saprochaete* | |
| *Colletosporium* | | *Scenomyces* | |
| *Cuticularia* | | *Sclerotium* | |
| *Dactuliophora* | 67 C | *Tretopileus* | 67 F |
| *Ectostroma* | | *Vasculomyces* | |
| *Himantia* | | | |

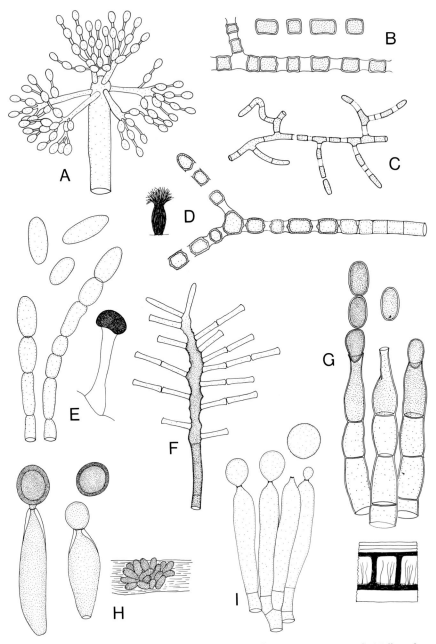

PLATE 1. A, *Amblyosporium spongiosum*; B, *Sporendonema purpurascens*; C, *Malbranchea pulchella*; D, *Coremiella ulmariae*; E, *Antromycopsis broussonetiae*; F, *Sympodiella acicola*; G, *Cryptostroma corticale*; H, *Stromatocrea cerebriforme*; I, *Tuberculina persicina*.

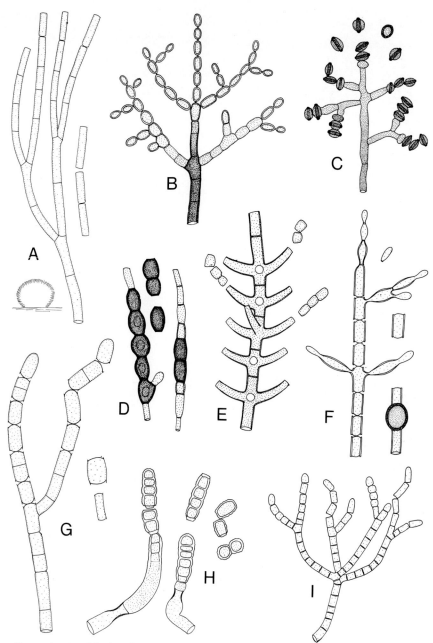

PLATE 2. A, *Cylindrocolla urticae*; B, *Oidiodendron* sp.; C, *Stephanosporium cereale*; D, *Scytalidium lignicola*; E, *Oncocladium flavum*; F, *Moniliella acetoabutans*; G, *Geotrichum candidum*; H, *Wallemia ichthyophaga*; I, *Arthrographis langeroni*.

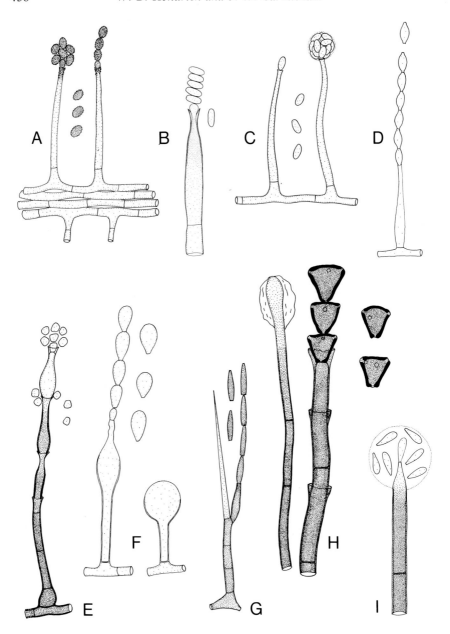

PLATE 3. A, *Gliomastix murorum*; B, *Monostachys transversalis*; C, *Acremonium* sp.; D, *Paecilomyces roseolus*; E, *Chloridium* state of *Chaetosphaeria myriocarpa*; F, *Torulomyces indica*; G, *Pleurocatena acicularis*; H, *Catenularia* state of *Chaetosphaeria novae-zelandiae*; I, *Sphaeromyces clavisporus*.

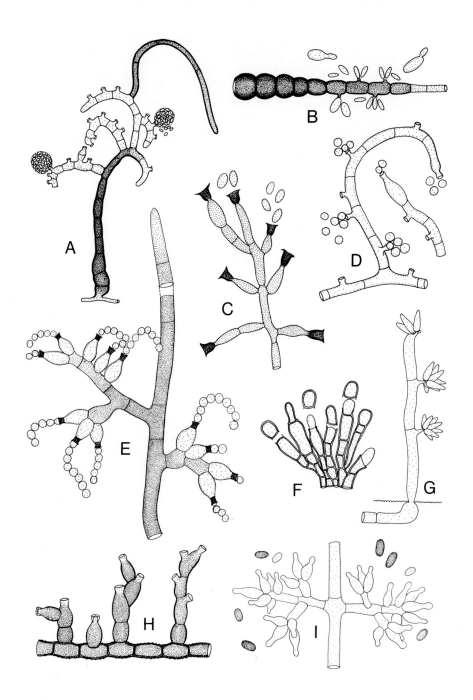

PLATE 4. A, *Bahupaathra samala*; B, *Aureobasidium pullulans*; C, *Phialophora verrucosa*; D, *Cladorrhinum foecundissimum*; E, *Angulimaya sundara*; F, *Phialophoropsis trypodendri*; G, *Meria coniospora*; H, *Capnophialophora* state of *Strigopodia resinae*; I, *Trichoderma koningii*.

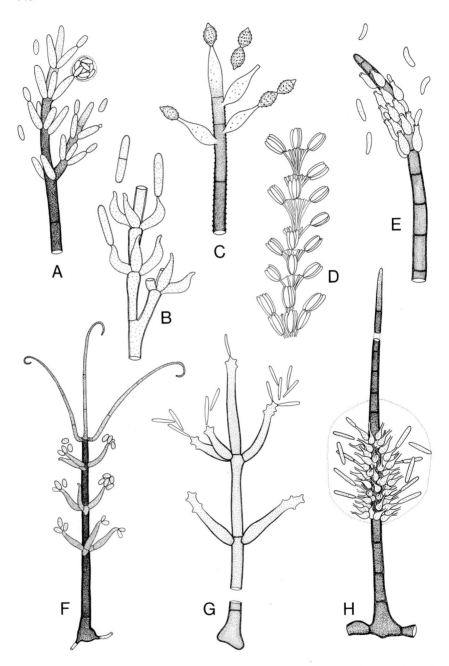

PLATE 5. A, *Stachylidium bicolor*; B, *Uncigera cordae*; C, *Acrophialophora nainiana*; D, *Articularia quercina*; E, *Zanclospora brevispora*; F, *Gonytrichum macrocladum*; G, *Selenosporella curvispora*; H, *Chaetopsina fulva*.

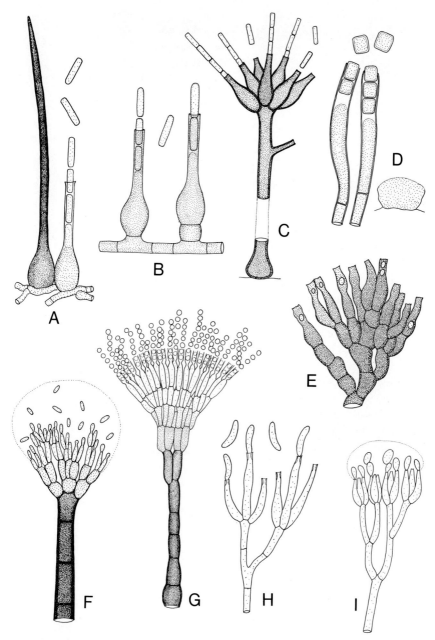

PLATE 6. A, *Chaetochalara bulbosa*; B, *Chalara fusidioides*; C, *Sporendocladia castaneae*; D, *Bloxamia truncata*; E, *Cystodendron dryophilum*; F, *Haplographium* state of *Hyaloscypha dematiicola*; G, *Phialocephala dimorphospora*; H, *Dendrodochium* state of *Nectria magnusiana*; I, *Gliocladium* sp.

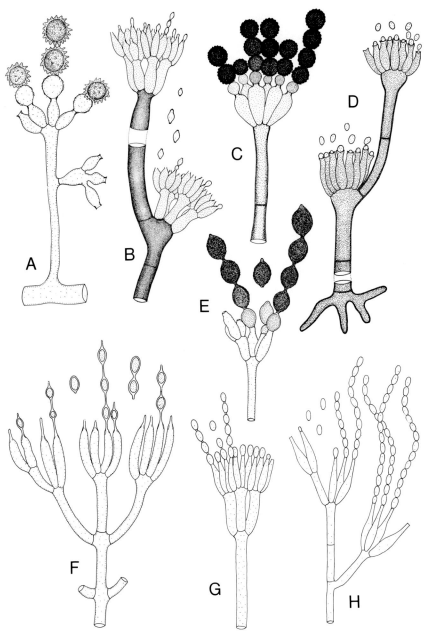

PLATE 7. A, *Eladia saccula*; B, *Thysanophora penicillioides*; C, *Stachybotrys* sp.; D, *Custingophora olivacea*; E, *Phialomyces macrosporus*; F, *Phialotubus microsporus*; G, *Penicillium* sp.; H, *Paecilomyces varioti*.

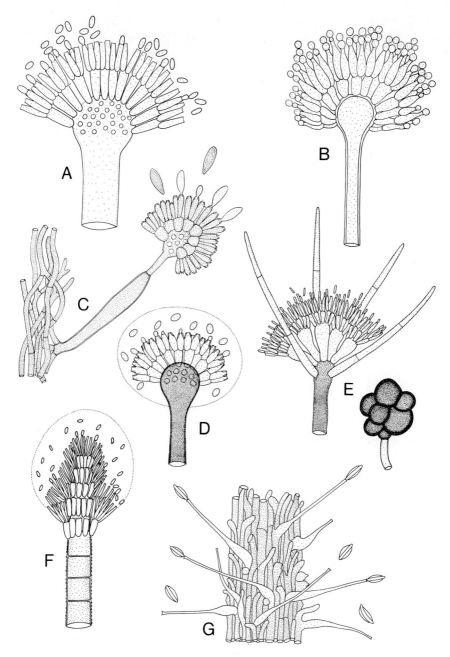

PLATE 8. A, *Gliocephalis hyalina*; B, *Aspergillus* sp.; C, *Gibellula suffulta*; D, *Goidanichiella* sp.; E, *Gliocephalotrichum bulbilium*; F, *Gloiosphaera clerciana*; G, *Hirsutella saussurei*.

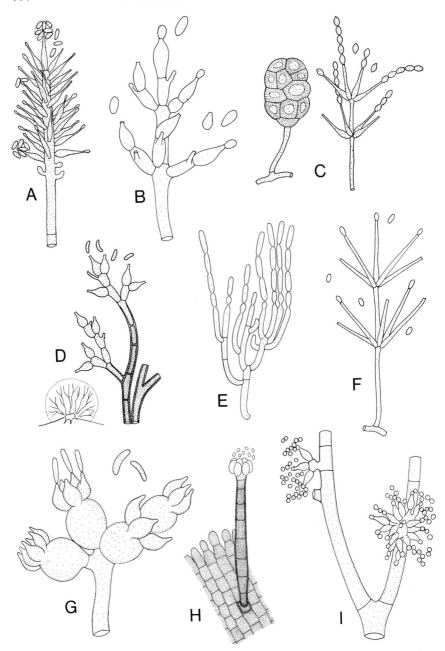

PLATE 9. A, *Harziella capitata*; B, *Sesquicillium candelabrum*; C, *Diheterospora catenulata*; D, *Agyriella nitida*; E, *Metarrhizium anisopliae*; F, *Verticillium dahliae*; G, *Endostilbum cerasi*; H, *Plokamidomyces* state of *Trichopeltheca asiatica*; I, *Myrioconium marina*.

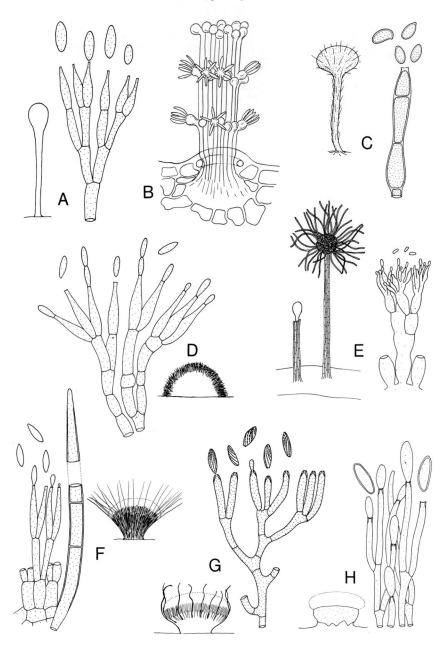

PLATE 10. A, *Akanthomyces* sp.; B, *Microstroma album*; C, *Agaricostilbum palmicola*; D, *Nalanthamala madreeya*; E, *Heterocephalum aurantiacum*; F, *Volutella* sp.; G, *Myrothecium striatisporum*; H, *Tubercularia vulgaris*.

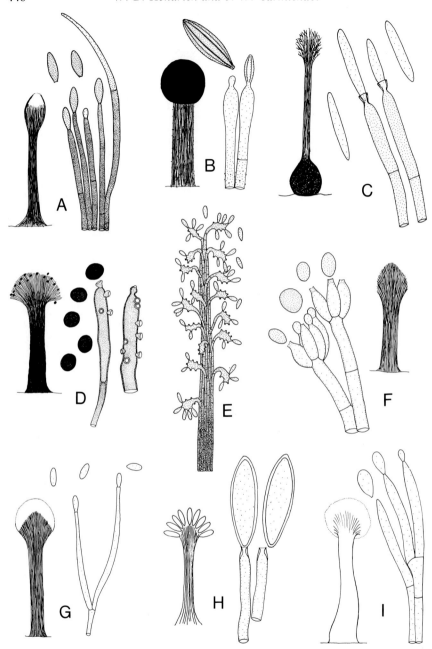

PLATE 11. A, *Saccardaea atra*; B, *Solheimia costaspora*; C, *Graphiothecium fresenii*; D, *Paathramaya sundara*; E, *Phaeoisaria clematidis*; F, *Phialostele scytopetali*; G, *Crinula caliciiformis*; H, *Ciliciopodium macrosporum*; I, *Stilbum* sp.

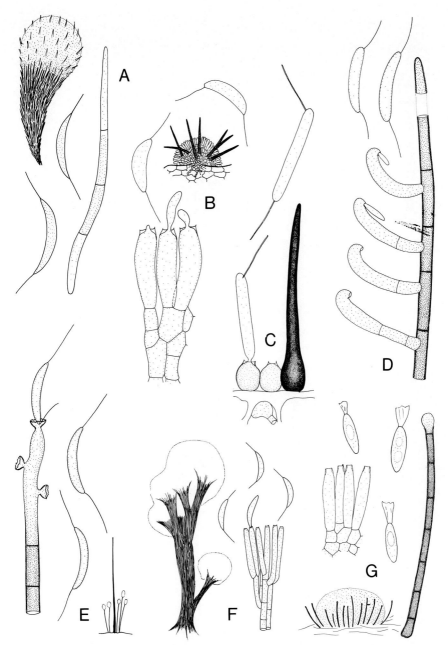

PLATE 12. A, *Neottiosporella radicata*; B, *Minimidochium setosum*; C, *Mahabalella acutisetosa*; D, *Menispora ciliata*; E, *Codinaea assamica*; F, *Thozetellopsis toklaiensis*; G, *Koorchaloma madreeya*.

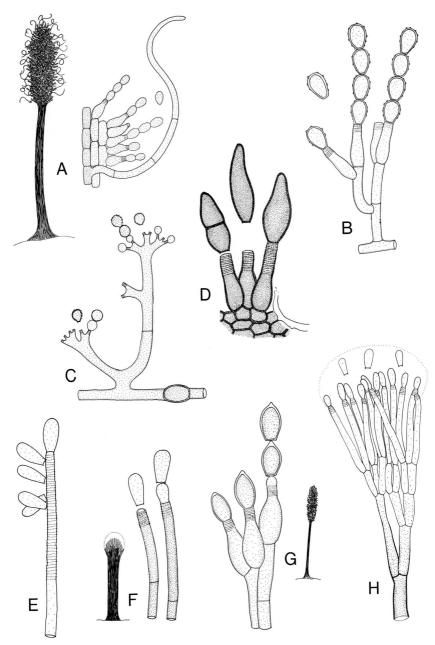

PLATE 13. A, *Trichurus spiralis*; B, *Scopulariopsis brevicaulis*; C, *Polypaecilum insolitum*; D, *Spilocaea pomi*; E, *Bainieria hyalina*; F, *Graphium penicillioides*; G, *Cephalotrichum stemonitis*; H, *Leptographium lundbergii*.

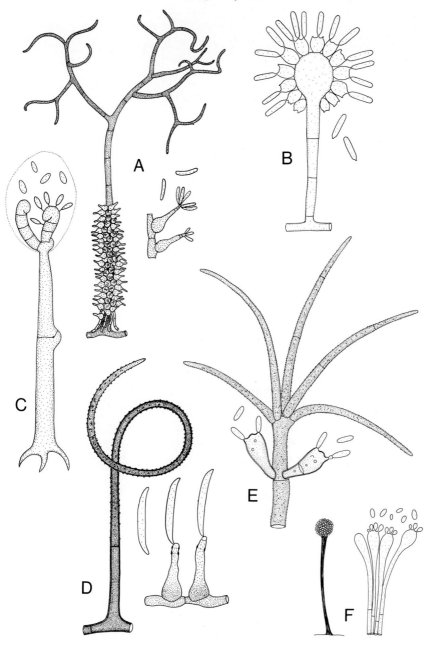

PLATE 14. A, *Ceratocladium microspermum*; B, *Basidiobotrys griseus*; C, *Jacobia conspicua*; D, *Circinotrichum maculiforme*; E, *Heimiodora verticillata*; F, *Pachnocybe ferruginea*.

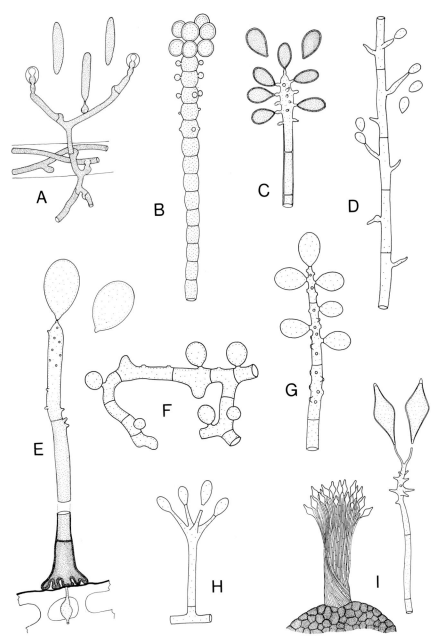

PLATE 15. A, *Nematoctonus haptocladus*; B, *Graphiola thaxteri*; C, *Olpitrichum macrosporum*; D, *Aphanocladium album*; E, *Hemibeltrania cinnamomi*; F, *Beniowskia sphaeroidea*; G, *Acladium conspersum*; H, *Dexhowardia tetraspora*; I, *Rhombostilbella rosea*.

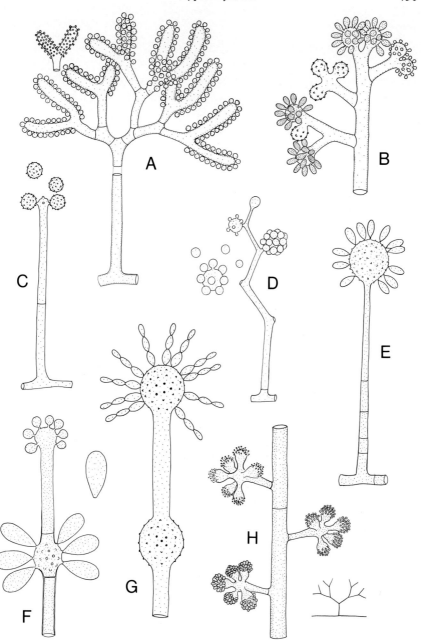

PLATE 16. A, *Ostracoderma* state of *Peziza ostracoderma*; B, *Botrytis cinerea*; C, *Sporocephalum peniophorae*; D, *Blastobotrys nivea*; E, *Oedocephalum* sp.; F, *Gonatobotrys simplex*; G, *Nematogonium parasiticum*; H, *Botryosporium pulchrum*.

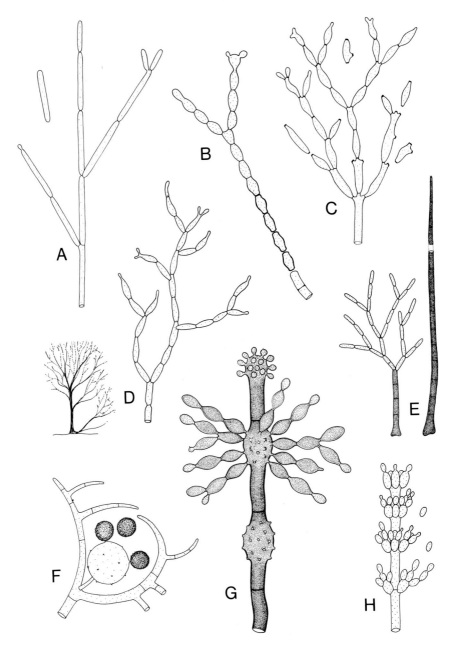

PLATE 17. A, *Fusidium griseum*; B, *Monilia cinerea*; C, *Hyalodendron* sp.; D. *Sphaeridium candidum*; E, *Hormiactella fusca*; F, *Sigmoideomyces dispiroides*; G, *Gonatobotryum apiculatum*; H, *Nomuraea prasina*.

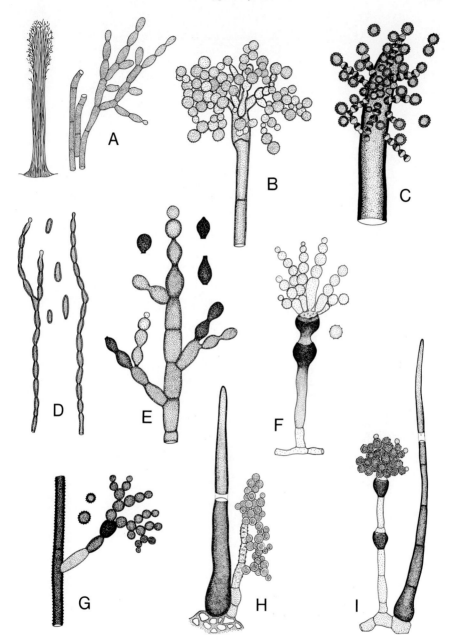

PLATE 18. A, *Sorocybe resinae*; B, *Periconia paludosa*; C, *Sadasivania girisa*; D, *Xylohypha nigrescens*; E, *Alysidium resinae*; F, *Haplobasidion thalictri*; G, *Trichobotrys effusa*; H, *Lacellina graminicola*; I, *Lacellinopsis sacchari*.

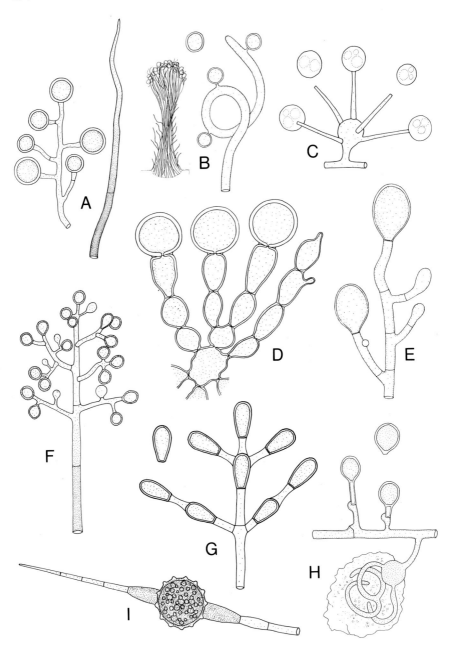

PLATE 19. A, *Botryotrichum piluliferum*; B, *Heydenia (Martindalia) spironema*; C, *Umbelopsis versiformis*; D, *Ambrosiella xylebori*; E, *Allescheriella crocea*; F, *Staphylotrichum coccosporum*; G, *Chrysosporium pannorum*; H, *Pagidospora amoebophila*; I, *Chaetoconidium arachnoideum*.

PLATE 20. A, *Pteroconium pterospermum*; B, *Arthrinium puccinioides* with conidia of (a) *A. caricicola*, (b) *A. luzulae*, (c) *A. cuspidatum*; C, *Cordella coniosporioides*; D, *Endocalyx melanoxanthus*; E, *Sepedonium chrysospermum*; F, *Glomopsis corni*; G, *Histoplasma capsulatum*; H, *Glomospora empetri*.

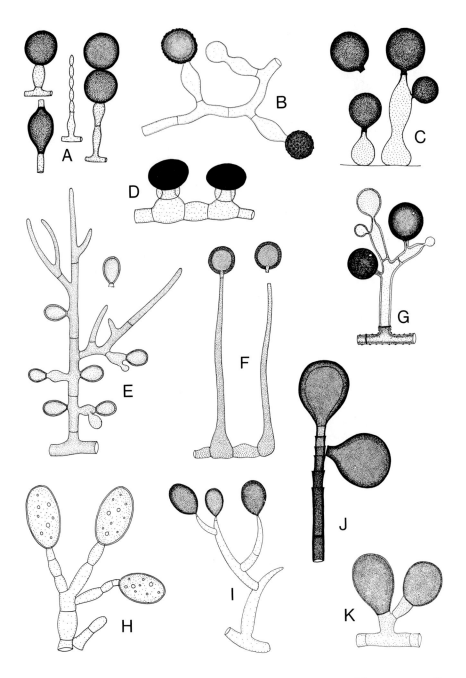

PLATE 21. A, *Humicola grisea*; B, *Thermomyces lanuginosus*; C, *Pilulina nigrospora*; D, *Nigrospora oryzae*; E, *Botryoderma lateritium*; F, *Domingoella asterinarum*; G, *Gilmaniella humicola*; H, *Bactridiopsis ulei*; I, *Acremoniella atra*; J, *Acrogenospora sphaerocephala*; K, *Acremoniula sarcinellae*.

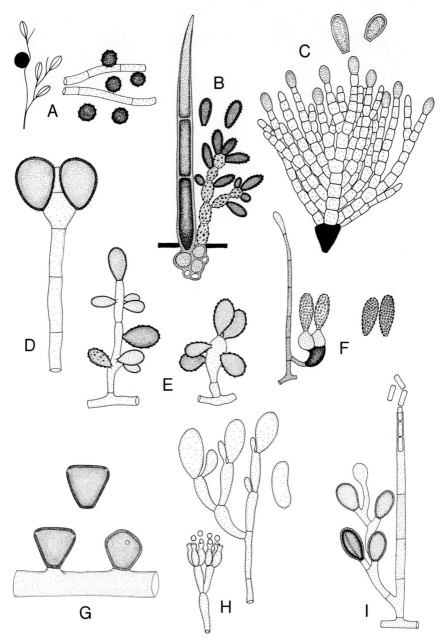

PLATE 22. A, *Ustilaginoidea virens*; B, *Herposira velutina*; C, *Chantransiopsis xantholini*; D, *Microclava bispora*; E, *Sporotrichum thermophile*; F, *Zygosporium geminatum*; G, *Rhinocladium coprogenum*; H, *Dimorphospora foliicola*; I, *Chalaropsis punctulata*.

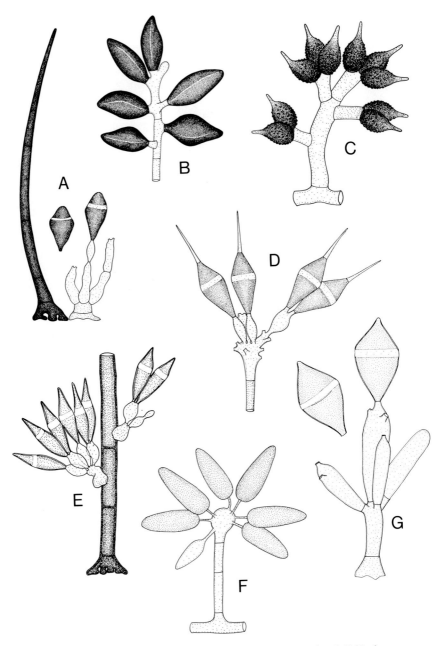

PLATE 23. A, *Beltraniella humicola*; B, *Mammaria echinobotryoides*; C, *Echinobotrym atrum*; D, *Beltrania rhombica*; E, *Beltraniopsis esenbeckiae*; F, *Asteromyces cruciatus*; G, *Pseudobeltrania cedrelae*.

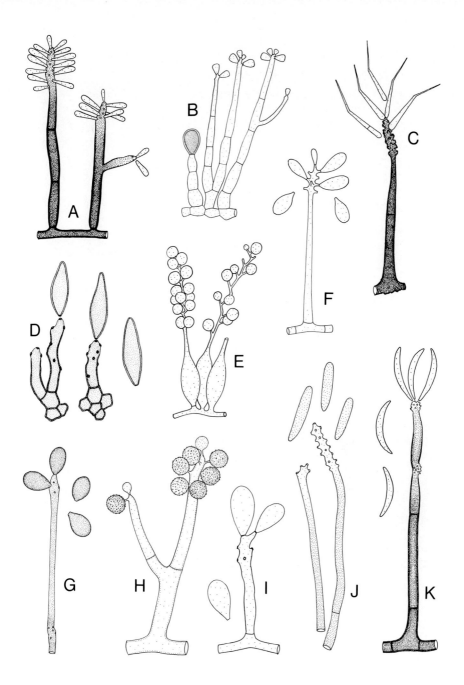

PLATE 24. A, *Rhinocladiella* state of *Dictyotrichiella mansonii*; B, *Raffaelea ambrosiae*; C, *Subulispora procurvata*; D, *Fusicladium* state of *Venturia pirina*; E, *Beauveria bassiana*; F, *Sporothrix* sp.; G, *Virgariella atra*; H, *Basifimbria aurea*; I, *Ovularia obliqua*; J, *Zasmidium cellare*; K, *Chloridiella* sp.

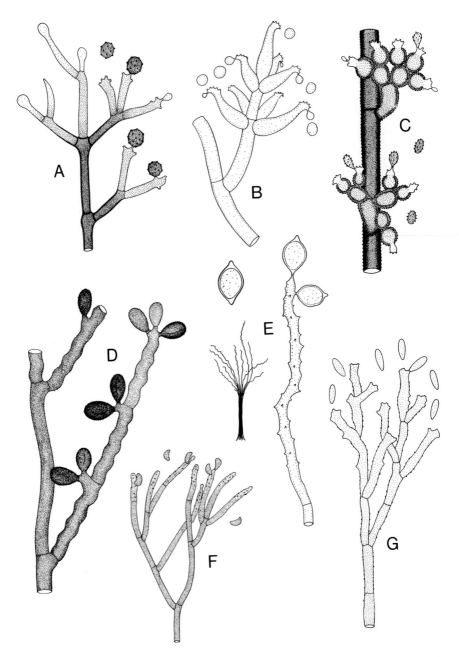

PLATE 25. A, *Dicyma* state of *Ascotricha chartarum*; B, *Costantinella micheneri*; C, *Kumanasamuha sundara*; D, *Conoplea sphaerica*; E, *Drumopama girisa*; F, *Virgaria nigra*; G, *Nodulisporium gregarium*.

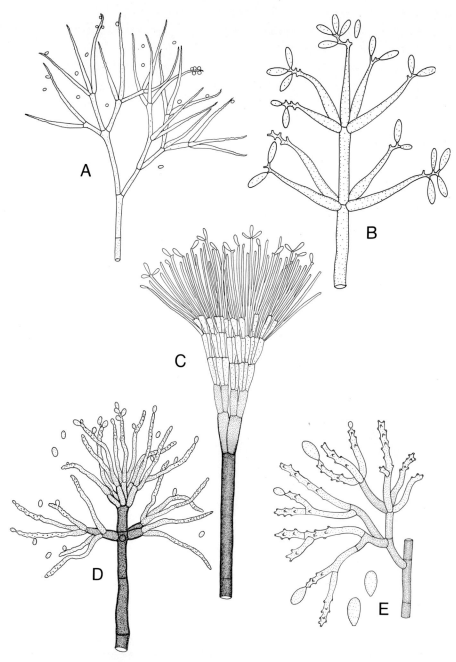

PLATE 26. A, *Tritirachium hydnicola*; B, *Calcarisporium arbuscula*; C, *Verticicladiella abietina*; D, *Verticicladium trifidum*; E, *Hansfordia ovalispora*.

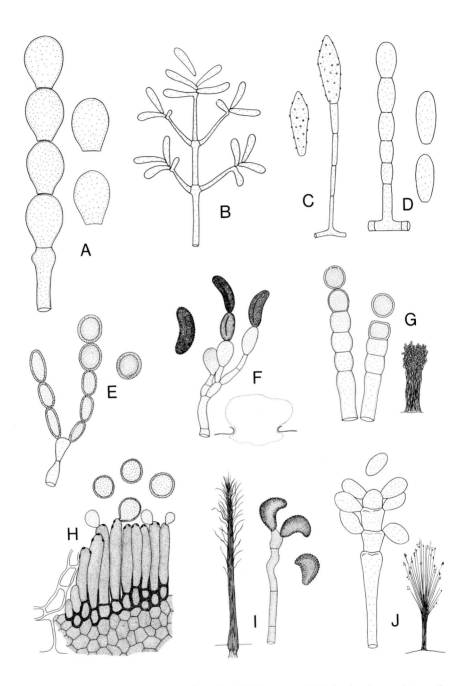

PLATE 27. A, *Basipetospora rubra*; B, *Cladobotryum* (*Cylindrophora*) *apiculatum*; C, *Ovulariopsis* state of *Phyllactinia* sp.; D, *Acrosporium* state of *Erysiphe graminis*; E, *Sphaerosporium lignatile*; F, *Megalodochium palmicola*; G, *Briosia ampelophaga*; H, *Hadrotrichum phragmitis*; I, *Melanographium selenioides*; J, *Stevensomyces palmae*.

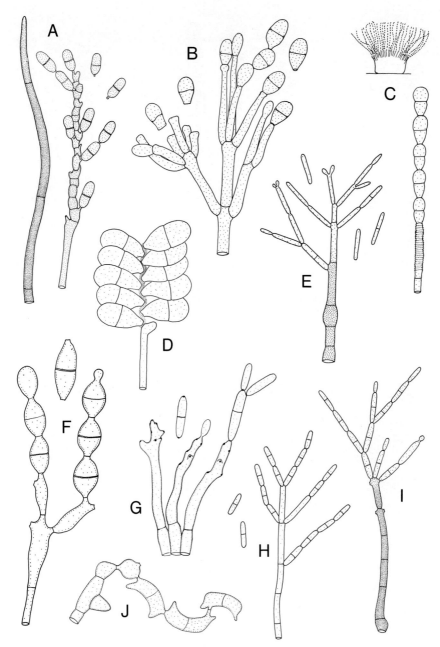

PLATE 28. A, *Paspalomyces aureus*; B, *Cladobotryum variospermum*; C, *Gymnodochium fimicolum*; D, *Trichothecium roseum*; E, *Polyscytalum foecundissimum*; F, *Diploospora rosae*; G, *Ramularia gei*; H, *Hormiactis alba*; I, *Polyscytalina grisea*; J, *Arxiella terrestris*.

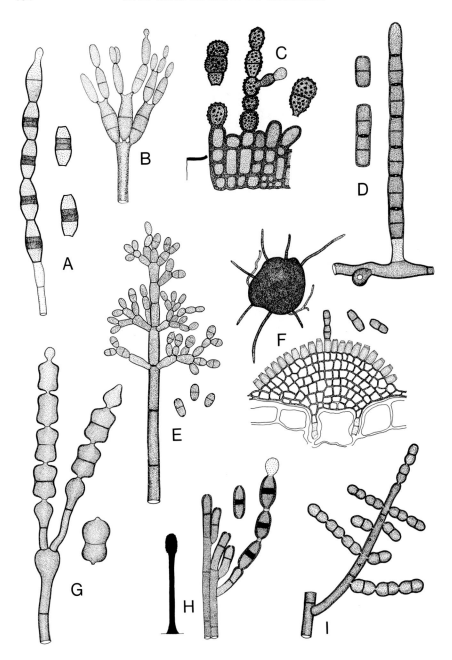

PLATE 29. A, *Bispora antennata*; B, *Cladosporium herbarum*; C, *Spilodochium vernoniae*; D, *Ampulliferina persimplex*; E, *Dendryphion (Dwayamala) prathilomaka*; F, *Trichodochium disseminatum*; G, *Oedemium didymum*; H, *Didymobotryum rigidum*; I, *Diplococcium spicatum*.

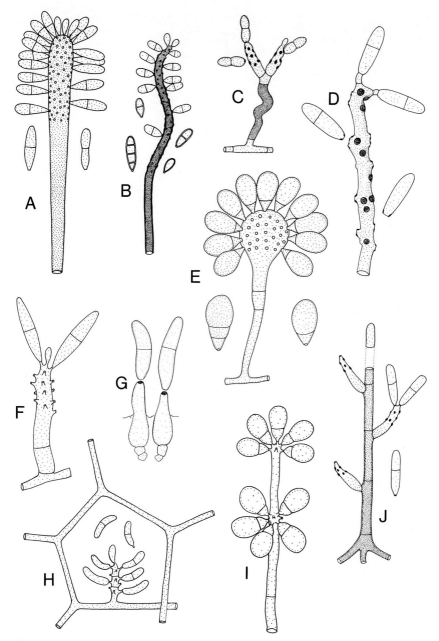

PLATE 30. A, *Blastophorella smithii*; B, *Veronaea botryosa*; C, *Zygophiala jamaicensis*; D, *Sceptrifera pulchra*; E, *Cephaliophora* (*Cephalomyces*) *nigricans*; F, *Dactylaria candidula*; G, *Fusicladiella pimpinellae*; H, *Neta patuxentica*; I, *Arthrobotrys superba*; J, *Haplariopsis fagicola*.

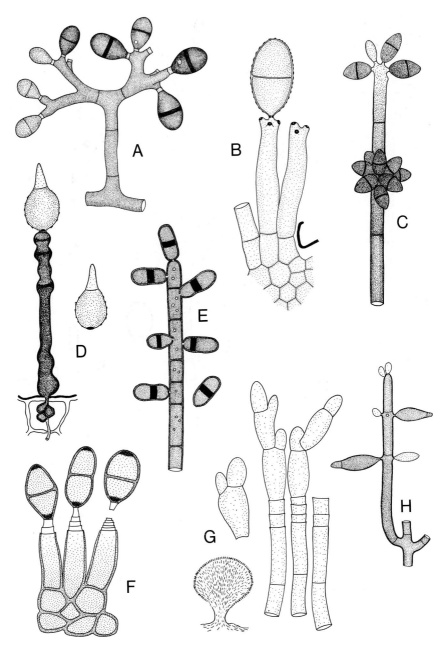

PLATE 31. A, *Balanium stygium*; B, *Asperisporium caricae*; C, *Cordana pauciseptata*; D, *Deightoniella africana*; E, *Spadicoides bina*; F, *Oedothea vismiae*; G, *Annellodochium ramulisporum*; H, *Didymotrichella quercina*.

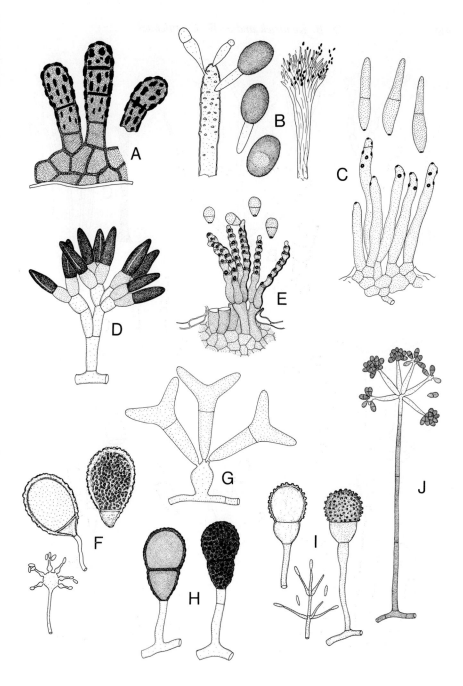

PLATE 32. A, *Hadronema orbiculare*; B, *Panchanania jaipurensis*; C, *Passalora depressa*; D, *Wardomyces humicola*; E, *Polythrincium trifolii*; F, *Chlamydomyces palmarum*; G, *Scolecobasidium terreum*; H, *Trichocladium asperum*; I, *Mycogone perniciosa*; J, *Pseudobotrytis bisbyi*.

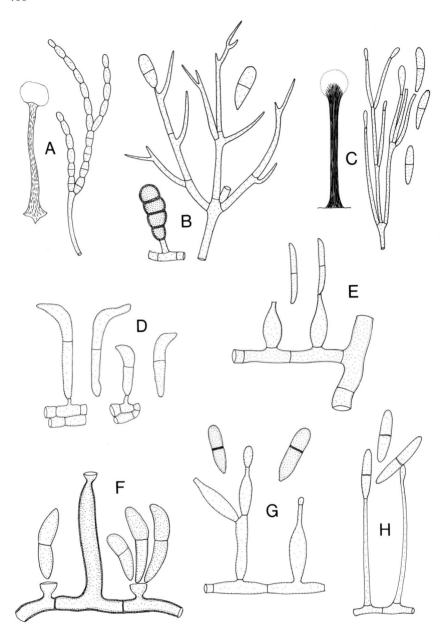

PLATE 33. A, *Antromyces copridis*; B, *Sympodiophora stereicola*; C, *Didymostilbe* sp.; D, *Rhynchosporium secalis*; E, *Cylindrodendrum album*; F, *Cyphellophora laciniata*; G, *Exophiala salmonis*; H, *Cephalosporiopsis* sp.

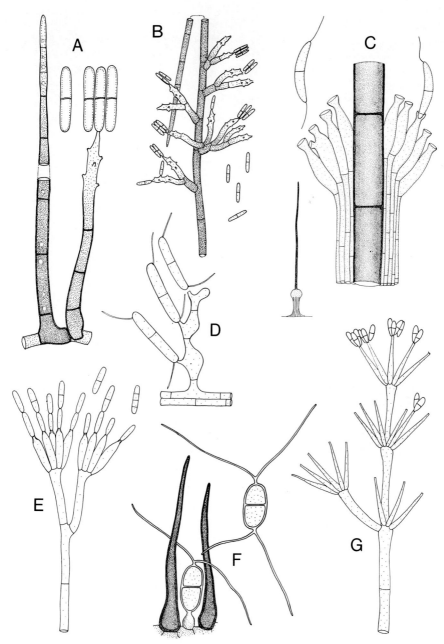

PLATE 34. A, *Cylindrotrichum oligospermum*; B, *Chaetopsis grisea*; C, *Menisporopsis novae-zelandiae*; D, *Leptodiscella africana*; E, *Penicillifer pulcher*; F, *Dwayaloma trina*; G, *Sibirina fungicola*.

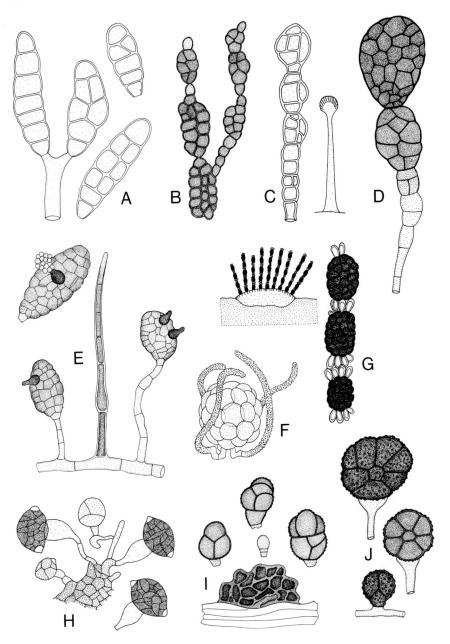

PLATE 35. A, *Hyalodictys degenerans*; B, *Peyronelia arbuscula*; C, *Dictyocatenulata alba*; D, *Coniosporium paradoxum*; E, *Septosporium bulbotrichum*; F, *Peyronelina glomerulata*; G, *Barnettella speciosa*; H, *Oncopodium antoniae*; I, *Epicoccum (Cerebellum) andropogonis*; J, *Epicoccum purpurascens*.

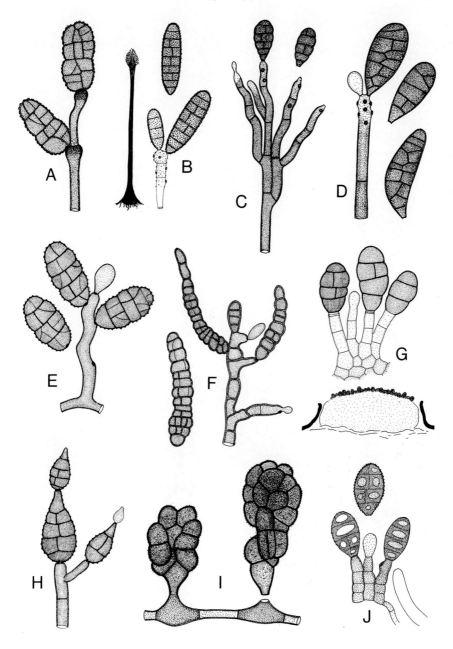

PLATE 36. A, *Stemphylium botryosum*; B, *Sclerographium aterrimum*; C, *Mystrosporiella litseae*; D, *Dactylosporium macropus*; E, *Ulocladium* sp.; F, *Sirosporium antenniforme*; G, *Stigmina lycii*; H, *Alternaria alternata*; I, *Chuppia sarcinifera*; J, *Thyrostromella myriana*.

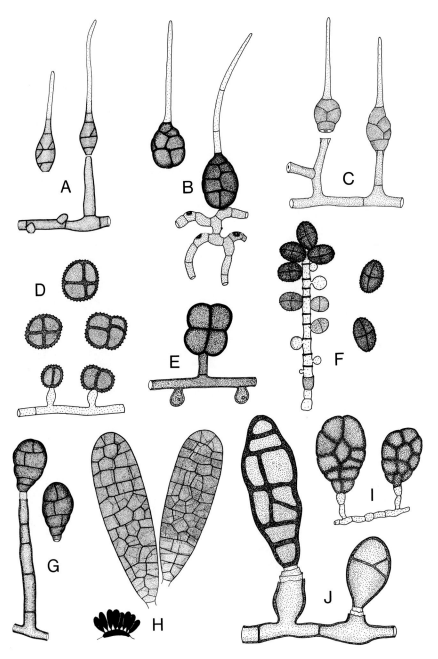

PLATE 37. A, *Hansfordiellopsis aburiensis*; B, *Piricauda paraguayense*; C, *Hansfordiella meliolae*; D, *Tetracoccosporium paxianum*; E, *Sarcinella fumosus*; F, *Dictyoarthrinium quadratum*; G, *Acrodictys bambusicola*; H, *Berkleasmium concinnum*; I, *Monodictys putredinis*; J, *Annellophorella faureae*.

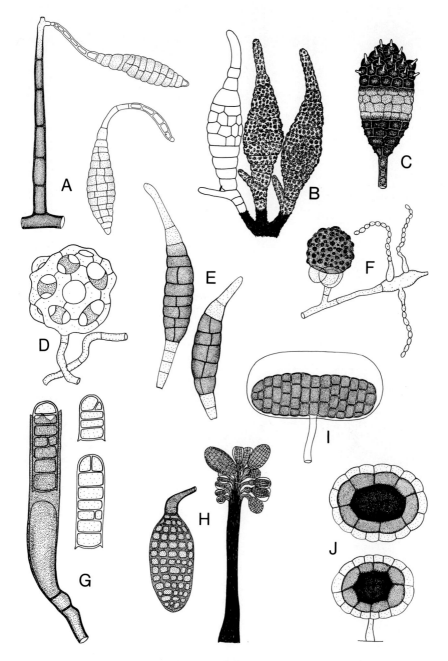

PLATE 38. A, *Pendulispora venezuelanica*; B, *Muiaria repens*; C, *Muiogone chromopteri*; D, *Clathrosphaerina zalewskii*; E, *Dictyodesmium ulmicola*; F, *Acrospeira mirabilis*; G, *Ascoconidium tsugae*; H, *Kostermansinda magna*; I, *Coleodictyospora cubensis*; J, *Hermatomyces tucumanensis*.

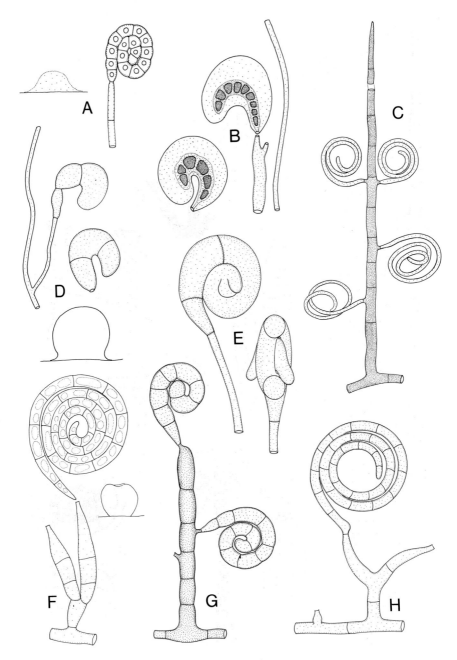

PLATE 39. A, *Everhartia hymenuloides*; B, *Drepanoconis larvaeformis*; C, *Helicosporium vegetum*; D, *Delortia palmicola*; E, *Diplorhynchus biloba*; F, *Vanbeverwijkia spirospora*; G, *Helicoma mulleri*; H, *Helicomyces roseus*.

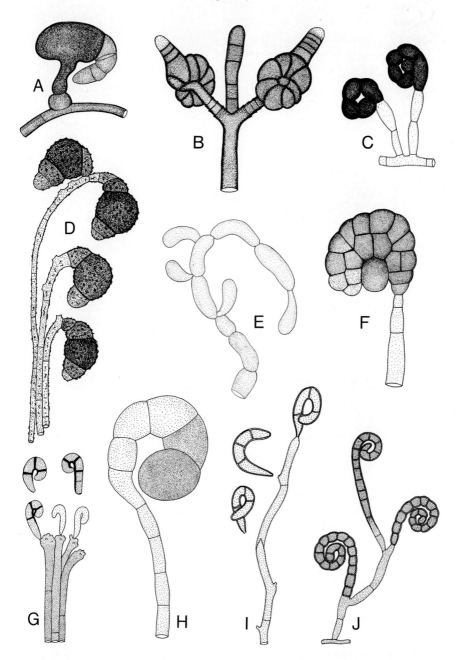

PLATE 40. A, *Curculiospora sydowii*; B, *Circinoconis paradoxa*; C, *Zalerion eistla*; D, *Curvulariopsis cymbisperma*; E, *Strumella olivatra*; F, *Xenosporium pleurococcum*; G, *Trochophora simplex*; H, *Cirrenalia pseudomacrocephala*; I, *Helicomina caperoniae*; J, *Troposporella fumosa*.

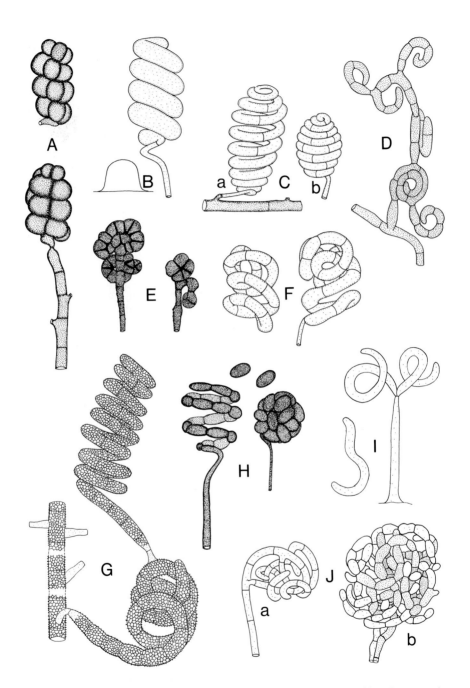

PLATE 41. A, *Helicorhoidion botryoideum*; B, *Troposporium album*; C, (a) *Helicoon sessile*, (b) *H. ellipticum*; D, *Helicodendron triglitziensis*; E, *Cheirospora botryospora*; F, *Hobsonia mirabilis*; G, *Hiospira* state of *Brooksia tropicalis*; H, *Helicocephalum sarcophilum*; I, *Brachyhelicoon xylogenum*; J, (a) *Spirosphaera beverwijkiana*, (b) *S. floriformis*.

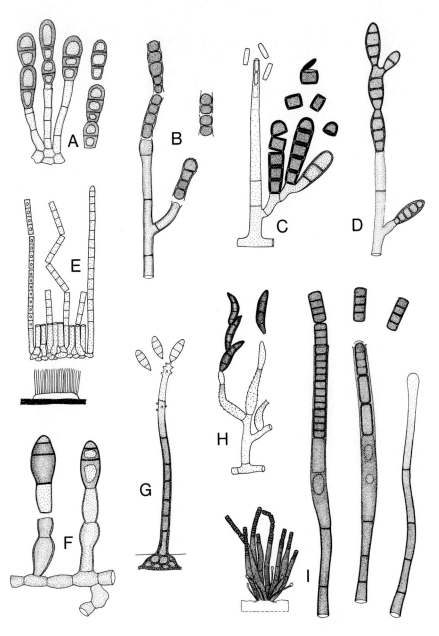

PLATE 42. A, *Bactrodesmiella masonii*; B, *Bahusakala olivaceo-nigra*; C, *Milowia (Thielaviopsis) basicola*; D, *Sporophiala prolifica*; E, *Septotrullula bacilligera*; F, *Phragmospathula phoenicis*; G, *Pleurophragmium simplex*; H, *Fusariella atrovirens*; I, *Sporoschisma mirabile*.

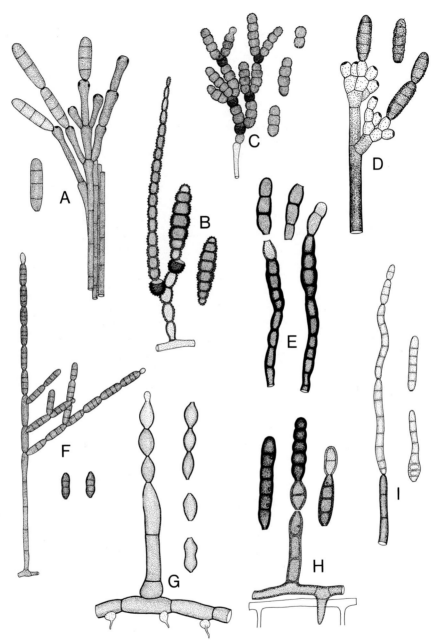

PLATE 43. A, *Dendrographium interseminatum*; B, *Dwayabeeja sundara*; C, *Torula herbarum*; D, *Dendryphion comosum*; E, *Taeniolella exilis*; F, *Septonema secedens*; G, *Ampullifera foliicola*; H, *Heteroconium cytharexyli*; I, *Ormathodium bombycinum*.

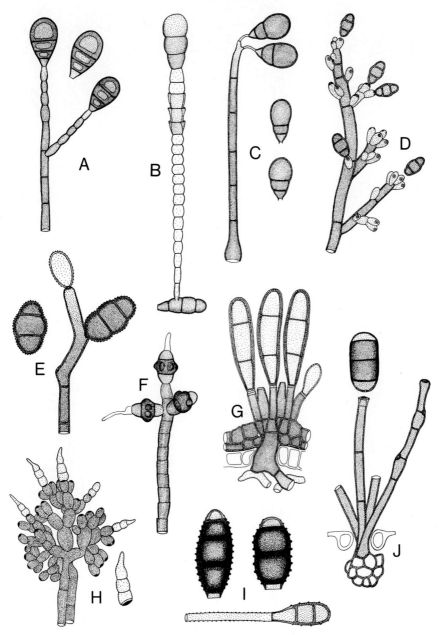

PLATE 44. A, *Brachysporiella gayana*; B, *Antennopsis gallica*; C, *Brachysporium obovatum*; D, *Spondylocladiella botrytioides*; E, *Acroconidiella tropaeoli*; F, *Acrophragmis coronata*; G, *Septoidium* state of *Parodiopsis hurae*; H, *Acarocybe hansfordii*; I, *Stemphyliomma valparadisiacum*; J, *Duosporium cyperi*.

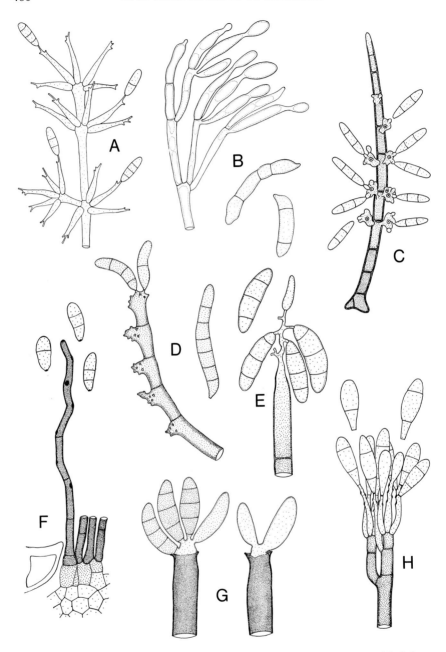

PLATE 45. A, *Eurasina bondarzewiae*; B. *Margaritispora aquatica*; C, *Spondylocladiopsis cupulicola*; D, *Gonatophragmium mori*; E, *Pleurothecium recurvatum*; F, *Cercosporidium chaetomium*; G, *Cacumisporium capitulatum*; H, *Sterigmatobotrys macrocarpa*.

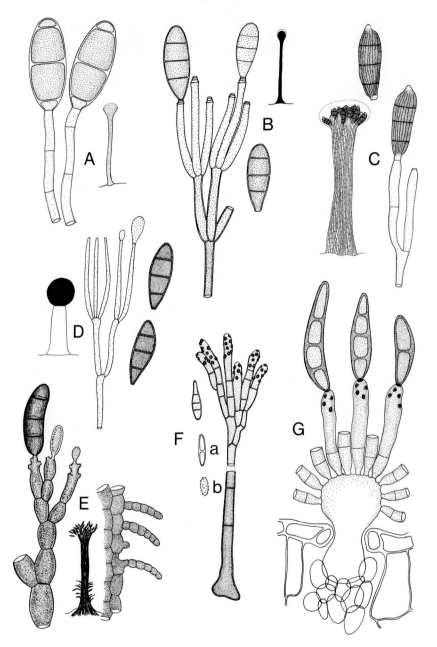

PLATE 46. A, *Calostilbella calostilbe*; B, *Arthrobotryum stilboideum*; C, *Virgatospora echinofibrosa*; D, *Albosynnema elegans*; E, *Capnocybe novae-zelandiae*; F, *Periconiella cyatheae*, with conidia of (a) *P. leptoderridis*, (b) *P. angusiana*; G, *Camptomeris albizziae*.

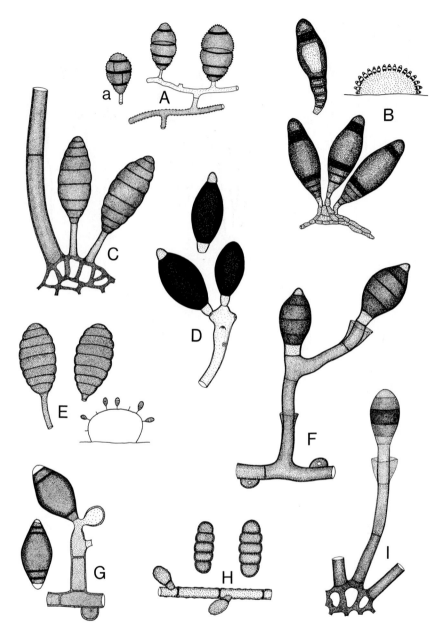

PLATE 47. A, *Pithomyces flavus* with conidium of (a) *P. cynodontis*; B, *Bactrodesmium abruptum*; C, *Excipularia narsapurensis*; D, *Brachydesmiella biseptata*; E, *Listeromyces insignis*; F, *Endophragmiopsis pirozynskii*; G, *Mitteriella ziziphina*; H, *Jainesia meliolicola*; I, *Endophragmia elliptica*.

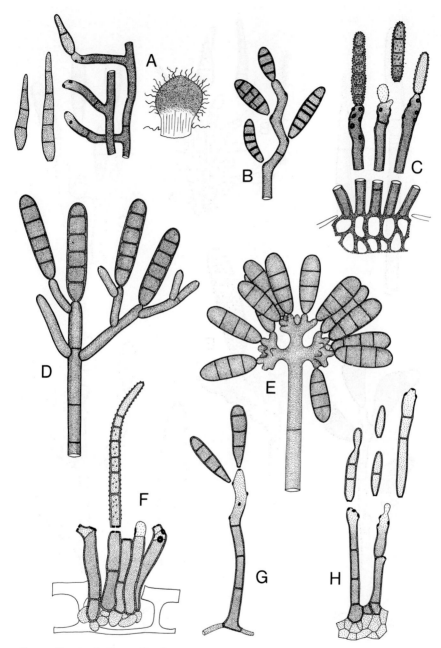

PLATE 48. A, *Podosporiella glomerata*; B, *Embellisia allii*; C, *Verrucispora proteacearum*; D, *Dendryphiopsis atra*; E, *Dichotomophthora indica*; F, *Stenellopsis fagraeae*; G, *Eriocercospora balladynae*; H, *Stenella araguata*.

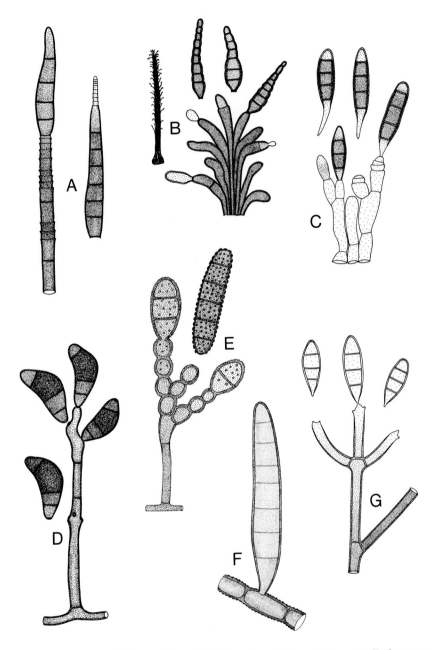

PLATE 49. A, *Annellophora africana*; B, *Podosporium rigidum*; C, *Exosporiella fungorum*; D, *Curvularia lunata*; E, *Hormocephalum ecuadorense*; F, *Antennatula atra*; G, *Phaeodactylium alpiniae*.

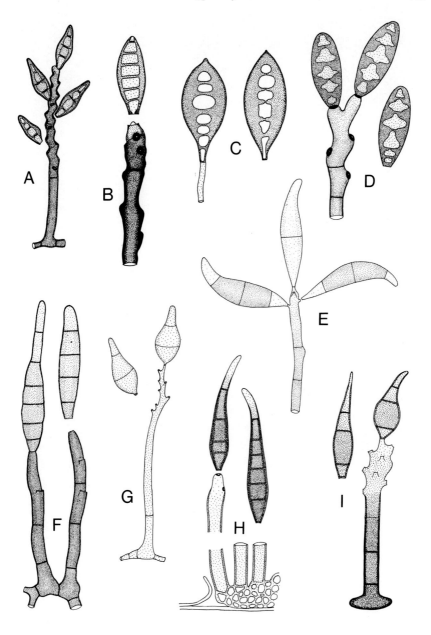

PLATE 50. A, *Spiropes guareicola*; B, *Pseudospiropes nodosus*; C, *Murogenella terrophila*; D, *Drechslera* sp.; E, *Nakataea* state of *Leptosphaeria salvinii*; F, *Tomenticola trematis*; G, *Pyricularia oryzae*; H, *Prathigada crataevae*; I, *Pyriculariopsis parasitica*.

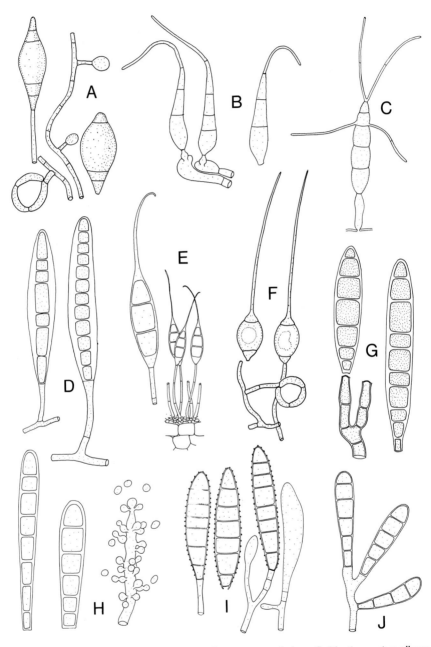

PLATE 51. A, *Dactylella ellipsospora*; B, *Spermospora subulata*; C, *Mastigosporium album*; D, *Keratinomyces ajelloi*; E, *Trichoconis caudata*; F, *Drechslerella acrochaeta*; G, *Bactridium flavum*; H, *Trichophyton mentagrophytes*; I, *Microsporum gypseum*; J, *Epidermophyton floccosum*.

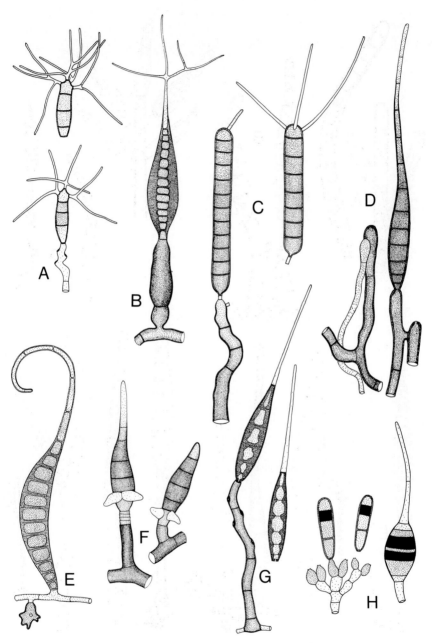

PLATE 52. A, *Pleiochaeta albizziae*; B, *Lomaantha pooga*; C, *Camposporium antennatum*; D, *Acarocybella jasminicola*; E, *Ceratophorum helicosporum*; F, *Teratosperma singulare*; G, *Phaeotrichoconis crotalariae*; H, *Triadelphia heterospora*.

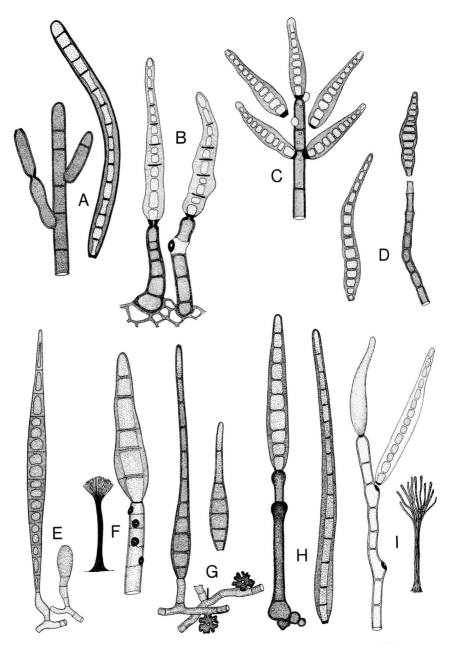

PLATE 53. A, *Corynesporella urticae*; B, *Exosporium tiliae*; C, *Helminthosporium velutinum*; D, *Sporidesmium bambusicola*; E, *Cercosperma subsessile*; F, *Annellophragmia coonoorense*, G, *Clasterosporium flagellatum*; H, *Corynespora cassiicola*; I, *Phaeoisariopsis armillata*.

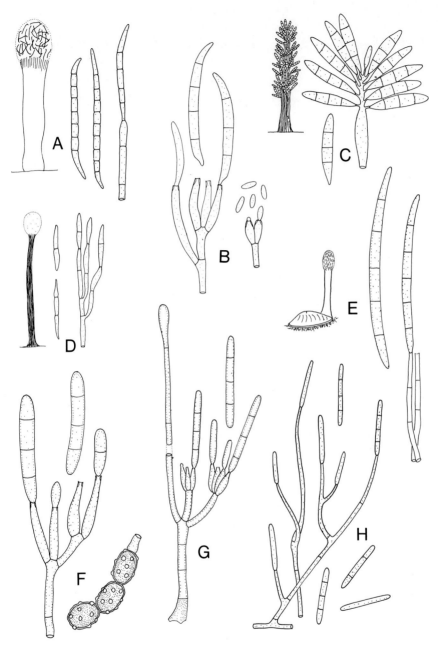

PLATE 54. A, *Morrisographium ulmi*; B, *Fusarium* state of *Nectria desmazierii*; C, *Arthrosporium candidum*; D, *Pseudographiella variaseptata*; E, *Atractium flammeum*; F, *Cylindrocarpon destructans*; G, *Cylindrocladium ilicicola*; H, *Bacillispora aquatica*.

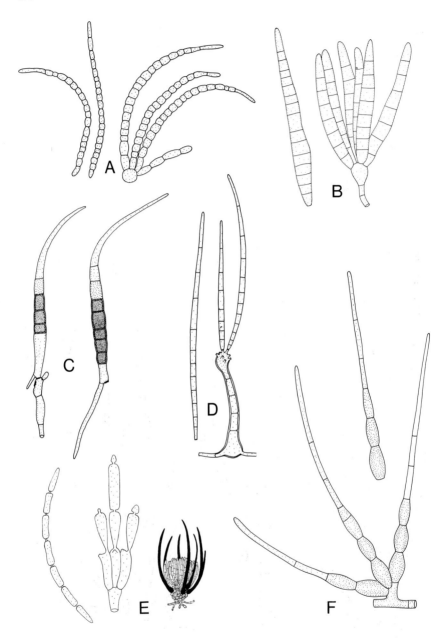

PLATE 55. A. *Petrakiopsis elegans*; B, *Tetracrium amphibium*; C, *Centrospora acerina*; D, *Mirandina corticola*; E, *Wiesneriomyces javanicus*; F, *Arthrocladium caudatum*.

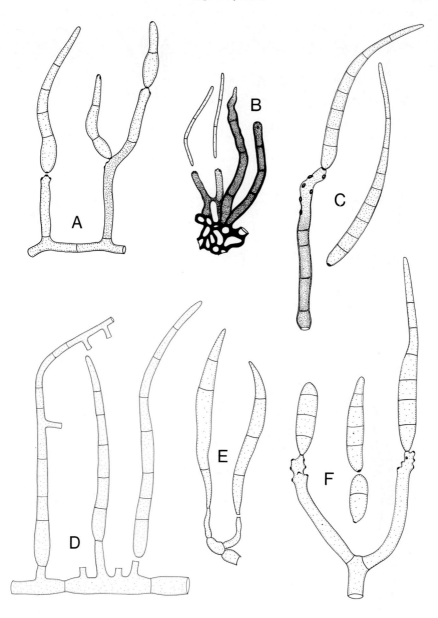

PLATE 56. A, *Cladosporiella uredinicola*; B, *Schizotrichum lobeliae*; C, *Cercospora elaeidis*; D, *Irpicomyces schiffnerulae*; E, *Cercosporella junci*; F, *Elletevera parasitica*.

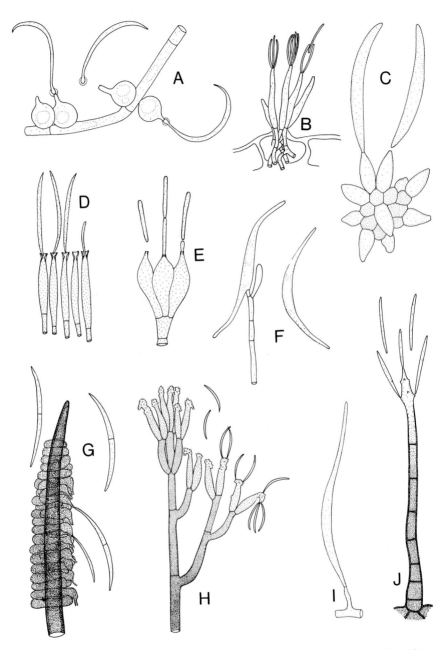

PLATE 57. A, *Harposporium oxycoracum*; B, *Kmetiopsis hymenaeae*; C, *Microdochium phragmitis*; D, *Kmetia exigua*; E, *Riclaretia urticae*; F, *Lunulospora curvula*; G, *Cryptophiale kakombensis*; H, *Harpographium* sp.; I, *Aciculariella lasiosphaeriae*; J, *Cercosporula corticola*.

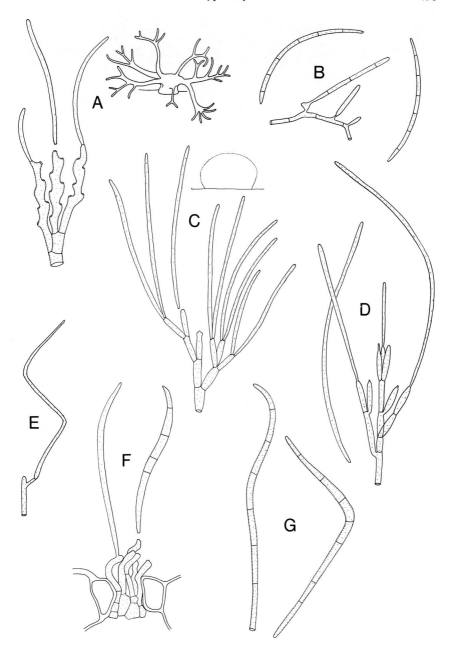

PLATE 58. A, *Pterulopsis dummeri*; B, *Sigmoidea prolifera*; C, *Linodochium hyalinum*; D, *Flagellospora curvula*; E, *Angulospora aquatica*; F, *Ramulispora sorghi*; G, *Anguillospora longissima*.

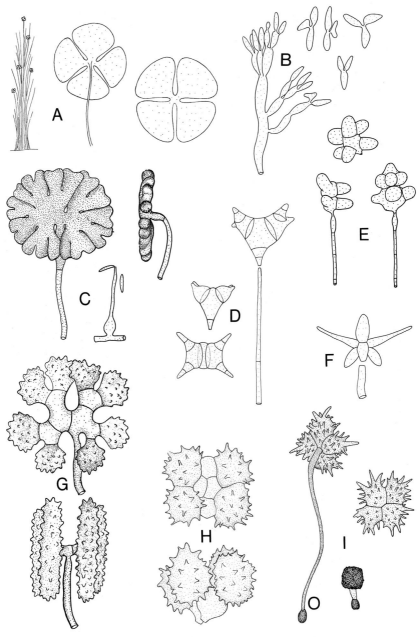

PLATE 59. A, *Riessia semiophora*; B, *Tricellula aquatica*; C, *Desmidiospora myrmecophila*; D, *Triposporina aphanophaga*; E, *Pyramidospora casuarinae*; F, *Titaea hemileiae*; G, *Candelabrum japonense*; H, *Isthmospora trochophila*; I, *Spegazzinia tessarthra*.

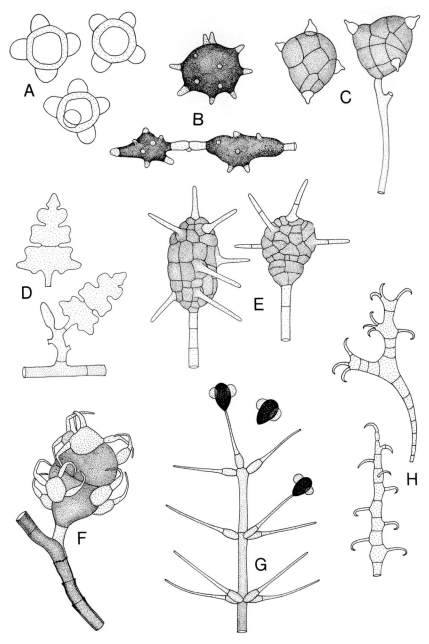

PLATE 60. A, *Stephanoma strigosum*; B, *Asterophora lycoperdoides*; C, *Oncopodiella trigonella*; D, *Dendrosporium lobatum*; E, *Petrakia echinata*; F, *Arachnophora fagicola*; G, *Physalidium elegans*; H, *Harpagomyces lomnicki*.

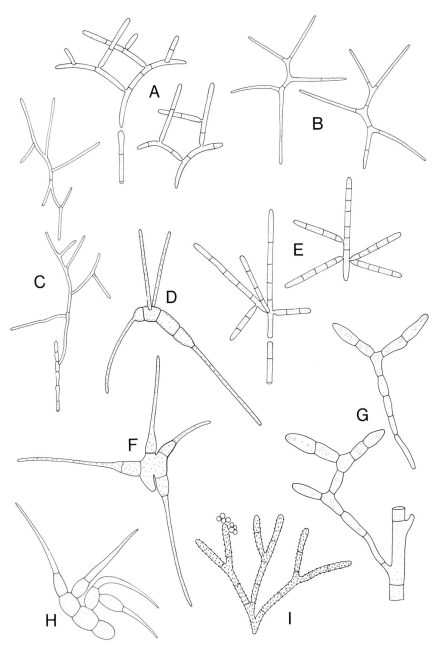

PLATE 61. A, *Varicosporium elodeae*; B, *Tricladium anomalum*; C, *Polycladium equiseti*; D, *Culicidospora aquatica*; E, *Dendrospora erecta*; F, *Campylospora chaetocladia*; G, *Varicosporina ramulosa*; H, *Gyoerffiella craginiformis*; I, *Thallospora aspera*.

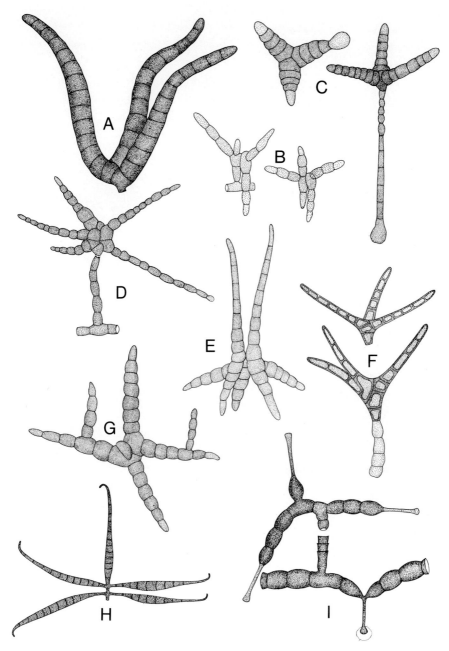

PLATE 62. A, *Ceratosporium fuscescens*; B, *Trispermum myrti*; C, *Triposporium elegans*; D, *Pentaposporium fourcroyae*; E, *Megaster longicornis*; F, *Actinocladium rhodosporum*; G, *Heptaster regnellianae*; H, *Casaresia sphagnorum*; I, *Grallomyces portoricensis*.

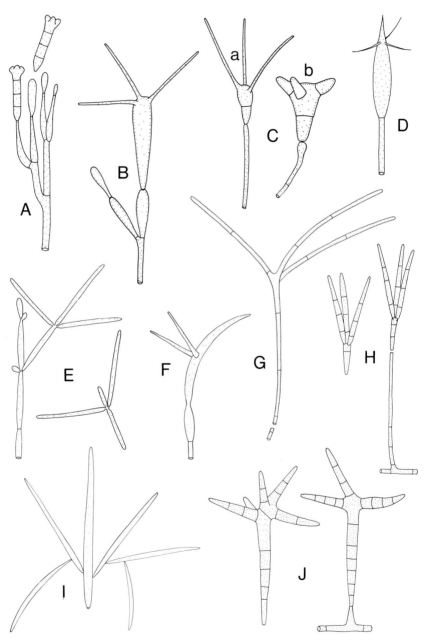

PLATE 63. A, *Heliscus lugdunensis*; B, *Clavatospora longibrachiata*; C, (a) *Clavariopsis aquatica*, (b) *C. brachycladia*; D, *Jaculispora submersa*; E, *Articulospora tetracladia*; F, *Alatospora acuminata*; G, *Tetrachaetum elegans*; H, *Tridentaria carnivora*; I, *Magdalaenaea monogramma*; J, *Trinacrium subtile*.

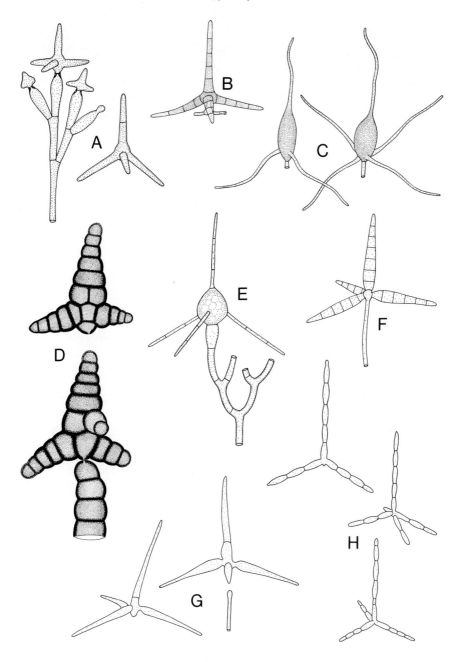

PLATE 64. A, *Lemonniera brachycladia*; B, *Tetraposporium asterinearum*; C, *Nia vibrissa*; D, *Hormiokrypsis* state of *Ophiocapnocoma phloiophilia*; E, *Actinospora megalospora*; F, *Flabellospora crassa*; G, *Triscelophorus monosporus*; H, *Trisulcosporium acerinum*.

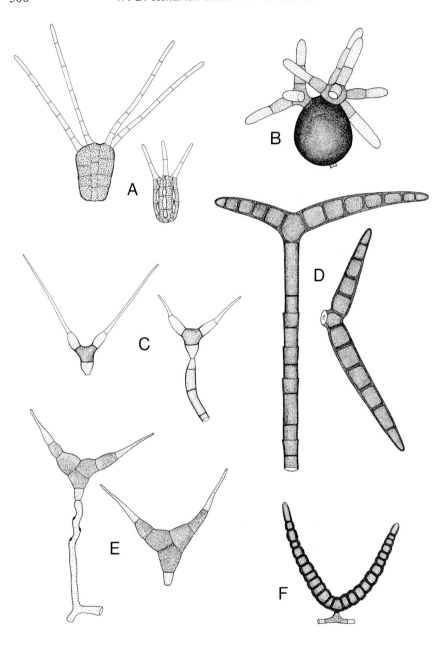

PLATE 65. A, *Tetraploa aristata*; B, *Orbimyces spectabilis*; C, *Iyengarina elegans*; D, *Ceratosporella bicornis*; E, *Diplocladiella scalaroides*; F, *Hirudinaria macrospora*.

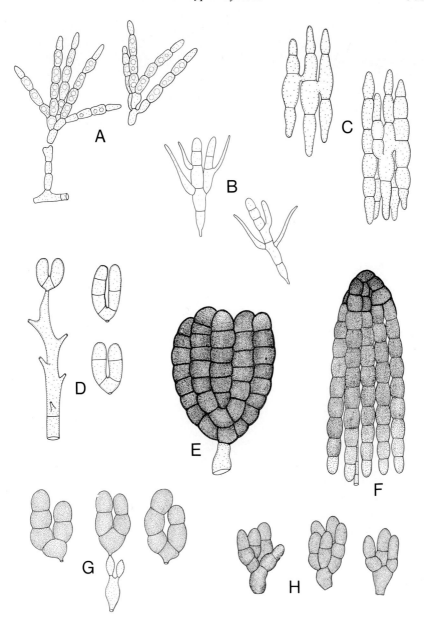

PLATE 66.  A, *Speiropsis pedatospora*; B, *Tetracladium setigerum*; C, *Amallospora dacrydion*; D, *Dicranidion fragile*; E, *Dictyosporium toruloides*; F, *Cryptocoryneum condensatum*; G, *Cheiromycella microscopica*; H, *Cheiromyces stellatus*.

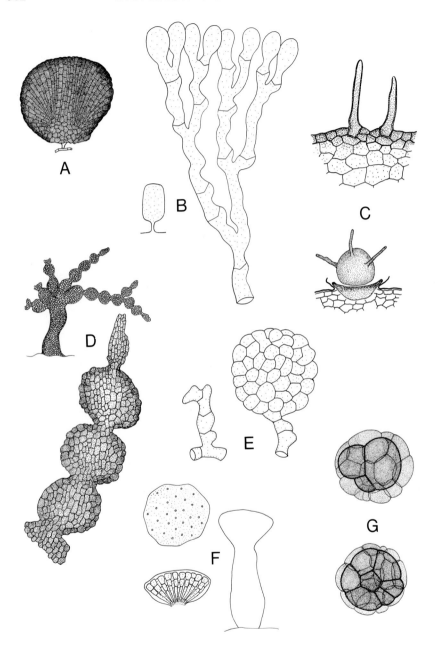

PLATE 67. A, *Mycoenterolobium platysporum*; B, *Aegerita candida*; C, *Dactuliophora tarrii*; D, *Arbuscula eugeniae*; E, *Burgoa verzuoliana*; F, *Tretopileus sphaerophorus*; G, *Papulaspora sepedonioides*.

## 10. Hyphomycetes 503

### INDEX TO PLATES

| GENERA | Plate | GENERA | Plate |
|---|---|---|---|
| *Acarocybe* | 44 H | *Anulohypha* | sterile |
| *Acarocybella* | 52 D | *Anulosporium* | sterile |
| *Aciculariella* | 57 I | *Aphanocladium* | 15 D |
| *Acinula* | sterile | *Arachnophora* | 60 F |
| *Acladium* | 15 G | *Arbuscula* | 67 D |
| *Acremoniella* | 21 I | *Arthrinium* | 20 B |
| *Acremoniula* | 21 K | *Arthrobotrys* | 30 I |
| *Acremonium* | 3 C | *Arthrobotryum* | 46 B |
| *Acroconidiella* | 44 E | *Arthrocladium* | 55 F |
| *Acroconidiellina* | see 44 E | *Arthrographis* | 2 I |
| *Acrodictys* | 37 G | *Arthrosporium* | 54 C |
| *Acrogenospora* | 21 J | *Articularia* | 5 D |
| *Acrophialophora* | 5 C | *Articulospora* | 63 E |
| *Acrophragmis* | 44 F | *Arxiella* | 28 J |
| *Acrospeira* | 38 F | *Ascoconidium* | 38 G |
| *Acrosporium* | 27 D | *Aspergillus* | 8 B |
| *Acrotamnium* | sterile | *Asperisporium* | 31 B |
| *Actinocladium* | 62 F | *Asteromyces* | 23 F |
| *Actinonema* | sterile | *Asterophora* | 60 B |
| *Actinospora* | 64 E | *Atractium* | 54 E |
| *Aegerita* | 67 B | *Aureobasidium* | 4 B |
| *Agaricostilbum* | 10 C | *Bacillispora* | 54 H |
| *Agyriella* | 9 D | *Bactridiopsis* | 21 H |
| *Akanthomyces* | 10 A | *Bactridium* | 51 G |
| *Akenomyces* | sterile | *Bactrodesmiella* | 42 A |
| *Alatospora* | 63 F | *Bactrodesmium* | 47 B |
| *Albosynnema* | 46 D | *Bahupaathra* | 4 A |
| *Allescheriella* | 19 E | *Bahusakala* | 42 B |
| *Alternaria* | 36 H | *Bainieria* | 13 E |
| *Alysidium* | 18 E | *Balanium* | 31 A |
| *Amallospora* | 66 C | *Barnettella* | 35 G |
| *Amblyosporium* | 1 A | *Bartheletia* | sterile |
| *Ambrosiella* | 19 D | *Basidiobotrys* | 14 B |
| *Ampullifera* | 43 B | *Basifimbria* | 24 H |
| *Ampulliferina* | 29 D | *Basipetospora* | 27 A |
| *Anguillospora* | 58 G | *Beauveria* | 24 E |
| *Angulimaya* | 4 E | *Beltrania* | 23 D |
| *Angulospora* | 58 E | *Beltraniella* | 23 A |
| *Annellodochium* | 31 G | *Beltraniopsis* | 23 E |
| *Annellophora* | 49 A | *Beniowskia* | 15 F |
| *Annellophorella* | 37 J | *Berkleasmium* | 37 H |
| *Annellophragmia* | 53 F | *Bispora* | 29 A |
| *Antennatula* | 49 F | *Blastobotrys* | 16 D |
| *Antennopsis* | 44 B | *Blastophorella* | 30 A |
| *Anthina* | sterile | *Blodgettia* | sterile |
| *Antromyces* | 33 A | *Bloxamia* | 6 D |
| *Antromycopsis* | 1 E | *Botryoderma* | 21 E |

| GENERA | Plate | GENERA | Plate |
|---|---|---|---|
| *Botryonipha* | sterile | *Chloridium* | 3 E |
| *Botryosporium* | 16 H | *Chrysosporium* | 19 G |
| *Botryotrichum* | 19 A | *Chuppia* | 36 I |
| *Botrytis* | 16 B | *Ciliciopodium* | 11 H |
| *Brachydesmiella* | 47 D | *Circinoconis* | 40 B |
| *Brachyhelicoon* | 41 I | *Circinotrichum* | 14 D |
| *Brachysporiella* | 44 A | *Cirrenalia* | 40 H |
| *Brachysporium* | 44 C | *Cladobotryum* | 27 B, 28 B |
| *Briosia* | 27 G | *Cladorrhinum* | 4 D |
| *Burgoa* | 67 E | *Cladosporiella* | 56 A |
| *Byssocladium* | sterile | *Cladosporium* | 29 B |
| *Cacumisporium* | 45 G | *Clasterosporium* | 53 G |
| *Calcarisporium* | 26 B | *Clathrosphaerina* | 38 D |
| *Calostilbella* | 46 A | *Clavariopsis* | 63 C |
| *Camposporium* | 52 C | *Clavatospora* | 63 B |
| *Camptomeris* | 46 G | *Clonostachys* | see 6 I |
| *Campylospora* | 61 F | *Coccidioides* | see 1 B |
| *Candelabrum* | 59 G | *Coccobotrys* | sterile |
| *Capnocybe* | 46 E | *Coccospora* | sterile |
| *Capnophialophora* | 4 H | *Codinaea* | 12 E |
| *Casaresia* | 62 H | *Coleodictyospora* | 38 I |
| *Catenularia* | 3 H | *Colletosporium* | sterile |
| *Cenococcum* | sterile | *Colletotrichum* | (coelomycete) |
| *Centrospora* | 55 C | *Coniosporium* | 35 D |
| *Cephaliophora* | 30 E | *Conoplea* | 25 D |
| *Cephalosporiopsis* | 33 H | *Cordana* | 31 C |
| *Cephalotrichum* | 13 G | *Cordella* | 20 C |
| *Ceratocladium* | 14 A | *Coremiella* | 1 D |
| *Ceratophorum* | 52 E | *Coronospora* | no illus. |
| *Ceratosporella* | 65 D | *Corynespora* | 53 H |
| *Ceratosporium* | 62 A | *Corynesporella* | 53 A |
| *Cercosperma* | 53 E | *Coryneum* | (coelomycete) |
| *Cercospora* | 56 C | *Costantinella* | 25 B |
| *Cercosporella* | 56 E | *Crinula* | 11 G |
| *Cercosporidium* | 45 F | *Cryptocoryneum* | 66 F |
| *Cercosporula* | 57 J | *Cryptophiale* | 57 G |
| *Chaetochalara* | 6 A | *Cryptostroma* | 1 G |
| *Chaetoconidium* | 19 I | *Culicidospora* | 61 D |
| *Chaetopsina* | 5 H | *Curculiospora* | 40 A |
| *Chaetopsis* | 34 B | *Curvularia* | 49 D |
| *Chaetospermum* | (coelomycete) | *Curvulariopsis* | 40 D |
| *Chalara* | 6 B | *Custingophora* | 7 D |
| *Chalaropsis* | 22 I | *Cuticularia* | sterile |
| *Chantransiopsis* | 22 C | *Cylindrocarpon* | 54 F |
| *Cheiromycella* | 66 G | *Cylindrocladium* | 54 G |
| *Cheiromyces* | 66 H | *Cylindrocolla* | 2 A |
| *Cheirospora* | 41 E | *Cylindrodendrum* | 33 E |
| *Chlamydomyces* | 32 F | *Cylindrotrichum* | 34 A |
| *Chloridiella* | 24 K | *Cyphellophora* | 33 F |

## 10. Hyphomycetes

| GENERA | Plate | GENERA | Plate |
|---|---|---|---|
| *Cystodendron* | 6 E | *Endostilbum* | 9 G |
| *Dactuliophora* | 67 C | *Epicoccum* | 35 I, 35 J |
| *Dactylaria* | 30 F | *Epidermophyton* | 51 J |
| *Dactylella* | 51 A | *Eriocercospora* | 48 G |
| *Dactylosporium* | 36 D | *Eurasina* | 45 A |
| *Deightoniella* | 31 D | *Everhartia* | 39 A |
| *Delortia* | 39 D | *Excipularia* | 47 C |
| *Dendrodochium* | 6 H | *Exophiala* | 33 G |
| *Dendrographium* | 43 A | *Exosporiella* | 49 C |
| *Dendrospora* | 61 E | *Exosporium* | 53 B |
| *Dendrosporium* | 60 D | *Flabellospora* | 64 F |
| *Dendryphion* | 29 E, 43 D | *Flagellospora* | 58 D |
| *Dendryphiopsis* | 48 D | *Fusariella* | 42 H |
| *Desmidiospora* | 59 C | *Fusarium* | 54 B |
| *Dexhowardia* | 15 H | *Fusicladiella* | 30 G |
| *Dichotomophthora* | 48 E | *Fusicladium* | 24 D |
| *Dichotomophthoropsis* | no illus. | *Fusidium* | 17 A |
| *Dicranidion* | 66 D | *Geotrichum* | 2 G |
| *Dictyoarthrinium* | 37 F | *Gibellula* | 8 C |
| *Dictyocatenulata* | 35 C | *Gilmaniella* | 21 G |
| *Dictyodesmium* | 38 E | *Gliocephalis* | 8 A |
| *Dictyosporium* | 66 E | *Gliocephalotrichum* | 8 E |
| *Dicyma* | 25 A | *Gliocladium* | 6 I |
| *Didymobotryum* | 29 H | *Gliomastix* | 3 A |
| *Didymostilbe* | 33 C | *Gloiosphaera* | 8 F |
| *Didymotrichella* | 31 H | *Glomopsis* | 20 F |
| *Diheterospora* | 9 C | *Glomospora* | 20 H |
| *Dimorphospora* | 22 H | *Goidanichiella* | 8 D |
| *Diplocladiella* | 65 E | *Gonatobotrys* | 16 F |
| *Diplococcium* | 29 I | *Gonatobotryum* | 17 G |
| *Diploospora* | 28 F | *Gonatophragmium* | 45 D |
| *Diplorhynchus* | 39 E | *Gonytrichum* | 5 F |
| *Domingoella* | 21 F | *Grallomyces* | 62 I |
| *Drechslera* | 50 D | *Graphiola* | 15 B |
| *Drechslerella* | 51 F | *Graphiothecium* | 11 C |
| *Drepanoconis* | 39 B | *Graphium* | 13 F |
| *Drumopama* | 25 E | *Gymnodochium* | 28 C |
| *Duosporium* | 44 J | *Gyoerffiella* | 61 H |
| *Dwayabeeja* | 43 B | *Gyrothrix* | see 14 D |
| *Dwayaloma* | 34 F | *Hadronema* | 32 A |
| *Echinobotryum* | 23 C | *Hadrotrichum* | 27 H |
| *Ectostroma* | sterile | *Hainesia* | (coelomycete) |
| *Eladia* | 7 A | *Hansfordia* | 26 E |
| *Elletevera* | 56 F | *Hansfordiella* | 37 C |
| *Embellisia* | 48 B | *Hansfordiellopsis* | 37 A |
| *Endocalyx* | 20 D | *Haplariopsis* | 30 J |
| *Endoconidium* | no illus. | *Haplobasidion* | 18 F |
| *Endophragmia* | 47 I | *Haplographium* | 6 F |
| *Endophragmiopsis* | 47 F | *Harpagomyces* | 60 H |

| GENERA | Plate | GENERA | Plate |
|---|---|---|---|
| *Harpographium* | 57 H | *Kmetiopsis* | 57 B |
| *Harposporium* | 57 A | *Koorchaloma* | 12 G |
| *Harziella* | 9 A | *Kostermansinda* | 38 H |
| *Heimiodora* | 14 E | *Kumanasamuha* | 25 C |
| *Helicocephalum* | 41 H | *Lacellina* | 18 H |
| *Helicodendron* | 41 D | *Lacellinopsis* | 18 I |
| *Helicoma* | 39 G | *Lemonniera* | 64 A |
| *Helicomina* | 40 I | *Leptodiscella* | 34 D |
| *Helicomyces* | 39 H | *Leptographium* | 13 H |
| *Helicoon* | 41 C | *Libertella* | (coelomycete) |
| *Helicorhoidion* | 41 A | *Linodochium* | 58 C |
| *Helicosporium* | 39 C | *Listeromyces* | 47 E |
| *Heliscus* | 63 A | *Loliomyces* | sterile |
| *Helminthosporium* | 53 C | *Lomaantha* | 52 B |
| *Hemibeltrania* | 15 E | *Lunulospora* | 57 F |
| *Heptaster* | 62 G | *Madurella* | sterile |
| *Hermatomyces* | 38 J | *Magdalaenaea* | 63 I |
| *Herposira* | 22 B | *Mahabalella* | 12 C |
| *Heterocephalum* | 10 E | *Malbranchea* | 1 C |
| *Heteroconium* | 43 H | *Mammaria* | 23 B |
| *Heydenia* | 19 B | *Margaritispora* | 45 B |
| *Himantia* | sterile | *Mastigosporium* | 51 C |
| *Hiospira* | 41 G | *Megalodochium* | 27 F |
| *Hirsutella* | 8 G | *Megaster* | 62 E |
| *Hirudinaria* | 65 F | *Melanconium* | (coelomycete) |
| *Histoplasma* | 20 G | *Melanographium* | 27 I |
| *Hobsonia* | 41 F | *Menispora* | 12 D |
| *Hormiactella* | 17 E | *Menisporopsis* | 34 C |
| *Hormiactis* | 28 H | *Meria* | 4 G |
| *Hormiokrypsis* | 64 D | *Metarrhizium* | 9 E |
| *Hormisciomyces* | see 9 H | *Microclava* | 22 D |
| *Hormocephalum* | 49 E | *Microdochium* | 57 C |
| *Hughesinia* | no illus. | *Microsporum* | 51 I |
| *Humicola* | 21 A | *Microstroma* | 10 B |
| *Hyalodendron* | 17 C | *Milowia* | 42 C |
| *Hyalodictys* | 35 A | *Minimidochium* | 12 B |
| *Hypha* | sterile | *Mirandina* | 55 D |
| *Hyphochlaena* | sterile | *Mitteriella* | 47 G |
| *Indiella* | sterile | *Monilia* | 17 B |
| *Irpicomyces* | 56 D | *Moniliella* | 2 F |
| *Isthmospora* | 59 H | *Moniliopsis* | sterile |
| *Itersonilia* | no illus. | *Monodictys* | 37 I |
| *Iyengarina* | 65 C | *Monostachys* | 3 B |
| *Jacobia* | 14 C | *Monotrichum* | see 51 E |
| *Jaculispora* | 63 D | *Morrisographium* | 54 A |
| *Jainesia* | 47 H | *Muiaria* | 38 B |
| *Johnstonia* | no illus. | *Muiogone* | 38 C |
| *Keratinomyces* | 51 D | *Murogenella* | 50 C |
| *Kmetia* | 57 D | *Mycoenterolobium* | 67 A |

## 10. Hyphomycetes

| GENERA | Plate | GENERA | Plate |
|---|---|---|---|
| *Mycogone* | 32 I | *Petrakiopsis* | 55 A |
| *Mycovellosiella* | see 29 B | *Peyronelia* | 35 B |
| *Myrioconium* | 9 I | *Peyronelina* | 35 F |
| *Myrothecium* | 10 G | *Phaeodactylium* | 49 G |
| *Mystrosporiella* | 36 C | *Phaeoisaria* | 11 E |
| *Myxotrichella* | sterile | *Phaeoisariopsis* | 53 I |
| *Nakataea* | 50 E | *Phaeoramularia* | see 29 B |
| *Nalanthamala* | 10 D | *Phaeotrichoconis* | 52 G |
| *Negeriella* | see 49 B | *Phialocephala* | 6 G |
| *Nematoctonus* | 15 A | *Phialomyces* | 7 E |
| *Nematogonium* | 16 G | *Phialophora* | 4 C |
| *Neottiosporella* | 12 A | *Phialophoropsis* | 4 F |
| *Neta* | 30 H | *Phialostele* | 11 F |
| *Nia* | 64 C | *Phialotubus* | 7 F |
| *Nigrospora* | 21 D | *Phloeoconis* | sterile |
| *Nodulisporium* | 25 G | *Phragmospathula* | 42 F |
| *Nomuraea* | 17 H | *Physalidium* | 60 G |
| *Oedemium* | 29 G | *Pilulina* | 21 C |
| *Oedocephalum* | 16 E | *Piricauda* | 37 B |
| *Oedothea* | 31 F | *Pithomyces* | 47 A |
| *Oidiodendron* | 2 B | *Plecotrichum* | sterile |
| *Olpitrichum* | 15 C | *Pleiochaeta* | 52 A |
| *Oncocladium* | 2 E | *Pleurocatena* | 3 G |
| *Oncopodiella* | 60 C | *Pleurophragmium* | 42 G |
| *Oncopodium* | 35 H | *Pleurothecium* | 45 E |
| *Orbimyces* | 65 B | *Plokamidomyces* | 9 H |
| *Ormathodium* | 43 I | *Podosporiella* | 48 A |
| *Ostracoderma* | 16 A | *Podosporium* | 49 B |
| *Overeemia* | sterile | *Polycladium* | 61 C |
| *Ovularia* | 24 I | *Polypaecilum* | 13 C |
| *Ovulariopsis* | 27 C | *Polyscytalina* | 28 I |
| *Paathramaya* | 11 D | *Polyscytalum* | 28 E |
| *Pachnocybe* | 14 F | *Polythrincium* | 32 E |
| *Paecilomyces* | 3 D, 7 H | *Prathigada* | 50 H |
| *Pagidospora* | 19 H | *Pseudobeltrania* | 23 G |
| *Panchanania* | 32 B | *Pseudobotrytis* | 32 J |
| *Papulaspora* | 67 G | *Pseudoepicoccum* | see 27 H |
| *Paracoccidioides* | see 19 G | *Pseudofusidium* | see 3 D, 7 H |
| *Paspalomyces* | 28 A | *Pseudographiella* | 54 D |
| *Passalora* | 32 C | *Pseudopetrakia* | see 60 E |
| *Pendulispora* | 38 A | *Pseudospiropes* | 50 B |
| *Penicillifer* | 34 E | *Pteroconium* | 20 A |
| *Penicillium* | 7 G | *Pterulopsis* | 58 A |
| *Pentaposporium* | 62 D | *Pyramidospora* | 59 E |
| *Periconia* | 18 B | *Pyricularia* | 50 G |
| *Periconiella* | 46 F | *Pyriculariopsis* | 50 I |
| *Pestalotia* | (coelomycete) | *Racodium* | sterile |
| *Pestalozziella* | (coelomycete) | *Raffaelea* | 24 B |
| *Petrakia* | 60 E | *Ramularia* | 28 G |

| GENERA | Plate | GENERA | Plate |
|---|---|---|---|
| Ramulispora | 58 F | Spondylocladiopsis | 45 C |
| Rhinocladiella | 24 A | Sporendocladia | 6 C |
| Rhinocladium | 22 G | Sporendonema | 1 B |
| Rhizoctonia | sterile | Sporidesmium | 53 D |
| Rhizohypha | sterile | Sporocephalum | 16 C |
| Rhizomorpha | sterile | Sporophiala | 42 D |
| Rhombostilbella | 15 I | Sporoschisma | 42 I |
| Rhynchosporium | 33 D | Sporothrix | 24 F |
| Riclaretia | 57 E | Sporotrichum | 22 E |
| Riessia | 59 A | Stachybotrys | 7 C |
| Saccardaea | 11 A | Stachylidium | 5 A |
| Sadasivania | 18 C | Staphylotrichum | 19 F |
| Saprochaete | sterile | Stemphyliomma | 44 I |
| Sarcinella | 37 E | Stemphylium | 36 A |
| Scenomyces | sterile | Stenella | 48 H |
| Sceptrifera | 30 D | Stenellopsis | 48 F |
| Schizotrichum | 56 B | Stephanoma | 60 A |
| Sclerographium | 36 B | Stephanosporium | 2 C |
| Sclerotium | sterile | Sterigmatobotrys | 45 H |
| Scolecobasidium | 32 G | Stevensomyces | 27 J |
| Scopulariopsis | 13 B | Stigmina | 36 G |
| Scytalidium | 2 D | Stilbum | 11 I |
| Seimatosporium | (coelomycete) | Stromatocrea | 1 H |
| Selenosporella | 5 G | Strumella | 40 E |
| Sepedonium | 20 E | Subulispora | 24 C |
| Septocylindrium | see 43 F | Sympodiella | 1 F |
| Septoidium | 44 G | Sympodiophora | 33 B |
| Septonema | 43 F | Synnematium | see 3 C |
| Septosporium | 35 E | Taeniolella | 43 E |
| Septotrullula | 42 E | Teratosperma | 52 F |
| Sesquicillium | 9 B | Tetrachaetum | 63 G |
| Sibirina | 34 G | Tetracladium | 66 B |
| Sigmoidea | 58 B | Tetracoccosporium | 37 D |
| Sigmoideomyces | 17 F | Tetracrium | 55 B |
| Sirosporium | 36 F | Tetraploa | 65 A |
| Solheimia | 11 B | Tetraposporium | 64 B |
| Sorocybe | 18 A | Thallospora | 61 I |
| Spadicoides | 31 E | Thermomyces | 21 B |
| Spegazzinia | 59 I | Thielaviopsis | see 6 B |
| Speiropsis | 66 A | Thozetellopsis | 12 F |
| Spermospora | 51 B | Thyrostromella | 36 J |
| Sphaeridium | 17 D | Thysanophora | 7 B |
| Sphaeromyces | 3 I | Titaea | 59 F |
| Sphaerosporium | 27 E | Tolypocladium | see 4 I |
| Spilocaea | 13 D | Tomenticola | 50 F |
| Spilodochium | 29 C | Torula | 43 C |
| Spiropes | 50 A | Torulomyces | 3 C |
| Spirosphaera | 41 J | Tretopileus | 67 F |
| Spondylocladiella | 44 D | Triadelphia | 52 H |

## 10. Hyphomycetes

| GENERA | Plate | GENERA | Plate |
|---|---|---|---|
| *Tricellula* | 59 B | *Ustilaginoidea* | 22 A |
| *Trichobotrys* | 18 G | *Vanbeverwijkia* | 39 F |
| *Trichocladium* | 32 H | *Varicosporina* | 61 G |
| *Trichoconis* | 51 E | *Varicosporium* | 61 A |
| *Trichoderma* | 4 I | *Vasculomyces* | sterile |
| *Trichodochium* | 29 F | *Veronaea* | 30 B |
| *Trichophyton* | 51 H | *Verrucispora* | 48 C |
| *Trichothecium* | 28 D | *Verticicladiella* | 26 C |
| *Trichurus* | 13 A | *Verticicladium* | 26 D |
| *Tricladium* | 61 B | *Verticillium* | 9 F |
| *Tridentaria* | 63 H | *Virgaria* | 25 F |
| *Trinacrium* | 63 J | *Virgariella* | 24 G |
| *Tripospermum* | 62 B | *Virgatospora* | 46 C |
| *Triposporina* | 59 D | *Volutella* | 10 F |
| *Triposporium* | 62 C | *Volutina* | no illus. |
| *Triscelophorus* | 64 G | *Wallemia* | 2 H |
| *Trisulcosporium* | 64 H | *Wardomyces* | 32 D |
| *Tritirachium* | 26 A | *Wiesneriomyces* | 55 E |
| *Trochophora* | 40 G | *Xenosporium* | 40 F |
| *Troposporella* | 40 J | *Xylohypha* | 18 D |
| *Troposporium* | 41 B | *Zalerion* | 40 C |
| *Tubercularia* | 10 H | *Zanclospora* | 5 E |
| *Tuberculina* | 1 I | *Zasmidium* | 24 J |
| *Ulocladium* | 36 E | *Zygophiala* | 30 C |
| *Umbelopsis* | 19 C | *Zygosporium* | 22 F |
| *Uncigera* | 5 B | | |

# Deuteromycotina
# Coelomycetes

MELANCONIALES, P. 555

SPHAEROPSIDALES, P. 560

# CHAPTER 11

# Coelomycetes

B. C. SUTTON

*Commonwealth Mycological Institute*
*Kew, Surrey, England*

---

## I. INTRODUCTION

The Fungi Imperfecti (Deuteromycotina) are an assemblage of fungi typically reproducing by spores which are formed without nuclear fusion followed by meiosis. This artificial subdivision embraces not only the imperfect, asexual, or conidial states of the Ascomycotina and Basidiomycotina but also those asexual fungi with which no perfect states have yet been correlated, if indeed any are in existence. Such artificiality is emphasized in the terminology "form-genera" and "form-species," the implication being that the names are proposed without a knowledge of the sexual states of the taxa they represent. In practice, however, the prefix "form-" is understood but usually omitted. The Deuteromycotina are comprised of three orders: the Sphaeropsidales with pycnidia, the Melanconiales with acervuli, and the Moniliales with superficial separate or aggregated conidiophores. Grove (1919) introduced the term Coelomycetes to accommodate in a single category the extremes in fructification type shown by the pycnidial and acervular genera *Phyllosticta*, *Phomopsis*, and *Phloeospora*. Later Grove (1935) recognized two classes in the Deuteromycotina. He characterized the Hyphomycetes by conidia borne on the exterior of the substrate bearing the fungi, and he adopted the term Coelomycetes for those fungi with conidia that are formed within a cavity of the substrate in which the fungus grows. This class now includes all immersed to superficial pycnidial and acervular fungi in which conidia are initiated within a cavity lined by either fungal or host tissue, or a combination of both.

More than 1100 generic names have been proposed in the Coelomycetes, of which approximately 400–500 are considered synonyms. Many are monotypic but several, such as *Phoma*, *Phyllosticta*, *Septoria*, and *Colletotrichum* contain binomials for more than 1000 species, and others like *Ascochyta*,

TABLE I

| Genera | Persoon (1801) | Link (1809) | Fries (1821) | Fries (1825) | Link (1833) |
|---|---|---|---|---|---|
| *Melanconium* | | C, Anandrae<br>O, Epiphytae | C, Coniomycetes<br>O, Entophytae<br>So, Stilbosporei | Scl, Sporomycetes<br>Coh, Coniomycetes<br>O, Entophyti<br>T, Stilbosporei | So, Epispori<br>S, Sporidochium spurium |
| *Phoma* | | | C, Gasteromycetes<br>O, Uterini veri<br>So, Pyrenomycetes Sphaeriacei | Scl, Ascomycetes<br>Coh, Pyrenomycetes<br>O, Cytisporei | |
| *Cytospora* | | | | Scl, Ascomycetes<br>Coh, Pyrenomycetes<br>O, Cytisporei | So, Sphaeriaceae |
| *Tubercularia* | C, Angiocarpi<br>O, Sclerocarpi | C, Anandrae<br>O, Gastromyci | C, Coniomycetes<br>O, Tuberculariae | Scl, Sporomycetes<br>Coh, Coniomycetes<br>O, Tubercularini<br>T, Tubercularei | So, Epispori<br>S, Sporidochium verum |
| *Gymnosporangium* | | | C, Coniomycetes<br>O, Sporodesmiae | Scl, Sporomycetes<br>Coh, Coniomycetes<br>O, Entophyti<br>T, Sporodesmii | So, Epispori<br>S, Sporidochium verum |
| "*Sphaeria*" | C, Angiocarpi<br>O, Sclerocarpi | | C, Gasteromycetes<br>O, Uterini veri<br>So, Pyrenomycetes Sphaeriacei | Scl, Ascomycetes<br>Coh, Pyrenomycetes<br>O, Sphaeriacei<br>So, Sphaerini | So, Sphaeriaceae |
| *Lycoperdon* | C, Angiocarpi<br>O, Dermatocarpi | C, Anandrae<br>O, Gastromyci | C, Gasteromycetes<br>O, Uterini veri<br>So, Trichospermi | Scl, Sporomycetes<br>Coh, Gasteromycetes<br>O, Trichospermi<br>So, Lycoperdinei | So, Gastromyci |

[a] *Abbreviations*: C, class; Coh, cohort; D, division; F, family; O, order; So, suborder; Scl, subclass; S, section; Sd, subdivision; T, tribe.

*Gloeosporium* auct., *Diplodia*, *Pestalotia*, *Hendersonia*, and *Stagonospora* are known to contain several hundred species names. Genera are worldwide in distribution, having been recorded from the tropics, temperate, and arctic regions. Although they are most commonly reported from living or dead plant material and from soil, they are known from a wide variety of substrates. Many species are facultative parasites causing minor leaf, stem, and root lesions and cankers or galls, while others are saprophytic and weakly pathogenic, being associated with secondary damage such as is found in diebacks and deteriorative conditions in shrubs and trees. A considerable number, however, are of economic importance as they cause serious agricul-

## 11. Coelomycetes

CLASSIFICATION OF SELECTED GENERA PRIOR TO 1869[a]

| Corda (1842) | Léveillé (1846) | Bonorden (1851) | Berkeley (1860) | Fuckel (1869) |
|---|---|---|---|---|
| O, Myelomycetes<br>So, Dermatogasteres<br>F, Melanconiaceae | | O, Cryptomycetes<br>F, Naemasporei | F, Coniomycetes<br>O, Melanconiei | Fungi Imperfecti<br>Hyphomycetes |
| O, Myelomycetes<br>So, Dermatogasteres<br>F, Sphaeronemeae | D, Clinosporés<br>Sd, Endoclinés<br>S, Spheronemés | | F, Coniomycetes<br>O, Sphaeronemei | Fungi Imperfecti<br>Phyllostictei |
| | D, Clinosporés<br>Sd, Endoclinés<br>S, Spheronemés | O, Cryptomycetes<br>F, Naemasporei | F, Coniomycetes<br>O, Sphaeronemei | Fungi Imperfecti<br>Cytosporacei |
| O, Hymenomycetes<br>So, Basidiophori<br>F, Tuberculariaceae | D, Clinosporés<br>Sd, Ectoclinés<br>T, Sarcopsidés<br>S, Tuberculariés | O, Mycetini<br>F, Tubercularini | F, Hyphomycetes<br>O, Stilbacei | Fungi Imperfecti<br>Gymnomycetes |
| | D, Clinosporés<br>Sd, Ectoclinés<br>T, Coniopsidés<br>S, Phragmidiés | | F, Coniomycetes<br>O, Pucciniei | Fungi Perfecti<br>Hypodermei<br>Uredinei |
| O, Myelomycetes<br>So, Dermatogasteres<br>F, Sphaeriacei | D, Thecosporés<br>Sd, Endothequés<br>T, Sphaeriacés | O, Pyrenomycetes<br>F, Sphaeriacei | F, Ascomycetes<br>O, Sphaeriacei | Fungi Perfecti<br>Ascomycetes<br>Pyrenomycetes |
| O, Myelomycetes<br>F, Lycoperdaceae | D, Basidiosporés<br>Sd, Entobasidés<br>T, Coniogastrés<br>S, Lycoperdés | O, Gasteromycetes<br>F, Lycoperdacei | F, Gasteromycetes<br>O, Trichogastres | Fungi Perfecti<br>Gasteromycetes<br>Lycoperdacei |

tural and forestry diseases. Damage may be caused by ubiquitous pathogens with wide host ranges such as *Macrophomina phaseolina*, *Botryodiplodia theobromae*, *Phyllostictina citricarpa*, *Colletotrichum gloeosporioides*, *Sclerophoma pythiophila*, and *Phoma exigua*. Within some of these, subspecific taxa are recognized. Other species are only capable of attacking single genera or species of host plant, such as *Colletotrichum musae*, *C. falcatum*, *Diplodia zeae*, *Septoria apiicola*, and *Phomopsis obscurans*.

Perfect states which have been correlated with Coelomycetes belong in the Ascomycotina. Although no Coelomycetes have been linked with members of the Hemiascomycetidae, there are numerous instances in the

Euascomycetidae, orders Sphaeriales, Diaporthales, Hypocreales, and Helotiales, and the Loculoascomycetidae, orders Myriangiales, Dothideales, and Pleosporales.

Advances in quality and resolution of optical instruments, developments in electron microscopy, and the diversity of applied techniques have led to an increase in number and complexity of classificatory criteria for different groups. The extent to which innovative techniques affect the spectrum of available criteria is often, but not always, dependent on the inherent characteristics of the organisms. The fewer the classificatory criteria readily available, as in yeasts and bacteria, the more likely that biochemical, serological and other physiological characteristics are used to differentiate taxa. The reverse may obtain in that the intransigence of some groups may result in a relatively static classification. Saccardo (1880, 1884) separated genera and species of the Coelomycetes by a few simple morphological features together with criteria based on proved or assumed, simple or spatial parasite/host relationships. In the majority of genera, means of separating species are heavily weighed in favour of host specialization because the morphological criteria available are relatively few and the extent to which techniques other than optical microscopy have been used with any success is limited. Consequently, the approach to classification has, with very few exceptions, remained static since Saccardo, and few major advances have materialized. Smith (1962) was led to observe that "satisfactory classification of the hyphomycetes... is, with the possible exception of the coelomycetes, in most urgent need of a comprehensive scheme to cover all genera." Since Hughes (1953) introduced his concepts on conidium and conidiophore ontogeny into hyphomycete classification, there has been a growing volume of work concerned with the application of his ideas to Coelomycetes. The ultimate aim is to derive a comprehensive classificatory scheme for the Deuteromycotina. This cannot be attained without an appraisal of conidium ontogeny in the Coelomycetes and its correlation with existing morphological data. The taxonomy and identification of the group is therefore in a transitional state as the new information is assimilated into the existing framework.

## II. SYSTEMS OF CLASSIFICATION

### A. Morphological Classifications

In reviewing the history of suprageneric taxa in the Coelomycetes, it is clear that the evolution of satisfactory systems for microfungi have been retarded for a combination of reasons. A lack of fundamental knowledge of life cycles has obscured the different origins of ascospores, basidiospores, and conidia. This, coupled with superficial similarities in the form of fructifications, such as exist between pycnidia and certain types of ascocarp, and between acervuli, sporodochia and telia, is the main contributory factor which has hindered progress. Genera of pycnidial and acervular fungi have been allied with members of the Uredinales, Sphaeriales, Moniliales, and even Gasteromycetes at the class, order, and family level. It was not until 1873 that Saccardo fully clarified their relationships with other groups. In Table I is shown a selection of genera from the Deuteromycotina and their classifications according to different authorities prior to 1869. The inclusion of *Gymnosporangium*, "*Sphaeria*", and *Lycoperdon* is designed to illustrate the relationships these fungi were thought to have with different members of the Deuteromycotina. In the following account, the spellings of suprageneric names as published by the authors are used to avoid the confusion which would result from the extensive correction of class, order, and family suffixes.

Persoon (1801) divided fungi into two classes, the Angiocarpi and Gymnocarpi, by the enclosed or exposed nature of the structures bearing spores. *Stilbospora* and *Naemospora*, both with acervuli, were included in the Angiocarpi, order Sclerocarpi, together with the ascomycetes *Sphaeria*, *Hysterium*, and *Xyloma* and the hyphomycete *Tubercularia*. Substantially the same criteria were used by Link (1809) but the fungi (class Anandrae) were divided into four orders based on the free or aggregated nature of spores and their relationships to the fructifications in which they were borne. These were the Epiphytae (leaf- and stem-inhabiting rusts, sporodochial hyphomycetes, and acervular fungi including *Stilbospora* and *Melanconium*), Mucedines (moniliaceous hyphomycetes), Gastromyci (gasteromycetes and myxomycetes), and Fungi (some homobasidiomycetes and ascomycetes).

Martius (1817) used the same ordinal separation, recognizing Coniomycetes (Epiphytae Link), Hyphomycetes (Mucedines Link), Gastromycetes, and Fungi. The Coniomycetes incorporated genera such as *Caeoma*, *Puccinia*, *Stilbospora*, *Seiridium*, *Aegerita*, *Fusarium*, *Melanconium*, *Epicoccum*, *Didymosporium*, *Exosporium*, *Tubercularia*, and *Gymnosporangium*. These were separated according to the degree of liberation of conidia from one another, and the types of host substrate upon which the fungi subsisted. *Naemospora* was placed with *Sphaeria*, *Hysterium*, and *Thelebolus* in the Fungi section Myelomycetes.

Nees von Esenbeck (1817) adopted a similar system to Martius and Link, differing in that the genera in the order Fungi were redistributed into three groups, the Fungi clavati et pileati (hymenomycetes), Fungi utrini (some gasteromycetes and *Peziza*), and Myelomyci in which *Naemospora* was again placed with various ascomycetes. The Protomyci, equivalent to the Coniomycetes of Martius and Epiphytae of Link, was again composed of rusts, acervular fungi (to which *Coryneum* had been added), and sporodochial hyphomycetes.

Fries (1821) included eleven pycnidial and acervular genera in his scheme which was the first to incorporate pycnidial fungi. He reverted to a four-class system, essentially similar to that of Link, but used the terms Coniomycetes, Hyphomycetes, Gasteromycetes, and Hymenomycetes. Acervular genera were again placed in orders and suborders within the Coniomycetes, together with sporodochial hyphomycetes and rusts. The pycnidial genera, including *Depazea*, *Actinothyrium*, *Leptostroma*, *Sphaeronaema*, *Phoma*, and *Bostrychia*, however, were placed with various ascomycetes in the suborder Pyrenomycetes, class Gasteromycetes, because of the superficially similar fructifications. The Pyrenomycetes were divided into four families, Xylomacei, Cytispori, Sphaeriacei, and Phacidiacei, on the basis of open or closed, superficial or erumpent fructifications and the situation of asci or spores. Later, Fries (1822) included all the pycnidial genera in the Cytisporei (to which *Cytispora* had been added) and the Xylomacei, reserving the Sphaeriacei and Phacidiacei for ascomycetes.

Fries (1825) produced a major reorganization of his classification in that the class Fungi was divided into two subclasses—Ascomycetes (with the cohorts Hymenomycetes and Pyrenomycetes) and Sporomycetes (with the cohorts Gasteromycetes and Coniomycetes). Significantly, all acervular fungi, hyphomycetes, and rusts were included in the Coniomycetes, while the Pyrenomycetes (including pycnidial fungi) were removed to a separate cohort from the Gasteromycetes. The breakdown within the Pyrenomycetes differed little from that of Fries (1821). The Coniomycetes were divided into four orders: the Tubercularini (including *Fusarium*, *Epicoccum*, and *Tubercularia*); Mucorini (with *Stilbum*, *Mucor*, *Thamnidium* and *Bactridium*); Mucedines which was equivalent to the Hyphomycetes of Martius and the Nematomyci of Nees; and the Entophyti which contained *Phragmotrichum*, *Coryneum*, *Seiridium*, *Asterosporium*, *Stilbospora*, *Didymosporium*, *Naemospora*, *Septoria*, *Melanconium*, *Sporidesmium*, *Exosporium*, *Papularia*, *Spilocaea*, *Gymnosporangium*, and other rusts. Within the Entophyti, four tribes were accepted: Sporodesmii, Stilbosporei, Naemosporei and Hypodermii. All the rusts were included in the Hypodermii, except *Gymnosporangium*, which was placed in the Sporidesmii.

Link (1833) favored the recognition of eight suborders within the order Fungi, an important addition being the separate category for Myxomycetes which had previously been grouped with the Gasteromycetes. Although the pycnidial genera were all grouped together with several ascomycetes in the suborder Sphaeriaceae, and the acervular genera maintained separately in the suborder Epispori, the inclusion of several rusts in the latter group was in some ways a retrograde step from that of Fries (1825).

Libert (1837) followed Fries (1822) in placing the eleven pycnidial genera she studied in the Ascochytacei and Cytosporei which were sections of the Pyrenomycetes. The sections Phacidiacei and Sphaeriacei were reserved for ascomycete genera.

Corda (1842) included 35 genera of pycnidial and acervular fungi in his scheme which consisted of four orders. These were named the Coniomycetes, Hyphomycetes, Myelomycetes, and Hymenomycetes and corresponded with the Epiphytae, Mucedines, Gastromyci, and Fungi of Link (1809). Acervular genera were scattered in each of these orders, *Colletotrichum* being assigned to the Psiloniaceae in the Hyphomycetes, and *Coryneum* to the Coryneaceae, suborder Basidiophori, in the Hymenomycetes. Several acervular genera again found their way into the Coniomycetes among rusts, but the important point was that some, including *Cryptosporium, Naemospora, Melanconium, Steganosporium,* and *Dilophospora*, were maintained in a separate, newly proposed family, the Melanconiaceae, in the Myelomycetes. Although pycnidial genera were scattered between the Sphaeronemeae and Sphaeriacei in the Myelomycetes, this probably represents the first significant attempt to separate pycnidial and acervular genera at the same taxonomic level within a single order.

Although overshadowed by later work, Léveillé (1845) placed most of the pycnidial genera considered in the Sphaeropsidei (Sphaeropsidaceae) class Stromatospori, from which originates the family name Sphaeropsidaceae. The significance of the origin and development of various types of spores was emphasized later by Léveillé (1846) who split fungi into six divisions, each based on a specific spore type. These were the Basidiosporés, Thecasporés, Clinosporés, Cystosporés, Trichosporés, and Arthrosporés. Pycnidial and acervular genera were only included in the Thecasporés, Clinosporés, and Arthrosporés. *Cheilaria, Cliostomum, Pilidium, Aposphaeria,* and *Depazea* were placed in the tribes Hysteriés, Cliostomés and Sphaeriacés of the Thecasporés (ascomycetes) and *Phragmotrichum* and *Cylindrosporium* in the Arthrosporés. The remainder of the pycnidial and acervular genera were placed in various sections of the Clinosporés. *Coryneum* was again included with *Phragmidium* in the rust section, but other attempts to place genera into natural sections displayed a degree of taxonomic intuitiveness that few contemporaries shared. These include the Excipulés (*Excipula, Dinemasporium,* and *Polynema*), Actinothyriés (*Actinothyrium, Leptostroma,* and *Leptothyrium*) and Pestalozziés (*Pestalotia, Robillarda, Discosia, Dilophospora, Neottiospora, Seiridium,* and *Prosthemium*).

Bonorden (1851) advocated major separation of the fungi into twelve orders including the Coniomycetes (moniliaceous hyphomycetes, rusts, *Stilbospora, Didymosporium*, and *Asterosporium*), Cryptomycetes (rusts, some ascomycetes, *Myxosporium, Libertella, Cryptosporium, Septoria, Blennoria, Steganosporium, Cytospora, Melanconium*, and *Ascochyta*), Mycetini (hyphomycetes and *Coryneum*), and Sphaeronemei (*Eurotium, Alphitomorpha*, ascomycetes, and pycnidial fungi). The scheme was noteworthy only in that the pycnidial fungi were removed from the Pyrenomycetes into a separate order, the Sphaeronemei. The acervular fungi remained distributed between as many as four different orders.

Berkeley (1860) pursued Bonorden's idea in removing pycnidial genera from the Pyrenomycetes to the Coniomycetes, Sphaeronemei. He also made a major contribution by separating acervular genera into the Coniomycetes, Melanconiei. Even though rusts were also placed in the Coniomycetes in the orders Pucciniaei and Aecidiacei, they were maintained separately from the pycnidial and acervular fungi. In addition, the family Hyphomycetes now contained sporodochial, moniliaceous, and synnematal forms. Thus, apart from the inclusion of some heterobasidiomycetes with the pycnidial and acervular fungi, Berkeley's system approached the modern classifications with a considerable degree of accuracy.

Fuckel (1869) produced a system similar to Berkeley's but with the innovation that the fungi were divided into two main groups named the Fungi Perfecti and the Fungi Imperfecti. The latter was further subdivided into several sections based on hyphomycetes, ascomycetes, rusts, jelly fungi, and coelomycetes. In fact, the composition of the Fungi Imperfecti became no more homogeneous than in previous systems.

Finally Saccardo (1873) adopted Fuckel's scheme in maintaining two classes, the Fungaceae perfectae and the Fungaceae imperfectae. The latter was divided into the Hyphomyceteae (including both *Stilbospora* and *Melanconium*), Gymnomyceteae (with *Fusarium, Polynema, Leptostroma, Coryneum, Epicoccum*), Phyllosticteae (with *Coniothyrium, Phoma, Depazea, Ascochyta, Phyllosticta, Septoria*), Sphaeropsideae (with *Pestalotia, Hendersonia, Diplodia, Actinothyrium, Sphaeropsis*), Dichaenaceae (with *Psilospora*), and the Mycelia Sterilia. Thus the Deuteromycetes, Sphaeropsidales, Melanconiales, and Moniliales became relatively homogeneous within their respective ordinal limits. Saccardo subsequently enlarged on these ideas. As a basis for separating three divisions in the Fungi Imperfecti, Saccardo (1880), used differences in fructification type. The Sphaeropsideae were characterized by pycnidia, the Melanconieae by acervuli and the Hyphomyceteae by loose webs of hyphae, sporodochia, or synnemata. Major differences in color, texture, and dehiscence, together with variations in the relationships of the fructification to the substrate, were employed by Saccardo (1884) to distinguish four families within the Sphaeropsideae and one in the Melanconieae. These were the Sphaerioideae, Nectrioideae, Leptostromaceae (equivalent to the dimidiato-scutatae of Saccardo, 1880), Excipulaceae [equivalent to Saccardo (1880) subcupulatae], and the Melanconiaceae. Below the family level, and in particular at the generic level, the means of separation was by the color, shape, septation, and size of conidia, together with more subtle differences between fructifications such as degree of aggregation or division of pycnidia, relationship to the substrate, shape and hairiness, etc. At the species level, Saccardo, together with most contemporaries, employed names based almost entirely on unproved simple or spatial host/parasite relationships. This approach was taken to its logical but impracticable conclusion by Tassi (1902)—of 24 genera considered to be related to *Phoma*, more than half were proposed as new on the basis of color, septation, and size of conidia, and the foliicolous or caulicolous nature of pycnidia.

Potebnia (1910) recognized five groups within the Deuteromycotina based on fructification structure. The Melanconiaceae was amalgamated with the Tuberculariaceae in the Acervulales, the Leptostromataceae, Excipulaceae and genera with incomplete pycnidia in the Pseudopycnidiales, and genera with normal pycnidia in the Pycnidiales. The remaining two groups contained hyphomycetes. A further modification was by von Höhnel (1911) who divided pycnidial and acervular fungi into five families. The Pycnidiaceae included both light- and dark-colored pycnidia with definite walls, and the Patelloidaceae contained forms with light- and dark-colored cupulate or discoid pycnidia. Thus the Nectrioideae, Excipulaceae, and Sphaerioideae of Saccardo (1884) were amalgamated into two families in which the color of the fructifications became secondary in importance to the overall form of the fructifications. The Pycnothyriaceae accommodated many of the genera placed by Saccardo in his Leptostromaceae. The Stromaceae, a newly conceived family, drew its members from the fungi in Saccardo's Sphaerioideae, Nectrioideae, and Excipulaceae in that they were stromatic with no true pycnidial wall, the conidia being formed in a simple locule. The concept of the Melanconiaceae differed little from that of Saccardo, although its division into the Eumelanconieae with irregular fructifications and Pseudosphaerioideae with pseudopycnidia lacking a definite wall was not entirely clear. Although Petch (1943) used von Höhnel's breakdown in his account of the British Nectrioideae, the scheme made no impression on subsequent workers. Von Höhnel (1923) later proposed a system of classification for Fungi Imperfecti based on more fundamental criteria and parts of his 1911 scheme were incorporated in it. Grove (1935, 1937) preferred to group the Melanconiales and Sphaeropsidales into the class Coelomycetes, bearing equivalent status to the Hyphomycetes. His approach to genera and species, however, was similar to that of Saccardo. So, despite the drawbacks of

Saccardo's system, namely the questionable reliability of the criteria used in delimiting genera and species, it remained the template upon which most of the work in the Fungi Imperfecti had been carried out prior to 1953. The general accounts and keys by Allescher (1901, 1903), Clements (1909), Diedicke (1912–1915), Clements and Shear (1931), and Bender (1934) are firmly grounded in Saccardo's principles.

## B. Ontogenetic Classifications

Early Saccardoan and pre-Saccardoan schemes of classification were primarily concerned with what were considered to be relatively static, invariable morphological criteria. From its inception by Costantin (1888) to the most recent developments, there has been an alternative approach based on the interacting complex of conidium and conidiophore ontogeny. The different methods by which conidia are formed and secede from the cells which bear them, and the ways in which those cells or groups of cells (conidiophores) behave before, during and after conidium formation are now considered to demonstrate fundamental similarities and differences more clearly. Most work in this area has been on hyphomycetes and the important contributions by Vuillemin (1910a,b, 1911, 1912), Mason (1933, 1937), Ingold (1942), and Hughes (1953) not only have been reviewed and placed in perspective, but have been modified, expanded and contracted by various authors including Goos (1956), Tubaki (1958, 1964), Subramanian (1962), Smith (1962), Luttrell (1963), Madelin (1966), Barron (1968), Cole and Kendrick (1968), and Kendrick and Cole (1969). Pycnidial and acervular fungi have not featured prominently in these developments because conidiophores are relatively small and details of conidium ontogeny consequently difficult to elucidate.

Costantin (1888) attempted to separate hyphomycetes by the different methods of insertion of conidia on parent hyphae. *Ellisiella*, an acervular fungus similar to *Colletotrichum*, was included with the hyphomycetes *Zygosporium*, *Beltrania* and *Circinotrichum* in a group characterized by long sterile setae and short fertile conidiophores. *Pestalotia* was equated with the hyphomycete *Mastigosporium* but placed in the Melanconiales. Vuillemin (1910a,b, 1911) classified hyphomycetes by differences in spore production but did not include pycnidial or acervular fungi. Von Höhnel (1923) introduced the new concept of endogenous spore formation. He separated genera into three groups, the Histiomyceten, Synnematomyceten and Hyphomyceten, of which only the first was extensively subdivided. The Histiomyceten contained pycnidial, acervular, and sporodochial fungi and was divided into two series. The Endogenosporae contained eighteen genera characterized by conidia formed within the inner cells of the fructification wall and liberated by histolysis of that wall. In the Exogenosporae, 530 genera were included with conidia formed by budding or division. Genera were distributed in the Pycnidiaceen and Stromaceen and separated into further categories by some of the criteria elucidated by von Höhnel (1911). Petrak (1925) criticised both the taxonomy and nomenclature of von Höhnel's system and only Klebahn (1933) and Goidanich and Ruggieri (1947) attempted to apply the scheme in practice.

Mason (1933) reevaluated the work of Vuillemin by clarifying terminology, particularly with reference to phialides, and introducing additional terms he considered necessary. Later Mason (1937) criticised the emphasis placed on sporodochia, acervuli, and pycnidia in classification, preferring to reduce them to secondary importance. He introduced a reorganization of the groups within the Fungi Imperfecti based on presence or absence of slime around the conidia. The two formal taxonomic categories, Gloiosporae and Xerosporae, were proposed by Wakefield and Bisby (1941) on the basis of Mason's arguments and adopted in the British List for Hyphomycetes. However Ingold (1942) considered these to be biological spore types, to which he added a third to accommodate aquatic hyphomycetes. Moreau (1953) adopted Vuillemin's system, amended by Langeron and Vanbreuseghem (1952), and included *Pestalotia* in the Thallospori, Aleuromycetales, and *Phialophorophoma* in the Conidiospori, Phialidales. Conidia in *Discosia* and *Seiridium* were also considered to be aleuriospores. However, the majority of pycnidial and acervular fungi were placed in families of the Conidiospori, Sporophorales together with sporodochial hyphomycetes. In fact the separation was little different from that of Saccardo (1884).

Hughes (1953) advanced an experimental classification for hyphomycetes in which eight sections were recognized. Most emphasis was given to different types of conidium development, and the Saccardoan criteria were relegated to secondary importance. Although based on and designed for hyphomycetes, Hughes suggested the scheme may be extended to Coelomycetes. He included the acervular genera *Alysisporium*, *Septotrullula*, and *Phragmotrichum* in Section V, characterized by meristem arthrospores which develop in gradually maturing basipetal series from conidiophores of indeterminate length. Some *Melanconium* species were referred to Section III typified by chlamydospores developing in succession from annellophores. The arthrospore state of *Hendersonula toruloidea* was placed in Section VII which was characterized by arthrospores formed from the basipetal fragmentation of conidiophores of determinate length. Subsequent work in the Sphaeropsidales and Melanconiales, stimulated by Hughes' approach, had shown his predictions to be correct, and that conidia are formed from definite conidiogenous cells in comparable ways to those outlined for the Moniliales (Dickinson and Morgan-Jones, 1966; Pirozynski and Morgan-Jones, 1968; Shoemaker, 1964; Shoemaker and Müller, 1964;

Sutton, 1963a, 1964a,b, 1967a, 1968a, 1971a; Sutton and Pirozynski, 1963, 1965; Sutton and Sellar, 1966; Walker, 1962). Subramanian (1962) attempted to place Hughes' scheme on a firm nomenclatural basis by recognizing and describing six families, Torulaceae, Bactridiaceae, Tuberculariaceae, Coniosporiaceae, Helminthosporiaceae and Geotrichaceae, while Barron (1968) accepted ten different series within the Hyphomycetes. No Coelomycetes were incorporated in either system.

## III. CRITERIA FOR CLASSIFICATION

### A. *Fructifications*

Fructifications are specialized pseudoparenchymatic structures within which, or on which, conidiogenous cells are produced. They may be separated into three recognizable categories—pycnidia, acervuli, and stromata—within which there is considerable variation. For fungi in which the fructifications start as pycnidia and then open out like acervuli, Kempton (1919) coined the term pseudoacervuli, and for pycnidia which only partly enclose the conidia Potebnia (1910) used the term pseudopycnidia. Saccardo (1884) proposed six suprageneric categories based on pigmented, hyaline, dimidiate, or cupulate pycnidia and acervuli, but Potebnia (1910) retained pycnidial and acervular genera (together with sporodochial forms) in two separate categories and placed the remaining miscellany in the Pseudopycnidiales. Von Höhnel (1923) recognized five types of fructification. These were pycnidia, acervuli, stromata, and cupulate and pycnothyrioid pycnidia. This scheme forms the basis of the following discussion, but no system has yet been advanced that will satisfactorily accommodate or distinguish the variability and diversity in form shown by coelomycete fructifications. Consequently there are many exceptions to, and divergences from, the general principles discussed.

Separate pycnidia are basically flask-shaped to globose, variably pigmented, with a papillate to rostrate apex and a single circular ostiole, generally glabrous but some genera with multicellular brown setae. In culture many isolates produce multiostiolate pycnidia. Pycnidia may be immersed, semi-immersed or superficial in relation to the substrate, often associated in the latter case with a subicle. The cavity is normally undivided and neither convoluted nor multiloculate; walls are pseudoparenchymatic and of varying degrees of complexity. The simplest type in *Scolecosporiella* is one to three cells thick, of very pale brown tissue, but darker near the ostiole. Conidiogenous cells are the inner undifferentiated wall cells. In *Chaetodiplodia* (Fig. 1A), a thicker, inner, hyaline, conidophore-bearing layer is distinct from the brown outer wall cells, whereas in *Botryosphaerostroma* (Fig. 1C) the

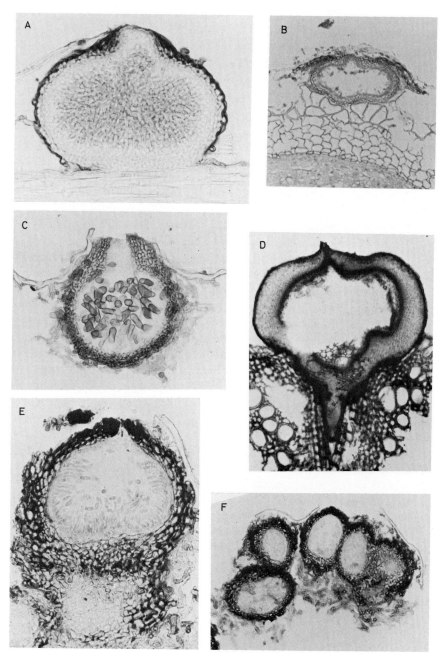

FIG. 1. (A) *Chaetodiplodia caulina*, v.s. pycnidium (× 125); (B) *Ascochytulina deflectens*, v.s. pycnidium (× 50); (C) *Botryosphaerostroma visci*, v.s. pycnidium (× 50); (D) *Plenodomus lingam*, v.s. pycnidium (× 50); (E) *Hendersonula toruloidea*, v.s. single pycnidium (× 125); (F) *Hendersonula toruloidea*, v.s. aggregated pycnidia (× 50).

dark brown wall-cells form a wider layer than the conidiogenous cell layer. In *Plenodomus* (Fig. 1D), the outer or peripheral layer is dark brown, thinner and quite distinct from the median layer of hyaline sclerotioid cells and the inner thin-walled conidiophore-bearing region. The outer wall cells in *Stilbophoma* are small, dark brown and sclerotioid and the thin-walled inner tissue is relegated to the base of the pycnidium which is immersed in the substrate. A clypeus is formed around the ostiole in *Ascochytulina* (Fig. 1B), and in *Cytonaema* the tissue organization is more complex because the clypeus is intimately associated with the rostrate beak. There is considerable variation within and between genera in all these features, particularly in culture. Such diversity in *Phoma* was emphasized by Dennis (1946).

Some pycnidia become loosely aggregated into linear or botryose groups without development of sterile stromatic tissue. Connecting tissue, if present, is prosenchymatous rather than pseudoparenchymatous. The genera in which this occurs are few. Aggregated pycnidia display almost the same range in complexity as separate pycnidia, from the *Scolecosporiella* type in *Darluca* to the *Botryosphaerostroma* type in *Botryodiplodia*, *Hendersonula toruloidea* (Fig. 1E,F), *Labridella*, and *Bleptosporium*.

The term stroma has come to mean a mass or matrix of vegetative hyphae, with or without tissue of the host, in which or on which fructifications are developed. Its use in Coelomycetes is rather vague, but the present concept excludes genera with separate or aggregated pycnidia. It embraces those with thick-walled, pulvinate, columnar, cylindrical, or clavate fructifications within which conidia are formed in simple or complex cavities. The simplest type of stroma is that most closely related to, and in some genera scarcely separable from, the pycnidium. It is consistently variable and irregular in overall shape, the cavity is convoluted or divided, the wall normally thick and of dark brown pseudoparenchyma, and the ostiole, if present, invariably singular. In *Phaeocytostroma* (Fig. 2D), *Mycohypallage*, and many *Phomopsis* species (Sutton, 1965; Hahn, 1928) the similarity to pycnidia is evident, but in *Cytospora*, *Operculella*, *Pleurocytospora* (Fig. 2B), and *Schizophoma* the differences are marked. In fact, the majority of genera here have relatively thick walls of even thin-walled pseudoparenchyma with almost no differentiation of specialized tissues. In more complex genera, the cavity in which conidia are formed may be lined by undifferentiated stromatic pseudoparenchyma, or, more specialized tissue appearing similar to pycnidial walls. In *Bellulicauda* (Fig. 2C), *Placonema*, and *Perizomella*, the convoluted conidiogenous cell-bearing layer is formed within a pseudostroma of host and fungus tissue, and in *Ceuthospora* several locules with pycnidiumlike walls are formed in a similar way. The linear pycnidiumlike locules of *Camarographium* and *Dilophospora* (Fig. 3D) are immersed in a eustroma entirely composed of fungal tissue. There is a wide range of stromata where two or more

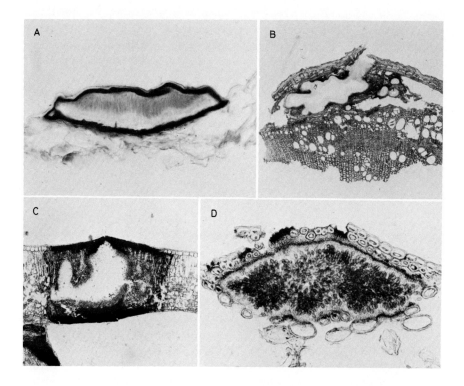

FIG. 2. (A) *Pseudothyrium polygonati*, v.s. flattened pycnidium (× 200); (B) *Pleurocytospora lycii*, v.s. stroma (× 20); (C) *Bellulicauda dialii*, v.s. stroma (× 20); (D) *Phaeocytostroma ambiguum*, v.s. stroma (× 50).

simple, convoluted, or divided locules are formed, presumably in similar ways to the Myriangiales and the Dothidiales. Clements and Shear (1931) described these types as valsoid, dothideoid, phyllachoroid, discoid, pulvinate, globoid, botryose, linear and effuse, and several genera in the Sphaeropsidales were separated by using such characters. In *Haplosporella*, *Chondrostroma* (Fig. 3C), and *Pleosphaeropsis* individual locules are at the same level in the stroma and separated by vertically arranged pseudoparenchyma. They dehisce by a single common apical ostiole, and the whole stroma has the external appearance of a single large pycnidium. In *Cytosphaera* (Fig. 3A), *Blennoria* and *Myxofusicoccum* structure is similar but conidia are released as the upper wall disintegrates. In *Strasseriopsis*, locules are peripheral in the stroma and, as in the nectrioidaceous genera, *Aschersonia* and *Diplozythia* each opens separately by an ostiole or a wide pore. Such regularly arranged locules may be contrasted with those in *Pycnidiella*,

*Camaropycnis* (Fig. 3B), *Scleropycnis*, and *Fuckelia* where they are convuluted and irregularly divided, and with *Pleurophomella* (Fig. 3E) where several stromata are formed from a common base, each having a single locule.

Cupulate fructifications are separate, astromatic, more or less superficial,

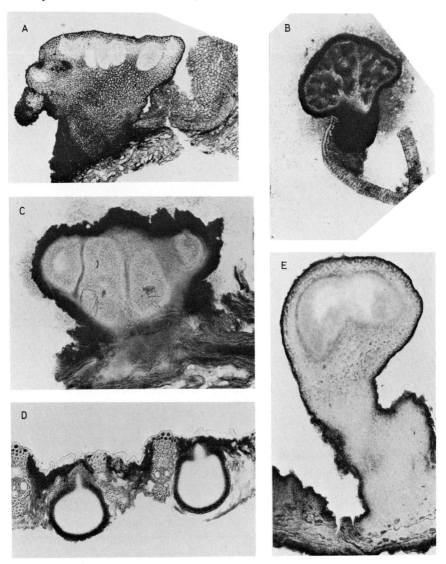

FIG. 3. (A) *Cytosphaera mangiferae*, v.s. stroma ($\times$ 20); (B) *Camaropycnis libocedri*, v.s. stroma ($\times$ 20); (C) *Chondrostroma laricis*, v.s. stroma ($\times$ 50); (D) *Dilophospora alopecuri*, v.s. stroma ($\times$ 50); (D) *Pleurophomella eumorpha*, v.s. stroma ($\times$ 100).

often setose and may be pigmented or hyaline. They are distinguished from individual pycnidia by their ontogeny, the absence of a definite ostiole and the complexity of wall structure. Initially globose and closed, they finally become cupulate with the conidiophore-bearing tissue fully exposed. Typical genera include *Phragmotrichum*, *Siroscyphella*, *Acarosporium*, *Hainesia*, *Dinemasporium*, *Heteropatella*, *Belaina*, and *Amphitiarospora*. In *Chaetopatella*, *Polynema*, and *Annellolacinia*, genera on monocotyledonous substrates, the cupulate features are reduced to such an extent that they may be justifiably interpreted as subcuticular acervuli.

Many so-called pycnidial fructifications do not fall satisfactorily into these categories. They mostly consist of genera belonging in the Leptostromataceae, Pycnothyriaceae, and Peltasterales and the segregates recognized by von Höhnel (1923) and Batista and Ciferri (1959). They comprise flat (*Colletotrichella*), elongated (*Leptothyrium*), hemispherical (*Pseudothyrium*, Fig. 2A), or irregular (*Pycnidiopeltis*) fructifications which are immersed (*Ciliochorella*), subcuticular (*Schizothyrella*), or superficial (*Actinotexis*). They dehisce in a variety of ways including by an ostiole (*Discosia*), slit (*Placella*), irregular tears (*Suttoniella*), or longitudinal fissures (*Leptostroma*). The arrangement of pseudoparenchymatic cells in the upper wall of the fructification may be radiate (*Pycnothyrium*), isodiametric (*Asterostromopsis*), or convoluted and irregular (*Plenocatenulis*). The locule may be simple or divided and the conidiogenous cell-bearing surface inverted on the upper wall (*Allothyriella*), on the lower surface (*Acarellina*), or both (*Schizothyra*). The relationships within this heterogeneous group have not been elucidated.

Acervuli display none of the complexities shown by pycnidia and stromata. There is a total lack of any definite wall structure, the fructification consisting of an immersed aggregation of pseudoparenchyma from the upper layers of which conidiophores are formed. The developing conidium mass ruptures covering host tissues and there is no specialized method of dehiscence. Fructifications may be subcuticular (*Titaeosporina*, Fig. 4A), epidermal (*Psammina*), subepidermal (*Dothistroma*, Fig. 4B), or subperidermal (*Asterosporium*), and in some genera, such as *Fominia* and *Colletotrichum*, dark brown setae are produced amongst the acervular cells.

Bauke (1876) differentiated two types of pycnidium primordium formation, and de Bary (1887) independently named these meristogenous (formed from the division of one or more cells of a single hypha) and symphogenous (formed from the intermingling of several hyphal branches). Von Tavel (1886) recognized a third type where the primordia are formed from the repeated division and growth of a single conidium. Kempton (1919) found the concepts applicable to the development of acervuli and sporodochia, but showed that few species develop strictly one way or the other. Mercer (1913) found pycnidia in *Phoma richardsiae* to be symphogenous, meristogenous, or formed from single conidia and Boerema (1964) and Punithalingam (1966) reported essentially the same for *Phoma herbarum* and *Septoria* spp., respectively. Chippindale (1929) described development in *Ascochyta gossypii* as simple or compound meristogenous. Whereas for *Plenodomus lingam*, Calvert and Pound (1949) considered development as symphogenous, Boerema and van Kesteren (1964) found it to be meristogenous. Differences in interpretation of such variable characteristics are inevitable; consequently the criteria have not been used extensively for classification.

Cavity formation within developing pycnidia was interpreted by de Bary (1887) as occurring by rupture of the inner cells of the primordium (schizogenous), while Baccarini (1890) suggested that in *Sphaeropsis malorum* both schizogenous and lysigenous factors were involved. Dodge (1923) reported cavity formation in *Phyllostictina carpogena* to be lysigenous, in *Schizoparme straminea* as schizogenous, while in *Sclerotiopsis concava* it is a combination of both. Archer (1926) in a study of several Sphaeropsidales maintained that cavity formation was lysigenous as did Chippindale (1929) in *Ascochyta gossypii*. Boerema and van Kesteren (1964) found the cavity in *Plenodomus lingam* to be lysigenous, but Harris (1935) and Punithalingam (1966) in *Septoria* and Boerema (1964) in *Phoma herbarum*, ascribed its formation to schizogenous and lysigenous factors. Dodge (1930) described an entirely different origin for *Chaetomella raphigera* where the cavity is formed by the upward and inward growth of the fructification wall.

Mechanisms of dehiscence are varied, the most common being by way of a more or less circular ostiole. The methods by which ostioles are considered to arise are diverse. The most frequently reported method is by lysis of the pycnidial wall in a restricted area. Archer (1926) described this for *Septoria lycopersici* and *Camarosporium robiniae* and Boerema and van Kesteren (1964) for *Plenodomus lingam*. Lysis combined with mechanical force exerted by the developing conidium mass was reported by Punithalingam (1966) for *Septoria* spp. and Archer (1926) for *Phomopsis arctii*, *Coniothyrium concentricum*, and *Septoria lycopersici*. In *Ascochyta gossypii*, Chippindale (1929) attributed it to lysis followed by growth of the inner wall cells to produce a papillate ostiole. Simple upward growth of similar cells, but without preceding lysis, was described by Archer (1926) for *Phoma betae*. In *Schizoparme straminea* Dodge (1923) reported buffer tissue formed to break the leaf epidermis, followed by perforation and disorganization of the pycnidial wall, and upward growth of inner wall-cells to produce the ostiole. Similar buffer tissue was also described in *Cytophoma pruinosa* and *Sphaeropsis conspersa* by Archer (1926). In the majority of cases, however, the ostiole is merely referred to as a pore in the wall. There is no special mechanism of dehiscence in acervuli, the conidial mass usually splitting the overlying host tissues irregularly. Occasionally discrete circular flaps of host tissue are pushed

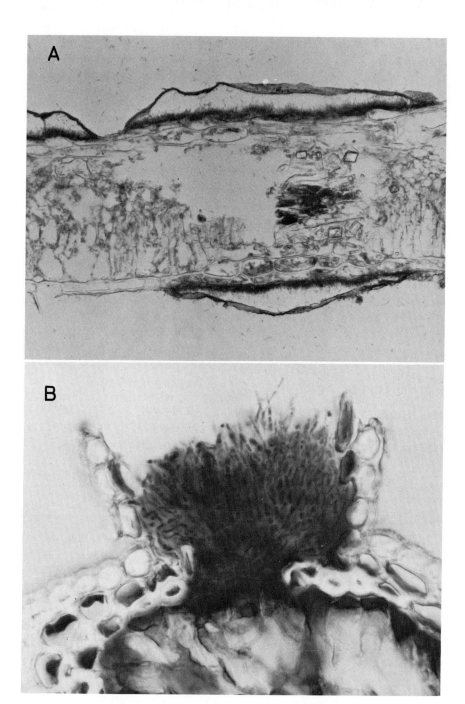

FIG. 4. (A) *Titaeosporina tremulae*, v.s. acervuli (× 100); (B) *Dothistroma pini*, v.s. acervulus (× 250).

aside as conidia emerge in *Dothistroma* and some *Septogloeum* species. Several genera in the Leptostromataceae and Pycnothyriaceae dehisce by a longitudinal opening. In pycnidial and stromatic fungi there are few genera with longitudinal lines of dehiscence and their ontogenies have not been elucidated. These include *Ajrekarella, Chaetodiscula, Hysterodiscula, Moralesia, Phaeolabrella, Psilospora,* and *Tetranacrium.* The development of *Chaetomella,* however, was described by Dodge (1930). The sequence of events in opening of cupulate fructifications is similarly obscure. Buddin and Wakefield (1926) report the young globose fructifications in *Heteropatella antirrhini* as entirely closed but with the central tissues in the upper wall of distinct structure. As the fructification enlarges, the central tissues rupture and the lateral walls open out to expose the conidiophores.

The relationship of the primordium, and sometimes the mature acervulus, to the host tissues is of importance in the separation of genera in the Melanconiales. Von Arx (1957a) differentiated between epidermal, subepidermal, and subcuticular acervuli in segregating genera related to *Gloeosporium* and Vassiljevsky and Karakulin (1950) separated graminicolous *Colletotrichum* species into *Dicladium* because of the subcuticular nature of the acervuli. The phenomenon has been reported by Wolf (1912), Dodge (1931), Stowell and Backus (1966), and Sutton (1966).

Although many genera are distinguished by setose fructifications, it has been shown by Frost (1964) that in some *Colletotrichum* species the presence or absence of setae can be controlled by varying the relative humidity. To what extent this is applicable to other genera in the Sphaeropsidales where setae are more common is unknown.

Paraphyses occur in a limited number of genera and they have been briefly discussed by Sutton and Sellar (1966). Most reports of supposedly sterile elements refer to immature conidiophores or conidiogenous cells.

Although fructifications fall into a series of recognizable groups, there is much variation between them, particularly under cultural conditions, and inevitably this leads to difficulties in the circumscription of generic limits. Duke (1928) showed that *Colletotrichum* and *Vermicularia* are identical even though *Colletotrichum* had been placed in the Melanconiales (Saccardo, 1884), Acervulales (Potebnia, 1910) and Stromataceae (von Höhnel, 1923), and *Vermicularia* in the Gymnomycetes (Fuckel, 1869), Sphaeropsidales (Saccardo, 1884), Pseudopycnidiales (Potebnia, 1910), Tuberculariaceae (von Höhnel, 1923), and the Melanconiales (Grove, 1937). A similar situation is found in *Pestalotia*, characterized by cupulate fructifications, and *Pestalotiopsis* with typical acervuli. Steyaert (1949) and Sutton (1970) emphasized the value of differences in fructifications (complemented by additional characters) in separating the two genera, whereas Servazzi (1953) and Guba (1961), on the basis of their own work and reports by other workers, claimed that variation in fructification type was sufficiently great to negate this criterion as being of any taxonomic value. A more significant consequence of variability in culture is the relationship between well-developed acervuli and sporodochia. Sporodochia are the typical fructifications of the Tuberculariaceae, in which the conidial mass is supported by an aggregation of short conidiophores. They are held to differ from acervuli in that conidiophores and conidia are formed above the surface of the substrate. In culture, however, it is impossible to distinguish between the two types of fructification. Mason (1937) viewed all fructifications in the Fungi Imperfecti as mere forms of a fungal stroma, and it is becoming increasingly obvious that there are more affinities between the Melanconiaceae and Tuberculariaceae than between the Melanconiaceae and other coelomycetes, especially truly pycnidial forms.

Differences in fructification type correlated with ontogenetic variations in conidiophores have only been elucidated for a few genera (Sutton, 1967a, 1968a), and it is too early to determine any general relationships. Such different types of conidium formation that are known appear to be distributed evenly between the families of the Coelomycetes.

## B. Conidiogenous Cells and Conidium Development

The two attributes conidiophores present for classification and identification are morphology and development, the latter encompassing the ontogenetic changes occurring before, during, and after conidium formation. Coelomycete literature abounds with references to conidiophores that are obsolete, lacking, indefinite, unknown, more or less present, none, absent, and distinctly absent (Petrak and Sydow, 1927; Clements and Shear, 1931; Bender, 1934; Grove, 1935, 1937; Sprague, 1950). The paucity of accurate information on conidiophores is attributable to the fact that the majority are unicellular and small, and frequently it is difficult to differentiate the conidiophore or conidiogenous cell from the individual vegetative cells comprising the fructification. Some, however, are long and filiform, producing terminal or lateral conidia, and these have been noted in individual descriptions and used in some instances to separate genera. However, most information on conidiophores and particularly their development is assumptive rather than factual. The majority of conidia have been thought to form by budding, and consequently development has generally been considered blastic. In the analysis of *Gloeosporium* and about 40 related genera, von Arx (1957a) described and figured the morphology of conidiophores but gave no details of differences in ontogeny. All were illustrated as blastic except in *Melanophora* where phialides were figured. Proven differences in conidium development would have facilitated separating some of the difficult genera. Similar examples may be cited in the work of Guba (1961), Zambettakis (1954), Biga et al. (1959), and Jørstad (1965, 1967). The presence of apical phialidic channels or annellides in Coelomycetes has rarely been described prior to 1953, and although some genera have been reported with catenate conidia, the acropetal or basipetal nature of the chains has rarely been elucidated. Chains are invariably assumed to be acropetal with the youngest conidium and therefore the meristem at the apex. That chains may have developed from phialides or annellides similar to those in *Penicillium* and *Scopulariopsis*, respectively, has not been considered. Therefore, it is not surprising that confusion has frequently arisen with the limited attempts to incorporate morphological or developmental features of the conidiophores in descriptions (Klebahn, 1933; Petrak and Sydow, 1927; Petrak, 1924; Zambettakis, 1954). The brief history of conidiophores in *Colletogloeum sissoo* illustrates a situation that is typical of many fungi. Sutton (1964b) showed that the distinct annellides had been referred to as with an apical scar (Chupp, 1954), polyblastic (Ahmad and Lodhi, 1953), and catenate (Petrak, 1953). There are several instances, particularly with appendaged conidia (Guba, 1961), where a lack of fundamental knowledge of conidium development has led to misrepresentation concerning the orientation of conidia on conidiophores (Sutton, 1963a, 1967a,b). There are, however, isolated examples in the literature where conidiophore morphology and conidium ontogeny have been fully elucidated and these are discussed in the following account of ontogenetic types.

Conidia in the Deuteromycotina are now recognized as originating in two basic ways (Kendrick, 1971). Blastic and thallic development are subdivided into categories based on fundamental differences in wall relationships between the conidium and the parent cell from which it is formed. Enterothallic conidia are sporangic, while holothallic conidia are arthric, meristematic, or solitary. Enteroblastic conidia are phialidic or tretic, while holoblastic conidia are annellidic, solitary, synchronous, sympodial, or retrogressive. The majority of coelomycete conidia are holoblastic or enteroblastic and a limited number are enterothallic or holothallic. Basauxic development was described by Hughes (1953) and Ellis (1965), and since it is unknown in Coelomycetes will not be considered further.

### 1. Thallic Development

Enterothallic ontogeny is present in the Sphaeropsidales inasmuch as the endogenous spermatia described by Higgins (1920, 1929, 1936) and Dring (1961) for the *Asteromella* states of *Mycosphaerella* species are formed entirely within the conidiogenous cells. Conidia are extruded through one or more projections on each cell but the full details of secession and extrusion are not known.

## 11. Coelomycetes

Holothallic ontogeny is typified by catenate and solitary conidia, both of which are distinguished from their holoblastic counterparts in being formed by separation of part of a preexisting element (the conidiophore or conidiogenous cell) and not from a newly formed element produced from the conidiophore or conidiogenous cell. Briefly, the conidium chains of Hughes' sections IA, IB, VA, VB, and VII are distinguished by the presence or absence of a meristem. If present, its location with respect to the developing conidum is of prime importance. Classificatory schemes have evolved to such complexity that unless these details are fully known the fungus cannot be assigned to its developmental section. In Coelomycetes this is often extremely difficult. In *Siropatella stenospora* the variably septate conidia merge into the ill-defined conidiophores, the acropetal or basipetal growth of which is unclear. Conversely in *Desmopatella salicis* (Fig. 5B) the long chains are of identical conidia but again with the meristematic region indistinct. The catenate conidia in *Acarosporium* species are in dichotomously branched chains and morphologically identical from the base to the apex of the chain. The presence of conidium appendages precludes considering the conidia as holothallic and arthric, but the absence of a clear meristem makes it similarly difficult to place them as meristematic holothallic or retrogressively holoblastic.

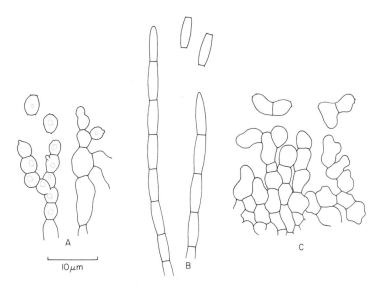

FIG. 5. (A) *Hormococcus conorum*, ontogeny holoblastic, solitary or synchronous, conidia catenate, solitary; (B) *Desmopatella salicis*, ontogeny holothallic, ? arthric, or ? meristematic; (C) *Neozythia handelii*, ontogeny holoblastic, solitary or synchronous, conidia catenate, solitary.

Meristematic holothallic development is known in *Phragmotrichum* (Hughes, 1953; Sutton and Sandhu, 1969a) and also *Septogloeum populiperdum*, but in this species chains are relatively short, conidia variable in shape, and conidiophores branched near the base.

Arthric holothallic development is more readily distinguished. Sydow (1913) described the basipetal fragmentation of determinate conidiophores in *Sclerographium induratum* and Bayliss Elliott and Stansfield (1923) illustrated a similar situation in *Patellina caesia* although details have not been confirmed. Under the name *Sirothecium lichenicola*, Grove (1937) figured arthroconidia in *Vouauxiella lichenicola*. Pirozynski and Morgan-Jones (1968), in describing *Trullula olivascens*, reported the conidiophores developing transverse septa at regular intervals in basipetal succession.

2. *Blastic Development*

Holoblastic development in which catenate conidia are formed from an apical meristem was considered to be rare in the Coelomycetes by Hughes (1953). However, it has been established in *Hormococcus conorum* (Fig. 5A) where chains of hyaline, globose acropetal conidia are formed from determinate elongated or doliiform conidiogenous cells. *Lemalis aurea*, described by Bayliss Elliott and Stansfield (1923) may also belong here. Several poorly described coelomycetes with similar catenate conidia may eventually be referred here once their conidiophore ontogeny is elucidated.

The simplest type of holoblastic ontogeny is represented by the conidiogenous cell, barely distinct from the inner cells of the pycnidial or acervular wall, which forms a solitary conidium as seen in *Cicinnobella* (Fig. 6A). The fungi that produce a few large conidia in simple pycnidia are solitarily holoblastic and include *Scolecosporiella* and *Kellermania* (Sutton, 1968a), *Aristastoma* (Sutton, 1964b), *Apiocarpella macrospora*, *Camarographium abietis*, *Tiarospora* (Fig. 6D), and *Prosthemium stellariae*. Dickinson and Morgan-Jones (1966) showed *Ascochytula* and *Microdiplodia*, genera with smaller conidia, to develop similarly. These conidiogenous cells display a transition to the next type which is elongated and distinct from the fructification wall cells. Again conidia are relatively large and complex. Fungi included in this group are *Septogloeum rhopaloideum* (Fig. 7E), *Mycohypallage* (Sutton, 1963a), *Labridella* (Shoemaker, 1963), *Toxosporiopsis* (Fig. 7B) (Sutton and Sellar, 1966), *Steganosporium muricatum*, *Asterosporium asterospermum* (Archer, 1924), *Entomosporium* (Fig. 7D), *Chaetospermum chaetosporum* (de Fonseka, 1960), *Dichomera* (Fig. 6B), and *Stevensonula*. Species of *Phyllostictina* probably belong here, with the *Ascochyta* and *Stagonospora* species described by Cunnell (1956, 1957b). In *Discosia* no more than a single conidium is formed from each elongated conidiogenous cell, but in related genera annellides are frequently produced. That some elongated

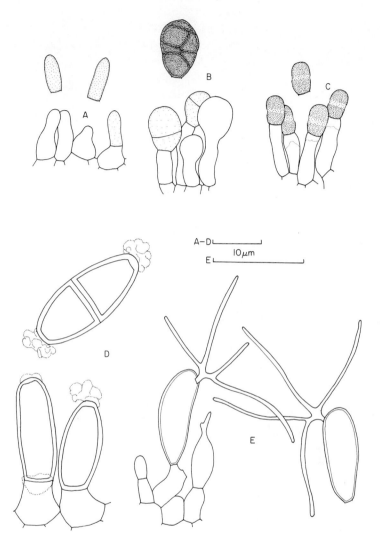

Fig. 6. (A) *Cicinnobella parodiellicola*, ontogeny holoblastic, solitary; (B) *Dichomera saubinetii*, ontogeny holoblastic, solitary; (C) *Perizomella inquinans*, ontogeny holoblastic, solitary; (D) *Tiarospora perforans*, ontogeny holoblastic, solitary; (E) *Pestalozziella subsessilis*, ontogeny holoblastic, solitary and sympodial or synchronous.

conidiogenous cells are the precursors of annellides seems most likely. A rare type of development is found in *Catenophora* (Luttrell, 1940) where conidiophores are septate. The individual cells function conidiogenously, each producing a single lateral holoblastic conidium (Fig. 8D).

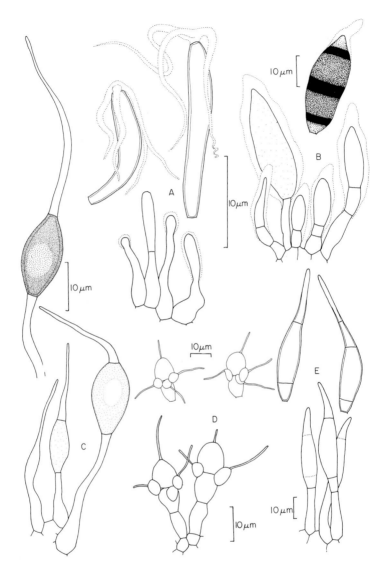

FIG. 7. (A) *Tiarosporella paludosa*, ontogeny holoblastic, solitary; (B) *Toxosporiopsis capitata*, ontogeny holoblastic, solitary; (C) *Harknessia caudata*, ontogeny holoblastic, solitary; (D) *Entomosporium maculatum*, ontogeny holoblastic, solitary; (E) *Septogloeum rhopaloideum*, ontogeny holoblastic, solitary.

Some holoblastic conidiogenous cells produce more than one conidium. This is accomplished by sympodial growth after each conidium secedes, by synchronous formation of several conidia at different points on the conidiogenous cell, or by annellidic growth of the conidiogenous cell after each conidium has seceded. The triradiate conidia in *Furcaspora* (Fig. 8B) are formed synchronously, but it is not often easy to separate this from sympodial development, particularly if the conidiogenous activity occurs in a restrict-

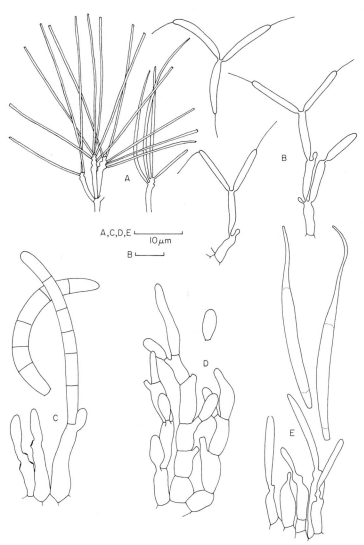

FIG. 8. (A) *Eriospora leucostoma*, ontogeny holoblastic, sympodial or synchronous; (B) *Furcaspora pinicola*, ontogeny holoblastic, sympodial or synchronous; (C) *Bothrodiscus berenice*, ontogeny holoblastic, sympodial; (D) *Catenophora pruni*, ontogeny holoblastic, solitary, conidiophore formed of several conidiogenous cells; (E) *Monochaetiella themedae*, ontogeny holoblastic, sympodial.

ed region at the tip of the conidiogenous cell. *Bothrodiscus* (Fig. 8C), *Alveophoma* (Sutton, 1964a), *Septoria didyma*, and *S. cytisi*, *Septocyta ramealis*, *Oramasia* (Sutton and Pirozynski, 1963), and *Marssonina delastrei* show restricted sympodial development, but in *Heteropatella antirrhini* (Buddin and Wakefield, 1926) and the conidial state of *Colpoma quercinum* (Twyman, 1947) the sympodial sequence is clearly seen. Several fungi bear conidiogenous cells which are nodular at the apex, possibly representing the thin scars left by successive conidia. These include *Monochaetiella cymbopogonis* (Punithalingam, 1969), *M. themedae* (Fig. 8E), *Calogloeum amentorum*, *Eriospora leucostoma* (Fig. 8A), *Corniculariella macrospora* and several *Septoria* species.

Annellidic development, as a result of the time-lapse photomicrography study of *Scopulariopsis brevicaulis* by Cole and Kendrick (1969a) and the electron microscopy of *Melanconium bicolor*, *M. apiocarpum*, and *Cryptosporiopsis* by Sutton and Sandhu (1969b), is more precisely defined and understood than most forms of conidium development. In Coelomycetes most annellides are hyaline or very pale brown with few apical annellations. The complex of acervular genera related to *Pestalotia* was shown by Sutton and Sellar (1966) to be characteristic of this group, while *Scolecosporium*, *Coryneum* (Sutton, 1963b), and *Steganosporium* form another natural group with similar wide elongated branched conidiophores and sparse annellations. There are many genera in the Sphaeropsidales and Melanconiales with comparable annellides, a selection of which includes *Ahmadia*, *Bleptosporium* (Sutton, 1963a), *Chondroplea* (Sutton, 1964a), *Colletogloeum* (Sutton, 1964b), *Camarosporium* (Fig. 9E), *Cryptosporium*, *Cryptocline*, *Cylindrosporium* auct., *Dichomera*, *Diplodia tumefaciens* (Zalasky, 1964), *Disculina*, *Monostichella*, *Neohendersonia*, *Lasmeniella* (Fig. 9C), *Lamproconium*, *Melanconium* (Corda, 1839; Kobayashi, 1968), *Titaeospora*, *Septogloeum* auct., *Phloeospora*, *Siroscyphella*, and *Sirococcus*. In *Idiocercus* (Sutton, 1967a) the annellations are flared, cupulate and widely spaced, but this is unusual. Some fungi with pigmented conidia often bear annellides which become dark brown towards the apices, occasionally with ornamentation similar to that of the conidia they produce. In *Didymosporina*, *Enerthidium* (Fig. 9D), *Obstipipilus* (Sutton, 1968a), and *Annellolacinia* (Sutton, 1964b), many closely spaced annellations are formed from smooth annellides of similar color to the conidia, but in *Lecanosticta* (Fig. 9A), *Cryptomela*, *Leptomelanconium*, *Melanconium* (Sutton, 1964c), *Gloeocoryneum*, *Schonbornia* (Bubák, 1906), *Stilbospora terminaliae*, and *Hendersonia eucalypti* (Walker, 1962), the annellides become rough-walled and deeply pigmented at the apices. Lability between annellidic and sympodial holoblastic ontogeny is illustrated by *Seimatosporium kriegerianum* (Sutton, 1964b) and *Ajrekarella* (Sutton, 1967b) where, in addition to simple annellides, the conidia are formed from laterally developed growing points.

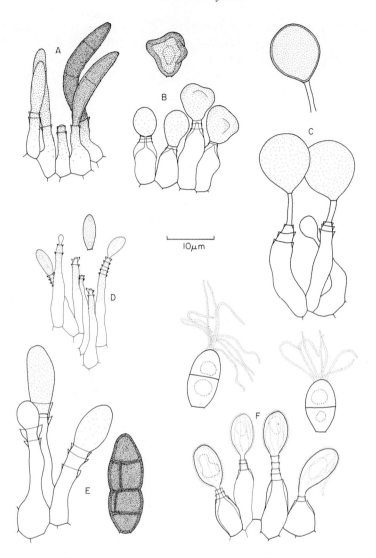

FIG. 9. (A) *Lecanosticta acicola*, ontogeny holoblastic, annellidic; (B) *Readeriella mirabilis*, ontogeny holoblastic, annellidic; (C) *Lasmeniella dendritica*, ontogeny holoblastic, annellidic; (D) *Enerthidium canarii*, ontogeny holoblastic, annellidic; (E) *Camarosporium quaternatum*, ontogeny holoblastic, annellidic; (F) *Comatospora suttonii*, ontogeny holoblastic, annellidic.

Enteroblastic development is of two main types. Tretic ontogeny (Hughes' Section VI, porospores) was claimed in *Phoma* by Boerema and van Kesteren (1964) but otherwise is unknown in Coelomycetes; however, phialidic ontogeny in its broadest sense (Hughes' Section IV) is widespread. At its extremes, phialidic development includes the types of development shown by *Bloxamia truncata* (Pirozynski and Morgan-Jones, 1968) and *Aposphaeria agminalis* (Chesters, 1938), which correspond respectively with subsections IVB and IVA of Tubaki (1958). Cole and Kendrick (1969b), working with *Phialophora*, *Penicillium*, and *Ceratocystis*, showed that the outer wall of the phialide apex does not become an integral part of the conidium, but that the entire outer wall of the first phialoconidium is newly formed. Each successive conidium is enclosed in a continuation of the secondary wall which encloses the first conidium. Successive conidia are separated by a double septum. The predominant type of conidiogenous cell which has been termed a phialide in the Coelomycetes by Chesters (1938), Linder (1944), Sutton (1964a), and Pirozynski and Morgan-Jones (1968), is doliiform with a truncate papillate apex and cytoplasm restricted near the apex to a channel which is often less than 1 μm wide. The channel is bordered or encircled by unstainable matter (possibly wall material), and occasionally there is an apical collarette. Chains of conidia from phialides are rarely seen in Coelomycetes, although *Discula junci* conforms to the phialidic concept of Cole and Kendrick (1969b). The majority of conidia formed from doliiform phialides are rounded or obtuse at the base when dispersed. There is limited evidence that initially they may be truncate at the base and secede by septum formation (Sutton and Sandhu, 1969b; Brewer and Boerema, 1965). Madelin (1966) has suggested that there may be two types of secession in phialides, budding and transverse septation, but the few electron microscopy studies of typically budding or phialidic organisms indicate only secession by septation to be operative (Zachariah and Fitz-James, 1967; Buckley *et al.*, 1969; Durrell, 1968; Agar and Douglas, 1955; Streiblova and Beran, 1965). Some conidia from simple phialides are distinctly flattened at the base even at maturity, e.g. *Microsphaeropsis callista*, *Cryptocline paradoxa*, *Oothyrium butyrospermi* (Fig. 11B), *Cryptosporiopsis* spp., *Erythrogloeum hymenaeae*, *Monostichella salicis*, *Hendersonula toruloidea*, *Marssonina* spp., and additional species. However, other species, such as *Readeriella mirabilis* (Fig. 9B), *Neomelanconium deightonii*, *Septogloeum carthusianum*, and *Discosporium deplanatum* (Fig. 11D), not only have basally truncate conidia and thin cytoplasmic channels in the conidiogenous cells but develop a few indistinct transverse annellations at the conidiogenous apices. The proliferation of phialides following conidium formation is not unusual. In the Hyphomycetes, genera such as *Catenularia* develop conidia at each of several successively higher points. To establish if conidiogenous cells in *Readeriella*, *Neomelanconium*, *Septogloeum*, and *Discosporium* are annellides or proliferating phialides, it will be necessary to determine the number of conidia produced at each level. In those cases where there appears to be no progressive growth of the conidiogenous cell, yet clearly the conidia are truncate, successive conidia may have seceded by separation of double septa at identical levels, or at retrogressive levels as demonstrated in *Cryptosporiopsis* by Sutton and Sandhu (1969b). If successive conidia secede at the same level, each by the formation of a double septum, then this ontogeny is fundamentally distinct from that described as phialidic by Cole and Kendrick (1969b) where there is wall continuity between successive conidia, and demonstrates a closer relationship to annellidic development.

The enteroblastic conidiogenous cells at present termed phialides and characteristic of Section IVA show parallel systems of arrangement to their holoblastic counterparts. The commonest type is indistinguishable from the inner cells of the pycnidial wall and is known in *Sclerophoma pythiophila* (Sutton, 1964a), *Plenodomus, Rhizosphaera, Microsphaeropsis, Camarosporium, Monostichella robergei, Placella* (Fig. 10A), *Monochaetiella hyparrheniae* (Fig. 10C), *Apiocarpella* spp., *Ciliochorella mangiferae* (Fig. 10E) (Sutton, 1964b), *Septogloeum oxysporum, Ascochytulina* (Fig. 10B), *Cytoplacosphaeria rimosa, Diplozythiella bambusina* (Fig. 10D), and *Wojnowicia*. Identical development has been demonstrated in several *Phoma* species as monopolar repetitive budding by Boerema and Dorenbosch (1965), Boerema, Dorenbosch, and van Kesteren (1965a), Boerema and Höwcler (1967), and Boerema and de Jong (1968). Wilson and Hahn (1928) and Urries (1956, 1957) figure similar phialides for *Sclerophoma magnusiana* and *Pycnothyrium microscopicum*, respectively.

A similar range of Coelomycetes bears elongated aseptate and unbranched conidiogenous cells which are distinct from the tissues that bear them. Chesters (1938) described such conidiophores in *Aposphaeria* species, and in *Cytoplea juglandis*, and they are now known in several genera and species, including *Phaeocytostroma* (Sutton, 1964b), *Pleurophomopsis, Myxosporina* (Fig. 11C), *Sirococcus strobilinus, Titaeosporina, Gloeosporidiella ribis, Cytoplea* (Fig. 11A), *Cryptocline, Cylindrogloeum, Discogloeum, Diplodia* spp., *Coniella* (Sutton, 1969), *Colletotrichum* (Sutton, 1962), and *Melanophora cercocarpi*. The group of genera with setose cupulate fructifications and setulate conidia including *Dinemasporium, Pseudolachnea, Chaetopatella* (Fig. 12F), *Stauronema*, and *Polynema* have similar conidiogenous cells.

In *Strasseria carpophila* (Fig. 11E) (Sutton, 1967a) and *Strasseriopsis* (Fig. 11F) (Sutton and Kobayashi, 1970) the appendaged conidia are produced from elongated phialides, with the body of the conidium formed first and the appendage last, whereas in the phialidic genera *Dilophospora, Neottiospora caricina* (Cunnell, 1957a), and *Ellisiella* the appendages and setulae are of varying origins and ontogeny. A more complex situation occurs where conidiophores are septate and the individual cells are functional conidiogenous cells with single phialidic apertures. *Eleutheromyces* (Fig. 12E) and *Eleutheromycella* are typical examples where development was originally thought to be blastic (Seeler, 1943) but is clearly phialidic. Linder

(1944) suggested the name *Phialophorophoma* for a *Phoma*-like fungus with acropleurogenous phialidic apertures. Similar development is now known for many genera and species including *Chaetomella, Pleurophomella, Discella acerina, Selenophoma* spp., *Pleurosticta lichenicola, Phacostromella, Pyreno-*

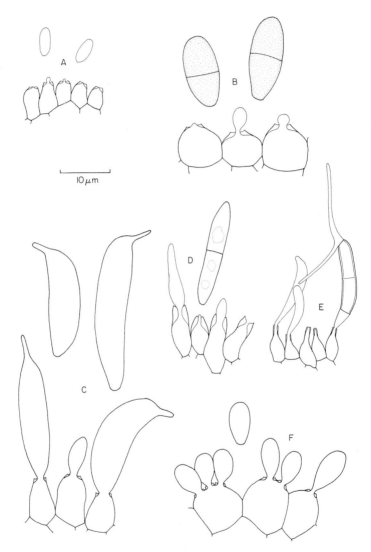

FIG. 10. (A) *Placella fraseriana*, ontogeny enteroblastic, monophialidic; (B) *Ascochytulina deflectens*, ontogeny enteroblastic, monophialidic; (C) *Monochaetiella hyparrheniae*, ontogeny enteroblastic, monophialidic; (D) *Diplozythiella bambusina*, enteroblastic, monophialidic; (E) *Ciliochorella mangiferae*, ontogeny enteroblastic, monophialidic; (F) *Sarcophoma miribelii*, ontogeny enteroblastic, polyphialidic.

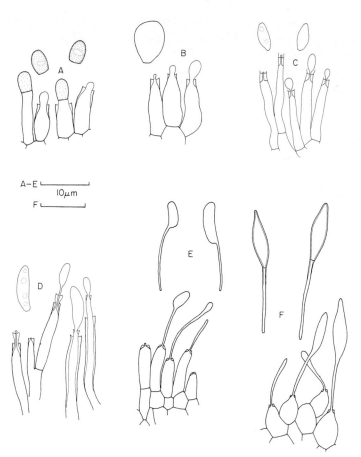

FIG. 11. (A) *Cytoplea juglandis*, ontogeny enteroblastic, monophialidic; (B) *Oothyrium butyrospermi*, ontogeny enteroblastic, monophialidic; (C) *Myxosporina subtecta*, ontogeny enteroblastic, monophialidic; (D) *Discosporium deplanatum*, ontogeny enteroblastic, monophialidic, percurrently proliferating; (E) *Strasseria carpophila*, ontogeny enteroblastic, monophialidic; (F) *Strasseriopsis tsugae*, ontogeny enteroblastic, monophialidic.

*chaeta* spp., *Pseudozythia* (Fig. 12D), *Subramanella*, and *Chaetoconis* (Sutton, 1968a). In addition to septate conidiophores some genera show basal and lateral branching of the conidiophores. Mason (1937) recognized phialidic development in *Cytospora*, a genus typified by well-branched conidiophores, and Sutton (1965) has demonstrated it in *Phomopsis*. Many genera and species are now known in this group, including *Volutella*, *Rhabdospora*, *Dendrophoma* auct., *Crocicreas atroviride*, *Cytogloeum tiliae*, *Cytonaema spinellum*, *Pseudopatellina conigena*, and *Cytosporella*. Polyphialides (Hughes, 1953) are present in *Suttoniella* which represents a phialidic analogue of the holoblastic *Furcaspora*, and in *Sphaerographium squarrosum*, *Ludwigiella*, *Kabatiella microsticta*, *Melanophora crataegi*, *Diachorella onobrychidis* (Fig. 12C) (Sutton, 1967a), *Sarcophoma miribelii* (Fig. 10F), and *Sporonema nigrificans*, but the phenomenon is not widespread in Coelomycetes. Enteroblastic development corresponding with Tubaki's section IVB is similarly infrequent. Berkeley and Broome (1854) described *Bloxamia truncata* with endogenously formed conidia. Trotter (1904) referred to a similar situation in *Cytosporella paradoxa*. Lehman (1918) gave a detailed account of endogenous conidia in *Sphaeronaema fimbriatum*, and Brierley (1915) related development in *Cytosporella*, *Sphaeronaema*, and *Bloxamia* to that found in *Thielaviopsis basicola*.

Whereas conidiophore and conidium development is of considerable value in the circumscription of generic limits, when similar principles are applied to an analysis of species in existing genera containing large numbers of species, the results highlight the enormity of their heterogeneity. In some genera, such as *Cytospora*, *Phomopsis*, and *Colletotrichum*, there is no difference in conidium ontogeny, and only a limited, yet predictable, range in conidiophore morphology between species. On the other hand, of 21 leguminous *Pyrenochaeta* species, all with simple setose pycnidia and small hyaline unicellular enteroblastic conidia, seventeen had simple doliiform phialides, one had elongated phialides and three had the elongated, septate type of conidiophore formed of individual conidiogenous cells. Similar situations obtain in *Monochaetiella* where the type species *M. hyparrheniae* is enteroblastic with simple phialides, but the two subsequent species, *M. cymbopogonis* and *M. themedae*, are holoblastic and sympodial; also in *Marssonina* where *M. brunnea*, *M. castagnei*, and *M. kriegeriana* are enteroblastic with simple phialides; *M. delastrei* is holoblastic sympodial and *M. potentillae* forma *fragariae* is holoblastic and solitary. One of the most complex situations is to be found in *Septoria* and similar genera such as *Phloeospora*, *Septogloeum* and *Cylindrosporium* auct. A selection of ten random *Septogloeum* species showed four as holoblastic and annellidic, one as holoblastic annellidic and sympodial, two solitarily holoblastic, one enteroblastic with simple phialides, one meristematic holothallic, and the type species with annellides! These are isolated examples of what is a relatively common phenomenon in large genera like *Ascochyta*, *Phyllosticta*, *Phoma*, *Hendersonia*, *Stagonospora*, *Camarosporium*, and *Diplodia*, etc. Consequently the fundamental relationships between such genera and their segregates cannot be elucidated without extensive studies on the conidium ontogeny in type or lectotype species. Until this is effected there can be neither satisfactory nor significant delimitation of general suprageneric categories for the Coelomycetes.

## C. Conidia

The size, shape, septation, color and ornamentation of conidia have been major, yet in many cases, irrational, sources of classificatory criteria. Conidia range in size from the spermatia of *Asteromella* (3–4 × 0.75–1 μm) to the macroconidia of *Coryneum* (60–70 × 12–14 μm), and between these two extremes there is diversity in both size and shape. Conidium length invariably fluctuates more than width within species; consequently the latter is often of more value for differentiating between taxa. Length/breadth ratios have not been used extensively, but some genera such as *Stagonospora* and *Rhabdospora*, *Phloeospora* and *Septogloeum*, and *Phoma* and *Macrophoma* are separated basically by this means. There is, however, a distinct correlation between conidium size and method of formation in that most large conidia produced in pycnidia are solitary and holoblastic. Conidium size and shape can be altered significantly by varying the nature of the artificial medium upon which isolates are grown (Sutton, 1964a; Brooks and Searle, 1921).

Diversity in conidium shape is not as marked as in Hyphomycetes because the pycnidial and acervular walls impose a degree of mechanical limitation. Nevertheless, the variety of conidium morphology is increased by ornamentation of conidium walls and the development of cellular and mucilaginous appendages and setulae. On natural substrata there is little variation in the actual shape of conidia within species and sometimes within genera (*Phomopsis*, *Cytospora*, *Phoma*, *Phyllosticta*), although in *Operculella* (Kheswalla, 1941) and *Neozythia* (Petrak, 1958) conidium shape and septation is labile (Fig. 5C). Most holoblastic, enterothallic, and holothallic conidia are truncate at the base, and at the apex if formed in chains. Enteroblastic conidia, however, are often rounded at the base and it is debatable as to whether this is the result of budding from the conidiogenous cell or rounding off following secession by septum formation.

Luttrell (1963) suggested that there are two basic types of septum: eusepta typical of *Sporidesmium tropicale* and distosepta (the pseudosepta of Ellis, 1960) typical of *Corynespora cassiicola*. Although the concepts were considered to be oversimplified by Campbell (1968), Sutton and Sandhu (1969b), and Sutton (1970) as far as ultrastructural morphology was concerned, at least two types can be recognized in Coelomycetes. Eusepta are most common, being found in both hyaline and brown conidia. Distosepta are rare and only found in a limited number of genera with brown conidia such as *Coryneum*, *Steganosporium*, *Pestalotia*, *Prosthemium*, and *Asterosporium*. Eusepta and distosepta may be transverse or longitudinal. In Hyphomycetes, differences in euseptate and distoseptate conidia do not necessarily coincide with generic limits. In Coelomycetes, however, most genera are homogeneous with respect to septation type. An exception was *Pestalotia* as maintained by Guba (1961) which included the type species, *P. pezizoides*, with distosepta and the rest of the accepted species with eusepta.

The number of septa in conidia has frequently been used to distinguish genera, but lability in septum formation and protraction in development, combined with tardiness in development of pigment, particularly in culture, have meant that fungi may be placed in different genera depending on the stage at which they are examined (Brooks and Searle, 1921; Sattar, 1934; Wollenweber and Hochapfel, 1936). A typical example is *Phoma seriata* for which Wollenweber and Hochapfel (1936) cited fourteen synonyms in six genera. The modern tendency, therefore, is to include in a single genus several species differing in degrees of septation (Sutton, 1964b, 1968a; Sutton and Pirozynski, 1965; Dorenbosch, 1970), an approach with many parallels in the Hyphomycetes.

Conidium pigment, or its absence, in Coelomycetes has led to a proliferation of described genera. This is in contrast to the situation in Hyphomycetes where differences in color have provided one of the major means of separating suprageneric categories. Development of pigment may occur as the conidium is formed on the conidiogenous cell or it may be a much slower process, taking place not only during development but after secession. Protracted pigment development often coincides with delay in septum formation. In *Botryodiplodia theobromae* conidia are aseptate and hyaline or pale brown for several weeks before the median septum is formed and the mature color achieved. Of the 48 *Botryodiplodia* species accepted by Petrak and Sydow (1927), 21 were originally described in *Macrophoma* and twelve in *Sphaeropsis*, respectively characterized by large hyaline aseptate conidia and large brown aseptate conidia. Other species with characteristically long periods of maturation are *Hendersonula toruloidea* (Nattrass, 1933) and *Diplodia tumefaciens* (Zalasky, 1964). *Macrophoma pinea*, initially hyaline and aseptate, ultimately becomes pale brown and from one- to two-septate, especially prior to germination. *Pestalotiopsis sydowiana* was shown by Sutton (1961) to take up to five months for conidia to mature, during which time conidia passed through three different morphological phases.

Conidia most frequently develop color evenly throughout, particularly when aseptate. The degree of pigment varies from some *Microsphaeropsis*, *Coniothyrium*, and *Ascochytula* species, where it is only detectable when conidia are viewed in mass, to *Melanconium* and its segregates, where individual conidia are dark brown and the mass is black. In *Lamproconium* conidia are bluish (Grove, 1937). In genera such as *Pestalotia*, *Pestalotiopsis*, *Bleptosporium*, *Seiridium*, *Monochaetia*, and *Seimatosporium* the pigment is not developed in the apical or basal cells, but is restricted to the median cells. Occasionally in *Pestalotiopsis* the median cells are pigmented to varying degrees, features useful in separation of species (Steyaert, 1949). In *Toxosporium, Toxosporiopsis*, and species allied to *Pestalotiopsis versicolor*, a dark black band is developed around septa, but in *Perizomella* (Fig. 6C) and *Asterostomopsis* (Batista and Ciferri, 1959), the aseptate conidia are brown, with a median hyaline band, possibly similar to those in *Beltrania* and *Beltraniella* (Pirozynski and Patil, 1970).

Hyaline conidia are usually smooth walled, but the surfaces of some pigmented conidia become punctate, granulate, tuberculate, or verrucate. The extent to which this occurs is unknown, particularly with very small conidia and the finer types of ornamentation. Little is known of the ultrastructure of walls and their surfaces, but variation in size and density of pigment granules and their relationships with the different wall layers shown for *Melanconium* spp. by Sutton and Sandhu (1969b) and for *Leptomelanconium* spp., *Gloeocoryneum*, and *Hendersonia pinicola* by Chao (1969) give indications of the complexity of the structural organization involved. The different surface characteristics of some *Coniothyrium* species have recently been described by Punithalingam and Jones (1970). In *Botryodiplodia theobromae* the conidium wall is puncticulate but longitudinal areas along which no pigment is produced give the conidia a striated appearance. In some *Harknessia* (Sutton, 1971b) and *Coniella* species (Sutton, 1969), the pigmented conidia are smoothwalled with a multiple or single longitudinal region paler in color. These may be germination slits or striations as in *B. theobromae*.

Many detached conidia bear apical, lateral, or basal attenuated structures which have been termed appendages, cilia, setae, setulae, and pedicels. One type originates from the body of the conidium with which cytoplasmic continuity is maintained at least during formation. These have been termed appendages by Sutton (1967a). The other type is extracellular and has no obvious cytoplasmic continuity with the conidium. These are setulae and mucilaginous attachments of various kinds. Apical appendages may be devoid of cytoplasmic contents, as, for example, in *Ciliochorella* (Subramanian and Ramakrishnan, 1956), *Seiridium*, and *Toxosporium*, or may retain their contents throughout development, as in *Scolecosporiella* (Sutton, 1968a), *Diachorella* (Ciccarone, 1963), *Heteropatella*, *Placonema* (Fig. 12A), and *Ellisiella*, but here the appendages are little more than the drawn-out apices of the terminal cells. Appendages are frequently branched, as in *Pestalozziella* (Fig. 6E), *Pestalotiopsis* and *Truncatella* (Steyaert, 1949), *Labridella* (Sutton, 1970), and *Mycohypallage* (Sutton, 1963a). Apical and basal setulae are filiform and typically of diameters of less than 1 $\mu$m. They are characteristic of *Dinemasporium* and related genera, *Shanoria* (Subramanian and Ramakrishnan, 1956) and several other genera. In *Stauronema* (Fig. 12B), *Polynema* (Sutton, 1968b), and *Obstipipilus* (Sutton, 1968a),

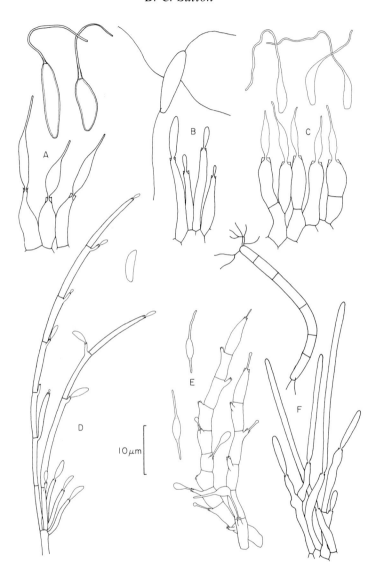

FIG. 12. (A) *Placonema napelii*, ontogeny enteroblastic, monophialidic; (B) *Stauronema sacchari*, ontogeny enteroblastic, monophialidic; (C) *Diachorella onobrychidis*, ontogeny enteroblastic, monophialidic; (D) *Pseudozythia pusilla*, ontogeny enteroblastic, monophialidic, conidiophore formed of several conidiogenous cells; (E) *Eleutheromyces subulatus*, ontogeny enteroblastic, monophialidic, conidiophore formed of several conidiogenous cells; (F) *Chaetopatella ryukyuensis*, ontogeny enteroblastic, monophialidic, conidiophore formed of several conidiogenous cells.

lateral setulae are present, and in *Dilophospora* and *Chaetopatella* (Hino and Katumoto, 1961) they are extensively branched. Extracellular attachments not included as setulae are the mucilaginous caps and brushlike structures described for many genera. They arise in a variety of ways. In *Sadasivanella* (Agnihothrudu, 1964) and *Amphitiarospora* (Agnihothrudu, 1963), *Samukuta* and *Sakireeta* (Subramanian and Ramakrishnan, 1957) their origin is not precisely known. Young conidia in some *Harknessia* spp., *Toxosporiopsis* (Sutton and Sellar, 1966), and *Kellermania* (Sutton, 1968a), however, are enclosed in a mucilaginous sheath which persists as an apical cap in *Toxosporiopsis*, disappears entirely in *Harknessia* spp., and remains as a stout yet filiform apical attachment in *Kellermania*. In *Neottiospora caricina*, Cunnell (1957a) described the mucilaginous envelope splitting from the conidium apex, becoming everted to produce a basal hollow cone, and in *Robillarda phragmitis* (Cunnell, 1958) there is a similar situation in which detached conidia were enclosed in a sheath, basal filiform mucilaginous attachments being formed by longitudinal splitting of the sheath from the apex toward the base. The reverse is known to occur in *Comatospora* (Fig. 9F) and *Tiarosporella* (Fig. 7A) where the sheath splits from the base and recurves over the apex as filiform attachments (Pirozynski and Shoemaker, 1971). Basal appendages are invariably devoid of cytoplasm and are formed in three ways. They may be the remains of the apical region of the conidiogenous cell carried away with the conidium following secession, varying from the minute basal frills on holoblastic conidia to the relatively long appendages found in *Lasmeniella*, *Harknessia* (Fig. 7C), *Bellulicauda*, and *Idiocercus*. They may be formed within the conidiogenous cell in *Pestalotiopsis* (Sutton, 1961) or as lateral outgrowths from the basal cell in *Seimatosporium* (Sutton, 1964b; Shoemaker, 1964). Occasionally, the appendage is formed first and the body of the conidium develops from the tip of the appendage as in *Strasseria* (Parmelee, 1958; Sutton, 1967a) and *Strasseriopsis* (Sutton and Kobayashi, 1970).

## D. Artificial Cultures

Although in some genera, such as *Sphaceloma*, *Phomopsis*, *Pestalotiopsis*, *Truncatella*, and *Cytospora*, cultural variation within genera is very limited, the majority of coelomycetes when isolated from their natural substrates and grown in artificial conditions show great diversity in cultural and morphological features, ability and extent of sporulation, and the formation of sclerotia, chlamydospores, and appressoria. References to the cultural attributes of coelomycetes and the factors governing them are numerous (Ali, 1962; Archer, 1926; Brewer, 1960; Burger, 1921; Coons, 1916; Dennis, 1946; Fournet *et al.*, 1970; Hoo and Wilcoxson, 1969; Klebahn, 1905; Leach,

1962; Leonian, 1924; MacNeill, 1950; Simmonds, 1965). Dennis (1946) and Boerema (1969) pointed out that in pycnidial genera related to *Phoma* there are few satisfactory morphological features available and since there are numerous specialized taxa with wide host ranges it is necessary to study behavior in culture and correlate it with established morphological criteria. This approach, combining pathological and cultural studies with the classical type method, is best illustrated by reference to recent work on *Phoma*. The foot rots and leaf spots of lucerne, red clover and pea were thought to be caused by three different fungi, variously referred to the genera *Phoma* and *Ascochyta*. Although there are apparently no morphological differences between them, and they vary in pathogenicity, Boerema *et al.* (1965b) were able to distinguish two varieties *in vitro* by differences in uniformity and variability in cultural form and their diverse ability to produce chlamydospores and crystals in culture. Another example is in *Peyronellaea*, which is distinguished from *Phoma* by an ability to form chains of dictyochlamydospores in artificial culture (Luedemann, 1959). Boerema *et al.* (1965a) and Chodat (1926) showed that some strains lose this ability, after which they are indistinguishable from *Phoma*. Correlating the work of Chodat (1926) and Lacoste (1955) who showed that the C/N ratio of the medium is a determining factor in dictyochlamydospore production, Boerema *et al.* (1965a) placed *Peyronellaea* in synonymy with *Phoma*. Species were separated according to the manner in which chlamydospores were produced and the presence or absence of single dictyochlamydospores. The use of chlamydospore production in culture to separate species is well known in *Fusarium* and has been used in other *Phoma* species by Dorenbosch (1970) and *Hendersonula toruloidea* by Nattrass (1933). A further example is the anthraquinone pigment production in culture by *Phoma exigua* var. *foveata*. This was used by Boerema (1967) to separate the variety from *P. exigua*, together with different temperature requirements and a greater pathogenicity to potato. Boerema and Höweler (1967) also separated four varieties of *P. exigua* by a combination of the metabolites which oxidize to blue-green and red pigments, mycelial characters and host ranges. Eveleigh (1961) separated *Phoma violacea* from other paint-inhabiting species by pigment formation but found a wide range in colony characters, spore mass color, colony pigment and growth among his isolates, as did Dennis (1946) who dealt with *Phoma* species from a variety of substrates.

Other features which have occasionally been used in separating species and genera in culture are appressoria in *Colletotrichum* (von Arx, 1957b; Sutton, 1962, 1968c), and sclerotia in *Macrophomina* (Ashby, 1927), *Phoma* (Dorenbosch, 1970), and *Colletotrichum* (von Arx, 1957b). The most notable example of successful cultural approach to coelomycetes is with *Colletotrichum* which was originally separated into several hundred species mainly according to host substrates. Von Arx (1957b), however, only accepted 20 species, with 600 synonyms for *C. gloeosporioides*, 90 for *C. dematium* and 35 for *C. graminicola*. As the synonymies are investigated, some taxa are inevitably recognized as distinct (Simmonds, 1965; Sutton, 1962, 1968c) with the result that the number of accepted species is slowly increasing and the taxa are being defined much more accurately by cultural and pathological behavior.

Gel electrophoresis has rarely been used in coelomycete taxonomy. Durbin (1966) found that protein patterns were inconclusive for distinguishing three graminicolous species of *Septoria*. It was postulated that species with different protein patterns arose in a similar manner to morphological and pathological variants.

Although the pathogenic, serological, chemical, and cultural characteristics of coelomycetes appear to be of increasing value in separation of species and subspecific taxa, it seems most unlikely that they will have any more than a minor influence on the circumscription of generic and suprageneric categories. The latter will ultimately be based on ontogenetic features of conidiogenous cells and conidia, criteria which seem to be genetically stable and subject to minimal variation under natural and artificial conditions.

## IV. KEYS

### A. Notes

It has already been explained why coelomycete classification and identification is in a transitional state. Consequently the following keys are an inevitable compromise between established suprageneric separations and those indicated by a consideration of conidium ontogeny. There are several obvious omissions with respect to family and generic names.

Within the two orders there is no separation into families for two reasons. (1) The new information on conidium ontogeny in the Deuteromycotina will bring about a radical suprageneric reorganization of taxa in the not- too- distant future which will materially affect traditional separations in the Coelomycetes. (2) The availability of legitimate and validly published family names for groups of genera separated according to fructification type means that the traditional format is untenable on a nomenclatural basis, notwithstanding the acknowledged deficiencies on taxonomic grounds. The earliest available family name for pycnidial fungi is Ascochytaceae Lib., 1837. This predates Sphaeropsidaceae Lév., 1845, which is illegitimate and not validly published, Sphaerioidaceae Sacc., 1899 (Sphaerioideae of Grove, 1935), which is also illegitimate and Phomaceae Clements, 1909, all three of which are commonly used. Despite the names Discellaceae Clem. and Shear, 1931, and Excipulaceae Bonord., 1851, having been extensively used to group fungi with discoid fructifications, neither is tenable because *Excipula = Pyrenopeziza* (Helotiales) and the fructification of *Discella* is not remotely discoid. No other family names are applicable. There appear to be no family names for the stromatic fungi as a whole, the Sclerophomaceae von Höhnel, 1916, and its type genus failing to provide the broadness in concept supplied by the illegitimate Stromataceae von Höhnel, 1911. In acervular fungi both the Stilbosporaceae Fr., 1832, and Coryneaceae Cda. 1839, predate the widely used and universally accepted Melanconiaceae Cda. 1842. Several later family names are known for the flattened stromatic fungi placed in the Leptostromataceae Sacc., 1884.

Many common, universally known genera such as *Ascochyta, Dothiorella, Diplodia, Phyllosticta, Fusicoccum, Septogloeum, Rhabdospora, Marssonina,* and *Aposphaeria* are not included in the key because the conidium ontogeny in type or lectotype (if designated) species is either unknown or questionable. Most of these genera are large and heterogeneous and no use would be served by keying them out either on the basis of a randomly chosen species, or alternatively, by keying them out to different positions in the key because of the several different methods of conidium formation inevitably found in a selection of species within a genus.

Two points concerning conidium ontogeny should be noted. The term conidiophore is used for a multicellular hyphal structure comprised of terminal conidiogenous cells supported by vegetative cells which are morphologically distinct from the pseudoparenchyma from which they arise.

Retrogressive annellides, hitherto only demonstrated by electron microscopy for *Cryptosporiopsis* by Sutton and Sandhu (1969b) are probably of wide occurrence in this group. With the limitations of optical microscopy it is impossible to distinguish this type of development from phialidic ontogeny; consequently *Cryptosporiopsis* and similar fungi are keyed out in the phialidic section.

## B. Coelomycetes

In this class, the thallus is mycelial, eucarpic, and septate; the conidia are formed in various ways from conidiogenous cells lining a cavity which is initially enclosed by fungal tissue (pycnidia and eustromata) or a combination of fungal and host tissue (acervuli and pseudostromata).

### 1. General Characteristics

These are primarily minute parasites and saprobes of terrestrial vascular plants, found commonly in soil, sewage, marine, and freshwater environments and as hyperparasites of other fungi. They are the asexual states of Ascomycotina.

Fructifications, when superficial or partly immersed, may be subtended by vegetative unspecialized ramifying hyphae or by hyphae arranged in subicula and anastomosing or radiating networks. Pycnidia are superficial or immersed, spherical, flattened or discoid, with a multicellular wall of isodiametric cells, pigmented yellow, orange, brown to black, separate or aggregated, usually with a dehiscent opening (ostiole). Acervuli are immersed, separate or confluent, consisting of a basal stroma lacking lateral and upper walls or specialized methods of dehiscence. Stromata may be convoluted, uni- or multilocular, superficial or immersed, of various shapes, with thick, multicellular walls of isodiametric cells, pigmented yellow, orange, brown or black, separate or aggregated, and with one or more dehiscent ostioles.

Conidiogenous cells are formed on the inner face of the fructification walls, sometimes restricted to the lateral or basal walls, unicellular, often grouped into complex structures (conidiophores), producing conidia asexually.

Conidia are enterothallic, holothallic, enteroblastic, or holoblastic, phialides and annellides being most common, deciduous, hyaline or pigmented, aseptate, euseptate, or distoseptate, of various shapes, and with cellular or extracellular appendages, setulae, and mucilaginous ornamentations.

## 2. *Important Literature*

von Arx (1957a); Grove (1935, 1937); Petrak and Sydow (1927, Sutton (1961, 1963a, 1964b, 1971a,b); Vassiljevsky and Karakulin (1950); Wehmeyer (1933; 1941). For details see References, pp. 574–581.

### KEY TO ORDERS OF COELOMYCETES

1. Fructifications acervuli . . . . . . . . . . . . . . . . . . . . **Melanconiales** p. 555
1'. Fructifications separate, aggregated, spherical, discoid or flattened pycnidia or stromata . . . . . . . . . . . . . . . . . . . . . . . . . . . . **Sphaeropsidales** p. 560

### KEY TO MELANCONIALES (MELANCONIACEAE)

Microscopic parasites or saprobes of plant material. Fructifications subcuticular, subepidermal or subperidermal, with or without subicle or setae; stromatic tissue restricted to the base of the fructification, forming conidia on the upper surface; dehiscing by regular or irregular splitting of the overlying host tissues. Consisting of a single family
. . . . . . . . . . . . . . . . . . . . . . . . . . . . . . . . . . . . . . . . **Melanconiaceae**
1. Conidiogenous cells enteroblastic, phialidic . . . . . . . . . . . . . . . . . . . 25
1'. Conidiogenous cells holoblastic . . . . . . . . . . . . . . . . . . . . . . . . . 2

2(1') Conidiogenous cells proliferating percurrently or sympodially ....... 9
2'(1') Conidiogenous cells not proliferating .................... 3
3(2') Multicellular conidiophores of several conidiogenous cells each forming a conidium laterally to the conidiophore; conidia hyaline, aseptate, oval-obovate .. **Catenophora**
C. pruni on Prunus.
3'(2') Conidiogenous cells separate, not comprising a multicellular conidiophore .... 1
    4(3') Conidia consisting of several branches from 1 or more basal cells ...... 5
    4'(3') Conidia unbranched ........................ 8
5(4) Conidia euseptate ............................ 6
5'(4) Conidia distoseptate ........................... 7
    6(5) Conidia and subtending septate conidiophores in mucilage, indehiscent; conidia dark brown, branches indistinct, compact, markedly constricted at the septa
.................................... **Thyrsidium**
    6'(5) Conidiogenous cells directly from acervular pseudoparenchyma; conidia palmate, hyaline, slightly constricted at the septa ................ **Psammina**
7(5') Conidia stellate with up to 14 divergent arms, each with a distinct basal cell; conidiogenous cells subtended by long septate conidiophores .......... **Prosthemium**
7'(5') Conidia stellate with 4 divergent arms, brown, not markedly constricted at the septa; conidiogenous cells subtended by long septate conidiophores ...... **Asterosporium**
    8(4') Conidiophores elongated, cylindrical, branched, septate only at the base, hyaline, becoming pigmented and tuberculate at the apex; conidia 3-euseptate, brown, clavate, base truncate, apex obtuse, tuberculate ......... **Gloeocoryneum**
G. cinereum on Pinus.
    8'(4') Conidiophores elongated, branched, septate, occasionally with terminal conidia from lateral branches (conidiogenous cells) immediately below septa; conidia 1–5-euseptate, hyaline, filiform ..................... **Dothistroma**
D. pini, red-band needle blight of pines. Ivory (1967) recognizes 3 varieties.
9(2) Conidiogenous cells sympodial ................... 10
9'(2) Conidiogenous cells annellidic .................... 11
    10(9) Conidiogenous cells stout, elongated, geniculations distinct but unthickened; conidia aseptate, hyaline, broadly navicular, base truncate ...... **Calogloeum**
C. amentorum on Salix cinerea.
    10'(9) Conidiophores septate and repeatedly branched, geniculations small, indistinct and unthickened; conidia lunate, hyaline, aseptate ........... **Libertella**
L. faginea on Fagus and L. betulina on Betula.
11(9') Conidia with apical, lateral, basal appendages or setulae (occasionally absent in Truncatella, Seimatosporium, and Pestalotiopsis) ................ 12
11'(9') Conidia lacking appendages or setulae ................. 17
    12(11) Annellides short but annellations distinct; conidia pale brown, 1-euseptate, clavate, truncate at the base, setula subapical, unbranched, extracellular
..................................... **Obstipipilus**
    12'(11) Annellides elongated, annellations indistinct; appendages cellular and lacking cytoplasm ............................. 13

## 11. Coelomycetes

13(12′) Acervuli setose, subcuticular; annellides cylindrical, pale brown with 1–8 annellations; conidia pale brown, aseptate, with 1 apical and 1 exogenous, basal, unbranched, cellular appendages ................................ **Annellolacinia**

13′(12′) Acervuli without setae; conidia 3-euseptate ................ 14

13″(12′) Acervuli without setae; conidia 4-euseptate ................ 15

13‴(12′) Acervuli without setae; conidia 5-euseptate ................ 16

 14(13′) Two median conidium cells pigmented, end cells hyaline, lacking cytoplasm, 0–1 basal, endogenous, cellular, unbranched appendages, 1–4 apical, cellular, simple or branched appendages; conidia and conidiogenous cells initially in mucilage ...................................... **Truncatella**
  Steyaert (1949) accepts 5 species. Guba (1961) maintains 34 spp. in *Pestalotia* sect. Quadriloculatae.

 14′(13′) Two median conidium cells pigmented, end cells hyaline, lacking cytoplasm, 0–1 basal, exogenous, cellular, simple or branched appendages, 0–1 apical, cellular, simple or branched appendages ................ **Seimatosporium**
  Synonyms accepted by Shoemaker (1964) and Sutton (1964b) include *Sporocadus*, *Cryptostictis*, *Dochmolopha*, *Amphichaeta*, *Disaeta*, *Seiridina*, *Basipilus*, *Monoceras*.

 15(13″) Three median cells pale brown, end cells hyaline, lacking cytoplasm, 1-apical and 1–3 subapical to lateral, cellular, simple appendages, basal appendage absent ...................................... **Pestalozzina**

 15′(13″) Three median cells pigmented, end cells hyaline, lacking cytoplasm, 2 or more apical, cellular, simple or branched appendages and 0–1 basal, endogenous, cellular, simple or branched appendages ................ **Pestalotiopsis**
  Steyaert (1949) accepts 45 spp. Guba (1961) maintains 182 spp. in *Pestalotia* sect. Quinqueloculatae.

 15″(13″) Three median cells pigmented, end cells hyaline, lacking cytoplasm, 1 apical, simple or branched, cellular appendage, 0–1 basal, endogenous, cellular, simple appendage ................ **Monochaetia**
  Guba (1961) accepts 41 spp.

 15‴(13″) Three median cells pigmented, end cells hyaline, 0–1 apical, cellular, simple or branched appendages, 1 basal, exogenous, cellular, simple or branched appendage ................ **Seimatosporium**
  See also 14′.

 16(13‴) Conidiophores developing in mucilage; 4 median cells pigmented, often longitudinally striate, end cells hyaline, lacking cytoplasm, 1 apical, simple, or branched, cellular appendage, 1 basal, endogenous, simple or branched, cellular appendage ...................................... **Seiridium**

 16′(13‴) Conidiophores lacking mucilage; 4 median cells pigmented with median septum encircled by a black band, end cells hyaline, lacking cytoplasm, 1 apical, simple, cellular appendage, 0–1 basal, simple, endogenous, cellular appendage ...................................... **Toxosporium**
  *T. camptospermum* on *Abies*.

 16″(13‴) Conidiophores lacking mucilage, four median cells pigmented, end cells hyaline and lacking cytoplasm, 0–1 apical, simple or branched, cellular appendage, 1 basal, simple or branched, exogenous, cellular appendage ... **Seimatosporium**
  See also 14′.

17(11′) Conidia distoseptate . . . . . . . . . . . . . . . . . . . . . . . . . . . . 18
17′(11′) Conidia euseptate . . . . . . . . . . . . . . . . . . . . . . . . . . . . . 19
17″(11′) Conidia aseptate . . . . . . . . . . . . . . . . . . . . . . . . . . . . . 21
    18(17) Annellations sparse; conidia brown, with longitudinal and transverse septa, cell lumina reduced, occasionally developing in mucilage . . . . . . . **Steganosporium**
    18′(17) Annellations sparse; conidia brown, with transverse septa, cell lumina reduced; conidiophore often persistent . . . . . . . . . . . . . . . . . . . . . **Coryneum**
        Correlated by Wehmeyer (1941) with *Pseudovalsa*.
19(17′) Annellides developing similar pigment as conidia, 1–4 distinct annellations; conidia unequally 1-septate, pale brown, smooth-walled, clavate, base truncate
. . . . . . . . . . . . . . . . . . . . . . . . . . . . . . . . . . . . . . **Didymosporina**
19′(17′) Annellides developing similar pigment and ornamentation as conidia, 1–2 annellations; conidia 1-septate, dark brown, tuberculate, clavate, base indistinctly truncate
. . . . . . . . . . . . . . . . . . . . . . . . . . . . . . . . . . . . . **Leptomelanconium**
    Sutton and Chao (1970) recognize 2 spp. on *Pinus* and *Picea*.
19″(17′) Conidia 2–4-septate . . . . . . . . . . . . . . . . . . . . . . . . . . . . 20
    20(19″) Annellides developing similar pigment and ornamentation as conidia, 1–2 annellations; conidia cylindrical, curved or irregular in shape, tuberculate, indistinctly 0–4 septate . . . . . . . . . . . . . . . . . . . . . . . . . . . **Lecanosticta**
        Brown spot needle blight of pines (Wolf and Barbour, 1941).
    20′(19″) Conidiophores septate, branched at the base; annellides hyaline; conidia 1–3 septate, hyaline, filiform . . . . . . . . . . . . . . . . . . . . . . . **Titaeospora**
    20″(19″) Conidiophores elongated, septate, branched, pale brown, annellations distinct; conidia hyaline, 1–4 septate, cylindrical, obclavate, fusoid to sigmoid
. . . . . . . . . . . . . . . . . . . . . . . . . . . . . . . . . . . . . . **Colletogloeum**
    20‴(19″) Annellides elongated, doliiform, pale brown, annellations distinct; conidia hyaline, 1–3 septate, irregularly cylindrical . . . . . . . . . . . . . . . **Ahmadia**
21(17″) Conidia rough-walled, brown . . . . . . . . . . . . . . . . . . . . . . . . 22
21′(17″) Conidia smooth-walled, brown . . . . . . . . . . . . . . . . . . . . . . . 23
21″(17″) Conidia hyaline . . . . . . . . . . . . . . . . . . . . . . . . . . . . . . 24
    22(21) Acervuli setose; conidiophores cylindrical, septate and branched at the base; annellides pigmented and rough-walled towards the apices, 1–3 annellations
. . . . . . . . . . . . . . . . . . . . . . . . . . . . . . . . . . . . . . **Phaeopolynema**
    22′(21) Acervuli lacking setae; annellides brown and rough-walled towards the apices, 1–2 annellations . . . . . . . . . . . . . . . . . . . . . . . . **Leptomelanconium**
        See also 19′.
23(21′) Acervuli with a central cushion of sterile tissue, ramicolous; conidiophores long, branched, hyaline and septate at the base, rarely pigmented at the apex; conidia obovate, ovate to ellipsoid, base truncate, guttulate or eguttulate . . . . . . . . . **Melanconium**
    Correlated by Wehmeyer (1941) with *Melanconis*.
23′(21′) Acervuli foliicolous, simple; conidiophores hyaline, elongated, sparingly branched and septate, annellations thickened, pale brown; conidia pale brown, biguttulate, base truncate, prominently thickened . . . . . . . . . . . . . . . . . . . . **Enerthidium**
23″(21′) Acervuli ramicolous, simple; paraphyses branched, septate, mucilaginous; coni-

## 11. Coelomycetes

diophores very long, branched, septate, annellations flared; conidia bluish, elliptical, thick-walled, base truncate, apex papillate . . . . . . . . . . . . . **Lamproconium**

23'''(21') Acervuli foliicolous, simple; conidiophores hyaline, elongated, septate and branched at the base, becoming dark brown towards the apex, with prominent annellations; conidia brown, guttulate, fusiform, base truncate . . . . . . . . . . **Cryptomela**
    *C. typhae* on *Typha*.

24(21'') Acervuli foliicolous; conidiophores shortly cylindrical; conidia curved, markedly so toward the apices, lunate . . . . . . . . . . . . . . . . . . . **Gloeosporidiella**
    Correlated by Rimpau (1962) with *Drepanopeziza*. Von Arx (1957a) accepts 5 spp.

24'(21'') Acervuli ramicolous; conidiophores elongated, branched, septate; conidia long, falcate . . . . . . . . . . . . . . . . . . . . . . . . . . . . . . **Disculina**
    *D. neesii* on *Alnus*.

25(1) Conidiogenous cells monophialidic . . . . . . . . . . . . . . . . . . . . . . 27
25'(1) Conidiogenous cells consistently polyphialidic . . . . . . . . . . . . . . . . 26

26(25') Acervuli foliicolous or caulicolous; conidiophores becoming brown with age; conidiogenous cells polyphialidic; conidia minute, oval . . . . . . . . **Sphaceloma**
    Conidial states of *Elsinoë*, causing leaf and stem scab or anthracnose of many hosts.

26'(25') Acervuli and phialides hyaline; conidia hyaline, aseptate, oval to pyriform, formed more or less synchronously from 2-8 apertures . . . . . . . . . **Kabatiella**
    Von Arx (1957a) recognizes 15 spp., including *K. lini* causing stem break and browning of flax.

27(25) Acervuli with a radiate hypomycelium; phialides cylindrical; conidia hyaline, aseptate, cylindrical to curved, tapered towards the base . . . . . . . . . . . . **Actinonemella**
27'(25) Acervuli separate, lacking hypomycelia, but ocasionally with setae . . . . . . . 28
28(27') Conidia aseptate . . . . . . . . . . . . . . . . . . . . . . . . . . . . . . 29

28'(27') Conidia medianly 1-septate, hyaline, oval; conidiophores elongated, branched and septate, individual cells functioning conidiogenously producing terminal conidia on small lateral branches . . . . . . . . . . . . . . . . . . . . . . **Discella**
    Synonyms accepted by von Arx (1963) and Boerema (1970) include *Cytodiplospora*, *Septomyxa*, and *Diplodina*. *Discella salicis* on *Salix* and *D. acerina* on *Acer*.

29(28) Conidia hyaline . . . . . . . . . . . . . . . . . . . . . . . . . . . . . . . 31
29'(28) Conidia pigmented . . . . . . . . . . . . . . . . . . . . . . . . . . . . . 30

30(29') Acervuli epidermal; conidiophores elongated, branched and septate at the base; conidia pale brown, rhomboid . . . . . . . . . . . . . . . . . . . **Vanderystiella**
    Syn. *Deightonia*.

30'(29') Acervuli subepidermal with supporting immersed stroma; phialides minute, hyaline, ampulliform; conidia dark brown, thick-walled, finely puncticulate, oval to pyriform . . . . . . . . . . . . . . . . . . . . . . . . . . . . . **Fairmaniella**

31(29) Acervuli ramicolous; conidiophores elongated, branched, septate, individual cells functioning conidiogenously producing terminal conidia on short lateral branches; conidia clavate to oblong . . . . . . . . . . . . . . . . . . . . . **Cytogloeum**

31'(29) Acervuli ramicolous; phialides often proliferating percurrently with flared collarettes formed on branched, septate conidiophores; conidia guttulate, oblong, tapered towards the base, ends obtuse . . . . . . . . . . . . . . . . . . . . . **Discosporium**
    *D. deplanatum* on *Carpinus*.

31″(29) Conidiophores neither septate nor branched, or only sparingly, and phialides not
proliferating percurrently . . . . . . . . . . . . . . . . . . . . . . . . . . . 32
    32(31″) Phialide channel relatively wide in relation to the width of the conidiogenous
    cell apex; conidia flattened at the base . . . . . . . . . . . . . . . . . . . . . 33
    32′(31″) Phialide channel relatively narrow in relation to the width of the conidiogen-
    ous cell apex; conidia obtuse at the base . . . . . . . . . . . . . . . . . . 37
33(32) Acervuli setose . . . . . . . . . . . . . . . . . . . . . . . . . . . . . . . . 34
33′(32) Setae absent . . . . . . . . . . . . . . . . . . . . . . . . . . . . . . . . . 35
    34(33) Acervuli subcuticular; conidia falcate with the apex prolonged into a simple
    cellular appendage . . . . . . . . . . . . . . . . . . . . . . . . . . . . **Ellisiella**
    34′(33) Acervuli subcuticular, epidermal or subepidermal; conidia cylindrical or falcate,
    lacking an apical appendage . . . . . . . . . . . . . . . . . . . . **Colletotrichum**
        von Arx (1957b) accepts 11 spp., including the ubiquitous plant pathogens *C.
        gloeosporioides, C. coccodes, C. graminicola,* and *C. dematium* causing anthracnoses.
35(33′) Acervuli subepidermal; conidia cylindrical or slightly curved, the apex prolonged
into a short cellular appendage . . . . . . . . . . . . . . . . . . . **Monochaetiella**
35′(33′) Conidia lacking an apical appendage . . . . . . . . . . . . . . . . . . . . 36
    36(35′) Phialides ampulliform to doliiform; conidia relatively large, broadly cylindrical
    with a protruding truncate base . . . . . . . . . . . . . . . . . . **Cryptosporiopsis**
        *C. malicorticis,* perennial canker of apple.
    36′(35′) Conidiophores elongated, aseptate and unbranched except at the base; conidia
    broadly navicular to elliptical, narrowed towards the apex and base
    . . . . . . . . . . . . . . . . . . . . . . . . . . . . . . . . . . . **Phacostroma**
37(32′) Acervuli foliicolous; conidiogenous cells tapered towards the apex; conidia
obovate . . . . . . . . . . . . . . . . . . . . . . . . . . . . . . . . **Erythrogloeum**
37′(32′) Acervuli foliicolous; conidiophores elongated, septate and branched at the base;
conidia broadly navicular, guttulate, flattened at the base . . . . . . . . **Myxosporina**
    *Rhodesia* is based on the same species.
37″(32′) Acervuli foliicolous, subcuticular; conidiophores irregularly branched to septate
with conidia formed from lateral branches; conidia cylindrical or slightly curved,
tapered towards the base . . . . . . . . . . . . . . . . . . . . . . . **Titaeosporina**
    *T. tremulae* on *Populus.*

## KEY TO SPHAEROPSIDALES

    This order consists of microscopic parasites or saprobes of plant material. Fructifications are superficial, semi-immersed or immersed, globose, discoid or hemispherical, eustromatic or pseudostromatic, unilocular, multilocular or convoluted, enclosing walls entirely formed of fungal pseudoparenchyma, forming conidia from the locular walls; dehiscing by a circular or longitudinal ostiole or by disintegration of the upper walls. The Sphaeropsidales is traditionally separated into the Sphaeropsidaceae, Discellaceae, Nectrioidaceae, and Leptostromataceae, but is undivided in the following treatment.

1. Pycnidia separate, spheroid, ampulliform or ovate, unilocular, ostiolate, pigmented or
hyaline; walls simple or several cells thick, outer layer forming a distinct specialized
wall . . . . . . . . . . . . . . . . . . . . . . . . . . . . . . . . . . . . . . . 2

## 11. Coelomycetes

1′. Pycnidia aggregated, spheroid, ampulliform or ovate, unilocular, ostiolate, pigmented or hyaline; walls simple or several cells thick, outer layer forming a distinct specialized wall; pseudoparenchymatic connecting tissue absent, plectenchyma sometimes present ................................................... 51

1″. Fructifications separate, rarely immersed, hyaline or pigmented, discoid, sometimes initially closed but opening to expose the conidiogenous layer ........... 58

1‴. Fructifications eustromatic or pseudostromatic, cavity simple convoluted or multilocular, uni- or multi-ostiolate; walls several cells thick, of unspecialized pseudoparenchyma, often with no definite outer wall layer ................... 71

    2(1) Conidiogenous cells holoblastic ........................ 3

    2′(1) Conidiogenous cells enteroblastic, phialidic ................ 24

    2″(1) Conidiogenous cells holothallic ...................... 50

    2‴(1) Conidiogenous cells enterothallic; conidia bacillar, hyaline, aseptate ................................................. **Asteromella**
        Spermatial states of Ascomycetes.

3(2) Conidiogenous cells annellidic ........................ 8

3′(2) Conidiogenous cells sympodial or polyblastic ................ 4

3″(2) Conidiogenous cells forming solitary conidia ................ 14

    4(3′) Pycnidia immersed; conidiogenous cells with 1–3 flat scars; conidia hyaline, thick-walled, medianly 1-septate, each end with a mucilaginous convoluted cap ................................................... **Tiarospora**

    4′(3′) Conidia lacking a mucilaginous cap .................. 5

5(4′) Conidia aseptate ............................... 6

5′(4′) Conidia euseptate ............................. 7

    6(5) Pycnidia thin-walled, immersed; conidiogenous cells cylindrical, unbranched, aseptate, 1–3 flat scars; conidia hyaline, oval, sometimes guttulate ................................................... **Alveophoma**

    6′(5) Pycnidia immersed, walls thick, of loose parenchyma; conidiogenous cells filiform, on branched, septate conidiophores, several nodular apical scars; conidia hyaline, filiform, frequently remaining attached to the conidiogenous cell ..... **Eriospora**

7(5′) Pycnidia semi-immersed, rostrate; conidiogenous cells elongated on branched, septate conidiophores, scars 1–4, prominent, protruding, flat; conidia navicular, hyaline, up to 6-septate ............................................ **Corniculariella**

7′(5′) Pycnidia immersed, sometimes papillate; conidiogenous cells doliiform, ampulliform or elongated, occasionally proliferating; conidia hyaline, filiform, multiseptate ................................................... **Septoria**
    Destructive leaf and stem pathogens of phanerogams; *S. apiicola* on *Apium*, *S. nodorum* on Gramineae, *S. lycopersici* on *Lycopersicon*.

    8(3) Conidia with a prominent basal frill or appendages ............ 9

    8′(3) Prominent basal frill and appendages absent ................ 10

9(8) Pycnidia with a dark brown papillate prominent ostiole; conidiogenous cells with several flared annellations; conidia hyaline, aseptate, guttulate with a basal frill or appendage derived from the annellide apex ................... **Idiocercus**

9′(8) Pycnidia thin-walled, immersed; annellides short, with indistinct annellations; conidia 4-septate, 3 median cells very pale brown, apical cell hyaline, devoid of cytoplasm, 1–3

apical, cellular, simple appendages, 1 endogenous, basal, cellular, simple
appendage . . . . . . . . . . . . . . . . . . . . . . . . . . . . . . . . . . . **Bartalinia**

   10(8′) Conidia aseptate . . . . . . . . . . . . . . . . . . . . . . . . . . . 11

   10′(8′) Conidia euseptate . . . . . . . . . . . . . . . . . . . . . . . . . . 12

11(10) Pycnidia superficial, thick-walled, dark brown; annellides pale brown, doliiform to ampulliform with 1–2 annellations; conidia spherical, dark brown, thick-walled, verrucate . . . . . . . . . . . . . . . . . . . . . . . . . . . . . . . . . **Coniothyrina**
      *C. agaves* on *Agave* spp.

11′(10) Pycnidia immersed; annellides hyaline, ampulliform to lageniform with a reduced apical cytoplasmic channel but distinct annellations; conidia deltoid, flattened at the apex, brown, thick-walled, smooth, base truncate . . . . . . . . . . . . . **Readeriella**
      *R. mirabilis* on *Eucalyptus* spp.

11″(10) Pycnidia semi-immersed to superficial; annellides cylindrical, hyaline with distinct annellations; conidia pyriform, dark brown, verrucate, base truncate . . . **Lichenoconium**
      *L. cladoniae* on *Cladonia*; several more spp. on lichens.

   12(10′) Conidia pigmented . . . . . . . . . . . . . . . . . . . . . . . . . . . 13

   12′(10′) Conidia hyaline, multiseptate, cylindrical, navicular, fusiform, truncate at the base; annellides doliiform, hyaline; pycnidia immersed, thin-walled
     . . . . . . . . . . . . . . . . . . . . . . . . . . . . . . . . . . . **Stagonospora**

13(12) Pycnidia immersed, dark brown; annellides doliiform, hyaline, 1–3 annellations; conidia 0–1-septate, cylindrical to oval, thick-walled, verrucate . . . . . **Coniothyrium**
      Biga, *et al.* (1959) accept 125 spp., Petrak and Sydow (1927) accept 89 spp. *C. palmarum* on Palmae; *C. fuckelii* with a wide host range.

13′(12) Pycnidia large, subperidermal, thick-walled, ostiole papillate; annellides ampulliform to cylindrical with 1–3 annellations; conidia obovoid, 2-septate, lower cell pale brown, upper cells darker, base truncate with a thickened scar, smooth-walled
   . . . . . . . . . . . . . . . . . . . . . . . . . . . . . . . . . . . **Neohendersonia**

13″(12) Pycnidia immersed, dark brown; annellides cylindrical with flared annellations; conidia muriform, irregular in shape, dark brown, thick-walled, smooth
   . . . . . . . . . . . . . . . . . . . . . . . . . . . . . . . . . . . **Camarosporium**

   14(3″) Conidiogenous cells similar to the inner cells of the pycnidial wall, ampulliform to doliiform, rarely shortly cylindrical . . . . . . . . . . . . . . . . . . . . 18

   14′(3″) Conidiogenous cells elongated, cylindrical, distinct from the inner cells of the pycnidial wall . . . . . . . . . . . . . . . . . . . . . . . . . . . . . . . 15

15(14′) Cellular appendages produced from the apical conidium cell . . . . . . . . . 16

15′(14′) Mucilaginous extracellular appendage(s) present at conidium apex . . . . . 17

   16(15) Pycnidia thin-walled, dark brown, becoming discoid, immersed; conidia cylindrical or curved, 3-septate, apical cell hyaline, devoid of cytoplasm, with 2–4 appendages formed from a common point, 3 lower cells very pale brown
     . . . . . . . . . . . . . . . . . . . . . . . . . . . . . . . . . . . **Hyalotiella**

   16′(15) Pycnidia formed from a basal stroma, dark brown; conidia cylindrical, 3-septate, 2 median cells pale brown, end cells lighter, with 2 separate dichotomously to irregularly branched appendages from the apical cell . . . . . . . . . . . . . **Hyalotiopsis**

17(15′) Pycnidia thick-walled, dark brown, immersed; conidia globose, subglobose to ovoid,

guttulate, aseptate, hyaline, with a single apical mucilaginous simple
appendage . . . . . . . . . . . . . . . . . . . . . . . . . . . . . . . . . . . . **Phyllostictina**
Boerema *et al.*, (1965b) place this genus in synonymy with *Phyllosticta*; *P. citricarpa*, black spot of citrus (Kiely, 1949).

17′(15′) Pycnidia thin-walled, clypeate; conidia clavate or curved, hyaline, aseptate, base truncate, developing in mucilage which splits from the base towards the apex, remaining as 1–4 apical everted appendages . . . . . . . . . . . . . . . . . . . . . . **Tiarosporella**

    18(14) Pycnidia thin-walled, superficial . . . . . . . . . . . . . . . . . . . . . 19
    18′(14) Pycnidia of various textures, immersed . . . . . . . . . . . . . . . . 20

19(18) Pycnidia setose; conidia muriform, pale brown, smooth-walled, base truncate
. . . . . . . . . . . . . . . . . . . . . . . . . . . . . . . . . . . . . . . . . . **Comocephalum**

19′(18) Pycnidia glabrous; conidia oblong, apex obtuse, base truncate, pale brown, smooth-walled, aseptate . . . . . . . . . . . . . . . . . . . . . . . . . . . . . . . **Cicinnobella**
Hyperparasites of superficial ascomycetes.

    20(18′) Conidia lacking cellular or extracellular appendages . . . . . . . . . . 22
    20′(18′) Conidia with apical mucilaginous extracellular appendage(s) . . . . . . 21
    20″(18′) Conidium apex drawn out into a filiform unbranched cellular appendage; conidia transversely and rarely longitudinally septate, brown, smooth-walled; pycnidia immersed, thin-walled . . . . . . . . . . . . . . . . . . . . . . . . . **Scolecosporiella**
    Sutton (1968a) accepts 3 spp.

21(20′) Pycnidia thick-walled, dark brown, immersed, almost clypeate; conidia 0–2-septate, hyaline, cylindrical, enclosed in a thin mucilaginous sheath which is drawn out into an apical appendage . . . . . . . . . . . . . . . . . . . . . . . . . . . **Kellermania**

21′(20) Pycnidia thick-walled, dark brown, immersed, ostiole lateral; conidia hyaline, aseptate, cylindrical to curved, developing in a mucilaginous sheath which splits from the base, remaining as 3–7 fine apical appendages . . . . . . . . . . . . . . . . . **Giulia**

    22(20) Pycnidia glabrous . . . . . . . . . . . . . . . . . . . . . . . . . . . . . . 23
    22′(20) Pycnidia setose, simple, immersed; conidia hyaline, more or less cylindrical, transversely and rarely longitudinally septate . . . . . . . . . . . . . **Aristastoma**
    Sutton (1964b) accepts 5 spp.; leaf pathogens of Leguminosae.

23(22) Pycnidia dark brown, thick-walled, immersed; conidia very large, pale brown, aseptate, smooth-walled, oval . . . . . . . . . . . . . . . . . . . **Botryosphaerostroma**

23′(22) Pycnidia brown, thick-walled, immersed; conidia obovoid, pale brown, smooth-walled, 1–2-septate . . . . . . . . . . . . . . . . . . . . . . . . . . . . . . . . **Ascochyta**

23″(22) Pycnidia of varying textures and pigment, immersed; conidia hyaline, filiform, multiseptate . . . . . . . . . . . . . . . . . . . . . . . . . . . . . . . . . . . **Septoria**
See also 7′.

    24(2′) Conidiogenous cells consistently polyphialidic . . . . . . . . . . . . . . 25
    24′(2′) Conidiogenous cells monophialidic . . . . . . . . . . . . . . . . . . . . 27

25(24) Conidia aseptate . . . . . . . . . . . . . . . . . . . . . . . . . . . . . . . . 26

25′(24) Conidia 1–3-septate, hyaline, falcate; conidiogenous cells cylindrical to obpyriform with 1–3 apical or lateral phialidic apertures; pycnidia superficial to semi-immersed, rostrate . . . . . . . . . . . . . . . . . . . . . . . . . . . . . . . . . **Sphaerographium**

    26(25) Pycnidia immersed, very pale brown to hyaline; conidiogenous cells doliiform,

hyaline, with 1–3 apical phialidic apertures; conidia hyaline, oval . . . **Sarcophoma**
*S. miribelii* on *Buxus*.

26′(25) Pycnidia immersed, medium to dark brown; conidiogenous cells cylindrical to obpyriform, hyaline, with 1–3 apical, subapical or lateral phialidic apertures; conidia hyaline, falcate . . . . . . . . . . . . . . . . . . . . . . **Ludwigiella**
Syn. *Selenophomopsis*.

27(24′) Phialides indistinguishable from the inner pycnidial wall cells . . . . . . . . 28

27′(24′) Phialides elongated but not associated with branched, septate conidiophores . . . . . . . . . . . . . . . . . . . . . . . . . . . . . . . . . . 37

27″(24′) Phialides associated with branched, septate conidiophores . . . . . . . . . 42

 28(27) Pycnidia setose . . . . . . . . . . . . . . . . . . . . . . . . . . . . 29

 28′(27) Pycnidia glabrous . . . . . . . . . . . . . . . . . . . . . . . . . . . . 31

29(28) Conidia pigmented . . . . . . . . . . . . . . . . . . . . . . . . . . . . . . 30

29′(28) Conidia hyaline, oblong, ends obtuse, aseptate, guttulate or eguttulate; pycnidia immersed, thin-walled . . . . . . . . . . . . . . . . . . . . . . . . **Pyrenochaeta**

 30(29) Conidia medianly 1-septate, pale brown, cylindrical, ends obtuse; phialides ampulliform, pale brown; pycnidia immersed, thin-walled . . . **Chaetodiplodia**

 30′(29) Conidia multiseptate, pale brown, falcate, apex almost corniform; phialides minute, ampulliform, hyaline; pycnidia immersed, thick-walled, rostrate to corniform . . . . . . . . . . . . . . . . . . . . . . . . . . . . . . . . . **Wojnowicia**
 *W. hirta* on Gramineae.

31(28′) Pycnidia globose to subglobose, papillate . . . . . . . . . . . . . . . . . . 32

31′(28′) Pycnidia clypeate, thin-walled, immersed; phialides ampulliform, hyaline; conidia medianly 1-septate, ellipsoid, pale brown . . . . . . . . . . . . . . . . **Ascochytulina**

31″(28′) Phialides pale brown, ampulliform; conidia oblong, apex obtuse, base truncate, dark brown, thick-walled, 3-septate, smooth-walled . . . . . . . . . . . **Ceratopycnis**

 32(31) Conidia pigmented . . . . . . . . . . . . . . . . . . . . . . . . . . . . 33

 32′(31) Conidia hyaline . . . . . . . . . . . . . . . . . . . . . . . . . . . . . 34

33(32) Pycnidia immersed or semi-immersed, thin-walled, brown; phialides ampulliform, hyaline; conidia aseptate, brown, smooth-walled, elliptical to oval
. . . . . . . . . . . . . . . . . . . . . . . . . . . . . . . . . **Microsphaeropsis**
*M. olivaceum*, wide host range.

33′(32) Pycnidia immersed, thin-walled, brown; phialides ampulliform, hyaline; conidia medianly 1-septate, pale brown, smooth-walled, oblong with obtuse ends
. . . . . . . . . . . . . . . . . . . . . . . . . . . . . . . . . . . **Ascochytella**

 34(32′) Pycnidia immersed . . . . . . . . . . . . . . . . . . . . . . . . . . . 35

 34′(32′) Pycnidia superficial . . . . . . . . . . . . . . . . . . . . . . . . . . 36

35(34) Pycnidia thin-walled, brown; phialides ampulliform, hyaline; conidia hyaline, aseptate, elliptical to oval . . . . . . . . . . . . . . . . . . . . . . . . . . . . . **Phoma**
 *P. herbarum* (Boerema, 1964); *P.* spp. in soil (Dorenbosch, 1970); *Peyronellaea* considered synonymous by Boerema *et al.* (1965a).

35′(34) Pycnidia thin-walled, hyaline; phialides ampulliform, hyaline; conidia hyaline, aseptate, globose, opaque . . . . . . . . . . . . . . . . . . . . . . . . . **Hapalosphaeria**

*H. deformans* causing sepal and anther hypertrophy of *Rubus* (Diedicke and Sydow, 1908).

36(34′) Pycnidia thin-walled, pale brown, irregular in shape; phialides ampulliform, hyaline; conidia hyaline, aseptate, elliptical to oval . . . . . . . . . **Cicinnobolus**
   *C. cesatii*, hyperparasite of Erysiphales.

36′(34′) Pycnidia formed through stomata, globose, furfuraceous, composed of large thin-walled cells; phialides brown, ampulliform; conidia hyaline, aseptate, oval, relatively large . . . . . . . . . . . . . . . . . . . . . . . . . **Rhizosphaera**
   Kobayashi (1967) recognizes 4 spp., associated with needle cast of Coniferae.

36″(34′) Pycnidia dark brown, thick-walled, sclerotioid; phialides ampulliform, hyaline; conidia hyaline, aseptate, cylindrical . . . . . . . . . . . **Plenodomus**
   *P. lingam*, dry rot of Cruciferae.

37(27′) Pycnidia rostrate, setose or both . . . . . . . . . . . . . . . . . . . . 38

37′(27′) Pycnidia simple, without setae . . . . . . . . . . . . . . . . . . . . 39

38(37) Pycnidia rostrate, without setae; phialides shortly cylindrical, hyaline; conidia hyaline, cylindrical to curved, medianly 1-septate . . . . . . . . . **Chaetopyrena**

38′(37) Pycnidia rostrate, setose; phialides shortly cylindrical, hyaline, conidia hyaline to olivaceous, cylindrical or slightly curved, medianly 1-septate
   . . . . . . . . . . . . . . . . . . . . . . . . . . . . . . **Chaetosphaeronema**

38″(37) Pycnidia immersed, not rostrate, thin-walled, setose; phialides shortly cylindrical, hyaline; conidia hyaline, oblong, ends obtuse, aseptate, guttulate or eguttulate . . . . . . . . . . . . . . . . . . . . . . . . . . . . **Pyrenochaeta**
   See also 29′.

38‴(37) Pycnidia semi-immersed to superficial, thin-walled, setose, indehiscent, collapsing; phialides long cylindrical, hyaline; paraphyses branched, septate, formed from lateral walls; conidia olivaceous, fusiform, aseptate . . . . . . . . . **Amerosporium**

39(37′) Conidia brown, aseptate, smooth, spherical to elliptical or almost falcate, sometimes with an evanescent laterally appressed, mucilaginous, extracellular appendage; phialides hyaline, cylindrical, restricted to the basal cushion of pseudoparenchyma . . . . **Coniella**
   Sutton (1969) accepts 5 spp., including *C. diplodiella*, causing white rot of *Vitis*.

39′(37′) Conidia hyaline . . . . . . . . . . . . . . . . . . . . . . . . . . 40

40(39′) Conidia without a mucilaginous appendage . . . . . . . . . . . . . . 41

40′(39′) Conidia aseptate, guttulate, fusiform, formed in a mucilaginous sheath which splits at the apex and becomes everted as a basal hollow cone . . . **Neottiospora**

41(40) Pycnidia immersed or semi-immersed, dark brown; phialides shortly cylindrical, hyaline; conidia subglobose, hyaline, aseptate . . . . . . . . . . . . . **Pleurophomopsis**

41′(40) Pycnidia immersed, hyaline, opening widely; phialides long, cylindrical, tapering to the apex and with a prominent flared collarette; conidia filiform, sigmoid, hyaline, aseptate, swollen and thickened at the base . . . . . . . . . . . . . . . . . . **Protostegia**

42(27″) Pycnidia superficial, setose, opening by a longitudinal raphe; conidiophores branched, septate, hyaline; conidia formed from apical and lateral phialides, hyaline, aseptate, cylindrical to fusiform . . . . . . . . . . . . . . . . . **Chaetomella**
   Stolk (1963) accepts 3 spp., common soil organisms.

42′(27″) Pycnidia opening by a circular ostiole . . . . . . . . . . . . . . . . 43

43(42′) Conidiophores branched, septate, producing phialospores at the apices of main and small lateral branches immediately below transverse septa . . . . . . . . . . . . 44

43′(42′) Conidiophores branched, septate, producing phialospores at the apices of main and long lateral branches . . . . . . . . . . . . . . . . . . . . . . . . . . . . . 46

    44(43) Pycnidia thin-walled, brown, simple, immersed . . . . . . . . . . . . 45

    44′(43) Pycnidia yellow-orange, superficial, opaque, rostrate; conidia hyaline, aseptate, elliptical, the apex and base drawn out into simple, filiform cellular appendages from which the contents have retracted into the conidium body . . . . . . . . . . . . . . . . . . . . . . . . . . . . . . . . . . . . . . . **Eleutheromyces**
        E. *subulatus*, hyperparasitic on Homobasidiomycetes.

    44″(43) Pycnidia indehiscent, consisting of a sclerotioid upper and lateral wall and a thin-walled basal region from which vertically aggregated conidiophores arise; conidia hyaline, aseptate, ellipsoid . . . . . . . . . . . . . . . . . . . . . . **Stilbophoma**

45(44) Pycnidia setose; conidia hyaline, aseptate, ellipsoid . . . . . . . . . **Pyrenochaeta**
    See also 29′.

45′(44) Pycnidia glabrous; conidia hyaline, aseptate, ellipsoid . . . . . **Phialophorophoma**
    P. *litoralis*, a marine organism.

    46(43′) Pycnidia entirely superficial, thick-walled, dark brown; conidia filiform, hyaline, aseptate . . . . . . . . . . . . . . . . . . . . . . . . . . . . . . **Cryptogenella**

    46′(43′) Pycnidia immersed . . . . . . . . . . . . . . . . . . . . . . . . . 47

47(46′) Pycnidia clypeate and rostrate, dark-brown, immersed; conidia hyaline, aseptate, ellipsoid, minute . . . . . . . . . . . . . . . . . . . . . . . . . . . . . **Cytonaema**

47′(46′) Pycnidia neither rostrate nor clypeate . . . . . . . . . . . . . . . . 48

    48(47′) Conidia septate, hyaline, apical cell drawn out into a simple cellular appendage; pycnidia thick-walled, occasionally divided . . . . . . . . . . . . . . **Chaetoconis**

    48′(47′) Conidia aseptate . . . . . . . . . . . . . . . . . . . . . . . . . . 49

49(48′) Pycnidia thin-walled, pale brown, immersed; conidia hyaline, falcate . . . . . . . . . . . . . . . . . . . . . . . . . . . . . . . . . . . . . . . . **Phlyctaena**
    P. *vagabunda*, saprobe and parasite of leaves and fruit.

49′(48′) Pycnidia thick-walled, dark-brown, immersed; conidiophores frequently collapsing (possibly paraphyses); conidia hyaline, cylindrical, apex obtuse, guttulate . . . . . . . . . . . . . . . . . . . . . . . . . . . . . . . . . . . . . . . **Coleophoma**

    50(2″) Pycnidia simple, immersed, brown; conidial chains unbranched, arthric; conidia olivaceous, cylindrical, both ends truncate, aseptate, biguttulate . . . **Vouauxiella**
    Hyperparasites of lichens.

    50′(2″) Pycnidia simple, immersed, pale brown; conidial chains branched irregularly, arthric; conidia hyaline, 0–1-septate, irregular in shape . . . . . . . . . **Neozythia**

    50″(2″) Pycnidia simple, immersed, brown, opening widely; conidial chains unbranched, arthric; conidia cylindrical to doliiform, aseptate, hyaline, both ends truncate . . . . . . . . . . . . . . . . . . . . . . . . . . . . . . . . . . . . . **Desmopatella**

51(1′) Conidiogenous cells holoblastic . . . . . . . . . . . . . . . . . . . . 52

51′(1′) Conidiogenous cells enteroblastic, phialidic . . . . . . . . . . . . . . . 55

    52(51) Conidiogenous cells annellidic . . . . . . . . . . . . . . . . . . . 53

    52′(51) Conidiogenous cells forming solitary conidia . . . . . . . . . . . . 54

53(52) Pycnidia aggregated in botryose clusters, thick-walled, dark brown; annellides doliiform

to shortly cylindrical, hyaline; conidia ellipsoid, 1-septate, dark brown, with longitudinal striations, lacking appendages . . . . . . . . . . . . . . . . . . . . **Botryodiplodia**
   *B. theobromae*, a common tropical pathogen of phanerogams.

53'(52) Pycnidia thick-walled, superficial, loosely aggregated; conidia 4-septate, 3 median cells dark brown, end cells hyaline, lacking cytoplasm, apical cell flattened with a single simple appendage inserted laterally . . . . . . . . . . . . . . . . . . . **Bleptosporium**

   54(52) Pycnidia thick-walled, superficial, loosely aggregated; conidia 5-septate, dark brown except the apical cell which is hyaline, lacking cytoplasm and prolonged into a cellular branched corniform appendage . . . . . . . . . . . . . . . **Labridella**
   *L. cornu-cervae*, conidial *Griphosphaerioma kansensis*.

   54'(52) Pycnidia thick-walled, linearly aggregated; conidia cylindrical to curved, pale-brown, guttulate, multiseptate . . . . . . . . . . . . . . . . . . . . **Septoriella**

55(51') Conidia olivaceous, aseptate, oval to cylindrical; phialides ampulliform to shortly cylindrical; pycnidia violaceous, linearly aggregated, immersed . . . . . . . **Calocline**

55'(51') Conidia 1-septate . . . . . . . . . . . . . . . . . . . . . . . . . . . 56

55"(51') Conidia 2-septate . . . . . . . . . . . . . . . . . . . . . . . . . . . 57

   56(55') Pycnidia brown, immersed, irregularly aggregated; phialides ampulliform, hyaline; conidia hyaline, fusiform, medianly septate . . . . . . . . . . . . . **Clypeopycnis**

   56'(55') Pycnidia brown, immersed to semi-immersed, linearly aggregated; phialides doliiform, hyaline; conidia hyaline, oval, medianly septate, with an evanescent, mucilaginous, extracellular appendage at each end . . . . . . . . . . . . . . . . . . . . **Darluca**
   *D. filum*, hyperparasitic on Uredinales; conidial *Eudarluca caricis* (Eriksson, 1966).

57(55") Pycnidia dark brown, immersed, clypeate, linearly aggregated; phialides ampulliform to shortly cylindrical; conidia hyaline, irregular in shape, cylindrical, sigmoid, curved
. . . . . . . . . . . . . . . . . . . . . . . . . . . . . . . . . . . **Cytoplacosphaeria**

57'(55") Pycnidia dark-brown, immersed, irregularly aggregated; phialides shortly cylindrical, hyaline; conidia at maturity 2-septate, median cell dark-brown, end cells hyaline, oval
. . . . . . . . . . . . . . . . . . . . . . . . . . . . . . . . . . . **Hendersonula**
   *H. toruloidea*, wide host range.

   58(1") Conidiogenous cells holoblastic . . . . . . . . . . . . . . . . . . . . . 59
   58'(1") Conidiogenous cells enteroblastic, phialidic . . . . . . . . . . . . . . 62
   58"(1") Conidiogenous cells holothallic . . . . . . . . . . . . . . . . . . . . 70

59(58) Conidiogenous cells annellidic . . . . . . . . . . . . . . . . . . . . . . 60

59'(58) Conidiogenous cells sympodial or polyblastic . . . . . . . . . . . . . . . 61

59"(58) Conidiogenous cells forming solitary conidia, doliiform, ampulliform to irregular, hyaline; conidia pale-brown, septate, smooth, branched dichotomously or irregularly; fructification dark-brown . . . . . . . . . . . . . . . . . . . . . . . **Digitosporium**
   *D. pinophilum*, conidial *Crumenulopsis sororia* (Groves, 1969).

   60(59) Conidiophores long, branched, septate; conidiogenous cells 1–3 annellate; conidia 5-distoseptate, 4 median cells brown, end cells hyaline, lacking cytoplasm, 3–9 apical, cellular, simple or dichotomously branched appendages, and 1 basal, endogenous, cellular, simple or branched appendage; fructification dark-brown . . . . **Pestalotia**
   *Pestalotia* sensu stricto, see Sutton (1970).

   60'(59) Conidiophores long, branched, septate, hyaline; conidiogenous cells 1–4 annellate; conidia hyaline, aseptate, cylindrical, apex obtuse, base truncate; fructifications yellow . . . . . . . . . . . . . . . . . . . . . . . . . . . . . . **Siroscyphella**

61(59') Fructifications dark-brown; conidiophores hyaline, septate, branched; conidia hyaline, 2–3-septate, falcate, the apical cell drawn out into a long unbranched cellular appendage, the basal cell occasionally with a short, excentric, unbranched, cellular appendage
................................................................ **Heteropatella**

61'(59') Fructifications dark-brown, conidiogenous activity restricted to undifferentiated locules separated by sterile tissue; conidiogenous cells cylindrical, hyaline with up to 3 flat scars; conidia lunate, apex obtuse, base truncate, hyaline, multiseptate, exuded in distinct white conidial balls ............................................ **Bothrodiscus**

    *B. berenice*, conidial *Ascocalyx abietis* (Groves, 1968).

62(58') Fructifications dark-brown; conidiophores cylindrical, pale-brown, septate, with a long terminal phialide within which conidia are formed in basipetal succession; conidia cuboid, truncate at each end, guttulate, smooth, hyaline ......... **Bloxamia**

62'(58') Phialides producing conidia from an apical meristem enclosed at the most in a short collarette ................................................................ 63

63(62') Conidia with setulae ............................................ 64

63'(62') Conidia lacking setulae ............................................ 69

64(63) Conidia septate ................................................ 65

64'(63) Conidia aseptate ................................................ 68

65(64) Conidia pale brown ............................................ 66

65'(64) Conidia hyaline ................................................ 67

66(65) Fructifications dark-brown, setose, almost reduced to subcuticular acervuli; conidiogenous cells hyaline, aseptate, cylindrical; conidia cylindrical to obclavate, 0–2-septate, verruculose, with 1 apical setula and 1–4 basal lateral setulae
................................................................ **Polynema**

66'(65) Fructifications dark brown, setose; conidiophores branched, septate, hyaline; conidia catenate, ellipsoid to cylindrical, thickened at each end, medianly 1-septate, smooth, with 1 apical and 1 basal branched setula .......... **Hoehneliella**

67(65') Fructifications dark-brown, setose; conidiophores branched, septate, hyaline; conidia falcate or slightly curved, medianly 1-septate, with a short unbranched setula at each end ................................................................ **Pseudolachnea**

67'(65') Fructifications dark brown, setose, almost reduced to subcuticular acervuli; conidiophores branched, septate, hyaline; conidia cylindrical or curved, multiseptate, guttulate, with several branched or simple setulae at each end ............ **Chaetopatella**

68(64') Fructifications dark-brown, setose; conidiophores branched, septate, hyaline or pale brown, with an unbranched setula at each end ........ **Dinemasporium**

68'(64') Fructifications dark-brown setose; conidiophores branched, septate, hyaline; conidia falcate or slightly curved, hyaline with 1 apical, 1 basal and several lateral setulae ................................................................ **Stauronema**

69(63') Fructifications olivaceous; conidiophores branched, septate, hyaline; conidia fusiform to ellipsoid, biguttulate, olivaceous, aseptate ................. **Crocicreas**

69'(63') Fructifications yellow to orange; conidiophores elongated, branched, septate; conidia formed from apical phialides on main and short lateral branches immediately below septa; hyaline, ellipsoid to falcate, aseptate ........................ **Hainesia**

70(58") Fructifications dark-brown; conidiophores long cylindrical, hyaline, septate, branched, with chains of conidia; conidia arthric, cylindrical, truncate at each end, aseptate, olivaceous, guttulate ................................. **Trullula**

## 11. Coelomycetes

70′(58″) Fructifications dark-brown; conidiophores short, cylindrical, hyaline, sometimes septate and branched, with chains of conidia; conidia arthric, doliiform, cylindrical or irregularly beaked in the middle, multiseptate to muriform, dark brown, truncate at each end . . . . . . . . . . . . . . . . . . . . . . . . . . . . . **Phragmotrichum**
    Sutton and Pirozynski (1965) and Sutton and Sandhu (1969a) restrict the genus to 3 spp.

71(1‴) Conidiogenous cells holoblastic . . . . . . . . . . . . . . . . . . . . 72
71′(1‴) Conidiogenous cells enteroblastic, phialidic . . . . . . . . . . . . . . . 89
71″(1‴) Conidiogenous cells holothallic; conidial chains branched, arthric; conidia cylindrical, aseptate, hyaline, both ends truncate; eustroma semi-immersed, dark brown, thick-walled, unilocular, simple, ostiolate . . . . . . . . . . . . . . . . . . . . . . **Siropatella**

    72(71) Conidiogenous cells proliferating . . . . . . . . . . . . . . . . . . 73
    72′(71) Conidiogenous cells not proliferating . . . . . . . . . . . . . . . . 80

73(72) Conidiogenous cells sympodial or polyblastic . . . . . . . . . . . . . . . 74
73′(72) Conidiogenous cells annellidic . . . . . . . . . . . . . . . . . . . . 75

    74(73) Eustromata immersed, pale-brown, thin-walled, unilocular, simple, ostiolate; conidiogenous cells cylindrical, hyaline, with 1–3 apical scars; conidia hyaline, aseptate, triradiate, with a short, simple, extracellular setula at the end of each branch . . . . . . . . . . . . . . . . . . . . . . . . . . . . . . . . **Furcaspora**

    74′(73) Eustromata immersed, brown, thin-walled, convoluted, uni-multilocular, uniostiolate; conidiogenous cells ampulliform to elongated, hyaline, with 1–2 flat apical scars; conidia hyaline, filiform to cylindrical, multiseptate . . . . . . . . . . **Septocyta**

75(73′) Eustromatic . . . . . . . . . . . . . . . . . . . . . . . . . . . . 77
75′(73′) Pseudostromatic . . . . . . . . . . . . . . . . . . . . . . . . . . 76

    76(75′) Stroma concentrated in the epidermis; locule convoluted, lined by a multicellular pseudoparenchymatic wall, ostiole single; annellides cylindrical, hyaline with 1–3 often widely spaced annellations; conidia clavate to ellipsoid, aseptate, pale brown, eguttulate, the remains of the conidiogenous cells persisting as a basal appendage . . . . . . . . . . . . . . . . . . . . . . . . . . . . . . . . **Bellulicauda**

    76′(75′) Stroma concentrated in the epidermis; locule convoluted or divided, ostiole single; annellides cylindrical, hyaline, with 1–3 annellations; conidia lenticular, circular in outline, aseptate, brown, guttulate, remains of the conidiogenous cell persisting as a basal appendage . . . . . . . . . . . . . . . . . . . . . . . **Lasmeniella**

77(75) Stroma superficial . . . . . . . . . . . . . . . . . . . . . . . . . . 78
77′(75) Stroma immersed . . . . . . . . . . . . . . . . . . . . . . . . . . 79

    78(77) Stroma hysteriiform, dark brown, thick-walled, setose, unilocular, ostiole longitudinal; conidiophores hyaline, septate, branched; annellides often sympodial, 1–2 annellations; conidia hyaline, aseptate, broadly navicular, with 1–2 simple cellular apical appendages and 2–6 exogenous, simple, cellular, basal appendages . . . . . . . . . . . . . . . . . . . . . . . . . . . . . . . . . **Ajrekarella**

    78′(77) Stroma hemispherical, dark brown, thick-walled, multilocular with a central core of sterile tissue, multiostiolate; conidiophores branched, septate, hyaline, 1–2 annellations; conidia very pale brown, doliiform, apex obtuse, base truncate, aseptate . . . . . . . . . . . . . . . . . . . . . . . . . . . . . . . . . . . **Lasmenia**

79(77′) Stroma brown, thin-walled, unilocular, simple, ostiolate; conidiophores hyaline,

branched, septate, with 1–2 annellations; conidia subglobose to obovate, hyaline, aseptate, thick-walled . . . . . . . . . . . . . . . . . . . . . . . . . . . . . . . . . **Chondroplea**
   *C. populea*, conidial *Cryptodiaporthe populea*, causing cankers on *Populus* (Butin, 1958).

79′(77′) Stroma subepidermal, irregularly multilocular, dark-brown, breaking down in upper tissues to release conidia; annellides thick-walled, hyaline, cylindrical, 1–2 annellations with reduced cytoplasmic channel; conidia dark-brown, subglobose, collapsing readily, truncate at the base . . . . . . . . . . . . . . . . . . . . . . . . . . . . . . . . **Neomelanconium**

   80(72′) Eustromatic . . . . . . . . . . . . . . . . . . . . . . . . . . . . . . 81

   80′(72′) Pseudostromata concentrated in the epidermis, multilocular or convoluted, ostiolate; conidiogenous cells cylindrical, hyaline; conidia oval, flattened at the apex and base, dark-brown with a median hyaline band, aseptate . . . . . **Perizomella**

81(80) Stromata regularly multilocular . . . . . . . . . . . . . . . . . . . . . 82
81′(80) Stromata with a single, simple or convoluted locule . . . . . . . . . . . . 84

   82(81) Conidia aseptate . . . . . . . . . . . . . . . . . . . . . . . . . . . 83

   82′(81) Stroma semi-immersed or immersed, dark-brown, locules each with a separate ostiole, restricted to the upper regions of the stroma; conidiogenous cells cylindrical, hyaline; conidia dark-brown, muriform . . . . . . . . . . . . . . . **Dichomera**

83(82) Stroma irregularly campanulate, dark-brown, locules each with a separate ostiole, restricted to the upper regions of the stroma; conidiogenous cells cylindrical, occasionally on septate branched conidiophores; conidia globose to subglobose, thick-walled, hyaline, aseptate; microconidia phialidic . . . . . . . . . . . . . . . . . . . . . . **Cytosphaera**

83′(82) Stroma ampulliform, dark-brown, locules with a single common ostiole; conidiogenous cells cylindrical, hyaline; paraphyses present; conidia oval, dark-brown, thick-walled, ornamentation reticulate . . . . . . . . . . . . . . . . . . . . . . . . **Haplosporella**
   Petrak and Sydow (1927) accept 80 spp.

   84(81′) Stromata with a longitudinal line of dehiscence; conidia lacking appendages
   . . . . . . . . . . . . . . . . . . . . . . . . . . . . . . . . . . . . . . 85

   84′(81′) Stromata opening by a circular ostiole, wide aperture or by rupture of the host tissues; conidia with cellular or extracellular appendages . . . . . . . . . . . 86

85(84) Stromata separate, dark brown, hysteriiform, thick-walled; conidiogenous cells hyaline, cylindrical; conidia tetraradiate, pale-brown, each arm 4-septate . . . . **Tetranacrium**

85′(84) Stromata separate or aggregated, dark-brown, hysteriiform, thick-walled; conidiophores hyaline or pale brown, cylindrical, septate, branched at the base; conidia hyaline, obovate, aseptate, thick-walled, with granular cytoplasm . . . . . . . . . . **Psilospora**

   86(84′) Stroma superficial, columnar, dark brown, thin-walled, with an ill-defined wide ostiole; conidiogenous cells long, clavate, pale-brown; conidia reniform, dark-brown, 1-septate, basal cell twice as long as apical cell, at the base and apex with tufts of simple extracellular setulae . . . . . . . . . . . . . . . . . . . . **Stevensonula**

   86′(84′) Stroma immersed or semi-immersed, pale brown to hyaline . . . . . . . 87

87(86′) Stroma convoluted, rupturing host tissues; conidiophores short, cylindrical, hyaline, branched and septate at the base; paraphyses present; conidia developing in mucilage which persists as an apical cap, fusiform, 3-septate, end cells brown, median cells dark-brown, septa encircled by black bands . . . . . . . . . . . . . . . . . . . . **Toxosporiopsis**

87′(86′) Stroma simple, pale-brown to hyaline; conidia with cellular appendages . . . . 88

88(87′) Conidia aseptate, hyaline, cylindrical, with several, separate, simple, cellular, apical and basal appendages .................... **Chaetospermum**
88′(87′) Stromata separate or aggregated; ostiole furfuraceous; conidiogenous cells long lageniform, hyaline; conidia brown, globose, subglobose, lenticular, gibbose, oblique-napiform, guttulate, with part of the conidiogenous cell persisting as a basal cellular appendage, some species with an apical short or filiform cellular appendage
................................................................... **Harknessia**
   Syn. *Mastigonetron*. Sutton (1971b) accepts 16 spp.
88″(87′) Stromata pale-brown, separate; conidiogenous cells cylindrical, hyaline; conidia clavate, 1-septate, brown, guttulate, smooth, with 1 branched, cellular hyaline apical appendage lacking cytoplasm ................. **Mycohypallage**
89(71′) Conidiogenous cells consistently polyphialidic ................ 90
89′(71′) Conidiogenous cells monophialidic ..................... 91
   90(89) Pseudostroma flattened, effuse, epidermal to subepidermal, dark-brown, opening by irregular slits; phialides cylindrical to obclavate with 1–2 apertures; conidia hyaline, aseptate, narrowly fusiform, the apex prolonged into a filiform simple cellular appendage ................................................ **Diachorella**
      Ciccarone (1963) accepts 3 formae of the type species, *D. onobrychidis*, on Leguminosae.
   90′(89) Eustroma flattened, subepidermal, brown, opening by irregular fissures; phialides doliiform, broadly oval to irregular, with 1–3 apertures at the apex; conidia triradiate, hyaline, aseptate ............................. **Suttoniella**
91(89′) Phialides associated with branched septate conidiophores ............ 92
91′(89′) Phialides indistinguishable from the inner cells of the stroma wall, or elongated but not associated with conidiophores ..................... 109
   92(91) Eustromatic ...................................... 93
   92′(91) Pseudostromatic, black, immersed, cylindrical between epidermises, with several separate locules opening by separate or individual furfuraceous ostioles; conidiophores hyaline, branched, septate; conidia hyaline, aseptate, cylindrical with an apical evanescent mucilaginous appendage .................... **Ceuthospora**
93(92) Stromata superficial or semi-immersed ................... 94
93′(92) Stromata immersed, epidermal, subepidermal, subperidermal, not becoming erumpent as separate pycnidiumlike fructifications ................. 98
   94(93) Stroma bright-colored (nectrioidaceous), with separate irregularly arranged locules each opening separately; conidia hyaline, fusiform, guttulate ..... **Aschersonia**
      Parasitic on scale insects; monographed by Petch (1921).
   94′(93) Stroma brown to black ............................ 95
95(94′) Locules regular, undivided, separated by vertically arranged pseudoparenchyma
   ........................................................... 96
95′(94′) Locules convoluted and irregular ..................... 97
   96(95) Stroma flattened, hemispherical; conidia cylindrical, hyaline, aseptate
      ............................................................. **Blennoria**
   96′(95) Stroma cuneiform in longitudinal section, ostiole central; conidia cylindrical, hyaline, aseptate ............................. **Chondrostroma**
97(95′) Stroma ampulliform, ostiole central; conidia cylindrical, hyaline, aseptate
   .............................................................. **Camaropycnis**

97'(95') Stroma consisting of a central column surmounted apically and laterally by pycnidium-like fructifications within which conidiogenous cells are formed from separate or convoluted locules; conidia medianly 1-septate, hyaline, guttulate, oval . . . . . . . . . **Fuckelia**

97"(95') Stroma consisting of a basal to central column surmounted by pycnidiumlike fructifications within which conidia are formed from single, multiple or occasionally convoluted locules; conidiophores forming conidia at the apices of main branches and small lateral branches immediately below transverse septa; conidia hyaline, aseptate, bacillar
. . . . . . . . . . . . . . . . . . . . . . . . . . . . . . . . . . . . . . . . . . **Pleurophomella**
>Conidial states of *Tympanis* (Groves, 1952).

>98(93') Conidiophores producing phialospores at the apices of main and small lateral branches immediately below transverse septa . . . . . . . . . . . . . . . . 99

>98'(93') Conidiophores producing phialospores at the apices of main and long lateral branches . . . . . . . . . . . . . . . . . . . . . . . . . . . . . . . . . . 100

99(98) Stroma brown, thick-walled, upper wall breaking down, cavity simple; conidia hyaline, ellipsoid, medianly 1-septate . . . . . . . . . . . . . . . . . . . . . . . **Hypocenia**

99'(98) Stroma dark brown, thick-walled, cavity convoluted; conidia hyaline, aseptate, allantoid . . . . . . . . . . . . . . . . . . . . . . . . . . . . . . **Pleurocytospora**

>100(98') Paraphyses present . . . . . . . . . . . . . . . . . . . . . . . . . 101
>100'(98') Paraphyses absent . . . . . . . . . . . . . . . . . . . . . . . . . 102

101(100) Stroma thick-walled, dark-brown, cavity convoluted, ostiole single; conidia aseptate, smooth, brown, cylindrical, elliptical to oval . . . . . . . . . . . **Phaeocytostroma**
>*P. sacchari*, rind disease of sugar cane (Sutton, 1964b).

101'(100) Stroma thick-walled, brown, cavity convoluted, ostiole single; conidia pale brown, aseptate, fusiform . . . . . . . . . . . . . . . . . . . . . . . . . . **Subramanella**

>102(100') Conidia septate . . . . . . . . . . . . . . . . . . . . . . . . . . 103
>102'(100') Conidia aseptate . . . . . . . . . . . . . . . . . . . . . . . . . 104

103(102) Stroma simple, thin-walled, brown, cavity undivided, ostiole wide; conidia acerose to fusiform, hyaline, medianly 1-septate . . . . . . . . . . . . . . . . . **Sirococcus**

103'(102) Stroma thick-walled, brown, cavity convoluted or divided, ostiole wide; conidia falcate, hyaline, 3–6-septate . . . . . . . . . . . . . . . . . . . . . . **Brunchorstia**
>*B. destruens*, conidial *Scleroderris lagerbergii*, causing defoliation and dieback of *Pinus*.

103"(102) Stroma thick-walled, brown, cavity convoluted and divided, ostioles numerous, furfuraceous; conidia multiseptate, hyaline, filiform, often curved irregularly
. . . . . . . . . . . . . . . . . . . . . . . . . . . . . . . . . . . . . . . . . . **Micropera**
>Conidial states of *Dermea* (Groves, 1946).

>104(102') Conidia filiform, hamate, sigmoid or lunate . . . . . . . . . . . . . 105
>104'(102') Conidia cylindrical or ellipsoid . . . . . . . . . . . . . . . . . . 106

>104"(102') Conidia allantoid, hyaline, aseptate; stroma complex, convoluted and divided but with a single furfuraceous ostiole . . . . . . . . . . . . . . . . **Cytospora**
>Conidial states of *Valsa*; associated with dieback conditions.

105(104) Stroma thick-walled, brown, simple or convoluted, with a single ostiole; conidia hyaline, filiform, hamate, sigmoid or curved; α-conidia present . . . . . . **Phomopsis**
>Conidial states of *Diaporthe*; associated with leaf spots and dieback conditions (Wehmeyer, 1933).

105′(104) Stroma thick-walled, brown, convoluted and divided, with a single ostiole; conidia hyaline, lunate . . . . . . . . . . . . . . . . . . . . . . . . . . . . . . . . . . . . . . **Cytosporina**

    106(104′) Conidia eguttulate . . . . . . . . . . . . . . . . . . . . . . . . . . . . . . . . 107

    106′(104′) Conidia regularly guttulate . . . . . . . . . . . . . . . . . . . . . . . . . 108

107(106) Stroma thin-walled, brown, simple or convoluted, with a single ostiole; conidia hyaline, short cylindrical . . . . . . . . . . . . . . . . . . . . . . . . . . . . . . **Sporonema**

107′(106) Stroma thick-walled, brown, convoluted and divided, often with a clypeus, ostiole single; conidia oval, hyaline, thick-walled . . . . . . . . . . . . . . . . . **Phacostromella**

    108(106′) Stroma thick-walled, brown, simple or convoluted, with a single ostiole; conidia hyaline, often biguttulate, fusiform; β-conidia present . . . . **Phomopsis**
    See also 105.

    108′(106′) Stroma thick-walled, brown, simple or convoluted, with a single ostiole; conidia hyaline, irregularly guttulate, irregularly rhomboid . . . . . . . . **Phacidiopycnis**
    Conidial states of *Potebniamyces*.

109(91′) Stromata immersed in the substrate . . . . . . . . . . . . . . . . . . . . . . . 110

109′(91′) Stromata superficial, brown, thick-walled, hypoxyloid, individual peripheral locules with separate ostioles; phialides hyaline, ampulliform; conidia hyaline, 1-septate, the basal cell filiform, the apical cell oval . . . . . . . . . . . . . . . . . . . . . . . . . **Strasseriopsis**

    110(109) Stroma distinctly flattened, superficial subcuticular or epidermal, dark-brown . . . . . . . . . . . . . . . . . . . . . . . . . . . . . . . . . . . . . . . . . . . 119

    110′(109) Stroma subepidermal or superidermal, unflattened . . . . . . . . . . . 111

111(110′) Eustromatic, simple, mostly unilocular, occasionally convoluted or multilocular, brown . . . . . . . . . . . . . . . . . . . . . . . . . . . . . . . . . . . . . . . . . . . . . . . 113

111′(110′) Pseudostromatic . . . . . . . . . . . . . . . . . . . . . . . . . . . . . . . . . . . 112

    112(111′) Stroma dark-brown, doliiform with a single central locule surrounded by fungus-host tissue; phialides hyaline, elongated; conidia pale-brown, oval, the apex prolonged into a cellular simple or branched appendage which is separated by a septum . . . . . . . . . . . . . . . . . . . . . . . . . . . . . . . . . . . . . . . . . **Placonema**

    112′(111′) Stroma black, effuse, with several separate locules, each with an ostiole arranged linearly below the main stroma tissue; phialides hyaline, ampulliform; conidia cylindrical, hyaline, 1–3-septate with several apical and basal branched, cellular appendages . . . . . . . . . . . . . . . . . . . . . . . . . . . . . . . . . . . . . . . . . **Dilophospora**
    *D. alopecuri* causing twist of Gramineae.

113(111) Ostiole surrounded by a superficial disk-like cap of sterile tissue; conidiophores hyaline, aseptate, unbranched, very long; conidia oval, subhyaline, aseptate, verruculose . . . . . . . . . . . . . . . . . . . . . . . . . . . . . . . . . . . . . . . . **Rabenhorstia**
    *R. tiliae*, conidial state of *Hercospora tiliae*.

113′(111) Ostiole, if present, simple . . . . . . . . . . . . . . . . . . . . . . . . . . . . . 114

    114(113′) Conidia aseptate . . . . . . . . . . . . . . . . . . . . . . . . . . . . . . . . . 115

    114′(113′) Conidia septate . . . . . . . . . . . . . . . . . . . . . . . . . . . . . . . . . 117

115(114) Apical or basal appendages absent . . . . . . . . . . . . . . . . . . . . . . . 116

115′(114) Conidia allantoid, hyaline, aseptate, with a basal, cellular, unbranched appendage; phialides cylindrical, hyaline; stroma simple or convoluted, brown . . . . . **Strasseria**

116(115) Phialides ampulliform, hyaline; conidia oval to cylindrical, guttulate, hyaline; stroma brown, simple . . . . . . . . . . . . . . . . . . . . . . . **Sclerophoma**

S. *pythiophila*, conidial *Sydowia polyspora*, associated with leaf blight and dieback of Coniferae.

116′(115) Phialides cylindrical, hyaline; conidia hyaline, filiform to cylindrical; stroma dark brown, simple or convoluted, clypeate . . . . . . . . . . . **Apomelasmia**

117(114′) Conidia hyaline . . . . . . . . . . . . . . . . . . . . . . . . . . . . . 118

117′(114′) Conidia pale-brown, guttulate, smooth, short-cylindrical, truncate at the base, obtuse at the apex; phialides hyaline, cylindrical; stroma dark-brown, simple or convoluted . . . . . . . . . . . . . . . . . . . . . . . . . . . . . . . . . . . . **Cytoplea**

118(117) Conidia 1-septate, irregular in shape, cylindrical, curved, sigmoid, guttulate; phialides ampulliform to cylindrical, hyaline; stroma dark-brown, simple . . . . . . . . . . . . . . . . . . . . . . . . . . . . . . . . . . . . . . . . . **Diplozythiella**

118′(117) Conidia 0–1-septate, oval or slightly curved; phialides minute, ampulliform, hyaline; stroma dark-brown, simple . . . . . . . . . . . . . . . . **Placonemina**

119(110) Stroma epidermal with multicellular upper and lower walls, unilocular, opening by radiating fissures; conidia 3-septate, median cells pale-brown, end cells hyaline, lacking cytoplasm, apical cell with 2 cellular, unbranched appendages, basal cells with 1 cellular unbranched appendage . . . . . . . . . . . . . . . . . . . . . . . . . **Ciliochorella**

119′(110) Conidia lacking appendages, hyaline, aseptate . . . . . . . . . . . . . . . 120

120(119′) Stroma subcuticular, upper wall of radiating hyphae, dark-brown, basal wall hyaline, opening by a circular pore; conidiophores cylindrical, sparsely septate and branched; conidia falcate to reniform . . . . . . . . . . . . . **Colletotrichella**

120′(119′) Stroma subcuticular upper and lower walls dark brown, 1 cell thick, opening by a longitudinal fissure; phialides ampulliform, hyaline; conidia irregularly cylindrical with a minute scar at each end . . . . . . . . . . . . . . . . . **Pseudothyrium**

120″(119′) Stroma subcuticular, upper and lower walls dark brown, 1 cell thick, opening by a circular pore; phialides cylindrical, hyaline; conidia cylindrical, multiguttulate . . . . . . . . . . . . . . . . . . . . . . . . . . . . . . . . . . . . . . **Crandallia**

## REFERENCES

Agar, H. D., and H. C. Douglas. (1955). Studies of budding and cell wall structure of yeast. Electron microscopy of thin sections. *J. Bacteriol.* **70**:427–434.

Agnihothrudu, V. (1963). Notes on fungi from north-east India. XIV. A new genus of Discellaceae from Assam. *Sydowia* **16**:73–76.

Agnihothrudu, V. (1964). Notes on fungi from north-east India. XXIV. *Sadasivanella*—A new genus of Sphaeropsidales from Assam. *Proc. Indian Acad. Sci. Sect. B* **60**:81–85.

Ahmad, S., and S. A. Lohdi. (1953). Some new or unreported fungi from West Pakistan. *Sydowia* **7**:266–269.

Ali, M. M. (1962). Comparison of the physiology of three isolates of *Colletotrichum graminicola*. *Mycopathol. Mycol. Appl.* **17**:261–268.

Allescher, A. (1901). "Rabenhorst's Kryptogamen-Flora. I. Die Pilze. VI. Fungi imperfecti: Hyalin-sporige Sphaerioideen" E. Kummer, Leipzig.

Allescher, A. (1903). "Rabenhorst's, Kryptogamen-Flora. I. Die Pilze. VII. Fungi imperfecti: Gefärbt-sporige Sphaerioideen sowie Nectrioideen, Leptostromaceen, Excipulaceen und der Ordnung der Melanconieen." E. Kummer, Leipzig.

## 11. Coelomycetes

Archer, W. A. (1924). The morphological development of *Asterosporium hoffmanni*. *Mycologia* **16**:220–232.
Archer, W. A. (1926). Morphological characters of some Sphaeropsidales in culture. *Ann. Mycol.* **24**:1–84.
Ashby, S. F. (1927). *Macrophomina phaseoli* (Maubl.) comb. nov. the pycnidial stage of *Rhizoctonia bataticola* (Taub.) Butl. *Trans. Brit. Mycol. Soc.* **12**:141–147.
Baccarini, P. (1890). Sullo sviluppo dei picnidii. *G. Ital. Bot.* **22**:150.
Barron, G. L. (1968). "The Genera of Hyphomycetes from Soil." Williams & Wilkins, Baltimore, Maryland.
Batista, A. C., and R. Ciferri. (1959). Sistematica dos fungos imperfeitos de picnostromas com himenio invertido (Peltasterales). *Mycopathol. Mycol. Appl.* **11**:1–102.
Bauke, H. (1876). Beiträge zur Kenntnis der Pycniden. I. *Nova Acta Leopold. Carol. Deut. Akad. Naturforsch.* **38**:443.
Bayliss Elliott, J. S., and O. P. Stansfield. (1923). Records of Fungi Imperfecti. *Trans. Brit. Mycol. Soc.* **8**:249–254.
Bender, H. B. (1934). "The Fungi Imperfecti: Order Sphaeropsidales." Tuttle, Morehouse & Taylor Co., New Haven, Connecticut.
Berkeley, M. J. (1860). "Outlines of British Fungology." Lorell Reeve, London.
Berkeley, M. J., and C. E. Broome. (1854). XLIII. Notices of British fungi. *Ann. Mag. Natur. Hist.* **13**:458–469.
Biga, M. L. B., R. Ciferri, and G. Bestagno. (1959). Ordinamento artificiale delle specie del genere *Coniothyrium* Corda. *Sydowia* **12**:258–320.
Boerema, G. H. (1964). *Phoma herbarum* Westend., the type species of the form-genus *Phoma* Sacc. *Persoonia (Leyden)* **3**:9–16.
Boerema, G. H. (1967). The *Phoma* organisms causing gangrene of potatoes. *Neth. J. Plant. Pathol.* **73**:190–192.
Boerema, G. H. (1969). The use of the term forma specialis for *Phoma*-like fungi. *Trans. Brit. Mycol. Soc.* **52**:509–513.
Boerema, G. H. (1970). Nomenclatural remarks on three plant parasitic fungi. *Neth. J. Plant. Pathol.* **76**:165–170.
Boerema, G. H., and C. B. de Jong. (1968). Über die samenburtigen *Phoma*-Arten von *Valerianella*. *Phytopathol. Z.* **61**:362–371.
Boerema, G. H., and M. M. J. Dorenbosch. (1965). *Phoma*-achtige schimmels in associatie met appelbladvlekken. *Versl. Meded. Plziektenk. Dienst. Wageningen* **142**:138–154.
Boerema, G. H., and L. H. Höweler. (1967). *Phoma exigua* Desm. and its varieties. *Persoonia (Leyden)* **5**:15–28.
Boerema, G. H., and H. A. van Kesteren. (1964). The nomenclature of two fungi parasitizing *Brassica*. *Persoonia (Leyden)* **3**:17–28.
Boerema, G. H., M. M. J. Dorenbosch, and H. A. van Kesteren. (1965a). Remarks on species of *Phoma* referred to *Peyronellaea*. *Persoonia (Leyden)* **4**:47–68.
Boerema, G. H., M. M. J. Dorenbosch, and L. Leffring. (1965b). A comparative study of the black stem fungi on lucerne and red clover and the footrot fungus on pea. *Neth. J. Plant. Pathol.* **71**:79–89.
Bonorden, H. F. (1851). "Handbuch der Allgemeinen Mykologie." Stuttgart.
Brewer, D. (1960). Studies in *Ascochyta pisi* Lib. *Can. J. Bot.* **38**:705–717.
Brewer, J. G., and G. H. Boerema. (1965). Electron microscope observations on the development of pycnidiospores in *Phoma* and *Ascochyta*. *Proc., Kon. Ned. Akad. Wetensch.* **68**:86–97.
Brierley, W. B. (1915). The "endoconidia" of *Thielavia basicola*, Zopf. *Ann. Bot. (London)* **39**:483–493.

Brooks, F. T., and G. O. Searle. (1921). An investigation of some tomato diseases. *Trans. Brit. Mycol. Soc.* **7**:173–197.
Bubák, F. (1906). Zweiter Beitrag zur Pilzflora von Montenegro. *Bull. Herb. Boissier* [2] **6**: 473–488.
Buckley, P. M., T. D. Wyllie, and J. E. deVay. (1969). Fine structure of conidia and conidium formation in *Verticillium alboatrum* and *V. nigrescens*. *Mycologia* **61**:240–250.
Buddin, W., and E. M. Wakefield. (1926). On the life-history of a fungus parasite on *Antirrhinum majus*, with some remarks on the genus *Heterosphaeria*. *Trans. Brit. Mycol. Soc.* **11**: 169–188.
Burger, O. F. (1921). Variations in *Colletotrichum gloeosporioides* Penz. *J. Agr. Res.* **20**:723–736.
Butin, H. (1958). Über die auf *Salix* und *Populus* vorkommenden Arten der Gattung *Cryptodiaporthe* Petr. *Phytopathol. Z.* **32**:399–415.
Calvert, O. H., and G. S. Pound. (1949). Stimulated pycnidium production and symphogenous pycnidia in *Phoma lingam*. *Phytopathology* **39**:848–857.
Campbell, R. (1968). An electron microscope study of spore structure and development in *Alternaria brassicicola*. *J. Gen. Microbiol.* **54**:381–392.
Chao, R. L. C. (1969). Conidium morphology and ontogeny in species of *Leptomelanconium*, *Gloeocoryneum*, and *Hendersonia*. M.Sc. Thesis, University of Manitoba.
Chesters, C. G. C. (1938). Studies on British Pyrenomycetes. II. A comparative study of *Melanomma pulvis-pyrius* (Pers.) Fuckel, *Melanomma fuscidulum* Sacc. and *Thyridaria rubronotata* (B.&Br.) Sacc. *Trans. Brit. Mycol. Soc.* **22**:116–150.
Chippindale, H. G. (1929). The development in culture of *Ascochyta gossypii* Syd. *Trans. Brit. Mycol. Soc.* **14**:201–215.
Chodat, F. (1926). Recherches expérimentales sur la mutation chez les champignons. *Bull. Soc. Bot. Genève* [2] **18**:41–144.
Chupp, C. (1954). "A Monograph of the Fungus Genus *Cercospora*." Ithaca, New York.
Ciccarone, A. (1963). Osservationi sistematiche su *Diachorella onobrychidis* (D.C.) v. Höhnel. *Phytopathol. Mediter.* **2**:239–250.
Clements, F. E. (1909). "Genera of Fungi." H. W. Wilson Co., Connecticut.
Clements, F. E., and C. L. Shear. (1931). "The Genera of Fungi." Hafner, New York.
Cole, G. T., and W. B. Kendrick. (1968). Conidium ontogeny in hyphomycetes. The imperfect state of *Monascus ruber* and its meristem arthrospores. *Can. J. Bot.* **46**:987–992.
Cole, G. T., and W. B. Kendrick. (1969a). Conidium ontogeny in hyphomycetes. The annellophores of *Scopulariopsis brevicaulis*. *Can. J. Bot.* **47**:925–929.
Cole, G. T., and W. B, Kendrick. (1969b). Conidium ontogeny in hyphomycetes. The phialides of *Phialophora*, *Penicillium* and *Ceratocystis*. *Can. J. Bot.* **4**:779–789.
Coons, H. C. (1916). Factors involved in the growth and the pycnidium formation of *Plenodomus fuscomaculans*. *J. Agr. Res.* **5**:713–769.
Corda, A. C. I. (1839). "Icones fungorum hucusque cognitorum," 3. Prague.
Corda, A. C. I. (1842). "Icones fungorum hucusque cognitorum," 5. Prague.
Costantin, J. (1888). "Les Mucédinés simples. Matériaux pour l'histoire des Champignons," 2. Librairie Paul Klincksieck, Paris.
Cunnell, G. J. (1956). Some pycnidial fungi on *Carex*. *Trans. Brit. Mycol. Soc.* **39**:21–47.
Cunnell, G. J. (1957a). On *Neottiospora caricina* (Desm.) Höhnel. *Trans. Brit. Mycol. Soc.* **40**: 433–442.
Cunnell, G. J. (1957b). *Stagonospora* spp. on *Phragmites communis* Trin. *Trans. Brit. Mycol. Soc.* **40**:443–455.
Cunnell, G. J. (1958). On *Robillarda Phragmitis* sp. nov. *Trans. Brit. Mycol. Soc.* **41**:405–412.
de Bary, A. (1887). "Comparative Morphology and Biology of the Fungi, Mycetozoa and Bacteria." Oxford Univ. Press (Clarendon), London and New York.

Dennis, R. W. G. (1946). Notes on some British fungi ascribed to *Phoma* and related genera. *Trans. Brit. Mycol. Soc.* **29**:11–42.

Dickinson, C. H., and G. Morgan-Jones. (1966). The mycoflora associated with *Halimione portulacoides*. IV. Observations on some species of Sphaeropsidales. *Trans. Brit. Mycol. Soc.* **49**:43–55.

Diedicke, H. (1912–1915). "Kryptogamenflora der Mark Brandenburg. IX. Pilze VII. Sphaeropsideae, Melanconieae." Borntraeger Bros, Leipzig.

Diedicke, H., and H. Sydow. (1908). Uber *Paepalopsis deformans* Syd. *Ann. Mycol.* **6**:301–305.

Dodge, B. O. (1923). Origin of the central and ostiolar cavities in pycnidia of certain fungous parasites of fruits. *J. Agr. Res.* **23**:743–759.

Dodge, B. O. (1930). Development of the asexual fructifications of *Chaetomella raphigera* and *Pezizella lythri*. *Mycologia* **22**:169–174.

Dodge, B. O. (1931). A further study of the morphology and life history of the rose black spot fungus. *Mycologia* **23**: 446–462.

Dorenbosch, M. M. J. (1970). Key to nine ubiquitous soil-borne *Phoma*-like fungi. *Persoonia (Leyden)* **6**:1–14.

Dring, D. M. (1961). Studies on *Mycosphaerella brassicicola* (Duby) Oudem. *Trans. Brit. Mycol. Soc.* **44**:253–264.

Duke, M. M. (1928). The genera *Vermicularia* Fr. and *Colletotrichum* Cda. *Trans. Brit. Mycol. Soc.* **13**:156–184.

Durbin, R. D. (1966). Comparative gel-electrophoretic investigation of the protein patterns of *Septoria* species. *Nature (London)* **210**:1186–1187.

Durrell, L. W. (1968). Studies of *Aureobasidium pullulans* (deBary) Arnaud. *Mycopathol. Mycol. Appl.* **35**:113–120.

Ellis, M. B. (1960). Dematiaceous Hyphomycetes. I. *Mycol. Pap.* **76**:1–36.

Ellis, M. B. (1965). Dematiaceous Hyphomycetes. VI. *Mycol. Pap.* **103**:1–46.

Eriksson, O. (1966). On *Eudarluca caricis* (Fr.) O. Eriks., comb. nov., a cosmopolitan uredinicolous Pyrenomycete. *Bot. Notis.* **119**:33–69.

Eveleigh, D. E. (1961). *Phoma* spp. associated with painted surfaces. *Trans. Brit. Mycol. Soc.* **44**:573–585.

de Fonseka, R. N. (1960). The morphology of *Chaetospermum chaetosporum*. *Trans. Brit. Mycol. Soc.* **43**:631–636.

Fournet, J., P. Pauvert, and F. Rapilly. (1970). Propriétés des geleés sporifères de quelques Sphaeropsidales et Melanconiales. *Ann. Phytopathol.* **2**:31–41.

Fries, E. M. (1821). "Systema Mycologicum," 1. Gryphiswald, Lund, Sweden.

Fries, E. M. (1822). "Systema Mycologicum," 2. Gryphiswald, Lund, Sweden.

Fries, E. M. (1825). "Systema Orbis Vegetabilis." Lund, Sweden.

Frost, R. R. (1964). Seta formation in *Colletotrichum*. *Nature (London)* **201**:730–731.

Fuckel, L. (1869). "Symbolae Mycologicae." J. Nieder, Wiesbaden.

Goidanich, G., and G. Ruggieri. (1947). Le Deuterophomaceae di Petri. *Ann. Sper. Agr.* [N.S.] **1**:431–448.

Goos, R. D. (1956). Classification of the Fungi Imperfecti. *Proc. Iowa Acad. Sci.* **63**:311–320.

Grove, W. B. (1919). Mycological notes. IV. *J. Bot. (London)* **57**:206–210.

Grove, W. B. (1935). "British Stem- and Leaf-Fungi (Coelomycetes)," I. Cambridge Univ. Press, London and New York.

Grove, W. B. (1937). "British Stem- and Leaf-Fungi (Coelomycetes)," II. Cambridge Univ. Press, London and New York.

Groves, J. W. (1946). North American species of *Dermea*. *Mycologia* **38**:351–431.

Groves, J. W. (1952). The genus *Tympanis*. *Can. J. Bot.* **30**:571–651.

Groves, J. W. (1968). Two new species of *Ascocalyx*. *Can. J. Bot.* **46**:1273–1278.

Groves, J. W. (1969). *Crumenulopsis*, a new name to replace *Crumenula* Rehm. *Can. J. Bot.* **47**:47–51.
Guba, E. F. (1961). "Monograph of *Monochaetia* and *Pestalotia*." Harvard Univ. Press, Cambridge, Massachusetts.
Hahn, G. G. (1928). *Phomopsis conorum* (Sacc.) Died.—an old fungus of the douglas fir and other conifers. *Trans. Brit. Mycol. Soc.* **13**:278–286.
Harris, H. A. (1935). Morphologic studies of *Septoria lycopersici*. *Phytopathology* **25**:790–799.
Higgins, B. B. (1920). Morphology and life history of some ascomycetes with special reference to the presence and function of spermatia. *Amer. J. Bot.* **7**:435–444.
Higgins, B. B. (1929). Morphology and life history of some ascomycetes. II. *Amer. J. Bot.* **16**:287–296.
Higgins, B. B. (1936). Morphology and life history of some ascomycetes. III. *Amer. J. Bot.* **23**:598–602.
Hino, I., and K. Katumoto. (1961). "Icones Bambusicolorum Japonicorum." The Fuji Bamboo Garden.
Hoo, S.-C., and R. D. Wilcoxson. (1969). Effect of conidial number and matrix on germination of conidia in *Phoma medicaginis*. *Phytopathology* **59**:440–442.
Hughes, S. J. (1953). Conidiophores, conidia and classification. *Can. J. Bot.* **31**:577–659.
Ingold, C. T. (1942). Aquatic Hyphomycetes of decaying alder leaves. *Trans. Brit. Mycol. Soc.* **25**:339–417.
Ivory, M. H. (1967). A new variety of *Dothistroma pini* in Kenya. *Trans. Brit. Mycol. Soc.* **50**:289–297.
Jørstad, I. (1965). *Septoria* and septorioid fungi on dicotyledons in Norway. *Skr. Nor. Videnskaps-Akad. Oslo, 1: Math.–Naturv. Kl.* **22**:1–110.
Jørstad, I. (1967). *Septoria* and septorioid fungi on Gramineae in Norway. *Skr. Nor. Videnskaps-Akad. Oslo, 1: Math.–Naturv. Kl.* **24**:1–63.
Kempton, F. E. (1919). Origin and development of the pycnidium. *Bot. Gaz.* **68**:233–261.
Kendrick, W. B., ed. (1971). "Taxonomy of Fungi Imperfecti." Univ. of Toronto Press, Toronto.
Kendrick. W. B., and G. T. Cole. (1969). Conidium ontogeny in hyphomycetes. *Trichothecium roseum* and its meristem arthrospores. *Can. J. Bot.* **47**:345–350.
Kheswalla, K. F. (1941). Foot-rot of Gram (*Cicer arietum* L.) caused by *Operculella Padwickii* Nov. Gen. Nov. Spec. *Indian. J. Agr. Sci.* **11**:316–318.
Kiely, T. B. (1949). Preliminary studies on *Guignardia citricarpa*, n. sp.: The ascigerous stage of *Phoma citricarpa* McAlp. and its relation to black spot of Citrus. *Proc. Linn. Soc. N.S.W.* **73**:249–292.
Klebahn, H. (1905). Untersuchungen über einige Fungi Imperfecti und die zugehorigen Ascomyceten- Formen. *Jahrb. Wiss. Bot.* **41**:485–560.
Klebahn, H. (1933). Über Bau und Konidien-bildung bei einigen stromatischen Sphaeropsideen. *Phytopathol. Z.* **6**:229–304.
Kobayashi, T. (1967). Critical revision of the genera *Rhizosphaera* Mangin et Hariot and *Rhizophoma* Petrak et Sydow, a little-known fungous group associated with needle disease of conifers. *Bull. Gov. Forest. Exp. Sta.* **204**:91–112.
Kobayashi, T. (1968). Notes on Japanese species of the genus *Melanconium*. *Trans. Mycol. Soc. Jap.* **9**:1–11.
Lacoste, L. (1955). De la morphologie et de la physiologie de *Peyronellaea stipae* nov. sp. *C. R. Acad. Sci.* **241**:818–820.
Langeron, M., and R. Vanbreuseghem. (1952). "Précis de Mycologie." Masson, Paris.
Leach, C. M. (1962). Sporulation of diverse species of fungi under near-ultraviolet radiation. *Can. J. Bot.* **40**:151–161.

## 11. Coelomycetes

Lehman, S. G. (1918). Conidial formation in *Sphaeronema fimbriatum*. *Mycologia* **10**:155–163.
Leonian, L. (1924). A study of factors promoting pycnidium formation in some Sphaeropsidales. *Amer. J. Bot.* **11**:19–50.
Léveillé, J. H. (1845). Champignons exotiques. *Ann. Sci. Natur. Bot.* [3] **3**:38–70.
Léveillé, J. H. (1846). Considérations mycologiques suives d'une nouvelle classification des champignons. Dict. Univ. d'Hist. natur., Paris.
Libert, A. (1837). Précis des observations sur la famille des Hypoxylons. *Ann. Sci. Natur. Bot.* [2] **7**:121–125.
Linder, D. H. (1944). I. Classification of the marine fungi. *Farlowia* **1**:401–433.
Link, H. F. (1809). Observationes in Ordines plantarum naturales. Dissertatio. I. Complectens Anandrarum ordines Epiphytas, Mucedines, Gastromycos et Fungos. *Mag. Ges. Natur. Freunde, Berlin* **3**:3–42.
Link, H. F. (1833). "Handbuch zur Erkennung der nutzbarsten und am häufigsten vorkommenden Gewächse," III. Berlin.
Luedemann, G. M. (1959). The dictyochlamydospores of *Peyronellaea glomerata* (Corda) Goidanich ex Togliani contrasted with the dictyoporospore of *Alternaria tenuis* Auct. *Mycologia* **51**:772–780.
Luttrell, E. S. (1940). An undescribed fungus on japanese cherry. *Mycologia* **32**:530–536.
Luttrell, E. S. (1963). Taxonomic criteria in *Helminthosporium*. *Mycologia* **55**:643–674.
MacNeill, B. H. (1950). Studies in *Septoria lycopersici* Speg. *Can. J. Res., Sect. C.* **28**:645–672.
Madelin, M. F. (1966). The genesis of spores in higher fungi. *In* "The Fungus Spore" (M.F. Madelin, ed.), p. 15–36.
Martius, C. F. P. (1817). "Flora cryptogamia Erlangensis. Norimbergae" J. L. Schrag.
Mason, E. W. (1933). Annotated account of fungi received at the Imperial Mycological Institute, List II (Fasc. 2). *Mycol. Pap.* **3**:1–67.
Mason, E. W. (1937). Annotated account of the fungi received at the Imperial Mycological Institute, List II (Fasc. 3, General part). *Mycol. Pap.* **4**:69–99.
Mercer, W. B. (1913). On the morphology and development of *Phoma Richardsiae* n.sp. *Mycol. Zentralbl.* **2**:244–253.
Moreau, F. (1953), "Les Champignons," Vol. 2, pp. 941–2120. Masson, Paris.
Nattrass, R. M. (1933). A new species of *Hendersonula* (*H. toruloidea*) on deciduous trees in Egypt. *Trans. Brit. Mycol. Soc.* **18**:189–198.
Nees von Esenbeck, C. G. (1817). "Das System der Pilze und Schwämme." Würzburg.
Parmelee, J. A. (1958). Some foliicolous fungi of the Pyrolaceae. *Can. J. Bot.* **36**:865–881.
Persoon, D. C. H. (1801). "Synopsis Methodica Fungorum." H. Dieterich, Göttingen.
Petch, T. (1921). Studies in entomogenous fungi. II. The genera *Hypocrella* and *Aschersonia*. *Ann. Roy. Bot. Gard., Peradeniya* **7**:167–278.
Petch, T. (1943). British Nectrioideae and allied genera. *Trans. Brit. Mycol. Soc.* **26**:53–70.
Petrak, F. (1924). Mykologische Notizen. VII. *Ann. Mycol.* **22**:1–182.
Petrak, F. (1925). Mykologische Notizen. Ueber v. Höhnel's neues System der Fungi Imperfecti. *Ann. Mycol.* **23**:1–11.
Petrak, F. (1953). *Colletogloeum* n. gen., eine neue Melanconieen-Gattung. *Sydowia* **7**:367–369.
Petrak, F. (1958). Mykologische Bemerkungen. *Sydowia* **11**:337–353.
Petrak, F., and H. Sydow, (1927). Die Gattungen der Pyrenomyzeten, Sphaeropsideen und Melanconieen. I. Tiel. Die phaeosporen Sphaeropsideen und die Gattung *Macrophoma*. *Rep. Spec. Nov. Regni. Veg. Beih.* **42**:1–551.
Pirozynski, K. A., and G. Morgan-Jones. (1968). Notes on microfungi. III. *Trans. Brit. Mycol. Soc.* **51**:185–206.
Pirozynski, K. A., and S. D. Patil, (1970). Some setose Hyphomycetes of leaf litter in south India. *Can. J. Bot.* **48**:567–581.

Pirozynski, K. A., and R. A. Shoemaker. (1971). Some Coelomycetes with appendaged conidia. *Can. J. Bot.* **49**:529–541.

Potebnia, A. (1910). Beiträge zur Micromycetenflora Mittel-Russlands. *Ann. Mycol.* **8**:42–93.

Punithalingam, E. (1966). Development of the pycnidium in *Septoria*. *Trans. Brit. Mycol. Soc.* **49**:19–25.

Punithalingam, E. (1969). New species of *Monochaetiella* and *Septoria*. *Trans. Brit. Mycol. Soc.* **53**:311–315.

Punithalingam, E. and D. Jones. (1970). Spore surface ornamentation in *Coniothyrium* species. *Trans. Brit. Mycol. Soc.* **55**:154–156.

Rimpau, R. H. (1962). Untersuchungen über die Gattung *Drepanopeziza* (Kleb.) v. Höhnel. *Phytopathol. Z.* **43**:257–306.

Saccardo, P. A. (1873). "Mycologiae Venetae Specimen." P. Prosperini. Padua.

Saccardo, P. A. (1880). Conspectus generum fungorum italiae inferiorum nempe ad Shaeropsideas, Melanconieas et Hyphomyceteas pertinentium, systemate sporologico dispositoru. "Michelia Commentarium Mycologicum," Vol. 2, pp. 1–38.

Saccardo, P. A. (1884). "Sylloge fungorum omnium hucusque cognitorum," Vol. 3. Pavia.

Sattar, A. (1934). A comparative study of the fungi associated with blight diseases of certain cultivated leguminous plants. *Trans. Brit. Mycol. Soc.* **18**:276–301.

Seeler, E. V. Jr. (1943). Several fungicolous fungi. *Farlowia* **1**:119–133.

Servazzi, O. (1953). *Pestalotia* o *Pestalotiopsis*? *G. Bot. Ital.* **60**:943–947.

Shoemaker, R. A. (1963). Generic correlations and concepts: *Griphosphaerioma* and *Labridella*. *Can. J. Bot.* **41**:1419–1423.

Shoemaker, R. A. (1964). *Seimatosporium* ( = *Cryptostictis*) parasites of *Rosa*, *Vitis*, and *Cornus*. *Can. J. Bot.* **42**:411–421.

Shoemaker, R. A. and E. Müller. (1964). Generic correlations and concepts: *Clathridium* (=*Griphosphaeria*) and *Seimatosporium* (=*Sporocadus*). *Can. J. Bot.* **42**:403–410.

Simmonds, J. H. (1965). A study of the species of *Colletotrichum* causing ripe fruit rots in Queensland. *Queensl. J. Agr. Sci.* **22**:437–459.

Smith, G. (1962). The morphological approach to the taxonomy of microfungi. *Sym. Soc. Gen. Microbiol.* **12**:111–118.

Sprague, R. (1950). "Diseases of Cereals and Grasses in North America (Fungi except Rusts and Smuts)." Ronald Press, New York.

Steyaert, R. L. (1949). Contribution a l'étude monographique de *Pestalotia* de Not. et *Monochaetia* Sacc. (*Truncatella* gen. nov. et *Pestalotiopsis* gen. nov.). *Bull. Jard. Bot. Brux.* **19**: 285–354.

Stolk, A. C. (1963). The genus *Chaetomella* Fuckel. *Trans. Brit. Mycol. Soc.* **46**:409–425.

Stowell, E. A., and M. P. Backus. (1966). Morphology and Cytology of *Diplocarpon maculatum* on *Crataegus* I. The *Entomosporium* stage. *Mycologia* **58**:949–959.

Streiblova, E., and K. Beran. (1965). On the question of vegetative reproduction in apiculate yeasts. *Folia Microbiol. (Prague)* **10**:352–356.

Subramanian, C. V. (1962). A classification of the Hyphomycetes. *Curr. Sci.* **31**:409–411.

Subramanian, C. V. and K. Ramakrishnan. (1956). *Ciliochorella* Sydow, *Plagionema* Subram. & Ramakr., and *Shanoria* gen. nov. *Trans. Brit. Mycol. Soc.* **29**:314–318.

Subramanian, C. V., and K. Ramakrishnan. (1957). *Neottiospora* Desm. and two new genera, *Samukuta* and *Sakireeta*. *J. Indian. Bot. Soc.* **36**:68–86.

Sutton, B. C. (1961). Coelomycetes. I. *Mycol. Pap.* **80**:1–16.

Sutton, B. C. (1962). *Colletotrichum dematium* (Pers. ex Fr.) Grove and *C. trichellum* (Fr. ex Fr.) Duke. *Trans. Brit. Mycol. Soc.* **45**: 222–232.

Sutton, B. C. (1963a). Coelomycetes. II. *Neobarclaya*, *Mycohypallage*, *Bleptosporium*, and *Cryptostictis*. *Mycol. Pap.* **88**:1–50.

Sutton, B. C. (1963b). Two new species of *Coryneum* Nees ex Fries. *Kew Bull.* **17**:309–314.

## 11. Coelomycetes

Sutton, B. C. (1964a). *Phoma* and related genera. *Trans. Brit. Mycol. Soc.* **47**:497–509.
Sutton, B. C. (1964b). Coelomycetes. III. *Annellolacinia* gen. nov., *Aristastoma, Phaeocytostroma, Seimatosporium* etc. *Mycol. Pap.* **97**:1–42.
Sutton, B. C. (1964c). *Melanconium* Link ex Fries. *Persoonia (Leyden)* **3**:193–198.
Sutton, B. C. (1965). Typification of *Dendrophoma* and a reassessment of *D. obscurans. Trans. Brit. Mycol. Soc.* **48**:611–616.
Sutton, B. C. (1966). Development of fructifications in *Colletotrichum graminicola* (Ces.) Wils. and related species. *Can. J. Bot.* **44**:887–897.
Sutton, B. C. (1967a). Two new genera of the Sphaeropsidales and their relationships with *Diachorella, Strasseria*, and *Plagiorhabdus. Can. J. Bot.* **45**:1249–1263.
Sutton, B. C. (1967b). Redescription of *Ajrekarella* Kamat & Kalani. *Mycopathol. Mycol. Appl.* **33**:76–80.
Sutton, B. C. (1968a). *Kellermania* and its generic segregates. *Can. J. Bot.* **46**:181–196.
Sutton, B. C. (1968b). *Polynema*, an earlier name for *Neobarclaya. Mycologia* **60**:201–203.
Sutton, B. C. (1968c). The appressoria of *Colletotrichum graminicola* and *C. falcatum. Can. J. Bot.* **46**:873–876.
Sutton, B. C. (1969). Type studies of *Coniella, Anthasthoopa* and *Cyclodomella. Can. J. Bot.* **47**:603–608.
Sutton, B. C. (1970). Forest microfungi. III. The heterogeneity of *Pestalotia* de Not. section sexloculatae Klebahn sensu Guba. *Can. J. Bot.* **47**:2083–2094.
Sutton, B. C. (1971a). Conidium ontogeny in pycnidial and acervular fungi. *In* "Taxonomy of Fungi Imperfecti" (W. B. Kendrick, ed.), pp. 263–278. Univ. of Toronto Press, Toronto.
Sutton, B. C. (1971b). Coelomycetes. IV. The genus *Harknessia*, and similar Fungi on *Eucalyptus. Mycol. Pap.* **123**: 1–46.
Sutton, B. C., and R. L. C. Chao. (1970). *Leptomelanconium. Trans. Brit. Mycol. Soc.* **55**:37–44.
Sutton, B. C., and T. Kobayashi. (1970). *Strasseriopsis* gen. nov., based on *Phellostroma tsugae* Kobayashi. *Mycologia* **61**:1066–1071.
Sutton, B. C., and K. A. Pirozynski. (1963). Notes on British microfungi. I. *Trans. Brit. Mycol. Soc.* **46**:505–522.
Sutton, B. C., and K. A. Pirozynski. (1965). Notes on Microfungi. II. *Trans. Brit. Mycol. Soc.* **48**:349–366.
Sutton, B. C., and D. K. Sandhu. (1969a). *Phragmotrichum pini* (W. B. Cooke) comb. nov. *Trans. Brit. Mycol. Soc.* **52**:67–71.
Sutton, B. C., and D. K. Sandhu. (1969b). Electron microscopy of conidium development and secession in *Cryptosporiopsis* sp., *Phoma fumosa, Melanconium bicolor*, and *M. apiocarpum. Can. J. Bot.* **47**:745–749.
Sutton, B. C., and P. W. Sellar. (1966). *Toxosporiopsis* n. gen., an unusual member of the Melanconiales. *Can. J. Bot.* **44**:1505–1513.
Sydow, H. and P. (1913). Novae fungorum species. IX. *Ann. Mycol.* **11**:54–65.
Tassi, F. (1902). I generi *Phyllosticta* Pers., *Phoma* Fr., *Macrophoma* (Sacc.) Berl. et Vogl. e i loro generi analoghi, giusta la legge d'analogia. *Bull. Lab. Ort. Bot., Siena* **5**:1–72.
Trotter, A. (1904), Notulae mycologicae. *Ann. Mycol.* **2**:533–538.
Tubaki, K. (1958). Studies on the Japanese hyphomycetes. V. Leaf and stem group with a discussion of the classification of hyphomycetes and their perfect stages. *J. Hattori Bot. Lab.* **20**:142–244.
Tubaki, K. (1964). Taxonomic study of hyphomycetes. *Annu. Rep. Inst. Ferment., Osaka* **1**: 1961/1962:25–54.
Twyman, E. S. (1947). Notes on the dieback of oak caused by *Colpoma quercinum* (Fr.) Wallr. *Trans. Brit. Mycol. Soc.* **29**:234–241.

Urries, M. J. (1956). Novedades micologicas de la Flora canaria. *Ann. Inst. Bot. A. J. Cavanillo* **14**:153–170.
Urries, M. J. (1957). Hongos microscopicos de Canarias. *Museo Canario* **57–64**:1–140.
Vassiljevsky, N. I., and B. P. Karakulin. (1950). "Fungi Imperfecti Parasitici," Pars II, Moscow.
von Arx, J. A. (1957a). Revision der zu *Gloeosporium* gestellten Pilze. *Proc., Kon. Ned. Akad. Wetensch.* **51**:1–153.
von Arx, J. A. (1957b). Die Arten der Gattung *Colletotrichum* Corda. *Phytopathol. Z.* **29**: 413–468.
von Arx, J. A. (1963). Revision der zu *Gloeosporium* gestellten Pilze. Nachträge und Berichtigungen. *Proc., Kon. Ned. Akad. Wetensch.* **66**:172–182.
von Höhnel, F. (1911). Zur Systematik der Sphaeropsideen und Melanconieen. *Ann. Mycol.* **9**:258–265.
von Höhnel, F. (1917). Fungi imperfecti. Beitrage zur Kenntnis derselben. *Hedwigia* **59**: 236–284.
von Höhnel, F. (1923). System der Fungi Imperfecti Fuckel. *Mykol. Unters.* **3**:301–369.
von Tavel, F. (1886). Beiträge zur Entwickelungsgeschichte der Pyrenomyceten. *Bot. Zentralbl.* **44**:825.
Vuillemin, P. (1910a). Matériaux pour une classification rationelle des Fungi Imperfecti. *C. R. Acad. Sci.* **150**:882–884.
Vuillemin, P. (1910b). Les Conidiosporés. *Bull. Soc. Sci. Nancy* [3] **11**:129–172.
Vuillemin, P. (1911). Les Aleuriosporés. *Bull. Soc. Sci. Nancy* [3] **12**:151–175.
Vuillemin, P. (1912). "Les Champignons. Essai de classification." Doin, Paris.
Wakefield, E. M. and G. R. Bisby. (1941). List of Hyphomycetes recorded for Britain. *Trans. Brit. Mycol. Soc.* **25**:49–126.
Walker, J. (1962). Notes on plant parasitic fungi. I. *Proc. Linn. Soc. N.S.W.* **87**:162–176.
Wehmeyer, L. E. (1933). The genus *Diaporthe* Nitschke and its segregates. *Univ. Mich. Stud.* **9**:1–349.
Wehmeyer, L. E. (1941). A revision of *Melanconis, Pseudovalsa, Prosthecium*, and *Titania*. *Univ. Mich. Stud.* **14**:1–161.
Wilson, M. M., and G. G. Hahn. (1928). The identity of *Phoma pitya* Sacc., *Phoma abietina* Hart., and their relation to *Phomopsis pseudotsugae* Wilson. *Trans. Brit. Mycol. Soc.* **13**:261–278.
Wolf, F. A. (1912). The perfect stage of *Actinonema rosae. Bot. Gaz.* **54**:218–234.
Wolf, F. A., and W. J. Barbour. (1941). Brown-spot needle disease of pines. *Phytopathology* **31**:61–74.
Wollenweber, H. W., and H. Hochapfel. (1936). Beitrage zur Kenntnis parasitarer und saprophytischer Pilze. I. *Phomopsis, Dendrophoma* und *Ascochyta* und ihre Bezeihung zur Fruchtfaule. *Z. Parasitenk.* **8**:561–605.
Zachariah, K. and P. C. Fitz-James. (1967). The structure of phialides in *Penicillium claviforme. Can. J. Microbiol.* **13**:249–256.
Zalasky, H. (1964). Nomenclature and description of *Diplodia tumefaciens* (Shear) Zalasky (= *Macrophoma tumefaciens* Shear apud Hubert). *Can. J. Bot.* **42**:1049–1055.
Zambettakis, C. E. (1954). Recherches sur la systématique des "Sphaeropsidales-Phaeodidymae." *Bull. Soc. Mycol. Fr.* **70**:219–350.

# Author Index

Numbers in italics refer to the pages on which the complete references are listed.

## A

Abadie, F., 19, 21 *28*
Abawi, G. S., 301(159), 303(159), *313*
Aebi, B., 292(1), *307*
Agar, H. D., 542, *574*
Agnihothrudu, V., 551, *574*
Ahearn, D. G., 27, *30*
Ahmad, S., 99, *130*, 174, 190, 191, 192, 193, 200, *217*, 534, *574*
Ainsworth, G. C., 4, *7*, 45, 46, 47, *64*, 71, 72, 81, *85*, *328*, 331
Ajello, L., *64*
Alexopoulos, C. J., 5, *7*, 11, *31*, 46, 47, *64*, 78, *85*, 233, *243*
Ali, M. M., 551, *574*
Allescher, A., 522, *574*
Ames, L. M., 99, *127*
Ammon, H. U., 142, *214*
Andrus, C. F., 46, *64*, *65*
Apinis, A. E., 48, 51, 55, *65*
Archer, W. A., 530, 536, 551, *575*
Arnaud, G., 71, *85*, 154, 155, 163, 165, 168, 169, 170, 173, 192, 203, 204, 205, 207, 208, 209, 211, 212, *214*
Arpin, N., 254, *307*
Arwidsson, T., 225, *243*
Ashby, S. F., 552, *575*

## B

Baccarini, P., 530, *575*
Backus, M. P., *65*, 295(264), *317*, 532, *580*
Bandoni, R. J., 23, *31*
Bánhegyi, J., 232, *243*
Banno, I., 22, 23, 27, *28*, 29
Baranetzky, J., 51, *65*
Barkley, F. A., 1, 3, *7*
Barnett, H. L., *327*
Barr, M. E., 119, 120, 127, 142, 154, 172, *214*, 273(32), *308*

Barron, G. L., 48, *65*, 100, *127*, *327*, 333, 522, 524, *575*
Bartnicki-Garcia, S., 12, *28*
Batista, A. C., 154, 168, 169, 171, 172, 173, 204, 206, 207, 208, 209, 210, 213, *214*, 529, 549, *575*
Batra, L. R., 15, *28*, 287(6, 8), 288(7), 305(228), *307*, *308*, *316*
Bauke, H., 529, *575*
Bayliss Elliott, J. S., 536, *575*
Beeli, M., 84, *85*
Bell, A., 183, 184, *215*
Bender, H. B., 522, 533, *575*
Benedek, T., 163, 164, 168, 169, 170, 171, 172, 173, 174, 175, 176, 178, 179, 180, 181, 182, 183, 184, 185, 186, 187, 188, 189, 190, 191, 192, 193, 196, 197, 198, 199, 203, 204, 206, 207, 208, 209, 210, 211, 212, 213, *218*
Benedix, E. H., 270(10), 285(9), 286(9), 301(9), 307(9), *308*
Benjamin, C. R., *65*
Benjamin, R. K., 51, 53, *65*, 224, 225, 226, 227, 231, 233, *243*
Beran, K., 542, *580*
Berkeley, M. J., 515, 520, 546, *575*
Berlese, A. N., 223, *243*
Bertault, R., 305(10a), *308*
Berthet, P., 254, 273(12), *308*
Bessey, E. A., 12, *28*, *65*, 71, 72, 76, *85*, 232, 233, *243*
Besson, M., 21, *28*
Bestagno, G., 534, 562, *575*
Bezerra, J. L., 104, *127*, 209, *214*
Biga, M. L. B., 534, 562, *575*
Biggs, R., 15, *28*
Bisby, G. R., *328*, 523, *582*
Blumer, S., 72, 73, 74, 77, 79, *85*
Boedijn, K. B., 18, *28*, *65*, 103, 123, *128*,

163, 165, *214*, 260(15), 265(13), 299(14), 301(14), *308*
Boelens, W. C., 224, *243*
Boerema, G. H., 530, 541, 542, 543, 552, 559, 563, 564, *575*
Boidin, J., 19, 21, *28*, *30*
Bonar, L., 284(16), *308*
Bonorden, H. F., 515, 519, *575*
Booth, C., *65*, 92, 99, 101, 102, 105, 113, 114, *128*, *131*, 187, *214*
Bose, S. K., 188, 190, 199, *214*
Boudier, E., *65*, 255(17), 299(18), *308*
Bowerman, C. A., 287(109), *311*
Bracker, C. E., 23, *28*
Brefeld, O., *65*
Breton, A., 105, *128*
Brewer, D., 551, *575*
Brewer, J. G., 542, *575*
Brierley, W. B., 546, *575*
Bro Larsen, E., 226, *243*
Brooks, F. T., 547, 548, *576*
Broome, C. E., 546, *575*
Brown, A. H. S., *65*
Brundza, K., 73, *85*
Bubák, F., 540, *576*
Buchwald, N. F., 286(22, 23), 287(22, 23), 288(22, 23), 289(22, 23), *308*
Buckley, H., 27, *31*
Buckley, P. M., 542, *576*
Buddin, W., 532, 540, *576*
Buecher, E. J., 20, *28*
Burdsall, H. H., Jr., 260(25), 274(24, 25), *308*
Burger, O. F., 551, *576*
Burton, K. A., 13, 20, *32*
Butin, H., 116, *128*, 176, *214*, 570, *576*
Butler, E. E., 16, *28*

**C**

Cailleux, R., 110, 111, *128*
Cain, R. F., 15, *28*, 45, 46, 53, 55, 57, 59, *65*, *66*, 98, 99, 100, 110, 111, *127*, *128*, *130*, *131*, 184, *214*, 267(141), 272(26, 142), 279(231), 280(231), 281(233), 284(233, 234), 295(232), *308*, *312*, *316*
Calvert, O. H., 530, *576*
Campbell, R., 547, *576*
Carmichael, J. W., *67*, 324, *328*
Carroll, G. C., 109, *128*, 232, 233, *243*
Casagrande, F., 188, 192, *214*, 284(199), *314*

Cash, E. K., 287(282a), 301(27), *308*, *318*
Chadefaud, M., 138, 154, *215*, 232, *243*
Chao, R. L. C., 549, 558, *576*, *581*
Chapman, B., 101, *129*
Chesters, C. G. C., 57, *65*, 183, 184, *215*, 542, 543, *576*
Chippindale, H. G., 530, *576*
Chodat, F., 552, *576*
Christiansen, M. P. 302(28), *308*
Christiansen, M. S., 282(29), *308*
Chupp, C., 534, *576*
Ciccarone, A., 107, *128*, 549, 571, *576*
Ciferri, R., 56, *65*, 122, *128*, 154, 168, 169, 170, 171, 172, 173, 204, *214*, *217*, 303(30), *308*, 529, 534, 549, 562, *575*
Clémencet, M., 259(31), *308*
Clements, F. E., 2, 7, 56, *65*, 325, *326*, *328*, 522, 527, 533, *576*
Cole, G. T., 48, *65*, *66*, 522, 540, 542, 543, *576*, *578*
Collart, A., 224, *243*
Cooke, J. C., 273(32), *308*
Cooke, W. B., 154, *215*
Cooney, D. G., 55, *65*
Coons, H. C., 551, *576*
Copeland, H. F., 1, 3, *7*
Corbaz, R., 182, *215*
Corda, A. C. I., 515, 519, 540, *576*
Corlett, M., 144, 172, *215*
Costantin, J., 522, *576*
Cribb, A. B., 117, *128*
Cribb, J. W., 117, *128*
Cunnell, G. J., 536, 543, 551, *576*
Cutter, V. M., 24, *30*

**D**

Dangeard, P. A., *65*
Darker, G. D., 197, *215*, 278(35), 279(34, 35), 280(35), 281(33), *308*
Da Silva Maia, H., 209, *214*
Davidson, R. W., 98, *131*
de Bary, A., *65*, 232, *243*, 529, 530, *576*
Défago, G., 291(36), 293(36, 192), *308*, *314*
de Fonseka, R. N., 536, *577*
Deighton, F. C., 172, *215*
Deinem, M. H., 27, *31*
de Jong, C. B., 543, *575*
Denison, W. C., 232, 233, *243*, 265(41),

Index 585

266(40, 42, 43), 267(43), 275(37, 39), 276(38), *309*
Dennis, R. W. G., 45, 46, 47, *65*, 88, 99, 121, 122, *128*, *130*, 154, 163, 168, 171, 173, 174, 175, 178, 180, 181, 182, 183, 184, 185, 186, 187, 188, 189, 190, 191, 192, 193, 194, 196, 197, 199, 200, 203, 204, 206, 207, 213, *215*, *217*, 256, 286(48), 286(48), 287(48), 288(48, 55, 56), 289(46, 48), 291(49), 292(53, 57a), 293(45, 47), 296(44, 54), 297(44, 46, 48, 50, 54), 298(44), 300(47, 48, 58), 301(48, 84), 302(48, 84), 303(48, 54), 304, 305(48, 56), 306(48), 307(48, 56), *309*, *310*, 526, 551, 552, *577*
Derx, H. G., 25, *28*
de Vay, J. E., 542, *576*
Devine, J., 27, *29*
Dharne, C. G., 296(59), *309*
Dickinson, C. H., 523, 536, *577*
Diedicke, H., 522, 565, *577*
Diehl, W. W., 126, *128*
Diesing, K. M., 223, *243*
Dingley, J. M., 103, 105, *129*
Dissing, H., 255(61a), 269(60, 61, 62, 63), *309*
Dixon, J. R., 293(63a), 301(63a), 304(63a), 305(63a), *310*
Dixon, P. A., 16, *28*
do Carmo-Sousa, L., 12, 14, 26, 27, *28*
Dodge, B. O., 164, *215*, 295(257), *317*, 530, 532, *577*
Dodge, C. W., 47, *65*, 229, 231, *243*, 259(64), *310*, *327*
Doguet, G., 99, *129*, 140, *215*
Doi, Y., 103, 104, *129*
Doidge, E. M., 81, *85*
Dorenbosch, M. M. J., 543, 548, 552, 563, 564, *575*, *577*
Douglas, H. C., 542, *574*
Drayton, F. L., 288(65), *310*
Drechsler, C., 187, *215*
Dring, D. M., 534, *577*
Duke, M. M., 532, *577*
Dumont, K. P., 273(241), 286(67, 160), 288(66, 66a), 289(68), *310*, *313*, *316*
Durand, E. J., 292(69), 298(70), 303(70), *310*
Durand, F., 48, *67*
Dubin, R. D., 553, *577*

Durrell, L. W., 542, *577*

E

Eckblad, F.-E., 263(72), 264(72), 265(72), 266(72), 267(72), 268(72), 271(72), 273(72), 277(72), 305(71), *310*
Egger, M. C., 280(73), *310*
Elliott, M. E., 286(74), 287(110), 288(75, 111), *310*, *311*
Ellis, J. B., 164, 172, 173, 175, 183, 184, 185, 188, 191, 192, 193, 205, *215*
Ellis, M. B., 324, *327*, *328*, 333, 534, 547, *577*
Emerson, R., 55, *65*
Emmons, C. W., *66*
Engler, A., 232, *243*
Erb, R. W., 272(76, 160a), *310*, *313*
Eriksson, O., 102, 119, *129*, 140, 182, 184, 186, 190, 191, 192, *215*, 282(77), *310*, *577*
Eveleigh, D. E., 552, *577*
Everhart, B. M., 164, 172, 173, 175, 183, 184, 185, 188, 191, 192, 193, 205, *215*

F

Farr, M. L., 179, 205, *215*
Farrow, W. M., 110, *128*
Faurel, L., 105, *128*, 267(78), *310*
Favre, J., 307(79), *310*
Fell, J. W., 22, 23, 24, 26, 27, *28*, *29*, 30
Fennell, D. I., *67*
Fischer, E., 53, 55, 56, 59, *66*, 259(80), 260(80), 261(80), *310*
Fitz-James, P. C., 542, *582*
Fitzpatrick, H. M., 91, 108, 109, *129*
Fournet, J. P., 551, *577*
Francke-Grosmann, H., 15, *28*
Fries, E. M., 36, *41*, 53, 56, *66*, 514, 518, 519, *577*
Frost, R. R., 532, *577*
Fuckel, L., 515, 520, 532, *577*
Fukazawa, Y., 12, *31*
Funk, A., 109, *129*, 279(81), 294(235), 303(235) *310*, *316*

G

Gäumann, E. A., 12, 15, *29*, 36, *41*, 45, 46, 47, *66*, 76, *85*, 229, 231, *243*
Galløe, O., 199, 200, 205, *215*
Gams, W., 101, 109, *129*, 132

Gamundi, I. J., 256, 258(83a), 271(83), 274(83, 83b), 275(82), 301(84), 302(84), *310*
Gareth-Jones, E. B., 104, *129*
Garnier de Souza, R., 209, *214*
Gayao, T. de J., 204, *214*
Gentles, J. C., 12, *29*
Gerlach, W., 105, *129*
Gerlagh, M., 101, *129*
Gilkey, H. M., 260(87, 88, 89, 90), 261(85, 86, 87, 88, 89, 90), *310*, *311*
Gilman, J. C., 100, *127*
Goidanich, G., 523, *577*
Goos, R. D., 522, *577*
Gordon, C. C., 90, *129*
Gordon, M. A., 27, *29*
Gorlenko, M. V., 26, *30*
Graddon, W. D., 282(91), *311*
Graham, J. H., 171, *215*
Gray, E., 287(294), *319*
Greenhalgh, G. N., 291(92), *311*
Gremmen, J., 263(93), 280(96), 281(94), 295(98), 301(97), 302(95, 281), *311*, *318*
Griffin, H. D., 98, *129*
Grove, W. B., 513, 521, 532, 533, 536, 548, 553, 555, *577*
Groves, J. W. 268(112), 286(120), 287(109, 110), 288(65, 111), 290(99, 100, 101), 294(102, 107, 108), 299(105, 113), 300(103, 104), 302(114), 303(108), 304(106), *310*, *311*, *312*, 567, 568, 572, *577*, *578*
Guba, E. F., 532, 533, 534, 548, 557, *578*
Guilliermond, A., 13, 14, 15, 16, *29*

**H**
Hahn, G. G., 526, 543, *578*, *582*
Hammarlund, C., 80, *85*
Hansford, C. G., 72, 77, 81, 82, 83, *85*, 90, 105, *129*, 168, 169, 170, 211, *215*
Harmaja, H., 269(115), 270(115), 271(116), *311*
Harr, J., 113, *130*
Harris, H. A., 530, *578*
Harrold, C. E., 16, *29*
Harter, L. L., 46, *64*, *65*
Hasegawa, T., 27, *29*
Hawker, L. E., 47, *66*
Hawksworth, D. L., 99, *129*, 154, *215*
Heald, F. D., 77, *85*

Hedjaroude, G. A., 154, 191, *215*
Heim, R., 268(118), 271(119), 275(117), *311*
Hennebert, G. L., 110, *132*, 268(119a) 286(119a, 120), 287(119a), *311*, *312*, 324, *328*
Heyting, C., 16, *31*
Higgins, B. B., 534, *578*
Hino, I., 101, *129*, 551, *578*
Hirata, K., 72, 73, 74, 79, 81, *85*
Hiura, U., 79, *85*
Hoare, S. C., 268(112), *311*
Hochapfel, H., 548, *582*
Höweler, L. H., 543, 552, *575*
Holm, L., 182, 186, 187, 188, 189, 190, 191, 192, *215*, *216*, 295(122a), *312*
Homma, Y., 73, 78, 79, *85*
Hoo, S.-C., 551, *578*
Hubbes, M., 113, *129*
Hütter, R., 292(193), 293(124), *312*, *314*
Hughes, S. J., 48, *66*, 168, 172, 173, *216*, 264(123), *312*, 324, *327*, *328*, 335, 516, 522, 523, 534, 536, 545, *578*
Hunt, J., 98, *129*
Hunter, I. L., 22, 23, 24, *29*, *30*

**I**
Imai, S., 259(125), 268(127), 269(127), 285(126, 128), 287(126), 301(126), 305(126, 128), 307(126), *312*
Ingold, C. T., 101, *129*, 522, 523, *578*
Ito, K., 115, *130*
Ivory, M. H., 556, *578*

**J**
Jacob, J. L., 19, *28*
Jenkins, A. E., 164, *218*
Jørstad, I., 534, *578*
Johnson, T. W., 116, 117, *129*
Johnston, H. W., 81, *85*
Jones, D., 549, *580*

**K**
Kamburov, S. S., 226, *244*
Kankainen, E., 286(129), 307(129), *312*
Kanouse, B. B., 265(130), 267(130), 268(133), 269(131), 272(132), 273(131), 274(132), 276(133), *312*
Kanzawa, S., 273(203), *315*
Kar, A. K., 183, 207, *216*
Karakulin, B. P., 532, 555, *582*

Karsten, H., 223, 229, *244*
Katumoto, K., 90, 101, 108, 114, *129*, 551, *578*
Kavina, K., 301(134), *312*
Kawakita, S., 12, *31*
Keddie, F. M., 26, *29*
Kempton, F. E., 524, 529, *578*
Kempton, P. E., 269(135), *312*
Kendrick, W. B., 48, *65*, *66*, 325, *327*, *328*, 522, 534, 540, 542, 543, *576*, *578*
Kenneth, R., 226, *244*
Kern, H., 113, *130*, 193, *216*
Kheswalla, K. F., 547, *578*
Kiely, T. B., 563, *578*
Kimbrough, T. W., 249, 267(136, 141), 272(26, 138, 140, 142), 273(140), *308*, *312*
Kirschstein, W., 284(144), 289(143), *312*
Klebahn, H., 532, 534, 551, *578*
Kluyver, A. J., 22, *29*
Knoch, Dr., 223, *244*
Kobayashi, T., 301(145), *312*, 540, 543, 551, 565, *578*, *581*
Kobayashi, Y., *66*, 115, *130*
Kobayasi, Y., 258(149), 260(146, 148), 274(147), *312*, *313*
Kohlmeyer, E., 116, *130*, 140, 144, 174, 182, 185, 186, 188, 189, 190, 191, 192, 199, *216*
Kohlmeyer, J., *66*, 116, *130*, 140, 144, 172, 182, 185, 186, 188, 189, 190, 191, 192, 199, *216*
Kolenati, F. A., *244*
Komagata, K., 27, *30*
Kominami, K., *66*
Korf, R. P., 15, 25, *29*, *31*, 250(153), 251, 253(156), 255(153, 158a), 263(161), 264(152), 265(152), 266(161), 268(151), 270(158), 271(185), 272(140, 160a), 273(12, 140, 155, 158, 241), 274(157), 276(157), 277(158), 279(200a), 281(156), 282(244), 284(156, 212), 285(158), 286(67, 160), 287(8, 158), 288(262), 289(68), 291(158), 296(150, 154), 297(156), 300(58, 158), 301(153, 159), 302(153, 158), 303(158, 159), 305(153, 164, 228), *308*, *309*, *310*, *312*, *314*, *315*, *316*, *317*
Kramer, C. L., 34, 40, *41*
Krassilnikov, N. A., 26, *30*
Kreger-van Rij, N. J. W., 19, 20, 21, 23, *29*
Kreisel, H., 3, *7*
Kubička, J., 266(274), 303(273), *318*

Kudriavzev, V. I., 12, *29*
Kuehn, H. H., 48, 53, *66*, *67*
Kursanova, L. I., 26, *30*
Kurtzman, C. P., 20, *32*

**L**

Lacoste, L., 552, *578*
Laffin, R. J., 24, *30*
Lange, M., 269(62), *309*
Langeron, M., 523, *578*
La Touche, C. J., 12, *29*
Leach, C. M., 551, *578*
Leffring, L., 552, 563, *575*
Le Gal, M., 255, 264, 265(166), 266(166), 267(166), 268(163), 270(165), 271(119, 165), 275(166, 171, 172, 172a), 276, (166, 167, 168, 170). 277(166), 303(170), *311*, *313*
Lehman, S. G., 546, *579*
Lehodey, Y., 19, 21, *28*
Leonian, L., 552, *579*
Lepesme, P., 224, *244*
Leth Back, A., 27, *66*
Léveillé, J. H., 72, *85*, 515, 519, *579*
Libert, A., 519, *579*
Lindau, G., *66*, *244*, 326
Lindenmayer, A., 2, *7*
Linder, D. H., 542, 543, *579*
Lindroth, C. H., 225, *244*
Link, H. F., 514, 517, 518, 519, *579*
Linnacus, C., 72, *85*
Lloyd, C. G., 303(173), *313*
Lodder, J., 12, 16, 19, 23, *30*
Loeffler, W., 175, *216*
Lohdi, S. A., 534, *574*
Luck-Allen, E. R., 184, *214*, 267(141), 272 (142), *312*
Ludwig, F., *66*
Luedemann, G. M., 552, *579*
Lundqvist, N., 109, 110, 111, *130*
Luttrell, E. S., 45, 46, 47, *66*, 140, 150, 171, 190, *215*, *216*, 232, *244*, 522, 537, 547, *579*

**M**

Maas Geesteranus, R. A., 251, 268(176), 276(177), 285(174, 177a), 286(174), 305 (175), 307(174), *313*, *314*
McKnight, K. H., 270(183), *314*
McLennan, E. I., 259(184), 261(184), *314*
MacNeill, B. H., 552, *579*

Madelin, M. F., 522, 542, *579*
Magasi, L. P., 280(205), *315*
Mains, E. B., 282(180), 284(178), 285(178, 179), 286(179), 301(180), 305(180), 307 (179), *314*
Maire, R., *66*, 224, 232, *244*
Maity, M. K., 183, 207, *216*
Malloch, D., 53, 55, 57, 59, *66*, 99, 100, *130*, 273(293), *319*
Manuel, J., 14, *30*
Marasas, W. F. O., 205, *216*
Marchionatto, J. B., 258(181), *314*
Margoliash, E., 2, *5*
Martin, G. W., 2, 5, *7*, 23, *30*, 36, *41*, 46, 47, *66*
Martin, P., 121, 122, 123, 124, *130*
Martius, C. F. P., 517, *579*
Mason, E. W., 323, 324, *328*, 332, 522, 523, 533, 545, *579*
Mattirolo, O., 259(182), 261(182), *314*
Mayr, G., 223, *244*
Menon, R., 182, *216*
Mercer, W. B., 529, *579*
Miller, J. H., 45, 46, 47, *66*, 124, *130*, 163, *216*
Miller, M. W., 12, 15, *30*
Mirza, F., 154, 192, *216*
Mirza, J. H., 110, *128*, *130*
Mix, A. J., 36, 40, *41*
Möller, A., 104, *130*
Moore, E. J., 271(185), *314*
Moravec, J., 267(187), 275(186), 276(187a), *314*
Moravec, Z., 127, *130*
Moreau, C., 110, *130*, 183, 198, 213, *216*
Moreau, F., 46, 47, *66*, 523, *579*
Moreau, M., 183, 198, 213, *216*
Morelet, M., 294(188), *314*
Morgan-Jones, G., 121, *131*, 194, *216*, 291 (92), *311*, 523, 536, 542, *577*, *579*
Mrak, E. M., 12, *30*
Müller, E., 46, 47, 48, *67*, *68*, 88, 91, 92, 99, 100, 101, 103, 105, 108, 111, 113, 114, 119, 120, 121, 122, *128*, *130*, *131*, *132*, 150, 152, 154, 163, 164, 168, 169, 170, 171, 173, 174, 175, 176, 178, 179, 180, 181, 182, 183, 184, 185, 186, 188, 189, 190, 191, 192, 193, 196, 197, 198, 199, 200, 203, 204, 205, 206, 207, 208, 209, 210, 211, 212, 213, *214*, *216*, *217*, *218*, 280(5, 190), 281(5), 284(199),

291(189), 293(192), *307*, *314*, 523, *580*
Mukerji, K. G., 100, *131*
Munk, A., 88, 102, 109, 111, 112, 113, 121, *128*, *130*, 140, 154, 174, 185, 186, 188, 189, 191, *217*
Muthappa, B. N., 174, *217*

N
Nadel, D. J., 226, *244*
Nakase, T., 27, *30*
Nannfeldt, J. A., 46, *67*, 140, 200, *217*, 269(63, 198), 273(196, 198), 274(198), 275(196), 284(197), 285(197), 289(195), 291(195), 292(195), 293(195), 294(195), 295(195), 302(195), 305(197), 307(194), *309*, *314*
Narasinham, M. J., 199, *218*
Nattrass, R. M., 548, 552, *579*
Naumov, N. A., 26, *30*
Nees von Esenbeck, C. G., 517, *579*
Neuwirth, F., 74, *85*
Newell, S. Y., 24, *29*, *30*
Nicot, J., 48, *67*
Nilsson, T., *68*
Nolan, C., 2, *7*
Nordin, I., 199, *217*
Novák, E. K., 12, *30*
Nüesch, J., 182, *217*
Nyland, G., 22, 23, 25, *30*

O
Oberwinkler, F., 284(199), *314*
Obrist, W., 120, *131*, 174, 189, *217*, 270(200), *315*
Ogimi, C., 279(200a), *315*
Olive, L. S., 23, 24, 25, *30*, 46, *67*
Orpurt, P. A., *65*
Orr, G. F., 53, *67*
Otani, Y., 273(202, 203), 297(201), *315*
Ou, S. H., 299(204), *315*
Ouellette, G. B., 280(205), *315*

P
Paden, J. W., 264(207), 265(206), *315*
Padhye, A. A., *67*
Palm, B. T., 258, *315*
Palmer, J. T., 286(208a), *315*
Pant, D. C., 264(209), 273(210), *315*
Pantidou, M. E., 284(211, 212), *315*
Parbery, D. G., 51, *67*, 107, *131*

# Index

Parguey-Leduc, A., 110, *131*, 154, *217*
Parker, A. K., 284(213), *315*
Parmelee, J. A., 551, *579*
Parmeter, J. R., Jr., 274(292), *318*
Patil, S. D., 549, *579*
Patouillard, N., 65, *67*
Pauvert, P., 551, *577*
Persoon, C. H., 514, 517, *579*
Petch, T., 521, 571, *579*
Petersen, L. J., 16, *28*
Peterson, G. A., 74, *85*
Petrak, F., 108, 111, *131*, 170, 175, 176, 183, 200, 204, 206, 213, *217*, 280(214), 281(217, 222), 282(219), 284(215), 292(221), 298(218), 300(220), 306(216), *315*, *327*, 523, 533, 534, 547, 548, 555, 562, 570, *579*
Peyritsch, J., 223, 232, *244*
Peyronel, B., 287(223), *315*
Pfister, D. H., 267(224b), 268(224, 224c), 273(224, 224c), 276(224a), 277(224), *315*, *316*
Phaff, H. J., 11, 12, 19, 20, 22, 23, 24, 25, 26, 27, *28*, *29*
Picard, F., 224, 232, *244*
Pignal, M. C., 19, 21, *28*, *30*
Pirozynski, K. A., 146, 172, 198, 199, 208, *215*, *217*, *218*, 278(236), 281(237), 297(264a), *316*, *317*, 325, *328*, 523, 524, 536, 540, 542, 548, 549, 551, 569, *579*, *580*, *581*
Pitt, J. I., 15, *30*
Pollack, F. G., 205, *215*
Potebnia, A., 521, 524, 532, *580*
Pound, G. S., 530, *576*
Powell, P. E., Jr., 279(224e), *316*
Pridham, T. G., 15, *30*
Punithalingam, E., 183, *217*, 530, 540, 549, *580*

## R

Rai, J. N., 100, *131*
Raitviir, A., 265(226), 269(225), 270(227a), 283, 297(227, 227b), *316*
Ramakrishnan, K., 549, 551, *580*
Ramamurthi, C. S., 305(228), *316*
Rao, R., 175, *218*
Raper, K. B., 15, *30*, 48, *67*
Rapilly, F., 551, *577*
Rawlings, G. B., 258(229), *316*

Redinger, K., 197, 200, *218*
Rehm, H., 279(230), 282(230), 290(230), *316*
Reid, J., 102, 113, 114, *131*, 146, 198, 199, *217*, *218*, 278(236), 279(231), 280(231), 281(233, 237), 284(213, 233, 234), 294(235), 295(232), 303(235), *315*, *316*
Reusser, F. A., 173, *218*
Reynolds, D. R., 154, 173, *218*
Richards, A. G., 225, 226, 227, *244*, *245*
Richle, R., 20, *30*
Rick, J., 266(238), *316*
Riedl, H., 193, 194, *218*
Rifai, M. A., 264(239), 265(239), 266(239), 268(239), 269(239), 270(239), 271(239), 274(239), 275(239), 276(239), 277(239), *316*
Rimpau, R. H., 295(240), *316*, 559, *580*
Robin, C., 223, 231, *245*
Robinson-Jeffrey, R. C., 98, *131*
Robison, B. M., 110, *131*
Rogers, J. K., 273(241), *316*
Rogerson, C. T., 103, 125, *131*, *328*
Rouget, A., *245*
Rouppert, C., 267(242), 270(242), *316*
Routien, J. B., 14, *31*
Ruggieri, G., 523, *577*

## S

Saccardo, P. A., *327*, 516, 520, 521, 523, 524, 532, *580*
Sadebeck, R., 36, *41*
Saho, H., 294(243), *316*
Sainclivier, M., 23, *31*
Salmon, E. S., 73, 80, *86*
Samson, R. A., 49, *67*
Sánchez, A., 282(244, 245), *316*
Sandhu, D. K., 536, 539, 540, 542, 543, 547, 549, 554, 569, *581*
Santesson, R., 197, 198, 200, *218*, 258(246), 306(247), *317*
Sattar, A., 548, *580*
Sawada, K., 73, *86*
Scheinpflug, H., 184, 185, *218*
Scheloske, H.-W., 226, *245*
Schippers-Lammertse, A. F., 16, *31*
Schläpfer-Bernard, E., 292(248), 294(248), 299(248), *317*
Scholer, H. J., 20, *30*
Schotter, G., 267(78), *310*
Schrantz, J. B., 116, 118, *131*

Schröter, J., 51, 53, *67*
Schüepp, H., 280(249), 295(249), *317*
Scott, DeB., 19, 26, *31*, *67*
Searle, G. O., 547, 548, *576*
Seaver, F. J., 256, 376(254), 287(253), 295(254), 302(252), 304(254), 305(254), 306(254), *317*
Seeler, E. V., Jr., 299(255), *317*, 543, *580*
Sellar, P. W., 524, 532, 536, 540, 551, *581*
Servazzi, O., 532, *580*
Setchell, W. A., 267(256), 270(256), *317*
Shadomy, H. J., 23, 27, *31*
Shanor, L., 224, 225, *243*, *245*
Shaw, C. G., 90, *129*
Shear, C. L., 2, 7, 56, *65*, 295(257), *317*, 325, *326*, *328*, 522, 527, 533, *576*
Shoemaker, R. A., 120, 121, *131*, 179, 208, *217*, *218*, 523, 536, 551, 557, *580*
Simmonds, J. H., 552, 553, *580*
Simmons, E. G., 324, *328*
Sivanesan, A., 144, 173, 211, *218*
Skou, J. P., 116, *131*
Slodki, M. E., 23, *31*
Slooff, W. C., 18, 19, 26, *31*
Smerlis, E., 280(258), 295(259), *317*
Smith, A. H., 268(133), 276(133), *312*
Smith, G., *65*, 77, *86*, 516, 522, *580*
Smith, M. N., 225, 226, 227, *244*, *245*
Sonck, C. E., 26, *31*
Sowell, G., 25, *31*
Sparrow, F. K., 3, 7, 116, 117, *129*
Spegazzini, C., 224, 232, *245*, 269(261), 271(260, 261), *317*
Spencer, J. F. T., 27, *30*, *31*
Spevak, M. B., 288(262), *317*
Spiltoir, C. F., 46, *67*
Sprague, R., 533, *580*
Stafleu, F. A., 1, *7*
Stansfield, O. P., 536, *575*
Starbäck, K., 251, *317*
Starkey, R. L., 19, *31*
Statzell, A. C., 22, 23, *29*
Stenderup A., 27, *31*
Stevens, F. L., 90, *131*
Steyaert, R. L., 532, 549, 557, *580*
Stodola, F. H., 27, *31*
Stolk, A. C., 49, *67*, 565, *580*
Storck, R., 11, *31*
Stowell, E. A., 295(264), *317*, 532, *580*
Streiblova, E., 542, *580*

Subramanian, C. V., 48, *67*, 522, 524, 549, 551, *580*
Sulmont, P., 113, *130*
Sutton, B. C., 297(264a), 317, 524, 526, 532, 533, 534, 536, 540, 542, 543, 545, 546, 547, 548, 549, 551, 552, 553, 554, 555, 557, 558, 563, 565, 567, 569, 571, 572, *580*, *581*
Svrček, M., 256, 266(274), 268(272), 271(265, 271), 274(265), 275(265), 276(265, 271), 277(271), 286(269), 289(266, 268), 297(270), 302(267), 303(273), *317*, *318*
Swinscow, T. D. V., 194, *216*
Sydow, H., 140, 154, 175, 205, 206, 209, 212, *218*, 533, 534, 536, 548, 555, 562, 565, 570, *577*, *579*, *581*
Sydow, P., 536, *581*

T

Takahashi, I., 294(243), *316*
Tassi, F., 521, *581*
Tavares, I. I., 227, 231, 233, *245*
Tehon, L. R., 280(275), *318*
Terrier, C.-A., 278(276), 279(276), 280(276), 281(276), *318*
Tewari, J. P., 100, *131*
Tewari, V. P., 264(209), 273(210), *315*
Thaxter, R., 223, 224, 227, 229, 231, 232, *245*, *246*, 258(278), 265(277), *318*
Theissen, F., 140, 154, 175, 205, 206, 209, 212, *218*
Thirumalachar, M. J., 127, *131*, 164, 199, *218*
Thom, C., *67*
Tilak, S. T., 175, *218*
Tim, S. K. M., 168, *218*
Ting, W. P., 187, *214*
Tommerup, I. C., 176, *218*
Toro, R. A., 154, *218*
Trappe, J. M., 259(278a), 260(278a), 261 (278a), *318*
Trotter, A., 546, *581*
Tsuchiya, T., 12, *31*
Tubaki, K., 25, *31*, 35, *41*, 48, *66*, *67*, 307(279), *318*, 522, 542, *581*
Tulasne, C., 74, *86*
Tulasne, L. R., 74, *86*
Turian, G., *68*
Twyman, E. S., 540, *581*
Tylutki, E. E., 264(207), 265(206), *315*

# Index

## U
Udagawa, S., 99, 104, *131*
Urries, M. J., 543, *582*

## V
Vanbreuseghem, R., 523, *578*
van Brummelen, J., 267(20, 21), 270(20), 271(21), 272(19), *308*
van der Walt, J. P., 16, 19, 20, 21, 22, 23, 24, 26, 27, *31*
van Kesteren, H. A., 530, 541, 543, 552, 564, *575*
van Niel, C. B., 22, *29*
van Uden, N., 15, 27, *30*, *31*, *32*
van Vloten, H., 302(281), *318*
Vassiljevsky, N. I., 532, 555, *582*
Veenhuis, M., 19, 23, *29*
Velenovský, J., 266(280), 298(280), 307(280), *318*
Verona, O., 163, 164, 168, 169, 170, 171, 172, 173, 174, 175, 176, 178, 179, 180, 181, 182, 183, 184, 185, 186, 187, 188, 189, 190, 191, 192, 193, 196, 197, 198, 199, 203, 204, 206, 207, 208, 209, 210, 211, 212, 213, *218*
Vey, A., 19, 21, *28*
Vidal-Leiria, M., 27, *32*
Viégas, J. H., 72, *86*
Vital, A. F., 204, *214*
Vlamis, J., 81, *86*
von Arx, J. A., 12, 20, *32*, 46, 47, 55, 59, 67, *68*, 88, 90, 91, 95, 99, 100, 102, 103, 105, 106, 108, 109, 110, 111, 114, 119, 120, 121, 122, *130*, *131*, *132*, 150, 152, 154, 163, 164, 165, 168, 169, 170, 171, 173, 174, 175, 176, 178, 179, 180, 181, 182, 183, 184, 185, 186, 188, 190, 196, 197, 198, 199, 200, 203, 204, 205, 206, 207, 208, 209, 210, 211, 212, 213, 214, *217*, *218*, 280(5), 281(5), 283(4), 291(2), 295(3), *307*, *327*, *333*, 532, 533, 552, 553, 555, 559, 560, *582*
von Büren, G., 35, *41*
von Höhnel, F., 102, *132*, 298(121), 299(122), *312*, 521, 522, 523, 524, 529, 532, *582*
von Keissler, K., 193, 194, *219*
von Tavel, F., 529, *582*
Vuillemin, P., 522, *582*

## W
Wakefield, E. M., 300(282), *318*, 523, 532, 540, *576*, *582*
Walker, J., 116, *132*, 524, 540, *582*
Walker, L. B., 15, *32*
Waraitch, K. S., 263(161), 266(161), *313*
Waterman, A. M., 174, *219*, 287(282a), *318*
Webster, J., 282(283), *318*, *328*
Weenhuis, M., 20, *29*
Wehmeyer, L. E., 102, 112, 113, 116, *132*, 154, 188, 192, *219*, 555, 558, 572, *582*
Weiss, F., 287(284), *318*
Wells, D. E., 16, *32*, 299(113), *311*
Wells, V. L., 269(135), *312*
Weresub, L. K., 98, *128*
Whetzel, H. H., 286(286), 287(286, 291), 288(285, 286, 288, 289), 289(286), *318*
Whisler, H., 225, 226, *246*
White, W. L., 15, *32*, 287(288, 291), 288(288, 290), 289(290), *318*
Whitney, H. S., 274(292), *318*
Whittaker, R. H., 1, *7*
Wickerham, L. J., 12, 13, 15, 18, 19, 20, 23, *31*, *32*
Wicklow, D. T., 273(293), *319*
Wieben, M., 40, *41*
Wilcoxson, R. D., 551, *578*
Wilson, D. E., 302(114), *311*
Wilson, M., 543, *582*
Wilson, M. M., 287(294), *319*
Windisch, S., 12, *32*
Winter, G., 232, *246*
Wolf, F. A., *68*, 174, *219*, 288(289), *318*, 532, 558, *582*
Wolf, F. T., *68*
Wollenweber, H. W., 548, *582*
Wyllie, T. D., 542, *576*

## Y
Yamauchi, S., 27, *29*
Yarrow, D., 26, *31*, *32*
Yarwood, C. E., 73, 74, 78, 80, 81, *86*

## Z
Zachariah, K., 542, *582*
Zalasky, H., 540, 548, *582*
Zambettakis, C. E., 533, 534, *582*
Zender, J., 13, *32*
Zogg, H., 154, 196, 197, *219*
Zsolt, J., 12, *30*
Zukal, H., 267(295), 299(296), *319*

# Index to Genera, Families, Orders, and Higher Taxa

An asterisk (*) after a page number indicates an illustration. (†) after a page number indicates a key. *Note*: Only the names of accepted genera of Hyphomycetes are included in this index where they are listed without authorities. For the full author citations and rejected and doubtful names of Hyphomycetes, see pp. 334–425. For new names see pp. 116 and 121.

## A

*Abrothallus* de Not., 199
*Acallomyces* Thaxter, 236
*Acantharia* Theiss. & Syd., 180
*Acanthonitschkia* Speg., 91
*Acanthophiobolus* Berl., 187
*Acanthorhynchus* Shear, 119
*Acanthosphaeria* Kirchst., 100
*Acarellina* Bat. & Maia, 529
*Acarocybe*, 432, 479*
*Acarocybella*, 432, 434, 487*
*Acarosporium* Bubák & Vleugel, 287, 529
*Acervus* Kanouse, 275
*Acetabula* Fuckel, 269
*Achaetobotrys* Bat. & Cif., 171
Achaetomiaceae, 98
*Achaetomiella* Arx, 99
*Achaetomium* Rai, Tewari & Mukerji, 100
*Achorodothis* Syd., 174
*Aciculariella*, 434, 492*
*Acinula*, 435
*Acladium*, 427, 450*
*Acompsomyces* Thaxter, 235, 237*
Acrasiomycetes, 2, 5
*Acremoniella*, 427, 456*
*Acremoniula*, 427, 456*
*Acremonium*, 64, 428, 438*
*Acroconidiella*, 433
*Acroconidiellina*, 430, 433, 479*
*Acrodictys*, 429, 472*
*Acrogenospora*, 427, 456*
*Acrogenotheca* Cif. & Bat., 172
*Acrogynomyces* Thaxter, 236
*Acrophialophora*, 425, 440*
*Acrophragmis*, 432, 435, 479*
*Acrospeira*, 429, 430, 473*
*Acrospermoides* Miller & Thompson, 127
*Acrospermum* Tode ex Fr., 127, 282

*Acrosporium*, 73, 425, 462*
*Acrotamnium*, 435
*Actidium* Fr., 197
*Actinocladium*, 435, 496*
*Actinocymbe* Höhn., 168
*Actinodothis* Syd., 82
*Actinomyxa* Syd., 205
*Actinonema*, 435
*Actinonemella* Höhn., 559
*Actinopeltis* Höhn., 168
*Actinoscypha* Karst., 291
*Actinospora*, 434, 499*
*Actinotextis* Arx, 529
*Actinothyrium* Kunze ex Fr., 518
*Aegerita*, 435, 502*, 517
*Aessosporon* van der Walt, 24
*Agaricostilbum*, 427, 445*
*Agonimia* Zahlbr., 125
*Agyriella*, 428, 444*
*Ahmadia* Syd., 558
*Ainsworthia* Bat., 169
*Aithalomyces* Woronich., 172
*Ajellomyces* McDonough & Lewis, 61
*Ajrekarella* Kamat & Kalani, 540, 569
*Akanthomyces*, 425, 445
*Akaropeltis* Bat. & Bezerra, 168
*Akenomyces*, 435
*Alatospora*, 434, 498*
*Albosynnema*, 433, 481*
*Albotricha* Raitviir, 283
*Aleuria* Fuckel, 268, 276
*Aleuria* (Sacc.) Sacc., 268
*Alina* Racib., 169
*Allescheriella*, 427, 454*
*Allonecte* Syd., 181
*Allophylaria* Karst., 306
*Allosoma* Syd., 170

*Allothyriella* Bat., Cif. & Nascim., 529
*Alphitomorpha* Wallr., 520
*Alternaria*, 429, 471*
*Alveophoma* Bausa Alcalde, 561
*Alysidium*, 425, 453*
*Alysisporium* Peyr., 523
*Amallospora*, 435, 501*
*Amauroascus* Schroet., 61
*Amazonia* Theiss., 82, 90
*Amazonotheca* Bat. & Maia, 213
*Ambrosiella*, 426, 454*
*Ambylosporium*, 425, 426, 436*
*Amerosporium* Speg., 565
*Amorphomyces* Thaxter, 227, 239, 240*
*Amorphotheca* Parberry, 57
Amorphothecaceae, 51, 57
*Amphichaeta* McAlp., 557
*Amphisphaerella* (Sacc.) Kirschst., 111, 119
*Amphisphaeria* Ces. & de Not., 120
Amphisphaeriaceae, 98, 118†
*Amphitiarospora* Agnihot., 529
*Amphobotrys* Hennebert, 286
*Amphoridium* Massal., 124
*Amphoroblastia* Serv., 125
*Ampullifera*, 426, 427, 430, 432, 478*
*Ampulliferina*, 429, 464*
*Amylascus* Trappe, 261
*Amylocarpus* Currey, 64
*Anariste* Syd., 208
*Angatia* Syd., 164
*Angelina* Fr., 292
*Anguillospora*, 434*, 493
*Angulimaya*, 425, 439*
*Angulospora*, 431, 434, 493*
*Anhellia* Racib., 164
*Anisochora* Theiss. & Syd., 107
*Anisomyces* Theiss. & Syd., 115
*Anisostomula* Höhn., 119
*Anixia* Fr., 84
*Anixiella* Saito & Minoura, 111
*Anixiopsis* Hansen, 54*, 62
*Annajenkinsia* Thirum. & Naras., 199
*Annellodochium*, 426, 430, 466*
*Annellolacinia* Sutton, 540, 557
*Annellophora*, 431, 432, 484*
*Annellophorella*, 429, 472*
*Annellophragmia*, 432, 433, 488*
*Anopodium* Lunqv., 110
*Antennatula*, 432, 484*
*Antennella* Theiss. & Syd., 172

*Antennellina* Mendoza, 172
*Antennellopsis* Mendoza, 172
*Antennopsis*, 432, 479*
*Antennospora* Mayers, 117
*Antennularia* Reichenb., 180
*Anthina*, 435
*Anthopeziza* Wettst., 265
*Anthostoma* Nitschke, 102
*Anthostomella* Sacc., 102, 122
*Anthrocobia* Boud., 276
*Antinoa* Vel., 307
*Antromyces*, 425, 429, 468*
*Antromycopsis*, 425, 436*
*Anulohypha*, 435
*Anulosporium*, 435
*Aparaphysaria* Speg., 271
*Apatomyces* Thaxter, 236
*Aphanoascus* Zukal, 54*, 62
*Aphanocladium*, 428, 450*
*Aphanopeltis* Syd., 209
*Aphanostigme* Syd., 178
*Apinisia* La Touche, 61
*Apiocarpella* Syd., 536
*Apiognomonia* Höhn., 115
*Apioporthe* Theiss. & Syd., 115
*Apioporthella* Petrak, 115
*Apiosordaria* Arx & Gams, 97*, 109
*Apiosphaeria* Höhn., 107
*Apiospora* Sacc., 120
*Apiosporina* Höhn., 139*, 152*, 180
*Apiosporopsis* Marian, 114
*Apiothyrium* Petrak, 119
*Apodospora* Mirza & Cain, 110
*Apomelasmia* Grove, 574
*Aposphaeria* Berk., 519, 542, 554
*Apostemidium* Karst., 282
*Appendiculella* Höhn., 82, 90
*Arachniotus* Schroet., 61
*Arachnocrea* Moravec, 127
*Arachnomyces* Massee & Salm., 54*, 62
*Arachnopeziza* Fuckel, 296
*Arachnophora*, 429, 434, 495*
*Arachnoscypha* Boud., 296
*Arachnotheca* Arx, 61
*Arbuscula*, 429, 435, 502*
*Arenaea* Penzig & Sacc., 297
*Arenariomyces* Höhnk, 117
*Aristastoma* Tehon, 563
*Armatella* Theiss. & Syd., 82, 90
*Armillaria* (Fr.) Staude, 16

*Arnaudiella* Petrak, 204
*Arnium* Nitschke apud Fuckel, 109
*Artallendaea* Bat. & Maia, 90
*Arthonia* Ach., 197
Arthoniaceae, 195, 197†
*Arthopyrenia* (Mass.) Müll. Arg., 193
*Arthrinium*, 427, 495*
*Arthrobotrys*, 430, 465*
*Arthrobotryum*, 433, 481*
*Arthrocladium*, 432, 434, 490*
*Arthroderma* Berk., 52*, 60
*Arthrographis*, 425, 437*
*Arthrorhynchus* Kolenati, 223, 236
*Arthrosporium*, 432, 489*
*Arthrothelium* Massal., 197
*Articularia*, 428, 440*
*Articulospora*, 434, 498*
*Arxiella*, 430, 463*
*Aschersonia* Mont., 571
Ascobolaceae, 263
*Ascobolus* Pers. ex Hook., 252*
*Ascocalvatia* Cain, 53
*Ascocalyx* Naumov, 294, 568
*Ascochyta* Lib., 530, 536, 554, 563
*Ascochytella* Tassi, 564
*Ascochytula* (Potebn.) Died., 536
*Ascochytulina* Petrak, 525* 544*, 564
*Ascoconidium*, 429, 473*
Ascocorticiaceae, 283
*Ascocorticium* Bref., 253
*Ascocoryne* Groves & Wilson, 302
*Ascocorynium* Ito & Imai, 285
*Ascodesmis* van Teigh., 61, 253, 270
*Ascoidea* Bref., 15
Ascoideaceae, 13, 15†
*Ascomyces* Mont. & Desm., 36
Ascomycetes, 3
Ascomycotina, 5, 6†, 9–319
*Ascophanella* Faurel & Schot., 267
*Ascophanella* Korf, 267
*Ascophanopsis* Faurel & Schot., 267
*Ascophanus* Boud., 267, 272, 273
*Ascopolyporous* Moeller, 126
*Ascoscleroderma* Clémencet, 259
*Ascosorus* P. Henn. & Ruhl. apud P. Henn. 253, 283
*Ascosparassis* Kobayasi, 274
*Ascosphaera*, 46
Ascosphaerales, 45
*Ascosporium* Berk., 36

*Ascotremella* Seaver, 302
*Ascotremellopsis* Teng & Ou apud Ou, 299
*Ascotricha* Berk., 99
*Ascozonus* (Renny) Hansen, 272
*Ashbya* Guill., 14
Aspergillales, 45
*Aspergillus*, 63, 64, 425, 443*
*Asperisporium*, 430, 466*
*Asteridiella* McAlp., 82, 83*, 90
*Asterina* Lév., 207
Asterinaceae, 202, 207†
*Asterinella* Theiss., 204
*Asterinema* Bat. & Gayao, 204
*Asterobolus*, 287
*Asterocalyx* Höhn., 299
*Astrocystis* Berk. & Br., 183
*Asterodothis* Theiss., 208
*Asterolibertia* Arnaud, 207
*Asteromassaria* Höhn., 188
*Asteromella* Pass. & Thüm., 534, 561
*Asteromyces*, 427, 458*
*Asterophora*, 426, 434, 495*
*Asterosporium* Kunze ex Wallr., 536, 556
*Asterostomopsis* Bat., Cif. & Maia, 529
*Asterotexis* Arx, 198
*Astomella* Thirum., 72
*Atichia* Flotow, 163
Atichiaceae, 163†
*Atksoniella* Diehl, 126
*Atopospora* Petrak, 182
*Atractium*, 433, 434, 489*
*Atropellis* Zeller & Goodd., 294
*Auerswaldia* Sacc., 173, 175
*Aulacostroma* Syd., 210
Aulographaceae, 202, 209†
*Aulographina* Arx & Müller, 209
*Aulographum* Lib., 210
*Aureobasidium*, 428, 439*
*Aurophora* Rifai, 266
*Autoicomyces* Thaxter, 235, 237*
*Autophagomyces* Thaxter, 235, 237*
*Auxarthron* Orr & Kuehn, 60

**B**

*Bacillispora*, 434, 489*
*Bactridiopsis*, 427, 486*
*Bactridium*, 432, 486*, 518
*Bactrodesmiella*, 431, 432, 477*
*Bactrodesmium*, 432, 482*
*Bagcheea* Miller & Menon, 114

## Index to Taxa 595

*Bagliettoa* Massal., 125
*Bagnisiella* Speg., 175
*Bagnisiopsis* Theiss. & Syd., 108
*Bahupaathra*, 428, 439*
*Bahusakala*, 428, 429, 431, 477*
*Bainieria*, 426, 448*
*Bakeromyces* Syd., 100
*Balanium*, 430, 466*
*Balansia* Speg., 126
*Balansiella* P. Henn., 126
*Balansina* Arnaud, 208
*Balansiopsis* Höhn., 126
*Balladyna* Racib., 82, 169, 170
*Balladynastrum* Hansf., 82, 83*
*Balsamia* Vittad., 261
*Banhegyia* Zeller & Toth, 199
*Barlaeina* Sacc. 270
*Barnettella*, 429, 470*
*Barssia* Gilkey, 261
*Bartalinia* Tassi, 562
*Bartheletia*, 435
*Barya* Fuckel, 127
*Basidiobotrys*, 426, 449*
Basidiomycetes, 3
Basidiomycetous yeasts, 22, 24†
Basidiomycotina, 5, 6†
*Basifimbria*, 427, 459*
*Basipetospora*, 61, 425, 462*
*Basipilus* Subram., 557
*Batistia* Cif., 122
*Batistinula* Arx, 208
*Beauveria*, 427, 459*
*Belaina* Bat. & Peres, 529
*Bellulicauda* Sutton, 527*, 569
*Belonidium* Mont. & Dur., 297
*Belonioscypha* Rehm, 305
*Belonium* Sacc., 283, 292, 305
*Belonopsis* (Sacc.) Rehm, 292, 293
*Beltrania*, 427, 458*, 522, 549
*Beltraniella*, 427, 458*, 549
*Beltraniopsis*, 427, 458*
*Benedekiella* Verona, 119
*Beniowskia*, 426, 450*
*Berkleasmium*, 429, 472*
*Berlesiella* Sacc., 188, 191
*Bertia* de Not., 92
*Biciliospora* Petrak, 91
*Bifusella* Höhn., 279
*Bifusepta* Darker, 279
*Bioporthe* Petrak, 112

*Biostictis* Petrak, 282
*Biotyle* Syd., 102
*Bipolaris*, 141, 187, 190, 305
*Bispora*, 430, 464*
*Bitancourtia* Thirum. & Jenkins, 164
*Blasdalea* Sacc., 205, 210
*Blastobotrys*, 426, 451*
*Blastocapnias* Cif. & Bat., 172
Blastomycetes, 7, 12, 24
*Blastophorella*, 430, 465*
*Blennoria* Fr., 571
*Bleptosporium* Steyaert, 540, 567
*Blogettia*, 435
*Blogiascospora* Shoem. & Müller, 121
*Bloxhamia*, 425, 441*, 541, 568
*Blumeriella* Arx, 279, 295
*Boedijnopeziza* Ito & Imai, 265
*Bolinia* Sacc., 123
*Bombardia* Fr., 97*, 109
*Bombardioidea* C. Moreau, 110
*Bonaria* Bat., 207
*Boothiella* Lodhi & Mirza, 100
*Bostrichia* Fr., 518
*Botryodiplodia* Sacc. 567
*Bothrodiscus* Shear, 294, 539*, 568
*Botryoderma*, 427, 456*
*Botryonipha*, 435
*Botryosphaeria* Ces. & de Not., 139*, 175, 183
Botryosphaeriaceae, 178, 183†
*Botryosphaerostroma* Petrak, 525*, 563
*Botryosporium*, 426, 451*
*Botryostroma* Höhn., 182
*Botryotinia* Whetzel, 286
*Botryotrichum*, 426, 454*
*Botrytis*, 274, 286, 427, 451*
*Boudiera* Cooke, 267
*Brachydesmiella*, 427, 482*
*Brachyhelicoon*, 431, 476*
*Brachysporiella*, 432, 479*
*Brachysoporium*, 433, 479*
*Brefeldiella* Speg., 210
Brefeldiellaceae, 202, 210†
*Bresadolella* Höhn., 100
*Brettanomyces* Kuffer. & van Laer, 26
*Briosia*, 426, 462*
*Brooksia* Hansf., 172
*Broomella* Sacc., 121
*Brunchorstia* Eriksson, 294, 572
*Buellia* de Not., 199

*Buerhenerula* Syd., 189
*Bulgaria* Fr., 264, 265, 300
*Bulgariastrum* Syd., 294
*Bulgariella* Karst., 300
*Bullera* Derx, 25
*Burgoa*, 435, 502*
*Butleria* Sacc., 164
*Byssoascus* Arx, 61
*Byssochlamys* Westl., 63
*Byssocladium*, 435
*Byssogene* Syd., 164
*Byssolophis* Clem., 183
*Byssonectria* Karst., 277

## C

*Caccobius* Kimbr. apud Kimbr. & Korf, 272
*Cacumisporium*, 432, 433, 480*
*Caenothyrium* Theiss. & Syd., 205
*Caeoma* Link, 517
*Cainia* Arx & Müller, 120
*Cainiella* E. Müller, 120
*Calcarisporium*, 427, 461*
*Caliciopsis* Peck, 109
*Calloria* Fr., 291
*Callorina* Korf, 291
*Calocline* Syd., 567
*Calogloeum* Syd., 556
*Calonectria* de Not., 105
*Calopora* Nitschke, 112
*Caloscypha* Boud., 275
*Calosphaeria* Tul., 102
*Calostilbella*, 433, 481*
*Calothyriopsis* Höhn., 204
*Calycella* Boud., 252*, 297, 301, 306
*Calycellina* Höhn., 297
*Calycina* Gray ex O. Kuntze, 307
*Calyculosphaeria* Fitzp., 91, 92*
*Calyptra* Theiss. & Syd., 171
*Camarographium* Bubák, 526
*Camarops* Theiss. & Syd., 124
*Camaropycnis* Cash, 528*, 571
*Camarosporium* Schulz., 541*, 562
*Camillea* Fr., 123
*Campoa* Speg., 212
*Camposporium*, 433, 487*
*Camptomeris*, 430, 432, 433, 481*
*Camptomyces* Thaxter, 241
*Campylospora*, 434, 496*
*Candelabrum*, 434, 494*

*Candida* Berkhout, 17*, 27
*Cantharomyces* Thaxter, 239, 240*
*Capillipes* Santesson, 306
*Capnobatista* Cif. & Leal, 171
*Capnociferria* Bat., 172
*Capnocrinum* Bat. & Cif., 172, 173
*Capnocybe*, 433, 481*
*Capnodaria* (Sacc.) Theiss. & Syd., 173
Capnodiaceae, 167, 171†
*Capnodium* Mont., 173
*Capnogonium* Bat. & Perez, 171
*Capnophaeum* Speg., 173
*Capnophialophora*, 428, 439*
*Capronia* Sacc., 188, 191
*Carbomyces* Gilkey, 259
*Carpenteles* Langeron, 64
*Caryospora* de Not., 185
*Casaresia*, 435, 497*
*Cashiella* Petrak, 292
*Castagnella* Arnaud, 103
*Catacauma* Theiss., 107
*Catenophora* Luttrell, 539*, 556
*Catenularia*, 101, 426, 438*
*Catinella* Boud., 291
*Caudella* Syd., 204
*Caulocarpa* Gilkey, 261
*Causalis* Theiss., 108
*Canangiopsis* Rehm, 302
*Cenangium* Fr., 302
*Cenococcum*, 435
*Centrospora*, 432–4, 490*
*Cephaliophora*, 432, 465*
*Cephaliosporopsis*, 430, 468*
*Cephaloascus* Hanawa, 16
*Cephalosporium*, 62, 64
*Cephalotheca* Fuckel, 63
Cephalothecaceae, 57, 63
*Cephalotrichum*, 426, 448*
*Ceratocladium*, 426, 449*
*Ceratocystis*, Ell. & Halst., 98, 542
*Ceratomyces* Thaxter, 235, 237*
Ceratomycetaceae, 234
*Ceratophacidium* J. Reid & Pirozynski, 278
*Ceratophorum*, 432, 487*
*Ceratopycnis* Höhn., 364
*Ceratosphaeria* Niessl, 112
*Ceratosporella*, 435, 500*
*Ceratosporium*, 435, 497*
*Ceratostomina* Hansf., 101
*Cercidospora* Körb., 184

## Index to Taxa

*Cercosperma*, 434, 488*
*Cercospora*, 432–4, 491*
*Cercosporella*, 432, 434, 491*
*Cercosporidium*, 430, 432, 433, 480*
*Cercosporula*, 434, 491*
*Ceriophora* Höhn., 110, 120
*Ceriospora* Niessl, 120
*Ceriosporopsis* Lindner, 117
*Cerodothis* Muthappa, 174
*Ceuthospora* Fr., 571
*Chaetapiospora* Petrak, 94*, 119
*Chaetochalara*, 425, 426, 428, 441*
*Chaetoconidium*, 426, 432, 454*
*Chaetoconis* Clem., 566
*Chaetodiplodia* Karst., 525*, 564
*Chaetodiscula* Bubák & Kabat, 532
*Chaetomelanops* Petrak, 183
*Chaetomella* Fuckel, 530, 565
Chaetomiaceae, 46, 98
*Chaetomidium* (Fuckel) Zopf, 99
*Chaetomium* Kunze, 46, 99
*Chaetonaevia* Arx, 290
*Chaetopatella* Hino & Katumoto, 550*, 568
*Chaetoplaca* Syd., 213
*Chaetopotius* Bat., 172
*Chaetopsina*, 428, 440*
*Chaetopsis*, 428, 430, 469*
*Chaetopyrena* Pass., 565
*Chaetoscutula* E. Müller, 198
*Chaetospermum* Sacc., 428, 536, 571
*Chaetosphaella* Booth & Müller, 10
*Chaetosphaeria* Tul., 101
*Chaetosphaeronema* Moesz, 565
Chaetothyriaceae, 167, 168†
*Chaetothyrina* Theiss., 207
*Chaetothyrium* Speg., 168
*Chalara*, 57, 127, 425, 426, 428–31, 433, 441*
*Chalaropsis*, 427, 457*
*Chalcosphaeria* Höhn., 115
*Chantransiopsis*, 426, 457*
*Charonectria* Sacc., 104
*Chaudefaudia* Feldmann, 98, 116
*Chaudefaudiella* Faurel & Schot., 62
*Cheilaria* Lib., 519
*Cheilodonta* Boud., 290
*Cheilymenia* Boud., 275
*Cheiromycella*, 435, 501*
*Cheiromyces*, 435, 501*
*Cheirospora*, 429, 476*
*Chevalieria* Arnaud, 169

*Chevalieropsis* Arnaud, 169
*Chiodecton* (Ach.) Müll.-Arg., 197
*Chitonomyces* Peyritsch, 241, 242*
*Chitonospora* Bomm., Rous., & Sacc., 121
*Chlamydomyces*, 426, 430, 467*
*Chlamydozyma* Wickerham, 15
*Chlorencoelia*, 305, 307
*Chloridiella*, 427, 459*
*Chloridium*, 101, 428, 429, 438*
*Chlorioactis* Kupfer ex Eckblad, 263
*Chlorociboria* Seaver ex Ramam., Korf & Batra, 289, 304, 305
*Chlorpscypha* Seaver, 300
*Chlorosplenium* Fr., 289, 293, 304, 305
*Choiromyces* Vitt., 261
*Chondroplea* Kleb., 540, 570
*Chondrostroma* Syd., 528*, 571
*Chromelosporium* Corda, 268
*Chrondroplea* Kleb., 116
*Chrysosporium*, 51, 60, 426, 427, 454*
*Chuppia*, 427, 429, 471*
Chytridiomycetes, 3, 5
*Ciboria* Fuckel, 287–9
*Ciboriella* Seaver, 307
*Ciborinia* Whetzel, 287
*Ciboriopsis* Dennis, 288
*Cicinnobella* P. Henn., 537*, 563
*Cicinnobolus* Ehrenb., 71, 565
*Ciferriomyces* Petrak, 108
*Ciferriotheca* Bat. & Lima, 213
*Ciferriusia* Bat., 168
*Ciliciopodium*, 427, 446*
*Ciliochorella* Syd., 543, 544*, 574
*Ciliomyces* Höhn., 105
*Circinoconis*, 431, 475
*Circinotrichum*, 434, 449*
*Circinotrichum* Nees, 522
*Cirrenalia*, 430, 475*
*Cirsosia* Arnaud, 209
*Cirsosiella* Arnaud, 209
*Cistella* Quél., 297
*Citeromyces* Santa Maria, 20
*Cladobotryum*, 127, 429, 430, 462*, 463*
*Cladorrhinum*, 428, 432, 439*
*Cladosporiella*, 434, 491*
*Cladosporium*, 57, 426, 430, 464*
*Clasterosporium*, 432, 488*
*Clathridium* Sacc., 121
*Clathrosphaerina*, 429, 473*
*Clathrospora* Rabenh., 192

*Claussenomyces* Kirschst., 301
*Clavaria* Vaill ex Fr., 285
*Clavariopsis*, 434, 498*
*Clavatospora*, 434, 498*
*Claviceps* Tul., 126
Clavicipitaceae, 98, 125†
Clavicipitales, 93
*Clavidisculum* Kirschst., 297
*Cleistosphaeria* Syd., 169
*Cliostomum* Fr., 519
*Clithris* (Fr.) Rehm, 279
*Clonophoromyces* Thaxter, 236
*Clonostachys*, 425, 428
*Clypeocarpus* Kirschst., 114
*Clypeolella* Höhn., 208
*Clypeopycnis* Petrak, 567
*Coccidiascus* Chatton, 15
*Coccidiella* Hara, 108
*Coccidioides*, 425
*Coccobotrys*, 435
*Coccochorella* Höhn., 107
*Coccodothis* Theiss. & Syd., 199
*Coccoidea* P. Henn., 181
*Coccoidella* Höhn., 175
*Coccomyces* de Not., 279, 295
*Cocconia* Sacc., 211
*Cocconiopsis* Arnaud, 211
*Coccophacidium* Rehm, 280
*Coccospora*, 435
*Coccostroma* Theiss. & Syd., 108
*Coccostromopsis* Plunkett, 108
*Cochliobolus* Drechs., 141*, 187
*Cochliomyces* Speg., 236, 238*
*Codinaea*, 428, 430, 433, 447
Coelomycetes, 7, 513–582, 555†
*Coelosphaeria* Sacc., 91
*Coleodictyospora*, 429, 473
*Colletotrichella* Höhn., 574
*Colletotrichum* Corda, 428, 532, 560
*Colmnosphaeria* Munk, 174
*Colphoma* Wallr, 279, 540
*Comatospora* Pirozy. & Shoem., 541*
*Comocephalum* Syd., 563
*Coniella* Höhn., 543, 549, 565
*Coniochaeta* (Sacc.) Massee, 109
*Coniosporium*, 429, 470*
*Coniothyrina* Syd., 562
*Coniothyrium* Corda, 191, 549, 562
*Conoplea*, 264, 427, 460*
*Cookeina* Kuntze, 265, 266

*Cookella* Sacc., 164
*Coprobia* Boud., 275
*Coprobolus* Cain & Kimbr., 272
*Copromyces* Lundq., 111
*Coprotinia* Whetzel, 288
*Coprotus* Korf & Kimbr. apud Kimbr. & Korf, 272
*Cordana*, 430, 466*
*Cordella*, 427, 455*
*Cordierites* Mont., 252*, 303
*Cordyceps* Link, 94*, 126
*Coremiella*, 425, 436*
*Coreomyces* Thaxter, 231, 236, 238*
*Corniculariella* Karst., 561
*Corollospora* Werderm., 117
*Coronellaria* Karsten, 292
Coronophoraceae, 89, 90, 91†
Coronophorales, 46, 91
*Coronospora*, 92, 435
*Coronosporella* Höhn, 91
*Coryne* Nees ex Gray, 302
*Corynelia* Fr., 108
Coryneliaceae, 96, 108†
*Coryneliopsis* Fitzp., 108
*Corynella* Boud., 301
*Corynespora*, 431, 432, 488*, 547
*Corynesporella*, 432, 488*
*Corynetes* Haszl., 285, 305
*Coryneum* Nees ex Fr., 113, 432, 540, 558
*Coscinaria* Ell. & Everh., 125
*Costantinella*, 427, 460*
*Crandallia* Ell. & Sacc., 574
*Crebrothecium* Routien, 14
*Creopus* Link, 103
*Crinula*, 303, 428, 446*
*Crocicreas* Fr., 568
*Crotone* Theiss. & Syd., 182
*Crumenula* de Not., 294, 299
*Crumenulopsis* Groves, 294, 567
*Cryptendoxyla* Malloch & Cain, 57
*Cryptocline* Petrak, 540
Cryptococcaceae, 25, 26†
*Cryptococcus* Kütz. em. Phaff & Spencer, 27
*Cryptocoryneum*, 435, 501*
*Cryptoderis* Auersw., 116
*Cryptodiaporthe* Petrak, 116, 570
*Cryptodiscus* Corda, 290
*Cryptogenella* Syd., 566
*Cryptoleptosphaeria* Petrak, 100

*Cryptomela* Sacc., 559
*Cryptomyces* Grev., 280
Cryptomycetaceae, 278
*Cryptomycina* Höhn., 281
*Cryptonectriella* (Höhn.) Weese, 104
*Cryptonectriopsis* Höhn., 114
*Cryptophiale*, 430, 434, 492*
*Cryptosphaerella* Sacc., 92
*Cryptosphaeria* Grev., 118
*Cryptospora* Tul. 111
*Cryptosporella* Sacc., 114
*Cryptosporiopsis* Bubák & Kabat, 289, 290, 540, 543, 554, 560
*Cryptosporium* Kunze ex Fr., 519, 520
*Cryptostictis* Fuckel, 557
*Cryptostroma*, 426, 436*
*Cryptothecium* Penz. & Sacc., 197
*Ctenomyces* Eidam, 52*, 60
*Cucurbidothis* Petrak, 192
*Cucurbitaria* Gray ex Grev., 192
*Cudonia* Fr., 252*, 304
*Cudoniella* Sacc., 252*, 307
*Cudoniopsis* Speg., 286
*Culicidospora*, 434, 496*
*Curculiospora*, 431, 475*
*Curvularia*, 433, 484*
*Curvulariopsis*, 431, 433, 475*
*Custingophora*, 429, 442*
*Cuticularia*, 435
*Cyathicula* de Not., 306
*Cyathipodia* Boud., 269
*Cycloschizon* P. Henn., 211
*Cyclotheca Theiss.*, 203
*Cylindrocarpon*, 105, 433, 489*
*Cylindrocladium*, 430, 433, 489*
*Cylindrocolla*, 291, 425, 437*
*Cylindrodendrum*, 430, 468*
*Cylindrogloeum* Petrak, 543
*Cylindrosporella* Höhn., 115
*Cylindrosporium* Grev., 519
*Cylindrosporium* Unger, 295
Cylindrotrichum, 430, 468*
*Cymadothea* Wolf, 147*, 174
*Cyphellophora*, 431, 433, 468*
*Cystodendron*, 428, 441*
*Cystotricha* Berk. & Br., 304
*Cytispora* Fr., 518
*Cytodiplospora* Oud., 559
*Cytogloeum* Petrak, 559
*Cytonaema* Höhn., 566

*Cytophoma* Höhn., 530
*Cytoplacosphaeria* Petrak, 567
*Cytoplea* Bizz. & Sacc., 545*, 574
*Cytosphaera* Died., 528*, 570
*Cytospora* Ehrenb. ex Fr., 113, 572
*Cytosporella* Sacc., 546
*Cytosporina* Sacc., 573
*Cyttaria* Berk., 253, 254*, 258
Cyttariaceae, 258
Cyttariales, 256, 257, 258
*Cyttariella* Palm, 258

**D**

*Dactuliophora*, 435, 502*
*Dactylaria*, 430, 432, 433, 465*
*Dactylobotrys*, 268
*Dactylella*, 432, 486*
*Dactylomyces* Sopp, 54*, 62
*Dactylosporium*, 429, 471*
*Daldinia* Ces. & de Not., 123
*Daleomyces* Setchell, 268
*Dangeardiella* Sacc. & Syd., 189
*Darluca* Cast., 567
*Dasyobolus* (Sacc.) Sacc., 267
*Dasyscypha* (Fr.) Fuckel, 297
*Dasyscyphella* Tranz., 297
*Davincia* Penzig & Sacc., 306
*Dasyscyphus* Gray, 283, 297
*Davisomycella* Darker, 280
*Debaryella* Höhn., 100, 105
*Debaryomyces* Lodder & Kreger-van Rij, 21
*Deightonia* Petrak, 559
*Deightoniella*, 430, 466*
*Dekkera* van der Walt, 19
*Dekkeromyces* Wickerham & Burton, 21
*Delastria* Tul. 260
*Delastriopsis* Mattir., 261
*Delitschia* Auersw., 184
*Delortia*, 431, 474*
*Delphinella* (Sacc.) Kuntze, 174
*Dencoeliopsis* Korf, 301
*Dendrodochium*, 427, 441*
*Dendrographium*, 433, 478*
*Dendrophoma* Sacc., 545
*Dendrosphaera* Pat., 62
*Dendrospora*, 434, 496*
*Dendrosporium*, 430, 435, 495*
*Dendrostibella* Höhn., 301
*Dendryphion*, 430, 432, 433, 464*, 478*
*Dendryphiopsis*, 432, 483*

*Densocarpa* Gilkey, 261
*Depazea* Fr., 518, 519
*Dermapteromyces* Thaxter, 236
*Dermatea* Fr., 294
Dermateaceae, 283
*Dermateopsis* Nannf., 294
*Dermatina* (Sacc.) Höhn., 194
*Dermatocarpon* Eschw., 124
*Dermea* Fr., 294, 527
*Desmazierella* Lib., 263
*Desmidiospora*, 435, 494*
*Desmopatella* Höhn., 535*, 566
Deuteromycotina, 4, 5, 7†, 321–519
Deuteromycetes, 3
*Dexhowardia*, 426, 450*
*Diachora* J. Müller, 107
*Diachlorella* Höhn., 107, 550*, 571
*Dialacenium* Syd., 82, 171
*Diandromyces* Thaxter, 241
*Diaphoromyces* Thaxter, 241
*Diapleella* Munk, 121
*Diaporthe* Nits., 116, 572
*Diaporthella* Petrak, 115
Diaporthiaceae, 98, 118†
*Diaporthopsis* Fabre, 114
*Diatractium* Syd., 116
Diatrypaceae, 98, 111†
*Diatrype* Fr., 94*, 118
*Dibeloniella* Nannf., 293
*Dichlaena* (Dur. & Mont.) Maire, 64
*Dichobotrys* Hennebert, 274
*Dichomera* Cooke, 537*, 570
*Dichomyces* Thaxter, 241, 242*
*Dichotomophthora*, 432, 433, 483*
*Dichotomophthoropsis*, 431
*Dichotomyces* Saito ex Scott, 58*, 64
*Dicladium* Cesati, 532
*Diclonomyces* Thaxter, 236
*Dicrandromyces* Thaxter, 239
*Dicranidion*, 289, 435, 501*
*Dictyoarthrinium*, 429, 472*
*Dictyocatenulata*, 429, 470
*Dictyodesmium*, 429, 473*
*Dictyodothis* Theiss. & Syd., 192
*Dictyonella* Höhn., 164
*Dictyoporthe* Petrak, 113
*Dictyosporium*, 429, 435, 501*
*Dictyostomiopelta* Viégas, 207
*Dictyothyrina* Theiss., 206
*Dictyotrichiella* Munk, 188, 191

*Dictyothyrium* Theiss., 206
*Dicyma*, 427, 460*
*Didymascella* Maire & Sacc. apud Maire et al., 284
*Didymella* Sacc., 182
*Didymobotryum*, 430, 464*
*Didymopleela* Munk, 186
*Didymosamarosporella* Johns. & Gold, 117
*Didymosphaeria* Fuckel, 184, 186
*Didymosporina* Höhn., 558
*Didymosporium* Nees ex Fr., 517
*Didymostilbe*, 430, 468*
*Didymotrichella*, 430, 466*
*Diehlia* Petrak, 300
*Diehliomyces* Gilkey, 261
*Digitosporium* Gremmen, 294, 567
*Diheterospora*, 429, 444*
*Dilophosphora* Desm., 528*, 573
Dimeriaceae, 177, 178†
*Dimeriella* Speg., 178, 179
*Dimerina* Theiss., 179
*Dimerium* Sacc. & Syd., 179
*Dimeromyces* Thaxter, 242*, 243
*Dimerosporium* Fuckel, 207
*Dimorphomyces* Thaxter, 242*, 243
*Dimorphospora*, 426, 428, 457*
*Dinemasporium* Lév., 568
*Dioicomyces* Thaxter, 228, 239, 240*
*Diplacella* Syd., 114
*Diplocarpa* Massee, 297
*Diplocarpon* Wolf, 295
*Diplocladiella*, 435, 500*
*Diplococcium*, 430, 464*
*Diplodia* Fr., 540, 554
*Diplodina* Westend., 559
*Diplogelasinospora* Cain, 111
*Diplomyces* Thaxter, 236
*Diploospora*, 430, 463*
*Diplorhynchus*, 430, 474*
*Diplotheca* Starb., 164
*Diplozythia* Bubák, 527
*Diplozythiella* Died., 544*, 574
Dipodascaceae, 15
*Dipodascus* Lagerh., 15
*Dipodomyces* Thaxter, 228
*Diporotheca* Gordon & Shaw, 90
*Disaeta* Bonar, 557
*Discella* Berk. & Br., 116, 559
Discellaceae, 554
*Discina* (Fr.) Fr., 252*, 264, 268–70

## Index to Taxa

*Discinella* Boud., 306
*Disciotis* Boud., 269
*Discocainia* J. Reid & Funk, 295
*Discocistella* Svrček, 297
*Discodiaporthe* Petrak, 112
Discogloeum Petrak, 543
*Discohainesia* Nannf., 295
Discomycetes, 6, 249–319, 256, 257†
*Discosia* Lib., 523, 529
*Discosporiella* Petrak, 289
*Discosporium* Höhn., 545*, 559
*Discostroma* Clem., 121
*Discostromella* Petrak, 115
*Discula* Sacc., 114, 542
*Disculina* Höhn., 559
*Distolomyces* Thaxter, 236
*Ditopella* de Not., 108
*Docholopha* Cooke, 557
*Dolabra* Booth & Ting, 187
*Domingoella*, 427, 456*
*Dothidasteroma* Höhn., 210, 211
*Dothidasteromella* Höhn., 208
*Dothidea* Fr, 144*, 175
Dothideaceae 168, 173†
Dothideales 136, 161, 165, 167†
*Dothidella* Speg., 203
*Dothidina* Theiss. & Syd. 104
*Dothiopeltis* E. Müller, 214
*Dothiora* Fr., 137, 144, 176
Dothioraceae, 167, 175†
*Dothiorella* Sacc., 554
*Dothiostroma* Hulbary, 531*, 556
*Drechslera*, 433, 485*
*Drechslera* Ito, 192
*Drechslerella*, 432, 486*
*Drepanoconis*, 431, 474*
*Drepanomyces* Thaxter, 229
*Drepanopeziza* (Kleb.) Höhn., 295, 559
*Drumopama*, 427, 460*
*Duebenia* Fr., 307
*Dubitatio* Speg., 104
*Duosporium*, 433, 479*
*Duplicaria* Fuckel, 279
*Durandiella* Seaver, 300
*Durandiomyces* Seaver, 268
*Durella* Tul., 304
*Dussiella* Pat., 126
*Dwayabeeja*, 430, 434, 478*
*Dwayaloma*, 432, 469*
*Dysrhynchis* Clem., 170

### E

*Echinobotryum*, 427, 458*
*Echinodothis* Atk., 126
*Echinopodospora* Robison, 110
*Echinothecium* Zopf, 178
*Echnidnodella* Theiss. & Syd., 209
*Ecteinomyces* Thaxter, 239, 240*
*Ectostroma*, 435
*Eidamella* Matr. & Dass., 52*, 60
*Einoa* Kohlmeyer, 64
*Eladia*, 426, 442*
*Elaphomyces* Nees ex Fr., 47, 259
Elaphomycetaceae, 18, 259
*Elderia* McLennan, 261
*Eleutherascus* Arx, 61
*Eleutheromycella* Höhn., 543
*Eleutheromyces* Fuckel, 550* 566
*Elletevera*, 433, 491*
*Ellisiella* Sacc., 522
*Ellisiodothis* Theiss., 137*, 152*, 203
*Elsinoë* Racib., 163, 559
*Elvela* L., 269, 270
*Elytroderma* Darker, 279
*Embellisia*, 429, 433, 483*
*Emericella* Berk. & Br., 58*, 63
*Emericellopsis* van Beyma, 58*, 64
*Enchnoa* Fr., 113
*Encoelia* (Fr.) Karst., 303
*Encoeliella* Höhn., 298
*Encoeliopsis* Nannf., 294, 303
*Endocalyx*, 427, 455*
*Endocoleroa* Petrak, 182
*Endoconidiophora* Münch, 98
*Endoconidium*, 425
*Endodothiora* Petrak, 176
*Endomyces* Rees, 16
Endomycetaceae, 13, 14, 16†
Endomycetales, 11–22, 13†
Endomycetes, 12
*Endomycopsella* Boedjn, 18
*Endomycopsis* Stelling-Dekker, 17*, 19
*Endostilbum*, 428, 444*
*Endothella* Theiss. & Syd., 107
*Endothia* Fr., 115
*Endoxyla* Fuckel, 111
*Endoxylina* Romell, 102
*Enerthidium* Syd., 541*, 558
*Englerodothis* Theiss. & Syd., 198
*Engleromyces* P. Henn., 123
*Englerula* P. Henn., 170

Englerulaceae, 167, 170†
*Englerulaster* Höhn., 170
*Enterographa* Fée, 198
*Entodesmium* Riess, 187, 190
*Entomosporium* Lév., 295, 536, 538*
*Entonaema* Moeller, 123
*Entopeltis* Höhn., 205, 210
*Endophragmia*, 432, 482*
*Endophragmiopsis*, 432, 482*
*Entosordaria* (Sacc.) Höhn., 119
*Ephelina* Sacc., 295
*Epibelonium* E. Müller, 199
*Epichloë* (Fr.) Tul., 125
*Epicoccum*, 429, 470*, 517
*Epicrea* Petrak, 100
*Epidermophyton*, 432, 456*
*Epiglia* Boud., 299
*Epipolaeum* Theiss. & Syd., 179
*Episphaerella* Petrak, 179
*Eremascus* Eidam, 16
*Eremothecium* Borzi, 14
*Erikssonia* Penz. & Sacc., 108
*Erinelle* Quél., 297
*Erinellina* Seaver, 296
*Eriocercospora*, 433, 434, 483*
*Eriopezia* (Sacc.) Rehm, 296
*Eriopeziza* Auct., 296
*Eriosphaeria* Sacc., 101
*Eriospora* Berk. & Br., 539*, 561
Erysiphaceae, 71, 72, 73†, 79† (conidial states)
Erysiphales, 47, 71–86, 89
*Erysiphe* Hedw. F ex Fr., 73, 75*, 77*, 79*, 80
*Erythrogloeum* Petrak, 560
*Eucantharomyces* Thaxter, 240*, 241
*Euacanthe* Theiss., 91
*Eudarluca* Speg., 182, 567
*Eudimeriolum* Speg., 179
*Eudimeromyces* Thaxter, 243
*Eumela* Syd., 178
*Eumonoicomyces* Thaxter, 239
Eumycota, 3, 4, 5†, 9–582
*Eupelta* Syd., 208
*Eupenicillium* Ludwig, 58*, 64
*Eurasina*, 432, 480*
*Europhium*, 47
*Europhium* A. K. Parker, 98
*Eupropolella* Höhn., 291
Eurotiaceae, 56, 63

Eurotiales, 50–64, 57†
*Eurotium* Link ex Fr., 58*, 63, 520
*Euryachora* Fuckel, 174, 182
*Eusynaptomyces* Thaxter, 235
*Eutryblidiella* (Rehm) Höhn., 199
Eutunicatae, 36
*Eutypella* (Nitschke) Sacc., 118
*Eutypha* Tul., 118
*Everhartia*, 431, 474*
*Evulla* Kavina, 301
*Exarmidium* Karst., 120
*Excipula* Fr., 554
Excipulaceae, 554
*Excipularia*, 432, 482*
*Exoascus* Fuckel, 36
*Exophiala*, 429, 431, 468*
*Exosporiella*, 432, 484*
*Exosporium*, 433, 488*, 517

# F

*Fabospora* Kudriavzev, 21
*Fabraea* Sacc., 295
*Febrella* Kirschst., 284
*Fairmaniella* Petrak & Syd., 559
*Fastigiella* Benedix, 270
*Farlowiella* Sacc., 196
*Fenestrella* Tul., 192
*Ferrarisia* Sacc., 210
Filobasidiaceae, 23
*Filobasidium* Olive, 24
*Fimaria* Vel., 272
*Fimetariella* Lundqv., 110
*Fischerula* Mattir., 261
*Fitzpatrickia* Cif., 91
*Flabellospora*, 434, 499*
*Flagellospora*, 434, 493*
*Fleischeria* Penz. & Sacc., 126
*Fomina* Girzitska, 529
*Furcaspora* Bonar, 539*, 569
*Fracchiaea* Sacc., 92
*Fragosphaeria* Shear, 64
*Fuckelia* Bon., 280, 572
Fungi, 4†
*Fusariella*, 431, 477*
*Fusarium*, 101, 433, 489*, 517
*Fusicladiella*, 430, 465*
*Fusicladium*, 427, 430, 459*
*Fusicoccum* Corda, 554
*Fusidium*, 282, 426, 452*

Index to Taxa 603

**G**

*Galactinia* (Cooke) Boud., 268
*Gaeumannia* Petrak, 101
*Gaeumannomyces* Arx & Olivier, 116
*Gaillardiella* Pat., 92
*Galiella* Nannf. & Korf apud Korf, 264, 265
Gasteromycetes, 3, 7
*Gelasinospora* Dowding, 111
*Gelatinodiscus* Kanouse & Smith, 268
*Geminospora* Pat., 107
*Genabea* Tul., 260
*Genea* Vitt., 260
Geneaceae, 259
*Geneosperma* Rifai, 275
*Geniculosporium* Chesters & Greenh., 124
*Geodina* Denison, 266
Geoglossaceae, 283
*Geoglossum* Pers. ex Fr., 284
*Geomorium* Speg., 269
*Geopora* Harkn, 252\*, 257
*Geoporella* Soehner, 260
*Geopyxis* (Pers.) Sacc., 271, 273
*Geotrichum*, 425, 428, 437\*
*Gibbera* Fr., 181
*Gibberella* Sacc., 105
*Gibberidea* Fuckel, 188
*Gibellina* Pass., 106
*Gibellula*, 425, 443\*
*Gilletiella* Sacc. & Syd., 180
*Gilmaniella*, 427, 456\*
*Guilia* Tassi, 563
*Gliocephalis*, 428, 443\*
*Gliocephalothrichm*, 428, 443\*
*Gliocladium*, 64, 103, 428, 429, 441\*
*Gliomastix*, 426, 429, 438\*
*Globulina* Speg., 126
*Gloeandromyces* Thaxter, 236
*Gloeocoryneum* Weindlmayr, 556
*Gloeoglassum* Durand, 284
*Gloeopeziza* Zukal, 299
*Gloeosporidiella* Petrak, 295
*Gloeosporidiella* Petrak, 559
*Gloeosporium* Auct., 532
*Gloeotinia* Wilson, Noble & Gray, 287
*Gloiosphaera*, 428, 443\*
*Glomerella* Spauld. & Schrenk, 106
*Glomopsis*, 427, 455\*
*Glomospora*, 426, 455\*
*Gloniella* Sacc., 196
*Gloniopsis* de Not., 196

*Glonium* Muhlenb. ex Fr., 196
*Glyphium* Nits., 197
*Gnomonia* Ces. & de Not., 115
*Gnomonina* Höhn., 114
*Gnomoniella* Sacc., 115
*Godronia* Moug. & Lév. apud Moug., 294, 299
*Goidanichiella*, 428, 443\*
*Gonatobotrys*, 426, 451\*
*Gonatobotryum*, 426, 452\*
*Gonatophragmium*, 433, 480\*
*Gonytrichum*, 428, 429, 440\*
*Gorgoniceps* Karst., 305
*Graddonia* Dennis, 293
*Grallomyces*, 435, 497\*
*Graphiola*, 426, 450\*
*Graphiothecium*, 426, 446\*
*Graphium*, 428, 448\*
*Gremmenia* Korf, 281
*Gremmeniella*, 294
*Griggsia* Stev. & Dalb., 108
*Griphosphaerella* Petrak, 121
*Griphosphaeria* Höhn., 105, 121
*Griphosphaerioma* Höhn., 120, 567
*Grovesia* Dennis, 304
*Grovesiella* Eriksson, 294
*Grovesiella* Morelet, 294
*Guignardia* Viala & Ravaz, 173
*Guilliermondella* Nadson & Krassil., 19
*Guttularia* Obermeyer, 99
Gymnoascaceae, 51, 59†
*Gymnoascopsis* C. & M. Moreau, 198
*Gymnoascus* Baranetzky, 52\*, 60
*Gymnocratera* P. Henn., 260
*Gymnodiscus* Zukal, 267
*Gymnodochium*, 429, 463\*
*Gymnomitrula* Imai, 307
*Gymnosporangium* Hedw. F., 514
*Gyoerffiella*, 434, 496\*
*Gyromitra* Fr., 252\*, 269
*Gyrothrix*, 434

**H**

*Habrostictis* Fuckel, 290
*Hadronema*, 430, 467\*
*Hadrotrichum*, 427, 462\*
*Haglundia* Nannf., 292
*Hainesia* Ell. & Sacc., 295, 428, 568
*Halbania* Racib., 209
*Halbaniella* Theiss., 205
*Haligena* Kohlmeyer, 118

*Halonectria* Gareth Jones, 104
*Halophiobolus* Cribb, 117
*Halosphaeria* Linder, 117
Halosphaeriaceae, 96, 116†
*Halosphaeropsis* Johnson, 117
*Halotthia* Kohlmeyer, 185
*Halstedia* Stev., 107
*Hamigera* Stolk & Sampson, 49, 63
*Hanseniaspora* Zikes, 19, 26
*Hansenula* Syd., 20
*Hansfordia*, 427, 461*
*Hansfordiella*, 429, 432, 472*
*Hansfordiellopsis*, 429, 432, 472*
*Hapalosphaeria* Syd., 564
*Haplariopsis*, 430, 465*
*Haplobasidion*, 426, 453*
*Haplographium*, 428, 441*
*Haplomyces* Thaxter, 240*, 241
*Haplosporella* Speg., 570
*Haplostroma* Syd., 108
*Hapsidospora* Malloch & Cain, 58*, 64
*Hariotula* Arnaud, 203
*Harknessia* Cooke, 538*, 549, 571
*Harpagomyces*, 432, 434, 495*
*Harpographium*, 428, 434, 492*
*Harposporium*, 434, 492*
*Harziella*, 428, 444*
*Heleococcum* Jörg., 104
*Helicocephalum*, 425, 428, 431, 476*
*Helicodendron*, 431, 476*
*Helicogonium* White, 15
*Helicoma*, 431, 474
*Helicomina*, 431, 433, 475*
*Helicomyces*, 431, 474*
*Helicoon*, 431, 476*
*Helicorhoidion*, 431, 476*
*Helicosporium*, 431, 474*
*Heliscus*, 432, 434, 498*
*Helminthocarpon* Fée, 197
*Helminthosphaeria* Fuckel, 110
*Helminthosporium*, 187, 192, 305, 433, 488*
Helotiales, 256, 257, 282, 283†
*Helotium* Pers. ex Gray, 297, 307
*Halvella* L. ex St.-Amans, 253, 269
Helvellaceae, 263
*Helvellella* Imai, 269
Hemiascomycetes, 6, 9–41
*Hemibeltrania*, 427, 450*
*Hemicarpenteles* Sarbhoy & Elphick, 64

*Hemiodora*, 428, 449*
Hemiphacidiaceae, 283
*Hemiphacidium* Korf, 284
Hemisphaeriales, 136, 161, 200, 202†
*Hendersonia* Sacc., 567
*Hendersonula* Speg., 525*
*Henningsia* Moeller, 122
*Henningsiella* Rehm, 198
*Heptaster*, 435, 497*
*Hercospora* Fr., 112, 573
*Hermatomyces*, 429, 472*
*Herpomyces* Thaxter, 228, 231, 239, 240*
*Herposira*, 427, 457*
*Herpotrichia* Fuckel, 187
*Herpotrichiella* Petrak, 188
Herpotrichiellaceae, 188
*Hesperomyces* Thaxter, 236, 237*
*Heterocephalum*, 428, 445*
*Heteroconium*, 432, 478*
*Heteropatella* Fuckel, 304, 532, 540, 568
*Heteropera* Theiss., 114
*Heterosphaeria* Grev., 304
*Heydenia*, 426, 427, 454*
*Heyderia* (Fr.) Lin, 252*, 285, 306
*Hiemsia* Svrček, 276
*Higginsia* Nannf., 295
*Himantia*, 435
*Hiospira*, 431, 476
*Hirsutella*, 427, 428, 443*
*Hirudinaria*, 435, 500*
*Histoplasma*, 427, 455*
*Hobsonia*, 431, 476*
*Hoehneliella* Bres. & Sacc., 568
*Holwaya* Sacc., 303
*Hormiactella*, 426, 452*
*Hormiactis*, 430, 463*
*Hormiokrypsis*, 435, 499*
*Hormisciomyces*, 428
*Hormocephalum*, 432, 484*
*Hormococcus* Robak, 535*, 536
*Hormotheca* Bon., 182, 206
*Hughesinia*, 435
*Humaria* Fuckel, 273, 274
*Humarina* Seaver, 268, 270–2, 274, 277
*Humicola*, 427, 456*
*Hyalinia* Boud., 289
*Hyalocapnias* Bat. & Cif., 172
*Hyalodendron*, 426, 452*
*Hyaloderma* Speg., 100
*Hyalodictys*, 429, 470*

*Hyalopeziza* Fuckel, 297
*Hyaloscypha* Boud., 297
Hyaloscyphaceae, 283
*Hyalotheles* Speg., 164
*Hyalotiella* Papendorf, 562
*Hyalotiopsis* Punithal., 562
*Hyalotricha* Dennis, 298
*Hydnobolites* Tul., 260, 261
*Hydnocystis* Tul. & Tul., 260
*Hydnoplicata* Gilkey, 261
*Hydnotrya* Berk. & Br., 260
*Hydnotryopsis* Gilkey, 261
*Hydraeomyces* Thaxter, 241
*Hydronectria* Kirschst., 104
Hymenomycetes, 7
*Hymenopleella* Munk, 121
*Hymenoscyphus* Gray, 305, 306, 307
*Hypha*, 435
*Hyphelia*, 268
*Hyphochlaena*, 435
Hyphochytridiomycetes, 3, 5
Hyphomycetes, 7, 323–509, 335–425 (list of accepted, rejected and doubtful generic names), 425†, 436*–502*, 503 (index to plates)
*Hypnotheca* Tommerup, 176, 213
*Hypocenia* Berk. & Curt., 572
*Hypocopra* Fuckel, 123
*Hypocrea* Fr., 94*, 103
Hypocreaceae, 97, 103†
Hypocreales, 45, 93
*Hypocreela* Sacc., 126
*Hypocreophis* Speg., 103
*Hypocreopsis* Karst., 126
*Hypoderma* DC. ex St.-Amans, 280
*Hypodermella* Tubeuf, 279, 280
*Hypomyces* (Fr.) Tul. 94*, 127
Hypomycetaceae, 98, 127†
*Hyponectria* Sacc., 119
*Hypospilina* (Sacc.) Trav., 115
*Hypoxylina* Starb., 123
*Hypoxylon* Bull. ex Fr., 124
*Hypoxylonopsis* P. Henn., 102
Hysteriaceae, 195, 196†
Hysteriales, 135, 162, 194, 195†
*Hysterium* Tode ex Fr., 196, 517
*Hysterocarina* Zogg, 196
*Hysterodiscula* Petrak, 532
*Hysterographium* Corda, 196
*Hysteropezizella* Höhn., 293

*Hysterostegiella* Höhn., 293
*Hysterostromella* Speg., 212

**I**

*Idiocercus* Sutton, 561
*Ijuhya* Höhn., 102
*Illosporium* Mart., 104
*Ilyomyces* Picard, 236
*Ilytheomyces* Thaxter, 236
*Incrupila* Raitviir, 233
*Indiella*, 435
*Inermisia* Rifai, 277
*Inocyclus* Theiss. & Syd., 212
Inophyta, 2
*Involucrocarpon* Serv., 124
*Involucrothele* Serv., 125
*Iodophanus* Korf apud Kimbr. & Korf, 267
*Ionomidotis* Durand, 303
*Irene* Theiss. & Syd., 90
*Irenopsis* Stev., 82, 83*, 90
*Irpicomyces*, 434, 491*
*Isothea* Fr., 107
*Isthmiella* Darker, 279
*Isthmospora*, 429, 435, 494*
*Itersonilia* Derx, 25, 426
*Iyengarina*, 435, 500

**J**

*Jacobia*, 428, 449*
*Jacobsonia* Boedijn, 301
*Jaculispora*, 427, 434, 498*
*Jaffuela* Speg., 169
*Jafnea* Korf, 269, 271, 273–4
*Jafneadelphus* Rifai, 271
*Jainesia*, 432, 482
*Johansonia* Sacc., 198
*Johnstonia*, 430
*Julella* Fabre, 193

**K**

*Kabatiella* Bubák, 559
*Karschia* Körb., 199
*Karstenella* Harmaja, 253, 271, 284
*Karstenula* Speg., 193
*Keissleriella* Höhn., 141*, 150*, 190
*Keisslerina* Petrak, 176
*Keithia* Sacc., 284
*Kellermania* Ell. & Ev., 563
*Keratinomyces*, 432, 486*
*Keratinophyton*, 62

*Kernia* Nieuwland, 47, 99, 100
*Kerniomyces* Toro, 213
*Klasterskya* Petrak, 101
*Kleidiomyces* Thaxter, 239
*Kloeckera* Janke, 26
*Kluyveromyces* van der Walt, 21, 22
*Kmetia*, 434, 492*
*Kmetiopsis*, 428, 492*
*Konradia* Rac., 126
*Koochaloma*, 428, 447*
*Korfia* J. Reid & Cain, 284
*Korfiella* Pant & Tewari, 264
*Kostermansinda*, 429, 473*
*Kotlabaea* Svrček, 277
*Kretzschmaria* Fr., 123
*Kriegeria* Rabenh., 301
*Kriegeriella* Höhn., 208
*Kubickia* Svrček, 302
*Kumanasamuha*, 427, 460*
*Kusanobotrys* P. Henn., 169
*Kusanotheca* Bat. & Cif., 171

L

*Laboulbenia* Mont. & Robin, 223, 225, 231, 236, 238*
Laboulbeniaceae, 235
Laboulbeniales, 223, 234†
Laboulbeniomycetes, 6, 223–246, 234†
*Labridella* Brenckle, 120, 536, 549, 567
*Labyrinthomyces* Boedijn, 260
Labyrinthulales, 5
*Lacellina*, 426, 453*
*Lacellinopsis*, 426, 453*
*Lachnaster* Höhn., 297
*Lachnella* Fr., 297
*Lachnellula* Karst., 295
*Lachnum* Retz. ex Karst., 297
*Lacunospora* Cailleux, 110
*Laestadia* Auers., 115
*Laetinaevia* Nannf., 290, 291
*Lagenulopsis* Fitzp., 109
*Lagerbergia*, 294
*Lambertella* Höhn., 288
*Lambro* Rac., 115
*Lamproconium* (Grove) Grove, 548, 559
*Lamprospora* de Not., 268, 270, 277
*Lanzia* Sacc., 288
*Larseniella* Munk., 100
*Lasiobelonium* (Sacc.) Sacc., 296
*Lasiobolus* Sacc., 272

*Lasiobotrys* Kunze ex Fr., 175
*Lasiosordaria* Chenantais, 110
*Lasiosphaeria* Ces. & de Not., 109
*Lasiostictis* Sacc., 284
*Lasmenia* Speg., 569
*Lasmeniella* Petrak & Syd., 541*, 569
*Laterotheca* Bat., 171
Lecanactidaceae, 196, 200†
*Lecanactis* Eschw., 200
*Lecanidion* Endlich., 139*, 200
*Lecanosticta* Syd., 541*, 558
*Lejosphaerella* Höhn., 120
*Lemalis* Fr., 536
*Lembosia* Lév., 209
*Lembosiellina* Bat. & Maia, 209
*Lembosina* Theiss., 209
*Lemonniera*, 435, 499*
*Lentoscopora* Linder, 117
*Leophloea* Gray, 193
*Leotia* Pers. ex Fr., 252*, 285, 301
Leotiaceae, 283
*Lepeutypa* Petrak, 121
*Leporina* Vel., 272
*Lepteutypella* Petrak, 121
*Leptodiscella*, 431, 469*
*Leptodothiora* Höhn., 176
*Leptographium*, 428, 448*
*Leptoguignardia* E. Müller, 189
*Leptomelanconium* Petrak, 558
Leptopeltidaceae, 203, 213†
*Leptopeltis* Höhn., 213
*Leptopeltopsis* Petrak, 137*, 213
*Leptophyma* Sacc., 213
*Leptopodia* Boud., 269
*Leptorhaphis* Körb., 193
*Leptosphaeria* Ces. & de Not., 139*, 186, 187, 189
*Leptosphaerulina* McAlp., 137*, 148*, 171
*Leptospora* Rabenh., 187
Leptostomataceae, 554
*Leptostroma* Fr., 518
*Leptothyrium* Kunze ex Wallr., 519
*Leptotrochila* Karst., 294–5
*Lespiaultinia* Zobel, 261
*Letendraea* Sacc., 137*, 180
*Leucocarpia* Vězda, 124
*Leucoconis* Theiss. & Syd., 72
*Leucoscypha* Boud., 274, 276
*Leucosporidium* Fell, Statz., Hunter & Phaff, 24

*Leucostoma* Nitschke, 113
*Leveillopsis* Stev., 108
*Leveillula* Arnaud, 73, 77*, 79*
*Libertella* Desm., 118, 434, 556
*Lichenoconium* Petrak & Syd., 562
*Licopolia* Sacc. & Syd., 180
*Lignicola* Höhnk, 117
*Lilliputia* Boud. & Pat., 64
*Limacinia* Neger, 172
*Limaciniella* Mendoza, 168
*Limacinula* (Sacc.) Höhn., 173
*Limnaiomyces* Thaxter, 241
*Lindra* Wilson, 118
*Linocarpon* Syd., 116
*Linodochium*, 434, 493*
*Linospora* Fuckel, 116
*Linostoma* Höhn., 98
*Linotexis* Syd., 82
*Lipomyces* Lodder & Kreger-van Rij, 19
Lipomycetoideae, 19
*Lirula* Darker, 279
*Listeromyces*, 432, 482*
*Lithomyces* Viala & Marsais, 99
*Lizonia* Ces. & de Not., 178
Loculoascomycetes, 6, 133–219, 155 (synoptic guide), 161†
*Lodderomyces* van der Walt, 22
*Loliomyces*, 435
*Lomaantha*, 432, 489*
*Lopadostoma* (Nitschke) Trav., 122
*Lophidiopsis* Berl. 184
*Lophidium* Sacc., 184
*Lophionema* Sacc., 183
*Lophiosphaeria* Trev., 183
*Lophiostoma* (Fr.) Ces. & de Not., 184
Lophiostomataceae, 178, 183†
*Lophiotrema* Sacc., 184
*Lophium* Sacc., 197
*Lophodermella* Höhn., 279
*Lophodermium* Chev., 286
*Lophomerum* Ouellette & Magasi, 280
*Lophophacidium* Lagerb., 281
*Lophotrichus* Benjam., 99
*Loramyces* Weston, 94*, 101
*Lowagia* Petrak, 107
*Ludwigiella* Petrak, 564
*Ludwigomyces* Kirschst., 284
*Lulworthia* Sutherland, 117
*Lunulospora*, 434, 492*
*Lycoperdon* Pers. 514

## M

*Macowaniella* Doidge, 208
*Macrodiaporthe* Petrak, 112
*Macrophoma* (Sacc.) Berl. & Vogl., 552
*Macrophomina* Petrak, 515
*Macropodia* Fuckel, 269, 273
*Macroscyphus* Nees ex Gray, 269
*Madurella*, 435
*Magdalaenaea*, 434, 498*
*Magnusiella* Sadebeck, 36
*Mahabalella*, 428, 447*
*Malassezia* Baillon, 26
*Malbranchea*, 425, 436*
*Mamiania* Ces. & de Not., 115
*Mamianiella* Höhn., 114
*Marcelleina* Brumm., Korf & Rifai apud Brumm., 270
*Margaritispora*, 427, 432, 480*
*Mammaria*, 427, 458
*Marssonuina* Magn., 554
*Martinia* Whetzel, 286
*Martininia* Dumont & Korf, 286
*Massaria* de Not., 188
*Massariella* Speg., 120
*Massarina* Sacc., 188
*Massariopsis* Niessl, 120
*Massariovalsa* Sacc., 112
Mastigomycotina, 5†
*Mastigonetron* Kleb., 571
*Mastigosporium*, 432, 433, 486*
*Mastigosporium* Riess, 522
*Mattirolomyces* Fischer, 260
*Maublancia* Arnaud, 204
*Maublancomyces* Herter, 270
*Maurodothella* Sacc. & Syd., 209
*Maurodothina* Arnaud ex Piroz. & Shoem., 208
*Mazosia* Mass., 197
*Mazzantia* Mont., 114
*Medeolaria* Thaxter, 258
Medeolariaceae, 258
Medeolariales, 256, 257
*Megalodochium*, 426, 462*
*Megaster*, 435, 497*
*Melachroia* Boud., 291
Melanconiaceae, 554, 555†
Melanconiales, 555†
*Melanconiella* Sacc., 112
*Melanconis* Tul., 94*, 112, 558
*Melanconium* Link ex Fr., 112, 428, 514, 540–558

*Melanobotrys* Rodway, 102
*Melanochaeta* Müller, Harr & Sulmont, 113
*Melanochlamys* Syd., 206
*Melanographium*, 427, 462*
*Melanomma* Nits. ex Fuckel, 190, 191
*Melanophora* Arx, 532, 543
*Melanopsamma* Niessl, 101
*Melanopsammella* Höhn., 100
*Melanospora* Corda, 99
Melanosporaceae, 96, 98†
*Melaspilea* Nyl., 200
*Melastiza* Boud., 275–6
*Melchioria* Penz. & Sacc. 101
*Meliola* Fr., 82, 83*, 84, 90
Meliolaceae, 71, 89
Meliolales, 89, 90†
*Meloderma* Darker, 280
*Melogramma* Fr., 113
*Melomastia* Nits., 101
*Melomastis* Nits., 189
*Mendoziopeltis* Bat., 207
*Menispora*, 428, 431, 433, 447*
*Menisporopsis*, 428, 431, 469*
*Meria*, 427, 439*
*Merostictis* Clem., 293
*Mesniera* Sacc. Syd., 183
Mesnieraceae, 178, 183†
*Mesophellia* Berk., 47, 259
*Metacapnodium* Speg., 172
*Metachora* Syd. & Butl., 107
*Metacoleroa* Petrak, 180
*Metarrhizium*, 425, 444*
*Metasphaeria* Sacc., 189
*Metathyriella* Syd., 213
*Metchnikowia* Kamienski, 15
*Micraspis* Darker, 281
Microascaceae, 98
Microascales, 45, 47
*Microascus* Zukal, 47, 100
*Microcallis* Syd., 168
*Microclava*, 427, 429, 457*
*Microcyclus* Sacc., 175
*Microdiplodia* Allesch., 526
*Microdochium*, 434, 492*
*Microdothella* Syd., 203, 206
*Microglaena* Koerb., 194
*Microglossum* Gill., 285
*Micronectriella* Höhn., 105
Micropeltidaceae, 202, 206
*Micropeltis* Mont., 206

*Micropeltopsis* Wain., 205
*Micropera* Lév., 294, 300, 572
*Micropyxis* Seeler, 299
*Microscypha* Syd., 297
*Microsomyces* Thaxter, 228
*Microsphaera* Lév., 73, 79*, 80
*Microsphaeropsis* Höhn., 564
*Microsporum*, 51, 60, 432, 486*
*Microstroma*, 265, 428, 445*
*Microthecium* Corda, 99
*Microthelia* Koerb., 185, 193
Microthyriaceae, 202, 203†
Microthyriales, 161, 200
*Microthyriella* Höhn., 213
*Microthyrium* Desm., 203
*Micula* Duby, 294
*Micularia* Boed., 165
*Midotis* Fr., 269
*Milowia*, 431, 432, 477*
*Minimidochium*, 428, 447*
*Mirandina*, 434, 490*
*Misgomyces* Thaxter, 236
*Mitrophora* Lév., 268
*Mitrula* Fr., 285, 286, 307
*Mitteriella*, 208, 433, 482*
*Mixia* Kramer, 34
*Miniaecia* Boud., 299
*Moelleriella* Bres., 126
*Moeszopeltis* Petrak, 176, 213
*Mollieriella* Wint., 165
*Mollisia* (Fr.) Karst., 292–3
*Mollisiella* (Phill.) Massee, 298
*Mollisina* Höhn., 292
Monascaceae, 53, 61
*Monascostroma* Höhn., 171
*Monascus* van Teigh., 54*, 61
*Monilia*, 111, 287, 426, 452*
*Moniliella*, 425, 426, 437*
*Monilinia* Honey, 287
*Moniliopsis*, 435
*Monoceras* Guba, 557
*Monochaetia* (Sacc.) Sacc., 557
*Monochaetiella* E. Castellani, 539*, 544*, 560
*Monocillium* Pers. ex Fr., 101
*Monodictys*, 429, 472*
*Monographella* Petrak, 121
*Monographus* Fuckel, 120
*Monoicomyces* Thaxter, 239, 240*
*Monostachys*, 425, 438*

# Index to Taxa

*Monostichella* Höhn., 542, 543
*Monotrichum*, 430
*Moralesia* Urries, 532
*Morchella* St.-Amans, 252*, 268
Morchellaceae, 263
*Morenoina* Theiss., 171
*Morfea* Roze em. Bat. & Cif., 172
*Morrisographium*, 433, 434, 489*
*Moschomyces* Thaxter, 228
*Mucor* Mich. ex Fr., 518
*Muellerites* Holm, 188
*Muiaria*, 429, 473*
*Muiogone*, 429, 473*
*Mukagomyces* Imai, 259
*Munkiella* Speg., 206
Munkiellaceae 202, 205†
*Munkiodothis* Theiss. & Syd., 107
*Muricopeltis* Viégas, 206
*Murogenella* 432, 485*
Mycetozoa 2
*Mycoangloisia* Arnaud, 168
*Mycoarachis* Malloch & Cain, 57
*Mycocitrus* Moeller, 126
*Mycoenterolobium*, 429, 435, 502*
*Mycogone*, 430. 467*
*Mycohypallage* Sutton, 536, 549, 571
*Mycolachnea* Maire, 274
*Mycomalus* Moeller, 109
*Mycomedusiospora* Munk & Carroll, 109
*Mycomelaspilea* Reinke, 200
*Mycomicrothelia* Keissler, 185
*Myconeesia* Kirschst., 122
*Mycophycophila* Kohlm., 116
Mycophyta, 2
Mycoporaceae, 178, 193†
*Mycoporellum* Müll.-Arg., 194
*Mycoporopsis* Müll.-Arg., 194
*Mycoporum* Flotow, 194
*Mycorhynchus* Sacc., 104
*Mycosphaerella* Johanson, 137*, 174, 534
*Mycothyridium* E. Müller, 121
*Mycovellosiella*, 426, 430, 432
*Myelosperma* Syd., 119
*Myiocopron* Speg., 203
Myriangiaceae, 153†
Myriangiales, 46, 135, 161, 162, 163†
*Myriangiella* Zimm., 213
*Myriangium* Mont. & Berk., 143*, 163
*Myrioconium*, 286, 425, 428, 444*
*Myriodiscus* Boedijn, 252

*Myriogonium* Cain, 15
*Myriosclerothinia* Buchw., 286
*Myriostigmella* Arnaud, 210
*Myrmaeciella* Lindau, 103
*Myrmaecium* Nitschke, 102
*Myrmecocystis* Harkn., 260
*Myrothecium*, 428, 429, 445*
*Mystrosporiella*, 429, 433, 471
*Mytilidium* Duby, 197
*Myxofusicoccum* Died., 527
Myxomycetes, 2, 5
Myxomycota, 3, 4, 5†
*Myxophacidiella* Höhn., 280
*Myxophacidium* Höhn., 280
*Myxosporina* Höhn., 545*, 560
*Myxosporium* Link ex Corda, 291, 520
*Myxotrichella* 435
*Myxotrichum* Kunze, 52*, 60

## N

*Nadsonia* Syd., 19
Nadsonioideae, 19
*Naemacyclus* Fuckel, 284
*Naemospora* Pers. ex Fr., 517
*Naetrocymbe* Körb., 173
*Naevia* Fr., 291
*Nakataea*, 433, 485*
*Nalanthamala*, 425, 445*
*Nannfeldtia* Petrak, 280
*Nannfeltiella* Eckblad, 264
*Nannizzia* Stockdale, 60
*Nanoscypha* Denison, 266
*Nanostictis* Christiansen, 282
*Narasimhella* Thirum. & Mathur, 61
*Nautisphaeria* E. B. G. Jones, 117
*Nectria* Fr., 94*, 105
*Nectriella* Nitschke, 104
*Nectriopsis* Maire, 105
*Negeriella*, 432
*Nematoctonus*, 427, 450*
*Nematogonium*, 426, 451*
*Nematospora* Peglion, 14
*Nematostigma* Syd., 170
*Neobulgaria* Petrak, 301–2
*Neocomospora* E. F. Smith, 104
*Neocudoniella* Imai, 252*, 301
*Neodeightonia* Booth apud Punithal., 183
*Neogodronia* Schlapfer, 294
*Neogymnomyces* Orr, 60

*Neogyromitra* Imai 269, 270
*Neohendersonia* Petrak, 562
*Neokeissleria* Petrak, 112
*Neolecta* Speg., 285
*Neomelanconium* Petrak, 570
*Neonaumovia* Schwarzman, 281
*Neoparodia* Petrak & Cif., 170
*Neopeltella* Petrak, 266
*Neophacidium* Petrak, 281
*Neorehmia* Höhn., 100
*Neostomella* Syd., 209
*Neotapesia* E. Müller & Hütter, 292
*Neotestudina* Segretain & Destombes, 47
*Neottiella* (Cooke) Sacc., 276
*Neottiospora* Desm., 543, 565
*Neottiosporella*, 428, 447*
*Neournula* Paden & Tylutki, 265
*Neozythia* Petrak, 535*, 566
*Neta*, 430, 465*
*Neurospora* Shear & Dodge, 111
*Nia*, 427, 434, 499*
*Niesslia* Auersw., 101
*Nigrosabulum* Malloch & Cain, 64
*Nigrospora*, 427, 456*
*Niptera* Fr., 292
*Nipterella* Starb. ex Dennis, 303
*Nitschkea* Otth, 91
*Nodulisporium*, 124, 427, 460*
*Nodulosphaeria* Rabenh., 187, 189
*Nomuraea*, 426, 452*
*Nothojafnea* Rifai, 274
*Nothomitra* Maas G., 285
*Nothophacidium* J. Reid & Cain, 295
*Nummularia* Tul., 124
*Nummulariola* House, 123
*Nycteromyces* Thaxter, 243
*Nymanomyces* P. Henn., 278

## O

*Obstipipilus* Sutton, 549, 556
*Ocellaria* (Tul.) Karst., 290
*Ocellariella* Petrak, 291
*Ochroglossum* Imai, 285
*Octospora* Hedw. ex Gray, 277
*Odontia*, 15
*Oedemium*, 430, 464*
*Oedocephalum*, 267, 268, 425, 451*
*Oedothea*, 430, 466*
*Oidiodendron*, 51, 425, 437*

*Oidium* Sacc., 73
*Olpitrichum*, 427, 450
*Ombrophila* Fr., 252*, 302
*Omphalospora* Theiss. & Syd., 174
*Oncocladium*, 425, 437*
*Oncopodiella*, 435, 495*
*Oncopodium*, 429, 470*
*Onygena* Pers. ex Fr., 54*, 62
Onygenaceae, 53, 61
Onygenales, 45
*Oomyces* Berk. & Br., 125
Oomycetes, 3, 5
*Oosporidium* Stautz, 27
*Opegrapha* Ach., 197
Opegraphaceae, 195, 197†
*Operculella* Kheswalla, 526, 547
*Ophiobolus* Riess, 116, 187, 189, 190
*Ophiocapnocoma* Bat. & Cif., 172
*Ophiocordyceps* Petch, 126
*Ophiodothella* (P. Henn.) Höhn., 107
*Ophiognomonia* Sacc., 116
*Ophionectria* Sacc., 105, 186
*Ophioparodia* Petrak & Cif., 170
*Ophiostoma* Syd., 46, 98
Ophiostomataceae, 46, 96, 98†
*Ophiostomella* Petrak, 98
*Oplotheciopsis* Bat. & Cif., 100
*Oplothecium* Syd., 100
*Oramasia* Urries, 540
*Orbilia* Fr., 289
Orbiliaceae, 283
*Orbiliaster* Dennis, 289
*Orbiliella* Kirschst., 289
*Orbiliopsis*, 305
*Orbimyces*, 435, 500*
*Orcadia* Sutherland, 105
*Ormanthodium*, 432, 434, 478*
*Ostracoderma*, 268, 426, 451*
*Ostreola* Darker, 197
*Ostropa* Fr., 282
Ostropales, 256, 257, 281, 282†
*Ostropella* (Sacc.) Höhn., 183
*Oswaldia* Rangel, 107
*Oswaldoa* Bat., 213
*Otidea* (Pers.) Bon., 269, 273
*Otidiella* Sacc., 275
*Otthia* Nits., 185
*Overeemia*, 435
*Ovularia*, 427, 459*
*Ovulariopsis*, 425, 462*

## Index to Taxa

*Ovulinia* Weiss, 287
*Ovulitis* Buchw., 287
*Oxydothis* Penz. & Sacc., 120

### P

*Paathramaya*, 427, 446*
*Pachnocybe*, 427, 449*
*Pachycudonia* Imai, 305,
*Pachyella* Boud., 268, 273
*Pachyphloeus* Tul., 261
*Pachysolen* Boidin & Adzet, 20
*Paecilomyces*, 62, 63, 425, 426, 438*, 442*
*Pagidospora*, 427, 454*
*Paidania* Rac., 108
*Palawania* Syd., 204
*Palawaniella* Doidge, 210, 212
*Palomyces* Höhnk, 117
*Panchanania*, 427, 430, 467*
*Papularia* Fr., 518
*Papulaspora*, 429, 435, 502*
*Paracapnodium* Speg., 173
*Paracesatiella* Petrak, 102
*Paracoccidioides*, 427, 454*
*Paraenglerula* Höhn., 82
*Paradiscina* Benedix, 270
*Paradoxa* Mattir., 259
*Paraliomyces* Kohlm., 186
*Paramazzantia* Petrak, 114
*Paranthostomella* Speg., 112
*Paraphaeosphaeria* O. Eriks., 191
*Paraplacidiopsis* Serv., 125
*Parastigmatea* Doidge, 206
*Parencoelia* Petrak, 298
*Parenglerula* Höhn., 170
*Parksia* Cash, 301
*Parmularia* Lév., 211
Parmulariaceae, 202, 210†
*Parmulariopsella* Sivanesan, 211
*Parmulariopsis* Petrak, 211
*Parmulina* Theiss. & Syd., 211
*Parodiella* (Speg.) Theiss & Syd., 181
Parodiopsidaceae, 167, 169†
*Parodiopsis* Maubl., 170
*Paryphydria* Zukal, 299
*Passalora*, 430, 433, 467*
*Paspalomyces*, 430, 463*
*Passarinula* Sacc., 104
*Passeriniella* Berl., 189
*Patella* Weber ex Morg., 271, 274–6
*Patellaria* Fr., 200

Patellariaceae, 196, 199†
*Patellariopsis* Dennis, 292
*Patellina* (Speg.) Speg.,
*Patinella* Sacc., 289
*Patinellaria* Karst., 304
*Patouillardina* Arnaud, 208
*Paxina* Kuntze, 264, 269, 273
*Peckiella* Sacc., 127
*Pectinotrichum* Vars. & Orr, 60
*Peloronectria* Moeller, 104
*Peloronectriella* Doi, 104
*Peltosphaeria* Berl., 193
*Pemphidium* Mont., 119
*Pendulispora*, 429, 473*
*Penicillifer*, 430, 431, 469*
*Penicillium*, 62, 63, 64, 425, 426, 442*, 534
*Penicillopsis* Ghosh, Orr & Kuehn, 54*, 62
*Pentaposporium*, 429, 435, 497*
*Penzigia* Sacc., 123
*Periaster* Theiss. & Syd., 108
*Periconia* 426, 453*
*Periconiella* 426, 427, 430, 432, 433, 481*
Pericystales, 46
*Peridoxylon* Shear, 123
*Periperidium* Darker, 281
*Perischizon* Syd., 212
Perisporiaceae, 71, 81, 82†
Perisporiales, 71
*Perisporiopsis* Stev., 169, 170
*Perithrichspora* Linder, 117
*Perizomella* Syd., 537*, 570
*Perrotia* Boud., 296
*Pestalopeziza* Seaver, 302
*Pestalotia* de Not., 121, 302, 433, 540, 548, 557, 567
*Pestalotiopsis* Steyaert, 556, 557
*Pestalozziella* Sacc. & Ell., 427, 537*
*Pestalozzina* (Sacc.) Sacc., 557
*Petaloporus*, 61
*Petchiomyces* Fischer & Mattir. apud Fischer, 260
*Petrakia*, 435, 495*
*Petrakina* Cif., 204
*Petrakiopsis*, 434, 490*
*Petriella* Curzi, 100
*Petriellidium* Malloch, 100
*Peyritschiella* Thaxter, 241
Peyritschiellaceae, 235
*Peyronelia*, 429, 470*
*Peyronelina*, 429, 470*

*Peyronellaea* Goid. ex Togliani, 564
*Pezicula* Tul., 290
*Peziza* L. ex St.-Amans, 268, 273, 275, 276, 517
Pezizaceae, 263
Pezizales, 45, 256, 257, 262†
*Pezizella* Fuckel, 295
*Pezizellaster* Höhn., 297
*Pezoloma* Clem., 300
Phacidiaceae, 278
Phacidiales, 256, 257, 277, 278†
*Phacidiella* Poteb., 280
*Phacidina* Höhn., 281
*Phacidiopycnis* Poteb., 280, 573
*Phacidiostroma* Höhn., 281
*Phacidium* Fr., 281
*Phacostroma* Petrak, 560
*Phacostromella* Petrak, 573
*Phaeangella* (Sacc.) Massee, 303
*Phaeangellina* Dennis, 300
*Phaeapsis* Petch, 112
*Phaedropeziza* Le Gal, 276
*Phaeoapiospora* Theiss. & Syd., 99
*Phaeobulgaria* Seaver, 300
*Phaeocapnias* Cif. & Bat., 172
*Phaeochora* Höhn., 99
*Phaeochorella* Theiss. & Syd., 108
*Phaeociboria* Höhn., 288
*Phaeocryptopus* Naumov, 181
*Phaeocytostroma* Petrak, 527\*, 543, 572
*Phaeodactylium*, 432, 484\*
*Phaeodiaporthe* Petrak, 112
*Phaeodimeriella* Speg., 179
*Phaeodiscus* Batra, 288
*Phaeodothiora* Petrak, 175
*Phaeohelotium* Kanouse, 305
*Phaeoisaria*, 427, 446\*
*Phaeoisariopsis*, 433, 488\*
*Phaeolabrella* Speg., 532
*Phaeomacropus* P. Henn., 269
*Phaeophomatospora* Speg., 122
*Phaeopolynema* Speg., 558
*Phaeoramularia*, 430, 432
*Phaeosaccardinula* P. Henn., 169
*Phaeosclerotinia* Hori apud Saski, 287
*Phaeosphaeria* Miyake, 190, 191
*Phaeospora* Hepp., 184
*Phaeostoma* Arx & Müller, 100
*Phaeotrichoconis*, 433, 487\*
*Pharcidia* Körb., 125

*Phaulomyces* Thaxter, 236
*Phurmomyces* Thaxter, 235
*Phialea* (Fr. ex Fr.) Gill., 288
*Phialina* Höhn., 297
*Phialocephala*, 428, 441\*
*Phialomyces*, 426, 442\*
*Phialophora* Medlar, 289, 292, 293, 428, 439\*, 542
*Phialophorophoma* Linder, 566
*Phialophoropsis*, 425, 439\*
*Phialostele*, 425, 446\*
*Phialotubus*, 426, 442\*
*Phibalis* Wallr., 303
*Philidium* Kunze ex Fr., 519
*Phillipsia* Berk., 266–7
*Phillipsiella* Cooke, 199
Phillipsiellaceae, 196, 198†
*Phloeoconis*, 435
*Phloeospora* Wallr., 547
*Phloeosporella* Höhn., 295
*Phlyctaena* Mont. & Desm., 566
*Phoenicostroma* Syd., 108
*Phoma* Sacc., 514, 529, 541, 552, 564
*Phomatospora* Sacc., 111
*Phomopsis* Sacc., 116, 280, 526, 545, 572, 573
*Phorcys* Niessl, 120
*Phragmidium* Link, 519
*Phragmocauma* Theiss. & Syd., 190
*Phragmodiscus* Hansf., 121
*Phragmonaevia* Rehm, 281
*Phragmospathula*, 430, 432, 477\*
*Phragmothyrium* Höhn., 205
*Phragmotrichum* Kunze ex Fr., 569
Phycomycetes, 2, 3
*Phycopsis* Mang. & Pat., 163
*Phylacia* Lév., 122
*Phylactinia* Lév., 73, 75\*, 77\*, 79\*, 80
*Phyllachora* Nitschke apud Fuckel, 107
*Phylleutypa* Petrak, 107
*Phyllocrea* Höhn., 108
*Phyllomyces* Lloyd, 303
*Phylloporthe* Syd., 115
*Phyllosticta* Pers. ex Desm., 554, 563
*Phyllostictina* Syd., 530, 563
*Phymatomyces* Kobayasi, 260
*Physalidium*, 427, 435, 495\*
*Physalospora* Niessl, 106, 109
*Physomitra* Boud., 269
*Piceomphale* Svrček, 289

## Index to Taxa

*Pichia* Hansen, 20, 21, 22
*Picoa* Vitt., 260
*Piedraia* Fonseca & Leão, 164
*Piersonia* Harkn., 261
*Pilatia* Vel., 298
*Pilgeriella* P. Henn., 169
*Pilidium* Kunze ex Fr., 295
*Pilulina*, 427, 456*
*Pindara* Vel., 266
*Piricauda*, 429, 472*
*Pirottaea* Sacc., 293
*Pithomyces* 429, 482*
*Pithya* Fuckel, 266
*Pithyella* Boud., 298
*Pityrosporum* Sab., 26
*Placella* Syd., 529
*Placidiopsis* Beltr., 125
*Placoasterella* Sacc., 208
*Placocrea* Syd., 180
*Placonema* (Sacc.) Petrak, 550, 573
*Placonemina* Petrak, 574
*Placuntium* Ehrenb. ex Höhn., 278
*Plagiosphaeria* Petrak, 116
*Plagiostigme* Syd., 112
*Plagiostoma* Fuckel, 115
*Plagiostomella* Höhn., 115
Plasmodiophoromycetes, 5
*Platychora* Petrak, 182
*Platypeltella* Petrak, 205
*Platyspora* Wehmeyer, 192
*Platystomum* Trev., 184
*Plectania* Fuckel, 264–7
Plectascales, 46
*Plectolitus*, 64
*Plectomyces* Thaxter, 235
Plectomycetes, 6, 43–68
*Plectomyriangium* C. & M. Moreau, 198, 213
*Plectosphaera* Theiss., 106
*Plectosphaerella* Kleb., 101
*Plectotrichum*, 435
*Pleiochaeta* 432, 433, 487*
*Plejobolus* (Bomm., Rouss., & Sacc.) O. Eriks., 186
*Plenocatenulis* Bat. & Cif., 529
*Plenodomus* Preuss, 525*, 530, 565
*Pleocryptospora* J. Reid & Booth, 102
*Pleomassaria* Speg., 193
*Pleonectria* Sacc., 105
*Pleosphaeropsis* Died., 527
*Pleospora* Rabenh., 192

Pleosporaceae, 178, 184†
Pleosporales, 136, 162, 176, 177†
*Pleosporopsis* Oerst., 122
*Pleurocatena*, 426, 438*
*Pleuroceras* Riess, 116
*Pleurocytospora* Petrak, 527*, 572
*Pleurophomella* Höhn., 528*, 572
*Pleurophomella* Höhn., 304
*Pleurophomopsis* Petrak, 565
*Pleurophragmium*, 433, 477*
*Pleurosticta* Petrak, 544
*Pleurothecium*, 432, 433, 480
*Plicaria* Fuckel, 270
*Plochmopeltis* Theiss., 198
*Ploettnera* P. Henn., 290–1
*Ploioderma* Darker, 280
*Plokamidimyces*, 428, 444*
*Plowrightia* Sacc., 175
*Plurisperma* Sivanesan, 173
*Pocillum* de Not., 304
*Poculum* Vel., 288
*Pododimeria* E. Müller, 179
*Podonectria* Petch, 126, 186
*Podophacidium* Niessl apud Rabenh., 291
*Podosordaria* Ell. & Holway, 123
*Podosphaera* Kunze ex Lév., 73, 79*, 80
*Podospora* Ces., 110
*Podosporiella*, 433, 483*
*Podosporium*, 432, 484*
*Podostroma* Karst., 103
*Poloniodiscus* Svreček & Kubická, 303
*Polyblastia* Lonnr., 125
*Polyblastiopsis* Zahlbr., 194
*Polycladium*, 434, 496*
*Polyclypeolina* Bat. & Lima, 207, 210
*Polycyclina* Theiss. & Syd., 211
*Polycyclinopsis* Bat., Vital & Lima, 204
*Polycyclus* Höhn., 211
*Polydesmia* Boud., 298
*Polydiscidium* Wakef., 252*, 300
*Polynema* Lév., 549, 568
*Polypaecilum*, 62, 426, 448*
*Polyrhizon* Theiss. & Syd., 180
*Polyscytalina*, 430, 463*
*Polyscytalum*, 426, 430, 463*
*Polystigma* DC, 106
Polystigmataceae, 97, 106†
*Polystigmella* Natalyna, 106
*Polystomella* Speg., 203
*Polythrincium*, 430, 467*

*Pontoporeia* Kohlm., 185
*Poronia* Willd., 123
*Porophoromyces* Thaxter, 241
*Potebniamyces* Smerlis, 280, 573
*Pragmopora* Massal., 304
*Prathigada*, 433, 485*
*Preussia* Fuckel, 184
*Prillieuxina* Arnaud, 209
*Pringsheimia* Schultz, 175
*Pronectria* Clem., 104
*Propolis* (Fr.) Fr., 290, 298
*Prosthemium* Kunze ex Fr., 556
*Prothecium* Fres, 113
*Protobaglieottoa* Serv., 124
*Protogenea* Kobayasi, 260
*Protomyces* Unger, 34*
Protomycetaceae, 33, 36
Protomycetales, 33†, 36
*Protoscypha* Syd., 199
*Protostegia* Cooke, 565
*Protothyrium* Arnaud, 212
Prototunicatae, 12, 36, 45
*Psammina* Rouss. & Sacc., 556
*Pseudoarachniotus* Kuehn,
*Pseudobalsamia* Fischer, 261
*Pseudobeltrania*, 427, 458*
*Pseudobotrytis*, 427, 430, 467*
*Pseudociboria* Kanouse, 307
*Pseudocollema* Kanouse & Smith, 276
*Pseudocryptosporella* J. Reid & Booth, 114
*Pseudodiaporthe* Petrak, 112
*Pseudodiscinella* Dennis, 300
*Pseudodiscus* Arx & Müller, 199
*Pseudoepicoccum*, 428
Pseudoeurotiaceae, 57, 62
*Pseudeurotium* van Beyma, 58*, 64
*Pseudofusidium*, 426
*Pseudographiella*, 431, 433, 489*
*Pseudoguignardia* Gutner, 119
*Pseudogymnoascus* Raillo, 60
*Pseudohelotium* Fuckel, 304
*Pseudolachnea* Ranojevic, 568
*Pseudolembosia* Theiss., 210
*Pseudomassaria* Jacz., 120
*Pseudombrophila* Boud., 272
*Pseudomeliola* Speg., 102, 108
*Pseudonectria* Seaver, 104
*Pseudoparodia* Theiss. & Syd., 180
*Pseudopatellina* Höhn., 545
*Pseudopetrakia*, 435

*Pseudopeziza* Fuckel, 293, 295
*Pseudophacidium* Karst., 280
*Pseudophysalospora* Höhn., 119
*Pseudopithyella* Seaver, 266
*Pseudoplea* Höhn., 171
*Pseudoplectania* Fuckel, 264, 275
*Pseudopyrenula* Müll.-Arg., 193
*Pseudorhizinia* Jaczew., 269
*Pseudorhytisma* Juel, 280
*Pseudoscypha* J. Reid & Piroz., 198
*Pseudosphaeria* Höhn., 152
Pseudosphaeriaceae, 167, 171†
*Pseudospiropes*, 433, 485*
*Pseudothia* Theiss. & Syd., 112
*Pseudothyridaria* Petrak, 102
*Pseudothyrium* Höhn., 527*, 574
*Pseudotis* (Boud.) Boud., 273, 275
*Pseudotrichia* Kirchst., 185
*Pseudotthiella* Petrak, 108
*Pseudovalsa* Ces. & de Not., 113
*Pseudovalsa* Ces. & de Not., 558
*Pseudovalsella* Höhn., 112
*Pseudozythia* Höhn., 550*
*Psilachnum* Höhn., 297
*Psiloglonium* (Höhn.) Petrak, 196
*Psilopezia* Berk., 268, 273, 276, 277
*Psilospora* Rabenh., 570
*Pteridospora* Penz. & Sacc., 186
*Pteroconium*, 427, 455*
*Pterulopsis* 434, 493*
*Ptychoverpa* Boud., 269
*Puscinia* Pers., 517
*Pulparia* (Karst.) Karst., 270
*Pulvinula* Boud., 273, 277
*Pustularia* Fuckel, 273
*Pustulina* Eckblad, 273
*Pycnidiella* Höhn., 527
*Pycnidiopeltis* Bat. & Costa, 529
*Pycnocarpon* Theiss., 210
*Pycnoderma* Syd., 212
*Pycnodermina* Petrak, 212
*Pycnopeltis* Syd., 212
*Pycnopeziza* White & Whetzel, 287
*Pycnothyrium* Died., 543
*Pycnothyrium* Died. em. Arx, 214
*Pyramidospora*, 434, 494*
*Pyrenochaeta* de Not., 564–6
Pyrenomycetes, 3, 6, 69–132, 87, 88 (outline classification), 89†
*Pyrenopeziza* Fuckel, 293, 554

## Index to Taxa 615

*Pyrenophora* Fr., 141*, 150*, 192
*Pyrenostigma* Syd., 183
*Pyricularia*, 430, 432, 433, 485*
*Pyriculariopsis*, 433, 485*
*Pyronema* Carus, 252*, 271
Pyronemataceae, 263
*Pyxidiophora* Bref. & Tavel, 127

**Q**

*Quaternaria* Tul., 118

**R**

*Rabenhorstia* Fr., 112, 573
*Racodium*, 435
*Radulum* Fr., 303
*Raffaelea*, 427, 459*
*Ramsbottomia* Buckley, 277
*Ramularia*, 430, 463*
*Ramulispora*, 434, 493*
*Readeriella* Syd., 541*, 562
*Rebentischia* Karst., 189
*Rechingerella* Petrak, 111
*Rehmiella* Winter, 114
*Rehmiellopsis* Bubák & Kabat, 174
*Rehmiodothis* Theiss. & Syd., 107
*Rehmiomycella* E. Müller, 102
*Remispora* Linder, 117
*Retinocyclus* Fuckel, 299
*Rhabdocline* Syd. apud Syd. & Petrak, 284
*Rhabdogloeum* Syd. apud Syd. & Petrak, 284
*Rhabdospora* (Sacc.) Sacc., 554
*Rhachomyces* Thaxter, 239, 240*
*Rhagodolobium* P. Henn. & Lindau, 212
*Rhagodostoma* Körb., 103
*Rhamphoria* Niessl, 112
*Rhinocladiella*, 427, 428, 459*
*Rhipidiomyces* Thaxter, 241
*Rhipidocarpon* Theiss. & Syd., 211
*Rhizina* Fr. ex Fr., 269
*Rhizoblepharia* Rifai, 272
*Rhizoctonia*, 274, 435
*Rhizogene* Syd., 175
*Rhizohypha*, 435
*Rhizomorpha*, 435
*Rhizomyces* Thaxter, 236
*Rhizosphaera* Mang. & Har., 565
*Rhodesia* Grove, 560
*Rhodosporidium* Banno, 17*, 24
*Rhodotorula* Harrison, 24, 27

*Rhombostilbella*, 427, 450*
*Rhopographus* Nits., 189
*Rhynchomeliola* Speg., 101
*Rhynchonectria* Höhn., 105
*Rhynchophoromyces* Thaxter, 235, 237*
*Rhynchosporium* 430, 468*
*Rhynchostoma* Karst., 101
*Rhyparobius* Boud., 272, 273
*Rhytidenglerula* Höhn., 170
*Rhytidhysterium* Speg., 143*, 200, 298
*Rhytisma* Fr., 278
Rhytismataceae, 278
*Rickia* Cav., 241
*Rickiella* Syd. apud Rick, 266
*Riclaretia*, 433, 492*
*Riessia*, 429, 434, 494*
*Rinia* Penz. & Sacc., 108
*Rizalia* Syd., 100
*Robergea* Desm., 282
*Robillarda* Sacc., 519, 551
*Roccella* DC, 197
Roccellaceae, 197
*Rollandina* Pat., 61
*Romanoa* Thirum., 127
*Rosellina* de Not., 122
*Rosenscheldia* Speg., 182, 186
*Rosenscheldiella* Theiss. & Syd., 182
*Rostrella* Zimmermann, 98
*Roussoëlla* Sacc., 120
*Rutstroemeria* Karst., 288
*Ryparobius* Boud., 272, 273

**S**

*Saccardaea*, 429, 446*
*Saccardia* Cooke, 164
Saccardiaceae, 165, 164†
*Saccardinula* Speg., 165
Saccardinulaceae, 163, 164†
*Saccardoella* Speg., 101
*Sacardomyces* P. Henn., 100
*Saccharomyces* Mayen em. Rees, 17*, 21, 22
Saccharomycetaceae, 13, 14, 18†
Saccharomycetales, 12
Saccharomycetoideae, 19
*Saccharomycodes* Hansen, 19
*Saccharomycopsis* Schiön., 20
*Saccobolus* Boud., 267
*Saccosoma* Casp. apud Rehm, 264, 265
*Saccothecium* Fr., 175
*Sadasivanella* Agnihot., 551

*Sadasivania*, 427, 453*
*Sakireeta* Subram. & Ramak., 551
*Samarosporella* Linder, 117
*Samukuta* Subram. & Ramakr., 551
*Sandersoniomyces* Benj., 236
*Saprochaete*, 435
*Sarawakus* Lloyd, 103
*Sarcinella*, 429, 472*
*Sarcinella* Sacc., 208
*Sarcoleotia* Ito & Imai apud Imai, 305
*Sarcophoma* Höhn., 544*, 564
*Sarcorhophalum* Rabenh., 36
*Sarcoscypha* (Fr.) Boud., 264–6
Sarcoscyphaceae, 263
Sarcosomataceae, 262
*Sarcosphaera* Auersw., 252*, 268
*Sarcotrochila* Höhn., 284
*Sarcostromella* Boedijn, 123
*Sarcoxylon* Cooke, 123
*Sartorya* Vuill. apud C. R. Benj., 58*, 63
*Scaphidiomyces* Thaxter, 236, 238*
*Scenomyces*, 435,
*Sceptrifera*, 428, 430, 465*
*Schiffnerula* Höhn., 171
*Schismatomma* Massal., 200
*Schistodes* Theiss., 71
*Schizoblastosporion* Cif., 26
*Schizochora* Syd., 107
*Schizoparme* Shear, 530
*Schizophoma* Kleb., 526
*Schizosaccharomyces* Lindner, 17*, 18
Schizosaccharomycoideae, 18
*Schizothyra* Bat. & Costa, 529
*Schizothyrella* Thüm., 529
Schizothyriaceae, 203, 213†
*Schizothyrina* Bat. & Lima, 213
*Schizothyrioma* Höhn., 295
*Schizothyrium* Desm., 137*, 213
*Schizotrichum*, 434, 491*
*Schizoxylon* Pers. ex Chev., 282
*Schonbornia* Bubák, 540
*Schweinitziella* Speg., 102, 108
*Scirrhia* Nits. 137*, 174
*Scleroderris* (Fr.) de Not. 299, 572
*Scleroglossum* Hara, 287
*Sclerographium*, 429, 471*, 526
*Scleromitrula* Imai, 287
*Sclerophoma* Höhn., 574
Sclerophomaceae, 554
*Scleropycnis* Syd., 528

*Sclerotinia* Fuckel, 286
Sclerotiniaceae, 283
*Sclerotiopsis* Speg., 530
*Sclerotium*, 435
*Scodellina* Mich. ex Gray, 269, 273
*Scolecobasidium*, 430, 433, 467
*Scoleccoccoidea* Stev., 108
*Scoleconectria* Stev. & Manter, 105, 108
*Scolecopeltidium* Stev. & Manter, 206
*Scolecosporiella* Petrak, 563
*Scolecosporium* Lib., 188, 540
*Scolionema* Theiss. & Syd., 170
*Scophanus*, 275
*Scopulariopsis* Bain., 425, 426, 448*, 534, 540
*Scorias* Fr., 171, 172
*Scortechinia* Sacc., 91
*Scortechiniella* Arx & Müller, 91
*Scutellinia* (Cooke) Lamb., 266, 275
*Scutula* Tul., 199
*Scytalidium*, 425, 429, 431, 437*
*Seaverinia* Whetzel, 287
*Seimatosporium* Corda, 121, 433, 540, 556, 557
*Seiridina* Höhn., 557
*Seiridium* Nees ex Fr., 121
*Selenaspora* Heim & Le Gal., 271
*Selenophoma* Marie, 544
*Selenophomopsis* Petrak, 564
*Selenosporella*, 428, 434, 440*
*Selenotila* Lagerh., 26
*Selinia* Karst., 103
*Semidelitschia* Cain & Luck-Allen, 184
*Sepedonium*, 127, 427, 455*
*Septocylindrium*, 431
*Septocyta* Petrak, 569
*Septogloeum* Sacc., 538*, 554
*Septoidium*, 432, 479*
*Septomyxa* Sacc., 559
*Septonema*, 432, 478*
*Septosporium*, 429, 470*
*Septoria* Sacc., 530, 561, 563
*Septoriella* Oud., 567
*Septotinia* Whetzel ex Groves & Elliott, 287
*Septotis* Buchw., 287
*Septotrullula*, 431, 433, 477*, 523
*Sepultaria* (Cooke) Boud., 274
*Serenomyces* Petrak, 99
*Sesquicillium*, 104, 428, 444*
*Setella* Syd., 172

# Index to Taxa 617

*Seynesia* Sacc., 120
*Seynesiella* Arnaud, 203
*Seynesiopeltis* Stev. & Ryan, 204
*Shanoria* Subram. & Ramakr., 549
*Shanorella* R. K. Benjamin, 60
*Shiraia* P. Henn., 104
*Shiraiella* Hara, 104
*Sibirina*, 431, 469*
*Sigmoidea*, 434, 493*
*Sigmoideomyces*, 426, 452*
*Sillia* Karst., 112
*Sirococcus* Preuss, 572
*Siropatella* Höhn., 569
*Siroscyphella* Höhn., 567
*Sirosporium*, 429, 433, 471*
*Sirothecium* Karst. 536
*Skoteinospora* Bat., 168
*Smardaea* Svrček, 271
*Soleella* Darker, 279
*Solheimia*, 429, 446*
*Sordaria* Ces. & de Not., 94*, 97*, 110
Sordariaceae, 97, 109†
*Sorokina* Sacc., 291
*Sowerbyella* Nannf., 275
*Spadicoides*, 428, 430, 433, 466*
*Spathularia* Pers ex Fr., 285
*Spathulariopsis*, 285
*Spegazzinia*, 429, 434, 435, 494*
*Spegazziniella* Bat. & Lima, 213
*Spegazzinula* Sacc., 104
*Speiropsis*, 435, 501*
*Spermophthora* Ashby & Nowell, 14
Spermophthoraceae, 13, 14†
*Spermospora*, 432, 433, 486*
*Sphaceloma* de Bary, 559
*Sphaeria* Fr., 116, 121, 514
Sphaeriaceae, 97, 100†
Sphaeriales, 47, 89, 93, 96†
*Sphaeridiobolus* Boud., 267
*Sphaeridium*, 426, 452*
*Sphaerodothis* Shear, 106
*Sphaerognomonia* Potn., 114
*Sphaerographium* Sacc., 563
*Sphaeromyces*, 428, 438*
*Sphaeronaemella* Karst., 98
*Sphaeronema* Fr., 518
Sphaeropsidales, 555, 560†
*Sphaeropsis* Sacc., 520, 530
*Sphaeropyxis* Bon., 122
*Sphaerosoma* Klotsch apud Dietr., 252*, 267, 270
*Sphaerospora* (Sacc.) Sacc., 274, 275
*Sphaerosporella* (Svrček) Svrček & Kubická, 274
*Sphaerosporium*, 426, 462*
*Sphaerotheca* Lév., 73, 75*, 78*, 79*
*Sphaerozone* Zobel apud Corda, 252*, 267, 270
*Sphaerulina* Sacc., 175
*Sphagnicola* Vel. 300
*Spilocaea*, 427, 430, 448*, 518
*Spilodochium*, 426, 430, 464*
*Spilopodia* Boud., 294
*Spiromastix* Kuehn & Orr, 61
*Spiropes*, 432, 433, 485*
*Spirosphaera*, 431, 476
*Spondylocladiella*, 432, 479*
*Spondylocladiopsis*, 432, 480*
Sporendocladia, 426, 441*
*Sporendonema*, 425, 436*
*Sporidesmium*, 432, 434, 488*, 518
*Sporidiobolus* Nyland, 24, 27
*Sporobolomyces* Kluyver & van Niel, 25
Sporobolomycetaceae, 24, 25†
*Sporocadus* Corda, 557
*Sporocephalum*, 426, 427, 451*
*Sporocybe*, 426, 453
*Sporomega* Corda, 279
*Sporonema* Desm., 295, 573
*Sporophiala*, 432, 477*
*Sporormia* de Not., 184
Sporormiaceae, 178, 184†
*Sporoschisma*, 113, 431, 477*
*Sporoschizon* H. Riedl, 193
*Sporothrix*, 62, 64, 427, 459*
*Sporotrichum*, 427, 457*
*Spragueola* Massee, 285
*Stachybotrys*, 101, 426, 428, 429, 442*
*Stachylidium*, 428, 429, 440*
*Stagonospora* (Sacc.) Sacc., 536, 562
*Stammaria* Fuckel, 305
*Staphylotrichum*, 427, 454*
*Stauronema* (Sacc.) Syd. & Butler, 550*, 568
*Steganopycnis* Syd., 120
*Steganosporium* Corda, 558
*Stegasphaeria* Syd., 183
*Stegastroma* Syd., 112
*Stegopeziza* Höhn., 297
*Stegopezizella* Syd., 284

*Stemmatomyces* Thaxter, 236
*Stemphyliomma*, 432, 479*
*Stemphylium*, 429, 471*
*Stenella* 432, 483*
*Stenellopsis*, 433, 483*
*Stephanoma*, 427, 434, 495*
*Stephanosporium*, 425, 437*
*Stephanotheca* Syd., 212
Stephanothecaceae, 203, 212†
*Stephensia* Tul., 261, 282
*Sterigmatöbotrys*, 433, 480*
*Sterigmatomyces* Fell, 26
*Stevensia* Trott., 164
*Stevensomyces*, 426, 462*
*Stevensonula* Petrak, 570
*Stichomyces* Thaxter, 239
Stictidiaceae, 281
*Stictis* Pers. ex Gray, 303
*Stigmatea* Fr., 182, 206
*Stigmatodothis* Syd., 206
*Stigmatomassaria* Munk, 188
*Stigmatomyces* Karst., 226, 231, 236, 237*
*Stigmina*, 429, 432, 471*
*Stigmochora* Theiss. & Syd., 93*, 107
*Stilbella*, 64
*Stilbocrea* Pat., 103
*Stilbohypoxylon* P. Henn., 122
*Stilbophoma* Petrak, 566
*Stilbospora* Pers. ex Mérat, 113, 517
*Stilbum*, 428, 446*, 518
*Stirtonia* A. L. Smith, 152*, 197
*Stomatogene* Theiss., 82, 169
*Stomatopeltis* Theiss., 137*, 207
*Stomiotheca* Bat., 206
*Strasseria* Bres. & Sacc., 545*, 573
*Strasseriopsis* Sutton & Kobayashi, 543, 545*, 573
*Streptobotrys* Henneb., 286
*Streptotheca* Vuill., 273
*Streptotinia* Whetzel, 286
*Stromatinia* (Boud.) Boud., 288
*Stromatocrea*, 426, 436*
*Strossmeyeria* Schulzer, 305
*Strumella* 264, 431, 475
*Succinaria* Syd., 108
*Subramanella* Srivastava, 572
*Subulispora*, 434, 459*
*Suttoniella* Ahmad, 571
*Svreckia* Kubická, 267
*Swanniomyces* Klöcker, 20

*Sydowia* Bres., 176, 574
*Sydowiella* Petrak, 112
*Sydowiellina* Bat. & Lima, 213
*Sydowinula* Petrak, 91
*Symphaeophyma* Speg., 212
*Symphaster* Theiss. & Syd., 207
*Symplectromyces* Thaxter, 236, 238
*Sympodiella*, 425, 436*
*Sympodiophora*, 430, 468*
*Synandromyces* Thaxter, 236
*Synaptomyces* Thaxter, 235
Synascomycetes, 36
*Synnematium*, 428
*Syringospora* Quinq., 27

**T**

*Taeniolella*, 432, 478
*Talaromyces* C. R. Benjamin, 49, 58*, 63
*Tapesia* Fuckel, 292
*Tapesina* Lamb., 296
*Taphridium* Lagerh. & Juel, 34
*Taphrina* Fr., 36, 37*, 39*
Taphrinaceae, 35, 36
Taphrinales, 35–40
*Tarzetta* (Cooke) Lamb., 271, 273
*Teichospora* Fuckel, 192
*Telimenia* Rac., 107
*Telimeniella* Petrak, 108
*Telimenopsis* Petrak, 107
Teliomycetes, 6
*Teratomyces* Thaxter, 236, 238*
*Teratonema* Syd., 91
*Teratosperma*, 432, 433, 487*
*Teratosphaeria* Syd., 182
*Terfezia* Tul., 260
Terfeziaceae, 259
*Tetrachaetum*, 434, 498*
*Tetracladium*, 434, 501*
*Tetracoccosporium*, 429, 472*
*Tetracrium*, 434, 490*
*Tetranacrium* Hudson & Sutton, 570
*Tetrandromyces* Thaxter, 239
*Tetraploa*, 435, 500*
*Tetraposporium*, 435, 499*
*Testudina*, 47
*Thalassoascus* Ollivier, 185
*Thallospora* 434, 496*
*Thamnidium* Link ex Wallr., 518
*Thamnomyces* Ehrenb. ex Spreng., 121
*Thaumasiomyces* Thaxter, 235

# Index to Taxa

*Thaxteria* Sacc., 92, 187
*Thecotheus* Boud., 252*, 267
*Thelebolus* Tode ex Fr., 252*, 272, 517
*Thelidium* Massal., 125
Thermoascaceae, 55, 62
*Thermoascus* Miehe, 54*, 62
*Thermomyces*, 427, 456*
*Therrya* Penzig & Sacc., 280
*Thielavia*, 47
*Thielavia* Zopf, 99
Thielaviaceae, 98
*Thielaviella* Arx & Taricq, 100
*Thielaviopsis*, 426, 441*, 546
*Thindia* Korf & Waraitch, 263, 266
*Thozetellopsis*, 428, 447*
*Thrauste* Theiss., 82, 170
*Thuemenella* Penz. & Sacc., 103
*Thuemenidium* Kuntze, 285
*Thyridaria* Fuckel, 102
*Thyridaria* Sacc., 188
*Thyridium* (Nits.) Sacc., 193
*Thyriopsis* Theiss. & Syd., 213
*Thyronectria* Sacc., 105
*Thyronectroidea* Seaver, 105
*Thyrostromella*, 429, 471*
*Thysanophora*, 426, 442*
*Thyrsidium* Mont., 556
*Tiarospora* Sacc. & March. 537*, 561
*Tiarosporella* Höhn., 538*, 563
*Tichothecium* Flotow, 125
*Tilletiopsis* Derx, 25
*Tirmania* Chatin, 259
*Titaea*, 434, 494*
*Titaeospora* Bubák, 558
*Titaeosporina* van Luiyk 529, 531*
*Tolypocladium*, 428, 439*
*Tomasellia* Massal., 185
*Tomenticola*, 433, 485*
*Tonduzia* Stev., 101
*Topospora* Fr., 299
*Torpedospora* Meyers, 117
*Torrubiella* Boud., 126
*Torula*, 431, 478*
*Torulaspora* Lindner, 21
*Torulomyces*, 426, 438*
*Torulopsis* Behrend, 27
*Toxosporiopsis* Sutton & Sellar, 536, 538*, 570
*Toxosporium* Vuill., 557
*Toxotrichum* Orr & Kuehn, 52*, 60

*Trabutia* Sacc. & Roum., 205
*Trabutiella* Stev., 107
*Trachyxylaria* Moeller, 102
*Trematomyces* Schrantz, 121
*Trematosphaeria* Fuckel 190, 191
*Tremella* Dill. ex Fr., 23
Tremellales, 23, 27
*Trenomyces* Chatton & Picard, 243
*Tretopileus*, 435, 502*
*Treubiomyces* Höhn., 169
*Triadelphia*, 432, 502*
*Triandromyces* Thaxter, 239
*Triangularia* Boedijn, 97*, 110
*Tricellula*, 434, 494*
*Tricharia* Boud., 271
*Tricharina* Eckblad, 271
*Trichasterina* Arnaud, 207
*Trichipezizella* Dennis ex Raitviir, 296
*Trichobelonium* (Sacc.) Rehm, 292
*Trichobolus* (Sacc.) Kimbr. & Cain apud Kimbr. & Korf, 253*, 272
*Trichobotrys*, 426, 453*
*Trichocladium*, 430, 433, 467*
Trichocomataceae, 48, 56, 62
*Trichoconis*, 432, 486*
*Trichodelitschia* Munk, 184
*Trichoderma*, 103, 428, 439*
*Trichodiscus* Kirschst., 296
*Trichodochium*, 430, 464*
*Trichodothella* Petrak, 181
*Trichodothis* Theiss. & Syd, 181
*Trichoglossum* Boud., 285, 299
*Trichoma* Jungh., 54*, 62
*Trichomerium* Speg., 172
*Trichometasphaeria* Munk, 141*, 150*, 190
Trichomycetes, 3, 6
*Trichopeltella* Höhn., 205
Trichopeltinaceae, 202, 205†
*Trichopeltheca* Bat., Costa & Cif., 173
*Trichopeltina* Theiss., 205
*Trichopeltum* Bat., Cif. & Costa, 205
*Trichophaea* Boud., 271, 274
*Trichophaeopsis* Korf & Erb apud Erb, 272, 274
*Trichophysalospora* Lebedova, 119
*Trichophyton*, 51, 60, 432, 486*
*Trichoscyphella* Nannf., 295
*Trichosphaerella* Bomm., Rouss. & Sacc.,100
*Trichosphaeria* Fuckel, 100
*Trichosporon* Behrend, 17*, 26

*Trichothecium*, 289, 429, 463*
Trichothyriaceae, 167, 168†
*Trichothyriella* Theiss, 168
*Trichothyrina* (Petrak) Petrak, 168
*Trichothyriomyces* Bat. & Maia, 168
*Trichothyrinula* Petrak, 168
*Trichothyriopsis* Theiss., 168
*Trichothyrium* Speg., 168
*Trichurus*, 426, 448*
*Tricladium* 434, 496*
*Tridentaria* 434, 498*
*Trigonopsis* Schachner, 26
*Trimmatothele* Norm., 124
*Trimmatothelopsis* Zschacke, 124
*Trinacrium*, 434, 498*
*Tripedotrichum* Orr & Kuehn 52*, 60
*Tripospermum* 435, 497*
*Tripospora* Sacc., 94*, 108
*Triposporina* 434, 494*
*Triposporium*, 435, 497*
*Tripterospora* Cain, 110
*Triscelophorus*, 434, 499*
*Trisulcosporium*, 434, 499*
*Tritirachium*, 427, 461*
*Trochila* Fr., 291
*Trochophora*, 431, 475*
*Troposporella*, 431, 475*
*Troposporium* 431, 476*
*Trullula* Ces., 536, 568
*Truncatella* Steyaert, 556, 557
*Tryblidaria* (Sacc.) Rehm, 200
*Tuber* Mich. ex Fr., 261
Tuberaceae, 259
Tuberales, 47, 256, 257
*Tubercularia*, 105, 428, 445*, 514
*Tuberculina*, 427, 436*
*Tubeufia* Penz. & Sacc., 186
Tulasnellaceae, 27
*Tympanis* Tode ex Fr., 299, 572
*Tympanopsis* Starb., 91
*Typhulochaeta* Ito & Hara, 78

## U

*Uleodothis* Theiss. & Syd., 181
*Uleomyces* P. Henn., 164
*Ulocladium*, 429, 471*
*Uloseia* Bat., 172
*Umbelopsis*, 426, 454*
*Uncigera*, 428, 431, 440*
*Uncinula* Lév., 73, 77*, 79*, 80

*Underwoodia* Peck, 269
*Unguicularia* Höhn., 298
*Unguiculella* Höhn., 297
*Unguilariopsis* Rehm, 298
Unitunicatae, 36
*Urceolella* Boud., 298
*Urnula* Fr., 263, 264
*Uropolystigma* Maubl., 107
*Urosporella* Atk., 119
Ustilaginales, 23
*Ustilaginoidea*, 126, 427, 457*
*Ustulina* Tul., 123

## V

*Valdensia* Peyron., 287
*Valdensinia* Peyron., 287
*Valetoniella* Höhn., 101
*Valsa* Fr., 113, 572
*Valsaria* Ces. & de Not., 102
*Valsella* Fuckel, 113
*Valseutypella* Höhn., 113
*Valsonectria* Speg., 103
*Vanbeverwijkia*, 431, 474*
*Vanderystiella* P. Henn., 559
*Varicosporina*, 434, 496*
*Varicosporium*, 434, 496*
*Vasculomyces*, 435
*Velutaria* Fuckel, 296
*Velutarina* Korf, 302
*Venturia* Sacc., 182
Venturiaceae, 178, 179†
*Vermicularia* Tode ex Fr., 532
*Veronaea*, 430, 465*
*Verpa* Swartz ex Pers., 252*, 269
*Verpatinia* Whetzel, 252*, 286
*Verrucaria* Schrad., 124
Verrucariaceae, 96, 124†
*Verrucispora*, 433, 483*
*Verrucobotrys*, 287
*Verticicladiella*, 428, 461*
*Verticicladium*, 263, 427, 461*
*Verticillium*, 428, 444*
*Vestergrenia* Rehm, 173
*Vialaea* Sacc., 118
*Vibrissea* Fr., 252*, 282, 285
*Viegasia* Bat., 208
*Virgaria*, 428, 460*
*Virgariella*, 428, 459*
*Virgatospora*, 433, 481*
*Virgella* Darker, 279

*Vizella* Sacc., 205
*Vleugelia* J. Reid & Booth, 113
*Volkartia* Marie, 34
*Volutella*, 428, 445*, 545
*Volutina*, 426
*Vouauxiella* Petrak & Syd., 536, 566

## W

*Wageria* Stev. & Dalby, 82
*Wallemia*, 425, 426, 437*
*Waltonia* Saho apud Saho & Takahashi, 294
*Wardomyces*, 427, 430, 467*
*Wawlia* Namyslowski, 122
*Wentiomyces* Koord., 179
*Wettsteinina* Höhn., 152, 171
*Whetzelinia* Korf & Dumont, 286
*Wickerhamia* Soneda, 19
*Wiesneriomyces*, 425, 426, 428, 431, 433, 434, 490*
*Wingea* van der Walt, 20
*Wojnowicia* Sacc., 564
*Wolfina* Seaver ex Eckblad, 265
*Wynnea* Berk. & Br. apud Berk. 253*, 265
*Wynnella* Boud., 269

## X

*Xenodium* Syd., 165
*Xenomeris* Syd., 181
*Xenonectriella* Weese, 104
*Xenosporium*, 429, 431, 475*
*Xenostigme* Syd., 82
*Xenostomella* Syd., 204
*Xylaria* Hill ex Grev., 94*, 97*, 122
Xylariaceae, 98, 121†
Xylariales, 93
*Xylobotryum* Pat., 102

*Xyloceras* Smith, 102
*Xylochora* Arx & Müller, 119
*Xylogone* Arx & Nilsson, 54*, 62
*Xylogramma* Wallr., 304
*Xylohypha*, 426, 453*
*Xyloma* Pers., 517
*Xyloschizon* Syd. apud Syd. & Petrak, 279
*Xylosphaera* Gray, 122
*Xylosphaera* Otth. em. Petrak, 193
*Xynophila* Cain, 53
*Xystozukalia* Theiss., 172

## Y

*Yatesula* Syd., 204
Yeasts, 11–32

## Z

*Zalerion*, 431, 475*
*Zanclospora*, 428, 440*
*Zasmidium*, 428, 459*
*Zignoella* Sacc., 112
*Zodiomyces* Thaxter, 234, 237*
*Zoellneria* Vel., 297
*Zopfiella* Winter, 110
*Zukalina* Kuntze, 267
*Zygofabospora* Kudriavzev, 21
*Zygolipomyces* Krassil., Babjeva & Meavahad, 19
Zygomycetes, 3, 6
Zygomycotina, 5, 6†
*Zygophiala*, 430, 465*
*Zygopleurage* Boedijn, 97*, 111
*Zygosaccharomyces* Barker, 21
*Zygospermella* Cain, 111
*Zygosporium*, 426, 427, 457*, 522
*Zythia* Fr., 115
*Zythiostroma* Höhn., 105